Advanced Calculus

Advanced Calculus

second edition

Angus E. Taylor

University of California

W. Robert Mann

University of North Carolina

John Wiley & Sons, Inc.

New York   Santa Barbara   London   Sydney   Toronto

CONSULTING EDITOR

George Springer
Indiana University

Second Edition Copyright © 1972 by John Wiley & Sons, Inc.
First Edition copyright © 1955 by Ginn and Company.
ISBN 0 471 00587 8
Library of Congress Catalog Card Number: 78-185259
Printed in the United States of America.

10 9 8 7 6 5 4

To
Our
Students

Preface to the Second Edition

The original edition of *Advanced Calculus* was published in 1955. This new edition retains most of the characteristic features of the original book, and many chapters are changed little or not at all, in deference to the expressed wishes of those who have found the book to be useful and instructive. The preface to the first edition is reproduced here separately, without change, since it expresses certain thoughts and views which are still cogent. In this preface to the new edition the authors wish to explain the principal ideas and motivations which have guided them in making the revision.

One of the important trends in modern mathematics is an effort to narrow the gap between the standard body of knowledge presented to students in courses, and the insights emerging from the researches of those who are working on the frontiers of the subject. These insights frequently come from interactions between different fields of mathematics. For a good many years it has been increasingly apparent that mathematical analysis, particularly in the realm of functions of several variables, is susceptible to great conceptual improvement and simplification through the systematic use of the methods of linear algebra and the closely related work on vector spaces in the theory of linear operators. Accordingly, we have made some revisions of the previous text, and added two new chapters, primarily to utilize the ideas of linear algebra and vector spaces in dealing with the differential calculus of functions of several variables. Our aim has been to retain what is still important and useful while adding new material. We wish to make sure that the student acquires facility in working with partial derivatives before being introduced to the much more general concept of differentiability of vector-valued functions.

In order to keep the size of the book within reasonable bounds we have included only those parts of linear algebra which are indispensable for a good understanding of the theory of differentiable functions from R^n to R^m. We have added a new Chapter XI, dealing with the algebra of linear spaces of finite dimensionality and linear transformations from one such space to another. This chapter also contains what is needed about metrics and norms to enable us to discuss continuity for linear transformations. The chapter builds up toward a fundamental theorem on the class of invertible linear mappings of R^n into itself.

The second new chapter, Chapter XII, is concerned with differentials and continuously differentiable transformations from R^n to R^m. The goal of the chapter is to generalize the inversion theorem of Chapter XI, which deals with linear mappings of R^n into itself, to continuously differentiable mappings of R^n into itself, where the condition for local invertibility is that the differential of the mapping at a particular point (which is a linear mapping of R^n into itself) be invertible. We then continue on to implicit functions, and show how to use the inversion theorem to obtain the classical implicit function theorem relating to q equations in p plus q variables; these equations are to be "solved" to determine q of the variables in terms of the remaining variables.

Some changes have been made in the discussion of the differential in Chapters VI and VII. Also, we have used the differential to motivate the discussion of surface area in Section (§) 14.6.

The last two chapters from the first edition have been dropped (they were on complex functions, and Fourier series and integrals, respectively). Other changes have been very minor.

The book is designed to be used in a variety of ways. Some parts of it can be used by the students purely for reference and for self-help. This may be particularly true of the first few chapters, which deal with material touched upon to some extent in courses in elementary calculus. The fundamental theory of differential calculus of functions of a single real variable is contained in the first four chapters. The elementary theory of integral calculus for functions of one variable is contained in a few sections of Chapter I and in the first part of Chapter XVIII. A major portion of the book is concerned with the differential and integral calculus of functions of several variables, and another portion of the book is concerned with infinite series and improper integrals.

In sending forth this revised edition we wish to pay tribute to all our students, from the teaching of whom we have learned much. We have enjoyed teaching, especially when we have seen our students growing in understanding and appreciation of mathematical analysis. The questions and comments of students have often led us to new insights as teachers. We hope that other students and other teachers will find that this book opens the doors to understanding and enjoyment.

We are happy to give credit to Mr. David Lane for reading the revised portions of the second edition and making many helpful suggestions, and we should like to acknowledge our gratitude to Mrs. Linda Leonard for her excellent typing under constant pressure.

ANGUS E. TAYLOR
W. ROBERT MANN

Preface to the First Edition

This book has grown out of my experience in teaching advanced calculus over a period of more than a dozen years. It has taken shape gradually as I observed the effect on my students of various presentations of theory and technique, and as I accumulated experience in the construction of problems and examination questions. Every part of the book has been planned to make the whole an effective instrument for imparting the fundamental principles and methods of analysis to students at the advanced calculus level. The book is aimed at the student reader; I have striven to arouse interest at every stage, to motivate the direction of the exposition, and to achieve clarity through ample illustrative examples and particular care in directing the course of the reasoning.

In many of the chapters the first section is devoted to one or more of the following: (1) setting forth in general terms the aim of the chapter, (2) supplying motivation for the subject matter, (3) explaining my point of view in fitting the chapter into the book as a whole.

Books on advanced calculus vary widely in choice of subject matter, in emphasis on particular topics, and in treatment of the relation between elementary and advanced calculus. These are matters on which no one book can fully please all users. For lack of space I have omitted treatment of some topics which I would have liked to include. The emphasis is on sound understanding of concepts, and on the basic principles of analysis: those properties of the real number system which support the theory of limits and continuity. But the thread of theoretical development is imbedded in an ample exposition of the methods and techniques which are needed by every student of advanced applied mathematics. There is a generous and rich supply of exercises and problems.

Learning in calculus is cumulative. It is also evolutionary. The student does not come all at once to a one and only correct understanding of new ideas. At each new level of his maturity he can gain a fresh appreciation of things he has already been taught. An advanced calculus should not ignore or discard all that a student already knows about calculus. Rather, it should build upon what he knows, and strengthen that knowledge by emphasis upon those aspects of elementary calculus which are given least attention in the usual freshman and sophomore courses, and which become increasingly important as a student progresses into more advanced analysis. Chapter I of the present book is designed to be used for building in this way.

The book is written on the assumption that students using it have normal skill in the formal aspects of elementary calculus, and that they can draw freely on the standard formulas of algebra, trigonometry, and calculus relative to the elementary functions. Some of the logical issues pertaining to the definition of logarithms, exponentials, and trigonometric functions are not fully met in elementary calculus, of course. But I prefer not to tackle these issues prematurely. They can be settled in due time, and in a variety of ways, once the student knows enough about definite integrals, infinite series, and uniform convergence. Meanwhile, the student is eager to get on to new ideas, new techniques, new applications.

A word about the system of numbering. Sections within each chapter are numbered in decimal order. Thus, in Chapter XIX, § 19.21 and § 19.22 follow each other between § 19.2 and § 19.3. The first section in each chapter has no digits after the decimal point. Formulas in each section are numbered with consecutive integers after a dash which follows the section number in which the formulas occur. Thus, formulas in § 1.11 range from (1.11–1) to (1.11–17). Theorems are numbered consecutively in Roman numerals, starting with Theorem I for the first theorem in each new chapter. References to theorems usually cite the theorem number and the section in which it occurs.

I have not attempted to include a bibliography. I have been influenced by many books, both American and European, but I cannot account for the influences in detail.

In sending the book forth I pay my respects to the memory of one of my teachers, Professor William Fogg Osgood, and I thank heartily all those students and colleagues who have taken an interest in seeing the book brought to completion.

ANGUS E. TAYLOR

Contents

8. Implicit-function theorems

9. Transformations and mappings

10. Vectors and vector fields

11. Linear transformations

15. Line and surface integrals

16. Point-set theory

17. Fundamental theorems on continuous functions

18. The theory of integration

19. Infinite series

20. Uniform convergence

21. Power series

22. Improper integrals

Advanced Calculus

Fundamentals of Elementary Calculus

1. Introduction

A course in advanced calculus must build upon the presumption that students studying the subject have already gained some knowledge of elementary calculus. We shall therefore begin by taking a backward look over those parts of calculus with which the reader of this book should have facility and a measure of understanding. Our object in such a retrospect is not to conduct a systematic review. The purpose is, rather, to establish a common point of view for students whose training in calculus, up to this point, must inevitably reflect a wide variety of practices in teaching, choice of subject matter, and distribution of emphasis between the acquisitions of problem-solving skills and mastery of fundamental theory. As we survey the field of elementary calculus we shall stress the conceptual aspect of the subject: fundamental definitions and processes which underlie all the applications. In a first course in calculus it is often the case that the fundamental notions are introduced through the medium of particular geometrical or physical applications. Thus, to the beginner, the derivative may be typified by, or even identified with, the speed of a moving object, while the integral is thought of as the area under a curve. We now seek to take a more general, or abstract, view. Differentiation and integration are processes which are carried out upon functions. We need to have a clear understanding of the definitions of these processes, quite apart from their applications.

Another aspect of our survey will be our concern with the logical unfolding of the fundamental principles of calculus. Here again we strive to take a more mature point of view. We wish to indicate in what respects it is desirable and necessary to look more deeply into the derivations of rules and proofs of theorems. There are places in elementary calculus, as usually taught to beginners, where the development is necessarily inadequate from the standpoint of logic. In many places the reasoning leans heavily on intuition or on one sort or another of plausibility argument. That this state of affairs persists is partly due to a deliberate placing of emphasis: we make our primary goal the attainment of skill in the manipulative techniques of calculus which lend themselves readily to applications at an elementary level in physics, engineering, and the like. This kind of skill (up to a certain point) can be imparted without paying much attention to questions of logical rigor. But it is also true that there are logical inadequacies in a first course in calculus which cannot be made good entirely within the customary time limits of such a course (two or three semesters), even where a reasonably heavy emphasis is laid upon "theory." At bottom the subject of calculus rests upon the real number system and the theory of limits. A full appreciation and understanding of this foundation material must come slowly, but the need for such understanding becomes more acute as we progress in learning. In advanced calculus we must make a deeper study of the real number system, of the theory of limits, and of the properties of continuous functions. In this way only can we proceed easily and with confidence to a mastery of many new concepts and processes of higher mathematics.

1.1 Functions

At the very outset we must discuss the mathematical concept of a *function*, for we shall constantly be talking about properties of functions and about processes which are applied to functions. The function concept has been very much generalized since the early development of calculus by Leibniz and Newton. At the present time the word "function" is used broadly to mean any determinate correspondence between two classes of objects.

Example 1. Consider the class of all plane polygons. If to each polygon we make correspond the number which is the perimeter of the polygon (in terms of some fixed unit of length), this correspondence is a function. Here the first class of objects is composed of certain geometrical figures, while the members of the second class are positive numbers.

To begin with, let us consider functions which are correspondences between sets of real numbers. Such functions are called *real functions of a real variable*. The first set of numbers is the *domain of definition* or simply the *domain* of the function. The second set, consisting of the values taken on by the function, is called the *range*. Once the domain, which we may call D, has been specified, the function is defined as soon as a definite rule of correspondence has been given, assigning to each number of D some corresponding number in the range. If x is a symbol which may be used to denote any member of D, we call x the *independent variable* of the function. In some situations it is very natural to have more than one number associated with a given value of x and to call such a correspondence a *multiple-valued function*. If each value of x corresponds to just one number in the range we have a *single-valued function*, which is what is properly meant by the term *function*. We usually find it possible to deal with multiple-valued functions by separating them into several (possibly infinitely many) single-valued functions. Hereafter we shall always assume that all functions referred to are single-valued, unless the situation explicitly indicates the contrary.

A function may be defined by an algebraic or trigonometric formula, but it need not be so.

Example 2. If x denotes any real number, let $[x]$ denote the algebraically largest integer which does not exceed x; e.g.,

$$[-2.3] = -3, [-1] = -1, [0] = 0, [3.5] = 3, [7] = 7, [7.2] = 7.$$

The correspondence between x and $[x]$ defines a function. If we use f to denote this function, then we would say that f is defined by $f(x) = [x]$.

Example 3. Another simple function is defined by associating with x its *absolute value* $|x|$. The definition of $|x|$ is:

$$|x| = x \text{ if } x \geqq 0,$$
$$|x| = -x \text{ if } x < 0.$$

Thus $|7| = 7, |0| = 0, |-5| = 5, |3 - 10| = 7$. If we think of x as a point on a number scale (the x-axis), then $|x|$ is the numerical distance (always nonnegative) between x and the origin.

The concept and the symbolism of absolute value are quite important. The student will need to get accustomed to reading sentences that contain inequalities and absolute values. Thus, for instance, $|7 - 5| = 2, |-16 - (-10)| = 6$, and, in general, $|x_1 - x_2|$ is the distance between points x_1 and x_2 on the x-axis. As another example, $|x - 5| < 2$ means that the distance between x and 5 is less than 2; this is equivalent to saying that x lies between 3 and 7. We can write this in the form $3 < x < 7$. A general statement of the same sort is that $|x - a| < b$ (where $b > 0$) is equivalent to $a - b < x < a + b$.

We regard functions as mathematical entities, and represent them by symbols. The commonly used symbols are the Latin letters f, g, h, F, G, H, and the Greek letters ϕ, ψ, Φ, Ψ, but in principle any symbol may be used. If f is the symbol for a particular function, we use $f(x)$ to represent the number which the function makes correspond to any particular value of the independent variable x; this is called the *value* of the function at x.

Example 4. Let f be the symbol for the function which makes correspond to a positive number the natural logarithm of that number. Then $f(x) = \log_e x$. (We shall normally drop the subscript e and write $\log x$ in place of $\log_e x$.)

There is some ambiguity in the use of functional notation, for $f(x)$ is frequently used as a symbol for the function itself, as well as for the value of the function. Thus, for example, we speak of the "function sin x," "the function $x^2 - 3x + 5$," or "the function $\phi(x)$." There is of course a difference between the function and the value of the function. If the symbol $f(x)$ appears, the context will usually make clear whether reference is being made to the function or to the value of the function. To avoid possible ambiguity we shall cultivate the practice of writing "the function f"

instead of "the function $f(x)$." This usage is in accord with prevalent practice in current literature, and the student will do well to become familiar with it.

If y is a symbol for the value of the function f at x, we can write $y = f(x)$. Here y is called the *dependent variable*; we say that y is a function of x. In elementary calculus most of the stress is upon functions which are defined by means of fairly simple formulas connecting the independent variable x and the dependent variable y. Here, however, we look toward understanding the principles of calculus as they apply to functions which are arbitrary except insofar as they are restricted by specified hypotheses.

We shall in due course have to deal with functions of more than one variable. The general notion of a function is still that of a correspondence. A real function F of two real variables x, y is a correspondence which assigns a number $F(x, y)$ as the value of the function corresponding to the pair of values x, y of the two independent variables. The use of functional notation and the designation of the function by the single letter F require no detailed comment, since the basic ideas are no different from those already explained.

The characteristic feature of calculus is its use of *limiting processes*. Differentiation and integration involve certain notions of passage to a limit. A fuller discussion of ideas about limits is presented later on in this chapter (§§ 1.6–1.64). Here we wish to touch on only one limit notion, that of the limit of a real function of one real variable. This notion is fundamental in the definition of a derivative.

Suppose f is a function which is defined for all values of x near the fixed value x_0, and possibly, though not necessarily, at x_0 as well. We wish to attach a clear meaning to the statement: $f(x)$ approaches A (or tends to the limit A) as x approaches x_0. The symbolic form of the statement is

$$(1.1\text{–}1) \qquad \lim_{x \to x_0} f(x) = A.$$

The symbol A is understood to stand for some particular real number. The arrow is used as a symbol for the word "approaches." Sometimes (1.1–1) is expressed in the form $f(x) \to A$ as $x \to x_0$. Here are three typical examples of statements of this kind: (a) $x^3 \to 8$ as $x \to 2$, (b) $(x - 1)^{1/2} \to 3$ as $x \to 10$, (c) $\log_{10} x \to 2$ as $x \to 100$.

DEFINITION. The assertion (1.1–1) means that the absolute value $|f(x) - A|$ can be made as small as we please merely by requiring that the absolute value $|x - x_0|$ be sufficiently small, and different from zero. This verbal statement is expressible in terms of inequalities as follows: Suppose ϵ is *any* positive number. Then there is *some* positive number δ such that

$$(1.1\text{–}2) \qquad |f(x) - A| < \epsilon \quad \text{if} \quad 0 < |x - x_0| < \delta.$$

Note that $0 < |x - x_0|$ is the same as $x \neq x_0$. Note also that $|f(x) - A|$

$< \epsilon$ is the same as $A - \epsilon < f(x) < A + \epsilon$, and $| x - x_0 | < \delta$ the same as $x_0 - \delta < x < x_0 + \delta$.

We can give a geometrical portrayal of the inequalities (1.1–2). Let the points (x, y) with $y = f(x)$ be located on a rectangular co-ordinate system; also locate the point (x_0, A). For any $\epsilon > 0$ draw the two horizontal lines $y = A \pm \epsilon$. Now (1.1–1) means that, by choosing δ small enough, those points of the graph of $y = f(x)$ which lie between the two vertical lines $x = x_0 \pm \delta$ and not on the line $x = x_0$ will also lie between the horizontal lines $y = A \pm \epsilon$. Fig. 1 shows a specimen of this situation. The diagram also shows how δ may have to be made smaller as ϵ becomes smaller.

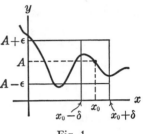

Fig. 1

It is to be emphasized that (1.1–1) places no restrictions whatever on the value of f at x_0, in case it is defined at that point.

Appreciation of the formal definition of the meaning of (1.1–1) takes time and experience. The formal definition is the basis for exact reasoning on matters involving the limit concept. But it is also quite important to develop an intuitive understanding of the notion of a limit. This may be done by considering a large number of illustrative examples and by observing the way in which the limit concept is used in the development of calculus. One needs to learn by example how a function $f(x)$ may fail to approach a limit as x approaches x_0.

The variable x may approach x_0 from either of two sides. Let us use $x \to x_0+$ to indicate that x approaches x_0 from the right, and $x \to x_0-$ to indicate approach from the left. The conditions for $\lim_{x \to x_0} f(x) = A$ are then that $f(x) \to A$ as $x \to x_0+$ and also $f(x) \to A$ as $x \to x_0-$. *In terms of inequalities the meaning of $f(x) \to A$ as $x \to x_0+$ is this: to any $\epsilon > 0$ corresponds some $\delta > 0$ such that $| f(x) - A | < \epsilon$ if $x_0 < x < x_0 + \delta$.* The meaning of $f(x) \to A$ as $x \to x_0 -$ may be expressed in a similar way.

Example 5. The limit of $f(x)$ as $x \to x_0$ may fail to exist because:

(a) The limits from right and left exist but are not equal. This is the case with

$$f(x) = 1 + \frac{|x|}{x},$$

where $f(x) \to 2$ as $x \to 0+$ and $f(x) \to 0$ as $x \to 0-$.

(b) The values of $f(x)$ may get larger and larger (tend to infinity) as $x \to x_0$ from one side or the other, or from both sides. This is the case with $f(x) = 1/x$ as $x \to 0$.

(c) The values of $f(x)$ may oscillate infinitely often, approaching no limit. This is the case with $f(x) = \sin(1/x)$ which oscillates infinitely often between -1 and $+1$ as $x \to 0$ from either side.

Example 6. If $f(x) = e^{-1/x^2}$, then $\lim_{x \to 0} f(x) = 0$. To "see" the correctness of this result, one must have clearly in mind the nature of the exponential function. When x is near zero, $-1/x^2$ is large and negative; now e raised to a large negative power is a small positive number. Hence e^{-1/x^2} is nearly 0 when x is nearly 0, and $f(x) \to 0$ as $x \to 0$. This is an example of a rough intuitive argument leading to a conclusion about a certain limit.

It is instructive to see how the intuitive argument is made precise by reference to the definition of a limit. The statement $\lim_{x \to 0} e^{-1/x^2} = 0$ means that to each $\epsilon > 0$ corresponds some δ such that

$$(1.1\text{–}3) \qquad |e^{-1/x^2} - 0| < \epsilon \text{ if } 0 < |x - 0| < \delta.$$

Let us see how we may find a suitable δ when ϵ is given. In doing this we take for granted the properties of the exponential and logarithmic functions.

Since e to any power is positive, $|e^{-1/x^2} - 0| < \epsilon$ is equivalent to $e^{-1/x^2} < \epsilon$. We rewrite this inequality in several successive equivalent forms:

$$\frac{1}{e^{1/x^2}} < \epsilon, \quad \frac{1}{\epsilon} < e^{1/x^2}, \quad \log\left(\frac{1}{\epsilon}\right) < \frac{1}{x^2}.$$

Let us suppose that $\epsilon < 1$, so that $\log(1/\epsilon) > 0$. Then further equivalent forms are

$$0 < x^2 < \frac{1}{\log(1/\epsilon)}, \quad 0 < |x| < \left\{\frac{1}{\log(1/\epsilon)}\right\}^{1/2}.$$

It now appears that, if $0 < \epsilon < 1$, we can choose

$$\delta = \left\{\frac{1}{\log(1/\epsilon)}\right\}^{1/2},$$

and then (1.1–3) will hold, as required.

Even in very obvious situations it is worth while to practice finding a δ corresponding to a given ϵ, just to drive home an appreciation of the meaning of the definition of a limit.

Example 7. Given $\epsilon > 0$, find δ so that $|f(x) - 4| < \epsilon$ if $0 < |x - 2| < \delta$, where $f(x) = x^3 - x^2$. This will show that $\lim_{x \to 2}(x^3 - x^2) = 4$.

We have $\qquad x^3 - x^2 - 4 = (x - 2)(x^2 + x + 2)$.

To begin with, let us consider only values of x such that $|x - 2| < 1$, or $1 < x < 3$. For such x we certainly have $4 < x^2 + x + 2 < 14$, and hence

$$|x^3 - x^2 - 4| \leq 14|x - 2|.$$

Now we see that $|x^3 - x^2 - 4| < \epsilon$ provided $14|x - 2| < \epsilon$, or $|x - 2| < \epsilon/14$. Hence we choose for δ any positive number such that both $\delta \leq 1$ and $\delta \leq \epsilon/14$. This choice meets the requirements.

Reasoning with limits is facilitated by various simple theorems. Among the most important such theorems are the following rules, which we state here informally:

Suppose that

$$\lim_{x \to x_0} f(x) = A \text{ and } \lim_{x \to x_0} g(x) = B; \text{ then}$$

(1.1–4)
$$\lim_{x \to x_0} [f(x) + g(x)] = A + B,$$

(1.1–5)
$$\lim_{x \to x_0} [f(x)g(x)] = AB,$$

(1.1–6)
$$\lim_{x \to x_0} \frac{f(x)}{g(x)} = \frac{A}{B}, \text{ provided } B \neq 0.$$

Formal proofs of the validity of these three rules are made in § 1.64. Meanwhile we accept them and use them.

Closely related to the limit concept is the concept of *continuity*.

DEFINITION. Suppose the function f is defined at x_0 and for all values of x near x_0. Then the function is said to be *continuous* at x_0 provided that

(1.1–7)
$$\lim_{x \to x_0} f(x) = f(x_0).$$

Most of the functions which we deal with in calculus are continuous; points of discontinuity are exceptional, but may occur. A function may fail to be continuous at x_0 either because $f(x)$ does not approach any limit at all as $x \to x_0$, or because it approaches a limit which is different from $f(x_0)$.

Example 8. The function $f(x) = [x]$ (see Example 2) is discontinuous at x_0 if x_0 is an integer, but is continuous at x_0 if x_0 is not an integer.

We observe in this case that $\lim_{x \to 2} f(x)$ does not exist, for when x is near 2, $f(x) = 1$ if $x < 2$ and $f(x) = 2$ if $x > 2$. The situation is similar at other integers.

Example 9. Suppose we define a function f by

$$f(x) = [x] + [2 - x] - 1.$$

Direct inspection shows the following:

$$f(0) = 1,$$
$$f(x) = 0 \text{ if } 0 < x < 1,$$

$$f(1) = 1,$$
$$f(x) = 0 \text{ if } 1 < x < 2,$$
$$f(2) = 1.$$

Consequently, $\lim_{x \to 1} f(x) = 0$; but $f(1) = 1$, and so f is not continuous at $x = 1$.

Example 10. Let us define $f(x) = (\sin x)/x$ if $x \neq 0$. This definition of $f(x)$ has no meaning if $x = 0$, since division by 0 is undefined. However, let us make the additional definition $f(0) = 1$. With this definition, f is continuous at $x = 0$. For, as we learn in elementary calculus,

$$(1.1\text{–}8) \qquad\qquad \lim_{x \to 0} \frac{\sin x}{x} = 1.$$

Since we have defined $f(0) = 1$, (1.1–8) shows that $\lim_{x \to 0} f(x) = f(0)$; therefore f is continuous at $x = 0$, by the definition.

We have based the concept of continuity directly upon the concept of a limit. A condition for continuity of a function may be given directly in terms of inequalities, just as we defined a limit in terms of inequalities. Thus, if f is defined throughout some interval containing x_0 and all points near x_0, f is continuous at x_0 if to each positive ϵ corresponds some positive δ such that

$$(1.1\text{–}9) \qquad |f(x) - f(x_0)| < \epsilon \quad \text{whenever} \quad |x - x_0| < \delta.$$

This form of the condition for continuity is equivalent to the original definition.

Many common words are used in mathematics in a specialized way. Usually the mathematical meaning of a word has some relation to the common meaning of the word; but mathematical meanings are precise, whereas common meanings are broad or variable. The adjective "continuous" is a word of this kind, with a restrictive and precise mathematical meaning. Experience shows that students tend to read more, in the way of preconceived notions about the meaning of the term, into the word "continuous" than is implied by the definition. In analytic geometry and calculus we become familiar with the graphs of many functions, and there is a tendency to associate the term "continuous function" with the picture of a smooth, unbroken curve. Now it is true that if f is continuous at each point of an interval, the corresponding part of the graph of $y = f(x)$ will be an unbroken curve. But it need not be smooth. Smoothness is related to differentiability; the more derivatives f has, the smoother is its graph. A function may be continuous without having a derivative. In that case the graph of $y = f(x)$ might be so crinkly, so devoid of smoothness, as to make correct visualization of it quite impossible.

EXERCISES

Where the square-bracket notation occurs in these exercises, $[f(x)]$ denotes the algebraically largest integer which is $\leq f(x)$ (see Example 2).

1. Find each of the limits indicated, using algebraic simplification and the rules (1.1–4)–(1.1–6).

(a) $\lim\limits_{x \to 4} \dfrac{x^2 - 16}{x - 4}$;

(b) $\lim\limits_{x \to 1} \dfrac{x^{10} - 1}{x - 1}$;

(c) $\lim\limits_{x \to 1} \dfrac{x^n - 1}{x - 1}$ (n a positive integer);

(d) $\lim\limits_{x \to 0} \dfrac{(x + 2)^2 - 4}{x}$;

(e) $\lim\limits_{x \to 0} \dfrac{1}{x} \left(\dfrac{1}{2 + x} - \dfrac{1}{2} \right)$;

(f) $\lim\limits_{x \to 0} \dfrac{1}{x} \left(\dfrac{1}{(4 + x)^2} - \dfrac{1}{16} \right)$;

(g) $\lim\limits_{x \to 2} \dfrac{x^3 + x^2 - 5x - 2}{x^2 - 4}$;

(h) $\lim\limits_{x \to -1} \dfrac{108(x^2 + 2x)(x + 1)^3}{(x^3 + 1)^3(x - 1)}$.

2. Find each of the following limits, using roughly quantitative arguments based on your knowledge of the various functions involved (somewhat as in the first paragraph of the discussion of Example 6).

(a) $\lim\limits_{x \to 0} 2^{\log x^2}$;

(b) $\lim\limits_{x \to 0} \cos (e^{-1/x^2})$;

(c) $\lim\limits_{x \to 0} \dfrac{1 + \cos x}{1 + (\log x^2)^2}$;

(d) $\lim\limits_{x \to \pi} \tan^{-1} \left(\tan^2 \dfrac{x}{2} \right)$, where the inverse tangent has its principal value, i.e., $v = \tan^{-1} u$ means $u = \tan v$ and $-\dfrac{\pi}{2} < v < \dfrac{\pi}{2}$;

(e) $\lim\limits_{x \to 0} \log \left(\dfrac{\sin x}{x} \right)$.

3. Draw the graphs for each of the following functions and then answer the questions:

(a) $f(x) = 2x - 1$ if $x \leq 1$, $f(x) = 6 - 5x$ if $x > 1$. Is f continuous at $x = 1$?

(b) $f(x) = (x^2/2) - 2$ if $0 < x < 2$, $f(x) = 2 - (8/x^2)$ if $2 < x$. Does $\lim_{x \to 2} f(x)$ exist? How should $f(2)$ be defined to make f continuous at $x = 2$?

(c) $f(x) = [1 - x^2]$. Consider only $-1 \leq x \leq 1$. Does $\lim_{x \to 0} f(x)$ exist? Is f continuous at $x = 0$?

(d) $f(x) = (x - 1)[x]$. Consider only $0 \leq x \leq 2$. Is f continuous at $x = 1$?

(e) $f(x) = x/|x|$ (undefined if $x = 0$). Does $\lim_{x \to 0} f(x)$ exist?

4. In each of the following cases f is defined by the given formula only if $x \neq 0$. How should $f(0)$ be defined to make f continuous at $x = 0$?

(a) $f(x) = \dfrac{\sin 2x}{x}$;

(b) $f(x) = \dfrac{\tan x}{x}$;

(c) $f(x) = \dfrac{(x + 2)^3 - 8}{x}$;

(d) $f(x) = 10^{-1/x^2}$.

Exercises 5–7 form a natural unit.

5. If $f(x) = cx$, where c is a constant, show that $\lim_{x \to x_0} f(x) = f(x_0)$ by

applying the definition of a limit as expressed by (1.1–2). If $c \neq 0$ what can you take δ to be in terms of ϵ and c?

6. If c is constant and n is a positive integer, show that $\lim_{x \to x_0} cx^n = cx_0^n$. Use Exercise 5, mathematical induction, and (1.1–5).

7. By a *polynomial* in x we mean a function defined by an expression

$$P(x) = a_0 x^n + a_1 x^{n-1} + \cdots + a_n$$

where the coefficients $a_0, a_1, \cdots, a_n$ are constants, and n is an integer ≥ 0. If $n = 0$, $P(x)$ is constant in value, and the degree of $P(x)$ is said to be zero. If $n \geq 1$ and $a_0 \neq 0$, we say that the degree of the polynomial is n. Prove that $P(x)$ is continuous at every point x_0. Use the result of Exercise 6. What other result about limits do you use?

8. By a rational function of x we mean a function defined by an expression

$$R(x) = \frac{p(x)}{P(x)},$$

where $p(x)$ and $P(x)$ are polynomials (see Exercise 7). The function is defined except when $P(x) = 0$. Show that it is continuous at x_0 if it is defined there. Use definition (1.1–7) and state exactly what appeal you make to facts about limits stated in the text or established in previous exercises.

9. If $f(x) = \sin \dfrac{1}{x}$, (a) find $f\left(\dfrac{1}{n\pi}\right)$, $n = 1, 2, \cdots$; (b) find $f\left(\dfrac{2}{n\pi}\right)$, $n = 1$,

5, 9, $\cdots$; (c) find $f\left(\dfrac{2}{n\pi}\right)$, $n = 3, 7, 11, \cdots$. (d) How does the derivative $f'(x)$ behave as $x \to 0$?

10. (a) How does $2^{-1/x}$ behave as $x \to 0+$? (b) as $x \to 0-$? (c) What can you say about $\lim\limits_{x \to 0} \dfrac{1}{1 + 2^{-1/x}}$?

11. Graph each of the following functions: (a) $|x|$, (b) $|x - 1|$, (c) $|x + 2|$, (d) $|x^3|$, (e) $|1 - x^2|$. Do any of these functions have any points of discontinuity?

12. Graph the function $x - [x]$ and discuss its discontinuities.

13. Which of the following functions is continuous at $x = 0$? (a) $[x^2 + 2]$, (b) $[4 - x^2]$, (c) $[x^2 - 1]$. Graph each function when $-1 \leq x \leq 1$.

14. If $f(x) = \dfrac{[\frac{3}{2} + x] - [\frac{3}{2}]}{x}$, (a) find $f(1)$, $f(-1)$, $f(\frac{1}{2})$, $f(-\frac{1}{2})$, $f(\frac{1}{4})$, $f(-\frac{1}{4})$.
(b) Without using the square brackets, write expressions for $f(x)$ if $0 < x < \frac{1}{2}$ and if $-\frac{1}{2} < x < 0$. (c) What is $\lim_{x \to 0} f(x)$?

15. If $f(x) = \dfrac{|x^2 - 1|}{x - 1}$, (a) find $f(\frac{1}{2})$, $f(\frac{3}{4})$, $f(\frac{7}{8})$, $f(\frac{9}{8})$, $f(\frac{5}{4})$, $f(\frac{9}{8})$. (b) Express $f(x)$ without absolute values if $x > 1$; if $0 < x < 1$. (c) What can you say about $\lim_{x \to 1} f(x)$?

16. If $f(x) = \dfrac{|2 + x| - |x| - 2}{x}$, (a) find $f(1)$, $f(-1)$, $f(-2)$, $f(2)$. (b) Write an expression for $f(x)$ without absolute values if $0 < x$; if $-2 < x < 0$. (c) What can you say about $\lim_{x \to 0} f(x)$?

17. If $f(x) = [7 x^2 - 14]$, (a) find $f(0)$, $f(1)$, $f(2)$, $f(\frac{1}{2})$, $f(\frac{3}{2})$, $f(\frac{5}{3})$. (b) Is f continuous at $x = 0$? (c) Is it continuous at $x = 1$? (d) at $x = \sqrt{2}$?

18. If $f(x) = [\sin x]$, (a) find $f(0)$, $f(\pi/2)$, $f(-\pi/2)$, $f(\pi/4)$, $f(-\pi/4)$. (b) Does $\lim_{x \to 0} f(x)$ exist? (c) What does $f(x)$ approach as $x \to (\pi/2) -$? (d) as $x \to 0 -$?

19. Prove that $\lim_{x \to 1} (x^2 + 2 x) = 3$ by finding δ in terms of a given positive ϵ so that $|x^2 + 2 x - 3| < \epsilon$ if $|x - 1| < \delta$.

20. Show that $\left| \dfrac{1}{x^2 + 16} - \dfrac{1}{25} \right| \leq \dfrac{7}{500} |x - 3|$ if $2 < x < 4$. Hence, for any $\epsilon > 0$, find δ so that $\left| \dfrac{1}{x^2 + 16} - \dfrac{1}{25} \right| < \epsilon$ if $|x - 3| < \delta$, thus proving directly that $\lim_{x \to 3} \dfrac{1}{x^2 + 16} = \dfrac{1}{25}$.

21. Show that $|(1 + x)^3 - 1| \leq 7 |x|$ if $-1 < x < 1$. Then prove by the definition (1.1–2) that $\lim_{x \to 0} (1 + x)^3 = 1$ (i.e., find a suitable δ for any given positive ϵ).

22. Show that $\left| \dfrac{1}{x} \left(\dfrac{1}{2 + x} - \dfrac{1}{2} \right) + \dfrac{1}{4} \right| < \epsilon$ if $0 < |x| < \delta$, where δ is the smaller of the numbers 1, 4ϵ (ϵ being positive). Translate this situation into a statement of the form $\lim_{x \to x_0} f(x) = A$, specifying what you take for f, x_0, and A.

23. Show that, if $0 < \epsilon < 1$, $10^{-1/x} < \epsilon$ when $0 < x < \left(\log_{10} \dfrac{1}{\epsilon} \right)^{-1}$. What is the corresponding statement about a limit?

24. Does $\lim_{x \to 0} \dfrac{2^{1/x} + 3}{2^{1/x} + 1}$ exist?

25. Suppose that a function f is defined by setting $f(x) = 1/n$ if $x = 1/2^n$, where $n = 1, 2, 3, \cdots$, and $f(x) = 0$ for all other values of x. Is f continuous at $x = 0$?

1.11   Derivatives

Elementary calculus deals with the processes of differentiation and integration, the techniques of these processes as they pertain to various common functions, and the applications of the processes to problems of geometry, physics, and other sciences. Let us examine the concept of the derivative.

Consider a function f, defined for values of the variable x in the interval $a < x < b$. Let x_0 be any fixed point of the interval, and consider the ratio

(1.11–1)
$$\frac{f(x) - f(x_0)}{x - x_0},$$

where $x \neq x_0$ and x is a variable point of the interval. The ratio (1.11–1) is called a difference quotient.

DEFINITION. If the difference quotient (1.11–1) approaches a limit as x approaches x_0, the limit is called the *derivative* of f at $x = x_0$, and is denoted by $f'(x_0)$. Thus, by definition,

$$(1.11-2) \qquad f'(x_0) = \lim_{x \to x_0} \frac{f(x) - f(x_0)}{x - x_0},$$

provided the limit exists.

Quite likely the student is familiar with another notation in connection with this definition. Sometimes we write $x = x_0 + \Delta x$, and then (1.11–2) takes the form

$$(1.11-3) \qquad f'(x_0) = \lim_{\Delta x \to 0} \frac{f(x_0 + \Delta x) - f(x_0)}{\Delta x}.$$

Here the symbol Δx denotes an independent variable. For many algebraic calculations it is convenient to use h in place of Δx. Also, we may drop the zero subscript; we then have the definition

$$(1.11-4) \qquad f'(x) = \lim_{h \to 0} \frac{f(x + h) - f(x)}{h},$$

provided the limit exists.

In addition to the notation $f'(x)$ for the derivative we frequently use the notation $\dfrac{d}{dx} f(x)$.

Example 1. Using form (1.11–2), calculate $f'(x_0)$ if $f(x) = x^2$. Here

$$f(x) - f(x_0) = x^2 - x_0{}^2 = (x - x_0)(x + x_0);$$
$$f'(x_0) = \lim_{x \to x_0} (x + x_0) = 2 x_0.$$

Example 2. Using form (1.11–4), calculate $f'(x)$ if $f(x) = 1/x$. Here

$$f(x + h) - f(x) = \frac{1}{x + h} - \frac{1}{x} = \frac{-h}{(x + h)x};$$
$$f'(x) = \lim_{h \to 0} \frac{-1}{(x + h)x} = -\frac{1}{x^2}.$$

DEFINITION. A function which has a derivative at a certain point is said to be *differentiable* at that point.

In the definition (1.11–2) we were assuming that f was defined in an interval extending some distance on each side of the point x_0. It is understood that x may approach x_0 from either side, and that the limit of the difference quotient is the same when x approaches x_0 from the left as when the approach is from the right.

It is useful to define *one-sided* derivatives. Using the notation for limits from the right and left, respectively, as explained just prior to Example 5 in § 1.1, we define the right-hand derivative $f'_+(x_0)$ and the left-hand derivative $f'_-(x_0)$ as follows:

$$(1.11-5) \qquad f'_+(x_0) = \lim_{x \to x_0+} \frac{f(x) - f(x_0)}{x - x_0},$$

$$(1.11-6) \qquad f'_-(x_0) = \lim_{x \to x_0-} \frac{f(x) - f(x_0)}{x - x_0},$$

provided the limits exist.

If in discussing a function we confine our attention wholly to an interval $a \leq x \leq b$, then we shall understand that $f'(a)$ means $f'_+(a)$, and that $f'(b)$ means $f'_-(b)$. If $a < x_0 < b$, however, and if the function is differentiable at x_0, then we must have $f'_+(x_0) = f'_-(x_0)$. The derivative $f'(x_0)$ is then the common value of the two one-sided derivatives.

Example 3. Let $f(x) = \sqrt{2(1 - \cos 2x)}$. Show that this function is not differentiable at the points $x = 0, \pm \pi, \pm 2\pi, \cdots$ and find the one-sided derivatives at these points.

We recall the trigonometric identity

$$1 - \cos 2x = 2 \sin^2 x.$$

Thus*

$$f(x) = \sqrt{4 \sin^2 x} = 2 |\sin x|.$$

This means that $f(x) = 2 \sin x$ when $\sin x \geq 0$; but $f(x) = -2 \sin x$ when $\sin x < 0$. The graph of $f(x)$ is shown in Fig. 2. The dotted portions repre-

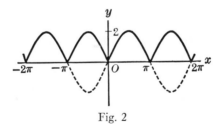

Fig. 2

sent the function $2 \sin x$ when $\sin x < 0$. From the symmetry of the figure we see that the situation at all the points $x = n\pi$ $(n = 0, \pm 1, \cdots)$ is the

---

*In this book we adhere to the standard convention that if $A \geq 0$, $\sqrt{A}$ *means the nonnegative square root of A.* According to this convention $\sqrt{a^2} = a$ if $a \geq 0$, but $\sqrt{a^2} = -a$ if $a < 0$. Both cases are covered by the formula $\sqrt{a^2} = |a|$. Finally, $A^{1/2}$ and $\sqrt{A}$ are merely different notations for the same thing.

same. The right-hand derivative at each of these points is 2, and the left-hand derivative is -2:

$$f'_+(0) = \lim_{x \to 0+} \frac{2 \sin x - 0}{x - 0} = 2,$$

$$f'_-(0) = \lim_{x \to 0-} \frac{-2 \sin x - 0}{x - 0} = -2.$$

We take it for granted that readers of this book are acquainted with the interpretation of the derivative $f'(x)$ as the slope of the curve $y = f(x)$ when we employ a graphical representation in rectangular co-ordinates with equal scales on the two axes (see Fig. 3). To say that f is differentiable at a point means geometrically that the curve $y = f(x)$ has at that point a unique tangent line which is not parallel to the y-axis.

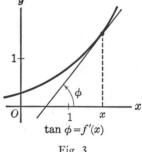

$$\tan \phi = f'(x)$$

Fig. 3

We also take it for granted that the student is familiar with the interpretation of the derivative as an instantaneous rate of change. Without going into detail we emphasize the fact that the concepts of velocity, acceleration, and all kinds of instantaneous rates of change find their precise mathematical formulation in terms of the notion of the derivative of a function.

Students beginning this book are expected to know the general rules of differentiation, including the rules for dealing with sums, products, and quotients.

It will be convenient to draw up a list of differentiation formulas.

GENERAL RULES

(1.11–7) $$\frac{d}{dx}[f(x) + g(x)] = f'(x) + g'(x).$$

(1.11–8) $$\frac{d}{dx}[cf(x)] = cf'(x) \ (c \text{ constant}).$$

(1.11–9) $$\frac{d}{dx}[f(x)g(x)] = f(x)g'(x) + f'(x)g(x).$$

(1.11–10) $$\frac{d}{dx}\left[\frac{f(x)}{g(x)}\right] = \frac{g(x)f'(x) - f(x)g'(x)}{[g(x)]^2}. \qquad [g(x)]^2 \neq 0$$

Each of these rules is in fact a theorem. We understand that the functions $f(x)$, $g(x)$ are defined on some interval $a < x < b$. Rule (1.11–7), when stated more fully as a theorem, may be expressed as follows: *If $f(x)$ and $g(x)$ are differentiable for a particular value of x, then their sum is also* *differentiable for this value of x, and* (1.11–7) *holds*. The student should amplify each of the rules (1.11–8)–(1.11–10) into a formally stated theorem in the same manner. What special provision must be made in connection with (1.11–10)?

We mentioned at the end of § 1.1 that a function may be continuous and yet not differentiable. For instance, the function of Example 3 is continuous for all values of x, but it is not differentiable at the points $n\pi, n = 0, \pm 1, \pm 2, \cdots$. However, differentiability *does* imply continuity as the following theorem shows.

THEOREM I. *If f is differentiable at x_0, it is continuous there.*

PROOF.   When $x \neq x_0$ we can write

$$f(x) = \frac{f(x) - f(x_0)}{x - x_0} (x - x_0) + f(x_0).$$

Then, by (1.1–4) and (1.1–5),

$$\lim_{x \to x_0} f(x) = f'(x_0) \cdot 0 + f(x_0) = f(x_0).$$

This completes the proof.

The student must already be accustomed to using the rule for differentiating a composite function (sometimes called the chain rule). By a *composite* function we mean a function formed by substitution of one function in place of the independent variable in another function:

$$(1.11–11) \qquad\qquad F(t) = f[g(t)].$$

To fix the ideas precisely, suppose that g is defined when $\alpha < t < \beta$, and that the values of the function satisfy the inequality $a < g(t) < b$. Suppose that f is defined when $a < x < b$. Then, replacing x by $g(t)$, we obtain the composite function F defined by (1.11–11).

THEOREM II. *Suppose g is differentiable at a point t_0 of the interval $\alpha < t < \beta$. Let $x_0 = g(t_0)$, and suppose that f is differentiable at x_0. Then the composite function F is differentiable at t_0, and*

$$(1.11–12) \qquad\qquad F'(t_0) = f'(x_0)g'(t_0).$$

In elementary calculus this theorem is often expressed symbolically in a different way, by writing

$$y = f(x), x = g(t).$$

Then $\qquad\qquad\qquad \dfrac{dy}{dt} = \dfrac{dy}{dx} \cdot \dfrac{dx}{dt} .$

Example 4.   Suppose $f(x) = x^{17}$, $g(t) = t - t^2$. Then, *if $F(t) = f(g(t))$*

$$F(t) = (t - t^2)^{17} \quad \text{and} \quad F'(t) = 17(t - t^2)^{16}(1 - 2t).$$

We accept Theorem II as known from elementary calculus. The proof is a somewhat delicate matter, however, and the student who wishes to study the proof will find a discussion of it in Exercise 26 at the end of this section.

Throughout calculus there are two aspects of the development of the subject. On the one hand we formulate concepts and rules applicable to *arbitrary* functions having certain properties. Theorems I and II are of this type. On the other hand there are the *particular* functions which we deal with as illustrations and in all practical applications, e.g.,

$$(1 - x^2)^{1/2}, \sin 2x, \tan^{-1} x, \log x, e^{-x^2/2},$$

and many others. We assume that the student knows the formulas for differentiation of the standard elementary functions, and in general we shall regard all such functions as being available for illustrative purposes.

In order to illustrate the possibility of various kinds of situations which do not ordinarily arise with the standard elementary functions, we sometimes resort to the contrivance of functions specifically defined so as to exhibit some peculiarity. Such specially contrived functions serve to help the student appreciate the generality of the concept of a function. They also teach him to be wary of tacitly assuming more than is implied in a given definition or hypothesis.

Example 5.   Let a function be defined as follows:

$$f(x) = x^2 \sin \frac{1}{x} \text{ if } x \neq 0,$$
$$f(0) = 0.$$

We shall see that this function is differentiable for all values of x, but that the calculation of its derivative at $x = 0$ requires special attention.

We note first of all that the formula by which $f(x)$ is defined when $x \neq 0$ cannot be used when $x = 0$, since $1/x$ is then undefined. Hence the assignment of the value of $f(0)$ may be made as we choose. The value $f(0) = 0$ is chosen because this makes the function continuous at $x = 0$; that is,

$$(1.11–13) \qquad\qquad \lim_{x \to 0} x^2 \sin \frac{1}{x} = 0.$$

The correctness of this result is seen from the inequality

$$(1.11–14) \qquad\qquad \left| x^2 \sin \frac{1}{x} \right| \leqq x^2,$$

which holds since the value of the sine function never exceeds unity.

To find the derivative, we follow standard procedures in writing

$$f'(x) = x^2 \cos \frac{1}{x} \cdot \left(\frac{-1}{x^2} \right) + 2 \, x \sin \frac{1}{x}$$

(1.11–15) $$f'(x) = - \cos \frac{1}{x} + 2 \, x \sin \frac{1}{x} \cdot$$

This result is correct when $x \neq 0$. If $x = 0$, however, the foregoing procedure for finding $f'(x)$ is not valid, for it is based on the rule for the derivative of the product of two functions, namely x^2 and $\sin (1/x)$; the second of these functions is not defined at $x = 0$, and cannot be defined there so as to be differentiable.

As yet, then, we do not know whether $f(x)$ is differentiable at $x = 0$. Now, by definition,

$$f'(0) = \lim_{x \to 0} \frac{f(x) - f(0)}{x - 0},$$

provided the limit exists. Since $f(0) = 0$,

$$\frac{f(x) - f(0)}{x - 0} = \frac{f(x)}{x} = x \sin \frac{1}{x},$$

and we see that

(1.11–16) $$f'(0) = \lim_{x \to 0} x \sin \frac{1}{x} = 0.$$

It is worth pointing out that $f'(x)$ is not continuous at $x = 0$; for from (1.11–15) we see that as $x \to 0$, $f'(x)$ approaches no limit but oscillates infinitely often from -1 to $+1$.

A graph is helpful in visualizing the nature of the function f. The student should construct such a graph, using the method of multiplication of ordinates. The curve $y = f(x)$ oscillates between the curves $y = x^2$, $y = - x^2$, crossing the axes at the points $\pm \dfrac{1}{\pi}, \pm \dfrac{1}{2 \pi}, \pm \dfrac{1}{3 \pi}, \cdots$.

In concluding this section we point out a certain principle of reasoning about limits which we used in arriving at (1.11–13) and (1.11–16). It is the following:

If two functions, F, G satisfy an inequality of the form

(1.11–17) $$A \leq F(x) \leq G(x),$$

where A is some fixed number, and if $\lim_{x \to x_0} G(x) = A$, *then* $\lim_{x \to x_0} F(x) = A$ *also.*

For example, in applying this principle to arrive at (1.11–16) we put $F(x) = | x \sin (1/x) |$, $G(x) = | x |$, $A = 0$, $x_0 = 0$. The principle just stated is a special case of Theorem XII, § 1.61. Its truth is an immediate consequence of the definition of a limit.

EXERCISES

1. (a) If $f(x) = x^n$, where n is a positive integer, compute $f'(x_0)$, using (1.11–2) and a factorization theorem. (b) Compute $f'(x)$, using (1.11–4) and the binomial theorem.

2. Suppose p and q are positive integers without a common factor. Let $f(x) = x^{p/q}$. Suppose x, x_0 are positive, and write $u = x^{1/q}$, $u_0 = x_0^{1/q}$. Verify that

$$\frac{f(x) - f(x_0)}{x - x_0} = \frac{u^p - u_0^p}{u^q - u_0^q}.$$

Proceed from here to show directly that $f'(x_0) = nx_0^{n-1}$, where $n = p/q$.

3. Let $f(x) = x^c$, where $x > 0$ and c is irrational. Since c cannot be expressed as a ratio of integers, the method of Exercise 2 is not available for calculating $f'(x)$. However, assuming as known the differentiability properties of the exponential and logarithmic functions, show that the formula $f'(x) = cx^{c-1}$ may be deduced from the fact that $f(x) = e^{c \log x}$.

4. Let $f(x) = \sin x$, $g(x) = \cos x$. Show that finding $f'(0)$ is the same as finding $\lim_{x \to 0} \dfrac{\sin x}{x}$, and that finding $g'(0)$ is the same as finding $\lim_{x \to 0} \dfrac{\cos x - 1}{x}$. Taking for granted that these limits have values 1, 0 respectively, deduce the formula $f'(x) = \cos x$, using (1.11–4) and the expansion formula for $\sin (x + h)$.

5. Show that the formula $g'(x) = -\sin x$ may be derived from the relations $f'(x) = \cos x$, $\cos x = \sin \left(\dfrac{\pi}{2} - x \right)$. ($f$ and g are defined in Exercise 4.) What theorem do you use?

6. Let $f(x) = \log_e x$. Using only the definition of the derivative and standard properties of the logarithm function, explain why $f'(1) = \lim_{h \to 0} \log (1 + h)^{1/h}$.

7. If $f(x) = \log x$, show that

$$\frac{f(x + h) - f(x)}{h} = \frac{1}{x} \frac{f(1 + t) - f(1)}{t},$$

where $t = h/x$. Hence explain why f is differentiable at x if it is differentiable at 1, and show that $f'(x) = \dfrac{f'(1)}{x}$.

8. Let $f(x) = e^x$. Explain why $f'(0) = \lim_{x \to 0} \dfrac{e^x - 1}{x}$. Show that, if $f'(0) = 1$ is known, one can deduce $f'(x) = e^x$ with the aid of the laws of exponents. Start from (1.11–4).

9. The radius of a sphere is being increased at a variable rate. This rate is 2 centimeters per second when the radius is 5 centimeters. Find the rate of change of the volume of the sphere, in cubic centimeters per second, at this particular moment.

10. A rocket is being launched straight upward from the earth. It burns liquid fuel at a variable rate, the rate being N gallons per mile when the rocket is 10 miles high. If the speed of the rocket at this time is 1000 miles per hour, what is the instantaneous rate of fuel consumption in gallons per hour? Let x be the altitude of the rocket t hours after launching. Suppose the rate of fuel consumption is $kx^{-1/2}$ gallons per mile and $3\,k(ct)^{1/2}$ gallons per hour where c and k are positive constants. Find a formula for the rate of rise of the rocket, and deduce the formula connecting x and t.

11. If f is the function of Example 5 and $F(t) = f(t^2 - 1)$, find $F'(1)$.

12. Show that $f(x) = |\,x\,|$ is not differentiable at $x = 0$. Is it continuous? Find $f'_+(0)$ and $f'_-(0)$.

13. Let $f(x) = e^{-|x|}$. **(a)** Graph this function. **(b)** Is it continuous at $x = 0$? **(c)** Is it differentiable there?

14. Let $f(x) = x\,|\,x\,|$. **(a)** Graph the function. **(b)** Find $f'(x)$ if $x > 0$; if $x = 0$; if $x < 0$. **(c)** Is the derivative f' differentiable at $x = 0$?

15. (a) If $f(x) = [x]$, compute $f'(\tfrac{3}{2})$ by (1.11–4). **(b)** How does Theorem I show that f is not differentiable at $x = 2$? **(c)** What is the value of $f'_+(2)$? **(d)** How does $\dfrac{f(2 + h) - f(2)}{h}$ behave as $h \to 0-$?

16. Discuss the continuity and differentiability at $x = 0$ of f, where $f(x) = x \sin (1/x)$ when $x \neq 0$, $f(0) = 0$.

17. Show that the function defined as $f(x) = x^3 \sin (1/x)$ if $x \neq 0$, $f(0) = 0$, has a derivative for all values of x, and that f' is continuous at $x = 0$ but not differentiable there.

18. (a) For what values of the exponent n (an integer) will $f'(x)$ exist at $x = 0$ if $f(x) = x^n \sin (1/x^2)$ when $x \neq 0$, and $f(0) = 0$? **(b)** For what values of n will f' be continuous at $x = 0$? **(c)** For what values of n will f' be differentiable at $x = 0$?

19. Let $f(x) = \dfrac{x}{1 + e^{1/x}}$ if $x \neq 0$, $f(0) = 0$. Does $f'(0)$ exist? Sketch the graph near $x = 0$, showing the directions from which a point approaches the origin along the curve.

20. Discuss the differentiability at $x = 1$ of $f(x) = (x - 1)[x]$. Draw the graph when $0 \leq x \leq 2$.

21. If $f(x) = [x] + (x - [x])^{1/2}$, sketch the graph when $0 \leq x \leq 3$. What can you say about continuity and differentiability of f at $x = 1$ and $x = 2$? Write a formula for $f(x)$ without square brackets when $0 < x < 1$, and use it to find $f'(\tfrac{1}{2})$.

22. Let f be a function which is defined for all x, with the properties (i) $f(a + b) = f(a)f(b)$, (ii) $f(0) = 1$, (iii) f is differentiable at $x = 0$. Show that f is differentiable for all values of x, and that $f'(x) = f'(0)f(x)$.

23. Let functions f and g be defined for all x and possess the following properties: (i) $f(x + y) = f(x)g(y) + f(y)g(x)$, (ii) f and g are differentiable at $x = 0$, with $f(0) = 0$, $f'(0) = 1$, $g(0) = 1$, $g'(0) = 0$. Show that f is differentiable for all values of x, with $f''(x) = g(x)$. If it is also known that $g(x + y) = g(x)g(y) - f(x)f(y)$, show that g is differentiable for all values of x, with $g'(x) = -f(x)$.

24. Suppose $f\left(\dfrac{1}{2^n}\right) = \dfrac{1}{2^{2n}}$, $n = 1, 2, \cdots$, and $f(x) = 0$ for all other values of x. Is f differentiable at $x = 0$? What is the situation if instead we define $\left(\dfrac{1}{2^n}\right) = \dfrac{1}{2^{n+1}}$?

25. Construct proofs for rules (1.11–7)–(1.11–10), using (1.11–4).

26. Consider the following remarks on the proof of Theorem II. If it were true that $g(t) \neq g(t_0)$ when $t \neq t_0$, we could prove (1.11–12) starting from the following formula:

$$\frac{F(t) - F(t_0)}{t - t_0} = \frac{f[g(t)] - f[g(t_0)]}{g(t) - g(t_0)} \cdot \frac{g(t) - g(t_0)}{t - t_0}.$$

Why does $g(t) \to g(t_0)$ as $t \to t_0$? What are the limits of the quotients on the right as $t \to t_0$? Refer to (1.1–5) and draw a conclusion.

This kind of argument is not adequate if $g(t) = g(t_0)$ for infinitely many values of t approaching t_0. To avoid the difficulty, define a new function in a somewhat artificial manner:

$$\phi(t) = \frac{f[g(t)] - f[g(t_0)]}{g(t) - g(t_0)} \text{ if } g(t) \neq g(t_0), \ \phi(t) = f'(x_0) \text{ if } g(t) = g(t_0).$$

Explain carefully why ϕ is continuous at t_0. Refer to (1.1–7) and explain how to use ϕ to prove (1.11–12).

1.12   Maxima and minima

One of the important things about the derivative is that it helps us to locate the relative maxima and minima of a function. Let us formulate exactly what can be said about such things.

It is necessary to say what we mean by an *open interval* of the x-axis. If a and b are numbers such that $a < b$, all numbers x such that $a < x < b$ form what is called the open interval from a to b, or more briefly, the open interval (a, b). The end points a, b do not belong to the open interval. By contrast, the set of all numbers x such that $a \leq x \leq b$ is called a *closed interval*; here the end points belong to the interval. We shall denote closed intervals by the use of square brackets: $[a, b]$; for open intervals we shall use ordinary parentheses: (a, b). By a *neighborhood* of x_0 we mean an open interval containing x_0.

DEFINITION.   Let f be a function which is defined on an open interval (a, b), and let x_0 be a point of the interval. We say that f has a relative maximum at x_0 if there is some neighborhood of x_0 (say (a_1, b_1)), where

$a_1 < x_0 < b_1$) contained in (a, b) and containing x_0 such that $f(x) \leq f(x_0)$ if $a_1 < x < b_1$. This means that $f(x_0)$ is *at least* as large (algebraically) as $f(x)$ at all points x for some distance *on either side* of x_0 (see Fig. 4).

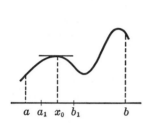

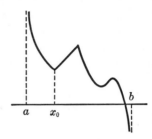

Fig. 4. *A relative maximum*   Fig. 5. *A relative minimum*

A similar definition is made for a relative minimum, the inequality being reversed: $f(x) \geq f(x_0)$ when x is near x_0.

Both a relative maximum and a relative minimum are covered by the term "a relative *extremum*."

THEOREM III. *Suppose that f has a relative extremum at the point x_0 of the open interval (a, b), and suppose that f is differentiable at x_0. Then $f'(x_0) = 0$.*

PROOF. For definiteness assume that f has a relative maximum at x_0. Consider the one-sided derivatives at x_0, and bear in mind that $f(x) \leq f(x_0)$ when x is sufficiently near x_0. Then $f(x) - f(x_0) \leq 0$; accordingly

$$\frac{f(x) - f(x_0)}{x - x_0} \leq 0 \text{ when } x > x_0$$

and so (see (1.11–5)) $f'_+(x_0) \leq 0.$

Likewise $\frac{f(x) - f(x_0)}{x - x_0} \geq 0 \text{ when } x < x_0,$

so that $f'_-(x_0) \geq 0$ (see (1.11–6)). But, since f is differentiable at x_0, we have $f'(x_0) = f'_+(x_0) = f'_-(x_0)$. Therefore $f'(x_0) = 0$, for it is neither positive nor negative. The proof for the case of a relative minimum is entirely similar.

It may happen, of course, that a function has a relative extremum at a point x_0, but is not differentiable there. This happens with $f(x) = 1 - x^{2/3}$ at $x = 0$, and with $f(x) = |x - 1|$ at $x = 1$. See Fig. 5, also.

The proof of Theorem III rests on the following reasoning about limits:

If a variable quantity is $\leqq 0$ and approaches the limit A, then $A \leqq 0$; likewise, if the variable quantity is $\geqq 0$ and approaches the limit B, then $B \geqq 0$. This principle is considered further in § 1.61 (Theorem X).

In defining a relative extremum we compared the value $f(x_0)$ with values $f(x)$ at points x *on both sides of* x_0. Theorem III applies only when x_0 is an *interior* point of the interval on which we are examining the values of f. By contrast, let us consider a function which is defined only when $a \leqq x \leqq b$, and suppose that $f(a)$ is greater than $f(x)$ when x

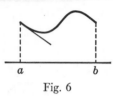

Fig. 6

is near a on the right. Then, if the right-hand derivative $f'_+(a)$ exists, we can infer that $f'_+(a) \leqq 0$, but not that $f'_+(a) = 0$ (see Fig. 6). We leave it for the student to draw the appropriate conclusion about $f'_-(b)$ if $f(x) \leqq f(b)$ when x is near b on the left.

Of course, the mere fact that $f'(x_0) = 0$ does not guarantee that f has a relative extremum at x_0. This is illustrated by $f(x) = x^3$ at $x = 0$, where the graph has a horizontal tangent but the function has neither a relative maximum nor a relative minimum. If f is differentiable at x_0, $f'(x_0) = 0$ is a *necessary*, but not a *sufficient*, condition for f to have a relative extremum at x_0.

There are tests which are sufficient, but not necessary, for a relative extremum at x_0, and which at the same time provide a means of distinguishing a relative maximum from a relative minimum. We postpone consideration of such tests to later sections. See, for instance, Example 8, § 1.2.

In many problems we are interested in a function which is defined over some given interval, and we wish to find the largest (or smallest) value which the function assumes on the given interval. The interval may be closed, e.g., $0 \leqq x \leqq 4$, or open, e.g., $1 < x < 3$, or neither. Examples of intervals which are neither open nor closed are furnished by inequalities such as $0 < x \leqq 10$ and $2 \leqq x < 8$ (open at one end and closed at the other). Also, we may be interested in finding the greatest or least value of $f(x)$ for an infinite range of values of x. For example, let x denote the altitude of a right circular cone circumscribed about a sphere of radius b, and let $f(x)$ denote the volume of the cone. One finds without much trouble that

$$f(x) = \frac{\pi}{3} \left(\frac{b^2 x^2}{x - 2b} \right).$$

The significant values of x are those for which $x > 2b$, since if $x \leqq 2b$ there can be no cone of altitude x circumscribed about the sphere. For further consideration of this problem see Exercise 10, page 26.

There may or may not be an absolute maximum or an absolute minimum in a given situation. This will depend on the nature of the function and the interval which are involved.

Example 1. $f(x) = x$; interval $0 < x \leq 10$.

Here there is no absolute minimum, since $f(x)$ can be as near 0 as we please, but never attains that value on the specified interval. There is an absolute maximum, occurring at $x = 10$.

Example 2. $f(x) = x^2$; interval $-1 \leq x < 2$.

Here there is an absolute minimum at $x = 0$; there is no absolute maximum on the interval, for $f(x)$ can approach but never reach the value 4 when x is restricted to the specified interval.

Example 3. $f(x) = \tan x$; interval $-\pi/2 < x < \pi/2$.

Here there is neither absolute maximum nor absolute minimum.

There is a very important theorem to the following effect:

If f is defined and continuous at each point of the finite closed interval $a \leq x \leq b$, then at some point of the interval $f(x)$ attains an absolute maximum value. Likewise, at some point of the interval $f(x)$ attains an absolute minimum value. This theorem is taken up carefully and proved in § 3.2. For the present we shall accept the theorem and strive to appreciate its usefulness. The requirement that the interval be finite and closed is quite essential, for in the absence of these limitations $f(x)$ might not attain any absolute extreme values, as we see by Examples 1–3.

In practice the functions we are interested in are usually differentiable at all points of the interval (there may sometimes be isolated exceptional points). If an *absolute* extreme value occurs at an interior point of the interval, it is also a *relative* extreme value in the sense of Theorem III, and therefore we must have $f'(x) = 0$ at the point, provided f is differentiable. We therefore have the following guiding principle in searching for points at which $f(x)$ can attain an absolute maximum or minimum value: *Suppose f is differentiable on the given interval, except perhaps at a finite number of points, and suppose it is known that an absolute maximum (or minimum) value is actually attained. Then the point of attainment is either*

                (a) a point where $f'(x) = 0$,
                (b) a point at one end of the interval,
or   *(c) a point where f is not differentiable.*

In the common type of problem studied in elementary calculus, the solution is usually found under *(a)*. In fact, it usually happens with physical or geometrical problems that there is only one interior point of the given interval where $f'(x) = 0$. Solutions under (b) do occur sometimes, even in physical problems, and a carefully reasoned solution should always take account of the situation at the ends of the interval, perhaps even before computing $f'(x)$. The situation (c) may occur also, but this will be more rare in common practice.

Example 4. Find a number x between 0 and 1 such that $f(x) = \dfrac{2}{x} + \dfrac{8}{1-x}$ is as small as possible.

We observe that $f(x) > 0$ when $0 < x < 1$; f is continuous in the specified open interval. Also, $f(x)$ becomes very large (in fact $f(x) \to +\infty$) as x approaches either end of the interval. We con-
clude that the graph of $y = f(x)$ near the ends of the interval has an appearance somewhat as shown in Fig. 7. It follows from this reasoning that if we choose a closed interval $a \leq x \leq b$, with $a > 0$ and very near 0, and $b < 1$ and very near 1, the function f will have smaller values in the interior of the interval $[a, b]$ than it has in the rest of the interval $(0, 1)$. Since f is continuous on the finite closed interval $[a, b]$, it must attain a value at some point of $[a, b]$ which is an ab-

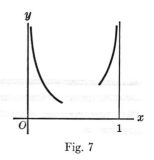

Fig. 7

solute minimum among all the values occurring on the interval. This absolute minimum will also be an absolute minimum among all the values of f occurring on the open interval $(0, 1)$. Now f is differentiable in $(0, 1)$; hence the required point of absolute minimum must be a point at which $f'(x) = 0$. We therefore proceed to compute the derivative and solve the equation $f'(x) = 0$:

$$f'(x) = \frac{-2}{x^2} + \frac{8}{(1-x)^2} = \frac{6x^2 + 4x - 2}{x^2(1-x)^2},$$

$$3x^2 + 2x - 1 = 0, \quad x = \tfrac{1}{3} \text{ or } x = -1.$$

The only solution in the interval $(0, 1)$ is $x = \tfrac{1}{3}$. We conclude that $f(x)$ attains its minimum value at $x = \tfrac{1}{3}$, the minimum value is 18. Observe that no test, by second derivatives or otherwise, is necessary to distinguish between a maximum and a minimum in this case, since we know that a minimum value exists, and there is only one point in the interval at which $f'(x) = 0$.

We emphasize that the assurance of the existence of an absolute minimum in Example 4 is based on use of the theorem cited following Example 3 on page 23. Likewise, in Example 5 (following), the existence of an absolute maximum is assured by the same theorem.

Example 5. If $f(x) = x(4 - x) \{(x - 2)^2 + 2\}$, find the absolute maximum value of $f(x)$ when $0 \leq x \leq 4$.

Here we see that f is differentiable (and continuous) for all values of x. Moreover, $f(0) = f(4) = 0$, and $f(x) > 0$ if $0 < x < 4$. There must be an absolute maximum for f somewhere on the closed interval $[0, 4]$, and it clearly does not occur at either end of the interval. Hence it must occur at some interior point where $f'(x) = 0$. A simple calculation shows that

$$f(x) = -x^4 + 8 x^3 - 22 x^2 + 24 x,$$
$$f'(x) = -4 x^3 + 24 x^2 - 44 x + 24,$$
$$f'(x) = -4(x - 1)(x - 2)(x - 3).$$

There are three points where $f'(x) = 0$: $x = 1, 2$, and 3. Calculation shows that

$$f(1) = 9, f(2) = 8, f(3) = 9.$$

Hence the absolute maximum value of f on $[0, 4]$ is 9, occurring at $x = 1$ and $x = 3$. All of this shows up clearly on a graph, which the student should construct for himself. It is noteworthy, however, that the reasoning is conclusive without the graph.

EXERCISES

1. (a) Find all the points of relative maxima and minima and sketch the graph of $f(x) = (x + 5)^2(x^3 - 10)$. (b) Find the absolute maximum and minimum values of $f(x)$ on the interval $-6 \leq x \leq -2$.

2. Find the algebraically largest and smallest values of $f(x) = 20 x^3 - 2700 x + 7000$ when $0 \leq x \leq 10$.

3. Find the absolute maximum of $f(x) = (x^2 - 75)/(x - 10)$ for $0 \leq x < 10$. Begin by sketching the graph enough to show why a maximum must be attained in this interval.

4. Consider $f(x) = x/(x^2 + a^2)^{3/2}$ for $x \geq 0$. Explain why $f(x)$ must attain an absolute maximum value at some point. Find the point.

5. Find the absolute maximum of $\sin^2 2\theta (1 + \cos 2\theta)$ for $0 \leq \theta \leq \pi/2$.

6. Discuss the possible absolute extrema of $f(x) = x^4 + (256/x^2)$ for $0 < x \leq 4$, and find any that exist.

7. Consider the function $(27/\sin x) + (64/\cos x)$, $0 < x < \pi/2$. Why must there be an absolute minimum in this open interval? Find where it occurs, and the corresponding value of the function.

8. Without drawing the graph, find the absolute maximum and minimum values of the function (a) $5 \cos^3 x - 3 \cos x$; (b) $2 \sin x - 1 + 2 \cos^2 x$.

9. A right circular cone of altitude x is inscribed in a sphere of radius c. Express the volume of the cone as a function of x. What open interval of values of x is of significance in considering nondegenerate cones? Explain why there must be an absolute maximum volume for some x in this interval, and find the x for which the maximum is attained.

10. Let V be the volume of a right circular cone of altitude x circumscribed about a sphere of radius b. Show that V attains its absolute minimum when $x = 4b$. Write the argument out fully and carefully after the manner of the discussion of Example 4.

11. A right circular cylinder with radius of base x is inscribed in a right circular cone with radius of base r and altitude h. (a) Express the total surface area of the cylinder (including ends) as a function of x. (b) What interval of x-values are of significance if a nondegenerate cylinder is wanted? (c) What inequality must be satisfied by r and h if there is to be a nondegenerate cylinder of maximum area? (d) What is the situation if $h = 6$, $r = 2$? if $h = 4$, $r = 2$? if $h = 4$, $r = 3$?

12. Consider $f(x) = \dfrac{\sin \pi x}{x(1 - x)}$ if $0 < x < 1$. (a) Complete the definition of f at $x = 0, 1$ so as to make f continuous on $[0, 1]$. (b) Find the absolute maximum and minimum of $f(x)$ on $[0, 1]$ after completing (a). (c) Sketch the graph of $y = f(x)$.

13. Let $f(x) = 5\sqrt{16 + x^2} + 4\sqrt{(3 - x)^2}$ (taking the positive square root in both cases, so that $\sqrt{(3 - x)^2} = |3 - x|$). Note that f is differentiable except when $x = 3$ and that $f(x) \to +\infty$ as $x \to +\infty$ or $x \to -\infty$. (a) How do you infer from this that f must attain an absolute minimum value? (b) Find formulas for $f'(x)$ according as $x < 3$ or $x > 3$, and show that $f'(x) < 0$ if $x < 3$, while $f'(x) > 0$ if $x > 3$. (c) What do you infer about the point of attainment of the minimum? (d) What is the minimum value of $f(x)$?

14. A spring is located at $(0, a)$, and a man's house is located at $(b, 0)$, where $a > 0$, $b > 0$. A pipeline is to be laid in two straight parts, the first part from the spring to $(x, 0)$, and the second part from $(x, 0)$ to the house. The two parts will cost c_1 and c_2 dollars per unit length, respectively.

(a) Show that the total cost of the pipeline, for *any* value of x, is

$$f(x) = c_1\sqrt{a^2 + x^2} + c_2 |b - x|.$$

(b) Show that f has its absolute minimum value for some x such that $0 < x \leq b$. HINT: Consider the sign of $f'(x)$ when $x \leq 0$ and $x > b$ respectively.

(c) Find the inequalities which must be satisfied by c_1, c_2, a, and b if f is to attain its minimum for an x such that $0 < x < b$.

Observe that Exercise 13 is a special case of this exercise. A contrasting special case is afforded by taking $c_1 = 5$, $c_2 = 4$, $a = 3$, $b = 5$. If one does these two special cases first, the general problem will be more interesting.

15. Discuss the intervals of definition of the function

$$f(x) = \{(16 - x^2)(x^2 - 9)\}^{1/2},$$

and find the absolute maximum of the function.

16. A man wishes to get from point A to point B, these points being diametrically opposite each other on the shores of a circular pond. The man can row $1\frac{1}{2}$ miles per hour and walk 5 miles per hour.

(a) If there is a boat available at A, what combination of rowing and walking will take him to B in the least possible time?

(b) Discuss the problem in case the rowing and walking speeds are, respectively, u and v miles per hour. Check your results carefully in the special case $u = 2$, $v = 4$.

17. Consider a and b as fixed, with $b < a$. Let c be a variable such that $b < c < a$. Let ϕ be the acute angle between the tangents to the circle $x^2 + y^2 = c^2$ and the ellipse $b^2x^2 + a^2y^2 = a^2b^2$ at a point of intersection. Find $\tan \phi$ when c is chosen so that ϕ is greatest.

18. Write out the proof of Theorem III for the case of a relative minimum. Show that, if f has a relative minimum at x_0, and $g(x) = -f(x)$, then g has a relative maximum at x_0. Hence deduce the proof for the case of a minimum from the facts already established for a maximum.

1.2 The law of the mean (The mean-value theorem for derivatives)

The theorem which goes by the name of *the law of the mean* is one of the most important theoretical results in the subject of differential calculus. It is used as a tool in many places in the later developments of calculus, both differential and integral, particularly in connection with proofs. We wish to emphasize very strongly that the student of advanced calculus needs to gain an appreciation of the power of the law of the mean as an instrument of systematic reasoning. The first step should be to become thoroughly familiar with the content of the law itself.

THEOREM IV. *(The law of the mean.) Let f be a function which is continuous at each point of the closed interval $a \leq x \leq b$, and let it have a derivative at each point of the open interval $a < x < b$. Then there is a point $x = X$ in the open interval $(a < X < b)$ such that*

(1.2–1) $$f(b) - f(a) = (b - a)f'(X).$$

The theorem has a geometrical interpretation. Represent the function graphically by the curve $y = f(x)$, and let A, B be the points on the curve corresponding to $x = a$, $x = b$, respectively. The formula (1.2–1) states that there is some point on the curve, with abscissa $x = X$, at which the tangent is parallel to the line AB. There may be more than one suitable value of X; the essential thing is that there is always at *least* one (see Fig. 8).

It is worth noting that (1.2–1) remains true if we exchange a and b, for both sides merely change sign when this is done. Thus, suppose x_1, x_2 are the end-points of an interval on which the conditions of the law of the mean are satisfied. Then we can write

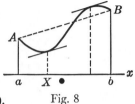

Fig. 8

$$(1.2\text{–}2) \qquad f(x_2) - f(x_1) = (x_2 - x_1)f'(\xi),$$

where $x = \xi$ is some point between x_1 and x_2. In writing this formula we do not need to know which of the numbers x_1, x_2 is the larger.

The geometrical interpretation of the law of the mean makes its truth plausible. A proof must be based on analytical reasoning, however. We follow the usual procedure of basing the proof on an auxiliary theorem named after the seventeenth-century mathematician Michel Rolle.

ROLLE'S THEOREM.   *Let g be a function which satisfies the conditions of [Theorem IV], and suppose further that g(a) = g(b) = 0. Then for some X such that a < X < b it is true that g'(X) = 0.*

PROOF.   We distinguish two cases: (1) the case in which $g(x)$ is zero on the whole interval, and (2) the case in which $g(x)$ assumes values other than 0 at some points of the interval. In case (1) the derivative $g'(x)$ is identically zero and the existence of X is assured. In case (2) let M and m be the maximum and minimum values, respectively, of $g(x)$ on the closed interval $[a, b]$. At least one of the values M, m must be different from 0, and must therefore occur at a point x of the *open* interval (a, b). By Theorem III (§ 1.12) we conclude that $g'(X) = 0$.

We have glossed over the main difficulty in this proof, namely the matter of the existence of the extreme values M and m. Here we appeal to the theorem which asserts that if a function is continuous on a finite closed interval, it actually attains its absolute extremal values at certain points of the interval. This theorem has been referred to before (see § 1.12, following Example 3); we treat it formally as Theorem III in § 3.2.

The law of the mean is deduced from Rolle's theorem by an artifice. The function f in Theorem IV need not vanish at $x = a$ and $x = b$. Suppose, however, that $y = F(x)$ is the equation of the straight line AB in Fig. 8, so that $F(a) = f(a)$, $F(b) = f(b)$. Let $g(x) = F(x) - f(x)$. Then g will be a function meeting the conditions of Rolle's theorem. The equation of the line in question is

$$y = \frac{f(b) - f(a)}{b - a}\,(x - a) + f(a).$$

Hence we set

$$g(x) = \frac{f(b) - f(a)}{b - a} (x - a) + f(a) - f(x).$$

Note that $g(a) = g(b) = 0$. The derivative is

$$g'(x) = \frac{f(b) - f(a)}{b - a} - f'(x).$$

The conclusion $g'(X) = 0$ of Rolle's theorem is now seen to be equivalent to the law of the mean in the form (1.2–1).

We now give some simple examples illustrating the law of the mean in particular instances.

Example 1. Suppose $f(x) = x^3$. Find a suitable value for X in (1.2–1) when $a = -1$ and $b = 2$.

Since $f'(x) = 3 x^2$, the law of the mean takes the form $b^3 - a^3 = (b - a)3 X^2$ in the present case. With $a = -1, b = 2$ we have $8 - (-1)^3 = (2 - (-1))3 X^2$, or $X^2 = 1$. Solving, we find $X = \pm 1$. We want a value of X such that $a < X < b$, i.e., $-1 < X < 2$. Hence $X = 1$ is the suitable value in question.

Example 2. If $f(x) = x^2$, show that the suitable value of X in the law of the mean is $X = (a + b)/2$.

We have $f'(x) = 2 x$. Hence the law of the mean becomes $b^2 - a^2 = (b - a)2X$, or $X = (a + b)/2$. Where is this point located in relation to a and b?

Example 3. If $f(x) = \sin x$ and $x_1 = 0$, $x_2 = 5 \pi/6$, find ξ such that $x_1 < \xi < x_2$ and (1.2–2) holds.

We have $f'(x) = \cos x$ and $\sin (5 \pi/6) = \frac{1}{2}$. Hence (1.2–2) takes the form

$$\frac{1}{2} - 0 = \sin \frac{5 \pi}{6} - \sin 0 = \frac{5 \pi}{6} \cos \xi$$

or

$$\cos \xi = \frac{3}{5 \pi} = 0.19099.$$

Since $0 < \xi < 5 \pi/6$, we find $\xi = \cos^{-1}(0.19099) = 1.37863$. This is somewhat less than $\pi/2$, which is to be expected, as may be seen from a carefully drawn graph.

In actual practice we are seldom interested in the exact value of the X occurring in (1.2–1); the important thing is that X lies between $x = a$ and $x = b$. This enables us to obtain inequalities for estimating the value of $f(x)$.

Example 4. Show that $\frac{1}{3} < \log 1.5 < \frac{1}{2}$.

This may be done as follows: We take $f(x) = \log x$, $a = 1$, $b = 1.5$; then $f'(x) = 1/x$, and by the law of the mean

$$\log 1.5 - \log 1 = \log 1.5 = (1.5 - 1)\frac{1}{X} = \frac{0.5}{X},$$

where $1 < X < 1.5$. It follows that

$$\frac{0.5}{1.5} < \frac{0.5}{X} < \frac{0.5}{1}, \text{ or } \frac{1}{3} < \frac{0.5}{X} < \frac{1}{2}.$$

This gives the required result.

Example 5. Use the law of the mean to show that

$$(1.2\text{-}3) \qquad\qquad \frac{\log x}{x} < \frac{\log a}{x} + \frac{1}{a}$$

if $0 < a < x$.

We use (1.2–1) with $f(x) = \log x$ and b replaced by x. Then

$$\log x - \log a = \frac{x - a}{X}, \, a < X < x.$$

Hence
$$\frac{\log x}{x} = \frac{\log a}{x} + \frac{x - a}{x} \cdot \frac{1}{X}.$$

Now
$$\frac{x - a}{x} \cdot \frac{1}{X} < \frac{1}{X} < \frac{1}{a},$$

and so (1.2–3) is seen to be correct.

A much used variant form of the law of the mean is obtained from (1.2–1) in the following way: Let $h = b - a$, so that $b = a + h$. Then X may be written $X = a + \theta h$, where $0 < \theta < 1$, because any number between a and $a + h$ can be expressed in this latter form. Thus we have

$$(1.2\text{-}4) \qquad\qquad f(a + h) = f(a) + hf'(a + \theta h), \, 0 < \theta < 1.$$

Example 6. Use (1.2–4) to show that

$$(1.2\text{-}5) \qquad\qquad (1 + h)^\alpha > 1 + \alpha h$$

if $h > 0$ and $\alpha > 1$.

We take $f(x) = x^\alpha$, $a = 1$. Then by (1.2–4),

$$(1 + h)^\alpha = 1 + h\alpha(1 + \theta h)^{\alpha-1}.$$

If $0 < h$ we have $(1 + \theta h)^{\alpha-1} > 1$, since $1 + \theta h > 1$ and $\alpha > 1$. The inequality (1.2–5) follows at once.

One very important consequence of the law of the mean will be formally stated here as a theorem.

THEOREM V. *Let f be differentiable at each point of the open interval* *a < x < b, and suppose that f'(x) = 0 at each such point. Then the value of the function is constant on the interval.*

PROOF. It follows by Theorem I, § 1.11, that f is continuous at each point of the open interval (a, b). Consider any two distinct points of the interval, say x_1, x_2, where $a < x_1 < x_2 < b$. We may apply the law of the mean, obtaining formula (1.2–2). But $f'(\xi) = 0$, by hypothesis, and so $f(x_1) = f(x_2)$. We have now proved the theorem, for we have shown that f has the same value at any two points of the interval.

Theorem V plays an essential role in the explanation of the relation between differentiation and integration, as we shall see when we come to the proof of Theorem VIII, § 1.53. Theorem V is also useful in dealing with the concept of the "general solution" of a certain elementary type of differential equation, as we shall see in § 1.4.

The following example also affords an important application of the law of the mean.

Example 7. Suppose that f satisfies the conditions of the law of the mean on the interval $a \leqq x \leqq b$, and that $f'(x) > 0$ when $a < x < b$. Show that $f(x)$ increases as x increases.

We are to show that $x_1 < x_2$ implies $f(x_1) < f(x_2)$ whenever $a \leqq x_1 < x_2 \leqq b$. The law of the mean tells us that there is some ξ such that $x_1 < \xi < x_2$ and $f(x_2) - f(x_1) = (x_2 - x_1)f'(\xi)$. Since $(x_2 - x_1) > 0$ and $f'(\xi) > 0$, we infer that $f(x_2) - f(x_1) > 0$; this is equivalent to $f(x_1) < f(x_2)$.

Example 8. Suppose that f is defined and differentiable on an open interval containing the point x_0. Suppose that $f'(x_0) = 0$ and that for all x sufficiently near x_0, $f'(x) > 0$ when $x < x_0$ and $f'(x) < 0$ when $x > x_0$. Show that these conditions are sufficient to guarantee that f has a relative maximum at x_0.

The argument is based on Example 7. As x increases, $f(x)$ increases when $f'(x) > 0$. By similar reasoning $f(x)$ decreases when x increases if $f'(x) < 0$. In the present case we see that the given conditions imply that for some small number h, $f(x)$ is increasing as x goes from $x_0 - h$ to x_0, and decreasing as x goes from x_0 to $x_0 + h$. Hence $f(x)$ must attain a relative maximum at x_0.

From this argument it will be apparent to the student how one may formulate sufficient conditions for a relative minimum at x_0.

EXERCISES

1. Use the law of the mean to show that $\frac{1}{9} < \sqrt{66} - 8 < \frac{1}{8}$.

2. Prove that there is no value of m such that $x^3 - 3x + m = 0$ has two distinct roots in the interval $0 \leqq x \leqq 1$. Use Rolle's theorem.

3. If $f(x) = x^3 - 3x^2 + 2x$, $a = 0$, $h = \frac{1}{2}$, find a suitable value of θ in the formula (1.2–4).

4. For what values of C is $Cx - \sin x$ an increasing function of x (for *all* x)?

5. Show that $2/\pi < (\sin \theta)/\theta < 1$ if $0 < \theta < \pi/2$.

HINT: Examine the sign of the derivative of $(\sin \theta)/\theta$.

6. Prove the following inequalities, using the law of the mean.
(a) $\sqrt{1 + x} < 4 + (x - 15)/8$ if $x > 15$.
(b) $\tan^{-1} x < (\pi/4) + (x - 1)/2$ if $1 < x$.
(c) $\dfrac{\pi}{4} - \dfrac{1 - x}{1 + x^2} < \tan^{-1} x < \dfrac{\pi}{4} - \dfrac{1 - x}{2}$ if $0 < x < 1$.
(d) $h/(1 + h^2) < \tan^{-1} h < h$ if $0 < h$.

7. Prove that the inequality (1.2–5) also holds if $-1 < h < 0$. Explain the reasoning about inequalities with care, noting that if $0 < A < 1$ and $B < 0$, then $AB > B$.

8. Prove the following inequalities:
(a) $\dfrac{x}{1 + x} < \log(1 + x) < x$, $-1 < x < 0$ or $0 < x$.

(b) $1 + \dfrac{x}{2\sqrt{1 + x}} < \sqrt{1 + x} < 1 + \frac{1}{2}x$, $-1 < x < 0$ or $0 < x$.

(c) $1 - \dfrac{x}{2} < \dfrac{1}{(1 + x)^{1/2}} < 1 - \dfrac{x}{2(1 + x)^{3/2}}$, $-1 < x < 0$ or $0 < x$.

(d) $p(x - 1) < x^p - 1 < px^{p-1}(x - 1)$, $1 < x$, $1 < p$.

(e) $\dfrac{m(x - 1)}{x^{1-m}} < x^m - 1 < m(x - 1)$, $0 < m < 1$, $1 < x$.

9. If $a > b > 0$ and m and n are positive numbers such that $m + n = 1$, show that $a^m b^n < ma + nb$. Use Exercise 8(e).

10. (a) Prove that $x^n + ax + b = 0$ (a, b real) has at most two distinct real roots if n is even, and at most three if n is odd.
(b) Prove that $x^n + ax^2 + b = 0$ (a, b real) has at most three distinct real roots if n is odd, and at most four if n is even.

11. Explain why $a_0 x^4 + a_1 x^3 + a_2 x^2 + a_3 x + a_4 = 0$ must have a root between 0 and 1 if $(a_0/5) + (a_1/4) + (a_2/3) + (a_3/2) + a_4 = 0$.

12. Let $F(x) = (f(x) - f(a))(g(b) - g(x))$, where f and g are continuous when $a \leq x \leq b$ and differentiable when $a < x < b$. Suppose further that $g'(x)$ is never zero. Show that there is a ξ between a and b such that

$$\frac{f'(\xi)}{g'(\xi)} = \frac{f(\xi) - f(a)}{g(b) - g(\xi)}.$$

13. Suppose $f''(x) > 0$ when $a \leq x \leq b$. Explain by an analytical argument why there can be at most one point of the interval at which $f'(x) = 0$. What is the geometrical interpretation as regards the curve $y = f(x)$?

14. Suppose f is continuous on the interval $a < x < b$, and that f is known to be differentiable on this interval except possibly at one point x_0. Suppose further that $\lim_{x \to x_0} f'(x)$ exists. Use the law of the mean to prove that f is differentiable at x_0 and that the derivative f' is continuous at that point.

15. Generalize the result of Illustrative Example 7 as follows: Suppose f is continuous on $a \leq x \leq b$ and differentiable on $a < x < b$. Suppose further that $f'(x) \geq 0$ when $a < x < b$ and that $f'(x) > 0$ for at least one value of x. Prove that $f(a) < f(b)$. [It is easy to see that $f(a) \leq f(b)$; what requires more care is to show that $f(a) \neq f(b)$.]

16. Suppose $f(x) = x^2 \sin(1/x) + (x/2)$ if $x \neq 0$, and define $f(0) = 0$. **(a)** Show that $f'(0) > 0$. **(b)** Show that, no matter how small the positive number h may be, there are infinitely many points on both sides of $x = 0$ and within distance h of $x = 0$ at which $f'(x) = \frac{3}{2}$ and also infinitely many at which $f'(x) = -\frac{1}{2}$. This shows that there is no interval containing $x = 0$ in which $f(x)$ is always increasing as x increases, in spite of the fact that the slope of the curve $y = f(x)$ is positive at $x = 0$.

17. Suppose the following things: that f is continuous when $0 \leq x \leq a$, differentiable when $0 < x < a$; that $f(0) = 0$, $f'(x) > 0$; and that $f'(x)$ increases as x increases. Show that $f(x)/x$ increases as x increases. (A suitable use of the law of the mean is indicated.) Examples are furnished by $f(x) = e^x - 1$ and by $f(x) = \tan x$, $0 \leq x \leq \pi/2$.

18. Let f be differentiable when $a < x < b$. Suppose x_1 and x_2 are distinct points of the interval, and let P_1 and P_2 be the corresponding points of the curve.

(a) Show that the condition for P_1 to lie above the line tangent to the curve at P_2 is

$$f(x_1) - f(x_2) - (x_1 - x_2) f'(x_2) > 0.$$

(b) If the condition in **(a)** holds for every pair of points on the curve, show that $f'(x)$ increases as x increases. In fact, if $x_1 < x_2$, show that

$$f'(x_1) < \frac{f(x_2) - f(x_1)}{x_2 - x_1} < f'(x_2).$$

(c) Show, conversely, that if $f'(x)$ increases as x increases, the condition in **(a)** is satisfied whenever $x_1 \neq x_2$. (Use the law of the mean.)

(d) Under the conditions in **(c)** show that the curve $y = f(x)$ between any two of its points lies entirely below the chord joining those points. Begin by showing that an analytic expression of this state of affairs is

$$\frac{f(x) - f(x_1)}{x - x_1} < \frac{f(x_2) - f(x)}{x_2 - x}$$

whenever $x_1 < x < x_2$.

1.3   Differentials

The notion of a differential is closely related to that of a derivative. For functions of a single independent variable this relationship is very close indeed, and very simple. For functions of several independent variables the relationship is less simple. At this point we are concerned only with functions of a single variable. We presume that the student is acquainted with differentials and their uses in the formal procedures of elementary calculus. Our purpose here is mainly to define differentials carefully and demonstrate the fundamental property upon which much of the usefulness of differentials depends.

Suppose that f is a function of the independent variable x, and let us assume that f is differentiable at x_0 (i.e., that $f'(x_0)$ exists).

DEFINITION. Let dx denote an independent variable which may take on any value whatsoever. Then the expression

$$f'(x_0)dx$$

is called the differential of f at x_0.

If we write $y = f(x)$, and if f is differentiable for a particular value of x, it is customary to write

$$(1.3–1) \qquad dy = f'(x)dx.$$

The differential of f is a homogeneous linear function of the independent variable dx; for a fixed x, dy is the value of the differential corresponding to any assigned value of dx. Thus dy is a dependent variable. The adjacent Figure 9 illustrates geometrically the functional dependence of dy on dx, as well as the relation to the function f itself. The xy-coordinate axes and the graph of the function $y = f(x)$ are shown in unbroken lines. A second co-ordinate system is shown with its origin at a typical point (x, y) of the curve $y = f(x)$. The axes in this system are scales for the measurement of the variables dx, dy. The equation (1.3–1) has as its graph a straight line of slope $f'(x)$. This line is, of course, the tangent to the curve $y = f(x)$ at the origin of the dx-dy coordinate system.

From (1.3–1) we have the quotient relation

$$(1.3–2) \qquad \frac{dy}{dx} = f'(x)$$

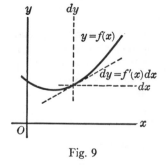

Fig. 9

whenever $dx \neq 0$. Actually the notation dy/dx for the derivative goes back to Leibniz, but the above presentation of separate meanings for dx and dy is of more modern origin. It is to be emphasized that there is no need for dx and dy to be small.

Example 1. If $y = f(x) = \sin x$, calculate the value of dy for $x = \pi/3$, $dx = \pi/6$. Here $f'(x) = \cos x$, so $dy = \cos x \, dx$. Evaluating, we obtain

$$dy = \left(\cos \frac{\pi}{3}\right) \frac{\pi}{6} = \frac{\pi}{12}.$$

Probably the most important feature of the formula (1.3–1) is that its truth is unaffected by the introduction of a new independent variable.

Example 2. Suppose $y = x^2$ and $x = t^3 + t$, so that $y = (t^3 + t)^2 = t^6 + 2 t^4 + t^2$. If we regard x as an independent variable, then $dy = 2 x \, dx$, by (1.3–1). Here dx is an independent variable. But if we regard t as an independent variable, then both x and y are dependent on t, and the notations dy, dx acquire new meanings:

$$x = t^3 + t, dx = (3 t^2 + 1)dt,$$
$$y = t^6 + 2 t^4 + t^2, dy = (6 t^5 + 8 t^3 + 2 t)dt.$$

But even with these new meanings, it is still true that $dy = 2 x \, dx$. We verify this by writing

$$2 x \, dx = 2(t^3 + t)(3 t^2 + 1)dt = (6 t^5 + 8 t^3 + 2 t)dt = dy.$$

What we have verified here in a particular case may be demonstrated in general by appealing to the rule for differentiating a composite function (Theorem II, § 1.11). Suppose $y = f(x)$ and $x = g(t)$, so that $y = F(t)$, where $F(t) = f(g(t))$. Then, with t as independent variable,

(1.3–3) $$dy = F'(t)dt, dx = g'(t)dt.$$

By Theorem II we have

(1.3–4) $$F'(t) = f'(x)g'(t).$$

Hence, combining (1.3–3) and (1.3–4),

$$dy = f'(x)g'(t)dt = f'(x)dx,$$

so that (1.3–1) holds, even though x and dx are no longer independent variables.

The use of differentials is a great convenience in algebraic manipulations which are incidental to much work in calculus. Using differentials rather than derivatives, one is often enabled to retain a desirable symmetry by not forcing a decision as to which variable is independent. The differential formula for arc length of a plane curve illustrates this point. The formula is

(1.3–5) $$ds^2 = dx^2 + dy^2;$$

the co-ordinates (x, y) of a point on the curve are functions of some parameter, and the arc length s, measured from some chosen initial point on the curve, likewise depends on the parameter. But the formula (1.3–5) holds (granted suitable conditions on the curve) no matter what the parameter may be.

The fact that $f'(x)$ is the ratio of dy to dx no matter what variable is independent is of great usefulness when we wish to compute the slope of a curve defined parametrically.

EXERCISES

1. (a) If $y = (1 - x^2)/(1 + x^2)$, compute dy when $x = 1$, $dx = 2$.

(b) If $x = \tan(t/2)$, compute dx when $t = \pi/2$, $dt = 2$.

(c) If x in (a) is replaced by its value in terms of t from (b), show that, on simplification, $y = \cos t$. From this formula compute dy when $t = \pi/2$, $dt = 2$. Compare the answers to (b) and (c) with your result in part (a).

2. From $x = r \cos \theta$, $y = r \sin \theta$, and $ds^2 = dx^2 + dy^2$ derive the formula $ds^2 = dr^2 + r^2 d\theta^2$.

3. (a) If $y = f(x)$, $y' = f'(x)$, and so on, why is $dy' = y'' dx$?

(b) What are dy'' and $d(y')^2$ when expressed with dx as a factor?

(c) Suppose that $x = f(t)$, $y = g(t)$, and write $x' = f'(t)$, $y' = g'(t)$, and so on. Show that

$$\frac{d}{dt}\left(\frac{dy}{dx}\right) = \frac{x'y'' - y'x''}{(x')^2}.$$

4. For a plane curve C, construct the tangent at a typical point $P(x, y)$, and let angles ϕ, ψ, θ be as indicated on Fig. 10, so that for the general case $\phi = \theta + \psi + n\pi$ where n is an integer ($n = 0$ in Fig. 10). From this equation and the relations

$$\frac{dy}{dx} = \tan \phi, \quad \frac{y}{x} = \tan \theta,$$

show that

$$\tan \psi = \frac{x\,dy - y\,dx}{x\,dx + y\,dy} = \frac{r\,d\theta}{dr}.$$

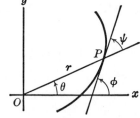

Fig. 10

5. From $\operatorname{ctn} \psi = \dfrac{dr}{r\,d\theta}$ (see Exercise 4) find $d\psi$ in terms of r, r', r'', and $d\theta$, where $r' = \dfrac{dr}{d\theta}$, and $r'' = \dfrac{dr'}{d\theta}$.

6. The curvature of a plane curve $y = f(x)$ is defined as $K = d\phi/ds$, where $dy/dx = \tan \phi$ and $ds^2 = dx^2 + dy^2$. Derive the formula

$$K = \pm \frac{y''}{(1 + y'^2)^{3/2}}.$$

7. Using the definition of K in Exercise 6, and the relation between ϕ, ψ, and θ in Exercise 4, derive the formula

$$K = \pm \frac{r^2 + 2r'^2 - rr''}{(r^2 + r'^2)^{3/2}}.$$

(Exercise 5 should be worked first.)

8. Let $y = f(x)$ be the equation of a plane curve C. The center of curvature of the curve corresponding to a particular point (x, y) on C is a point (X, Y) a distance R from (x, y) along the normal to C at (x, y), in the direction toward the concave side of the curve; here R is the radius of curvature. It may be shown that

$$X = x - \frac{y'(1 + y'^2)}{y''},$$

$$Y = y + \frac{1 + y'^2}{y''}.$$

It is assumed, of course, that $y'' \neq 0$.

The locus of (X, Y), as (x, y) moves along C, is a curve called the *evolute* of C. We may regard the above equations as parametric equations of the evolute, with x as parameter. Show that $dY/dX = -1/y'$. This proves that the normal to C at (x, y) is tangent to the evolute at (X, Y).

1.4 The inverse of differentiation

Many applications of calculus, some of them quite elementary, require the determination of a function from the knowledge of its first or second derivative, together with supplementary data about the function for particular values of the independent variable. The first and simplest general problem of this kind may be put as follows:

PROBLEM. *Given a continuous function $f(x)$, defined on a certain interval, find all functions defined on this interval and having $f(x)$ as derivative. In symbols, find y as a function of x such that*

$$(1.4\text{--}1) \qquad \frac{dy}{dx} = f(x).$$

Directly out of his experience with differentiation, the student is able to solve many problems of this type. The process is one of using standard differentiation formulas in reverse. Let us examine the reasoning carefully in a typical case.

Example. Find y as a function of x such that $y = 1$ when $x = 0$ and

$$(1.4\text{--}2) \qquad \frac{dy}{dx} = \frac{x}{1 + x^2}.$$

Here $f(x)$ is defined and continuous for all values of x, so we want the solution to be defined for all values of x.

The normal procedure is to write

$$(1.4\text{--}3) \qquad dy = \frac{x\,dx}{1 + x^2} = \frac{1}{2}\frac{d(1 + x^2)}{1 + x^2};$$

$$(1.4\text{--}4) \qquad \begin{aligned} y &= \tfrac{1}{2} \log (1 + x^2) + C; \\ 1 &= \tfrac{1}{2} \log (1 + 0) + C, \quad C = 1; \end{aligned}$$

(1.4–5) $y = \frac{1}{2} \log (1 + x^2) + 1.$

The supporting argument (usually not made explicit in practice) runs as follows: *If* there is a function y satisfying (1.4–2), then it also satisfies (1.4–3). In this latter form, as finally written, we recognize that a y satisfying (1.4–2) is furnished by (1.4–4), where C is an arbitrary constant. The proper determination of C is made by substituting the given matched values for x and y. Thus we obtain (1.4–5), which is actually a solution of the problem, as may be checked.

One question remains: Is (1.4–5) the *only* solution to the problem? Once we obtain (1.4–4) it is clear that C is uniquely determined by the condition that $y = 1$ when $x = 0$. The question is then: Does (1.4–4) give *all* the functions y satisfying (1.4–2)? The answer is affirmative, and is supplied by Theorem V, § 1.2. For let y be *any* function which is differentiable for all values of x and satisfies (1.4–2). Then the derivative of the function

$$y - \frac{1}{2} \log (1 + x^2)$$

is zero for all values of x; hence, by Theorem V, this function is constant, so that y is given by (1.4–4) for *some* value of C.

The equation (1.4–1) is the very simplest type of first-order differential equation. The problem which we have posed in connection with this equation may be called the problem of finding the "general solution" of (1.4–1). The general solution is the family of all functions $y = y(x)$ satisfying the equation. Any member of this family may be called a "particular solution." By appeal to Theorem V, § 1.2, we obtain the following conclusion:

If $y_1(x)$ and $y_2(x)$ are any two particular solutions of (1.4–1), they differ by a constant. Thus, if $y_1(x)$ is any particular solution, the general solution is given by

(1.4–6) $y = y_1(x) + C,$

where C is an arbitrary constant. In all this we assume, of course, that all the solutions considered are differentiable on the interval where $f(x)$ is defined.

The main problem is thus reduced to the finding of any one particular solution of (1.4–1). In many important simple problems such a particular solution may be found either by direct inspection or by various ingenious devices, all of which depend upon extensive familiarity with formulas of differentiation and manipulation of differentials. But it is not difficult to give examples in which no solution is forthcoming from the class of functions which a student meets and learns to differentiate in elementary calculus. Thus, for example, the equations

$$\frac{dy}{dx} = e^{-x^2}, \frac{dy}{dx} = \sqrt{1 + x^3}$$

have no solution within this class.

It is well to pause at this point and reflect upon the meaning of the word "function." In elementary differential calculus practically all our experience is with functions of a few basic types: algebraic, trigonometric and inverse trigonometric, exponential and logarithmic, and rather simple compounding of these types. It turns out that within this class of "elementary" functions, differentiation always leads to functions which are again in the class. Such is not the case with the inverse of differentiation, however; there are elementary functions which are not derivatives of elementary functions, e.g., e^{-x^2} and $\sqrt{1 + x^3}$. Now the general theorems of calculus deal with functions which are arbitrary except for requirements of differentiability or continuity, and which certainly need not be elementary in the sense of the first part of this paragraph. Once we have rejected the limitation of our considerations to "elementary" functions, we may well ask: What nonelementary functions do we know? If e^{-x^2} is not the derivative of any elementary function, how are we to find solutions of the equation $dy/dx = e^{-x^2}$? Evidently it is necessary in some fashion to acquire a supply of nonelementary functions of which we know the derivatives.

There are several very important methods for building such a supply. One method is that of integration of known functions. Starting with a given continuous function $f(x)$ defined on some interval, we form

(1.4–7) $$F(x) = \int_a^x f(t)dt,$$

where a and x belong to the interval, and a is kept fixed. Another method is that of forming infinite series whose terms are given functions of x:

$$F(x) = u_1(x) + u_2(x) + u_3(x) + \cdots.$$

We shall later be able to show that

$$\int_0^x \sqrt{1 + t^3}\, dt$$

is a function $F(x)$ such that $F'(x) = \sqrt{1 + x^3}$, and that

$$x - \frac{x^3}{3 \cdot 1!} + \frac{x^5}{5 \cdot 2!} - \frac{x^7}{7 \cdot 3!} + \cdots$$

is a function $F(x)$ such that $F'(x) = e^{-x^2}$.

The study of functions defined by infinite series will concern us in a later chapter of this book. Our immediate interest will be confined to functions of the type (1.4–7) defined by integration. We shall presently learn that if $f(x)$ is continuous on a given interval $a \leq x \leq b$, the general solution of the equation $dy/dx = f(x)$ on that interval is $y = F(x) + C$, where C is an arbitrary constant and $F(x)$ is defined by (1.4–7). The integral is defined by a limiting process involving sums.

Every student who has come this far in his study of calculus knows that there is a very close connection between differentiation and integration. But elementary calculus books vary widely in their discussions of integration and in their treatment of the link between differentiation and integration. Thus we shall not assume any uniformity of knowledge on this subject by the readers of this book. The next sections will be devoted to the subject of integration and its relation to differentiation. Our aim will be to lay out a precise logical pattern of definitions and theorems which will provide a common understanding of the integration concept and of the connection between differentiation and integration.

EXERCISES

1. What is the logical objection to the following procedure? Let the symbol $\int_0^t e^{-s^2} ds$ be *defined* as that function $F(t)$ such that $F'(t) = e^{-t^2}$ and $F(0) = 0$. Then $\dfrac{d}{dt}(F(t) + C) = F'(t) = e^{-t^2}$, and therefore $v = \int_0^t e^{-s^2} ds + C$ is the general solution of the equation $dv/dt = e^{-t^2}$.

2. What is the logical objection to the following procedure? Let $f(x)$ be a given function defined when $a \leqq x \leqq b$. Let $F(x)$ be any differentiable function such that $F'(x) = f(x)$ for each x of the given interval. Then define $\int_a^b f(x)dx = F(b) - F(a)$.

3. Try to find a continuous $F(x)$ such that $F(0) = 0$ and $F'(x) = [x]$ when $0 \leqq x \leqq 3$ (where $[x]$ denotes the greatest integer $\leqq x$). Do you succeed fully? Where is the difficulty?

1.5   Definite integrals

We are going to define what we mean by the definite integral of a function. We start with a function f, which we suppose to be defined and continuous on a closed interval $a \leqq x \leqq b$. These things being given, the definite integral of f over the interval is a certain number, which we denote by

$$(1.5–1) \qquad\qquad \int_a^b f(x)dx,$$

and which we arrive at by a defining process which we shall outline in four steps, as follows:

Step 1. Choose an integer $n \geqq 1$ and subdivide the interval $[a, b]$ into n subintervals by choosing points $x_0, x_1, \cdots, x_n$ such that $a = x_0 < x_1 < \cdots < x_n = b$. We adopt the notation

$$\Delta x_i = x_i - x_{i-1}, \quad i = 1, \cdots, n.$$

Step 2. In each subinterval choose an arbitrary point, denoting the point in the ith subinterval by x_i', so that

$$x_{i-1} \leqq x_i' \leqq x_i.$$

Step 3. Find the value of the function at each of the points chosen in Step 2, and form the sum

$$(1.5\text{–}2) \qquad f(x_1')\Delta x_1 + f(x_2')\Delta x_2 + \cdots + f(x_n')\Delta x_n.$$

This is called an approximating sum.

Step 4. Find the limit of the sums (1.5–2) as n is increased and the maximum of the numbers $\Delta x_1, \cdots, \Delta x_n$ is made to approach zero. This limit is, by definition, the definite integral (1.5–1), so that

$$(1.5\text{–}3) \qquad \int_a^b f(x)dx = \lim \sum_{i=1}^n f(x_i')\Delta x_i.$$

In a later chapter (XVIII) we shall study the theory of integration systematically. At that time we shall prove that the sums (1.5–2) do actually approach a limit in the case of any continuous function. For the present we take for granted the existence of this limit, and its uniqueness. A further discussion of the limit concept associated with the integral will be found in § 1.63.

A definite integral is thus defined as the limit of a certain kind of sum associated with the function. A geometrical interpretation of the integral can be made in terms of the area under the curve $y = f(x)$ from $x = a$ to $x = b$. Each term in the approximating sum (1.5–2) is the area of one of the shaded rectangles in Fig. 11. The area under the curve is the limit of the sum of the areas of these rectangles. We assume that the student is already familiar with this geometrical interpretation of the integral, and with the extension of

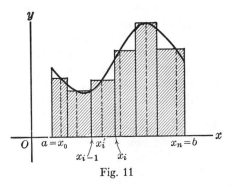

Fig. 11

the interpretation (by the concept of negative area) to those situations where the curve $y = f(x)$ goes below the x-axis. We emphasize, however, that we do not *define* the definite integral as the area under the curve. The area interpretation is merely a convenient method of bringing our intuition into play to aid us in grasping the nature of the definition (1.5–3). We "feel" that the area exists, and that a good approximation to it can be obtained by the sums (1.5–2), provided we take all the subintervals short enough. Actually, the area is defined as being equal to the limit in (1.5–3).

Area is only one of many geometrical and physical concepts whose exact quantitative measurement is furnished by a definite integral. The student is no doubt already familiar with such applications of integration as the finding of volumes (including volumes of solids of revolution) by slicing them into thin slabs or thin cylindrical shells, the calculation of arc-lengths of curves, the location of centroids and centers of gravity, and the reckoning of force due to water pressure on submerged plane surfaces. In every one of these applications the integral as the limit of a sum is the fundamental concept. Now we wish to emphasize that the concept is applicable to *any* continuous function, and that the concept itself does not depend on any geometrical or physical interpretation.

Suppose that m_i is the smallest value of $f(x)$ in the subinterval $x_{i-1} \leqq x \leqq x_i$. If we choose x_i' as a point at which $f(x)$ takes on the value m_i, the sum (1.5–2) becomes

$$(1.5\text{–}4) \qquad m_1\,\Delta x_1 + m_2\,\Delta x_2 + \cdots + m_n\,\Delta x_n.$$

This particular approximating sum is called a *lower sum*. Likewise we define an *upper sum* by choosing x_i' to be a point of the subinterval $x_{i-1} \leqq x \leqq x_i$ at which $f(x)$ takes on its greatest value M_i for that subinterval. The upper sum is

$$(1.5\text{–}5) \qquad M_1\,\Delta x_1 + M_2\,\Delta x_2 + \cdots + M_n\,\Delta x_n.$$

The upper and lower sums have the property that

$$(1.5\text{–}6) \qquad\qquad s \leqq \int_a^b f(x)dx \leqq S,$$

where s represents the lower sum and S represents the upper sum. This system of inequalities makes it possible to obtain numerical estimates of the value of an integral. Upper and lower sums are discussed at greater length in § 18.1.

Example 1.   Estimate the value of $\int_2^6 (75\,x - x^3)dx$ by lower and upper sums, using four equal subintervals.

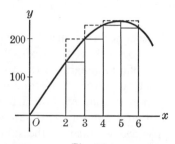

Fig. 12

We take $f(x) = 75 x - x^3$, $x_0 = 2$, $x_1 = 3$, $x_2 = 4$, $x_3 = 5$, $x_4 = 6$. Each $\Delta x_i = 1$. Since $f'(x) = 75 - 3 x^2$, it appears that $f(x)$ increases from $x = 2$ to $x = 5$, and decreases from $x = 5$ to $x = 6$. From the accompanying table of values we can then read off the values of m_i and M_i:

x	$f(x)$
2	142
3	198
4	236
5	250
6	234

i	m_i	M_i
1	142	198
2	198	236
3	236	250
4	234	250

Thus the lower and upper sums are

$$s = 142 + 198 + 236 + 234 = 810,$$
$$S = 198 + 236 + 250 + 250 = 934.$$

Therefore

$$810 < \int_2^6 (75 x - x^3)dx < 934.$$

A better estimate could be obtained by using more and smaller subintervals.

One of the simple but very important facts about integrals is expressed by the formula

(1.5–7)
$$\int_a^c f(x)dx = \int_a^b f(x)dx + \int_b^c f(x)dx,$$

where $a < b < c$, and f is continuous on the interval $a \leq x \leq c$. The analytical proof of this formula runs as follows: Divide the whole interval $[a, c]$ into parts in such a way that $x = b$ is always one of the points of subdivision. Suppose that $[a, b]$ is divided into m parts and $[b, c]$ into n parts:

$$a = x_0 < x_1 < \cdots < x_m = b = \xi_0 < \xi_1 < \cdots < \xi_n = c.$$

Choose points x_i' and ξ_j' such that $x_{i-1} \leq x_i' \leq x_i$, $i = 1, \cdots, m$, and $\xi_{j-1} \leq \xi_j' \leq \xi_j$, $j = 1, \cdots, n$. Then, as $m \to \infty$ and $n \to \infty$, and as the greatest of the differences Δx_i, $\Delta \xi_j$ approaches zero, we have

$$\int_a^b f(x)dx = \lim \sum_{i=1}^m f(x_i')\Delta x_i,$$

$$\int_b^c f(x)dx = \lim \sum_{j=1}^n f(\xi_j')\Delta \xi_j.$$

But also,

$$\int_a^c f(x)dx = \lim \left[\sum_{i=1}^m f(x_i')\Delta x_i + \sum_{j=1}^n f(\xi_j')\Delta \xi_j \right].$$

Formula (1.5–7) then follows from the principle that the limit of a sum is equal to the sum of the limits. We apply the principle to the sum of the two expressions

$$\sum_{i=1}^{m} f(x_i')\Delta x_i \quad \text{and} \quad \sum_{j=1}^{n} f(\xi_j')\Delta\xi_j.$$

There are various limiting processes used in calculus: limits of functions, limits of sequences, and the type of limit (of approximating sums) used in defining an integral. For each of these limiting processes there is the principle that the limit of a sum is equal to the sum of the limits. For limits of functions this principle was formulated in the rule (1.1–4). A proof of the validity of the rule (1.1–4) is given in § 1.64. With slight formal modifications, the idea of this proof applies equally well to sequences and to limits of approximating sums. Basically, all these various forms of the principle are covered by the theorem that the process of addition is a continuous function of the things which are added (see Theorem VI, § 17.5).

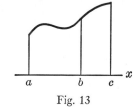

Fig. 13

Of course, the geometric interpretation of (1.5–7) is very simple and obvious: *The area between the curve $y = f(x)$ and the x-axis from a to c is the algebraic sum of the partial areas from a to b and from b to c* (see Fig. 13).

We do not in practice compute the precise values of definite integrals by direct application of the definition (1.5–3). In some simple cases, however, we may be able to find the precise value of the limit of the approximating sums by a direct examination of the sums. Usually these direct procedures involve subdivisions of the interval into equal parts.

Example 2. Find the value of $\int_a^b x^2\,dx$, assuming $0 < a < b$.

For convenience we write (by (1.5–7))

$$\int_a^b x^2\,dx = \int_0^b x^2\,dx - \int_0^a x^2\,dx.$$

We concentrate on finding the value of $\int_0^b x^2\,dx$.

Dividing $[0, b]$ into n equal parts, we write

$$x_0 = 0, \; x_1 = \frac{b}{n}, \; x_2 = \frac{2b}{n}, \; \cdots, \; x_n = \frac{nb}{n} = b.$$

Here each $\Delta x_i = \dfrac{b}{n}$. We choose $x_i' = x_i$. Then

$$\sum_{i=1}^{n} f(x_i')\Delta x_i = \sum_{i=1}^{n} x_i^2\,\Delta x_i$$

$$= \left[\left(\frac{b}{n} \right)^2 + \left(\frac{2 b}{n} \right)^2 + \cdots + \left(\frac{nb}{n} \right)^2 \right] \frac{b}{n}$$

$$= \frac{b^3}{n^3} [1^2 + 2^2 + \cdots + n^2].$$

There is a convenient formula for the sum of the squares of the integers from 1 to n (see Exercise 6):

(1.5–8) $$1^2 + 2^2 + \cdots + n^2 = \frac{n(n + 1)(2 n + 1)}{6}.$$

Combining the foregoing observations, we see that

(1.5–9) $$\int_0^b x^2 \, dx = \lim_{n \to \infty} \frac{n(n + 1)(2 n + 1)b^3}{6 n^3}.$$

Now

(1.5–10) $$\frac{n(n + 1)(2 n + 1)}{n^3} = 2 + \frac{3}{n} + \frac{1}{n^2},$$

so that, as $n \to \infty$, the expression on the right in (1.5–10) approaches 2 as a limit. Hence, from (1.5–9),

$$\int_0^b x^2 \, dx = \frac{2 b^3}{6} = \frac{b^3}{3}.$$

Since b was arbitrary, it follows that

$$\int_0^a x^2 \, dx = \frac{a^3}{3}.$$

Therefore

$$\int_a^b x^2 \, dx = \frac{b^3 - a^3}{3}.$$

EXERCISES

1. (a) Using four equal subintervals, calculate upper and lower sums for the integral $\int_{-1}^3 (x^3 - 3 x^2 + 3)dx$.

(b) Repeat (a), using eight equal subintervals.

(c) Calculate the value of the approximating sum (1.5–2), using four equal subintervals, and taking x_k' to be the midpoint of the kth subinterval.

2. (a) Calculate the value of the approximating sum (1.5–2) for the integral $\int_1^3 \frac{dx}{x}$, using six equal subintervals and taking x_k' to be the midpoint of the kth subinterval.

(b) Calculate upper and lower sums for the integral in (a), using six equal subintervals. A table of reciprocals will be found convenient for this exercise.

3. Follow the instructions of Exercise 1 as applied to the integral $\int_0^4 (4 x^2 - 12 x + 10)dx$.

4. Apply the definition (1.5–3) to find the value of the integral $\int_a^b f(x)dx$ if $f(x) = c$, where c is a constant.

5. (a) Let $A_1(n) = 1 + 2 + \cdots + n$. Noting that $A_1(n) = n + (n - 1) + \cdots + 1$, show that $2 A_1(n) = n(n + 1)$.

(b) Using the formula for $A_1(n)$ found in (a), calculate $\int_a^b x \, dx$ by a method like that used in Example 2.

6. Let $A_2(n) = 1^2 + 2^2 + \cdots + n^2$. Obtain a formula for $A_2(n)$ as follows: Start with

$$(p + 1)^3 - p^3 = 3 p^2 + 3 p + 1.$$

Write this out with $p = 0, 1, \cdots, n$, putting the results in order as shown.

$$1^3 - 0^3 = 3 \cdot 0^2 + 3 \cdot 0 + 1$$
$$2^3 - 1^3 = 3 \cdot 1^2 + 3 \cdot 1 + 1$$
$$\cdots\cdots\cdots\cdots\cdots\cdots\cdots$$
$$(n + 1)^3 - n^3 = 3 \cdot n^2 + 3 \cdot n + 1.$$

Now add, noting the cancellations on the left, to obtain

$$(n + 1)^3 = 3 A_2(n) + 3 A_1(n) + (n + 1).$$

Use the formula for $A_1(n)$ from Exercise 5(a) to solve for $A_2(n)$, obtaining

$$A_2(n) = \frac{n(n + 1)(2 n + 1)}{6}.$$

Imitate the foregoing procedure **(a)** to find the formula for $A_1(n)$ in this new way; **(b)** to obtain a formula for $A_3(n) = 1^3 + 2^3 + \cdots + n^3$. Then use your result to calculate $\int_a^b x^3 \, dx$ by the method of Example 2.

7. Let $f(x) = e^x$ and assume $b > 0$. Using n equal subintervals and a method somewhat like that of Example 2, show that

$$\int_0^b e^x \, dx = \lim_{n \to \infty} (e^b - 1) \left(\frac{e^h - 1}{h} \right)^{-1},$$

where $h = b/n$. Use the definition of $f'(0)$ to calculate the value of the limit, and so find the value of the integral.

8. Taking $f(x) = x$, and choosing $x_i' = (x_{i-1} + x_i)/2$, show that the approximating sum (1.5–2) has a value which is independent of n and the choice of the points $x_1, x_2, \cdots, x_{n-1}$. From this result calculate $\int_a^b x \, dx$, using the definition (1.5–3).

9. The function $f(x)$ defined as $(\sin x)/x$ when $x \neq 0$ and $f(0) = 1$ is continuous for all values of x. Why? Show that the value of $f(x)$ decreases steadily as x increases from 0 to $\pi/2$. Below is given a table of values of $\sin x$. Use it to obtain high and low estimates of the value of $\int_0^{\pi/2} f(x) \, dx$.

n	1	2	3	4	5	6	7
$\sin \dfrac{n\pi}{16}$	0.1951	0.3827	0.5556	0.7071	0.8315	0.9239	0.9808

10. Calculate $\int_1^2 x^p\,dx$, where p is a positive integer, by the following procedure. Divide the interval $[1, 2]$ into n parts by the points $x_0 = 1$, $x_1 = h$, $x_2 = h^2, \cdots$, $x_n = h^n$, where $h = 2^{1/n}$, so that $h^n = 2$. This does not give subintervals of equal lengths, but of lengths in geometric progression. Choose $x_k' = x_{k-1}$, and show that (using (1.5-3))

$$\int_1^2 x^p\,dx = (2^{p+1} - 1) \lim_{n \to \infty} \frac{1 - 2^{1/n}}{1 - 2^{(p+1)/n}}.$$

Use the formula for the sum of a geometric progression on $(1 - h)/(1 - h^{p+1})$, and thus show that the limit in the above formula has the value $1/(p + 1)$. Make use of the fact that $\lim_{n \to \infty} C^{1/n} = 1$ if $C > 0$. This is proved in § 1.62, Exercise 16.

11. After careful study of Exercise 10, adapt the method to find the value of $\int_a^b x^p\,dx$ where $0 < a < b$ and p is a positive integer. Start with $x_0 = a$, $x_1 = ah$, $x_2 = ah^2$, etc., where $h = (b/a)^{1/n}$.

1.51   The mean-value theorem for integrals

The theorem which we prove in this section has an importance much like that of the law of the mean (Theorem IV, § 1.2), in that it is valuable as a tool in systematic reasoning in calculus. We are introducing it here because of the use we shall have for it in our program of outlining the relation between differentiation and integration.

THEOREM VI. *(The mean-value theorem for integrals.) Let f be continuous on the closed interval $[a, b]$. Then there is some number X such that $a \leq X \leq b$ and*

(1.51–1) $$\int_a^b f(x)dx = (b - a)f(X).$$

PROOF. Let m and M denote the minimum and maximum values of f on the interval. Consider the approximating sums (1.5–2) of Step 3 in the definition of the integral (§ 1.5).

We have $m \leq f(x_i') \leq M.$

Therefore $m \sum_{i=1}^{n} \Delta x_i \leq \sum_{i=1}^{n} f(x_i')\Delta x_i \leq M \sum_{i=1}^{n} \Delta x_i.$

But evidently $\sum_{i=1}^{n} \Delta x_i = b - a.$

Hence the approximating sum (1.5–2) lies between $m(b - a)$ and $M(b - a)$. Consequently the definite integral, being the limit of the approximating sums, must also lie between these two numbers; that is,

$$m(b - a) \leq \int_a^b f(x)dx \leq M(b - a),$$

or

(1.51–2)
$$m \leq \frac{1}{b-a} \int_a^b f(x)dx \leq M.$$

Let us set

(1.51–3)
$$\mu = \frac{1}{b-a} \int_a^b f(x)dx.$$

We now reason that $f(x)$ must take on the value μ at some point $x = X$, $a \leq X \leq b$, since by (1.51–2) μ lies between the smallest and largest values of $f(x)$ on the interval. Once this argument is accepted, the proof is complete, for the equation $f(X) = \mu$ is equivalent to (1.51–1), in view of the definition of μ.

The existence of X such that $f(X) = \mu$ depends upon the hypothesis that f is continuous. Such existence is made plausible by intuitive consideration of the variation in value of a continuous function. An indubitable proof of the existence of X must await our systematic consideration of the properties of continuous functions (in Chapter III). The remarks which we made about the existence of m and M in connection with the proof of Rolle's theorem (§ 1.2) apply equally here to the existence of X.

The number μ defined by (1.51–3) is called the *average value* of the function $f(x)$ on the interval $[a, b]$. The sense in which this concept of average value is an extension of the simple notion of the arithmetic mean as an average value is indicated in Exercise 1.

EXERCISES

1. Let $[a, b]$ be divided into n equal parts, and let y_i be the value of $f(x)$ at the midpoint of the ith subinterval. The arithmetic mean of $y_1, \cdots, y_n$ is

$$A_n = \frac{y_1 + \cdots + y_n}{n}.$$

Show that $\mu = \lim_{n \to \infty} A_n$.

2. By interpreting the integral as an area, calculate the average value of $f(x) = \sqrt{a^2 - x^2}$ on the interval $-a \leq x \leq a$.

3. A right circular cone of altitude H and radius of base R has its axis along the x-axis. For a given value of x let $A(x)$ denote the area of cross section of the cone by a plane perpendicular to the x-axis at that point. What is the average value of $A(x)$, x ranging over all values for which the plane cuts the cone?

4. In Theorem VI it was asserted that an X can be found on the closed interval $[a, b]$ such that (1.51–1) holds. If $f(x)$ is constant on $[a, b]$, say $f(x) \equiv C$, then X may be taken as *any* point of the closed interval, for in that case $\int_a^b f(x) \, dx = C(b - a)$, and $C = f(X)$, no matter how we choose X. Hence certainly we can choose X so that $a < X < b$. Show that this can also be done if $f(x)$ is not constant on $[a, b]$. State precisely what you are taking for granted about continuous functions.

1.52   Variable limits of integration

Before coming to the main subject of this section it will be well to consider a matter of notation. In the symbolic expression

$$\int_a^b f(x)dx$$

we refer to x as the *variable of integration*. The value of the integral does
not depend upon the letter which is used for the variable of integration.
For example,

$$\int_0^2 x^3\, dx = \int_0^2 t^3\, dt = \int_0^2 u^3\, du.$$

In cases where the limits of integration are literal symbols it is important
to avoid using the same letter for a limit of integration and also for the
variable of integration. The danger of confusion is made apparent by
asking: What is the value of the expression

$$\int_0^x x^3\, dx$$

when $x = 2$?

Let us now suppose that we have a function $f(t)$ which is continuous on
the interval $a \leq t \leq b$. Regarding x as a variable, consider the function
$F(x)$ defined by the integral

$$(1.52\text{–}1) \qquad\qquad F(x) = \int_a^x f(t)dt.$$

It is clear that the integral does actually define a function of the upper
limit x provided $a \leq x \leq b$, for once the function $f(t)$ and the lower limit
$t = a$ are chosen and fixed, the integral has a definite numerical value
which depends only on x. If we resort to the interpretation of the integral
as an area, we may obtain a geometrical representation of the function
$F(x)$ as the area under the curve $y = f(t)$ from $t = a$ to $t = x$ (see Fig. 14).

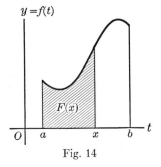

Fig. 14

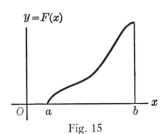

Fig. 15

It is natural to complete the definition (1.52–1) by setting $F(a) = 0$. As a matter of fact, it is usual to define

(1.52–2)
$$\int_a^a f(t)dt = 0$$

and

(1.52–3)
$$\int_b^a f(t)dt = -\int_a^b f(t)dt, \quad a < b.$$

These formalities are convenient when dealing with integrals as functions of the limits of integration. With these two formulas available it is not difficult to see that formula (1.5–7) is valid for any positions of a, b, c on an interval where f is continuous.

A graph of $y = F(x)$, corresponding to a graph of $y = f(t)$ as shown in Fig. 14, would appear somewhat as in Fig. 15.

We are going to be primarily interested in the derivative of the function $F(x)$ defined by the integral (1.52–1).

THEOREM VII.   *Let $f(t)$ be continuous, $a \leq t \leq b$, and define*

$$F(x) = \int_a^x f(t)dt, \quad a \leq x \leq b.$$

Then $F(x)$ is differentiable, with derivative

(1.52–4)
$$F'(x) = f(x).$$

The formula (1.52–4) may be put verbally in the form: the derivative of the definite integral of a continuous function with respect to the upper limit of integration is equal to the value of the integrand function at this upper limit.

PROOF.   We consider two points x, $x + \Delta x$ of the interval $[a, b]$. Then

$$F(x + \Delta x) - F(x) = \int_a^{x+\Delta x} f(t)dt - \int_a^x f(t)dt$$
$$= \int_x^{x+\Delta x} f(t)dt.$$

By the mean-value theorem for integrals (Theorem VI, § 1.51) we have

$$\int_x^{x+\Delta x} f(t)dt = \Delta x \cdot f(X),$$

where X is some number between x and $x + \Delta x$. Thus we see that

$$\frac{F(x + \Delta x) - F(x)}{\Delta x} = f(X).$$

In this equation we now hold x fixed and make Δx approach zero. Then
X approaches x, and $f(X)$ approaches $f(x)$.

Therefore
$$\lim_{\Delta x \to 0} \frac{F(x + \Delta x) - F(x)}{\Delta x} = f(x).$$

The limit on the left is $F'(x)$, by definition. Thus (1.52–4) is established. It
is clear from the proof that if x is at one end of the interval $[a, b]$, Δx must
be restricted to have but one sign (e.g., $\Delta x > 0$ if $x = a$). In this case
$F'(x)$ is a one-sided derivative.

Theorem VII furnished us with a complete solution to the problem
raised in § 1.4 in connection with (1.4–1). If $f(x)$ is a given function, continuous on the closed interval $[a, b]$, then the general solution of the equation

$$\frac{dy}{dx} = f(x)$$

on this interval is
$$y = \int_a^{x} f(t)dt + C,$$

where C is an arbitrary constant. This follows at once from Theorem VII
and the italicized statement accompanying (1.4–6).

EXERCISES

1. The functions f, g, h are assumed to be continuous for all values of their
independent variables. Complete each of the following equations:

(a) $\dfrac{d}{du} \displaystyle\int_2^{u} f(x)\, dx =$ (b) $\dfrac{d}{dx} \displaystyle\int_1^{x} g(s)\, ds =$ (c) $G(y) = \displaystyle\int_{y_0}^{y} h(t)\, dt,\ G'(y) =$

2. If $\phi(t) = \int_0^t (3 + x^3)^{-1/2}\, dx$, find (a) $\phi'(0)$, (b) $\phi'(1)$, (c) $\phi'(s)$.

3. If $F(x) = \displaystyle\int_1^{x} \dfrac{\sin y}{y}\, dy$, find (a) $F'\left(\dfrac{\pi}{6}\right)$, (b) $F'\left(\dfrac{\pi}{2}\right)$, (c) $F'(\pi)$.

4. If $G(x) = \int_{-2}^{x} |s|\, ds$, find (a) $G'(-1)$, (b) $G'(0)$, (c) $G'(2)$, (d) $G'(a)$.

5. (a) If $F(x) = \int_0^x t(t - 1)e^{-t^2}\, dt$, find the points of relative maxima and
minima of $F(x)$. (b) What is the value of $F(0)$? (c) For what values of x is $F'(x) > 0$,
and for what values of x is $F'(x) < 0$?

6. If $F(x) = \int_0^x t^3 e^{-t^3}\, dt$, find the absolute minimum value of $F(x)$.

1.53   The integral of a derivative

The theorem which we shall prove in this section is fundamental, for it
establishes the standard technique whereby definite integrals are calculated
in practice. The four-step defining process of arriving at a definite integral,

as set forth in § 1.5, is difficult to apply. For a large and important class of integrands the following theorem provides a convenient method of finding the value of the integral.

THEOREM VIII. *Let f be a given function continuous on the closed interval [a, b]. Suppose that F is any differentiable function such that $F'(x) = f(x)$ when $a \leq x \leq b$. Then*

(1.53–1) $$\int_a^b f(x)dx = F(b) - F(a).$$

PROOF. We are by hypothesis given a function $F(x)$ whose derivative is $f(x)$. By Theorem VII (§ 1.52) we know another function with this same derivative, namely

$$\int_a^x f(t)dt.$$

Thus the function $$G(x) = F(x) - \int_a^x f(t)dt$$

is constant, by Theorem V (§ 1.2), since its derivative is zero.

Now $$G(a) = F(a) - 0,$$

by (1.52–2). Also, $$G(b) = F(b) - \int_a^b f(t)dt.$$

Since $G(x)$ is constant we have $G(b) = G(a)$,

or $$F(b) - \int_a^b f(t)dt = F(a).$$

This result is equivalent to (1.53–1), so the proof is complete.

Example 1.   Find the value of the integral $\int_0^{\pi/2} \sin x \, dx$.

Applying Theorem VIII, we seek a function of x whose derivative is $\sin x$. Such a function is $-\cos x$. Therefore

$$\int_0^{\pi/2} \sin x \, dx = -\cos \frac{\pi}{2} + \cos 0 = 1.$$

We are now in a position to see clearly the connection between differentiation and integration. As concepts, by their definition, these processes are quite independent of each other. It turns out, however, that each process is in a certain sense inverse to the other. The two aspects of this mutual inverseness are displayed by Theorems VII and VIII. If we want a function defined when $a \leq x \leq b$ and having as its derivative a certain given continuous function $f(x)$, the class of all functions satisfying our want is the family $\int_a^x f(t)dt + C$. If, on the other hand, we wish to integrate a given continuous function $f(x)$, we can do so by the formula

$$\int_a^b f(x)dx = F(b) - F(a)$$

provided we can find a function $F(x)$ having $f(x)$ as its derivative at all points of the interval $[a, b]$.

Example 2.   Evaluate the integral $\int_{-10}^{-2} \dfrac{dx}{x}$:

We seek a function whose derivative is $1/x$ when $-10 \leqq x \leqq -2$. The familiar formula

(1.53–2) $\dfrac{d}{dx} \log x = \dfrac{1}{x}$

will not quite do, for $\log x$ is not defined if $x < 0$. But, if $x < 0$, $\log(-x)$ is defined, and

(1.53–3) $\dfrac{d}{dx} \log(-x) = \dfrac{1}{-x}(-1) = \dfrac{1}{x}$.

Hence, by Theorem VIII, with $f(x) = 1/x$, $F(x) = \log(-x)$, we have

$$\int_{-10}^{-2} \frac{dx}{x} = \log(-x)\Big|_{-10}^{-2} = \log 2 - \log 10 = \log \tfrac{1}{5}.$$

The formulas (1.53–2) and (1.53–3) can be combined in the single formula

(1.53–4) $\dfrac{d}{dx} \log|x| = \dfrac{1}{x}$ if $x \neq 0$.

We now consider a theorem which is much used in transforming definite integrals by substitution.

THEOREM IX.   *Suppose the following conditions are fulfilled:*
   (1) *$f(x)$ is continuous, $a \leqq x \leqq b$;*
   (2) *$\phi(t)$ and $\phi'(t)$ are continuous, and $a \leqq \phi(t) \leqq b$ when $\alpha \leqq t \leqq \beta$;*
   (3) *$\phi(\alpha) = a$, $\phi(\beta) = b$.*
   Let $g(t) = f(\phi(t))\phi'(t)$. Then

(1.53–5) $\displaystyle\int_a^b f(x)dx = \int_\alpha^\beta g(t)dt.$

We note that the formula (1.53–5) is easily remembered in the following way: when $x = \phi(t)$, $dx = \phi'(t)dt$, so that $f(x)dx = g(t)dt$. The limits $x = a$,

$x = b$ correspond to the limits $t = \alpha$, $t = \beta$ by (3). The proof of the theorem is left to the student (Exercise 10).

EXERCISES

1. What is wrong with the following equations?

(a) $\displaystyle\int_{-1}^{2} \frac{dx}{x^2} = -\frac{1}{x} \Big|_{-1}^{2} = -\frac{1}{2} - 1 = -\frac{3}{2}.$

(b) $\displaystyle\int_{0}^{\pi} \sec^2 x \, dx = \tan x \Big|_{0}^{\pi} = 0.$

2. Show that $\dfrac{d}{dx} \log |x - c| = \dfrac{1}{x - c}.$ Under what restrictions on a, b, c is

the formula $\displaystyle\int_{a}^{b} \frac{dx}{x - c} = \log \left| \frac{b - c}{a - c} \right|$

correct?

3. Find the values of the following integrals:

(a) $\displaystyle\int_{4}^{10} \frac{dx}{x - 2}$;

(b) $\displaystyle\int_{-4}^{-10} \frac{dx}{x - 2}$;

(c) $\displaystyle\int_{3}^{4} \frac{x \, dx}{x^2 - 25}$;

(d) $\displaystyle\int_{-4}^{-3} \frac{x \, dx}{x^2 - 25}$;

(e) $\displaystyle\int_{-1}^{2} \frac{dx}{9 - x^2}$;

(f) $\displaystyle\int_{4}^{10} \frac{dx}{9 - x^2}.$

4. Show that $\dfrac{d}{dx} \log \left| \dfrac{a + x}{a - x} \right| = \dfrac{2a}{a^2 - x^2}$ if $x^2 \neq a^2,$

and hence that

$$\int_{x_1}^{x_2} \frac{dx}{a^2 - x^2} = \frac{1}{2a} \log \left\{ \left| \frac{a + x_2}{a - x_2} \right| \left| \frac{a - x_1}{a + x_1} \right| \right\}$$

provided that x_1 and x_2 are not separated by either of the points $x = a$, $x = -a$.

5. The student will need to recall that, by the standard conventions about principal values of the inverse sine and inverse tangent, $y = \sin^{-1} x$ is the unique y such that $x = \sin y$ and $-\pi/2 \leq y \leq \pi/2$, while $y = \tan^{-1} x$ is the unique y such that $x = \tan y$ and $-\pi/2 < y < \pi/2$. Find the values of the following integrals:

(a) $\displaystyle\int_{-1/2}^{1/2} \frac{dx}{\sqrt{1 - x^2}}$

(c) $\displaystyle\int_{-\sqrt{3}}^{1} \frac{dx}{3 + x^2}$

(b) $\displaystyle\int_{-1}^{\sqrt{3}} \frac{dx}{1 + x^2}$

(d) $\displaystyle\int_{-2}^{2\sqrt{3}} \frac{dx}{\sqrt{16 - x^2}}.$

6. Show that $\tan^{-1} x = \displaystyle\int_{0}^{x} \frac{dt}{1 + t^2}$;

$\sin^{-1} x = \displaystyle\int_{0}^{x} \frac{dt}{\sqrt{1 - t^2}},$ $-1 < x < 1.$

7. Show that

$$\int_{x_1}^{x_2} \tan x \, dx = \log \left| \frac{\cos x_1}{\cos x_2} \right|, \quad -\frac{\pi}{2} < x_1 \leq x_2 < \frac{\pi}{2}.$$

8. Standard tables of integrals list the formula

$$\int \frac{dx}{a + b \cos x} = \frac{2}{\sqrt{a^2 - b^2}} \tan^{-1} \left[\frac{\sqrt{a^2 - b^2} \tan \dfrac{x}{2}}{a + b} \right], \quad a^2 > b^2.$$

(a) Let
$$F(x) = \frac{2}{\sqrt{a^2 - b^2}} \tan^{-1} \left[\frac{\sqrt{a^2 - b^2} \tan \dfrac{x}{2}}{a + b} \right]$$

and verify that
$$F'(x) = \frac{1}{a + b \cos x}$$

whenever x is not an odd multiple of π. What can be said about $F(x)$ for these exceptional values of x?

(b) Assuming $0 < b < a$, find the limit of $F(x)$ as $x \to \pi$ from the left; as $x \to \pi$ from the right.

(c) Consider $F(x)$ as defined at π by its limit as $x \to \pi$ from the left, use Theorem VIII to show that

$$\int_0^\pi \frac{dx}{a + b \cos x} = \frac{\pi}{\sqrt{a^2 - b^2}} \quad \text{if } a > b > 0.$$

(d) Find the value of $\displaystyle\int_0^{3\pi/2} \frac{dx}{5 - 3 \cos x}$.

9. (a) Discuss critically the integration formula

$$\int \frac{dx}{a^2 \cos^2 x + b^2 \sin^2 x} = \frac{1}{ab} \tan^{-1} \left(\frac{b}{a} \tan x \right),$$

assuming that a and b are positive. Compare with Exercise 8 **(a)**, **(b)**.

(b) Explain the reason for the apparent failure of Theorem VIII in the obviously false result

$$\int_0^\pi \frac{dx}{\cos^2 x + 4 \sin^2 x} = \tfrac{1}{2} \tan^{-1} (2 \tan x) \Big|_0^\pi$$

$$= \tfrac{1}{2} \tan^{-1} (0) - \tfrac{1}{2} \tan^{-1} (0) = 0.$$

(c) Show that $\displaystyle\int_0^{\pi/2} \frac{dx}{a^2 \cos^2 x + b^2 \sin^2 x} = \frac{\pi}{2ab}.$

10. Prove Theorem IX with the aid of the following suggestions: Define $F(x) = \int_a^x f(s) \, ds$, $H(t) = F(\phi(t))$. Show that $H'(t) = g(t)$ (by what theorem?). Then $H(\beta) - H(\alpha) = \int_\alpha^\beta g(t) dt$ (by what theorem?). Now complete the proof of Theorem IX.

11. Show that $\int_0^{\pi/2} \sin^n x \, dx = \int_0^{\pi/2} \cos^n y \, dy$ by an appropriate substitution.

12. Show that $\int_0^1 x^m (1 - x)^n \, dx = \int_0^1 x^n (1 - x)^m \, dx$ by an appropriate substitution.

13. If f is continuous on $[a, b]$, show that $\int_a^b f(x)\, dx = \int_a^b f(a + b - x)\, dx$.

14. If ϕ is a continuous function on $[0, 1]$, show that $\int_0^{\pi/2} \phi(\sin x)\, dx = \int_{\pi/2}^\pi \phi(\sin x)\, dx$, and hence that $\int_0^\pi \phi(\sin x)\, dx = 2\int_0^{\pi/2} \phi(\sin x)\, dx$.

15. Show that the substitution $x = a \cos^2 t + b \sin^2 t$ fulfills condition (2) of Theorem IX for the integral $\int_a^b \sqrt{(x - a)(b - x)}\, dx$. Use the substitution to find the value of the integral.

16. Show that $\int_{-a}^a f(x)\, dx = \int_0^a [f(x) + f(-x)]\, dx$ if f is continuous on $[-a, a]$. What do you conclude about the value of the integral if f is an odd function? an even function?

17. Show that $\int_a^b xf''(x)\, dx = bf'(b) - f(b) + f(a) - af'(a)$.

18. Show that, if f is continuous on $[0, 1]$, $\int_0^\pi xf(\sin x)\, dx = (\pi/2)\int_0^\pi f(\sin x)\, dx$.

Use this result to find the value of $\displaystyle\int_0^\pi \frac{x \sin x}{1 + \cos^2 x}\, dx$.

1.6 Limits

Most students of calculus get their first extensive experience with the limit concept in the course of learning about differentiation. In the process of working out formulas for the derivatives of x^n, $\sin x$, and other elementary functions, as well as in the establishment of the rules for differentiating sums, products, and quotients, students are taught to use some of the fundamental theorems about limits, such as:

The limit of a sum is equal to the sum of the limits.
The limit of a product is equal to the product of the limits.
The limit of a quotient is equal to the quotient of the limits, provided the limit of the denominator is not zero.

These are not fully and precisely formulated theorems in the form here given; nevertheless, each statement conveys the central idea of an important theorem about limits. No doubt most students accept the truth of these three propositions as being intuitively evident. Probably the most that can be expected, perhaps all that is desirable, in an elementary course in calculus, is the cultivation in the student of an awareness that these propositions exist and that it is necessary to appeal to them in building up the structure of calculus. As we proceed to a more advanced level, however, it becomes more important for us to analyze the limit concept carefully, and to see how the whole theory of limiting processes is developed. As with any part of mathematics, we cannot build a clear and precise theory of limits unless we formulate our basic definitions in terms sharp enough for use in the giving of clean-cut proofs.

There are at least three recognizably distinct limit concepts in elementary calculus. We describe them briefly in turn.

(1) The limit of a function of a continuous variable: $\lim_{x \to x_0} f(x)$. As examples of this kind of limit we cite

$$(a)\ \lim_{x \to 1} \frac{6\,x}{x^2 + 2} = 2,\ (b)\ \lim_{x \to 0} \frac{\sin x}{x} = 1,\ (c)\ \lim_{x \to e} \log_e x = 1.$$

The limits (1.11–2)–(1.11–4) defining a derivative are also of this type. The definition of this kind of limit was given in § 1.1 (see 1.1–2).

(2) The limit of a sequence of numbers: $\lim_{n \to \infty} s_n$. As examples we may choose

$$(a)\ \lim_{n \to \infty} \frac{2\,n}{n + 1} = 2, \qquad\qquad (b)\ \lim_{n \to \infty} \frac{\sin (n\pi/2)}{n} = 0,$$

$$(c)\ \lim_{n \to \infty} 2^{1/n} = 1, \qquad\qquad (d)\ \lim_{n \to \infty} \left(1 + \frac{1}{n}\right)^n = e.$$

Here the variable n is *discrete*, running through the natural numbers 1, 2, 3, $\cdots$. We have not yet defined this kind of limit formally.

(3) The type of limit occurring in the definition of an integral:

$$\lim \sum_{i=1}^{n} f(x'_i)\Delta x_i = \int_a^b f(x)dx$$

(in the notation of § 1.5). Here the variable quantity is an approximating sum; it does not depend merely on n, nor does it depend on a single continuous variable x. This kind of limit is therefore different from either of the types (1) and (2).

We shall now discuss these three types of limit in more detail.

1.61 Limits of functions of a continuous variable

In speaking of $\lim_{x \to x_0} f(x)$ it is usually understood that there is some open interval (a, b) such that $a < x_0 < b$, and such that f is defined at each point of the interval except possibly at x_0 itself. We repeat the definition of a limit from § 1.1:

DEFINITION. The function f has the limit A (a certain real number) as $x \to x_0$ provided that to each positive number ϵ there corresponds some positive number δ such that

(1.61–1) $|f(x) - A| < \epsilon$ if $0 < |x - x_0| < \delta$

and if f is defined at x.

In certain cases f may be defined only on one side of x_0; then we speak of one-sided limits.

In all these cases we speak of x as a *continuous variable*, because it is free to assume all values on certain intervals of the real-number scale. The adjective "continuous" here contrasts with "discrete."

The foregoing definition of limit was first used systematically in the foundations of calculus by the French mathematician Augustin-Louis Cauchy (1789–1857). Prior to Cauchy there had been various attacks on the problem of putting the fundamental concepts of the calculus on a sound basis. During the period from Newton and Leibniz to the work of Cauchy (roughly 1665–1821) a good part of the formal side of calculus and the technique of applying it to physics and geometry had been developed, but the reasoning was often hazy and dependent upon intuition rather than logic. The work of clarifying the fundamentals and establishing a satisfactory standard of logical rigor did not end with Cauchy, of course. A more adequate understanding of the real number system had yet to come.

We have put in this brief reference to mathematical history in order to draw a parallel. The student's understanding of calculus will normally pass through stages of development not unlike the historical ones, but with a difference. The student need not embrace all or even many of the misconceptions which have been nourished about calculus in the long evolution of the subject since the time of Newton and Leibniz, provided he will put aside preconceived notions of fundamental mathematical concepts and base his understanding on careful study of modern definitions and theorems. Intuition and experience must play their part in learning, of course.

The notation for a limit is frequently employed in conjunction with the use of the symbols $+\infty$, $-\infty$. These symbols are associated with the words "infinity" and "infinite," with which the student no doubt already has some familiarity. Our present object is to discuss the meanings of such symbolic assertions as

$$\lim_{x \to x_0} f(x) = +\infty, \quad \lim_{x \to x_0} f(x) = -\infty.$$

These statements are given various verbal renderings. The first one may be put in the form "$f(x)$ becomes positively infinite (or approaches plus infinity) as $x \to x_0$." For the second statement "positively" is replaced by "negatively," and "plus" by "minus." The meanings of the statements are contained in the following definitions:

DEFINITION. We write $\lim_{x \to x_0} f(x) = +\infty$ if to every $M > 0$ there corresponds some $\delta > 0$ such that $M < f(x)$ whenever $0 < |x - x_0| < \delta$. We write $\lim_{x \to x_0} f(x) = -\infty$ if to every $M > 0$ there corresponds some $\delta > 0$ such that $f(x) < -M$ whenever $0 < |x - x_0| < \delta$.

The definitions are modified in obvious ways for one-sided approach of x to x_0.

Example 1. $\lim_{x \to 0} x^{-2} = + \infty$.

To obtain this result from the definition, we consider the inequality $M < 1/x^2$, where M is any positive number. An equivalent inequality is $x^2 < 1/M$, or $|x| < M^{-1/2}$. Thus we may choose $\delta = M^{-1/2}$ and the assertion stated in Example 1 is seen to be true by definition.

A possible misconception of the definition may be allayed by the following example:

Example 2. Consider $f(x) = (1/x^2) \sin^2 (\pi/x)$ as $x \to 0$.

In this case it is *not* true that $\lim_{x \to 0} f(x) = + \infty$, in spite of the fact that f takes on values as large as we please as near 0 as we please. For, considering the definition, choose $M = 1$. The inequality $(1/x^2) \sin^2 (\pi/x) > 1$ is true for *some* values of x very close to 0, e.g., $x = \frac{2}{3}, \frac{2}{5}, \frac{2}{7}, \frac{2}{9}, \cdots$; but for certain other values of x near 0 it is not true, e.g., $x = \frac{1}{2}, \frac{1}{3}, \cdots,$ $1/n, \cdots$.

The symbols $+ \infty, - \infty$ are also used in connection with the independent variable. We use $x \to + \infty$ as a symbolic equivalent of the phrase "x tends to plus infinity" (or "x becomes positively infinite"). This occurs in such contexts as $\lim_{x \to + \infty} f(x) = A$, which by definition means "to each $\epsilon > 0$ there corresponds an $M > 0$ such that $|f(x) - A| < \epsilon$ whenever $M < x$."

Example 3. $\lim_{x \to + \infty} (2x - 1)/(x - 3) = 2$.

To verify this assertion by direct application of the definition, we proceed as follows:

$$\frac{2x - 1}{x - 3} - 2 = \frac{2x - 1 - 2x + 6}{x - 3} = \frac{5}{x - 3}.$$

Therefore, if $x > 3$ we have

$$\left| \frac{2x - 1}{x - 3} - 2 \right| = \frac{5}{x - 3} < \epsilon$$

provided $(5/\epsilon) < x - 3$, or $3 + (5/\epsilon) < x$. This shows that, for given $\epsilon > 0$, the conditions of the definition are fulfilled with $M = 3 + (5/\epsilon)$.

The following theorem states explicitly an important principle which is frequently used in mathematical arguments.

THEOREM X.   *Let $f(x)$ be defined at all points except $x = a$ of some open interval containing that point, and suppose that, as x tends to a, $f(x)$ approaches a limit which is positive. Then $f(x)$ is positive when x is sufficiently near the point a.*

PROOF.   Let the limit of $f(x)$ be A. Then, according to the definition, if any positive number ϵ is given, there is some corresponding positive number δ such that $|f(x) - A| < \epsilon$ when $0 < |x - a| < \delta$. Now $A > 0$, by hypothesis. Suppose that we take $A/2$ for the ϵ. Then there is a certain δ such that $|f(x) - A| < A/2$ if $0 < |x - a| < \delta$. Now $|f(x) - A| < A/2$ is equivalent to the double inequality $-A/2 < f(x) - A < A/2$, as the student will easily see. Using the left one of these last two inequalities, and transposing, we see that $-A/2 + A < f(x)$ when $0 < |x - a| < \delta$. Certainly then $f(x)$ is positive, for $A/2 > 0$. This completes the proof.

Example 4. As an illustration of the use of Theorem X, we shall prove the following statement: *Let f be defined in some open interval containing* $x = a$, *and suppose that* $f'(a)$ *exists and is positive. Then, for points* x_1, x_2 *sufficiently close to* $x = a$, $x_1 < a < x_2$ *implies*

$$(1.61\text{--}2) \qquad f(x_1) < f(a) < f(x_2).$$

To prove this assertion, consider the definition

$$f'(a) = \lim_{x \to a} \frac{f(x) - f(a)}{x - a}.$$

We are assuming that $f'(a) > 0$; consequently, by Theorem X,

$$(1.61\text{--}3) \qquad \frac{f(x) - f(a)}{x - a} > 0$$

when x is sufficiently close to a. But from (1.61–3) we infer that $f(x) > f(a)$ if $x > a$, and $f(x) < f(a)$ if $x < a$. Thus the assertion about the inequalities (1.61–2) is seen to be true.

We shall state two further theorems which, like Theorem X, are frequently used in the kind of reasoning with limits which occurs regularly in calculus.

THEOREM XI. *Let f be defined at all points except* $x = a$ *of some open interval containing that point. Suppose that there is a number* M *such that* $f(x) \leqq M$ *when* x *is sufficiently near* a. *Further suppose that* $\lim_{x \to a} f(x) = A$. *Then* $A \leqq M$.

THEOREM XII. *Let f, g, h be functions of x defined at all points except* $x = a$ *of some open interval containing that point. Suppose that* $f(x) \leqq g(x) \leqq h(x)$, *and suppose that the limits* $\lim_{x \to a} f(x)$, $\lim_{x \to a} h(x)$ *exist and are equal. Then* $\lim_{x \to a} g(x)$ *exists also, and all three limits are equal.*

We leave the proofs of these theorems as exercises for the student, but we shall use the theorems whenever the need arises. These theorems have analogues for other kinds of limits, e.g., limits of sequences and the limits defining definite integrals. For example, the proof of (1.51–2) employed a principle similar to that of Theorem XI, as applied to the integral as the limit of approximating sums. One of the standard proofs that $\lim_{x \to 0} (\sin x)/x = 1$ uses the principle of Theorem XII, with $f(x) = \cos x$, $g(x) = (\sin x)/x$, $h(x) = 1/\cos x$.

EXERCISES

1. Show that $\lim_{x \to 0} (\cos x)/x^2 = +\infty$. SUGGESTION: Note that $\cos x \geq \frac{1}{2}$ if $|x| \leq \pi/3$. If $M > 0$ is given, choose δ as the lesser of the numbers $\pi/3$, $(2 M)^{-1/2}$, and explain why $M < (\cos x)/x^2$ if $|x| < \delta$.

2. Show that $\lim_{x \to 0} (1 + \sin^2 x)/x^2 = +\infty$.

3. What is $\lim_{x \to 0} (\sin^2 x)/x^4$? Justify your answer.

4. Suppose $f(x) \geq m > 0$ and $g(x) > 0$ when x is any point in an interval $a < x < b$ containing x_0. Also suppose $g(x) \to 0$ as $x \to x_0$. Show that

$$\lim_{x \to x_0} \frac{f(x)}{g(x)} = +\infty.$$

5. Show that $\lim_{x \to +\infty} (2 + \sin x)/e^{-x} = +\infty$.

6. Given $\epsilon > 0$, find M so that

$$\left| \frac{9 x + 4}{3 x - 1} - 3 \right| < \epsilon \text{ if } M < x.$$

What statement about a limit does this prove?

7. Suppose that, as $x \to +\infty$ (or as $x \to x_0$), $|f(x)| \leq M$, where M is a constant, and that $|g(x)| \to +\infty$. Show that $\dfrac{f(x)}{g(x)} \to 0$.

8. Write $f(x) = \dfrac{4 x^2 - x + 1}{x^2 + 2 x - 3} = \dfrac{4 - (1/x) - (1/x^2)}{1 + (2/x) - (3/x^2)}$.

Then find $\lim_{x \to +\infty} f(x)$, using the rules for limits of sums and quotients.

9. If $P(x)$ and $Q(x)$ are polynomials of degrees m and n respectively, discuss $\lim_{x \to +\infty} \dfrac{P(x)}{Q(x)}$ according as $m > n$, $m = n$, $m < n$. Show that the results when $x \to -\infty$ are the same as when $x \to +\infty$, if $m \leq n$ or if $m - n$ is even.

10. In the following cases discuss the behavior of the given function $f(x)$ as $x \to +\infty$. Does $f(x)$ approach a number as limit? Does $f(x)$ tend to $+\infty$ or to $-\infty$? Or does it do none of these things?
(a) $\sin x$; (b) $\sin (1/x)$; (c) $(\sin x)/x$; (d) $x \sin x$; (e) $x^2 + x \sin x$; (f) $x^2 + x^2 \cos x$; (g) $x + x^2 \sin x$.

11. If $\lim_{h \to 0} \dfrac{f(x + h) - f(x)}{h} = +\infty$ we say that $f'(x) = +\infty$.
Likewise we define the meaning of $f'(x) = -\infty$. (a) If $f(x) = x^{1/3}$, what is $f'(0)$? (b) If $f(x) = x^{2/3}$, what about $f'(0)$? What about $f'_+(0)$ and $f'_-(0)$? (c) If $f(x) = 0$

when $x < 0$, $f(0) = 1$ and $f(x) = 2$ when $x > 0$, show that $f'(0) = +\infty$. This shows that $f'(x_0) = +\infty$ does not imply that f is continuous at x_0.

12. Prove that the law of the mean is true with the following hypotheses, which are weaker than those imposed in § 1.2: f is continuous when $a \le x \le b$, and, for each x of the open interval $a < x < b$, f is either differentiable (i.e., $f'(x)$ exists as a finite limit) or $f'(x)$ is $+\infty$ or $-\infty$ as defined in Exercise 11.

13. Prove Theorem XI and Theorem XII. For Theorem XI begin by supposing $M < A$, and show that this leads to a contradiction. For Theorem XII let $f(x)$ and $h(x)$ approach A as limit, and note that, if $\epsilon > 0$, the inequalities $A - \epsilon < f(x) < A + \epsilon$, $A - \epsilon < h(x) < A + \epsilon$ must hold when $|x - a|$ is sufficiently small.

14. State and prove a theorem similar to Theorem XI in which the inequalities are reversed.

1.62   Limits of sequences

A sequence is an ordered set of numbers in one-to-one correspondence with the positive integers. This correspondence may be shown by numbering the terms of the sequence in order:

$$s_1, s_2, s_3, \cdots, s_n, \cdots .$$

The sequence is then denoted symbolically by $\{s_n\}$. As examples we cite:

(a) $2, 4, 6, 8, \cdots, 2n, \cdots$;

(b) $\frac{1}{2}, \frac{1}{4}, \frac{1}{8}, \cdots, 1/(2n), \cdots$;

(c) $\frac{1}{2}, \frac{2}{5}, \frac{3}{10}, \frac{4}{17}, \cdots, n/(n^2 + 1), \cdots$;

(d) $1, 1 + \frac{1}{2}, 1 + \frac{1}{2} + \frac{1}{3}, \cdots, 1 + \frac{1}{2} + \cdots + (1/n), \cdots$;

(e) $1, 0, 1, 0, \cdots \left(s_n = \dfrac{1 - (-1)^n}{2}, n = 1, 2, \cdots \right).$

A sequence is in fact a particular kind of function, a function whose independent variable n ranges over the set of positive integers. We could (and occasionally do) use the functional notation $f(n)$ for a sequence, but notations such as $\{x_n\}$, $\{s_n\}$, $\{a_n\}$ are more common. Observe that $\{s_n\}$ is the symbol for the sequence (function) as a whole, whereas s_n is the symbol for the nth term (the value of the function).

Sometimes it is convenient to have a notation in which the terms are numbered $0, 1, 2, \cdots$ instead of $1, 2, 3, \cdots$. For instance, if the sequence is $1, 2, 4, 8, 16, 32, \cdots$, it is convenient to denote it by $s_0, s_1, s_2, \cdots, s_n$, so that $s_n = 2^n$ (note that $s_0 = 2^0 = 1$).

The definition of the limit of a sequence is quite similar to the definition of the limit of a function of a continuous variable, as given in § 1.61.

Thinking of the sequence $\{s_n\}$ as a function, let us compare $\{s_n\}$ with a function $f(x)$ of the continuous variable x. The definition of $\lim_{n \to \infty} s_n$ is then very similar to the definition of $\lim_{x \to +\infty} f(x)$ (just preceding Example 3 in § 1.61).

DEFINITION. We say that $\lim_{n \to \infty} s_n = A$ if for each positive ϵ there is some integer N depending on ϵ such that $| s_n - A | < \epsilon$ whenever $N \leq n$. In this case we say that the sequence $\{s_n\}$ is *convergent* and that it has the limit A.

Example 1.   We shall show that $\lim_{n \to \infty} (\frac{2}{3})^n = 0$.

Suppose $\epsilon > 0$ is given. We wish to find N so that $N \leq n$ will insure $(\frac{2}{3})^n < \epsilon$, or, what is equivalent, $1/\epsilon < (\frac{3}{2})^n$. Now, in the sequence $\{(\frac{3}{2})^n\}$, each term is half again as large as its predecessor, and the first term is $\frac{3}{2}$. Hence certainly $(\frac{3}{2})^n > n(\frac{1}{2})$. Thus, to get $1/\epsilon < (\frac{3}{2})^n$, it is amply sufficient to have $1/\epsilon < n(\frac{1}{2})$, or $2/\epsilon < n$. Therefore we take N as the first integer which is greater than $2/\epsilon$, and then certainly $(\frac{2}{3})^n < \epsilon$ if $N \leq n$. This is proof that $\lim_{n \to \infty} (\frac{2}{3})^n = 0$.

Example 2.   We now show that $\lim_{n \to \infty} a^n = 0$ if $0 < a < 1$. This includes Example 1 as a special case.

We can express a in the form

$$a = \frac{1}{1 + h}, \quad h > 0.$$

This is because $0 < a < 1$; h is given by

$$h = \frac{1 - a}{a} .$$

We now use (1.2–5), with $\alpha = n$:

$$a^{-n} = (1 + h)^n \geq 1 + nh.$$

This implies $$a^n \leq \frac{1}{1 + nh} .$$

Suppose now that $\epsilon > 0$ is given. We wish to find an N such that $N \leq n$ implies $a^n < \epsilon$. It is sufficient to have

$$\frac{1}{1 + nh} < \epsilon, \quad \text{or} \quad \frac{1}{\epsilon} < 1 + nh, \quad \text{or} \quad \frac{1 - \epsilon}{\epsilon} < nh.$$

Since $h > 0$, some multiple of h will exceed $(1 - \epsilon)/\epsilon$. Let N be a positive integer such that $(1 - \epsilon)/\epsilon < Nh$. Then $N \leq n$ implies $a_n < \epsilon$, and the proof is complete.

Theorems X, XI, and XII have analogues for sequences. The student can easily formulate these analogues for himself. The general theorems about sums, products, and quotients (see § 1.6) also apply to sequences.

Example 3.   Find $\lim\limits_{n \to \infty} \dfrac{2\,n^3 + n^2 - 7\,n}{n^3 + 2\,n + 2}$.

We write the general term of the sequence in the form

$$\frac{2 + (1/n) - (7/n^2)}{1 + (2/n^2) + (2/n^3)} \, .$$

By the theorem on sums,

$$\lim_{n \to \infty} \left(2 + \frac{1}{n} - \frac{7}{n^2} \right) = 2 \quad \text{and} \quad \lim_{n \to \infty} \left(1 + \frac{2}{n^2} + \frac{2}{n^3} \right) = 1.$$

By the theorem on quotients,

$$\lim_{n \to \infty} \frac{2 + (1/n) - (7/n^2)}{1 + (2/n^2) + (2/n^3)} = \frac{2}{1} \, .$$

Hence the required limit is 2.

Many important sequences have a property which is described by the word *monotonic*. A sequence $\{s_n\}$ is called *nondecreasing* if $s_1 \leqq s_2 \leqq s_3 \leqq \cdots$, i.e., if $s_n \leqq s_{n+1}$ for every positive integer n. It is called *strictly increasing* (or just *increasing*) if $s_1 < s_2 < s_3 < \cdots$. For example, the sequence

$$1, 1, 2, 2, 3, 3, \cdots$$

is nondecreasing. The nth term of this sequence is given by $\left[\dfrac{n+1}{2} \right]$ $\left(\text{the greatest integer} \leqq \dfrac{n+1}{2} \right)$. The sequence $\{s_n\}$, where $s_n = \dfrac{n}{n+1}$, is strictly increasing:

$$\tfrac{1}{2}, \tfrac{2}{3}, \tfrac{3}{4}, \cdots .$$

Likewise, we call a sequence *nonincreasing* if $s_n \geqq s_{n+1}$ for every n, and (*strictly*) *decreasing* if $s_n > s_{n+1}$ for every n. A sequence of any one of these four types is called *monotonic*.

One very important type of nondecreasing sequence is the sequence of decimal approximations to a positive number. For example, let s_n be the number obtained by writing the first n decimal places of the number $\tfrac{1}{3}$:

$$s_1 = 0.3, \ s_2 = 0.33, \ s_3 = 0.333, \ \cdots .$$

The general formula for s_n may be written

$$s_n = \frac{3}{10} + \frac{3}{10^2} + \cdots + \frac{3}{10^n} \, .$$

Using the formula

$$(1.62\text{--}1) \qquad a + ar + ar^2 + \cdots + ar^{n-1} = a\left(\frac{1 - r^n}{1 - r}\right)$$

for the sum of a geometric progression, we find

$$s_n = \tfrac{3}{10}\,\frac{1 - (\tfrac{1}{10})^n}{1 - \tfrac{1}{10}} = \tfrac{1}{3}[1 - (\tfrac{1}{10})^n].$$

We see from this that

$$(1.62\text{--}2) \qquad\qquad \lim_{n \to \infty} s_n = \tfrac{1}{3}.$$

When we write

$$\tfrac{1}{3} = 0.333 \cdots,$$

the meaning is exactly that expressed by (1.62–2).

Similar remarks apply to all nonterminating decimal representations, e.g.,

$$\tfrac{2}{3} = \lim_{n \to \infty} s_n, \quad s_n = \underbrace{0.66 \cdots 6}_{n}, \quad \sqrt{3} = \lim_{n \to \infty} x_n,$$

where $x_1 = 1.7$, $x_2 = 1.73$, $x_3 = 1.732$, $x_4 = 1.7321$, etc.

Suppose that $\{s_n\}$ is a nondecreasing sequence. There are just two possibilities: either (1) there is some number M such that $s_n \leqq M$ for every n, or (2) $s_n \to +\infty$ as $n \to \infty$ (this means that no matter what M is chosen, $M < s_n$ for all sufficiently large values of n). In case (1) we say that the sequence is *bounded above*. Likewise, for a nonincreasing sequence $\{s_n\}$ there are just two possibilities: either (1) the sequence is *bounded below*, i.e., there is some number M such that $M \leqq s_n$ for every n, or (2) $s_n \to -\infty$ as $n \to \infty$.

A sequence is called *bounded* if it is bounded both above and below. Observe that an increasing or nondecreasing sequence is always bounded *below*, since $s_1 \leqq s_n$ for every n. Likewise, a decreasing or nonincreasing sequence is bounded above, since $s_n \leqq s_1$.

There is a very important theorem which asserts that a monotonic sequence is convergent (i.e., has a limit) provided it is bounded.

THEOREM XIII. *Suppose that either*

$$(a) \quad s_n \leqq s_{n+1} \quad and \quad s_n \leqq M, n = 1, 2, \cdots,$$
$$or \qquad (b) \quad s_{n+1} \leqq s_n \quad and \quad M \leqq s_n, n = 1, 2, \cdots,$$

where M is a constant. Then $\lim_{n \to \infty} s_n$ *exists.*

The great importance of this theorem lies in the fact that by using it we can be sure that certain sequences are convergent without knowing precisely what the limits are. The theorem is proved in Chapter II (Theorems III and IV). The proof is based upon a careful discussion of the nature of the real number system.

Example 4. Let $s_n = \dfrac{1 \cdot 3 \cdot 5 \cdots (2n-1)}{2 \cdot 4 \cdot 6 \cdots (2n)}$. Show that this sequence is monotonic and bounded, and therefore convergent.

In order to make sure the notation is understood, let us write a few terms of the sequence, by substituting successively $n = 1, 2, 3, \cdots$. We have $\quad s_1 = \dfrac{1}{2}, s_2 = \dfrac{1 \cdot 3}{2 \cdot 4} = \dfrac{3}{8}, s_3 = \dfrac{1 \cdot 3 \cdot 5}{2 \cdot 4 \cdot 6} = \dfrac{5}{16}$, etc.

The sequence is decreasing. For,

$$s_2 = \tfrac{3}{4} s_1, \quad s_3 = \tfrac{5}{6} s_2, \quad s_4 = \tfrac{7}{8} s_3,$$

and in general

$$s_{n+1} = \frac{2n+1}{2n+2} s_n < s_n.$$

All the terms are positive; so that $0 < s_n \le \tfrac{1}{2}$. Thus the sequence is bounded. Hence it must have a limit. This argument does not show what the limit may be.

Example 5. Consider the sequence defined by

$$s_n = \frac{1}{1!} + \frac{1}{2!} + \frac{1}{3!} + \cdots + \frac{1}{n!},$$

the first few terms of which are

$$s_1 = 1, \; s_2 = 1 + \frac{1}{1 \cdot 2} = \frac{3}{2}, \; s_3 = 1 + \frac{1}{1 \cdot 2} + \frac{1}{1 \cdot 2 \cdot 3} = \frac{5}{3}.$$

This sequence is plainly monotonic, for

$$s_{n+1} = s_n + \frac{1}{(n+1)!}, \; s_n < s_{n+1}.$$

We shall show that the sequence is bounded above. Now, if $n > 2$,

$$\frac{1}{n!} = \frac{1}{1 \cdot 2 \cdot 3 \cdots n} < \underbrace{\frac{1}{2 \cdot 2 \cdots 2}}_{n-1} = \frac{1}{2^{n-1}}.$$

Therefore, if $n > 2$,

$$s_n < 1 + \frac{1}{2} + \frac{1}{2^2} + \cdots + \frac{1}{2^{n-1}}.$$

But, by (1.62–1), taking $a = 1$, $r = \frac{1}{2}$, we have

$$1 + \frac{1}{2} + \frac{1}{2^2} + \cdots + \frac{1}{2^{n-1}} = 2[1 - (\tfrac{1}{2})^n] < 2.$$

Therefore $s_n < 2$ for all n. Consequently, by Theorem XIII, the sequence $\{s_n\}$ is convergent. The theorem does not tell us the exact value of the limit, though of course we can see that it is not larger than 2.

Example 6.   Let $\{s_n\}$ be the sequence

$$s_n = \left(1 + \frac{1}{n}\right)^n.$$

We shall show that the sequence is increasing and bounded above. First we write the binomial expansion

$$(1 + a)^n = 1 + na + \frac{n(n-1)}{1 \cdot 2} a^2 + \frac{n(n-1)(n-2)}{1 \cdot 2 \cdot 3} a^3 + \cdots + a^n.$$

The coefficient of a^k $(1 \leq k \leq n)$ is

$$\frac{n(n-1)(n-2) \cdots (n-k+1)}{k!}.$$

Putting $1/n$ in place of a, we have

$$\left(1 + \frac{1}{n}\right)^n = 1 + n\left(\frac{1}{n}\right) + \frac{n(n-1)}{1 \cdot 2}\left(\frac{1}{n}\right)^2 + \cdots + \left(\frac{1}{n}\right)^n.$$

The expression on the right has $n + 1$ terms, a typical one of which is

$$\frac{n(n-1)(n-2) \cdots (n-k+1)}{k! \, n^k} = $$
$$\frac{1}{k!} \cdot 1 \cdot \left(1 - \frac{1}{n}\right)\left(1 - \frac{2}{n}\right) \cdots \left(1 - \frac{k-1}{n}\right).$$

Thus

$$(1.62–3) \quad \left(1 + \frac{1}{n}\right)^n = 1 + 1 + \frac{1 - \dfrac{1}{n}}{2!} + \frac{\left(1 - \dfrac{1}{n}\right)\left(1 - \dfrac{2}{n}\right)}{3!}$$
$$+ \cdots + \frac{\left(1 - \dfrac{1}{n}\right) \cdots \left(1 - \dfrac{n-1}{n}\right)}{n!}.$$

Now suppose that n is replaced by $n + 1$, so as to form the corresponding formula for the expression

$$\left(1 + \frac{1}{n+1}\right)^{n+1}.$$

We see from (1.62–3) that each of the numerators after the first two terms on the right in (1.62–3) is increased when n is replaced by $n + 1$. Moreover, the total number of terms on the right is increased from $n + 1$ to $n + 2$. Hence it is certainly the case that

$$\left(1 + \frac{1}{n}\right)^n < \left(1 + \frac{1}{n+1}\right)^{n+1}.$$

This shows that our sequence is increasing. Moreover, from (1.62–3) it is clear that

$$\left(1 + \frac{1}{n}\right)^n < 1 + 1 + \frac{1}{2!} + \frac{1}{3!} + \cdots + \frac{1}{n!}.$$

In Example 5, we saw that

$$1 + \frac{1}{2!} + \frac{1}{3!} + \cdots + \frac{1}{n!} < 2.$$

Therefore

(1.62–4)
$$\left(1 + \frac{1}{n}\right)^n < 3.$$

This shows that our sequence is bounded above. It therefore has a limit. The limit is denoted by the letter e:

(1.62–5)
$$e = \lim_{n \to \infty} \left(1 + \frac{1}{n}\right)^n.$$

This number e is taken as the base in defining natural logarithms, of which we assume the student already has a working knowledge.

The definition of the limit of a sequence enables us (theoretically, at least) to decide whether any specified number A is or is not the limit of a given sequence $\{s_n\}$. To make the decision in a given case we must work with inequalities to form an estimate of the magnitude of the difference $s_n - A$ as n becomes larger and larger. In practice we often find the limits of sequences by using the theorem on limits of sums, products, and quotients. In this procedure the given sequence is expressed in terms of other sequences whose limits we already know.

There are many cases, however, in which we cannot find the limit of a sequence by direct use of the definition or by using the theorem on limits of sums, products, and quotients. It is very important to be able to recognize, by intrinsic characteristics of the sequence itself, that the sequence is convergent. Then we can say with certainty: "By virtue of such and such a characteristic possessed by this sequence $\{s_n\}$, *there must exist* a number A such that $\lim_{n \to \infty} s_n = A$." One such characteristic feature is the property of being monotonic and bounded. Not all convergent sequences possess this characteristic, of course, but it is nevertheless of great importance.

In higher analysis there are many situations in which we assert the *existence* of something having specified properties. A problem leading to such an assertion may be called an *existence problem*. Many important concepts are introduced via existence problems. Definite integrals and functions defined by infinite series are examples.

EXERCISES

1. Which of the following sequences is convergent? (a) $\{(-1)^n\}$, (b) $\left\{\dfrac{(-1)^n}{n}\right\}$, (c) $\{n(-1)^n\}$, (d) $\{n[1 - (-1)^n]\}$, (e) $\{(-1)^{2n+1}\}$, (f) $\left\{\dfrac{\sin n}{n}\right\}$.

2. If $s_n = \dfrac{n^2 + n - 1}{3 n^2 + 1}$ and $\epsilon > 0$, find N so that $\mid s_n - \frac{1}{3} \mid < \epsilon$ if $N \leq n$.

3. If $s_n = \dfrac{6 n^3 + 2 n + 1}{n^3 + n^2}$, find A so that if $0 < \epsilon < 6$, $\mid s_n - A \mid < \epsilon$ provided $n > \dfrac{6 - \epsilon}{\epsilon}$. What does this show about $\lim_{n \to \infty} s_n$?

4. If $s_n = \dfrac{10^n}{n!}$, show that $s_n \leq \dfrac{10^{10}}{10!} \cdot \dfrac{10}{n}$ if $n > 10$. Hence, if $\epsilon > 0$, find N so that $\mid s_n \mid < \epsilon$ if $N \leq n$. From what value of n onward is $s_{n+1} < s_n$?

5. Let $s_1 = \dfrac{2}{1} \cdot \dfrac{1}{1^2}$, $s_2 = \dfrac{2 \cdot 4}{1 \cdot 3} \cdot \dfrac{1}{2^2}$, $s_3 = \dfrac{2 \cdot 4 \cdot 6}{1 \cdot 3 \cdot 5} \cdot \dfrac{1}{3^2}$, etc. Write the general expression for s_n and show that $s_n < \dfrac{2}{n}$. What do you conclude about $\lim_{n \to \infty} s_n$? Show that $s_{n+1} < s_n$.

6. Let $P(x) = a_0 x^p + a_1 x^{p-1} + \cdots$ and $Q(x) = b_0 x^q + b_1 x^{q-1} \cdots$ be polynomials, with $a_0 b_0 \neq 0$. Discuss $\lim_{n \to \infty} P(n)/Q(n)$ according as (a) $p < q$, (b) $p = q$, (c) $p > q$. (Compare with Example 3.)

7. (a) If $P(x)$ is a polynomial of degree $r > 0$, show that $\lim_{n \to \infty} P(n + 1)/P(n) = 1$. (b) What is $\lim_{n \to \infty} P(2 n)/P(n)$?

8. Find
$$\lim_{n \to \infty} \frac{1 - \left(1 - \dfrac{1}{n}\right)^4}{1 - \left(1 - \dfrac{1}{n}\right)^3}.$$

9. Using rationalization techniques, find the limits of the following sequences: (a) $\{\sqrt{n + 1} - \sqrt{n}\}$; (b) $\{\sqrt{n}(\sqrt{n + 1} - \sqrt{n})\}$.

10. If $\lim_{n \to \infty} \dfrac{x_n - x}{x_n + x} = 0$, show that $\lim_{n \to \infty} x_n = x$. (Write $y_n = \dfrac{x_n - x}{x_n + x}$ and solve for x_n.)

11. Let $f(x)$ be defined as $f(x) = \lim\limits_{n \to \infty} \dfrac{1}{x^n + x^{-n}}$, where $x > 0$. Find the value of $f(x)$ for each positive x.

12. Consider $\lim\limits_{n \to \infty} \left(\dfrac{1 - x^2}{1 + x^2} \right)^n$. For what values of x does the limit exist? Classify the values of x according to the value of the limit.

13. Let $f(x) = x/|x|$ if $x \neq 0$, and define $f(0) = 0$. Show that $f(x) = \lim_{n \to \infty} (2/\pi) \tan^{-1}(nx)$.

14. Find each of the following limits:

(a) $\lim\limits_{n \to \infty} \left(\dfrac{1}{n^2} + \dfrac{2}{n^2} + \cdots + \dfrac{n}{n^2} \right)$;

(b) $\lim\limits_{n \to \infty} \left(\dfrac{1^2}{n^3} + \dfrac{2^2}{n^3} + \cdots + \dfrac{n^2}{n^3} \right)$.

15. If $s_n = \dfrac{1}{\sqrt{n^2 + 1}} + \dfrac{1}{\sqrt{n^2 + 2}} + \cdots + \dfrac{1}{\sqrt{n^2 + n}}$, show that $\left(1 + \dfrac{1}{n} \right)^{-1/2} < s_n < 1$, and hence find $\lim_{n \to \infty} s_n$.

16. (a) Prove that, if $C > 1$, $\lim_{n \to \infty} C^{1/n} = 1$, using the following suggestions. First explain why $C^{1/n} > 1$. Then let $x_n = C^{1/n} - 1$ and use (1.2–5) to show that $C > 1 + nx_n$. It then follows that $\lim_{n \to \infty} x_n = 0$. (Why?) Why then does $C^{1/n} \to 1$? (b) If $0 < C < 1$, show that $\lim_{n \to \infty} C^{1/n} = 1$ by applying the *result* (not the method) of (a).

17. Let $a_n = n^{1/2n}$. For $n > 1$ write $a_n = 1 + x_n$. Use (1.2–5) to prove that $x_n \leq \dfrac{\sqrt{n} - 1}{n} < \dfrac{1}{\sqrt{n}}$. Use this to deduce that $n^{1/n} < 1 + \dfrac{2}{\sqrt{n}} + \dfrac{1}{n}$, and hence that $\lim_{n \to \infty} n^{1/n} = 1$.

18. Let $s_n = n/C^n$, where $C > 1$. Write $\sqrt{C} = 1 + x$ and use (1.2–5) to show that $\sqrt{s_n} < 1/(x\sqrt{n})$. What do you conclude about $\lim_{n \to \infty} s_n$?

19. Let $s_n = \left(1 + \dfrac{1}{n} \right)^{n+1}$. Show that $\dfrac{s_{n-1}}{s_n} = \dfrac{n-1}{n} \left(1 + \dfrac{1}{n^2 - 1} \right)^{n+1}$, and then use (1.2–5) to prove that $s_{n-1} > s_n$. Show that $\lim_{n \to \infty} s_n = e$ (see (1.62–5)).

20. (a) Show that $\lim\limits_{n \to \infty} \left(1 + \dfrac{1}{2n} \right)^n = \sqrt{e}$.

(b) Observe that $\left(1 + \dfrac{2}{n} \right) = \left(1 + \dfrac{1}{n+1} \right) \left(1 + \dfrac{1}{n} \right)$. Now find $\lim\limits_{n \to \infty} \left(1 + \dfrac{2}{n} \right)^n$.

(c) Find $\lim\limits_{n \to \infty} \left(1 + \dfrac{3}{n} \right)^n$.

21. Suppose $0 < s_n$ and $s_{n+1} < rs_n$, where r is a constant such that $0 < r < 1$. Show that $\lim_{n \to \infty} s_n = 0$.

22. Suppose $0 < s_n$ and $s_{n+1} \geq r s_n$, where r is a constant such that $r > 1$. Show that $\lim_{n \to \infty} s_n = +\infty$.

23. Suppose $C > 1$ and $s_n = C^{1/n}$. Show that $1 < s_n$ and $s_{n+1} < s_n$ (assume the contrary in each case, and deduce a contradiction). Hence, by Theorem XIII, $\lim_{n \to \infty} s_n$ exists. Denote the limit by r. Why is $r \geq 1$? Prove that $r = 1$ by showing that $r > 1$ leads to $1/C < (1/r)^n$. This contradicts the result of Example 2. (Why?)

24. Let $a_n = 1 + \dfrac{1}{1!} + \dfrac{1}{2!} + \cdots + \dfrac{1}{n!}$, $b_n = \left(1 + \dfrac{1}{n}\right)^n$. In the course of Example 6 it was proved that $b_n < a_n$. Let $a = \lim_{n \to \infty} a_n$ (the existence of the limit follows from Example 5). By definition, $e = \lim_{n \to \infty} b_n$ (see (1.62–5)). Show that $a = e$, using the following suggestions: First explain why you know that $e \leq a$. Then refer to (1.62–3) and explain why

$$1 + \frac{1}{1!} + \frac{\left(1 - \dfrac{1}{n}\right)}{2!} + \frac{\left(1 - \dfrac{1}{n}\right)\left(1 - \dfrac{2}{n}\right)}{3!} + \cdots + \frac{\left(1 - \dfrac{1}{n}\right) \cdots \left(1 - \dfrac{p-1}{n}\right)}{p!} < e$$

if $1 < p < n$. In this result let $n \to \infty$ and obtain $a_p \leq e$. Now complete the argument as to why $a = e$.

1.63   The limit defining a definite integral

Here we have to do with a limit different in kind from the limit of a function of a continuous variable and the limit of a sequence. Our discussion will be couched in terms of the notation used in the definition of a definite integral in § 1.5. Consider the approximating sums

(1.63–1)
$$\sum_{i=1}^{n} f(x_i')\Delta x_i$$

associated with a given fixed function f which is defined on the interval $a \leq x \leq b$. Although the index n appears here, these sums do not form a sequence. A particular sum depends not merely on n, but on the points of subdivision $x_1, x_2, \cdots, x_{n-1}$ and on the intermediate points $x_1', \cdots, x_n'$. We may say that, the function and the interval being fixed, the approximating sum (1.63–1) is a function of n and of the points $x_1, x_1, \cdots, x_{n-1}$, $x_1', \cdots, x_n'$. When we say that

(1.63–2)
$$\lim \sum_{i=1}^{n} f(x_i')\Delta x_i = \int_a^b f(x)dx,$$

we mean that to each positive number ϵ corresponds another positive number δ such that

$$\left| \sum_{i=1}^{n} f(x_i')\Delta x_i - \int_a^b f(x)dx \right| < \epsilon$$

for all choices of n and the points x_i, x_i' such that the greatest of the numbers $\Delta x_1, \cdots, \Delta x_n$ is less than δ.

We give this definition here for comparison with the definitions of the two kinds of limits already discussed. The problem of showing that the sums (1.63–1) do actually have a limit (when f is a continuous function) is another example of an "existence problem" for a real number. As with most such problems, we cannot obtain a solution without having at our command a systematic knowledge of the fundamentals of the real number system.

1.64 The theorem on limits of sums, products, and quotients

In § 1.6 we stated three fundamental theorems about limits (see also (1.1–4)–(1.1–6)). The student has used these theorems from the very beginning of his study of calculus. We are now going to give formal statement and proof of the propositions as they apply to limits of functions of a continuous variable. First, however, it will be convenient to consider certain rules governing the use of absolute values. The absolute value $|A|$ of a number A was defined in Example 3, § 1.1. Now for any two numbers A, B, it is always true that

(1.64–1) $$|A + B| \leq |A| + |B|.$$

There are four cases to consider: (1) A and B both positive, (2) A and B both negative, (3) $A = 0$ or $B = 0$, (4) A and B of opposite sign, and neither of them equal to 0. In the first three cases it is easily seen that $|A + B| = |A| + |B|$, while in case (4) $|A + B| < |A| + |B|$. Thus (1.64–1) is true in all cases.

We now turn to the limit theorem.

THEOREM XIV.   *Let f and g be defined in an interval containing $x = x_0$, but not necessarily at the point $x = x_0$ itself. Suppose that the limits*

$$\lim_{x \to x_0} f(x), \quad \lim_{x \to x_0} g(x)$$

exist. Then the sum $f(x) + g(x)$ and the product $f(x)g(x)$ approach limits as x tends to x_0, given by

(1.64–2) $$\lim_{x \to x_0} \{f(x) + g(x)\} = \lim_{x \to x_0} f(x) + \lim_{x \to x_0} g(x),$$

(1.64–3) $$\lim_{x \to x_0} \{f(x)g(x)\} = \{\lim_{x \to x_0} f(x)\} \{\lim_{x \to x_0} g(x)\}.$$

Furthermore, if $\lim_{x \to x_0} g(x) \neq 0$, the quotient $\dfrac{f(x)}{g(x)}$ has a limit given by

$$(1.64\text{-}4) \qquad \lim_{x \to x_0} \frac{f(x)}{g(x)} = \frac{\lim_{x \to x_0} f(x)}{\lim_{x \to x_0} g(x)}.$$

PROOF.   Let the limits of $f(x)$ and $g(x)$ be denoted by A and B respectively. We shall prove (1.64–2) and (1.64–4), leaving the proof of (1.64–3) as an exercise for the student. To prove (1.64–2) we must show that, if a positive ϵ is given, we can choose a positive δ such that

$(1.64\text{-}5) \quad |(f(x) + g(x)) - (A + B)| < \epsilon \text{ if } 0 < |x - x_0| < \delta.$

Now $(f(x) + g(x)) - (A + B) = (f(x) - A) + (g(x) - B),$

and therefore, by an application of (1.64–1),

$(1.64\text{-}6)$
$$|(f(x) + g(x)) - (A + B)| \leq |f(x) - A| + |g(x) - B|.$$

Now by hypothesis we can make $|f(x) - A|$ and $|g(x) - B|$ as small as we like by restricting x to lie sufficiently near x_0. In particular, there are positive numbers δ_1 and δ_2 such that

$(1.64\text{-}7) \qquad |f(x) - A| < \dfrac{\epsilon}{2} \text{ if } 0 < |x - x_0| < \delta_1$

and

$(1.64\text{-}8) \qquad |g(x) - B| < \dfrac{\epsilon}{2} \text{ if } 0 < |x - x_0| < \delta_2.$

Let δ be the smaller of the numbers δ_1, δ_2. It then follows from (1.64–6), (1.64–7), and (1.64–8) that (1.64–5) holds. This completes the proof of (1.64–2).

In proving (1.64–4) we first of all observe that, since $\lim_{x \to x_0} g(x) = B$ and $B \neq 0$, we can be sure that $|g(x)| > \frac{1}{2}|B|$ if we require x to be sufficiently near x_0. It suffices to choose δ_0 so that

$$|g(x) - B| < \tfrac{1}{2}|B| \text{ if } 0 < |x - x_0| < \delta_0,$$

for then $B = (B - g(x)) + g(x),$

and by (1.64–1) we have

$$|B| \leq |B - g(x)| + |g(x)| < \tfrac{1}{2}|B| + |g(x)|,$$

and consequently

$(1.64\text{-}9) \qquad \tfrac{1}{2}|B| < |g(x)| \text{ if } 0 < |x - x_0| < \delta_0.$

Now we can write

$$\frac{f(x)}{g(x)} - \frac{A}{B} = \frac{Bf(x) - Ag(x)}{g(x)B} = \frac{B[f(x) - A] + A[B - g(x)]}{g(x)B},$$

and so

(1.64–10)

$$\left| \frac{f(x)}{g(x)} - \frac{A}{B} \right| \leqq \frac{|B| |f(x) - A| + |A| |B - g(x)|}{|g(x)| |B|}.$$

(Here we have used the fact that the absolute value of a product is the product of the absolute values.) Now let a positive ϵ be assigned arbitrarily. We wish to choose a positive δ so that the left member of (1.64–10) is less than ϵ if $0 < |x - x_0| < \delta$. Let

$$\epsilon_1 = \frac{|B|^2 \epsilon}{2(|B| + |A|)}.$$

Choose δ_1 and δ_2 so that

(1.64–11) $|f(x) - A| < \epsilon_1$ if $0 < |x - x_0| < \delta_1$,

(1.64–12) $|g(x) - B| < \epsilon_1$ if $0 < |x - x_0| < \delta_2$.

At the same time we make sure that $\delta_2 < \delta_0$ so that (1.64–9) will hold. Now, making use of (1.64–9), (1.64–11), and (1.64–12) we see that the right member of (1.64–10) is less than

$$\frac{|B| \epsilon_1 + |A| \epsilon_1}{\frac{1}{2} |B|^2} = \epsilon$$

if $0 < |x - x_0| < \delta$, where δ is the smaller of the numbers δ_1, δ_2. This completes the proof of (1.64–4).

EXERCISES

1. Prove part (1.64–3), using the following suggestions: Show that $|g(x)| < |B| + 1$ if $|g(x) - B| < 1$. Write

$$f(x)g(x) - AB = (f(x) - A)g(x) + (g(x) - B)A,$$

and use (1.64–1). If $\epsilon > 0$ is given, let

$$\epsilon_1 = \frac{\epsilon}{|A| + |B| + 1}.$$

Now show that $|f(x)g(x) - AB| < \epsilon$ provided $|f(x) - A| < \epsilon_1, |g(x) - B| < 1$ and $|g(x) - B| < \epsilon_1$. Explain how these last inequalities are to be guaranteed by appropriate restrictions on x. Write out the entire proof in full detail.

2. Formulate and prove the counterpart of Theorem XIV for limits of sequences.

MISCELLANEOUS EXERCISES

1. Suppose that f is continuous on $[a, b]$, that $f'(x)$ and $f''(x)$ exist when $a < x < b$, that $f(a) = f(b) = 0$, and that there is a number c such that $a < c < b$ and $f(c) > 0$. Prove that there is a number ξ between a and b such that $f''(\xi) < 0$.

2. Let $f(x) = \log_a x$ $(0 < a, a \neq 1)$. Deduce that $f'(x) = \dfrac{f'(1)}{x}$, using nothing about logarithms other than: (1) the assumption that $f'(x)$ exists when $x > 0$, and (2) the property $\log_a (xu) = \log_a x + \log_a u$ when $x > 0$ and $u > 0$.

3. Find $$\lim_{x \to \infty} \{ \sqrt{(x + a)(x + b)} - x \}.$$

4. In each of the following cases, investigate the one-sided limits at the point indicated, and decide whether or not $\lim_{x \to x_0} f(x)$ exists.

(a) $f(x) = [x] + [3 - x]$, $x_0 = 2$;

(b) $f(x) = \dfrac{e^{\tan x} - 1}{e^{\tan x} + 1}$, $x_0 = \dfrac{\pi}{2}$;

(c) $f(x) = 2^{-1/x} \sin \dfrac{1}{x}$, $x_0 = 0$.

5. Let $s_1 = \sqrt{2}$, $s_{n+1} = \sqrt{2 s_n}$, $n = 1, 2, \cdots$; find $\lim_{n \to \infty} s_n$.

6. Suppose $0 < a_1 \leq a_2 \leq \cdots \leq a_k$, where the a's are fixed numbers. Let $b_n = (a_1{}^n + a_2{}^n + \cdots + a_k{}^n)^{1/n}$. Show that $\lim_{n \to \infty} b_n = a_k$.

7. (a) Show that $$\lim_{n \to \infty} \sum_{p=0}^{n-1} \frac{n}{n^2 + p^2} = \frac{\pi}{4}.$$

$\left(\text{Bring in consideration of } \displaystyle\int_0^1 \frac{dx}{1 + x^2} \cdot \right)$

(b) Find $\lim_{n \to \infty} \dfrac{1}{n} \displaystyle\sum_{p=1}^{n} \sin \dfrac{p \pi \alpha}{n}$, where $\alpha > 0$.

8. In (1.2–4) take $f(x) = x^3$, $a \neq 0$. Obtain a formula connecting a, θ, and h, and show that $| \theta - \frac{1}{2} | \leq \dfrac{2}{3} \dfrac{|h|}{|a|} \cdot$ Hence conclude that $\lim_{h \to 0} \theta = \frac{1}{2}$.

9. Let (x_0, y_0) be a point of the ellipse $b^2 x^2 + a^2 y^2 = a^2 b^2$. Let the tangent to the ellipse at (x_0, y_0) intersect the x-axis at A and the y-axis at B. Find the minimum possible value of the distance AB.

10. Define $f(x) = x \tan^{-1}(1/x)$ if $x \neq 0$, and $f(0) = 0$. Is f continuous at $x = 0$? Is it differentiable there?

11. Show, using the law of the mean, that
$$| x \log x | < | x \log \xi | + \xi$$
if $0 < x < \xi$. From this deduce that $\lim_{x \to 0+} x \log x = 0$. Specify carefully how to choose δ so that $0 < x < \delta$ will imply $| x \log x | < \epsilon$, where ϵ is given in advance. Begin by choosing ξ in terms of ϵ in a suitable manner.

12. Suppose $f(x) = x^2 + 2x$, and define g by $g(t) = t^2 \sin (\pi/t) + t$ if $t \neq 0$, $g(0) = 0$. Let $F(t) = f(g(t))$. Find $F'(0)$.

13. Find the absolute maximum and minimum of $f(x) = x^2(1 - x)^3$ on the interval $-1 \leq x \leq 2$ *without using a graph.*

14. Consider the functions

$$f_0(x) = \cos x - 1$$
$$f_1(x) = \sin x - x$$
$$f_2(x) = \cos x - 1 + \tfrac{1}{2} x^2$$
$$f_3(x) = \sin x - x + \tfrac{1}{6} x^3$$
$$f_4(x) = \cos x - 1 + \tfrac{1}{2} x^2 - \tfrac{1}{24} x^4$$
$$f_5(x) = \sin x - x + \tfrac{1}{6} x^3 - \tfrac{1}{120} x^5.$$

Note that $f'_1(x) = f_0(x)$ and that $f_0(x) \leqq 0$ if $x > 0$. Thus $f_1(x)$ decreases as x increases, and since $f_1(0) = 0$, we conclude that $f_1(x) < 0$ when $x > 0$. Next note that $f'_2(x) = -f_1(x)$. Explain how you conclude that $f_2(x) > 0$ when $x > 0$. Continuing in this way, show that for $x > 0$,

$$x - \tfrac{1}{6} x^3 < \sin x < x - \tfrac{1}{6} x^3 + \tfrac{1}{120} x^5$$

and $\qquad 1 - \tfrac{1}{2} x^2 + \tfrac{1}{24} x^4 - \tfrac{1}{120} x^6 < \cos x < 1 - \tfrac{1}{2} x^2 + \tfrac{1}{24} x^4.$

What is the generalization?

15. Suppose that f satisfies the hypotheses: f is defined and continuous when $a \leqq x < b$, differentiable when $a < x < b$, $f(a) = 0$, and $f(x) > 0$ if $a < x < b$. Prove that there cannot be a positive constant M such that $0 \leqq \dfrac{f'(x)}{f(x)} \leqq M$ when $a < x < b$.

The Real Number System   **2**

2. Numbers

Our experience with numbers begins with the positive integers (also called the whole numbers, or natural numbers). Next we become acquainted with zero, and in due course we become familiar with negative integers and with rational fractions (ratios of positive and negative integers). At some stage we learn the adjective *irrational* for numbers such as $\sqrt{2}$, $\sqrt[3]{5}$, π. In algebra we meet the equation $x^2 = -1$ and are told that it gives rise to a new number i. Numbers such as i, $2\,i$, $-7\,i$ are called *pure imaginary*, and numbers such as $3 + 5\,i$ are called *complex*. Our learning, with numbers as with everything else, proceeds mainly by particular cases and illustrative examples. But in due time it is possible to reduce our knowledge to order and to give it logical coherence by a systematic study of number and number systems. We make a beginning on such a systematic study in this chapter.

The integers, the rational fractions, and the irrational numbers compose the number system which lies at the foundation of calculus and of all analysis. The numbers of this system are called the *real* numbers, and the system of all such numbers is called the real number system.

The adjectives "real" and "imaginary," as applied to numbers, have entirely conventional technical meanings. They are not meant to convey, and the student should not let them convey, any implications whatsoever of a philosophical nature about existence or nonexistence as genuine entities.

We shall not attempt to define the concept of number or to build up the concepts of rational and irrational numbers from the concept of the integers. Instead, we shall take the real number system as something known (though somewhat imperfectly and unsystematically) by the student, and shall make specific the algebraic laws governing the real numbers. The student is already familiar with most of these laws, but it is now very important to know a complete set of properties of the real number system. From such a complete set we may deduce every property of the system.

2.1   The field of real numbers

Addition and multiplication are the fundamental operations with ordinary numbers. These operations conform to the following laws:

	ADDITION	MULTIPLICATION
The commutative law:	$a + b = b + a$	$ab = ba$
The associative law:	$a + (b + c) = (a + b) + c$	$a(bc) = (ab)c$
The distributive law:		$a(b + c) = ab + ac$

Throughout this section the word "number" will be understood to mean "real number," and symbols $a, b, c, x, \cdots$ will stand for real numbers.

There are two numbers with special properties, namely 0 and 1. The special properties are expressed by the laws

(2.1–1) $a + 0 = a$ and $a \cdot 1 = a$

for every number a.

The number 0 is special for addition, while 1 is special for multiplication. The operations of subtraction and division may be defined with the aid of these special numbers in the following way: To every number a corresponds its negative, $- a$, which is the "additive inverse" of a. By this we mean that $x = - a$ satisfies the equation

(2.1–2) $a + x = 0.$

Likewise every number a except 0 has a multiplicative inverse, denoted by a^{-1}. That is, if $a \neq 0$, $x = a^{-1}$ satisfies the equation

(2.1–3) $ax = 1.$

We then *define* the subtraction of b from a by the equation

(2.1–4) $a - b = a + (- b).$

Similarly, we define the division of a by b as

(2.1–5) $\dfrac{a}{b} = a(b^{-1}).$

The properties of the real numbers which we have just been discussing are summed up briefly in the language of modern algebra by saying that the real numbers form a *field*. The word "field" here has a special technical meaning. *When we say that a system of numbers F constitutes a field we mean the following:*

(1) *If a and b are in F, then a + b and ab are in F.*
(2) *The commutative, associative, and distributive laws hold.*
(3) *F contains distinct special numbers 0 and 1 with the properties (2.1–1).*
(4) *Equation (2.1–2) has a solution in F for each a, and (2.1–3) has a solution in F for each a $\neq$ 0.*

In higher algebra it is shown how the other familiar laws of elementary algebra are deducible from the laws governing a field. We shall not undertake any systematic deductions of this kind. We mention, however, the rule:

(2.1–6)   If $ab = 0$ and $b \neq 0$, then $a = 0.$

This is proved as follows: Since $b \neq 0$, there is a number b^{-1} such that

$bb^{-1} = 1$. From $ab = 0$ we conclude that $a(bb^{-1}) = 0 \cdot b^{-1}$, or $a \cdot 1 = 0$, or $a = 0$. Such rules as $a \cdot 0 = 0$ and $(-a)(-b) = ab$ are also readily deducible.

It may be noted that the system of integers (positive, negative, and zero) is *not* a field, although it fails to be one only through the fact that a^{-1} need not be an integer when a is. The rational numbers *do* form a field. Thus there are fields other than the field of real numbers.

2.2   Inequalities. Absolute value

One of the important properties of the real numbers is that they are *ordered*; that is, there is a notion "a is less than b" expressed by the inequality $a < b$; of any two numbers a, b one and only one of the following three things is true:

$$a < b \quad \text{or} \quad a = b \quad \text{or} \quad b < a.$$

The properties of order can be stated simply in terms of properties of positive numbers. We express the fact that a is positive by the symbols $0 < a$. The basic laws governing the positive numbers are three in number:

(2.2–1)   If $0 < a$ and $0 < b$, then $0 < a + b$.

(2.2–2)   If $0 < a$ and $0 < b$, then $0 < ab$.

(2.2–3)   For each a, one and only one of the following relations is true:
$$0 < a \quad \text{or} \quad 0 = a \quad \text{or} \quad 0 < -a.$$

In terms of positivity we lay down the following definitions:

If $b - a$ is positive, we say that a is less than b and write $a < b$. Under the same conditions we also say that b is greater than a and write $b > a$.

Other properties of order, or rules for manipulating inequalities, are deducible from the above definitions and the three basic laws. Among the important rules are the following:

(2.2–4) $0 < a^2$ if $a \neq 0$.

(2.2–5)   If $a < b$ and $b < c$, then $a < c$.

(2.2–6)   If $a < b$, then for any c, $a + c < b + c$.

(2.2–7)   If $a < b$ and $0 < c$, then $ac < bc$.

A field is said to be ordered if certain of its members are distinguished by calling them positive and if this notion of "being positive" satisfies the laws (2.2–1), (2.2–2), and (2.2–3).

For many purposes it is convenient to introduce the symbolism $a \leq b$, meaning that either $a < b$ or $a = b$. The symbolism $b \geq a$ has the same meaning.

The notion of the *absolute value* of a real number is defined as follows (the absolute value of a is denoted by $|a|$):

$$|a| = a \text{ if } 0 < a; \quad |a| = -a \text{ if } a < 0; \quad |0| = 0.$$

In actual calculations with absolute values we rely largely upon the two rules:

(2.2–8) $$|ab| = |a||b|,$$
(2.2–9) $$|a + b| \leq |a| + |b|.$$

(See § 1.64 for a brief discussion of (2.2–9)).
The further rule

(2.2–10) $$\left| |a| - |b| \right| \leq |a - b|$$

is also convenient; it is deducible with the aid of (2.2–9).

We observe in passing that the inequality $|x| \leq \epsilon$ (where $\epsilon > 0$) is equivalent to the inequalities $x \leq \epsilon$ and $-\epsilon \leq x$, which we write as a double inequality $-\epsilon \leq x \leq \epsilon$. By setting $x = b - a$ and doing a little transposing we see that $|a - b| \leq \epsilon$ is equivalent to $b - \epsilon \leq a \leq b + \epsilon$.

EXERCISES

1. Prove (2.2–4). Note first that $a^2 = a \cdot a$ and $a^2 = (-a) \cdot (-a)$. Then use (2.2–2) and (2.2–3).

2. Prove (2.2–5). Appeal to (2.2–1) and the fact that $x < y$ is equivalent to $0 < y - x$.

3. Prove (2.2–6).

4. Prove (2.2–7).

5. Write $a = (a - b) + b$ and apply the rule (2.2/9) to obtain $|a| \leq |a - b| + |b|$, whence $|a| - |b| \leq |a - b|$. Why is it also true that $|b| - |a| \leq |a - b|$? Explain now why (2.2–10) is correct.

2.3 The principle of mathematical induction

The natural numbers 1, 2, 3, $\cdots$ occupy a position of especial importance in the field of real numbers. To single out the class of all natural numbers from the rest of the real numbers we may make the following assertion: The totality of natural numbers forms the *smallest* class of real numbers possessing the two properties

(1) The number 1 is a member of the class.
(2) If x is a member of the class, so is $x + 1$.

The statement that the natural numbers form the smallest class having these properties means that any collection of real numbers having these two properties must include all the natural numbers.

The characteristic feature of the totality of natural numbers, as just described, is logically equivalent to the principle of mathematical induction (also called complete induction). We formulate the principle as follows:

Let A (n) denote a proposition (e.g., a verbal statement, or a formula) associated with the natural number n. Suppose it is possible to show that the proposition is true if n = 1, and suppose also that, for each particular n, we can prove the truth of A(n + 1) if we assume the truth of A(n). Then A(n) is true for every natural number n.

To see the validity of the principle, let S be the class of natural numbers n such that $A(n)$ is true. The assumption that $A(1)$ is true means that 1 belongs to S. Also, the fact that assumption of the truth of $A(n)$ allows us to deduce the truth of $A(n + 1)$ means that if n is in S, so is $n + 1$. Hence, by the remarks made in the first paragraph of this section, S must contain all the natural numbers. That is, $A(n)$ must be true for every n.

We shall not analyze in detail the logical position of the principle of mathematical induction in the description of the natural numbers. If the real numbers are thought of as having been built up from the numbers 1, 2, 3, $\cdots$, we need the principle of mathematical induction as an *axiom* about the natural numbers. If, on the other hand, we start from the assumption that the real numbers are somehow presented to us as a particular sort of ordered field, we may *define* the natural numbers as the smallest class of real numbers which contains the number 1, and which contains $x + 1$ if it contains x. The theory of classes or sets, into which we prefer not to venture at this stage, permits us to *prove* that there *is* such a smallest class. The principle of mathematical induction is, from this latter point of view, a *theorem* about the class of natural numbers.

By way of illustrating the principle of mathematical induction, we shall prove formally some things about natural numbers.

Example 1.   Every natural number is positive.

In the first place, $0 < 1$. This follows from (2.2–4), since $1 \neq 0$ and $1 = 1^2$. To complete the proof of the general assertion, suppose $0 < n$ for a particular n. Then $0 < n$ and $0 < 1$ imply $0 < n + 1$, by (2.2–1). Thus $0 < n$ for every natural number n, by the induction principle.

From now on we shall usually refer to "the natural numbers" as "the positive integers." The unmodified term "integers" refers to the class consisting of the positive integers 1, 2, 3, $\cdots$, their negatives -1, -2, -3, $\cdots$, and the number 0.

We shall take for granted without formal proof such familiar facts as that *there is no integer between n and n + 1, if n is an integer.*

Example 2. If S is a class of positive integers containing at least one member, it contains a smallest number.

Most students will feel inclined to accept this as true without demonstration. A proof is logically necessary, however. Observe that the assertion ceases to be true if the word "integers" is replaced by "real numbers"; for instance, there is no smallest member of the class of positive rational numbers.

The proof of the assertion in Example 2 uses the characteristic properties of the class of natural numbers in an interesting way. Observe, in the first place, that if 1 is in S, 1 *is* the smallest member of S, since there is no positive integer less than 1. Hence, we need consider only the case in which 1 does not belong to S. We now let T be the class of positive integers p such that $p < n$ for every n in S. This class T contains 1 since 1 is not in S. On the other hand, T does not contain *all* positive integers, since *some* positive integers belong to S. Hence, it cannot be that T contains $p + 1$ whenever it contains p, for in that case T *would* contain all positive integers, by the characteristic properties of the class of all such integers. There must therefore be some integer p_0 in T such that $p_0 + 1$ is not in T. Then, by the way in which T is defined, there must be an integer n_0 in S such that $n_0 \leq p_0 + 1$. We assert that n_0 is the smallest integer in S. For, if n is any member of S, we have $p_0 < n$. Let $m = n - p_0$, or $p_0 + m = n$. Here m is a positive integer. Therefore, $n_0 \leq p_0 + 1 \leq p_0 + m = n$, or $n_0 \leq n$. This completes the argument.

EXERCISES

1. Prove by induction that, for every natural number n, either $1 = n$ or $1 < n$,

2. Prove the validity of the following form of the principle of mathematical induction, resting your argument on the form enunciated in the text. *Let B(n) denote a proposition associated with the integer n. Suppose B(n) is known (or can be shown) to be true when $n = n_0$, and suppose the truth of B(n + 1) can be deduced if the truth of B(n) is assumed. Then B(n) is true for every integer n such that $n_0 \leq n$.* SUGGESTION: Let $A(n)$ be the proposition $B(n_0 + n - 1)$.

2.4   The axiom of continuity

The facts expressed in the statement that the real numbers form an ordered field are quite familiar. We are now going to discuss a much less familiar property of the real number system. Most students beginning a course in advanced calculus will have had no experience in making use of this property, and quite possibly may never have heard of it. We call it the *axiom of continuity.*

THE AXIOM OF CONTINUITY. *Suppose that all real numbers are separated into two collections, which we denote by L and R, in such a way that*

(1) *every number is either in L or in R,*
(2) *each collection contains at least one number,*
(3) *if a is in L and b is in R, then a < b.*

Then there is a number c such that all numbers less than c are in L and all numbers greater than c are in R. (The number c itself may belong either to L or to R, depending on the particular way in which L and R are formed.)

It is convenient to have a name for a separation of all real numbers into collections L and R meeting the specifications (1)–(3). We call such a separation a *cut*; the number c is then called the *cut number*. The cut number corresponding to a particular cut is unique. For suppose a given cut has the distinct cut numbers c_1 and c_2. One of them is the greater, say $c_1 < c_2$. Consider the number

$$b = \frac{c_1 + c_2}{2},$$

which lies halfway between c_1 and c_2: $c_1 < b < c_2$. Now $c_1 < b$ implies that b is in R, by one of the properties of the cut number c_1. Likewise $b < c_2$ implies that b is in L. Hence b is in both L and R. This is impossible, however, for by the specification (3) L and R cannot have any members in common. The assumption of distinct cut numbers has led to a contradiction. Therefore, we conclude that any cut has but one cut number.

It is clear that if c belongs to L then L consists of all numbers x such that $x \leqq c$, while R consists of all numbers x such that $c < x$. If c belongs to R, then L consists of all x such that $x < c$ and R consists of all x such that $c \leqq x$.

The idea of a cut was originated by the German mathematician Dedekind for use in his theory of the structure of the real number system (published in 1872 under the title *Continuity and Irrational Numbers*). If the number system is built up by stages from the integers, it is possible to arrange the exposition in such a way that what we have called the axiom of continuity is provable as a theorem about the real numbers. But since we are taking the point of view that the real numbers are somehow given to us to work with, we are listing our assumptions about them. The axiom of continuity is one of our assumptions. It is a far-reaching assumption, for with it we can deal satisfactorily with the existence problems for real numbers referred to near the end of each of §§ 1.62 and 1.63.

No further assumptions need be made about the real number system.

The system is described with logical completeness by saying that it is an ordered field satisfying the axiom of continuity.

At this point it is convenient to introduce a theorem which expresses what is called the *Archimedean law* of real numbers. Its proof is a good illustration of arguments using the axiom of continuity.

THEOREM I.   *Let a and b be positive real numbers. There exists a positive integer n such that $b < na$.*

PROOF.   Suppose the theorem false, so that $na \leq b$ for every positive integer n. We shall define a cut as follows: Let L consist of all numbers x such that $x < na$ for *some n*, and let R consist of all numbers not in L, i.e. all numbers y such that $na \leq y$ for every n. We must verify that the three specifications for a cut are fulfilled:

(1) Every number is in either L or R, by definition.

(2) b is in R by our initial supposition, and a is in L, since $a < na$ if $n = 2$. Thus neither L nor R is without members.

(3) If x is in L and y is in R, we have $x < na \leq y$ for some n, and hence $x < y$.

Now let c be the cut number. We observe that all the numbers na are in L, since $na < (n + 1)a$. Therefore $na \leq c$, for $c < na$ would mean that na is in R, by one of the properties of a cut number. Hence, also $(n + 1)a \leq c$. But this implies $na \leq c - a$. This being true for all n, we conclude that $c - a$ is in R. But $c - a < c$, and hence $c - a$ is in L. We have now reached a contradiction, and the proof is complete.

2.5   Rational and irrational numbers

Numbers of the form p/q, where p and q are integers, are called *rational*. Real numbers which are not rational are called *irrational*. The theory of the nature of irrational numbers began with the ancient Greeks. It has been known since the time of Pythagoras that $\sqrt{2}$ is not a rational number. (With a little knowledge about expressing integers as products of their prime factors the student may easily prove this fact for himself.) The theory of incommensurable ratios, developed by Eudoxus, is in essence a geometrical treatment of irrational numbers. It was not, however, until the nineteenth century that mathematicians (Dedekind with his cuts important among them) arrived at the understanding we have today of the real number system and of the position of irrationals in it.

Between any two real numbers there are both rational and irrational numbers. If a and b are given real numbers with $a < b$, we obtain a rational number r such that $a < r < b$ by the following argument: Since

$0 < 1$ and $b - a > 0$ there exists (by Theorem I, § 2.4) a positive integer n such that $1 < n(b - a)$. Let m be the smallest integer such that $m > na$. Then $(m - 1) \leq na$, and therefore

$$m \leq na + 1 < na + n(b - a) = nb.$$

Consequently

$$a < \frac{m}{n} < b,$$

so that $r = m/n$ is a rational number of the required sort. Finding an irrational number between a and b is left as an exercise for the student (see Exercise 2).

It follows from the previous paragraph that if x is a real number, rational or irrational, we can find a rational number r as near it as we please. That is, if ϵ is any positive number, we can find r so that $x - \epsilon < r < x + \epsilon$. We do in fact, in any actual computations or measurements, use rational approximations to real numbers. For many purposes we even limit ourselves to special kinds of rational numbers, namely decimal fractions. A decimal fraction is simply a rational number of the form $m/10^n$, where m and n are integers and $n \geq 0$. There is a number of this form between any two real numbers, as may be readily shown from the fact that if $a < b$, then $1 < 10^n(b - a)$ for a suitably chosen n.

EXERCISES

1. Assume that $\sqrt{2} = m/n$, where m and n are integers and the fraction is reduced to its lowest terms. Then $m^2 = 2n^2$. Now deduce a contradiction, and hence prove that $\sqrt{2}$ is irrational. Use the fact that if the product of two integers is even, then at least one of them is even.

2. Show that if $a < b$ there is an irrational number x between a and b. SUGGESTION: Let y be a positive irrational number, e.g., $\sqrt{2}$. Then show that there are integers m, n, with $n > 0$, such that $a < (m/n)y < b$. Why is $x = (m/n)y$ irrational?

3. Prove by induction that $n < 10^n$ if n is a positive integer. Then prove that if $a < b$, there are integers m, n, with $n > 0$, such that $a < m/10^n < b$.

2.6   The axis of reals

It is customary and convenient to use geometric language a good deal in speaking about the number system. On a given straight line we take an arbitrary point as an origin, an arbitrary direction as positive, and an arbitrary unit of length. We then mark off segments of unit length on either side of the origin, thus obtaining the points which we label as shown in Fig. 16. There is a one-to-one correspondence

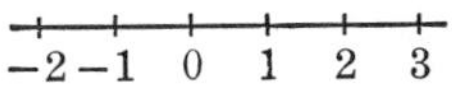

Fig. 16

between the real numbers and the points on the line. This entitles us, for brevity, to speak of "the point a" instead of "the point corresponding to the number a." The inequality $a < b$ has the geometrical interpretation that b lies in the positive direction from a along the line.

We call the line, thus regarded as a geometrical representation of the real number system, *the axis of reals,* or *the real number scale.*

2.7   Least upper bounds

By a *set* of real numbers we mean an aggregate or class of numbers. It may be formed according to any rule, and the number of its members may be finite or infinite. If the conditions laid down for determining the set are such that no number satisfies them, the set is said to be empty. It is very convenient to have a brief symbolism to indicate that a number belongs to a given set. The statement that the number s belongs to the set S is expressed symbolically in the form $s \in S$ (read s is a member of S). The symbolic form of the statement that s does *not* belong to S is $s \notin S$. Thus, if S is the set of prime positive integers, $3 \in S$ and $8 \notin S$.

If S is a set of numbers, and if M is a number such that $s \le M$ for each $s \in S$, we say that M is an *upper bound* of S. If A is an upper bound of S and if there is no number smaller than A which is also an upper bound for S, we call A the *least upper bound* of S. Obviously a set cannot have more than one least upper bound. The following theorem is of fundamental importance:

THEOREM II.   *If S is a set of real numbers which is not empty and which has an upper bound, then it has a least upper bound.*

PROOF.   We appeal to the axiom of continuity. Let L be the set of all numbers x such that $x < s$ for some s in S, and let R be the set of all numbers y such that $s \le y$ for every s in S. Clearly L and R together comprise all real numbers. If $s \in S$, then $s - 1 \in L$; $A \in R$ if A is an upper bound of S. Thus neither L nor R is empty. If $x \in L$ and $y \in R$, we have $x < s$ for some s in S. But $s \le y$, and therefore $x < y$. We have therefore defined a cut. Let c be the cut number. We shall prove that c is the least upper bound of S. It certainly is *an* upper bound. For if we suppose $c < s$ for some s in S, we can choose a number z between c and s. Then $z \in R$ since $c < z$, and $z \in L$ since $z < s$; thus we have a contradiction, for a number cannot belong to both L and R. If b is any number smaller than c, the properties of the cut number insure that $b \in L$, and hence $b < s$ for some s in S. Thus b cannot be an upper bound of S. The proof that c is the least upper bound of S is now complete.

Theorem II expresses a property of the real number system which is a direct consequence of the axiom of continuity. It is easily demonstrated that if the statement of Theorem II is taken as an axiom concerning the real numbers, the truth of the axiom of continuity may be deduced (making it a theorem instead of an axiom). For a key to this demonstration see Exercise 1. Thus the axiom of continuity and the existence of least upper bounds as stated in Theorem II are equivalent propositions. Hereafter, in arguments where we could lean equally well either on the axiom of continuity or on Theorem II, we shall usually appeal to the latter.

As an immediate application we shall prove Theorem XIII of Chapter I (§ 1.62). We reword it slightly.

THEOREM III.   *Let $\{x_n\}$ be a sequence such that $x_1 \leq x_2 \leq \cdots \leq x_n \leq x_{n+1} \leq \cdots$, and suppose that the set of numbers x_n has an upper bound: $x_n \leq M$ for every n. Then the sequence is convergent, its limit being the least upper bound of the numbers x_n.*

PROOF. Let A be the least upper bound of the numbers x_n. Then if $\epsilon > 0$ we have $A - \epsilon < x_n$ for *some n*, say $n = N$, and $x_n \leq A$ for every *n*. Since $x_N \leq x_n$ for every $n \geq N$ (by virtue of the assumption that $x_n \leq x_{n+1}$), we see that $A - \epsilon < x_n \leq A$ if $N \leq n$. Thus by definition $\lim_{n \to \infty} x_n = A$. This proves the theorem.

The notion of *lower bound* of a set, and of the *greatest lower bound*, are defined in exactly the same way as upper bound and least upper bound, except that the notions of "less than" and "least" are replaced throughout by "greater than" and "greatest." We may summarize the defining properties of the least upper bound and greatest lower bound as follows:

The set S has the least upper bound A if $s \leq A$ for every s in S and if, ϵ being any positive number, $A - \epsilon < s$ for at least one s in S.

The set S has the greatest lower bound B if $B \leq s$ for every s in S and, ϵ being any positive number, $s < B + \epsilon$ for at least one s in S.

THEOREM IV.   *If S is a set of real numbers which is not empty and which has a lower bound, then it has a greatest lower bound.*

PROOF. We could give a proof similar to that of Theorem II. Instead, we base the proof on Theorem II itself. Define a set T as consisting of all numbers $t = -s$, where $s \in S$. If M is a lower bound of S, we have $M \leq s$, or $-s \leq -M$ for each s in S. Thus T has an upper bound $-M$. By Theorem II, T has a least upper bound, say A. We leave it for the student to show that $-A$ is the greatest lower bound of S, thus completing the proof.

Just as Theorem IV matches Theorem II, so there is a theorem which matches Theorem III.

THEOREM V. *Let $\{x_n\}$ be a sequence such that $x_1 \geq x_2 \geq \cdots$ (in general $x_n \geq x_{n+1}$), and suppose that the set of numbers x_n has a lower bound: $x_n \geq M$ for every n. Then the sequence is convergent, its limit being the greatest lower bound of the numbers x_n.*

We leave it for the student to base a proof of this theorem on Theorem IV or to deduce the proof from Theorem III by considering the sequence $\{y_n\} = \{-x_n\}$.

EXERCISES

1. Let L and R be sets of real numbers satisfying the conditions of the axiom of continuity. Assuming the truth of Theorem II, show that the set L has a least upper bound c and that this least upper bound has the properties asserted for the cut number in the axiom of continuity.

2. Prove Theorem IV by an argument similar to that used in proving Theorem II, using the axiom of continuity.

3. Carry out the proof of Theorem V in each of two ways, as suggested in the text.

4. Prove Theorem I (§ 2.4), starting as follows: Suppose the assertion in the theorem false, so that $na \leq b$ for all positive integers n. Use Theorem II and the defining property of a least upper bound to arrive at a contradiction.

5. Let $x_1 = \sqrt{2}, x_{n+1} = \sqrt{2 + x_n}$. Use mathematical induction to show that $x_n < x_{n+1}$. Next show that if $x_n \geq 2$ (where $n > 1$), then $x_{n-1} \geq 2$ also. How do you conclude from this that $x_n < 2$ for all n? State why $\lim_{n \to \infty} x_n$ exists, and find the limit.

6. Suppose $c > 0$, and let $x_1 = \sqrt{c}, x_{n+1} = \sqrt{c + x_n}$. Show that $x_n < x_{n+1}$ and $x_n < 1 + \sqrt{c}$ (see Exercise 5). State why $\lim_{n \to \infty} x_n$ exists, and find the limit.

2.8   Nested intervals

The theorem which we are going to discuss in this section is an immediate consequence of Theorems III and V of § 2.7. We are going to use the language of geometry rather than the language of numbers; the theorem is about closed intervals on the real axis. It will be convenient to denote intervals by single letters, such as I, I_1, I_2, and so on. If I_1 and I_2 are closed intervals, and if the end points of I_2 lie in I_1, we say that I_2 is contained in I_1.

Now suppose that we are given a sequence of closed intervals, I_1, I_2,

$I_3, \cdots, I_n, \cdots$ with the property that I_2 is contained in I_1, I_3 is contained in I_2, and so on; that is, we assume that each interval contains the one which follows it in the sequence. Let us denote the length of the interval I_n by the symbol l_n, and let us assume that $\lim_{n \to \infty} l_n = 0$. In these circumstances we shall say that $\{I_n\}$ is a sequence of *nested* intervals, or that the sequence forms a *nest*.

THEOREM VI. *If $\{I_n\}$ is a nest of closed intervals, there is exactly one point which is common to all the intervals.*

PROOF. Let the interval I_n be described by the inequalities $a_n \leqq x \leqq b_n$, $(a_n < b_n)$. The fact that $\{I_n\}$ is a nest is then described as follows. The inequalities

(2.8–1) $a_n \leqq a_{n+1}, \; b_{n+1} \leqq b_n, \quad n = 1, 2, \cdots$

correspond to the fact that I_n contains I_{n+1}. Also

(2.8–2) $\lim_{n \to \infty} (b_n - a_n) = 0,$

since $l_n = b_n - a_n$. Now $a_n < b_n \leqq b_1$, and $a_1 \leqq a_n < b_n$. It now follows by Theorem III of § 2.7 that the sequence $\{a_n\}$ has a limit; likewise, by Theorem V, $\{b_n\}$ has a limit. Let us write

$$\lim_{n \to \infty} a_n = a, \quad \lim_{n \to \infty} b_n = b.$$

These two limits must coincide, that is $a = b$, by virtue of (2.8–2). Now $a_n \leqq a = b$, for a is the least upper bound of the sequence $\{a_n\}$; likewise $a = b \leqq b_n$. Hence the point a belongs to every one of the intervals I_n. Obviously there cannot be more than one such point common to all the intervals. For if two such points existed, say at a positive distance h apart, we could choose n so large that $l_n = b_n - a_n < h$, and under this condition the two points could not both lie in I_n.

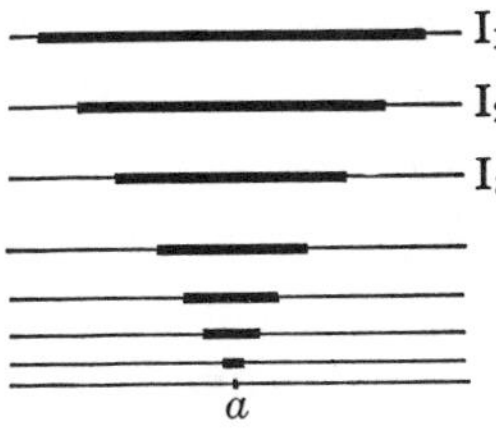

Fig. 17

Theorem VI is illustrated in Fig. 17. Uses of Theorem VI will appear in the next chapter.

MISCELLANEOUS EXERCISES

1. Suppose f is defined as follows: $f(x) = 1$ if x is a rational number, $f(x) = 0$ if x is an irrational number. For what values of x (if any) is f continuous?

2. Show that $\lim_{n \to \infty} \dfrac{n^n}{n! \, e^n}$ exists without finding the limit.

3. Show that $\lim\limits_{n \to \infty} \dfrac{1}{n} \left\{ \dfrac{2 \cdot 4 \cdots (2n)}{1 \cdot 3 \cdots (2n-1)} \right\}^2$ exists without finding the limit.

4. If $x_n = (\log n)/n$, show that $x_n > x_{n+1}$ when $n \geq N$, where N is a certain fixed integer. What is N? What do you conclude about the sequence $\{x_n\}$?

5. Let S be the set of all rational numbers r such that $r^2 < 2$. What is the least upper bound of S? What is the greatest lower bound of S?

6. Let S be the point set consisting of all the points $x_n = (-1)^n \left[2 - \dfrac{4}{2^n} \right]$, $n = 1, 2, \cdots$. Find the least upper bound and greatest lower bound of S.

7. The same as Exercise 6, except that $x_n = (-1)^n + (1/n)$.

8. Let a sequence of numbers a_1, a_2, $\cdots$ be such that $(2 - a_n) a_{n+1} = 1$. (a) Show that $\lim_{n \to \infty} a_n$ exists. Consider two cases: either $a_n < 2$ for all n, or $a_n > 2$ for some n. (b) Find the limit of the sequence.

9. Let x_1 and y_1 be given with $x_1 > y_1 > 0$, and define

$$x_{n+1} = \frac{x_n + y_n}{2} \,, \quad y_{n+1} = \sqrt{x_n y_n}, \quad n = 1, 2, \cdots.$$

Show (a) that $y_n < y_{n+1} < x_1$, (b) that $y_1 < x_{n+1} < x_n$, (c) that $0 < x_{n+1} - y_{n+1} < (x_1 - y_1)/2^n$. Explain why the sequences $\{x_n\}$ and $\{y_n\}$ are convergent and have the same limit.

10. The number $\sqrt{3}$ is defined as the positive number c such that $c^2 = 3$. That there *is* such a number may be proved as follows: Let S be the set of all positive numbers x such that $x^2 < 3$. This set is not empty, since $1 \in S$. It is a bounded set, for, if $x \in S$, either $x \leq 1$ or $x > 1$, and in the latter case $x < x^2 < 3$. By Theorem II, S has a least upper bound, which we shall denote by c. It remains to prove that $c^2 = 3$. (Why must c be positive?) If $\epsilon > 0$, there is some x in S such that $c - \epsilon < x$. Hence $(c - \epsilon)^2 < 3$, or $c^2 - 2c\epsilon + \epsilon^2 < 3$, since $x^2 < 3$. Now let $\epsilon \to 0$, and by Theorem XI, § 1.61, we conclude that $c^2 \leq 3$. Finally, we show that $c^2 < 3$ is impossible. For, if $c^2 < 3$, the fact that $\lim_{h \to 0} (c^2 + 2ch + h^2) = c^2$ shows that $(c + h)^2 < 3$ if h is sufficiently small. This means that $c + h$ is in S if h is any sufficiently small positive number. Since c is the least upper bound of S, this is a contradiction. Therefore $c^2 = 3$.

Now let the student develop a similar argument to show that if $A > 0$ and n is a positive integer, there is a positive number c such that $c^n = A$.

Continuous Functions 3

3. Continuity

In § 1.12 and § 1.2 we pointed out the need to know that if a function is continuous at all points of a finite closed interval it actually attains an absolute maximum and an absolute minimum at points of the interval. Again, in § 1.51, we saw that another property of continuous functions occupies a key position in the proof of the mean-value theorem for integrals. After our study of the real number system in Chapter II we are prepared to prove that continuous functions do in fact possess the properties referred to above.

The definition of a continuous function was given in § 1.1. We repeat the definition.

DEFINITION. Let f be a function which is defined in some interval containing the point x_0 either inside or at one end. We say that f is continuous at x_0 provided that $\lim_{x \to x_0} f(x) = f(x_0)$. If x_0 is at one end of the interval, x must approach x_0 from one side only. We say that f is continuous on an interval if it is continuous at each point of the interval.

Often it is convenient to express the definition of continuity in an alternative but equivalent way, using inequalities: f is continuous at x_0 provided that to each positive number ϵ corresponds some positive number δ such that $|f(x) - f(x_0)| < \epsilon$ whenever $|x - x_0| < \delta$ and x is in the interval on which f is defined. Observe that $|f(x) - f(x_0)| < \epsilon$ is equivalent to the double inequality $f(x_0) - \epsilon < f(x) < f(x_0) + \epsilon$. The choice of the number δ will as a rule depend both on ϵ and on x_0 (and of course on the particular function f).

Among the important theorems about continuity is the following assertion about the continuity of sums, products, and quotients:

THEOREM I. *Let f and g be functions defined on the same interval. If $f(x)$ and $g(x)$ are continuous at a point $x = x_0$, so are $f(x) + g(x)$ and $f(x) \cdot g(x)$. If $g(x_0) \neq 0$, the quotient $\dfrac{f(x)}{g(x)}$ is also continuous at $x = x_0$.*

The proof stems directly from the fundamental limit theorem (Theorem XIV, § 1.64). We have, for example,

$$\lim_{x \to x_0} \frac{f(x)}{g(x)} = \frac{\lim\limits_{x \to x_0} f(x)}{\lim\limits_{x \to x_0} g(x)} = \frac{f(x_0)}{g(x_0)} ,$$

provided $g(x_0) \neq 0$. This is precisely the statement that the quotient $\dfrac{f(x)}{g(x)}$

is continuous at $x = x_0$. The other parts of the theorem are proved in the same way.

EXERCISES

1. Assuming it is known that the functions sin x, cos x are continuous for all values of x, discuss the continuity of tan x and ctn x. Is either of these functions discontinuous at a point where it is defined?

2. If f is defined and continuous on $a \leqq x \leqq b$, is $1/f(x)$ defined on the same interval? What about continuity of $1/f(x)$? May it be discontinuous at a point where it is defined?

3. May a product $f(x) \cdot g(x)$ be continuous at a point x_0 where g is discontinuous? Support your answer. May the product be continuous at a point where both f and g are discontinuous?

4. May a sum $f(x) + g(x)$ be continuous at a point where f is continuous and g discontinuous? May the sum be continuous at a point where both f and g are discontinuous?

5. Suppose $f(x)$, $g(x)$, and $\dfrac{f(x)}{g(x)}$ are defined in some open interval containing x_0, and suppose $\dfrac{f(x)}{g(x)}$ and $g(x)$ are continuous at x_0. May f be discontinuous at x_0?

3.1   Bounded functions

A set of real numbers is said to be *bounded* if the set has both an upper and a lower bound. Consider a function f, defined on a given interval. We say that the function is *bounded on the interval* if the set of all the values of the function is a bounded set. This means that there is some number A such that $|f(x)| \leqq A$ for all x on the interval; or, alternatively, there are two numbers m, M such that $m \leqq f(x) \leqq M$ for all x on the interval. We make this definition whether or not the interval on which the function is defined is closed.

Example 1. The function $f(x) = \sin x$ is bounded on the interval $0 \leqq x \leqq 2\pi$, because $-1 \leqq \sin x \leqq 1$. Actually, the function is bounded on *every* interval in this case.

Example 2. The function $f(x) = 1/x$ is not bounded on the interval $0 < x \leqq 1$, for there is no upper bound to its values. We can make $f(x)$ as large as we please by taking x sufficiently near zero.

Example 3. The function $f(x) = x \sin x$ is not bounded on the infinite interval $0 \leqq x$. There is neither an upper nor a lower bound to the values of the function, since $f(n\,\pi/2) = n\,\pi/2$ if $n = 1, 5, 9, 13, \cdots$ and $f(n\,\pi/2) = -n\,\pi/2$ if $n = 3, 7, 11, 15, \cdots$.

THEOREM II. *If a function is continuous on a finite closed interval, it is bounded on that interval.*

FIRST PROOF.   Let the function f be continuous on the interval $a \leqq x \leqq b$. We observe in the first place that if x_0 is any point of the interval $[a, b]$ there is *some* subinterval containing x_0 in which f is bounded. For, if we take $\epsilon = 1$ in the criterion for continuity as stated in § 3, and denote the corresponding δ by δ_1, we have $f(x_0) - 1 < f(x) < f(x_0) + 1$ provided that x is a point of the interval $[a, b]$ such that $x_0 - \delta_1 < x < x_0 + \delta_1$. Note that this subinterval extends on both sides of x_0 if $a < x_0 < b$, and on one side of x_0 if $x_0 = a$ or $x_0 = b$. To show the dependence of δ_1 on x_0 we shall write $\delta_1 = \delta_1(x_0)$. A second observation of importance is this: If f is bounded on each of two abutting or overlapping subintervals, then it is bounded on the single interval which consists of all points which belong to either one or both of the original subintervals. For example, if $|f(x)| \leqq 10$ when $0 \leqq x \leqq 3/2$ and $|f(x)| \leqq 15$ when $1 \leqq x \leqq 3$, then certainly $|f(x)| \leqq 15$ when $0 \leqq x \leqq 3$.

Now let us define T as the set of all numbers t for which $a < t \leqq b$ and such that f is bounded on the interval $a \leqq x \leqq t$. The set T is not empty, for by the previous paragraph it certainly contains t if $a < t < a + \delta_1(a)$ (see Fig. 18). Furthermore, T has the upper bound b. Therefore, by Theorem II, § 2.7, T has a least upper bound, say c. To complete the proof we shall prove two things: (1) that c belongs to T, so that f is bounded on $[a, c]$, and (2) that $c = b$. By the definition of c we know that either c belongs to T, or there are points of T as near c as we please on the left

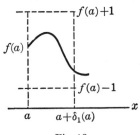

Fig. 18

of c. Now, as we observed at the outset, f is bounded in *some* subinterval containing c and extending at least a distance $\delta_1(c)$ to the left of c. There will certainly be a point of T, say t_1, such that $c - \delta_1(c) < t_1 \leqq c$. If $t_1 = c$, (1) is clear; if $t_1 < c$, f is bounded on each of the overlapping intervals $[a, t_1]$, $[c - \delta_1(c), c]$, and hence on the single interval $[a, c]$. In either case we see that c belongs to T. This proves (1).

To prove (2) we assume $c < b$ and deduce a contradiction. Once more we use the fact that f is bounded in *some* subinterval containing c; since $c < b$ this subinterval will extend somewhat to the right of c, say as far as $c + \delta_1(c)$. But now we know that f is bounded on each of the abutting intervals $[a, c]$, $[c, c + \delta_1(c)]$, and hence on the single interval $[a, c + \delta_1(c)]$. This means, however, that $c + \delta_1(c)$ is in the set T, by definition. Here we have a contradiction, since $c + \delta_1(c)$ is greater than the least upper bound of T. We have thus proved (2), and thereby completed the proof of the theorem.

SECOND PROOF. This proof uses the theorem of nested intervals (Theorem VI, § 2.8). As in the first proof, we observe that if a function is bounded on each of two abutting closed intervals, it is bounded on the interval which is obtained by combining the two intervals into one. Now suppose the theorem to be proved were false; that is, suppose f is not bounded on $[a, b]$. Denote the interval $[a, b]$ by I_1. Consider the closed intervals $[a, (a + b)/2]$, $[(a + b)/2, b]$ obtained by bisecting I_1. On at least one of these closed subintervals (denote such a one by I_2) f must fail to be bounded. We proceed to bisect I_2, obtaining a new subinterval I_3 on which f fails to be bounded. By repetition of this process we generate a sequence I_n of closed intervals on each of which f is not bounded. The length of I_n is $(b - a)/2^{n-1}$. Hence it is clear that I_n is a nest, as defined in § 2.8. By Theorem VI of § 2.8 there is a single point, say $x = c$, which is in each of the intervals I_n, and hence in the interval $[a, b]$. Now, as shown at the beginning of our first proof, f is bounded on *some* interval containing the point c. Denote such an interval by J. Since the length of I_n tends to zero as n increases, and since c is in I_n, it is clear that J must contain I_n when n is sufficiently large. But this involves a contradiction, for f is *not* bounded on I_n and it *is* bounded on J. Because of this contradiction, our initial assumption that the theorem is false must be rejected. We have thus completed the proof.

EXERCISES

1. Let $f(x) = 2x \sin(1/x) - \cos(1/x)$. Is this function bounded on the interval $0 < x \leq 1$?

2. Consider the function $\tan^{-1} x$, defined for all values of x. Is it bounded?

3. Which of the following functions are bounded on the indicated intervals?

(a) $\dfrac{1-x}{1+x}$, $-1 < x < 1$;

(c) $\dfrac{1}{\sqrt{5} - 2\sin x}$, $0 \leq x \leq 2\pi$;

(b) $\dfrac{|x|}{x}$, $0 < x < 1$;

(d) $\dfrac{\sin x}{x}$, $0 < x \leq \dfrac{\pi}{2}$;

(e) $\dfrac{1}{x} \sin \dfrac{\pi}{x}$, $0 < x \leq 1$.

4. Without attempting to find exact absolute maxima, find numbers M such that $|f(x)| \leq M$ on the intervals indicated in each of the following cases:

(a) $f(x) = x^{17} - 6x^{13} + 5x^2 - 2$, $-1 \leq x \leq 1$;

(b) $f(x) = 3\sin^2 x - 2\cos x - \sin \dfrac{x}{2} \cos \dfrac{x}{3}$, $0 \leq x \leq 2\pi$;

(c) $f(x) = \dfrac{x^3 - x^2 + 1}{1 + x^4}$, $-1 \leq x \leq 2$.

3.2   The attainment of extreme values

Suppose we are given a function f, and suppose we know that the function is bounded on a certain given interval. Let m and M be the greatest lower bound and least upper bound, respectively, of the values of $f(x)$ on the given interval. Is it necessarily the case that $f(x)$ actually takes on the values m and M on the interval? Examples show that the answer to this question is negative.

Example 1. Suppose we define $f(x) = x^2$ if $0 \leq x < 1, f(x) = 0$ if $x = 1$ (see Fig. 19). This function has $M = 1$ for the interval $0 \leq x \leq 1$, but there is no x on the interval such that $f(x) = 1$. Note, however, that the function is not continuous at $x = 1$.

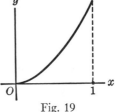

Fig. 19

Example 2. Let $f(x) = \dfrac{1}{x+1} \sin \dfrac{1}{x}$, $0 < x$. This function is continuous for all positive values of x. We have

$$\left| \frac{1}{x+1} \sin \frac{1}{x} \right| \leq \frac{1}{1+x} < 1$$

and hence $-1 < f(x) < 1$ when $x > 0$. Actually $m = -1$ and $M = 1$ for this function on any interval $0 < x \leq a$, where $a > 0$. As x approaches 0, $f(x)$ oscillates in value between $1/(x + 1)$ and $-1/(x + 1)$ an infinite number of times (see Fig. 20). If N is any number such that $-1 < N < 1$, there are infinitely many values of x as near 0 as we please such that $f(x) = N$. But the values ± 1 are never attained.

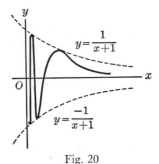

Fig. 20

THEOREM III. *Let f be continuous when $a \leq x \leq b$, and let m and M be the greatest lower bound and least upper bound of the values $f(x)$ on this interval. Then $f(x)$ assumes each of the values m and M at least once in the interval.*

PROOF.   Suppose the value M is never attained, so that $M - f(x) > 0$ for all x in the interval. Consider the function

$$g(x) = \frac{1}{M - f(x)} \cdot$$

It is a continuous function, since the denominator never vanishes (see Theorem I, § 3). Hence, by Theorem II, § 3.1, $g(x)$ is bounded. Let A be an upper bound of $g(x)$ (A is necessarily positive): *since* $M - f(x) > 0$.

$$\frac{1}{M - f(x)} \leqq A.$$

This inequality may be transformed successively into

$$\frac{1}{A} \leqq M - f(x), f(x) \leqq M - \frac{1}{A} \cdot$$

But we now have the result that $M - (1/A)$ is an upper bound for $f(x)$. This is a contradiction, however, for $M - (1/A) < M$, and M is the *least* upper bound of $f(x)$. We must then conclude that $M - f(x)$ vanishes for some value of x. This completes the proof so far as M is concerned. A similar proof can be given that the value m is assumed. Alternatively, one can consider the function $- f(x)$, whose least upper bound is $-m$. Then, by what has already been proved, $- f(x) = - m$ must hold for some x. This is equivalent to $f(x) = m$, of course.

EXERCISES

1. Let $f(x) = 3 x$ if $0 < x < 1$, and define $f(0) = 1, f(1) = 2$. Find m and M for f on the interval $[0, 1]$. Are these extreme values attained? What can you say about the continuity of f?

2. Let $f(x) = x - [x]$, where $[x]$ denotes the greatest integer less than or equal to x. Find m and M for f on $[0, 2]$. Are these extreme values attained? Graph the function.

3. Find m and M for $f(x) = \tan^{-1} x$, where x can range over all values. Are these extreme values attained?

4. Let $f(x) = \dfrac{x^2 + x^4}{2} + \dfrac{x^2 - x^4}{2} \sin \dfrac{\pi}{x}, 0 < x \leqq 1$.

Find m and M for this function on the indicated interval. Are these extreme values attained?

5. Given a point Q and a circle in the xy-plane, with Q not on the circle, explain with the aid of Theorem III how you know that there is a point P_1 on the circle which is at least as near to Q as any other point of the circle is; also explain why there is a point P_2 on the circle which is at least as far from Q as any other point on the circle is.

6. Given a point Q and a parabola in the xy-plane, with Q not on the parabola, explain why there is a point P of the parabola at least as close to Q as any other point of the parabola is? Why, in contrast to the case of the circle in Exercise 5, is there no point of the parabola at maximum distance from Q? If you wish, you may choose a co-ordinate system in which the parabola has a very simple equation.

3.3   The intermediate-value theorem

THEOREM IV. *Suppose that f is continuous on the closed interval $a \leqq x \leqq b$, and that $f(a) \neq f(b)$. Then, as x varies from a to b, $f(x)$ takes on every value between $f(a)$ and $f(b)$.*

This theorem expresses a property of continuous functions which has a simple geometric interpretation. Suppose, for example, that $f(a) < f(b)$, and that k is a number between $f(a)$ and $f(b)$.

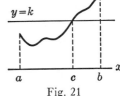
Fig. 21

Consider the graph of $y = f(x)$. It is a continuous curve joining the points $(a, f(a))$ and $(b, f(b))$. These points are on opposite sides of the line $y = k$; the theorem asserts that the curve $y = f(x)$ intersects the line $y = k$ at some point $x = c$ between a and b (see Fig. 21).

With this geometric interpretation before him, the student may be strongly tempted to say that the theorem is obviously true and requires no further proof. If he does so, however, he is not relying upon our definition of continuity given in § 3, but upon intuitive assumptions about the geometrical meaning of the term "continuous curve." We wish to show that the theorem can be proved on the basis of the definition.

Before proving Theorem IV we shall find it convenient to prove the following proposition:

THEOREM V. *Let f be defined on an interval containing the point c, and let f be continuous at $x = c$. Then, if $f(c) \neq 0$, there is a subinterval containing c throughout which $f(x)$ has the same sign as $f(c)$.*

Suppose, for example, that $f(x)$ is defined when $c - h < x < c + h$, where $h > 0$, and that $f(c) > 0$. Then if f is continuous at $x = c$ the theorem

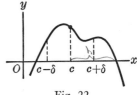
Fig. 22

asserts that we can choose δ, $0 < \delta < h$, so that $f(x) > 0$ when $|x - c| < \delta$ (see Fig. 22). The proof is an immediate consequence of Theorem X, § 1.61, since $\lim_{x \to c} f(x) = f(c) > 0$. The proof when $f(c) < 0$ is left to the student. If c is at one end of the interval on which f is defined, the subinterval on which f is of constant sign will extend on one side only of c.

PROOF OF THEOREM IV.   Consider the function $g(x) = f(x) - k$, where k is between $f(a)$ and $f(b)$. Let us assume for definiteness that $f(a) < k < f(b)$. We shall assume the theorem false and deduce a contradiction, thus completing the proof. Thus we assume that $g(x)$ is never zero on the interval $[a, b]$.

Now $g(a) < 0$ and $g(b) > 0$. Bisect the interval $[a, b]$. At the midpoint $g(x)$ is not zero, and hence is either positive or negative. Choose that half interval for which $g(x)$ is positive at one end and negative at the other, denoting it by I_2. The interval $[a, b]$ we denote by I_1. We now repeat the bisection process, successively obtaining intervals $I_1, I_2, I_3, \cdots$ such that $g(x)$ is negative at the left end of I_n and positive at the right end. These intervals obviously form a nest, and so by Theorem VI, § 2.8, close down on a unique point c. Now $g(c) \neq 0$, and so by Theorem V $g(x)$ is of one sign throughout some interval J containing c. But such an interval will contain I_n when n is sufficiently large, because the length of I_n approaches zero as n increases. Thus $g(x)$ takes on both positive and negative values in J. We have now reached a contradiction, and the proof is complete.

EXERCISES

1. Suppose f is defined as $f(x) = |x| \log |x|$ if $x \neq 0$, $f(0) = e$. Without investigating the limit of $f(x)$ as $x \to 0$, explain why it is certain that f is not continuous at $x = 0$.

2. A point Q is inside a circle C. The point of C nearest Q is a distance d from Q, and the point of C furthest from Q is a distance D from Q. Explain how you know that there is a point P on C whose distance from Q is $\frac{1}{2}(d + D)$.

3. If f is continuous and $f(x) \geqq 0$ on $[a, b]$, while $f(x) > 0$ for at least one x of the interval $[a, b]$, prove that $\int_a^b f(x)\, dx > 0$. Give a carefully reasoned proof, not merely an intuitive argument based on geometric plausibility.

4. Let P be a point on an ellipse. Consider rays (half-lines) emanating from P. Explain how you know that there is one such ray which divides the area enclosed by the ellipse into two equal parts.

5. Suppose f is continuous on $[a, b]$ and let λ be a number such that $0 < \lambda < 1$. Show that there is an x such that $a < x < b$ and $\int_a^x f(t)\, dt = \lambda \int_a^b f(t)\, dt$, provided that the condition $\int_a^b f(t)\, dt \neq 0$ is fulfilled.

MISCELLANEOUS EXERCISES

1. Six functions are defined below, each by a certain formula. For each function answer the questions: (a) For what values of x is the function defined? (b) For what values of x is the function continuous? (c) Are there any values of x where the function is not already defined, but may be defined so as to make the function continuous?

(i) $f(x) = \dfrac{x^3 - 1}{x^2 + x + 1}$,

(iii) $h(x) = \dfrac{2^{1/x} + 3}{2^{1/x} + 1}$,

(ii) $g(x) = \dfrac{x - 1}{x^2 - 1}$,

(iv) $F(x) = \dfrac{x^2 - 16}{x + 6 - 5\sqrt{x}}$,

(v) $G(x) = \dfrac{x^2 - 5x - 50}{(x^2 - 8x - 20)\sqrt{x^2 - 25}}$,

(vi) $H(x) = \dfrac{3x + 5}{\sqrt{2x - 3} - \sqrt{5x - 6} + \sqrt{3x - 5}}$.

2. Given that f is defined for all x, continuous at $x = 0$, and that, for all x and y, $f(x + y) = f(x)f(y)$, show that f is continuous for all values of x.

3. Given that f is defined for all x, continuous at $x = 0$, and that, for all x and y, $f(x + y) = f(x) + f(y)$, show that f is continuous for all values of x.

4. A function f is defined and continuous on the interval $a \leq x \leq b$, and $f(x) = 0$ when x is rational. Using Theorem V, explain why $f(x) = 0$ at *all* points of the interval.

5. Show that $|f(x) - f(x_0)| < |x - x_0|$ if $f(x) = \sqrt{4 + x^2}$ and $x \neq x_0$. What does this prove about f?

6. Let $P(x)$ be a polynomial of odd degree with real coefficients. Then the equation $P(x) = 0$ has at least one real root. Prove this by use of Theorem IV.

7. Let $P(x) = x^n + a_1 x^{n-1} + \cdots + a_n$, where n is an even positive integer, the a's are real, and $a_n < 0$. Show that the equation $P(x) = 0$ has at least two real roots. What more can you say about them?

8. Explain why $\dfrac{x^2 + 1}{x + 2} + \dfrac{x^4 + 1}{x - 3} = 0$ has at least one root between -2 and 3.

9. Let $f(x) = \dfrac{A}{a^2 + x} + \dfrac{B}{b^2 + x} + \dfrac{C}{c^2 + x} - 1$, where A, B, and C are positive and $a > b > c > 0$. Discuss the nature of the graph of $y = f(x)$ and explain why the equation $f(x) = 0$ has exactly three roots x_1, x_2, x_3 satisfying the inequalities $-a^2 < x_1 < -b^2 < x_2 < -c^2 < x_3$.

10. Suppose that f is continuous for all values of x, that $\lim_{x \to -\infty} f(x) = -1$, and $\lim_{x \to +\infty} f(x) = 10$. Explain how you use Theorem IV to show that there is at least one value of x such that $f(x) = 0$.

11. Suppose that f is given continuous for all x, with $f(x) < 0$ when $x < x_1$ and $f(x) > 0$ when $x > x_2$, where $x_1 < x_2$. (a) Define a cut (see § 2.4) in such a way that the cut number c satisfies the conditions $f(c) = 0$, $f(x) < 0$ if $x < c$. (b) Define a cut in such a way that the cut number satisfies the conditions $f(c) = 0$, $f(x) > 0$ if $x > c$.

12. Prove Theorem III by using a repeated bisection method, as in the second proof of Theorem II. At each bisection, retain a half interval on which the least upper bound of f is M. Let $\{I_n\}$ be the resulting nest of intervals. Choose a point x_n in I_n such that $f(x_n) > M - (1/n)$. Why is this possible? What happens as $n \to \infty$? Write out the whole argument carefully, showing how you are led to a point c such that $f(c) = M$.

13. Consider the following theorem: *If f is continuous when $a \leqq x \leqq b$ and if $f(a)f(b) < 0$ (i.e., if $f(a)$ and $f(b)$ are of unlike signs), then there is some point c between a and b at which $f(c) = 0$.* As we saw in § 3.3, the proof of Theorem IV may be made very easily once this theorem is proved. Give a proof, using the existence of a least upper bound, as guaranteed in Theorem II,§2.7. Start by defining S to be the set of all x such that $a \leqq x < b$ and $f(x)f(a) > 0$. Why does S have an upper bound? Let c be this upper bound and prove that $f(c) = 0$. Write out your whole argument clearly, with specific justification for each step in the reasoning.

14. A function f is defined and continuous when $0 \leqq x \leqq 1$, and $f(1) = 2$. The function has the further property that the value $f(x)$ is always a rational number. Find $f(0)$.

15. A function f is defined when $0 \leqq x \leqq 2$, with at most one point of discontinuity. Furthermore, the value $f(x)$ is rational if $0 \leqq x < 1$ and irrational if $1 < x \leqq 2$. Why must f have *exactly* one point of discontinuity? What is that point?

16. Let f be defined on $[0, 1]$ as follows: $f(x) = x$ if x is rational, $f(x) = 1 - x$ if x is irrational. (**a**) For what values of x is f continuous? (**b**) In spite of the fact that f does not satisfy the hypothesis of Theorem IV on $[0, 1]$, show that as x varies from 0 to 1, $f(x)$ takes on every value between $f(0)$ and $f(1)$. What is x, for example, if $f(x) = \sqrt{2}/2$?

17. Let f be defined for all x by the conditions: $f(x) = 0$ if x is irrational or if $x = 0$, $f(x) = 1/n$ if x is the rational number m/n, where $m \neq 0$, $n > 0$, and the fraction is reduced to its lowest terms. Explain carefully (**a**) why f is discontinuous at x_0 if x_0 is rational and not zero, (**b**) why f is continuous at x_0 if x_0 is irrational. (**c**) Is f continuous at $x = 0$?

18. Suppose that f is defined and has a continuous derivative when $a \leqq x \leqq b$, and suppose that $f(a) = f(b) = 0$, but that $f(x)$ is not 0 for every x. Let λ be a constant different from zero, and let $g(x) = f'(x) + \lambda f(x)$. Prove that there is some number ξ such that $a < \xi < b$ and $g(\xi) = 0$. This is fairly hard. There are two cases: Case 1, in which $f(x)$ assumes both positive and negative values, and Case 2, in which $f(x)$ takes on values of one sign only. The proof in Case 1 is easy. For Case 2 it is fairly easy to prove that $g(\xi) = 0$ for some ξ such that $a \leqq \xi \leqq b$, but it is much more difficult to prove the result as originally stated, with $a < \xi < b$.

19. A function f is continuous on $a \leqq x \leqq b$ and differentiable on $a < x < b$. Furthermore, $f(a) = a$ and $f(b) = b$. Show that there are two points x_1, x_2 such that $a < x_1 < x_2 < b$ and $\dfrac{1}{f'(x_1)} + \dfrac{1}{f'(x_2)} = 2$.

20. Suppose that f is defined and differentiable on the closed interval $[a, b]$, with $f'(a) > 0$, $f'(b) < 0$. Prove that $f'(\xi) = 0$ for some ξ such that $a < \xi < b$. Show also that the same conclusion may be reached if $f'(a) < 0$, $f'(b) > 0$.

21. There is a counterpart of Theorem IV for functions which need not be continuous, but are known to be derivatives of continuous functions. The theorem reads as follows: *Suppose that f is defined and differentiable on $[a, b]$, with $f'(a) \neq f'(b)$. Let k be a number between $f'(a)$ and $f'(b)$. Then $f'(\xi) = k$ at some point ξ, where $a < \xi < b$.* This theorem is known as Darboux's theorem. Prove it by setting $g(x) = f(x) - kx$ and using the results of Exercise 20.

22. Let f be a function which is defined for all x, continuous at $x = 0$, and such that $f(x + y) = f(x) + f(y)$ for all values of x and y. Show that $f(x) = Cx$, where $C = f(1)$. Begin by proving **(a)** that $f(m/n) = (m/n) f(1)$ if m and n are positive integers, **(b)** that $f(-x) = -f(x)$, and **(c)** that $f(0) = 0$. Then note Exercise 3 and apply Exercise 4 to the function $f(x) - xf(1)$.

Extensions of the Law of the Mean

4

4. Introduction

In this chapter we consider various generalizations of the law of the mean (Theorem IV, § 1.2), and related topics. The most important extension of the law of the mean is Taylor's formula with remainder. In elementary calculus Taylor's formula is often closely associated with the study of expansions of functions in power series. There is indeed an important connection between Taylor's formula and expansions in power series, but the formula is not important solely because of that connection. The chapter closes with a discussion of l'Hospital's rule.

4.1 Cauchy's generalized law of the mean

Cauchy's generalization of the law of the mean deals with two functions instead of with just one.

THEOREM I. *Let $F(x)$ and $G(x)$ be continuous on the closed interval $a \leqq x \leqq b$, and differentiable on the open interval $a < x < b$. Assume further that $G(a) \neq G(b)$, and that $F'(x)$ and $G'(x)$ never vanish simultaneously. Then for some value $x = X$ such that $a < X < b$ we have*

$$(4.1\text{--}1) \qquad \frac{F(b) - F(a)}{G(b) - G(a)} = \frac{F'(X)}{G'(X)}.$$

As a special case, we obtain the ordinary law of the mean (Theorem IV, § 1.2) if we take $G(x) = x$.

PROOF. As with the proof of the ordinary law of the mean, we appeal to Rolle's theorem (§ 1.2). Let us set

$$(4.1\text{--}2) \qquad k = \frac{F(b) - F(a)}{G(b) - G(a)},$$

and define $\qquad \phi(x) = F(x) - F(a) - k[G(x) - G(a)].$

Observe that $\phi(a) = 0$. Also, because of the definition of k, we see that $\phi(b) = 0$.

Now $\qquad \phi'(x) = F'(x) - kG'(x).$

The function $\phi(x)$ satisfies the conditions of Rolle's theorem; therefore $\phi'(X) = 0$ for some X, $a < X < b$. That is,

$$0 = F'(X) - kG'(X).$$

In this formula $G'(X) \neq 0$. For if $G'(X) = 0$ then $F'(X) = 0$ also, whereas we assumed that these derivatives never vanish simultaneously. Thus we can write

$$k = \frac{F'(X)}{G'(X)},$$

a formula which is equivalent to (4.1–1) because of (4.1–2). This completes the proof.

A geometrical interpretation of (4.1–1) may be given as follows: Let a plane curve be represented parametrically by equations

(4.1–3) $y = F(t), \ x = G(t), \ a \leqq t \leqq b.$

The slope of the curve for a given t is

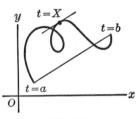

(4.1–4) $\dfrac{dy}{dx} = \dfrac{F'(t)}{G'(t)}.$

The constant k in (4.1–2) is the slope of the straight line joining the points on the curve corresponding to $t = a$ and $t = b$, respectively. The theorem says that the two slopes (4.1–2) and (4.1–4) are equal for at least one value of t between a and b (see Fig. 23).

Fig. 23

The uses of Theorem I are largely in proving other theorems. Its most conspicuous use is in the proof of the rule of l'Hospital (§ 4.5).

EXERCISE

Cauchy's generalized mean-value formula (4.1–1) may be written in the form

$$\begin{vmatrix} F(a) & G(a) & 1 \\ F(b) & G(b) & 1 \\ F'(X) & G'(X) & 0 \end{vmatrix} = 0.$$

This suggests the following more general theorem:

Suppose that F, G, H are continuous when $a \leqq x \leqq b$ and differentiable when $a < x < b$. Then there is a value X such that $a < X < b$ and

$$\begin{vmatrix} F(a) & G(a) & H(a) \\ F(b) & G(b) & H(b) \\ F'(X) & G'(X) & H'(X) \end{vmatrix} = 0.$$

Prove this theorem by considering

$$\phi(x) = \begin{vmatrix} F(a) & G(a) & H(a) \\ F(b) & G(b) & H(b) \\ F(x) & G(x) & H(x) \end{vmatrix}.$$

4.2   Taylor's formula with integral remainder

Consider a polynomial $P(x)$ of degree n:

(4.2–1) $P(x) = b_0 x^n + b_1 x^{n-1} + \cdots + b_n, \ n \geqq 0,$

where $b_0 \neq 0$. If we choose any particular value of x, say $x = a$, it is possible to express $P(x)$ as a sum of powers of $(x - a)$, the highest power being n:

$$(4.2\text{--}2) \qquad P(x) = c_0(x - a)^n + c_1(x - a)^{n-1} + \cdots + c_n.$$

That this is the case may be seen as follows: It is clearly true if $n = 0$, for in that case the two expressions for $P(x)$ are identical in form. For $n = 1$ we have a linear function $P(x) = b_0 x + b_1$, and we wish to express it in the form $c_0(x - a) + c_1$. Choosing $c_0 = b_0$, we have

$$P(x) - b_0(x - a) = b_1 + ab_0,$$

so that $P(x) = b_0(x - a) + c_1$ with $c_1 = b_1 + ab_0$. In general, we proceed by induction, assuming that the desired type of representation is possible with polynomials of degree $\leq n - 1$, where $n \geq 1$. Then for the polynomial (4.2–1) we choose $c_0 = b_0$, so that $P(x) - b_0(x - a)^n$ is a polynomial of degree at most $n - 1$. Hence we can express this polynomial as a sum of powers of $(x - a)$:

$$P(x) - b_0(x - a)^n = c_1(x - a)^{n-1} + \cdots + c_n.$$

This is equivalent to (4.2–2), and completes the induction proof.

Once we know that the representation (4.2–2) is possible, it is very easy to find convenient formulas for $c_0, \cdots, c_n$. Let us differentiate (4.2–2) k times, where $0 \leq k \leq n$ (if $k = 0$, $P^{(k)}(x)$ means $P(x)$). After doing this we set $x = a$. In this process the only term of $P(x)$ which leads to a nonzero result is $c_{n-k}(x - a)^k$. Therefore

$$P^{(k)}(a) = \left\{ \frac{d^k}{dx^k} [c_{n-k}(x - a)^k] \right\}_{x = a} = k! \, c_{n-k},$$

where, according to the usual convention, $0! = 1$.

Consequently

$$c_0 = \frac{P^{(n)}(a)}{n!}, \; c_1 = \frac{P^{(n-1)}(a)}{(n-1)!}, \cdots, c_n = P(a),$$

so that (4.2–2) may be written in the form

$$(4.2\text{--}3) \quad P(x) = P(a) + P'(a)(x - a) + \frac{P''(a)}{2!} (x - a)^2 + \cdots$$

$$+ \frac{P^{(n)}(a)}{n!} (x - a)^n,$$

where we have reversed the order of the terms to suit our convenience.

Now let us ask whether there is any counterpart of this formula when the polynomial $P(x)$ is replaced by a function $f(x)$ which is not a polynomial. That is, let us ask what relation the expression

$$f(a) + f'(a)(x - a) + \frac{f''(a)}{2!} (x - a)^2 + \cdots + \frac{f^{(n)}(a)}{n!} (x - a)^n$$

bears to $f(x)$. Naturally we assume that f can be differentiated n times at $x = a$. Let us assume more than that, however. A convenient assumption is that f and its first $n + 1$ derivatives are continuous in a closed interval containing $x = a$.

Now by Theorem VIII, § 1.53, we know that

$$\int_a^x f'(t)dt = f(x) - f(a).$$

Let us write this in the form

(4.2–4) $$f(x) = f(a) + \int_a^x f'(t)dt.$$

We transform the integral by integration by parts, taking

$$u = f'(t), \qquad dv = dt,$$
$$du = f''(t)dt, \qquad v = -(x - t).$$

Thus $$\int_a^x f'(t)dt = -f'(t)(x - t)\Big|_a^x + \int_a^x f''(t)(x - t)dt,$$

and so

(4.2–5) $$f(x) = f(a) + f'(a)(x - a) + \int_a^x f''(t)(x - t)dt.$$

We can now integrate by parts again, taking

$$u = f''(t), \qquad dv = (x - t)dt,$$
$$du = f^{(3)}(t)dt, \qquad v = \frac{-(x - t)^2}{2!}.$$

The integral in (4.2–5) now becomes

$$\int_a^x f''(t)(x - t)dt = -f''(t)\frac{(x - t)^2}{2!}\Big|_a^x + \int_a^x f^{(3)}(t)\frac{(x - t)^2}{2!}\,dt,$$

and so

$$f(x) = f(a) + f'(a)(x - a) + f''(a)\frac{(x - a)^2}{2!} + \frac{1}{2!}\int_a^x f^{(3)}(t)(x - t)^2\,dt.$$

It is now clear that repeated integration by parts will lead to the formula

(4.2–6) $$f(x) = f(a) + f'(a)(x - a) + \cdots$$
$$+ \frac{f^{(n)}(a)}{n!}(x - a)^n + \frac{1}{n!}\int_a^x (x - t)^n f^{(n+1)}(t)dt.$$

This is the generalization of (4.2–3) which we have been seeking. The function $f(x)$ has been expressed as a polynomial of degree n in $(x - a)$, *plus a remainder term*.

We state our findings in a formal theorem.

THEOREM II.   *Let $f(x)$ and its first $n + 1$ derivatives ($n \geq 0$) be continuous in a closed interval containing $x = a$ (either inside or at one end). Let x be any point of this interval. Then*

$$(4.2\text{-}7) \quad f(x) = f(a) + f'(a)(x - a) + \cdots + \frac{f^{(n)}(a)}{n!}(x - a)^n + R_{n+1},$$

the remainder being given by

$$(4.2\text{-}8) \qquad R_{n+1} = \frac{1}{n!} \int_a^x (x - t)^n f^{(n+1)}(t)dt.$$

We have indicated the procedure for proving the theorem by successive integration by parts, starting from (4.2–4). If one wishes, he may give the proof more formally by mathematical induction.

Formula (4.2–7) is called *Taylor's formula with remainder*. Various formulas for the remainder may be given, as we shall see in § 4.3.

The size of the remainder may sometimes be estimated from (4.2–8). Thus, for example, if $|f^{(n+1)}(t)| \leq M$ when $a \leq t \leq x$, we can see that

$$|R_{n+1}| \leq \frac{M}{n!} \int_a^x (x - t)^n \, dt = \frac{M(x - a)^{n+1}}{(n + 1)!}.$$

4.3   Other forms of the remainder

It is possible to obtain (4.2–7) with a different formula for R_{n+1}, under slightly less stringent assumptions. It will suffice to assume merely that $f^{(n+1)}(x)$ *exists* on an interval, without necessarily being continuous.

THEOREM III.   *Let f and its first n derivatives be continuous when $a \leq x \leq b$ (where $a < b$), and let the $(n + 1)$st derivative $f^{(n+1)}(x)$ exist when $a < x < b$. Then there is a value $x = X, a < X < b$, such that*

$$(4.3\text{-}1) \quad f(b) = f(a) + f'(a)(b - a) + \cdots + \frac{f^n(a)}{n!}(b - a)^n$$

$$+ \frac{f^{(n+1)}(X)}{(n + 1)!}(b - a)^{n+1}.$$

The same formula holds in case $b < a$, all the inequalities then being reversed.

This is a generalization of the law of the mean, and actually coincides with the law of the mean in the special case $n = 0$.

It is difficult to give a proof of (4.3–1) which will seem well motivated and free from artifice. The following proof has been discovered as a result of careful study and a certain amount of trial and error.

We define two functions

$$(4.3\text{–}2) \quad F(x) = f(b) - f(x) - f'(x)(b - x) - \cdots - \frac{f^{(n)}(x)}{n!}(b - x)^n,$$

and

$$(4.3\text{–}3) \qquad\qquad G(x) = \frac{(b - x)^{n+1}}{(n + 1)!}.$$

Observe that $F(b) = G(b) = 0$. In calculating the derivative of $F(x)$ we find that a great deal of cancellation occurs between terms arising from the differentiation of the right member of (4.3–2). Thus

$$\frac{d}{dx}[-f'(x)(b - x)] = -f''(x)(b - x) + f'(x).$$

The $f'(x)$ here cancels the derivative of the previous term, $-f(x)$; the term $-f''(x)(b - x)$ is canceled by one of the terms coming from the differentiation of $-\dfrac{f''(x)}{2!}(b - x)^2$. The final result, which the student should verify for himself, is

$$(4.3\text{–}4) \qquad\qquad F'(x) = -\frac{f^{(n+1)}(x)}{n!}(b - x)^n.$$

We also have

$$(4.3\text{–}5) \qquad\qquad G'(x) = -\frac{(b - x)^n}{n!}.$$

Let us now apply Cauchy's form of the law of the mean (4.1–1). Since $F(b)$ and $G(b)$ are zero, it reads

$$\frac{F(a)}{G(a)} = \frac{F'(X)}{G'(X)}, \quad \text{or} \quad F(a) = \frac{F'(X)}{G'(X)}G(a).$$

Taking account of (4.3–3), (4.3–4), and (4.3–5), this may be written

$$(4.3\text{–}6) \qquad\qquad F(a) = f^{(n+1)}(X)\frac{(b - a)^{n+1}}{(n + 1)!}.$$

If we now put $x = a$ in (4.3–2) and use (4.3–6), we obtain the desired formula (4.3–1). This proof is valid whether $a < b$ or $b < a$, because Cauchy's formula (4.1–1) is unaffected by an interchange of a and b.

The formula in Theorem III is written in a variety of different ways by changes in notation. One important form commonly occurring in the literature is obtained by putting $b = a + h$, where h may be either positive or negative. The number X between a and $a + h$ may then be written in the form $X = a + \theta h$, where θ is some number such that $0 < \theta < 1$. Thus we have

$$(4.3\text{-}7) \qquad f(a + h) = f(a) + f'(a)h + \cdots + \frac{f^{(n)}(a)}{n!}\, h^n$$

$$+ \frac{f^{(n+1)}(a + \theta h)}{(n + 1)!}\, h^{n+1}.$$

Another form results by putting $b = x$ in (4.3–1). In this form we may write

$$(4.3\text{-}8) \qquad f(x) = f(a) + f'(a)(x - a) + \cdots + \frac{f^{(n)}(a)}{n!}\, (x - a)^n + R_{n+1},$$

with

$$(4.3\text{-}9) \qquad R_{n+1} = \frac{f^{(n+1)}(X)}{(n + 1)!}\, (x - a)^{n+1},$$

where X lies between x and a.

Formula (4.3–9) is called *Lagrange's form* of the remainder.

Brook Taylor (English) published in 1715 the form of the infinite series (4.3–10) which bears his name today. The formulas (4.3–8) and (4.3–9) were derived by J. L. Lagrange (French) in 1797. Lagrange's proof utilized integrals. The first proof using the law of the mean appears to have been by Ampère in 1806.

The student should distinguish between (4.3–8) and Taylor's *series*, which is the infinite-series formula

$$(4.3\text{-}10) \qquad f(x) = f(a) + f'(a)(x - a) + \cdots + \frac{f^{(n)}(a)}{n!}\, (x - a)^n + \cdots .$$

This expansion of a function in powers of $(x - a)$ is valid under certain circumstances. Proofs of such validity in particular cases may be made with the aid of Taylor's formula and one or the other of the forms of the remainder. Power-series developments are considered systematically in Chapter XXI.

Example 1. Write Taylor's formula for $f(x) = 1/(2 + x)$ with $a = -1$ and $n = 2$, using Lagrange's form of the remainder.

We have $f(-1) = 1$ and

$$f'(x) = \frac{-1}{(2 + x)^2}\, , \qquad\qquad f'(-1) = -1,$$

$$f''(x) = \frac{2}{(2+x)^3}, \qquad\qquad f''(-1) = 2,$$

$$f^{(3)}(x) = \frac{-6}{(2+x)^4}, \qquad\qquad f^{(3)}(-1) = -6.$$

Thus (4.3–8) becomes

$$\frac{1}{2+x} = 1 - (x+1) + (x+1)^2 + R_3$$

with $$R_3 = \frac{-6}{(2+X)^4}\frac{(x+1)^3}{3!} = -\frac{(x+1)^3}{(2+X)^4},$$

where X is between x and -1. If we wish to estimate R_3 we observe that $1/(2+X)$ lies between $1/(2+x)$ and 1, and hence that $|R_3|$ lies between $\dfrac{|x+1|^3}{(2+x)^4}$ and $|x+1|^3$. For example, if $x = -0.9$, R_3 is negative, and in absolute value between $0.001/(1.1)^4$ and 0.001.

Example 2. Write Taylor's formula for $f(x) = \log(1+x)$ with $a = 0$ and a general value of n. Estimate R_{n+1} if $0 \leqq x \leqq \frac{1}{2}$.

We have $f(0) = 0$ and

$$f'(x) = \frac{1}{1+x}, \qquad\qquad f'(0) = 1,$$

$$f''(x) = \frac{-1}{(1+x)^2}, \qquad\qquad f''(0) = -1,$$

$$\cdots\cdots\cdots\cdots\cdots \qquad\qquad \cdots\cdots\cdots\cdots$$

$$f^{(n)}(x) = \frac{(-1)^{n-1}(n-1)!}{(1+x)^n}, \qquad f^{(n)}(0) = (-1)^{n-1}(n-1)!$$

The general formula for $f^{(n)}(x)$ may be surmised after the first few instances, and verified by induction. Thus

$$(4.3\text{–}11)\quad \log(1+x) = x - \tfrac{1}{2}x^2 + \tfrac{1}{3}x^3 - \cdots + (-1)^{n-1}\frac{1}{n}x^n + R_{n+1},$$

$$(4.3\text{–}12)\qquad\qquad R_{n+1} = (-1)^n\frac{1}{n+1}\left(\frac{x}{1+X}\right)^{n+1},$$

where X is between 0 and x. Of course we must have $0 < 1 + x$, or $-1 < x$. If $0 \leqq x \leqq \frac{1}{2}$, then

$$0 \leqq \frac{x}{1+X} \leqq \frac{1}{2},$$

and so $$|R_{n-1}| \leqq \frac{1}{n+1}\cdot\frac{1}{2^{n+1}}.$$

There are many other possible formulas for R_{n+1} besides that of Lagrange. Probably the most important of these other forms is that due to Cauchy. It appears in formula (4.3–13), which follows.

THEOREM IV. *An alternative form of the remainder in* (4.3–1), *with hypotheses as in Theorem III, is*

(4.3–13) $$R_{n+1} = \frac{f^{(n+1)}(X)}{n!}(b - X)^n(b - a),$$

where X is some number between a and b (in general different from the X of (4.3–1)). *With $b = a + h$, $X = a + \theta h$ $(0 < \theta < 1)$, this takes the form*

(4.3–14) $$R_{n+1} = \frac{f^{(n+1)}(a + \theta h)}{n!}\, h^{n+1}(1 - \theta)^n.$$

This last form is an alternative to the last term in (4.3–7), *but the θ's in the two forms are usually different.*

PROOF. We start with $F(x)$ as defined by (4.3–2), but instead of using (4.3–3) we define $G(x) = b - x$. Then, following the method of proof of Theorem III, we have (4.3–4) and the formula $G'(x) = -1$ in place of (4.3–5).

Just as before we have, for some X between a and b,

$$F(a) = \frac{F'(X)}{G'(X)}\, G(a),$$

or $$F(a) = \frac{f^{(n+1)}(X)}{n!}(b - X)^n(b - a).$$

This is equivalent to (4.3–13), for comparison of (4.3–2) and (4.3–1) shows that the remainder term is $R_{n+1} = F(a)$.

Other forms of the remainder, as well as other proofs of Theorems III and IV, are indicated in the exercises.

Example 3. Find Cauchy's form of R_{n+1} for $f(x) = \log(1 + x)$ with $a = 0$.

Using the work of Example 2, and putting $h = x$, $a = 0$ in (4.3–14), we find

(4.3–15) $$R_{n+1} = (-1)^n\, \frac{x^{n+1}(1 - \theta)^n}{(1 + \theta x)^{n+1}}.$$

Example 4. Show that, in Taylor's formula (4.3–11) for $\log(1 + x)$, the following estimates of the remainder hold:

(4.3–16) $$|R_{n+1}| < \frac{|x|^{n+1}}{1 + x} \quad \text{if } -1 < x < 0,$$

(4.3–17) $$|R_{n+1}| < \frac{x^{n+1}}{n+1} \text{ if } 0 < x \le 1.$$

We use Cauchy's form of the remainder to get (4.3–16), and Lagrange's form to get (4.3–17). If $-1 < x < 0$ we write (4.3–15) in the form

$$R_{n+1} = (-1)^n \frac{x^{n+1}}{1+\theta x} \left(\frac{1-\theta}{1+\theta x}\right)^n,$$

and observe that $1 + \theta x > 1 + x$ and

$$0 < \frac{1-\theta}{1+\theta x} < 1.$$

The inequality (4.3–16) is then seen to be correct. On the other hand, if $0 < x \le 1$, we observe that $0 < X < x$ and hence $1 + X > 1$ in (4.3–12). The inequality (4.3–17) is then seen to be correct.

We observe from the foregoing inequalities that $R_{n+1} \to 0$ as $n \to \infty$ if $-1 < x < 0$ or if $0 < x \le 1$. Of course $R_{n+1} = 0$ if $x = 0$. It follows that Taylor's formula, neglecting the remainder, gives a better and better approximation to $\log(1 + x)$ when n is increased, provided that $-1 < x \le 1$. Accordingly, we have the Taylor's series expansion, valid if $-1 < x \le 1$:

(4.3–18) $$\log(1+x) = x - \tfrac{1}{2}x^2 + \cdots + (-1)^{n-1}\frac{x^n}{n} + \cdots.$$

EXERCISES

1. Arrange x^4 in powers of $(x - 3)$.

2. Write Taylor's formula with Lagrange's remainder in the case of $f(x) = \dfrac{1}{x^2 + 3}$ with $a = 1, n = 2$.

3. Write Taylor's formula with Lagrange's remainder in each of the following cases:

(a) $f(x) = \sin^2 x, a = 0, n = 3$;
(b) $f(x) = \tan x, a = 0, n = 3$;
(c) $f(x) = e^{-x^2}, a = 0, n = 3$;
(d) $f(x) = \log(1 + e^x), a = 0, n = 2$;
(e) $f(x) = \dfrac{\log(1-x)}{1-x}, a = 0, n = 1$.

4. Find R_5 in (4.3–9) with $f(x) = \sin x, a = \pi/2$.

5. Write Taylor's formula for $f(x) = \sin x$ with $a = 0$. Show that

$$|R_{n+1}| \le \frac{|x|^{n+1}}{(n+1)!}.$$

Show that the same inequality holds for $f(x) = \cos x, a = 0$.

6. Show that

$$e^x = 1 + x + \frac{x^2}{2!} + \cdots + \frac{x^n}{n!} + R_{n+1},$$

with

$$0 < R_{n+1} < e^x \frac{x^{n+1}}{(n+1)!} \quad \text{if } 0 < x$$

and

$$|R_{n+1}| < \frac{|x|^{n+1}}{(n+1)!} \quad \text{if } x < 0.$$

7. Show that

$$(1-x)^{-1/2} = 1 + \tfrac{1}{2}x + \frac{1 \cdot 3}{2 \cdot 4}x^2 + \cdots + \frac{1 \cdot 3 \cdots (2n-1)}{2 \cdot 4 \cdots (2n)}x^n + R_{n+1},$$

and write both Lagrange's and Cauchy's form of the remainder. Show that

$$|R_{n+1}| < \frac{1 \cdot 3 \cdots (2n+1)}{2 \cdot 4 \cdots (2n+2)}|x|^{n+1} \quad \text{if } -1 < x < 0$$

and

$$|R_{n+1}| < \frac{1 \cdot 3 \cdots (2n+1)}{2^{n+1}n!}\frac{x^{n+1}}{(1-x)^{3/2}} \quad \text{if } 0 < x < 1.$$

8. Observe that the curves $y = \sin x$, $y = \lambda x$ intersect near $x = \pi$ if λ is small. Let $f(x) = \sin x - \lambda x$ and apply Taylor's formula with $a = \pi$, $n = 2$, assuming that x is near π and neglecting R_3. Use this result to show that an approximate solution of $\sin x = \lambda x$ is $x = \pi/(1 + \lambda)$.

9. Proceed as in Exercise 8 to find an approximate formula for the solution of $\operatorname{ctn} x = \lambda x$ near $x = \pi/2$ (assuming that λ is small).

10. Suppose that f is twice differentiable in the interval $a < x < b$ and that $f''(x) \geqq 0$ at each point. If $a < x_0 < b$ and $y_0 = f(x_0)$, show that in the given interval no point of the curve $y = f(x)$ is below the line which is tangent to the curve at (x_0, y_0). In particular, if $f'(x_0) = 0$, the function has a minimum value at x_0.

11. In the method of proof for Theorem III let the function $G(x)$ in (4.3–3) be replaced by $G(x) = (x - b)^p$. Then carry on the method and show that the expression for the last term in (4.3–7) is replaced by

$$R_{n+1} = \frac{f^{(n+1)}(a + \theta h)}{n!\,p} h^{n+1}(1 - \theta)^{n+1-p}, \qquad 0 < \theta < 1.$$

This is called Schlömilch's form of the remainder. Lagrange's form is the special case $p = n + 1$, and Cauchy's form is the special case $p = 1$.

12. The following suggestions provide a method of proving Taylor's formula with remainder without appeal to Cauchy's generalized law of the mean. Suppose that g is continuous on the closed interval $[a, a + h]$ and differentiable on the open interval $(a, a + h)$. Assume further that g has the properties (i) $g(a) = 1$, (ii) $g(a + h) = 0$, (iii) $g'(x) \neq 0$ if $a < x < a + h$. Define $F(x)$ by (4.3–2) with $a + h$ in place of b. Define $\phi(x) = F(x) - g(x)F(a)$. Apply Rolle's theorem to ϕ and hence obtain a formula for $R_{n+1} = F(a)$. The result is

$$f(a + h) = f(a) + f'(a)h + \cdots + \frac{f^{(n)}(a)}{n!}h^n - \frac{f^{(n+1)}(a + \theta h)h^n(1 - \theta)^n}{n!\,g'(a + \theta h)}.$$

If we choose
$$g(x) = \left(\frac{a + h - x}{h} \right)^{n+1}$$

we obtain (4.3–7) (Lagrange's form). The choice $g(x) = \dfrac{a + h - x}{h}$ leads to Cauchy's form (4.3–14).

4.4 An extension of the mean-value theorem for integrals

Theorem VI of § 1.51 may be generalized somewhat as follows:

THEOREM V. *Let $f(x)$ and $p(x)$ be continuous functions defined on the closed interval $a \leq x \leq b$, and suppose that $p(x) \geq 0$ for every x of the interval. Then there is some number X such that $a \leq X \leq b$ and*

(4.4–1)
$$\int_a^b f(x)p(x)dx = f(X) \int_a^b p(x)dx.$$

PROOF. Let m and M be the minimum and maximum values of $f(x)$ on the interval $[a, b]$. Then

$$mp(x) \leq f(x)p(x) \leq Mp(x)$$

and so

(4.4–2)
$$m\int_a^b p(x)dx \leq \int_a^b f(x)p(x)dx \leq M \int_a^b p(x)dx.$$

There are two cases to consider:

$$(1)\int_a^b p(x)dx > 0; \quad (2)\int_a^b p(x)dx = 0.$$

In case (1) we have

$$m \leq \frac{\displaystyle\int_a^b f(x)p(x)dx}{\displaystyle\int_a^b p(x)dx} \leq M.$$

Since the quotient of the integrals lies between the extreme values attained by $f(x)$ on the interval, it follows by Theorem IV of § 3.3 that there must be some point $x = X$ on the interval such that

$$f(X) = \frac{\displaystyle\int_a^b f(x)p(x)dx}{\displaystyle\int_a^b p(x)dx}.$$

This is equivalent to (4.4–1).

In case (2) we see by (4.4–2) that the left member of (4.4–1) is zero. The right member is also zero, no matter how we choose X. Thus the proof is complete.

As one application of Theorem V we shall show how Lagrange's form of the remainder in Taylor's formula (see (4.3–9)) may be deduced from the integral form (4.2–8) of the remainder. Suppose first that $a < x$. Then the function $(x - t)^n$ is ≥ 0 when $a \leq t \leq x$. Hence we can apply Theorem V to the integral in (4.2–8), taking $p(t) = (x - t)^n$. Thus

$$\int_a^x (x - t)^n f^{(n+1)}(t)dt = f^{(n+1)}(X)\int_a^x (x - t)^n \, dt$$

$$= f^{(n+1)}(X)\left[\frac{-(x - t)^{n+1}}{n + 1}\right]_a^x = f^{(n+1)}(X)\frac{(x - a)^{n+1}}{n + 1},$$

where X is some number such that $a \leq X \leq x$. When this result is put back into (4.2–8) we obtain the Lagrange form (4.3–9). If $x < a$ we write (4.2–8) in the form

$$R_{n+1} = \frac{(-1)^{n+1}}{n!}\int_x^a (t - x)^n f^{(n+1)}(t)dt.$$

If we take $p(t) = (t - x)^n$ in this integral, then $p(t) \geq 0$, and we can apply Theorem V. We leave it for the student to carry out the details of showing that we are once more led to formula (4.3–9).

It should be pointed out that in the present section we are assuming the *continuity* of $f^{(n+1)}(x)$, whereas in § 4.3 we merely assumed the existence of $f^{(n+1)}(x)$.

4.5   L'Hospital's rule

Theorem XIV of § 1.64 states that

$$\lim_{x \to c} \frac{f(x)}{g(x)} = \frac{\lim_{x \to c} f(x)}{\lim_{x \to c} g(x)},$$

provided that $\lim_{x \to c} f(x)$ and $\lim_{x \to c} g(x)$ both exist and $\lim_{x \to c} g(x) \neq 0$. There are many instances in which these conditions are not fulfilled, however, e.g.,

$$\lim_{x \to 0} \frac{e^x - \cos x}{\tan x} = ? \qquad\qquad \lim_{x \to +\infty} \frac{\log x}{x} = ?$$

$$\lim_{x \to +\infty} \frac{e^x}{x^{10}} = ? \qquad\qquad \lim_{x \to 0} \frac{e^{-1/x^2}}{x^5} = ?$$

The important cases not covered by the theorem of § 1.64, and not otherwise easily disposed of, are of two kinds. The first kind is that in which

$f(x)$ and $g(x)$ both approach 0 as $x \to c$. The second kind is that in which $|g(x)| \to \infty$ as $x \to c$. In both cases it may be that $x \to c$ from one side only; it may also happen that $x \to c$ is replaced by $x \to +\infty$ or $x \to -\infty$.

There is a useful rule for finding the limit of a quotient in the most commonly occurring instances of the cases just mentioned. This rule was popularized by the Marquis de l'Hospital in his textbook on calculus published in 1696, and is generally known as l'Hospital's rule. The exact statement of the rule actually amounts to the statement of two theorems, or of one theorem with two alternative assumptions leading to the same conclusion. Before coming to the formal statement of the theorem, we need to make a few preliminary remarks. We shall have two functions $f(x)$ and $g(x)$ to consider. Let c denote either a real number or one of the symbols $+\infty$, $-\infty$. We assume that f and g are defined on a portion of the x-axis which we denote by I, and we assume that, if c is a real number, I is a finite open interval with c as one of its end-points, while if c is $+\infty$ or $-\infty$, I is a semi-infinite open interval, extending indefinitely in the positive direction if $c = +\infty$, and indefinitely in the negative direction if $c = -\infty$. We then talk about limits as $x \to c$, it being understood that x is to range over I, and to approach c from one side only. We furthermore assume that the derivatives $f'(x)$, $g'(x)$ exist at each point of I, and that $g(x)$ and $g'(x)$ are never equal to zero:

THEOREM VI.   (*L'Hospital's rule.*) *Suppose either that*

     CASE 1. $f(x) \to 0$ *and* $g(x) \to 0$ *as* $x \to c$,
or that

     CASE 2. $|g(x)| \to \infty$ *as* $x \to c$.

Let A *denote either a real number or one of the symbols* $+\infty$, $-\infty$. *Suppose that*

(4.5–1)
$$\lim_{x \to c} \frac{f'(x)}{g'(x)} = A.$$

Then it is also true that

(4.5–2)
$$\lim_{x \to c} \frac{f(x)}{g(x)} = A.$$

PROOF.   The proof of Theorem VI is easiest when I is a finite interval and Case 1 is assumed. Indications as to the procedure to be followed are given in Exercise 8. The situation is less simple if c is $+\infty$ or $-\infty$, or if we assume Case 2. We shall give the proof in such a way that the argument does not depend on whether or not I is a finite interval, and also so that the arguments for Case 1 and Case 2 are very much alike.

We take x and y to be any distinct points of I, with y between x and c. Then, by Cauchy's generalized law of the mean (Theorem I, § 4.1), there is some point X between x and y such that

$$(4.5\text{-}3) \qquad \frac{f(x) - f(y)}{g(x) - g(y)} = \frac{f'(X)}{g'(X)}.$$

Observe that X also is between x and c. We are assured that $g(x) \neq g(y)$ by the ordinary law of the mean, since $g'(x)$ is never zero, by hypothesis.

Now consider Case 1, and rewrite (4.5-3) in the form

$$(4.5\text{-}4) \qquad \frac{\dfrac{f(x)}{g(x)} - \dfrac{f(y)}{g(x)}}{1 - \dfrac{g(y)}{g(x)}} = \frac{f'(X)}{g'(X)}.$$

Now suppose, for definiteness, that A is not $+\infty$ or $-\infty$. Suppose $\epsilon > 0$. The meaning of (4.5-1) is that there is some number x_0 such that, if x is between x_0 and c,

$$(4.5\text{-}5) \qquad A - \epsilon < \frac{f'(x)}{g'(x)} < A + \epsilon.$$

If x is such a number, so is X, and hence, no matter how y is chosen, it follows from (4.5-4) that

$$A - \epsilon < \frac{\dfrac{f(x)}{g(x)} - \dfrac{f(y)}{g(x)}}{1 - \dfrac{g(y)}{g(x)}} < A + \epsilon.$$

Now make $y \to c$. Then, since we are assuming Case 1, we conclude that

$$A - \epsilon \leqq \frac{f(x)}{g(x)} \leqq A + \epsilon.$$

This holds if x is between x_0 and c, where x_0 may depend on ϵ. But this means that (4.5-2) is true.

The argument is essentially the same if A is $+\infty$ or $-\infty$. For example, if $A = +\infty$, we take any number M, and choose x_0 so that

$$M < \frac{f'(x)}{g'(x)}$$

if x is between x_0 and c. Then we find that, for such values of x,

$$M \leqq \frac{f(x)}{g(x)}.$$

So we get (4.5-2) from (4.5-1) in this case also. The student may wish to review the definitions of limits involving the symbols $+ \infty$, $- \infty$; these definitions are found in § 1.61.

To treat Case 2, we return to (4.5–3) and write it in the form

$$(4.5\text{–}6) \qquad \frac{\dfrac{f(y)}{g(y)} - \dfrac{f(x)}{g(y)}}{1 - \dfrac{g(x)}{g(y)}} = \frac{f'(X)}{g'(X)} .$$

As in the previous argument, we may suppose the assumption (4.5–1) expressed in the form (4.5–5), with the consequence from (4.5–6) that

$$A - \epsilon < \frac{\dfrac{f(y)}{g(y)} - \dfrac{f(x)}{g(y)}}{1 - \dfrac{g(x)}{g(y)}} < A + \epsilon$$

for all x between x_0 and c and all y between x and c. We shall now let $y \to c$, keeping x fixed. Since $|g(y)| \to \infty$, we may safely assume that

$$1 - \frac{g(x)}{g(y)} > 0.$$

We then see that

$$(A - \epsilon)\left(1 - \frac{g(x)}{g(y)}\right) + \frac{f(x)}{g(y)} < \frac{f(y)}{g(y)} < \frac{f(x)}{g(y)} + (A + \epsilon)\left(1 - \frac{g(x)}{g(y)}\right).$$

As $y \to c$, the left and right members of this set of inequalities approach $A - \epsilon$ and $A + \epsilon$, respectively. Certainly then it will be true that, for y beyond a certain point,

$$A - 2\epsilon < \frac{f(y)}{g(y)} < A + 2\epsilon.$$

This, however, is equivalent to saying that

$$\lim_{y \to c} \frac{f(y)}{g(y)} = A,$$

which is what we set out to prove. As in Case 1, the argument is not very different if A is $+ \infty$ or $- \infty$.

Example 1. Take $f(x) = e^x - \cos x$ and $g(x) = \tan x$ with $c = 0$. The conditions of Theorem VI, Case 1, are fulfilled with x approaching zero from either side. Thus

$$\lim_{x \to 0} \frac{e^x - \cos x}{\tan x} = \lim_{x \to 0} \frac{e^x + \sin x}{\sec^2 x} = \frac{1 + 0}{1} = 1.$$

If, in attempting to apply Theorem VI, it should turn out that $\lim_{x \to c}$

$f'(x) = 0$ and $\lim_{x \to c} g'(x) = 0$, or that $\lim_{x \to c} |\, g'(x)\,| \to \infty$, it may be that the rule can be applied a second time.

Example 2. Find $\lim\limits_{x \to 0} \dfrac{e^x + e^{-x} - 2}{3\,x^2}$.

Here one application of Theorem VI gives

$$\lim_{x \to 0} \frac{e^x + e^{-x} - 2}{3\,x^2} = \lim_{x \to 0} \frac{e^x - e^{-x}}{6\,x}.$$

A second application gives

$$\lim_{x \to 0} \frac{e^x - e^{-x}}{6\,x} = \lim_{x \to 0} \frac{e^x + e^{-x}}{6} = \tfrac{2}{6} = \tfrac{1}{3}.$$

Hence the original limit is equal to $\tfrac{1}{3}$.

The effect of Theorem VI is to replace the original problem of finding $\lim\limits_{x \to c} \dfrac{f(x)}{g(x)}$ by the new problem of finding $\lim\limits_{x \to c} \dfrac{f'(x)}{g'(x)}$, provided this latter limit exists or is $+\infty$ or $-\infty$. Any legitimate methods may be brought to bear on the new problem, including algebraic or trigonometric transformations of the problem, breaking the quotient up into a product of simpler quotients, repeated use of Theorem VI, etc.

Example 3. Find $\lim\limits_{x \to 0} \dfrac{(e^x - 1)\sin x}{\cos x - \cos^2 x}$.

We first apply Theorem VI:

$$\lim_{x \to 0} \frac{(e^x - 1)\sin x}{\cos x - \cos^2 x} = \lim_{x \to 0} \frac{(e^x - 1)\cos x + e^x \sin x}{-\sin x + 2\cos x \sin x}.$$

Next we simplify the new quotient:

$$\frac{(e^x - 1)\cos x + e^x \sin x}{-\sin x + 2\cos x \sin x} = \frac{\cos x}{2\cos x - 1} \cdot \frac{e^x - 1}{\sin x} + \frac{e^x}{2\cos x - 1}.$$

Now

$$\lim_{x \to 0} \frac{\cos x}{2\cos x - 1} = \frac{1}{2 - 1} = 1, \quad \lim_{x \to 0} \frac{e^x}{2\cos x - 1} = \frac{1}{2 - 1} = 1,$$

and

$$\lim_{x \to 0} \frac{e^x - 1}{\sin x} = \lim_{x \to 0} \frac{e^x}{\cos x} = \frac{1}{1} = 1.$$

Hence the original limit is equal to 2.

The next example illustrates Case 1 with $c = +\infty$.

Example 4. Find $\lim\limits_{x \to +\infty} x \log \dfrac{x + 1}{x - 1}$.

Here we write
$$x \log \frac{x+1}{x-1} = \frac{\log \dfrac{x+1}{x-1}}{\dfrac{1}{x}},$$

taking $f(x) = \log \dfrac{x+1}{x-1}$, $g(x) = \dfrac{1}{x}$. It is clear that Theorem VI is applicable. Thus

$$\lim_{x \to +\infty} \frac{\log \dfrac{x+1}{x-1}}{1/x} = \lim_{x \to +\infty} \frac{\log(x+1) - \log(x-1)}{1/x},$$

$$= \lim_{x \to +\infty} \frac{\dfrac{1}{x+1} - \dfrac{1}{x-1}}{-1/x^2}.$$

On simplification, the new limit takes the form

$$\lim_{x \to +\infty} \frac{2x^2}{x^2 - 1} = 2,$$

so that the original limit has the value 2.

Next we illustrate Case 2 of Theorem VI.

Example 5. Find $\lim\limits_{x \to +\infty} \dfrac{e^x}{x^{10}}$.

Case 2 of Theorem VI is applicable, but it must be applied ten times before we reach a result. We have

$$\lim_{x \to +\infty} \frac{e^x}{x^{10}} = \lim_{x \to +\infty} \frac{e^x}{10\,x^9} = \lim_{x \to +\infty} \frac{e^x}{10 \cdot 9\,x^8} = \cdots$$

$$= \lim_{x \to +\infty} \frac{e^x}{10!} = +\infty.$$

It might at first sight appear as though we could solve the problem of Example 5 by use of Case 1. For we can write

$$\frac{e^x}{x^{10}} = \frac{x^{-10}}{e^{-x}},$$

and if $f(x) = x^{-10}$, $g(x) = e^{-x}$, the conditions of Case 1 are satisfied. But when we differentiate, things get worse instead of better:

$$\lim_{x \to +\infty} \frac{x^{-10}}{e^{-x}} = \lim_{x \to +\infty} \frac{-10\,x^{-11}}{-e^{-x}} = \lim_{x \to +\infty} \frac{11 \cdot 10\,x^{-12}}{e^{-x}} = \cdots.$$

As one of the very interesting applications of l'Hospital's rule, let us discuss the function e^{-1/x^2}.

Example 6. Let us define $f(x) = e^{-1/x^2}$ if $x \neq 0$, and $f(0) = 0$. Then all the derivatives of $f(x)$ have the value 0 at $x = 0$, as we shall now show.

In the first place, f is continuous at $x = 0$, for $f(x) \to 0$ as $x \to 0$ (see Example 6, § 1.1). Now, by definition,

$$f'(0) = \lim_{x \to 0} \frac{f(x) - f(0)}{x - 0} = \lim_{x \to 0} \frac{e^{-1/x^2}}{x}.$$

This appears to be a place to use Case 1 of Theorem VI, but the application of the method gives

$$\lim_{x \to 0} \frac{e^{-1/x^2}}{x} = \lim_{x \to 0} \frac{2\,x^{-3}e^{-1/x^2}}{1} = \lim_{x \to 0} \frac{2\,e^{-1/x^2}}{x^3},$$

and the new limit is harder to deal with, rather than easier. We may, however, use Case 2 of Theorem VI, as follows:

$$\lim_{x \to 0} \frac{e^{-1/x^2}}{x} = \lim_{x \to 0} \frac{1/x}{e^{1/x^2}} = \lim_{x \to 0} \frac{-1/x^2}{-\dfrac{2}{x^3}\,e^{1/x^2}} = \lim_{x \to 0} \tfrac{1}{2}\,xe^{-1/x^2} = 0.$$

This shows that $f'(0) = 0$.

To deal with higher derivatives we observe that, if $x \neq 0$,

$$f'(x) = 2\,x^{-3}e^{-1/x^2},$$
$$f''(x) = 4\,x^{-6}e^{-1/x^2} - 6\,x^{-4}e^{-1/x^2}.$$

It is easy to see, by induction, that $f^{(n)}(x)$ is a linear combination of terms of the form

$$e^{-1/x^2}/x^m$$

with $0 < m \leqq 3\,n$. Consequently, to see that $f^{(n)}(0) = 0$, as well as to see that $f^{(n)}(x)$ is continuous at $x = 0$, it will be sufficient to show that

(4.5–7)
$$\lim_{x \to 0} \frac{e^{-1/x^2}}{x^m} = 0$$

for all positive integers m. We prove (4.5–7) by repeated use of Case 2 of Theorem VI:

$$\lim_{x \to 0} \frac{x^{-m}}{e^{1/x^2}} = \lim_{x \to 0} \frac{-mx^{-m-1}}{-2\,x^{-3}e^{1/x^2}} = \frac{m}{2} \lim_{x \to 0} \frac{x^{-m+2}}{e^{1/x^2}};$$

after a finite number of steps the exponent in the numerator will be positive, and then the limit is seen to be 0.

The graph of $y = e^{-1/x^2}$ is indicated in Fig. 24. It is very flat (but of course not perfectly straight) in the neighborhood of the origin.

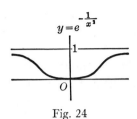

$$y = e^{-\frac{1}{x^2}}$$

Fig. 24

There are various kinds of functions whose limiting values may be found by using suitable devices to bring the problem to a form where l'Hospital's rule is applicable. The principal types of problems and the appropriate devices are indicated in Exercises 3 and 4.

EXERCISES

1. Evaluate the following limits:

(a) $\lim\limits_{x \to 0+} \dfrac{\sin 3x}{1 - \cos 4x}$.

(b) $\lim\limits_{x \to 0} \dfrac{\tan x - \sin x}{x^3}$.

(c) $\lim\limits_{x \to 0} \dfrac{1}{x} \log \dfrac{1+x}{1-x}$.

(d) $\lim\limits_{x \to 1} \dfrac{x - 6x^6 + 5x^7}{(1-x)^2}$.

(e) $\lim\limits_{x \to 0} \dfrac{x^3 \sin x}{(1 - \cos x)^2}$.

(f) $\lim\limits_{x \to 0} \dfrac{\tan^{-1} x - x}{x^3}$.

(g) $\lim\limits_{x \to +\infty} \dfrac{\log (1 + e^{-x})}{e^{-x}}$.

(h) $\lim\limits_{x \to 0} \dfrac{10^x - 5^x}{x}$.

(i) $\lim\limits_{x \to 0} \dfrac{\sin (\pi \cos x)}{x \sin x}$.

2. Evaluate the following limits:

(a) $\lim\limits_{x \to +\infty} \dfrac{x^n}{e^x}$, n a positive integer.

(b) $\lim\limits_{x \to +\infty} \dfrac{(\log x)^n}{x}$, n a positive integer.

(c) $\lim\limits_{x \to +\infty} \dfrac{\log x}{\sqrt{1 + x^3}}$.

(d) $\lim\limits_{x \to +\infty} \dfrac{x^3 (\log x)^2}{e^x}$.

(e) $\lim\limits_{x \to +\infty} \dfrac{\log (1 + e^{2x})}{x}$.

(f) $\lim\limits_{x \to +\infty} \dfrac{\log (1 + xe^{2x})}{x^2}$.

(g) $\lim\limits_{x \to +\infty} \dfrac{\log (1 + x^{-1} e^{2x})}{\sqrt{x}}$.

(h) $\lim\limits_{x \to 0+} \dfrac{\log (\tan 5x)}{\log (\tan 3x)}$.

(i) $\lim\limits_{x \to 0+} \dfrac{\log (1 - \cos 2x)}{\log \tan 2x}$.

3. One sometimes wants to find the limit of a function of the form $y = [F(x)]^{G(x)}$ as x approaches some limit x_0, or as $x \to \infty$. In such problems $F(x)$ is positive. If $F(x)$ and $G(x)$ both approach nonzero limits, the problem presents no unusual difficulty. There are, however, three cases in which the limit of y is not apparent without some investigation:

CASE 1. $F(x) \to 1$ and $G(x) \to \infty$;
CASE 2. $F(x) \to \infty$ and $G(x) \to 0$;
CASE 3. $F(x) \to 0$ and $G(x) \to 0$.

In these cases it is usually appropriate to investigate the limit of the logarithm of y:

$$\log y = G(x) \log F(x) = \frac{\log F(x)}{1/G(x)} .$$

The rule of l'Hospital may then be applicable. If $\log y \to b$, then $y \to e^b$.

Use this procedure to evaluate the following limits:

(a) $\lim\limits_{x \to 0+} x^x$.

(b) $\lim\limits_{x \to +\infty} x^{1/x}$.

(c) $\lim\limits_{x \to 0} \left(\dfrac{\sin x}{x} \right)^{1/x^2}$.

(d) $\lim\limits_{x \to \infty} \left(\cos \dfrac{2}{x} \right)^{x^2}$.

(e) $\lim\limits_{x \to +\infty} \left(1 + \dfrac{1}{2x} \right)^{x^2}$.

(f) $\lim\limits_{x \to 0+} (1 - 2^{-x})^{-x}$.

(g) $\lim\limits_{x \to +\infty} (e^x + e^{2x})^{1/x}$.

(h) $\lim\limits_{x \to +\infty} (e^{-x} + e^{-2x})^{1/x}$.

(i) $\lim\limits_{x \to 0+} \left(\log \dfrac{1}{x} \right)^x$.

(j) $\lim\limits_{x \to +\infty} \left(1 + \dfrac{1}{2x} \right)^{3x + \log x}$.

4. Sometimes a limit is not easy to determine because it is of the type $\lim_{x \to x_0} (f(x) - g(x))$, where $f(x)$ and $g(x)$ both become infinite as $x \to x_0$ (or as $x \to \infty$). In practice the best plan for such cases is usually to bring the expression $f(x) - g(x)$ to the form of a single quotient, e.g.,

$$\lim_{x \to 0} \left(\frac{1}{x} - \frac{1}{\sin x} \right) = \lim_{x \to 0} \frac{\sin x - x}{x \sin x} .$$

The limit may then be treated by l'Hospital's rule, or by devices of algebraic or trigonometric reduction. In some cases an algebraic device alone will be sufficient, e.g.,

$$\lim_{x \to +\infty} (x - \sqrt{x^2 + 1}) = \lim_{x \to +\infty} \frac{(x - \sqrt{x^2 + 1})(x + \sqrt{x^2 + 1})}{x + \sqrt{x^2 + 1}}$$

$$= \lim_{x \to +\infty} \frac{-1}{x + \sqrt{x^2 + 1}} = 0.$$

Evaluate the following limits:

(a) $\lim\limits_{x \to 0} \left(\dfrac{1}{x} - \dfrac{1}{\sin x} \right)$.

(b) $\lim\limits_{x \to 0} \left(\dfrac{1}{x \sin x} - \dfrac{1}{x^2} \right)$.

(c) $\lim\limits_{x \to 0+} \left(\dfrac{1}{x} - \log \dfrac{1}{x} \right)$.

(d) $\lim\limits_{x \to 0} \left(\dfrac{1}{\log (1 + x)} - \dfrac{1}{x} \right)$.

(e) $\lim\limits_{x \to 0} \left(\dfrac{1}{x^2} - \dfrac{1}{\tan^2 x} \right)$.

(f) $\lim\limits_{x \to +\infty} x(\sqrt{x^2 + a^2} - x)$.

(g) $\lim\limits_{x \to 2} \left(\dfrac{1}{x - 2} - \dfrac{5}{x^2 + x - 6} \right)$.

(h) $\lim\limits_{x \to \infty} (x^2 - \sqrt{x^4 - x^2 + 2})$.

(i) $\lim\limits_{x \to 0} \left(\operatorname{ctn} x - \dfrac{1}{x} \right)$.

(j) $\lim\limits_{x \to 0} \left(\dfrac{e^{-x}}{x} - \dfrac{1}{e^x - 1} \right)$.

5. Suppose that f and g have continuous derivatives of the first n orders in a closed interval $a \leq x \leq b$. Furthermore, assume that $f(a) = f'(a) = \cdots = f^{(n-1)}(a) = 0, g(a) = g'(a) = \cdots = g^{(n-1)}(a) = 0$, and that $g^{(n)}(a) \neq 0$. Without using l'Hospital's rule, show that

$$\lim_{x \to a+} \frac{f(x)}{g(x)} = \frac{f^{(n)}(a)}{g^{(n)}(a)}.$$

Use Taylor's formula with remainder.

6. What can you conclude in Theorem VI, if $\lim \dfrac{f'(x)}{g'(x)}$ fails to exist either as a definite numerical limit or as $+\infty$ or $-\infty$? Give your answer after a careful examination of the limits

$$\lim_{x \to 0} \frac{x^2 \sin \dfrac{1}{x}}{\sin x} \quad \text{and} \quad \lim_{x \to \infty} \frac{x - \sin x}{x}.$$

7. Evaluate each of the following limits:

(a) $\lim\limits_{x \to \infty} x \sin \dfrac{1}{x}.$

(b) $\lim\limits_{x \to +\infty} \dfrac{\log (\log x)}{\log (x - \log x)}.$

(c) $\lim\limits_{x \to +\infty} \dfrac{1}{x} \displaystyle\int_1^x \dfrac{\log t}{1 + t}\, dt.$

(d) $\lim\limits_{x \to +\infty} \dfrac{1}{x^2} \displaystyle\int_1^x \sin^2 t\, dt.$

(e) $\lim\limits_{x \to 0} \dfrac{e - (1 + x)^{1/x}}{x}.$

(f) $\lim\limits_{x \to 0+} \left(\dfrac{1}{x^2} - \dfrac{1}{x \log x} \right).$

(g) $\lim\limits_{x \to +\infty} (\log (1 + e^x) - x).$

8. Give a proof of Case 1 of Theorem VI, assuming that c is a real number, and utilizing the following suggestions: Let b be a point of the interval I, and consider the functions f, g on the *closed* interval with end-points b and c, after defining $f(c) = g(c) = 0$. Now apply Theorem I, § 4.1, and make the deduction of (4.5–2) from (4.5–1). Explain the argument carefully. Why do we define $f(c) = g(c) = 0$?

MISCELLANEOUS EXERCISES

1. Suppose that f is defined and differentiable in an interval containing $x = a$ (the point $x = a$ may be at one end of the interval, in which case derivatives at $x = a$ are to be considered as one-sided limits). Suppose also that $f''(a)$ exists, but assume nothing else about second derivatives. Show that

$$f''(a) = \lim_{x \to a} \frac{f(x) - f(a) - (x - a)f'(a)}{\dfrac{(x - a)^2}{2!}}.$$

2. Suppose that f satisfies the conditions of Exercise 1, and that $x = a$ is an interior point of the interval in question. Show that, if f has a relative minimum at $x = a$, then $f''(a) \geq 0$, while $f''(a) \leq 0$ if f has a relative maximum at $x = a$.

These are *necessary* conditions for a relative extreme at $x = a$. Now assume that $f'(a) = 0$ and $f''(a) > 0$, and prove that f must have a relative minimum at $x = a$. These are *sufficient* conditions for a relative minimum. State a set of sufficient conditions for a relative maximum at $x = a$, and prove the sufficiency of the conditions.

3. Generalize the result of Exercise 1, obtaining

$$f^{(n)}(a) = \lim_{x \to a} \frac{f(x) - f(a) - (x - a)f'(a) - \cdots - \dfrac{(x - a)^{n-1}}{(n - 1)!} f^{(n-1)}(a)}{\dfrac{(x - a)^n}{n!}} ,$$

with the assumptions that f is defined and has derivatives of orders $1, \cdots, n - 1$ in an interval including $x = a$, while $f^{(n)}(a)$ exists. Suggestion: Apply Theorem VI $(n - 1)$ times. Use this result to show that there is a function $E_n(x)$ defined in the interval, except at $x = a$, such that

$$f(x) = f(a) + (x - a)f'(a) + \cdots + \frac{(x - a)^{n-1}}{(n - 1)!} f^{(n-1)}(a)$$

$$+ \frac{(x - a)^n}{n!} \{ f^{(n)}(a) + E_n(x) \}$$

and $\lim_{x \to a} E_n(x) = 0$. The function $E_n(x)$ depends upon the function $f(x)$, of course. Find $E_2(x)$ explicitly if $f(x)$ is defined as $x^4 \sin (1/x)$ when $x \neq 0$, and $f(0) = 0$; take $a = 0$.

4. With the conditions of Exercise 3 on f suppose that $f'(a) = \cdots = f^{(n-1)}(a) = 0$ and that f has a relative minimum at $x = a$. Also suppose that n is even. Show that $f^{(n)}(a) \geq 0$. What if there is a relative maximum at $x = a$? What about sufficient conditions for a relative maximum or minimum when $f'(a) = \cdots = f^{(n-1)}(a) = 0$? What can you say if n is odd, $f'(a) = \cdots = f^{(n-1)}(a) = 0$ and $f^{(n)}(a) \neq 0$? (Assume that $x = a$ is an interior point of the interval.)

5. Use the result of Exercise 3 to prove the following theorem:

Let f and g be defined and have derivatives of orders $1, \cdots, n - 1$ in an interval including $x = a$, while $f^{(n)}(a)$ and $g^{(n)}(a)$ exist and $g^{(n)}(a) \neq 0$. Suppose that both f and g, as well as their first $n - 1$ derivatives, have the value 0 at $x = a$. Then

$$\lim_{x \to a} \frac{f(x)}{g(x)} = \frac{f^{(n)}(a)}{g^{(n)}(a)} .$$

This theorem holds for $n \geq 1$. It may give a result when Theorem VI is inapplicable, e.g., in the case $f(x) = x^2 \sin (1/x)$, $f(0) = 0$, $g(x) = x$.

6. Calculate the following limits:

(a) $\displaystyle \lim_{x \to 0} \frac{\cos \left(\dfrac{\pi}{2} \cos x \right)}{\sin^2 x} .$

(b) $\displaystyle \lim_{x \to 0+} x(\log x)^n$, n a positive integer.

(c) $\lim\limits_{x \to +\infty} \dfrac{x^{3/2} \log x}{\sqrt{1 + x^4}}$.

(e) $\lim\limits_{x \to 1} x^{1/(1-x)}$.

(d) $\lim\limits_{x \to +\infty} x^n \log (1 + e^{-x})$, n a positive integer.

(f) $\lim\limits_{x \to 1} \dfrac{x^x - x}{1 - x + \log x}$.

7. Discuss $\lim_{x \to +\infty} (a^x + b^x)^{1/x}$, assuming $0 < a \leq b$. Generalize to the case of $\lim_{x \to +\infty} (a_1{}^x + a_2{}^x + \cdots + a_n{}^x)^{1/x}$, where $0 < a_1 \leq a_2 \leq \cdots \leq a_n$.

8. Find $\lim\limits_{x \to 0} \dfrac{x \int_0^x e^{-t^2}\, dt}{1 - e^{-x^2}}$.

9. If $f(x) = e^{-1/x^2}$, show that $f^{(n)}(x) = e^{-1/x^2} P_n(1/x)$, where $P_n(t)$ is a polynomial of degree $3\,n$ in t, with leading term $2^n t^{3n}$. Show also that $P_{n+1}(t) = 2\,t^3 P_n(t) - t^2 P_n'(t)$.

10. Show that to each positive integer n corresponds a number θ_n, $0 < \theta_n < 1$, such that $\log \left(1 + \dfrac{1}{n}\right) = \dfrac{1}{n} - \dfrac{\theta_n}{2\,n^2}$.

11. Apply (4.3–7) to $f(x) = 1/(1 - x)$ with $a = 0$, $h = x$, and find the form of the remainder term. Then compare with the algebraic identity

$$\frac{1}{1 - x} = 1 + x + \cdots + x^n + \frac{x^{n+1}}{1 - x},$$

and so find that $\theta = \dfrac{1 - (1 - x)^{1/(n+2)}}{x}$ (where $x < 1$). Now find $\lim\limits_{x \to 0} \theta$.

12. Suppose that f is differentiable in an interval of which $x = a$ is an interior point, and assume that $f''(a)$ exists. Show that

$$f''(a) = \lim_{h \to 0} \frac{f(a + 2\,h) - 2\,f(a + h) + f(a)}{h^2}.$$

This may be proved using the result of Exercise 1. If one assumes that the second derivative exists on the whole interval and is continuous at $x = a$, the proof may be given by using (4.3–7) in a suitable way.

Functions of Several Variables 5

5. Functions and their regions of definition

Elementary calculus deals mainly with functions of a single independent variable. But we do not go far in either pure or applied mathematics
until we have occasion to consider functions of two or more variables.
We assume that the student has some familiarity with the concept of a
function of several independent variables.

One of the first things that claims our attention when we begin to study
functions of several variables is the nature of the region of definition of
such a function. The functions of one variable which we study in calculus
are usually defined on intervals of the real axis. There are only a few
different types of intervals. If the interval is finite, it may contain both its
end-points, or just one, or neither. If the interval is infinite, but is not the
entire axis, it has just one end-point, and this may or may not be counted
as belonging to the interval. There is much more variety in the case of
functions of several variables. We shall give some illustrative examples,
taking the number of independent variables to be two.

Example 1. $f(x, y) = \log (1 - x^2 - y^2)$.

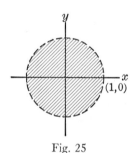

Fig. 25

The function is defined only when $x^2 + y^2 < 1$, since otherwise the
logarithm is undefined. The region of definition is the interior of the
unit circle with center at the origin. In Fig. 25 the circle is dashed to indicate that the boundary of the circular area does not belong to the region
of definition.

Example 2. $F(x, y) = \sqrt{x^2 + y^2 - 1} + \log (4 - x^2 - y^2)$.

Here we must have $x^2 + y^2 \geqq 1$ in order for the square root to be real,
while we must have $x^2 + y^2 < 4$ for the logarithm to be defined. The
region of definition of $F(x, y)$ is the annular region between the circles

126

$x^2 + y^2 = 1$ and $x^2 + y^2 = 4$. The inner circumference is part of the region of definition, while the outer circumference is not (see Fig. 26).

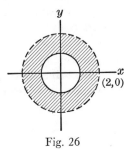

Fig. 26

Example 3. $g(x, y) = \dfrac{x}{y^2 - 4x}$.

The function is defined except when the denominator is zero, that is, everywhere except at the points of the parabola $y^2 = 4x$ (see Fig. 27).

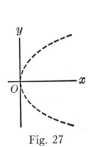

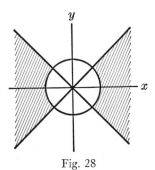

Fig. 27   Fig. 28

Example 4. $G(x, y) = \sqrt{x^2 - y^2} + \sqrt{x^2 + y^2 - 1}$.

The region of definition here is defined by the inequalities $x^2 \geqq y^2$, $x^2 + y^2 \geqq 1$. The lines $x - y = 0$, $x + y = 0$ divide the plane into four quadrants. The inequality $x^2 \geqq y^2$ states that the point (x, y) lies in (or on the edge of) one of those two of the four quadrants which contain the x-axis. The other inequality states that (x, y) lies outside or on the circle $x^2 + y^2 = 1$. Hence the region of definition of $G(x, y)$ is that part of the xy-plane which is shaded in Fig. 28.

Similar examples might be given for functions of three independent variables. The region of definition might be the interior of a cube, the interior and boundary of an ellipsoid, the space between two concentric spheres, or the interior of a surface formed like the inner tube of an automobile tire.

Because of the great variety of possible regions of definition of a function of two or more variables, it is desirable to devote some attention to matters of terminology about configurations of points in the plane. Not only will this make it easier for us to state things clearly, but it will eventually become absolutely indispensable in developing parts of our subject. We shall sometimes use the word "domain" for "region of definition," and by the range of a function we shall mean the set of values which the function takes on.

5.1   Point sets

In § 2.7 we explained the meaning of the phrase "a set of real numbers." Since we identify real numbers with points on the axis of reals, we may equally well speak of sets of points on a line. We are now going to talk about sets of points, or point sets, in the plane or in space of three dimensions. For the sake of simplicity and definiteness we shall speak principally of point sets in the plane.

In dealing with intervals we found it necessary to introduce the notions of closed and open intervals. Likewise, in our discussion of regions of definition of a function of several variables, we find it necessary to introduce and use the concepts *open set* and *closed set*. Furthermore, it is well to give a precise definition of the *boundary* of a set. It is much more difficult to define these concepts carefully than one might at first suppose, because of the great variety of configurations of points which are available for consideration. We proceed as follows: First let us define the phrase *a circular neighborhood* of the point (x_0, y_0) to mean the set of all points (x, y) lying inside some circle with center at (x_0, y_0). That is, if $\delta > 0$, the set of all (x, y) such that

$$(x - x_0)^2 + (y - y_0)^2 < \delta^2$$

constitutes a circular neighborhood of (x_0, y_0) (see Fig. 29). A neighborhood may have any positive number as its radius. The further concepts which we are now going to introduce are built upon the concept of a neighborhood. It is convenient to abandon the coordinate notation for the time being, and denote points by such symbols as $P, Q, P_1, P_2, \cdots$.

Fig. 29

DEFINITION.   A set S is called *open* if each point P of S has some circular neighborhood which belongs entirely to the set S.

Example 1.   A circular neighborhood, for instance the set of all points inside the circle $x^2 + y^2 = 1$, is an open set. For, if P is inside the unit circle, say at a distance r from the center (where $r < 1$), the circular neighborhood of P of radius δ also lies inside the unit circle provided δ is chosen so that $0 < \delta < 1 - r$ (see Fig. 30).

Example 2. The set of all points in the plane *not* on the parabola $y^2 = 4x$ is an open set. The student should verify this for himself.

Fig. 30

Example 3. Let S consist of all points on or inside the circle $x^2 + y^2 = 1$. This set is not open, for if P is on the circle there is no neighborhood of P which belongs entirely to S.

DEFINITION. If S is any point set, the set of all points of the plane which are not in S is called the *complement* of S. On occasion it is convenient to use the notation $C(S)$ for the complement of S.

To illustrate this notion, consider the sets defined in the foregoing examples. For the S of Example 1, $C(S)$ is the set of all points outside or on the circle $x^2 + y^2 = 1$, i.e., all points for which $x^2 + y^2 \geqq 1$. In Example 2, $C(S)$ consists of the points on the parabola $y^2 = 4x$. In Example 3, $C(S)$ is the exterior of the circle, i.e., all points such that $x^2 + y^2 > 1$.

DEFINITION. A set S is called *closed* if its complement is open.

The set of Example 3 is closed (the student should verify this to his own satisfaction); the sets of Examples 1 and 2 are not closed. A set consisting of any finite number of points is closed. A set may be neither open nor closed, as we see in the next example.

Example 4. Let S be the set of all points for which $1 \leqq x^2 + y^2 < 4$. This set is the region of definition of the function $F(x, y)$ of Example 2, § 5. It is not open, because a point of the circle $x^2 + y^2 = 1$ has no circular neighborhood which belongs entirely to S (see Fig. 26). The complement of S has two parts: the set of all points for which $x^2 + y^2 < 1$, and the set of all points for which $x^2 + y^2 \geqq 4$. It is easily seen that $C(S)$ is not open, for a point on the circle $x^2 + y^2 = 4$ has no circular neighborhood which belongs entirely to $C(S)$. Therefore S is not closed. The set $C(S)$ is shown in Fig. 31.

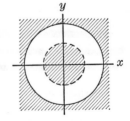

Fig. 31

If S is a set, the complement of $C(S)$ is S itself. Hence, by definition, $C(S)$ is closed if S is open. Thus, if one of the two sets S, $C(S)$ is open, the other is closed.

DEFINITION. If S is a point set, a point P is called a *boundary point* of

S if every neighborhood of P contains at least one point of S and one point of the complement $C(S)$. The collection of all boundary points of S is called the *boundary* of S. We denote it by $B(S)$.

The sets introduced in Examples 1–4 have the following boundaries:

Example 1. $B(S)$ is the circle $x^2 + y^2 = 1$.

Example 2. $B(S)$ is the parabola $y^2 = 4x$.

Example 3. $B(S)$ is the circle $x^2 + y^2 = 1$.

Example 4. $B(S)$ consists of the two circles $x^2 + y^2 = 1$ and $x^2 + y^2 = 4$.

It is clear from the definition of boundary that a set S and its complement $C(S)$ have the same boundary. If a set S is open, no boundary point of S is actually in S. If S is closed, $B(S)$ is part of S. These statements may be verified by referring to the definitions. In Example 4, $B(S)$ is partly in S and partly in $C(S)$.

DEFINITION.   A point P of a set S is called an *interior* point of S if there is some circular neighborhood of P which belongs entirely to S. The interior of a set S is the set consisting of all interior points of S.

An open set consists entirely of interior points. A set may have no interior points.

We are going to be most interested in sets of points which are of one of the following kinds:

   (*a*) open;
   (*b*) a closed set formed by an open set together with its boundary;
   (*c*) the boundary of an open set.

Furthermore, the sets which we consider will usually have boundaries which are composed of one or more curves or segments of curves (straight lines included), or possibly of isolated points.

DEFINITION.   By a *region* we shall mean a set of points which is either a nonempty open set or such a set together with some or all of the points forming its boundary.

All the sets occurring as domains of definition of the functions in Examples 1–4 of § 5 are regions in the sense just defined. Of these regions, those of Examples 1 and 3 are open; that of Example 4 is closed (i.e., the entire boundary is part of the region); that of Example 2 is neither open nor closed (it contains only one of the two circles forming the boundary).

If we wish to deal with point sets in three-dimensional space we define *spherical* neighborhoods instead of *circular* neighborhoods. Once this is

done we can define open set, closed set, and boundary just as in the case of point sets in the plane.
These concepts also apply to point sets on a line.
In that case we define a neighborhood of x_0 as an
open interval $x_0 - \delta < x < x_0 + \delta$, where δ is any
positive number.

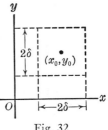

Fig. 32

Finally, we remark that we could use square
neighborhoods instead of circular neighborhoods
without affecting the whole development of the con-
cepts of open sets, closed sets, boundary of a set, and so forth. A square
neighborhood of (x_0, y_0) is the set of all points inside a square with center
(x_0, y_0) (see Fig. 32). If the square has sides of length 2 δ the neighborhood
is defined by the inequalities

$$| x - x_0 | < \delta, | y - y_0 | < \delta.$$

In the future we shall often speak of neighborhoods without the adjectives *circular* or *square*. It will usually be immaterial which is meant
or which the student chooses to think of on any particular occasion.

EXERCISES

1. The set S consists of all points (x, y) such that $x^2 + y^2 < 1$ and $x < 0$ if
$y = 0$. Describe S in geometrical language, with the aid of a figure. Is S open,
closed, or neither? What is the boundary of S?

2. The set S consists of all points (x, y) such that either $x^2 + y^2 = 1$ or $y = 0$
and $0 \leq x \leq 1$. Does this set have any interior points? Is it closed?

3. The set S consists of all points (x, y) such that $y \geq x^2$ and $y \leq 1$. Draw a
figure, and describe S in geometrical language. Is S open, closed, or neither?
What is the boundary of S?

4. The set S consists of all points (x, y) such that $0 < xy \leq 1$ and $x > 0$. Is
this set open, closed, or neither? What is $B(S)$?

5. The set S consists of all points (x, y) such that $y = \sin(1/x)$ and $x > 0$. Does
this set have any interior points? Is it closed? What is $B(S)$?

6. The set S consists of all points (x, y) for which $x^2 + y^2 < 4$ and $y > 0$
except for the points with $0 < y \leq 1$ and $x = 1/n, n = 1, 2, \cdots$, i.e., except for
the points of a certain infinite sequence of line segments each one unit long.
Is this set S open? What is the boundary of S? Are there any points of $B(S)$ which
are also in S? Is S a region? Is its complementary set $C(S)$ a region?

7. Let $f(x, y) = \left(y - \sin \dfrac{1}{x} \right)^{-1}$, the function being defined whenever this expression has a meaning, but not otherwise. Is the set of points where f is defined
a region? What is the boundary of the set?

8. Let $f(x, y) = \log \sin x + y^{-1/2}$, the function being defined whenever this expression has a meaning (real numbers only are to be considered). Describe, with the aid of a diagram, the set of points (x, y) where f is defined. Is the set open, closed, or neither? Of what does its boundary consist?

5.2   Limits

We wish to define what is meant by the statement "$f(x, y)$ approaches A as a limit when the point (x, y) approaches (x_0, y_0)." The statement is written in the form

$$(5.2\text{-}1) \qquad\qquad \lim_{(x, y) \to (x_0, y_0)} f(x, y) = A.$$

In giving the definition we shall assume that the function f is defined in a region R and that (x_0, y_0) is either an interior point of R or on the boundary of R. The point (x_0, y_0) *may*, but *need not*, belong to R. If it is in R, the meaning of (5.2–1) has nothing whatever to do with the value $f(x_0, y_0)$ at the point (x_0, y_0). The statement (5.2–1) is now defined to mean that if ϵ is any positive number, there is some neighborhood of (x_0, y_0) such that if (x, y) is in the neighborhood, in R, and different from (x_0, y_0), then $|f(x, y) - A| < \epsilon$. This definition may be compared with that for functions of one variable in §§ 1.1, 1.61. The limit notion can be expressed verbally as follows: The meaning of (5.2–1) is that $f(x, y)$ is in a prescribed neighborhood of A on the real axis provided (x, y) is any point other than (x_0, y_0) in a suitably chosen (sufficiently small) neighborhood of (x_0, y_0) in the plane.

With the adoption of the term *neighborhood* we obtain a unification of the limit concept for functions of one, two, or three independent variables. The extension to more than three variables causes no trouble and involves no new principle. We continue to use geometric language; the meaning of a "spherical neighborhood" in a space of four variables is made clear by the inequality

$$(x - x_0)^2 + (y - y_0)^2 + (z - z_0)^2 + (w - w_0)^2 < \delta^2.$$

The fundamental theorems about limits carry over to functions of several variables. We cite particularly Theorems X (§ 1.61) and XIV (§ 1.64).

When we say that $f(x, y) \to A$ as $(x, y) \to (x_0, y_0)$, it must be stressed that *the limit must exist and be the same, no matter how (x, y) approaches (x_0, y_0).* The student will recall that, for a function of one variable, $f(x) \to A$ as $x \to x_0$ means that $f(x) \to A$ as $x \to x_0+$ and also as $x \to x_0-$. But in the case of two variables, (x, y) can approach (x_0, y_0) in an infinite number of ways. If it is possible to find two different modes of approach to (x_0, y_0) such that $f(x, y)$ approaches different limits in the two cases, or no limit at all in at least one of the cases, then we can say that $\lim_{(x, y) \to (x_0, y_0)} f(x, y)$ does not exist.

Example 1. Let $f(x, y) = \dfrac{x^2 - y^2}{x^2 + y^2}$. This function is defined except at the origin. Let us show that the limit of $f(x, y)$ as $(x, y) \to (0, 0)$ does not exist. If $(x, y) \to (0, 0)$ along the x-axis, we have $f(x, 0) = 1$ $(x \neq 0)$. If $(x, y) \to (0, 0)$ along the y-axis, we have $f(0, y) = -1$ $(y \neq 0)$. Thus the limits for the two modes of approach are 1 and -1 respectively. This shows that $f(x, y)$ has no limit as $(x, y) \to (0, 0)$.

To prove directly that a certain function approaches a certain limit as $(x, y) \to (x_0, y_0)$, we have to work with inequalities. The following example will illustrate the technique. It is not our intent, at this stage of a student's training, to have him cultivate extensively the technique of working exercises of the type represented by the example. The purpose is merely to make clearer the essential content of the definition of a limit.

Example 2. Show that

$$(5.2\text{--}2) \qquad \lim_{(x, y) \to (0, 0)} \frac{2x^3 - y^3}{x^2 + y^2} = 0.$$

In terms of inequalities, this means that if ϵ is any positive number, we have to show that another positive number δ (depending on ϵ) can be found, such that

$$(5.2\text{--}3) \qquad \left| \frac{2x^3 - y^3}{x^2 + y^2} \right| < \epsilon \text{ if } 0 < (x^2 + y^2)^{1/2} < \delta;$$

in other words, denoting the function under consideration by $f(x, y)$, we have to show that, if $\epsilon > 0$, there is *some* circular neighborhood of the origin (whose radius we denote by δ) such that $|f(x, y) - 0| < \epsilon$ if (x, y) is in the specified neighborhood of the origin but not actually at the origin. We proceed to find such a number δ, considering ϵ as given.

Now $\qquad |2x^3 - y^3| \leq 2|x|^3 + |y|^3 = 2|x|\,x^2 + |y|\,y^2.$

Also, $\qquad |x| \leq (x^2 + y^2)^{1/2}$ and $|y| \leq (x^2 + y^2)^{1/2}.$

Therefore $\quad |2x^3 - y^3| \leq (x^2 + y^2)^{1/2}(2x^2 + y^2) \leq 2(x^2 + y^2)^{3/2},$

and $\qquad \left| \dfrac{2x^3 - y^3}{x^2 + y^2} \right| \leq 2(x^2 + y^2)^{1/2}$ if $0 < x^2 + y^2.$

It is now clear that (5.2–3) will hold if δ is chosen in any manner such that $0 < \delta \leq \epsilon/2$. Thus (5.2–2) is proved.

In work of this kind the student will find the simple inequalities

$$(5.2\text{--}4) \qquad 2|ab| \leq a^2 + b^2,$$

$$(5.2\text{--}5) \qquad |a| + |b| \leq \sqrt{2}(a^2 + b^2)^{1/2}$$

quite useful. See Exercise 6 for remarks about these inequalities.

EXERCISES

1. Find the limit of $\dfrac{x^2 - y^2}{x^2 + y^2}$ as (x, y) approaches $(0, 0)$ along the line $y = x$; along the line $y = mx$.

2. Does $\displaystyle\lim_{(x, y) \to (0, 0)} \dfrac{xy}{x^2 + y^2}$ exist? Give reasons.

3. Examine the behavior of $\dfrac{x^4 y^4}{(x^2 + y^4)^3}$ as (x, y) approaches $(0, 0)$ along various straight lines. Then consider what happens for approach to the origin along the curve $y^2 = x$. Is there a limit as $(x, y) \to (0, 0)$ without restriction?

4. Show in each case that the given function does not approach a limit as $(x, y) \to (0, 0)$, by examining the behavior of the function for at least two modes of approach.

(a) $\dfrac{x - y}{x^2 + y^2}$.

(c) $\dfrac{x^2 + y}{(x^2 + y^2)^{1/2}}$.

(b) $\dfrac{xy^2}{x^2 + y^4}$.

(d) $\dfrac{x^4 + 3 x^2 y^2 + 2 xy^3}{(x^2 + y^2)^2}$.

5. Define a function by setting $f(x, y) = 0$ if $y \leq 0$ or if $y \geq x^2$, and $f(x, y) = 1$ if $0 < y < x^2$. Show that $f(x, y) \to 0$ as $(x, y) \to (0, 0)$ along any straight line through the origin. Find a curve through the origin along which $f(x, y) = 1$ (except at the origin).

6. If A and B are nonnegative numbers, the inequality $A \leq B$ is equivalent to $A^2 \leq B^2$. Use this fact to prove the correctness of (5.2–4); then show that (5.2–5) is correct.

7. Let $f(x, y) = xy \left(\dfrac{x^2 - y^2}{x^2 + y^2} \right)$. Show that $| f(x, y) | \leq \frac{1}{2}(x^2 + y^2)$, and hence prove that $f(x, y)$ approaches a limit as $(x, y) \to (0, 0)$.

8. Let $f(x, y) = \dfrac{x^2 y^2}{x^2 + y^2}$. If $\epsilon > 0$, find δ so that $0 < (x^2 + y^2)^{1/2} < \delta$ implies $| f(x, y) | < \epsilon$.

9. If $\epsilon > 0$, show that $| 2 x^2 - 6 xy + 5 y^2 | < \epsilon$ when $(x^2 + y^2)^{1/2} < (\epsilon/13)^{1/2}$.

10. Show that $\dfrac{x^4 + y^4}{x^2 + y^2} < \epsilon$ if $0 < x^2 + y^2 < \delta^2$, for a suitably chosen δ depending on ϵ.

11. Show that $| x^3 - y^3 | \leq (x^2 + y^2)^{3/2}$.

12. Does $\dfrac{2 x^5 + 2 y^3(2 x^2 - y^2)}{(x^2 + y^2)^2}$ approach a limit as $(x, y) \to (0, 0)$?

5.3 Continuity

The notion of continuity depends on the notion of limit, as was pointed out at the beginning of Chapter III.

DEFINITION.   Let $f(x, y)$ be defined in a region R, and let (x_0, y_0) be a point of R. We say that f is continuous at this point if

$$\lim_{(x, y) \to (x_0, y_0)} f(x, y) = f(x_0, y_0).$$

If (x_0, y_0) is an interior point of R, the mode of approach of (x, y) to (x_0, y_0) is unrestricted in this definition. But, if (x_0, y_0) is a boundary point of R, there is the restriction that (x, y) must remain in R. We say that f is continuous in R if it is continuous at each point of R.

If f and g are defined in the same region R, and each is continuous at a point (x_0, y_0) of R, then the sum and product functions

$$f(x, y) + g(x, y), \quad f(x, y)g(x, y)$$

are also continuous at (x_0, y_0). The quotient function

$$\frac{f(x, y)}{g(x, y)}$$

is continuous at (x_0, y_0) provided $g(x_0, y_0) \neq 0$. These assertions are direct generalizations of Theorem I, § 3. They may be extended to functions of more than two variables.

The theorems of Chapter III all have important analogues for functions of several independent variables. We do not wish at this point to prove all these analogous theorems, but we shall discuss the statements of certain theorems which will be used in the chapters immediately following.

In dealing with the analogues of Theorems II (§ 3.1) and III (§ 3.2) of Chapter III it is necessary to introduce the concept of a *bounded point set*.

DEFINITION.   A point set S in the plane is called bounded if all its points are inside some sufficiently large circle. For a point set in space the definition is similar; we write "sphere" instead of "circle."

Examples.   The interior and boundary of a triangle form a bounded point set. The set of all points between the lines $y = 0$, $y = 1$ is not a bounded point set.

We now state two important theorems.

THEOREM I.   *If a function is continuous at each point of a closed and bounded region R, the function is bounded on the region (i.e., the values of the function form a bounded set of real numbers).*

THEOREM II. *Let f be continuous on a closed and bounded region R. Let m and M be the greatest lower bound and least upper bound of the values of f on R. Then f takes on each of the values m, M at least once in R.*

Proofs of these two theorems will be considered later, in §§ 17.2, 17.3. Theorem V of Chapter III (§ 3.3) has the following analogue. We state it for the case of two independent variables.

THEOREM III. *Let f be defined in an open set containing the point (x_0, y_0). Suppose that f is continuous at that point and that $f(x_0, y_0) \neq 0$. Then there is a neighborhood of (x_0, y_0) throughout which $f(x, y)$ has the same sign as at (x_0, y_0).*

The proof is left to the student.

There is a feature of the continuity of a function $f(x, y)$ which deserves notice. If we fix y, say $y = y_0$, $f(x, y_0)$ is a function of x alone. Likewise $f(x_0, y)$ is a function of y alone. It can happen that each of these functions of a single variable is continuous, and yet that $f(x, y)$ is not continuous. An illustration of this possibility is given in Exercise 3.

There is another theorem which will be needed later. It deals with composite functions, and may be roughly stated in the form: *A continuous function of continuous functions is continuous.* The number of variables is immaterial.

Examples. $F(z) = \sin z$ is a continuous function of z, and $f(x, y) = (1 + xy)^2$ is a continuous function of x, y. Therefore

$$F(f(x, y)) = \sin (1 + xy)^2$$

is a continuous function of x, y. Or, again,

$$F(x, y, z) = x^2 + y^2 + z^2 \quad \text{and} \quad f(x, y) = x(1 + x^2 + y^2)^{-3/2}$$

are continuous functions of x, y, z and x, y, respectively. Therefore

$$F(x, y, f(x, y)) = x^2 + y^2 + \frac{x^2}{(1 + x^2 + y^2)^3}$$

is a continuous function of x, y.

We formalize one such theorem about composite functions.

THEOREM IV. *Let $F(x, y, z)$ be continuous in an open set B of space. Let $f(x, y)$ be continuous in an open set R of the xy-plane. Writing $z = f(x, y)$, suppose that the point (x, y, z) is in B when (x, y) is in R. Then $F(x, y, f(x, y))$ is continuous in R.*

We shall not give a proof here. This theorem is a special case of the Theorem II which is provided in § 11.7.

EXERCISES

1. If $f(x, y) = \dfrac{\sin (x^2 + y^2)}{x^2 + y^2}$ when $x^2 + y^2 \neq 0$, how must $f(0, 0)$ be defined so as to make f continuous at $(0, 0)$?

2. Let us define $f(x, y) = \dfrac{\sin (xy)}{x}$ if $x \neq 0$, and $f(x, y) = y$ if $x = 0$. Does f have any points of discontinuity?

3. If we define $f(x, y) = xy/(x^2 + y^2)$ when $x^2 + y^2 \neq 0$, and $f(0, 0) = 0$, show that f is discontinuous at $(0, 0)$, but also that $f(x, 0)$ and $f(0, y)$ are continuous functions of x and y, respectively, with no exceptions.

4. Let $f(x, y) = (x^2 + y^2) \tan^{-1} (y/x)$ if $x \neq 0$, and define $f(0, 0) = 0$, but do not consider f defined if $x = 0$ and $y \neq 0$. **(a)** Is f continuous at $(0, 0)$ according to the definition in the text? **(b)** Is it possible to define f at the one additional point $(0, 1)$ so as to make it continuous there?

5. Let $f(x, y) = xy \log (xy)$ if $xy > 0$, and define $f(x, y) = 0$ if $xy = 0$. Where, if at all, is f discontinuous?

6. Let $f(x, y) = (5 x + y)/(x - y)$. Show directly by the definition that f is continuous at $(4, 1)$ by proving that, if $\epsilon > 0$, $| f(x, y) - f(4, 1) | < \epsilon$ provided (x, y) is in a sufficiently small neighborhood of $(4, 1)$. Start by showing that

$$| f(x, y) - f(4, 1) | \leqq 2 | x - 4 | + 8 | y - 1 |$$

at the points of the square $3 < x < 5, 0 < y < 2$.

7. If $f(x, y) = e^{-1/|x-y|}$ when $x \neq y$, how must f be defined when $x = y$ so as to make it continuous at all points of the plane?

8. Let us define $f(x, y) = 0$ if $y \leqq 0$ or if $x^2 \leqq y$, and $f(x, y) = 4 y(x^2 - y)/x^4$ if $0 < y < x^2$.
(a) Is f continuous at $(0, 0)$? **(b)** Discuss possible discontinuity at other points of the line $y = 0$ or the curve $y = x^2$.

9. Let $f(x, y) = x(1 - x^2 - y^2)^{-1/2}$, the region R of definition being defined by $x^2 + y^2 < 1$. Is it possible to add the single point $(0, 1)$ to R and define f so as to make it continuous at that point? Consider values of f on the circle $x^2 + y^2 - y = 0$, and also at other points in R near $(0, 1)$.

10. If $f(x, y, z) = xyz/(x^2 + y^2 + z^2)$ when $x^2 + y^2 + z^2 \neq 0$, is it possible to define $f(0, 0, 0)$ so as to make f continuous at the origin?

5.4   Modes of representing a function

The standard method of representing a function of one variable is by graphing in rectangular co-ordinates. We write $y = f(x)$ and plot the points (x, y). If f is continuous on an interval, the graph will be a curve in the plane.

The corresponding procedure for the case of two independent variables is familiar. We write $z = f(x, y)$, and plot the points (x, y, z). If f is continuous in a region R of the xy-plane we obtain a surface in space (see Fig. 33).

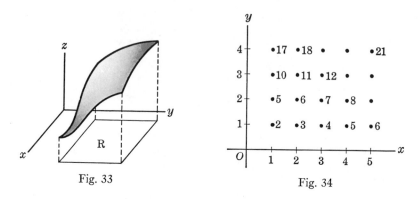

Fig. 33

Fig. 34

When we go to three independent variables there is no satisfactory analogue of the foregoing methods of graphical representation, for we cannot draw upon any familiar geometric intuition to visualize $w = f(x, y, z)$ as defining a configuration in space of four dimensions. There is, however, another mode of representation which is helpful. It is available as well in the case of two independent variables, and since the figures are easier to draw, we begin with that case.

When f is defined in a region R, we can think of each point of R as being given a label, namely, the value $f(x, y)$ at that point. A good example is obtained by thinking of the xy-plane as a map on which elevations above sea level are marked at various locations, $f(x, y)$ being the elevation in feet at (x, y) (see Fig. 34). To carry this example further, imagine that the map is a topographic map with *contour lines* drawn in, showing lines of equal elevation. Each line is labeled; there is a line for 500 feet above sea level, others for 400, 600, and so on. In the aggregate, the configuration of these lines, together with their numbering, gives us a good visual representation of the elevation as a function of x and y.

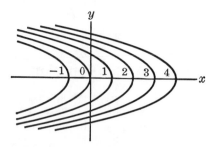

Fig. 35. *Level curves of* $f(x, y) = x + y^2$

This "topographic map" idea can be carried over to any function $f(x, y)$. Instead of contour lines we consider curves along which $f(x, y)$ is constant in value. Such a curve is called a *level curve* of the function. If the constant value is C, the equation of the level curve is $f(x, y) = C$. Isobars (curves of equal atmospheric pressure) on a meteorological chart furnish another good example of level curves of a function.

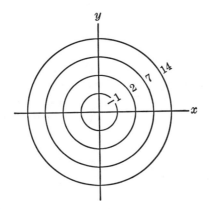

Fig. 36.　*Level curves of* $f(x, y) = x^2 + y^2 - 2$

The three-dimensional analogue of this mode of representation is now easily grasped. Instead of level curves we shall have *level surfaces* $f(x, y, z) = C$. Common physical examples of functions of three variables which are conveniently visualized in this way are density and temperature in a gas or other medium.

The Elements of Partial Differentiation **6**

6. Partial derivatives

In this chapter the main objective is the exposition of the fundamentals of differential calculus for functions of two or more variables, with emphasis on the formal procedures which are used in dealing with various kinds of particular problems. The subject is developed from the beginning. Many students will already know some things about partial differentiation from courses in elementary calculus. For them the earlier parts of this chapter will serve as a review. By concentrating on procedures and techniques in this chapter, and putting most of the theoretical considerations in later chapters, the authors have hoped to make the treatment of partial differentiation adaptable to the needs of students of varying degrees of preparation. For further expression of the authors' intentions in the organization of Chapters VI, VII, and VIII, see the introductory sections of Chapters VII and VIII.

Let $f(x, y)$ be defined in a region R of the xy-plane. If we think of y as fixed and x as variable, the derivative of $f(x, y)$ with respect to x is called the *partial derivative* with respect to x. This partial derivative is denoted by $\dfrac{\partial f}{\partial x}$. If we write $u = f(x, y)$, the partial derivative is also denoted by $\dfrac{\partial u}{\partial x}$.

Likewise, the partial derivative with respect to y, $\dfrac{\partial f}{\partial y}$ or $\dfrac{\partial u}{\partial y}$, is the derivative of $f(x, y)$ with respect to y when x is regarded as a constant.

Example 1. If
$$u = x^2 y + e^{-xy^3},$$
$$\frac{\partial u}{\partial x} = 2\, xy - y^3 e^{-xy^3},$$
$$\frac{\partial u}{\partial y} = x^2 - 3\, xy^2 e^{-xy^3}.$$

Similar definitions and notations apply in dealing with functions of three or more independent variables.

The partial derivatives of $\dfrac{\partial f}{\partial x}$ and $\dfrac{\partial f}{\partial y}$ are called the second partial derivatives of $f(x, y)$. There are in all four second derivatives of $f(x, y)$. The notations for these derivatives, if we write $u = f(x, y)$, are the following:

$$\frac{\partial}{\partial x}\left(\frac{\partial u}{\partial x}\right) = \frac{\partial^2 u}{\partial x^2}, \qquad\qquad \frac{\partial}{\partial y}\left(\frac{\partial u}{\partial x}\right) = \frac{\partial^2 u}{\partial y\,\partial x},$$

140

$$\frac{\partial}{\partial x}\left(\frac{\partial u}{\partial y}\right) = \frac{\partial^2 u}{\partial x\,\partial y}, \qquad\qquad \frac{\partial}{\partial y}\left(\frac{\partial u}{\partial y}\right) = \frac{\partial^2 u}{\partial y^2}.$$

Example 2. For the function of Example 1 we have

$$\frac{\partial^2 u}{\partial x^2} = 2\,y + y^6 e^{-xy^3},$$

$$\frac{\partial^2 u}{\partial y\,\partial x} = 2\,x + 3\,xy^5 e^{-xy^3} - 3\,y^2 e^{-xy^3},$$

$$\frac{\partial^2 u}{\partial x\,\partial y} = 2\,x + 3\,xy^5 e^{-xy^3} - 3\,y^2 e^{-xy^3},$$

$$\frac{\partial^2 u}{\partial y^2} = 9\,x^2 y^4 e^{-xy^3} - 6\,xy e^{-xy^3}.$$

We observe that

(6-1)
$$\frac{\partial^2 u}{\partial y\,\partial x} = \frac{\partial^2 u}{\partial x\,\partial y}$$

in this example. We shall ordinarily find that the relation (6-1) holds true for the functions we meet in practice, for the relation is valid at a point provided both the second derivatives are defined in a neighborhood of the point and continuous at the point. This will be proved in § 7.2 (Theorem III).

A partial derivative of $f(x, y)$ is again a function of x, y. To denote the value of $\dfrac{\partial f}{\partial x}$ at the point (x_0, y_0) we may use one of the expressions

$$\left(\frac{\partial f}{\partial x}\right)_{(x_0,\ y_0)}, \qquad \left.\frac{\partial f}{\partial x}\right|_{(x_0,\ y_0)}.$$

These notations are rather awkward, however; it is desirable to have a standard functional notation for partial derivatives. For a function $f(x, y)$ of two independent variables we shall write

$$f_1(x, y) = \frac{\partial f}{\partial x}, \qquad f_2(x, y) = \frac{\partial f}{\partial y}.$$

For the value of a partial derivative at a point we then have expressions such as

$$f_1(x_0, y_0) = \left(\frac{\partial f}{\partial x}\right)_{(x_0,\ y_0)}, \qquad f_2(a, b) = \left(\frac{\partial f}{\partial y}\right)_{(a,\ b)}.$$

For second derivatives we use the notations

$$f_{11}(x, y) = \frac{\partial}{\partial x}\left(\frac{\partial f}{\partial x}\right), \qquad f_{12}(x, y) = \frac{\partial}{\partial y}\left(\frac{\partial f}{\partial x}\right),$$

$$f_{21}(x, y) = \frac{\partial}{\partial x}\left(\frac{\partial f}{\partial y}\right), \qquad f_{22}(x, y) = \frac{\partial}{\partial y}\left(\frac{\partial f}{\partial y}\right).$$

Observe the ordering of the numerical subscripts in relation to the order of carrying out the differentiations; 1 refers to x and 2 refers to y.

The notation is extended in an obvious way to derivatives of order higher than the second, and also to functions of more than two independent variables. Thus for example,

$$f_{122}(x, y) = \frac{\partial^2}{\partial y^2} \left(\frac{\partial f}{\partial x} \right),$$

and $$g_3(x, y, z) = \frac{\partial g}{\partial z}, \qquad g_{123}(x, y, z) = \frac{\partial}{\partial z} \left[\frac{\partial}{\partial y} \left(\frac{\partial g}{\partial x} \right) \right].$$

6.1 Implicit functions

We often deal with functions which are defined implicitly as the solution of certain equations. In ordinary practice we can find the partial derivatives of such a function by the same procedures which we learn in elementary calculus.

Example 1. Find $\dfrac{\partial z}{\partial x}$ from the equation

(6.1–1) $$\frac{x^2}{16} + \frac{y^2}{12} + \frac{z^2}{9} = 1,$$

on the understanding that z is dependent and x, y are independent.

We have $$\frac{2 x}{16} + \frac{2 z}{9} \frac{\partial z}{\partial x} = 0,$$

and so

(6.1–2) $$\frac{\partial z}{\partial x} = - \frac{9 x}{16 z}.$$

The equation (6.1–1) actually defines two functions of (x, y), corresponding to the two choices of sign in

(6.1–3) $$z = \pm 3 \left(1 - \frac{x^2}{16} - \frac{y^2}{9} \right)^{1/2}.$$

By substituting (6.1–3) in (6.1–2) we obtain the partial derivative for each of these two functions:

(6.1–4) $$\frac{\partial z}{\partial x} = \mp \frac{3 x}{16} \left(1 - \frac{x^2}{16} - \frac{y^2}{9} \right)^{-1/2}.$$

The result (6.1–4) could also have been obtained by differentiating (6.1–3) directly.

The procedure can also be applied in the case of functions defined by simultaneous equations.

Example 2. If u and v are defined as functions of x, y by the equations

(6.1–5)
$$u \cos v - x = 0$$
$$u \sin v - y = 0,$$

find the partial derivatives $\dfrac{\partial u}{\partial x}$, $\dfrac{\partial v}{\partial x}$.

Method I. One method of procedure is to attempt to solve for u, v in terms of x, y. If this can be accomplished, we can then calculate the required partial derivatives directly.

From (6.1–5) we have

$$u^2 \cos^2 v = x^2, \; u^2 \sin^2 v = y^2.$$

Now add these equations and use a familiar trigonometric identity. The result is $u^2 \cos^2 v + u^2 \sin^2 v = x^2 + y^2$

(6.1–6)
$$u^2 = x^2 + y^2, \text{ or } u = \pm \sqrt{x^2 + y^2}.$$

Next, going back to (6.1–5), we substitute the value just found for u. We find

(6.1–7)
$$\cos v = \frac{x}{\pm \sqrt{x^2 + y^2}}, \quad \sin v = \frac{y}{\pm \sqrt{x^2 + y^2}},$$

the same sign being taken before the radical in both cases. We might also write

(6.1–8)
$$\tan v = \frac{y}{x}$$

in cases $x \neq 0$. We see that there are in general two possible determinations of u from (6.1–6); for v there are an infinite number of possible determinations from (6.1–7), differing by multiples of 2π. The derivatives of u may be found from (6.1–6):

$$\frac{\partial u}{\partial x} = \pm \frac{x}{\sqrt{x^2 + y^2}}.$$

In finding the derivatives of v it is easier to work from (6.1–8). We have

$$\sec^2 v \frac{\partial v}{\partial x} = \frac{-y}{x^2}, \; \sec^2 v = 1 + \tan^2 v = 1 + \frac{y^2}{x^2},$$

$$\frac{\partial v}{\partial x} = -\frac{y}{x^2 \sec^2 v} = \frac{-y}{x^2 + y^2}.$$

This result could also have been obtained by expressing v as an inverse tangent and then differentiating. It should, however, be noted that v is not necessarily the *principal value* of the inverse tangent of y/x.

Method II. As an alternative to the first method we may proceed directly to differentiate the equations (6.1–5). In doing so we must bear in mind that x, y are independent variables and that u, v are dependent variables. Since we are seeking partial derivatives with respect to x, we think of y as a constant. Then

(6.1–9)
$$\cos v \frac{\partial u}{\partial x} - u \sin v \frac{\partial v}{\partial x} - 1 = 0,$$
$$\sin v \frac{\partial u}{\partial x} + u \cos v \frac{\partial v}{\partial x} - 0 = 0.$$

These simultaneous equations are now solved for $\dfrac{\partial u}{\partial x}$ and $\dfrac{\partial v}{\partial x}$. The solution may be achieved by elimination, or by determinants and Cramer's rule. We leave it for the student to carry out the solution and find

(6.1–10)
$$\frac{\partial u}{\partial x} = \cos v, \qquad \frac{\partial v}{\partial x} = \frac{-\sin v}{u}.$$

To reconcile these answers with the answers as found by Method I, observe that, by (6.1–5) and (6.1–6),

$$\cos v = \frac{x}{u} = \frac{x}{\pm \sqrt{x^2 + y^2}},$$
$$\frac{-\sin v}{u} = -\frac{y}{u^2} = \frac{-y}{x^2 + y^2}.$$

One of the things to be noted in connection with problems like that of Example 2 is that the solution by Method II can be carried out even if the explicit solution for u and v in terms of x, y is impracticable.

In all the procedures which have been illustrated here it has been taken for granted that the given equations *do* implicitly define certain functions, and that these functions *do have* partial derivatives. The deeper theoretical questions on these matters are considered fully in Chapter VIII.

EXERCISES

1. Find $\dfrac{\partial u}{\partial y}$ and $\dfrac{\partial v}{\partial y}$ from Example 2. Use both of the methods given in the text, and reconcile your answers.

2. Suppose that u and v are defined in terms of x, y by the equations $u^2 - v^2 = xy$, $uv = x^2 + y^2$. Find the first partial derivatives of u and v with respect to x and y, respectively. Use the implicit function procedure (Method II). See how far you can go with Method I in this case.

3. Assuming that z is defined as a function of x and y by the equation $4 \sin^2 x + 2 \cos(y + z) = 2$, find $\partial z/\partial x$ and $\partial z/\partial y$ when $x = \pi/6$, $y = \pi/3$, $z = 0$.

4. Find $\partial z/\partial x$ and $\partial z/\partial y$ from the equation $x^3 + y^3 + z^3 - 3xyz = 0$.

6.2   Geometrical significance of partial derivatives

Just as the ordinary derivative of a function of one variable has its geometric realization in the slope of a line which is tangent to a curve, so the partial derivatives of a function of two variables have a geometrical significance in connection with a plane which is tangent to a surface. Our purpose in this section is to show how the partial derivatives $\dfrac{\partial f}{\partial x}$ and $\dfrac{\partial f}{\partial y}$, when $x = a$ and $y = b$, are related to the plane which is tangent to the surface $z = f(x, y)$ at the point (a, b, c). In this section we shall not give a formal definition of the tangent plane. We reserve the full discussion of this matter to § 6.4, because the concept of the tangent plane is the geometrical counterpart of the concept of the differential of a function of two variables.

Let S be the surface $z = f(x, y)$, and let (a, b, c) be a point on S. Then $c = f(a, b)$. Consider the line through the point (a, b, c) parallel to the z-axis. Let us visualize various planes containing this line, each such plane cutting the surface S in a curve. One such plane, cutting the y-axis perpendicularly at $y = b$, is shown in Fig. 37. In the diagram, the curve of the intersection of S and this plane, $y = b$, is represented as having a tangent line L at the point (a, b, c). One can also imagine a plane $x = a$, passing through (a, b, c) and intersecting the x-axis perpendicularly at $(a, 0, 0)$. More generally, one can imagine a plane different from either of these two, but so placed that it passes through (a, b, c) and is parallel to the

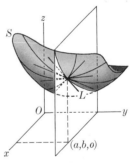

Fig. 37

z-axis. The student should construct for himself a diagram similar to Fig. 37, with a plane through (a, b, c), parallel to the z-axis but cutting neither the x-axis nor the y-axis at right angles. This exercise in geometrical visualization will be helpful for the following discussion.

Suppose that each plane through (a, b, c) and parallel to the z-axis cuts the surface S in a curve which has at the point (a, b, c) a tangent line which is not parallel to the z-axis. Suppose further that all of these tangent lines, corresponding to the different planes of the type described, lie in a single plane. Then this single plane must surely be the tangent plane to the surface S at (a, b, c), if indeed there is such a tangent plane. Fig. 37 shows four different curves on S, together with their tangent lines, all intersecting at (a, b, c).

Assuming now that there is a tangent plane to S at (a, b, c) not parallel to the z-axis, let us see how to find its equation. If $\cos \alpha$, $\cos \beta$, $\cos \gamma$ are the direction cosines of a line which is normal (perpendicular) to this plane, the equation of the plane can be written

$$(\cos \alpha)(x - a) + (\cos \beta)(y - b) + (\cos \gamma)(z - c) = 0.$$

Since the plane is not parallel to the z-axis, we know that $\cos \gamma \neq 0$; we can therefore solve for $z - c$ by dividing by $\cos \gamma$, thus obtaining an equation of the form

$$(6.2\text{--}1) \qquad z - c = A(x - a) + B(y - b).$$

Our problem is to find A and B. If we put $y = b$ in this equation, we find the relation between x and z on the line in which the tangent plane (6.2–1) cuts the plane $y = b$. This line of intersection must be the line L which is tangent to the curve in which the surface S is cut by the plane $y = b$ (see Fig. 37). The relation between x and z on L is obtained by putting $y = b$ in (6.2–1); that is,

$$z - c = A(x - a).$$

The relation between x and z along the curve in which S is cut by the plane $y = b$ is

$$z = f(x, b).$$

Hence, according to the usual relationship between slopes and derivatives, A must be the derivative of z with respect to x when y is held constantly equal to b:

$$(6.2\text{--}2) \qquad A = \frac{\partial}{\partial x} f(x, b) \bigg|_{x=a} = f_1(a, b).$$

In the same way, by putting $x = a$ in (6.2–1) and interpreting B as a slope in the relation between y and z, with x held constantly equal to a, we see that

$$(6.2\text{--}3) \qquad B = \frac{\partial}{\partial y} f(a, y) \bigg|_{y=b} = f_2(a, b).$$

Consequently, we see that the equation of the plane tangent to S at (a, b, c) can be written in the form

$$(6.2\text{--}4) \qquad z - c = f_1(a, b)(x - a) + f_2(a, b)(y - b),$$

or

$$(6.2\text{--}5) \qquad z = f(a, b) + f_1(a, b)(x - a) + f_2(a, b)(y - b).$$

The foregoing discussion of the tangent plane is not complete because of the fact that no full and formal definition of the tangent plane has been given. However, the discussion here is wholly consistent with the definition to be given in § 6.4. It is shown there that when S has a tangent plane at (a, b, c), not parallel to the z-axis, then f has partial derivatives when $x = a$ and $y = b$. The arguments we have just given can then be used to derive the equation (6.2-4) for the tangent plane. Observe that the assumption that f has partial derivatives at $x = a$, $y = b$ rules out the possibility that the tangent plane might be parallel to the z-axis.

The line through (a, b, c) perpendicular to the plane (6.2-5) is called the *normal* to the surface S at that point. As we see from the equation (6.2-5), the direction of the normal is determined by the ratios

$$f_1(a, b) : f_2(a, b) : -1,$$

for the direction cosines of the normal to a plane are proportional to the coefficients of x, y, and z respectively, in the equation of the plane. Here is a result to be remembered:

The line normal to the surface $z = f(x, y)$ at a given point has direction ratios $\dfrac{\partial z}{\partial x} : \dfrac{\partial z}{\partial y} : -1$, the partial derivatives being evaluated at the point in question.

Example 1. (a) Find the equation of the plane tangent to the paraboloid $48 z = 2 x^2 + 3 y^2$ at the point $(3, 2, \frac{5}{8})$. (b) Find the direction cosines of the normal to the surface at the point.

(a) We have $48 \dfrac{\partial z}{\partial x} = 4 x, \qquad 48 \dfrac{\partial z}{\partial y} = 6 y;$

at the point in question, therefore, $\dfrac{\partial z}{\partial x} = \dfrac{1}{4}, \dfrac{\partial z}{\partial y} = \dfrac{1}{4}$, and the equation of the tangent plane is

$$z - \tfrac{5}{8} = \tfrac{1}{4}(x - 3) + \tfrac{1}{4}(y - 2), \text{ or } 2 x + 2 y - 8 z = 5.$$

(b) To obtain the direction cosines from the ratios $\frac{1}{4} : \frac{1}{4} : -1$, we first compute $[(\tfrac{1}{4})^2 + (\tfrac{1}{4})^2 + (-1)^2]^{1/2} = \dfrac{3\sqrt{2}}{4}.$

The direction cosines are found by dividing $\frac{1}{4}, \frac{1}{4}, -1$ by $\dfrac{3\sqrt{2}}{4}$. They are, accordingly, $\dfrac{1}{3\sqrt{2}}, \dfrac{1}{3\sqrt{2}}, \dfrac{-4}{3\sqrt{2}}.$

Example 2. Show that at every point of intersection of the two surfaces $z = 2(x^2 + y^2)$, $8 z = 17 - (x^2 + y^2)$, the normals to the two surfaces are perpendicular. (Because of this we say that the surfaces intersect *orthogonally*.)

The student will readily find that the surfaces are paraboloids of revolution intersecting each other all along the circle $x^2 + y^2 = 1$ in the plane $z = 2$. There are no other intersections. Now, at a point of the first paraboloid the direction ratios of the normal to the surface are

$$4\,x : 4\,y : -1;$$

for the second paraboloid the direction ratios of the normal are found to be

$$-\frac{x}{4} : -\frac{y}{4} : -1.$$

The condition for perpendicularity of these two normals at a point which is common to the two surfaces is therefore

$$4\,x\left(\frac{-x}{4}\right) + 4\,y\left(\frac{-y}{4}\right) + 1 = 0,$$

or $-(x^2 + y^2) + 1 = 0$. Since this equation is satisfied along the intersection of the surfaces, the demonstration of perpendicularity is complete.

EXERCISES

1. Find the equation of the tangent plane to the surface $z = e^{-x} \sin y$
(a) at $x = 0$, $y = \pi/2$, **(b)** at $x = 0$, $y = \pi$, **(c)** at $x = 0$, $y = 0$. **(d)** Make as good a diagram as you can of the surface for $0 \leqq y \leqq \pi$, $x > 0$.

2. Find the equation of the plane tangent to the surface $x^3 + 2\,xy^2 - 7\,z^3 + 3\,y + 1 = 0$ at $(1, 1, 1)$.

3. Prove that the plane tangent to the surface $z = x^2 - y^2$ at the point (a, b, c) is pierced by the z-axis at the point for which $z = -c$.

4. Find the points of the paraboloid $z = x^2 + y^2 - 1$ at which the normal to the surface coincides with the line joining the origin to the point. What is the acute angle between the normal and the z-axis at these points?

5. If $a^2 \neq b^2$, prove that no normal to the surface $z = (x^2/a^2) + (y^2/b^2) - c$, at a point for which $x \neq 0$ and $y \neq 0$, can pass through the origin.

6. Prove that the spheres $x^2 + y^2 + z^2 = 16$, $x^2 + (y - 5)^2 + z^2 = 9$ intersect orthogonally, using the method of Example 2.

6.3 Maxima and minima

We sometimes have occasion to inquire about the largest or smallest value attained by a function under specified circumstances. In speaking about maximum (or minimum) values it is very important to distinguish between a *relative* maximum and an *absolute* maximum. Suppose we are dealing with a function $f(x, y)$ defined in a region R of the xy-plane.

DEFINITION. We say that the function f has a *relative* maximum at the point (a, b) if there is some neighborhood of (a, b) such that $f(x, y) \leqq f(a, b)$ for all points (x, y) of R which are in this neighborhood. We may express the definition otherwise by saying that the value of f at (a, b) is at least as big as at any of the points (x, y) around (a, b) and not too far away.

Thus, for instance, within a given range of mountains, the elevation of the land surface above sea level attains a relative maximum at the summit of any particular peak in the range.

DEFINITION.   Let f be defined in a region R, and let S be any part of R (i.e., any point-set in R). In particular, S might be all of R. Suppose there is in S a point (a, b) such that $f(x, y) \leqq f(a, b)$ for all points (x, y) in S. We then say that on the set S the function f has an *absolute* maximum at (a, b).

Observe that, on a given set S, f can have a relative maximum which is not an absolute maximum. Observe also that a function may fail to have an absolute maximum on a given set (think of the function $1/(xy)$ in the first quadrant).

Similar definitions are made for relative and absolute minima of a function.

In problems where we have to find the absolute maximum of a function on a given set we usually find that it is convenient to begin by looking for relative maxima. If there are only a few of the latter we may be able easily to select one which furnishes an absolute maximum. Hence it is useful to have criteria for locating relative extrema.

THEOREM I.   *Let f be defined on a region R, and let the function have a relative extreme (maximum or minimum) at the point (a, b) of R. Suppose further that (a, b) is an interior point of R (not on the boundary), and that f has first partial derivatives at (a, b). Then these derivatives are zero at that point:*

$$(6.3\text{--}1) \qquad\qquad f_1(a, b) = 0, \qquad f_2(a, b) = 0.$$

PROOF.   This theorem should be compared with Theorem III of § 1.12. The proof is based on this earlier theorem. Consider $f(x, b)$; this is a function of the single variable x, its values being those of the function $f(x, y)$ along the line $y = b$. As a function of x, $f(x, b)$ has a relative extreme at $x = a$. Moreover, the derivative of $f(x, b)$ at $x = a$ is $f_1(a, b)$. Therefore, by Theorem III, § 1.12, we conclude that $f_1(a, b) = 0$. In the same way, applying this earlier theorem to the function $f(a, y)$ of the single variable y, we conclude that $f_2(a, b) = 0$.

The hypothesis that (a, b) is an interior point of R is essential. A relative extreme can occur at a boundary point of R, and in that case equations (6.3–1) may not hold.

It is important to realize that, under the conditions stated in Theorem I, the vanishing of the first partial derivatives is a necessary, but not sufficient, condition for a relative extreme. If the surface $z = f(x, y)$ has at the point $x = a$, $y = b$ a tangent plane which is parallel to the xy-plane, then equations (6.3–1) hold; but z need not be a relative extreme at such a point. A "saddle-point" of a surface is an illustration of such a situation.

The foregoing definitions and Theorem I extend to functions of three or more variables in an obvious manner.

As in elementary calculus, sufficient conditions for a relative maximum or minimum can be formulated by adding to (6.3–1) certain conditions on the second derivatives of f at the point (a, b). We discuss such conditions in § 7.6. For the present, however, we proceed to illustrate some uses of Theorem I.

Example 1. Find the point of the plane $2\,x - 3\,y - 4\,z = 25$ which is nearest to the point $(3, 2, 1)$.

If D is the distance from the point (x, y, z) of the plane to $(3, 2, 1)$, we have $D^2 = (x - 3)^2 + (y - 2)^2 + (z - 1)^2$ and $z = \frac{1}{4}(2\,x - 3\,y - 25)$. Hence, eliminating z,

$$D^2 = (x - 3)^2 + (y - 2)^2 + (\tfrac{1}{2}\,x - \tfrac{3}{4}\,y - \tfrac{29}{4})^2.$$

We seek the minimum value of D^2 as x, y range through all possible values. In this case *all* points are interior points of the region (namely the whole xy-plane), and D^2 has partial derivatives at all points. We therefore look for points at which $\dfrac{\partial(D^2)}{\partial x} = \dfrac{\partial(D^2)}{\partial y} = 0$. The equations to be considered are

$$2(x - 3) + 2(\tfrac{1}{2}\,x - \tfrac{3}{4}\,y - \tfrac{29}{4}) \cdot \tfrac{1}{2} = 0,$$
$$2(y - 2) + 2(\tfrac{1}{2}\,x - \tfrac{3}{4}\,y - \tfrac{29}{4}) \cdot (\tfrac{-3}{4}) = 0.$$

On simplifying, we obtain

$$10\,x - 3\,y = 53,$$
$$-\,6\,x + 25\,y = -\,55.$$

The solution is found to be $x = 5$, $y = -1$. Substituting in the equation of the plane, we find $z = -3$. We now argue as follows: The function D^2 certainly has an absolute minimum (from the geometrical nature of the problem). This absolute minimum is also a relative minimum, and the conditions of Theorem I apply. But we obtain a unique point at which the two first partial derivatives vanish. Hence, this point must furnish the desired absolute minimum.

Example 2. Locate the points which might furnish relative maxima and minima of the function

$$f(x, y) = 2 xy - (1 - x^2 - y^2)^{3/2}$$

in the closed region $x^2 + y^2 \leq 1$ (which is the region of definition of the function). Hence, find the absolute maximum and minimum values of the function.

We first apply the criterion of Theorem I. We have

$$\frac{\partial f}{\partial x} = 2 y + 3 x(1 - x^2 - y^2)^{1/2},$$

$$\frac{\partial f}{\partial y} = 2 x + 3 y(1 - x^2 - y^2)^{1/2}.$$

The interior points of the region are those for which $x^2 + y^2 < 1$. The interior points which might furnish a relative maximum or minimum are among those which we find by solving the equations

$$
\begin{aligned}
3 \, x(1 - x^2 - y^2)^{1/2} &= -2 \, y, \\
3 \, y(1 - x^2 - y^2)^{1/2} &= -2 \, x.
\end{aligned}
$$

(6.3–2)

An obvious solution of these equations is $x = 0$, $y = 0$. If neither x nor y is zero we may divide one equation by the other and obtain the result

$$\frac{x}{y} = \frac{y}{x} \text{, or } x^2 = y^2.$$

Hence, substituting back in the first equation of (6.3–2) after squaring both sides, we obtain

$$9 \, x^2(1 - 2 \, x^2) = 4 \, x^2, \text{ or } 9 - 18 \, x^2 = 4.$$

Thus we find $x^2 = y^2 = \frac{5}{18}$. Going back again to (6.3–2) we have $(1 - x^2 - y^2)^{1/2} = \frac{2}{3}$, and hence $3 \, x(\frac{2}{3}) = -2 \, y$, or $x = -y$. Note that $x = y$ is ruled out. There are therefore three points in all which satisfy (6.3–2). They are

$$P_0 = (0, 0), \ P_1 = (\tfrac{1}{3}\sqrt{\tfrac{5}{2}}, -\tfrac{1}{3}\sqrt{\tfrac{5}{2}}), \ P_2 = (-\tfrac{1}{3}\sqrt{\tfrac{5}{2}}, \tfrac{1}{3}\sqrt{\tfrac{5}{2}}).$$

The table of values at the right may now be constructed:

Point	Value of f
P_0	-1
P_1 and P_2	$-23/27$

We emphasize that Theorem I does not assert that the function has relative extrema at all three of these points; it only states that if any relative extrema occur at interior points, such extrema are to be found among these three points. Before drawing any conclusions about absolute extrema we must investigate the behavior of the function on the boundary of the region. On the boundary we have $x^2 + y^2 = 1$ and therefore $f(x, y) = 2xy$. To look for extreme values of f on the boundary we might solve for y: $y = \pm\sqrt{1 - x^2}$, and look for the extremes of $\pm 2x\sqrt{1 - x^2}$ by the methods of elementary calculus. We find such extremes when $x^2 = \frac{1}{2}$. A more elegant procedure is to introduce the parametric equations $x = \cos\theta$, $y = \sin\theta$ for the boundary circle (here θ is the usual angle of polar co-ordinates). Then $2xy = 2\cos\theta\sin\theta = \sin 2\theta$, and we see that the values range between

$$-1 \left(\text{at } \theta = \frac{3\pi}{4} \text{ or } \frac{7\pi}{4} \right) \text{ and } +1 \left(\text{at } \theta = \frac{\pi}{4} \text{ or } \frac{5\pi}{4} \right).$$

We have now found four more points which must be considered along with the original three when we look for the absolute minimum and maximum values of the function in the closed region. When we compare the values ± 1 with the values at the points listed in the table, we see that the function has the absolute maximum value $+1$, and the absolute minimum value -1. The maximum occurs at the two boundary points $(\sqrt{2}/2, \sqrt{2}/2)$, $(-\sqrt{2}/2, -\sqrt{2}/2)$. The minimum occurs at the interior point P_0 and at the two boundary points $(\sqrt{2}/2, -\sqrt{2}/2)$, $(-\sqrt{2}/2, \sqrt{2}/2)$. Our work has not settled the question as to whether the interior points P_1, P_2 are points of relative extrema or saddle points. They are in fact saddle points, as may be shown by an examination of the function in polar co-ordinates.

Example 3. A shelter for use at the beach is to be built in the form of a box-like space with canvas covering on the top, back, and ends. If 96 square feet of canvas are available, what should be the dimensions of the shelter to give it maximum cubic content?

Let the shelter be y feet between ends, x feet from front to back, and z feet high. Its volume is $V = xyz$. The area to be covered by canvas is

$$A = 2xz + xy + yz.$$

Since $A = 96$, we can use this last equation to eliminate one variable, say y:

$$y = \frac{96 - 2xz}{x + z}, \quad V = 2xz\frac{48 - xz}{x + z}.$$

Here V is expressed in terms of the independent variables x, z; we can set $\frac{\partial V}{\partial x} = \frac{\partial V}{\partial z} = 0$ and solve for x and z. An alternative procedure which is in some ways preferable is the following: Differentiate both of the equations

$$V = xyz, \quad A = 2xz + xy + yz$$

with respect to x and z, regarding y as a function of x and z. Differentiation with respect to x gives

$$0 = \frac{\partial V}{\partial x} = yz + xz\frac{\partial y}{\partial x}, \quad 0 = 2z + x\frac{\partial y}{\partial x} + y + z\frac{\partial y}{\partial x}.$$

We now eliminate $\dfrac{\partial y}{\partial x}$ between these two equations.

$$\frac{\partial y}{\partial x} = -\frac{y}{x}, \; 0 = 2z + x\left(-\frac{y}{x}\right) + y + z\left(-\frac{y}{x}\right),$$

$$2z - y + y - \frac{yz}{x} = 0, \; 2x = y.$$

Since x and z enter symmetrically, we infer that $2z = y$ also, and hence that $x = z$. To get the values of x, y, z we return to the formula for A. We now have

$$96 = 2x(x) + x(2x) + (2x)x = 6x^2.$$

Hence $x^2 = 16$, $x = z = 4$, $y = 8$. The volume of the shelter of these dimensions is 128 cubic feet.

One logical issue still remains to be settled in the foregoing "solution" of the problem posed in Example 3. How do we know that we really found the dimensions which yield maximum volume? Our method was based on two assumptions: First, that there is a shelter of maximum volume under the given conditions, and second, that when V is expressed as a function of the independent variables x, z, the equations $\dfrac{\partial V}{\partial x} = \dfrac{\partial V}{\partial z}$ $= 0$ are satisfied when V attains its maximum. If we can justify these two assumptions, our solution will be fully established. Let us then consider V as a function of x and z. The formula is

(6.3–3) $$V = 2xz\,\frac{48 - xz}{x + z} \; ;$$

we do not consider all values of x and z, however, but only those which have a meaning for the problem under consideration. Thus we must have $x \geq 0$, $z \geq 0$. We must have $xz \leq 48$ also, since a negative volume would have no meaning. The region R in which we consider the values of V is thus composed of all points of the xz-plane for which $x \geq 0$, $z \geq 0$, $xz \leq 48$, except the one point $x = 0$, $z = 0$. This point is ruled out, since V is not defined there. The region R is shown in Fig. 38.

Let us now establish the fact that among all possible values of V in the region R, there is an absolute maximum value, and that this maximum occurs at an interior point of R. Observe that V is positive in the interior of R and that $V = 0$ at all points of the boundary of R except the origin (where V is not defined). Now, from the fact that $xz \leq 48$ in R we see that

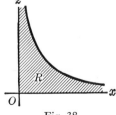

Fig. 38

$$V \leqq 96 \frac{48}{x + z},$$

and hence that $V \to 0$ as a variable point (x, z) of R moves in such a way that either $x \to \infty$ or $z \to \infty$. Finally, $V \to 0$ as $(x, z) \to (0, 0)$. For, $2\,xz \leqq x^2 + z^2$ and $x + z \geqq \sqrt{x^2 + z^2}$ are true inequalities (the latter when x and z are non-negative), and therefore

$$0 \leq V \leqq \frac{96(x^2 + z^2)}{\sqrt{x^2 + z^2}} = 96\sqrt{x^2 + z^2},$$

so that $V \to 0$ as $(x, z) \to (0, 0)$. From the foregoing arguments it is now clear that if we form a new region R_0 by excluding from R the points for which $x^2 + z^2 < \delta^2$ and $x^2 + z^2 > 1/\delta^2$, where δ is sufficiently small, the values of V at these excluded points will all be smaller than *some* of the values of V in the remaining region R_0. Since R_0 is a bounded and closed region in which V is continuous, V must have a maximum value in R_0 (Theorem II, § 5.3). The maximum value of V in R_0 will also be a maximum value of V in relation to all points of the larger region R. Since this maximum is positive, it must occur at an interior point of R.

We now apply Theorem I to draw the conclusion that $\dfrac{\partial V}{\partial x} = \dfrac{\partial V}{\partial z} = 0$ at the point where V is a maximum. Since there turned out to be only one point in R, namely $x = z = 4$, at which these conditions are satisfied, this point must be the point where V is a maximum.

EXERCISES

1. Find the point of the plane $x + 4\,y + 4\,z = 39$ nearest the point $(2, 0, 1)$.

2. Find the greatest value of the function $xy(c - x - y)$ in the closed triangular region with vertices $(0, 0)$, $(c, 0)$, and $(0, c)$. Assume $c > 0$.

3. Find the absolute maximum of $144\,x^3y^2(1 - x - y)$ in the first quadrant of the xy-plane.

4. Does $f(x, y) = x^2 + \frac{2}{3}xy + y^2 + (576/x) + (576/y)$ have any absolute extrema in the region $x > 0$, $y > 0$? If so, find where such extrema occur, and the type (maximum or minimum). Give all the supporting details of your argument.

5. Find the absolute minimum value of

$$f(x, y) = x^2 + y^2 + \left(\frac{2\,A - ax - by}{c}\right)^2$$

where A, a, b, and c are positive constants. All values of x and y are admitted. How do you know that a minimum exists?

6. Find the absolute extreme values of the function $f(x, y) = 2\,xy + (1 - x^2 - y^2)^{1/2}$ in the region $x^2 + y^2 \leqq 1$.

7. Find the absolute extreme values of the function $f(x, y) = xy - (1 - x^2 - y^2)^{3/2}$ in the region $x^2 + y^2 \leqq 1$.

8. (a) Introducing polar co-ordinates, show that the function of Example 2 becomes $f(x, y) = F(r, \theta) = r^2 \sin 2\theta - (1 - r^2)^{3/2}$. (b) Find the extreme values of F for $-1 \leqq r \leqq 1$, θ unrestricted, considering F as defined in a region of the $r\theta$ plane. (c) At the points of the $r\theta$-plane which correspond to the points P_1, P_2 of Example 2, show that $\dfrac{\partial F}{\partial r} = 0$, $\dfrac{\partial F}{\partial \theta} = 0$, $\dfrac{\partial^2 F}{\partial r^2} < 0$, $\dfrac{\partial^2 F}{\partial \theta^2} > 0$. From these facts explain why these points are saddle points and not points of relative extrema.

9. Solve Exercises 6 and 7 by the introduction of polar co-ordinates as independent variables.

10. Find the greatest value of the function $\sin x \sin y \sin (x + y)$ in the closed triangular region with vertices $(0, 0)$, $(\pi, 0)$, $(0, \pi)$.

11. Find the absolute maximum value of the function $(x^2 + 2 y^2)e^{-(x^2+y^2)}$, considering all possible values of x and y.

12. Find the absolute extrema of the function $3 x^2 - 8 xy - 4 y^2 + 2 x + 16 y$ in the square $0 \leqq x \leqq 2$, $0 \leqq y \leqq 2$.

13. Find the absolute extrema of the function $x^3 + y^3 + 3 xy^2 - 15 x - 15 y$ in the square $0 \leqq x \leqq 3$, $0 \leqq y \leqq 3$.

14. Find the minimum value of the function $(12/x) + (18/y) + xy$ in the first quadrant. How do you know there is a minimum?

15. Find the maximum value of the function $(xy - 4 y - 8 x)/x^2y^2$ in the first quadrant. How do you know there is a maximum?

16. A rectangular box without a top has length x, width y, and depth z. The combined area of the sides and bottom is fixed as S square feet. Express the volume V of the box as a function of x, y, and show that V is greatest when $x = y = (S/3)^{1/2}$, $z = x/2$. Justify your solution completely.

17. Consider the function $\sqrt{x^2 + y^2} + \sqrt{(x - 1)^2 + y^2}$. (a) Explain carefully why this function must have an absolute minimum at some point of the plane. (b) What is the minimum and where does it occur? (c) Are all the minimum points found by setting the partial derivatives equal to zero?

18. Consider the function $f(x, y) = |y| + \sqrt{x^2 + (y - 1)^2}$.
(a) At what points do one or both of the first partial derivatives of f fail to exist? (b) Find all points where they both exist and are equal to zero. (c) What is the absolute minimum value of f, and at what points does it occur?

19. For what position of the point (x, y) is the sum of the distance from (x, y) to the x-axis and twice the distance from (x, y) to the point $(0, 1)$ a minimum?

20. Let $f(x, y, z)$ be the sum of the three distances: from (x, y, z) to the y-axis, from (x, y, z) to the z-axis, and from (x, y, z) to the point $(1, 0, 0)$. Find the absolute minimum value of f, and where it occurs.

6.4 Differentials

In § 1.3 we defined the differential of a function of one independent variable. Let us recall the definition. If f has a derivative at the point x on the axis of real numbers, the differential of f at x, which we now denote by df, is by definition given by

$$df = f'(x)dx.$$

Observe that this is a very simple function of the independent variable dx, namely, a multiple of dx. The coefficient of dx is the derivative of f at x. We may also regard the differential as a function of x, and as such, it is defined at each point where the derivative of f exists.

When we come to consider the concept of the differential of a function of two variables, it is natural to look back to the one-variable case for guidance. For a function f of x and y it is plausible to think of defining df as a linear combination of dx and dy, where dx and dy are independent variables. Moreover, by analogy with the one-variable case, we might consider defining df at (x, y) by the formula

$$df = f_1(x, y)dx + f_2(x, y)dy,$$

where, as in § 6.2, we write

$$f_1(x, y) = \frac{\partial f}{\partial x}, \quad f_2(x, y) = \frac{\partial f}{\partial y}.$$

This does turn out to be the right formula to use for the calculation of df, but the formula alone does not give us all the insight we need in understanding the nature of the differential of a function of two variables. We can write this particular linear combination of dx and dy whenever the partial derivatives f_1 and f_2 both exist at the point (x, y). But, as it turns out, the mere existence of these derivatives is not an adequate condition on which to base the definition of the differential. We need something more.

Let us recall some things about differentiability in the one-variable case. Suppose $y = f(x)$. If f is differentiable at x_0, then f is continuous at x_0 (Theorem I, § 1.11). We would like this situation to carry over to the two-variable case. That is, we want a function of x and y to be continuous at (x_0, y_0) if it is differentiable at (x_0, y_0). But the mere fact that $f_1(x_0, y_0)$ and $f_2(x_0, y_0)$ exist is not sufficient to insure that f is continuous at (x_0, y_0). For an illustration of this see Exercise 2.

Going back again to the one-variable case, let us look more closely at the situation. The key to a deeper understanding lies in comparing Δf and df, where

$$\Delta f = f(x + dx) - f(x).$$

We can see that Δf and df both approach 0 as $dx \to 0$. Then, of course, $\Delta f - df$ also approaches 0. But we can say much more than this. As $dx \to 0$, the difference $\Delta f - df$ becomes so small in comparison with dx that the ratio of $|\Delta f - df|$ to $|dx|$ approaches 0. That is,

$$\lim_{dx \to 0} \frac{|\Delta f - df|}{|dx|} = 0.$$

To prove this we write

$$\frac{\Delta f - df}{dx} = \frac{f(x + dx) - f(x) - f'(x)dx}{dx}$$

$$\frac{|\Delta f - df|}{|dx|} = \left| \frac{\Delta f - df}{dx} \right| = \left| \frac{f(x + dx) - f(x)}{dx} - f'(x) \right| .$$

The limit of the last expression, as $dx \to 0$, is clearly 0, by the definition of $f'(x)$.

This aspect of the differential turns out to be crucial. We can restate it as follows: The function f is differentiable at x provided there is some multiple of dx, e.g. $A dx$, such that

$$\lim_{dx \to 0} \frac{|\Delta f - A dx|}{|dx|} = 0.$$

This turns out to be the case if and only if f has a derivative at x, and then we find that the multiplier A must be $f'(x)$.

The foregoing discussion of the one-variable case provides us with the motivation which we need for the definition of the differential of a function of two variables. We shall look for coefficients A, B in the expression

$$A dx + B dy$$

which will, if possible, make the difference

$$\Delta f - (A dx + B dy)$$

small in comparison with $\sqrt{(dx)^2 + (dy)^2}$ when dx and dy are both small.

Here

(6.4–1) $$\Delta f = f(x + dx, y + dy) - f(x, y).$$

We make the definition formally.

DEFINITION. Let f be a function of two variables, defined in some neighborhood of the point (x, y). We call f differentiable at (x, y) if there exist numbers A, B such that

(6.4–2) $$\lim_{(dx, dy) \to (0, 0)} \frac{|\Delta f - (A dx + B dy)|}{\sqrt{dx^2 + dy^2}} = 0.$$

If such numbers A, B do exist, the expression

(6.4-3) $$df = A\,dx + B\,dy$$

is called the differential of f at (x, y).

It is not always easy to tell when numbers A, B satisfying (6.4-2) do exist, but it is not difficult to find out what they must be if they exist. For convenience let us write (h, k) in place of (dx, dy). Then (6.4-2) becomes

$$\lim_{(h,\,k)\to(0,\,0)} \frac{|\,f(x+h, y+k) - f(x, y) - Ah - Bk\,|}{\sqrt{h^2 + k^2}} = 0.$$

Let us see what happens if we choose $k = 0$, $h \neq 0$. We obtain

$$\lim_{h\to 0} \left| \frac{f(x+h, y) - f(x, y) - Ah}{h} \right| = 0,$$

which is the same as saying that

$$\lim_{h\to 0} \frac{f(x+h, y) - f(x, y)}{h} = f_1(x, y)$$

must exist and be equal to A. Similarly, by starting with $h = 0$, $k \neq 0$, we see that $f_2(x, y)$ must exist and be equal to B.

We have now demonstrated that, if f is differentiable at (x, y), the first partial derivatives $f_1(x, y)$, $f_2(x, y)$ exist and the differential of f is given by the formula

(6.4-4) $$df = f_1\,dx + f_2\,dy.$$

Observe that df is a linear combination of dx and dy. We may also regard df as a function of (x, y) if (x, y) varies over some set, at each point of which f is differentiable. We should emphasize that although the differential of f may exist only at certain points (x, y) in the plane, at each point where it does exist it is defined for *all* values of dx and dy. There are functions for which f_1 and f_2 both exist at a point (x, y) and yet the function fails to have a differential there. This means of course that

$$\lim_{(dx,\,dy)\to(0,\,0)} \frac{|\,f(x+dx, y+dy) - f(x, y) - f_1(x, y)dx - f_2(x, y)dy\,|}{\sqrt{(dx)^2 + (dy)^2}}$$

is not zero—in fact, the limit doesn't even exist (Exercise 8). Simple examples of such functions are found in Exercises 2 and 7.

It will be proved later that if the first partial derivatives are defined throughout a neighborhood of the point (a, b), and are continuous at that point, then f is differentiable at (a, b). (See Theorem II, § 7.1.) This simple criterion assures us that most of the functions we ordinarily encounter are differentiable with only isolated exceptional points.

In working with differentials it is important to recognize the equivalence of several different notations. In particular, the differential of f at the point (x_0, y_0) is frequently regarded as a function $A\,dx + B\,dy$ such that

$$\lim_{(dx,\,dy)\to(0,\,0)} \frac{|f(x_0 + dx, y_0 + dy) - f(x_0, y_0) - A\,dx - B\,dy|}{\sqrt{(dx)^2 + (dy)^2}} = 0$$

or as a function $A(x - x_0) + B(y - y_0)$ of $x - x_0$ and $y - y_0$ such that

$$\lim_{(x,\,y)\to(x_0,\,y_0)} \frac{|f(x, y) - f(x_0, y_0) - A(x - x_0) - B(y - y_0)|}{\sqrt{(x - x_0)^2 + (y - y_0)^2}} = 0.$$

The equivalence of these two forms is seen by recognizing x as $x_0 + dx$ and y as $y_0 + dy$. Further variations are encountered where (h, k) or $(\Delta x, \Delta y)$ are used instead of (dx, dy). The real meaning of the differential lies in its close approximation to the increment in a function, and once this is understood, the variations in notation will cease to be confusing.

The approximating property which we used in defining differentials is also the key to extending the notion of tangency from curves to surfaces. The equation of the line tangent to the curve $y = f(x)$ at x_0 is given by

$$y = f(x_0) + f'(x_0)(x - x_0).$$

Notice that $f'(x_0)(x - x_0)$ is the differential of f at x_0, corresponding to the increment $dx = x - x_0$ in the independent variable. The tangent line approximates the curve so closely near $x = x_0$ that not only does

$$|f(x) - [f(x_0) + f'(x_0)(x - x_0)]| \to 0 \text{ as } x \to x_0,$$

but

$$\frac{|f(x) - [f(x_0) + f'(x_0)(x - x_0)]|}{|x - x_0|} \to 0 \text{ as } x \to x_0.$$

This suggests another way of defining what we mean by a tangent line, which is completely equivalent to the usual way. Notice that every non-vertical line through the point $(x_0, f(x_0))$ has the equation

$$y = f(x_0) + c(x - x_0)$$

for some constant c. We could say that the curve $y = f(x)$ has a nonvertical tangent line at $(x_0, f(x_0))$ if and only if there is some number, say c, such that

$$\frac{|f(x) - [f(x_0) + c(x - x_0)]|}{|x - x_0|} \to 0 \text{ as } x \to x_0.$$

If there is such a c, it is unique and must be $f'(x_0)$, so the tangent line given by this definition is just the usual one.

The reason for the foregoing discussion is that it suggests what turns out to be the best way of defining a tangent plane to a surface $z = f(x, y)$

at a point (x_0, y_0). Notice that every nonvertical plane through $(x_0, y_0, f(x_0, y_0))$ (i.e., every plane not parallel to the z-axis) has the equation

$$z = f(x_0, y_0) + A(x - x_0) + B(y - y_0)$$

for proper choice of A and B. We now define the surface to have a nonvertical tangent plane at the point $(x_0, y_0, f(x_0, y_0))$ if and only if there exist numbers, A and B, such that

$$\frac{|f(x, y) - [f(x_0, y_0) + A(x - x_0) + B(y - y_0)]|}{\sqrt{(x - x_0)^2 + (y - y_0)^2}} \to 0$$

as $(x, y) \to (x_0, y_0)$. But this just means that $A(x - x_0) + B(y - y_0)$ has to be the differential of f at (x_0, y_0), so if any such A and B exist we have already shown that they must be $f_1(x_0, y_0)$ and $f_2(x_0, y_0)$, respectively.

Thus, to say that the surface $z = f(x, y)$ has a nonvertical tangent plane at $(x_0, y_0, f(x_0, y_0))$ means the same as saying that the function $f(x, y)$ has a differential at (x_0, y_0), and the equation of the tangent plane can be written as

$$z = f(x_0, y_0) + df,$$

or

$$z = f(x_0, y_0) + f_1(x_0, y_0)(x - x_0) + f_2(x_0, y_0)(y - y_0),$$

just as the equation of the tangent line to the curve $y = f(x)$ at x_0 can be written as

$$y = f(x_0) + df,$$

or

$$y = f(x_0) + f'(x_0)(x - x_0).$$

In each case, of course, the differential is to be evaluated at the point of tangency, as indicated.

In view of the equivalence between f being differentiable and the surface $z = f(x, y)$ having a nonvertical tangent plane, it is not very difficult to understand how a function $f(x, y)$ can possess partial derivatives $f_1(a, b)$, $f_2(a, b)$ and yet fail to be differentiable at (a, b). All that is necessary is to have a surface $z = f(x, y)$ such that it has no tangent plane at $x = a, y = b$, and yet such that the surface is cut by the two planes $x = a, y = b$ in curves which have tangents at the point in question. The cone $z = \sqrt{x^2 + y^2}$ does not have a tangent plane at its vertex (the origin). The paraboloid $z = x^2 + y^2$ has the tangent plane $z = 0$ at the origin. We can readily imagine a surface which coincides with the paraboloid where the latter intersects the planes $x = 0, y = 0$, and which coincides with the cone where the cone intersects the planes $y = \pm x$. Such a surface $z = f(x, y)$ will have $f_1(0, 0) = 0$, $f_2(0, 0) = 0$, but it will not have a tangent plane at the origin (see Fig. 39).

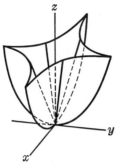

Fig. 39

One such surface is defined by the equation

$$z = (\,|\,x\,| - |\,y\,|\,)^2 + \frac{2\,|\,xy\,|}{\sqrt{x^2 + y^2}} \qquad \text{(with } z = 0 \text{ when } x = y = 0\text{)}.$$

Equivalent forms of the differentiability condition

If we let

$$\frac{f(x_0 + h, y_0 + k) - f(x_0, y_0) - Ah - Bk}{\sqrt{h^2 + k^2}} = \eta(h, k),$$

then to say that $Ah + Bk$ is the differential of f at (x_0, y_0) is the same as saying that

$$\lim_{(h,\,k)\,\to\,(0,\,0)} \eta(h, k) = 0.$$

Therefore, to say that f is differentiable at (x_0, y_0) is the same as saying that there exists a function η of h and k such that

(6.4–5)
$$f(x_0 + h, y_0 + k) - f(x_0, y_0) - f_1(x_0, y_0)h - f_2(x_0, y_0)k = \eta(h, k)\sqrt{h^2 + k^2}$$

and such that

$$\lim_{(h,\,k)\,\to\,(0,\,0)} \eta(h, k) = 0.$$

There is a useful alternative but equivalent form of the differentiability condition, employing $|\,h\,| + |\,k\,|$ instead of $\sqrt{h^2 + k^2}$. First, we should observe the following inequalities (see § 5.2, Exercise 6):

$$\sqrt{h^2 + k^2} \leqq |\,h\,| + |\,k\,| \leqq \sqrt{2}\,\sqrt{h^2 + k^2}\,.$$

It then follows that

$$1 \leqq \frac{|\,h\,| + |\,k\,|}{\sqrt{h^2 + k^2}} \leqq \sqrt{2},$$

and

$$\frac{1}{\sqrt{2}} \leqq \frac{\sqrt{h^2 + k^2}}{|h| + |k|} \leqq 1,$$

provided that $(h, k) \neq (0, 0)$.

Now let us define a function $\epsilon(h, k)$ in terms of $\eta(h, k)$ as follows:

$$\epsilon(h, k) = \eta(h, k) \frac{\sqrt{h^2 + k^2}}{|h| + |k|}.$$

It is clear from the foregoing inequalities that $\epsilon(h, k) \to 0$ as $(h, k) \to (0, 0)$ if and only if $\eta(h, k) \to 0$ as $(h, k) \to 0$. Consequently, in defining the differential of $f(x, y)$ at (x_0, y_0) it is immaterial whether we consider the ratio

$$\frac{f(x_0 + h, y_0 + k) - f(x_0, y_0) - Ah - Bk}{\sqrt{h^2 + k^2}} = \eta(h, k),$$

or the ratio

$$\frac{f(x_0 + h, y_0 + k) - f(x_0, y_0) - Ah - Bk}{|h| + |k|} = \epsilon(h, k).$$

An assertion equivalent to that made in connection with (6.4–5) is the following: f is differentiable at (x_0, y_0) provided there exists a function $\epsilon(h, k)$ and numbers A, B such that

(6.4–6)

$$f(x_0 + h, y_0 + k) - f(x_0, y_0) - Ah - Bk = \epsilon(h, k)(|h| + |k|)$$

when $(h, k) \neq (0, 0)$, and such that

$$\lim_{(h, k) \to (0, 0)} \epsilon(h, k) = 0.$$

In both (6.4–5) and (6.4–6) we can, if we wish, write Δx in place of h and Δy in place of k. As remarked earlier, we know that we must have

$$A = f_1(x_0, y_0), \quad B = f_2(x_0, y_0).$$

If we put $z = f(x, y)$ and $\Delta z = f(a + \Delta x, b + \Delta y) - f(a, b)$, then the preceding results imply the following frequently useful formulas:

(6.4–7) $$\lim_{(\Delta x, \Delta y) \to (0, 0)} \frac{\Delta z - [f_1(a, b) \Delta x + f_2(a, b) \Delta y]}{|\Delta x| + |\Delta y|} = 0,$$

and

(6.4–8) $$\Delta z = \frac{\partial z}{\partial x} \Delta x + \frac{\partial z}{\partial y} \Delta y + \epsilon(|\Delta x| + |\Delta y|),$$

where $\epsilon \to 0$ as $(\Delta x, \Delta y) \to (0, 0)$.

Example 1.   Verify (6.4–7) for $f(x, y) = x^2 y^3$, $a = 3$, $b = 2$. Here $f(a, b) = 72$,

$$f_1(a, b) = (2\ xy^3)_{(3,\ 2)} = 48,$$
$$f_2(a, b) = (3\ x^2 y^2)_{(3,\ 2)} = 108,$$
$$\Delta z = (3 + \Delta x)^2 (2 + \Delta y)^3 - 72.$$

Upon expanding the expression for Δz we find that

$$\Delta z = 48\ \Delta x + 108\ \Delta y + R,$$

where the remainder term R is composed of terms such as $72\ \Delta x\ \Delta y$, $8(\Delta x)^2$, $54(\Delta y)^2$, and terms of still higher degree in Δx and Δy. It is thus intuitively evident (and not difficult to prove rigorously) that

$$\lim_{(\Delta x,\, \Delta y) \to (0,\, 0)} \frac{R}{|\Delta x| + |\Delta y|} = 0;$$

this is precisely the relation (6.4–7) for the present situation.

The extension of the foregoing definitions to functions of three or more independent variables is easily comprehended. For $u = f(x, y, z)$ we have

$$du = \frac{\partial u}{\partial x}\ dx + \frac{\partial u}{\partial y}\ dy + \frac{\partial u}{\partial z}\ dz.$$

In the condition analogous to (6.4–7) we have $|\Delta x| + |\Delta y| + |\Delta z|$ in the denominator; we might also use $[(\Delta x)^2 + (\Delta y)^2 + (\Delta z)^2]^{1/2}$. The formula analogous to (6.4–8) is

(6.4–9)

$$\Delta u = \frac{\partial u}{\partial x}\ \Delta x + \frac{\partial u}{\partial y}\ \Delta y + \frac{\partial u}{\partial z}\ \Delta z + \epsilon(\,|\Delta x| + |\Delta y| + |\Delta z|\,),$$

where $\epsilon \to 0$ as the quantity in the last parenthesis approaches zero.

As a matter of technique in using differentials, it is important to know the standard formulas

$$d(u + v) = du + dv,$$
$$d(uv) = u\ dv + v\ du,$$
$$d\left(\frac{u}{v}\right) = \frac{v\ du - u\ dv}{v^2}.$$

Here u and v may represent differentiable functions of several independent variables. Suppose, for example, that

$$u = f(x, y), \quad v = g(x, y).$$

Then, by definition,

$$du = \frac{\partial u}{\partial x}\ dx + \frac{\partial u}{\partial y}\ dy,$$

$$dv = \frac{\partial v}{\partial x} \, dx + \frac{\partial v}{\partial y} \, dy,$$

$$d(uv) = \frac{\partial(uv)}{\partial x} \, dx + \frac{\partial(uv)}{\partial y} \, dy.$$

But

$$\frac{\partial(uv)}{\partial x} = u \frac{\partial v}{\partial x} + v \frac{\partial u}{\partial x},$$

with a similar formula for $\dfrac{\partial(uv)}{\partial y}$. With these relations before him the student can readily write out the work necessary to verify the relation $d(uv) = u \, dv + v \, du$.

We likewise have such formulas as

$$de^u = e^u \, du,$$

$$d \sin u = \cos u \, du,$$

$$d \tan^{-1} u = \frac{du}{1 + u^2},$$

where u is any differentiable function of several variables. Since these are all exactly the same in appearance as the formulas of elementary calculus, we refrain from presenting a formal list.

All the formulas just considered afford us special instances of the following principle: *A differentiable function of differentiable functions is differentiable.* A fully precise statement and proof of the theorem embodying this principle will be given later (see § 7.3). For the present we observe that it covers such statements as the following:

Let $v = f(x, y)$ and $v = g(x, y)$ be differentiable functions of x, y. Let $F(u, v)$ be a differentiable function of u, v, and let $w = F(f(x, y), g(x, y))$. Then w is a differentiable function of x, y, and

$$dw = \frac{\partial F}{\partial u} \, du + \frac{\partial F}{\partial v} \, dv$$

when the right side of this last equation is expressed entirely in terms of x, y, dx, and dy.

A particular consequence of the principle enunciated in the previous paragraph is that the form of the relation

$$df(x, y) = \frac{\partial f}{\partial x} \, dx + \frac{\partial f}{\partial y} \, dy$$

is valid even when x and y are not independent variables, but are differentiable functions of other variables. This invariance of form is one of the very important properties of differentials.

Example 2. If $z = \log(x^2 + y^2)$, find dz.
We use the formula

$$d \log u = \frac{du}{u}$$

with $u = x^2 + y^2$. The result is

$$dz = \frac{d(x^2 + y^2)}{x^2 + y^2} = \frac{2\,x\,dx + 2\,y\,dy}{x^2 + y^2}\,.$$

The coefficient of dx here is $\dfrac{\partial z}{\partial x}$, and that of dy is $\dfrac{\partial z}{\partial y}$, as the student should verify for himself.

Example 3. Let a, b, c be the sides of a triangle, and let θ be the angle opposite the side c. Regarding c as a function of a, b, and θ, find the differential dc. Use the result to find c approximately when $a = 6.20$, $b = 5.90$, and $\theta = 58°$.

By the law of cosines we have

$$c^2 = a^2 + b^2 - 2\,ab \cos \theta.$$

Hence (using radian measure for θ),

(6.4–10)
$$2\,c\,dc = 2\,a\,da + 2\,b\,db + 2\,ab \sin \theta\,d\theta - 2\,a \cos \theta\,db - 2\,b \cos \theta\,da.$$

This equation permits us to calculate dc when a, b, θ, da, db, $d\theta$ are known. To solve the numerical problem proposed in the example we start from a triangle with $a = 6$, $b = 6$, $\theta = \pi/3$ (60°). Then $\cos \theta = \frac{1}{2}$ and $c = 6$. We are interested in Δc when $\Delta a = 0.20$, $\Delta b = -0.10$, and $\Delta \theta = -\pi/90$ radians (the equivalent of $-2°$). Hence we set $da = 0.20$, $db = -0.10$, $d\theta = -\pi/90$, and use dc as an approximation for Δc. We obtain from (6.4–10), after dividing by 2,

$$6\,dc = 1.20 - 0.60 + 36\left(\frac{\sqrt{3}}{2}\right)\left(-\frac{\pi}{90}\right) + 0.30 - 0.60,$$

$$6\,dc = -0.79, \quad dc = -0.13.$$

Hence the new value of c is approximately $6 - 0.13 = 5.87$. This result does in fact agree with the exact value of c to two decimal places.

We conclude this section with a theorem which we shall find useful later.

THEOREM II. *If a function is differentiable at a point, it is continuous there.*

This theorem is true whether the number of independent variables is one or several. The proof (say for the case of two variables) is immediate from (6.4–8). We see from that formula that $\Delta z \to 0$ as Δx and $\Delta y \to 0$. Taking into account the meaning of Δz, we see that if f is differentiable at (a, b), then

$$\lim_{(\Delta x,\, \Delta y) \to (0,\, 0)} f(a + \Delta x, b + \Delta y) = f(a, b);$$

this is precisely the condition for continuity at (a, b).

EXERCISES

1. Find the differential of $f(x, y)$ in each of the following cases.
(a) $x^4 + y^4 + 6\,xy$ (c) $(x^2 + y^2)/xy$
(b) $e^x y$ (d) $(x^2 + y^2)^{3/2}$

2. Suppose $f(x, y) = \dfrac{xy}{x^2 + y^2}$ for all (x, y) except $(0, 0)$, and let $f(0, 0) = 0$. Show that $f_1(0, 0)$ and $f_2(0, 0)$ both exist with values equal to 0 but that f is discontinuous at $(0, 0)$.

3. (a) Show that

$$x\, d\left(\frac{x}{\sqrt{x^2 + y^2}}\right) + y\, d\left(\frac{y}{\sqrt{x^2 + y^2}}\right) = 0$$

for all values of x and y such that $x^2 + y^2 > 0$, and for all values of dx and dy.
(b) Generalize (a) by proving that $\sum_{k=1}^{n} x_k d\,(x_k/r) = 0$ if $r = (x_1^2 + \cdots + x_n^2)^{1/2}$ and $x_1, \cdots, x_n$ are independent variables.

4. Find dz at $x = 1$, $y = \pi/2$ if $z = x^2y + e^x \sin y$. What is the resulting value of dz if $dx = \pi - e$, $dy = e^2$?

5. If $f(x, y) = (50 - x^2 - y^2)^{1/2}$, find an approximate value of the difference $f(3, 4) - f(2.9, 4.1)$ by use of differentials.

6. If u and v are differentiable functions of x and y, prove the formula

$$d\left(\frac{u}{v}\right) = \frac{v\, du - u\, dv}{v^2},$$

assuming that $v \neq 0$.

7. Let $f(x, y) = \sqrt{|\,xy\,|}$. (a) Verify that $f_1(0, 0) = f_2(0, 0) = 0$. (b) Does the surface $z = f(x, y)$ have a tangent plane at $x = 0$, $y = 0$? Consider the section of the surface made by the plane $x = y$. (c) Verify that (6.4–7) is not satisfied at the point $a = 0$, $b = 0$, and hence that this function is not differentiable at the origin.

8. Assuming that f_1 and f_2 both exist at a point (x, y), prove that either

$$\frac{f(x + dx, y + dy) - f(x, y) - f_1(x, y)\, dx - f_2(x, y)\, dy}{\sqrt{(dx)^2 + (dy)^2}}$$

approaches zero as $(dx, dy) \to (0, 0)$, or else the limit does not exist.

6.5   Composite functions and the chain rule

We use the term *composite function* for a function which is obtained from a given function of one or more variables by substituting other functions in place of the variables in the first function. As an example, consider the function

$$F(x, y) = \sqrt{x^2 + y^2} \ .$$

It is a composite function which may be built up as follows: Let

$$f(u) = u^{1/2}, \ \phi(x, y) = x^2 + y^2.$$

On replacing u by $\phi(x, y)$ we get

$$f(\phi(x, y)) = F(x, y).$$

As another example, consider the function

$$f(x, y) = xy + \log \frac{y}{x} \ .$$

In this function let us replace x and y by certain functions of new variables r, θ, as follows:

$$x = r \cos \theta, \ y = r \sin \theta.$$

The result is a function of r and θ:

$$F(r, \theta) = r^2 \sin \theta \cos \theta + \log \tan \theta.$$

Note that we may also write

$$F(r, \theta) = f(r \cos \theta, r \sin \theta).$$

Most of the functions we deal with are built up as composite functions. We meet the concept of a composite function very early in the study of elementary differential calculus; it is there that we learn the very important rule of differentiation embodied in Theorem II, § 1.11. We are now going to be concerned with extensions of this rule for functions of several variables.

Let us first consider a function of three variables, say

$$(6.5\text{–}1) \qquad\qquad u = F(x, y, z).$$

We are going to suppose that the variables x, y, z are made to depend upon a variable t; let the notation for this dependence be

$$(6.5\text{–}2) \qquad\qquad x = f(t), \ y = g(t), \ z = h(t).$$

On substituting these functions for x, y, z in the function $u = F(x, y, z)$, we obtain u as a composite function of t; if we denote this function by $G(t)$, we have

$$(6.5\text{–}3) \qquad\qquad u = F(f(t), g(t), h(t)) = G(t).$$

The formula of differentiation for this composite function is

(6.5–4) $$\frac{du}{dt} = \frac{\partial u}{\partial x}\frac{dx}{dt} + \frac{\partial u}{\partial y}\frac{dy}{dt} + \frac{\partial u}{\partial z}\frac{dz}{dt}.$$

We shall presently prove this formula under suitable hypotheses. It is very important for the student to learn the structure of this formula; as a part of such learning, he must grasp clearly the role of the variables in the notation of each term in (6.5–4). In the term $\frac{\partial u}{\partial x}$, u and x are related by (6.5–1); u is dependent, and x is one of three independent variables x, y, z. In the term $\frac{dx}{dt}$, x and t are related by the first of equations (6.5–2); x is dependent and t is independent. In the term $\frac{du}{dt}$, u and t are related by (6.5–3); u is dependent and t is independent. An alternative notation for (6.5–4) is

(6.5–5) $$\frac{dG}{dt} = \frac{\partial F}{\partial x}\frac{df}{dt} + \frac{\partial F}{\partial y}\frac{dg}{dt} + \frac{\partial F}{\partial z}\frac{dh}{dt}.$$

This notation is clearer and less subject to misunderstanding. However, both methods of writing the formula are widely used, and the student should become familiar with both of them.

There are other varieties of composite functions in addition to the type presented in (6.5–3). The function F may depend on a different number of variables (e.g., 2, or 4, 5, $\cdots$). Also, the variables x, y, z may depend upon more than one variable. Suppose, for example, that we have

(6.5–6) $$u = F(x, y),$$
$$x = f(s, t), \quad y = g(s, t).$$

Then, under suitable differentiability assumptions, if we write $G(s, t) = F(f(s, t), g(s, t))$, we have the differentiation formulas

(6.5–7) $$\frac{\partial G}{\partial s} = \frac{\partial F}{\partial x}\frac{\partial f}{\partial s} + \frac{\partial F}{\partial y}\frac{\partial g}{\partial s},$$
$$\frac{\partial G}{\partial t} = \frac{\partial F}{\partial x}\frac{\partial f}{\partial t} + \frac{\partial F}{\partial y}\frac{\partial g}{\partial t}.$$

The general rule covering all formulas such as (6.5–5) and (6.5–7) is often called the *chain rule*, or the *composite-function rule*.

To describe the situation generally, let us use the term *first-class variables* for the independent variables on which F depends, and the term *second-class variables* for the variables on which G depends. Observe that G is formed by replacing each first-class variable in F by a function of the second-class variables. In the differentiation formulas such as (6.5–7) we have as many different formulas as there are variables of the *second* class;

each formula has as many terms as there are variables of the *first* class.

We shall now formulate and prove a theorem about formulas such as (6.5–5) or (6.5–7). For simplicity we assume the situation is that represented in (6.5–6).

THEOREM III. *Let $F(x, y)$ be defined in some region R of the xy-plane having the point (a, b) as an interior point, and let F be differentiable at (a, b). Let $f(s, t)$, $g(s, t)$ be defined in some neighborhood of the point (s_0, t_0) in the st-plane. Let these functions admit first partial derivatives at (s_0, t_0), and suppose further that*

$$(6.5–8) \qquad a = f(s_0, t_0), \; b = g(s_0, t_0).$$

Then the composite function $G(s, t) = F(f(s, t), g(s, t))$ has first partial derivatives at (s_0, t_0), given by formulas (6.5–7), where $\dfrac{\partial F}{\partial x}$, $\dfrac{\partial F}{\partial y}$ are evaluated at (a, b), and the partial derivatives with respect to s, t are evaluated at s_0, t_0.

PROOF. We shall prove the formula for $\dfrac{\partial G}{\partial s}$. Since we deal with only one second-class variable at a time, the same reasoning will prove the assertion of the theorem as regards $\dfrac{\partial G}{\partial t}$.

We fix the value of t; then $G(s, t_0)$ is a function of s alone, and

$$G_1(s_0, t_0) = \lim_{\Delta s \to 0} \frac{G(s_0 + \Delta s, t_0) - G(s_0, t_0)}{\Delta s}.$$

For arbitrary Δs let us write

$$\Delta x = f(s_0 + \Delta s, t_0) - f(s_0, t_0),$$
$$\Delta y = g(s_0 + \Delta s, t_0) - g(s_0, t_0).$$

Then by definition,

$$(6.5–9) \qquad \lim_{\Delta s \to 0} \frac{\Delta x}{\Delta s} = f_1(s_0, t_0), \; \lim_{\Delta s \to 0} \frac{\Delta y}{\Delta s} = g_1(s_0, t_0).$$

Finally, let us write

$$\Delta u = F(a + \Delta x, b + \Delta y) - F(a, b).$$

It then follows from (6.5–8) that

$$\Delta u = G(s_0 + \Delta s, t_0) - G(s_0, t_0),$$

and so

$$(6.5–10) \qquad \lim_{\Delta s \to 0} \frac{\Delta u}{\Delta s} = G_1(s_0, t_0).$$

Now, since F is differentiable at (a, b), we can write (see (6.4–8))

$$(6.5\text{–}11) \qquad \Delta u = \frac{\partial F}{\partial x} \Delta x + \frac{\partial F}{\partial y} \Delta y + \epsilon(|\Delta x| + |\Delta y|),$$

it being understood that here

$$\frac{\partial F}{\partial x} = F_1(a, b), \frac{\partial F}{\partial y} = F_2(a, b).$$

We now divide (6.5–11) by Δs and take the limit as $\Delta s \to 0$. Since $\Delta x \to 0$ and $\Delta y \to 0$, it follows from the definition of a differential that $\epsilon \to 0$. Hence in view of (6.5–9) and (6.5–10) we get

$$G_1(s_0, t_0) = F_1(a, b) f_1(s_0, t_0) + F_2(a, b) g_1(s_0, t_0).$$

This is the formula we set out to prove; it is the first of the formulas (6.5–7) in a different notation.

Example 1. If $F(x, y)$ is a given differentiable function of x, y, and if we introduce polar co-ordinates r, θ by writing $x = r \cos \theta$, $y = r \sin \theta$, then F is transformed into a function G of r and θ:

$$(6.5\text{–}12) \qquad G(r, \theta) = F(r \cos \theta, r \sin \theta).$$

Use the chain rule to find $\dfrac{\partial G}{\partial r}$ and $\dfrac{\partial G}{\partial \theta}$ in terms of $\dfrac{\partial F}{\partial x}$ and $\dfrac{\partial F}{\partial y}$.

Here the first-class variables are x, y and the second-class variables are r, θ. We have

$$\frac{\partial x}{\partial r} = \cos \theta, \qquad \frac{\partial y}{\partial r} = \sin \theta,$$

$$\frac{\partial x}{\partial \theta} = -r \sin \theta, \qquad \frac{\partial y}{\partial \theta} = r \cos \theta.$$

The chain rule, in the form (6.5–7) with r, θ replacing s, t, then gives

$$(6.5\text{–}13) \qquad \begin{aligned} \frac{\partial G}{\partial r} &= \frac{\partial F}{\partial x} \cos \theta + \frac{\partial F}{\partial y} \sin \theta, \\ \frac{\partial G}{\partial \theta} &= -\frac{\partial F}{\partial x} r \sin \theta + \frac{\partial F}{\partial y} r \cos \theta. \end{aligned}$$

To emphasize the meaning of these formulas, let us evaluate each part of them for $r = 1$, $\theta = \pi/6$. The corresponding values of x, y are $x = \sqrt{3}/2$, $y = \frac{1}{2}$. Then, using the subscript notation for partial derivatives, equations (6.5–13) become

$$(6.5\text{–}14) \qquad \begin{aligned} G_1\left(1, \frac{\pi}{6}\right) &= F_1\left(\frac{\sqrt{3}}{2}, \frac{1}{2}\right) \frac{\sqrt{3}}{2} + F_2\left(\frac{\sqrt{3}}{2}, \frac{1}{2}\right) \frac{1}{2}, \\ G_2\left(1, \frac{\pi}{6}\right) &= -F_1\left(\frac{\sqrt{3}}{2}, \frac{1}{2}\right) \frac{1}{2} + F_2\left(\frac{\sqrt{3}}{2}, \frac{1}{2}\right) \frac{\sqrt{3}}{2}. \end{aligned}$$

Thorough familiarity with the chain rule is of great importance to any student who wants to read books and journals in which mathematical methods are used extensively. Adequate training in the use of the chain rule must include, among other things, stress on the idea of transforming a given function of one set of variables into a new function of another set of variables, with a resulting set of formulas relating the two sets of partial derivatives of these functions. Ideas of this sort are constantly used in the theory of partial differential equations, and such ideas are at the very root of tensor analysis, which is an important subject beyond the scope of this book. Tensor analysis is an outgrowth and generalization of vector analysis. A brief discussion of some of the fundamental concepts of Euclidean vector analysis is given in Chapter X. The chain rule plays an important part in the discussion of such concepts as *gradient*, *divergence*, and *curl* in vector analysis.

Matters of notation play a considerable role in connection with the chain rule. Wide varieties of usage exist in mathematical writing where the chain rule is concerned. Consider the situation expressed in (6.5–6). Instead of writing the chain rule here in the form (6.5–7), it is frequently written

$$(6.5\text{–}15) \qquad \frac{\partial u}{\partial s} = \frac{\partial u}{\partial x}\frac{\partial x}{\partial s} + \frac{\partial u}{\partial y}\frac{\partial y}{\partial s},$$

with a similar formula in which t replaces s. The difference between (6.5–7) and (6.5–15) is that in (6.5–15) we have used *dependent variables* in place of *functional symbols*. The student must bear in mind some of the subtleties of distinction between a dependent variable and a function. If we write $u = F(x, y)$, u denotes the value of F at the point (x, y). If $x = f(s, t)$ and $y = g(s, t)$, and if $F(x, y)$ is thereby transformed into $G(s, t)$, we may also write $u = G(s, t)$, since the value of G at (s, t) is the same as the value of F at (x, y) when x and y are the values of f and g, respectively, at (s, t). But though we may write $u = F(x, y) = G(s, t)$, this does not mean that F and G are the same function; in general they are not.

Example 2.   If F is a differentiable function of two variables, and

$$u = F(s^2 - t^2, t^2 - s^2),$$

show that

$$(6.5\text{–}16) \qquad t\,\frac{\partial u}{\partial s} + s\,\frac{\partial u}{\partial t} = 0.$$

One method of handling this problem is to introduce new variables

$$x = s^2 - t^2,\, y = t^2 - s^2,$$

so that $u = F(x, y)$. Then, by the chain rule,

$$\frac{\partial u}{\partial s} = \frac{\partial u}{\partial x}(2\,s) + \frac{\partial u}{\partial y}(-\,2\,s),$$

$$\frac{\partial u}{\partial t} = \frac{\partial u}{\partial x}(-\,2\,t) + \frac{\partial u}{\partial y}(2\,t).$$

The correctness of (6.5–16) is now readily deduced.

It sometimes happens that a problem involves several variables and several relations between them. Consider, for example, the four variables V, S, r, h connected by the two equations

(6.5–17) $V = \pi r^2 h, \; S = 2\,\pi r h + 2\,\pi r^2.$

These formulas arise if we consider a right circular cylinder of height h, radius of base r, volume V, and total surface area S. Of the four variables, just two are independent. Ordinarily we most naturally think of r and h as independent, but other choices are legitimate. We may choose r and S as independent. Then

(6.5–18) $h = \dfrac{S}{2\,\pi r} - r, \; V = \tfrac{1}{2}\,rS - \pi r^3.$

In view of what has been said, it is evident that a notation such as $\dfrac{\partial V}{\partial r}$ is ambiguous, for if we calculate from (6.5–17) we have $\dfrac{\partial V}{\partial r} = 2\,\pi r h$, while if we calculate from (6.5–18) we have $\dfrac{\partial V}{\partial r} = \tfrac{1}{2}\,S - 3\,\pi r^2$, and these two results are not in agreement. What is needed is a notation that makes clear the choice of the independent variables. A customary notation employs subscripts. According to this practice,

$$\left(\frac{\partial V}{\partial r}\right)_h$$

indicates that we are regarding V as a function of the independent variables r, h, with h held constant in the differentiation. With this notation we have

$$\left(\frac{\partial V}{\partial h}\right)_r = \pi r^2, \; \left(\frac{\partial V}{\partial S}\right)_r = \tfrac{1}{2}\,r, \; \left(\frac{\partial S}{\partial h}\right)_r = 2\,\pi r,$$

and so on.

Situations like this arise in thermodynamics, for example, with different variables, of course, and different relations between them.

EXERCISES

In all these exercises it is assumed without explicit reference that all the functions introduced are differentiable.

1. If w is a function of p, q, r and each of these latter variables is a function of s, write the chain-rule formula for $\dfrac{dw}{ds}$.

2. If $u = x^2 - y^2$ and $v = 2\,xy$ transform $F(u, v)$ into $G(x, y)$, find $\dfrac{\partial G}{\partial x}$ and $\dfrac{\partial G}{\partial y}$ in terms of $\dfrac{\partial F}{\partial u}$ and $\dfrac{\partial F}{\partial v}$.

3. Define the composite function G and write the formula or formulas of the chain rule in each of the following situations. Identify the first- and second-class variables in each case.
(a) $u = F(x, y, z)$, $x = f(p, q)$, $y = g(p, q)$, $z = h(p, q)$.
(b) $u = F(w)$, $w = \phi(x, y)$.
(c) $u = F(x, y, z, \theta)$, $x = f(t)$, $y = g(t)$, $z = h(t)$, $\theta = t$.

4. If $x = u$, $y = u + v$ and $z = u + v + w$ transform $F(x, y, z)$ into $G(u, v, w)$, find each of the first partial derivatives of G in terms of the first partial derivatives of F.

5. Suppose u depends on x_1, x_2, x_3 and the x's depend on ξ_1, ξ_2 as follows:
$$x_1 = a_{11}\xi_1 + a_{12}\xi_2$$
$$x_2 = a_{21}\xi_1 + a_{22}\xi_2$$
$$x_3 = a_{31}\xi_1 + a_{32}\xi_2$$
where the a's are constants. Write the formulas connecting $\dfrac{\partial u}{\partial \xi_1}$ and $\dfrac{\partial u}{\partial \xi_2}$ with $\dfrac{\partial u}{\partial x_1}$, $\dfrac{\partial u}{\partial x_2}$, and $\dfrac{\partial u}{\partial x_3}$. Generalize for the case of n x's and m ξ's.

6. If $w = F(a + \alpha t, b + \beta t, c + \gamma t)$, where a, b, c and α, β, γ are constants, write a formula for $\dfrac{dw}{dt}$ involving $F_1(a + \alpha t, b + \beta t, c + \gamma t)$ and other similar expressions.

7. If $u = f(x - y, y - x)$, prove that $\dfrac{\partial u}{\partial x} + \dfrac{\partial u}{\partial y} = 0$.

8. If $u = F\left(\dfrac{y - x}{xy}, \dfrac{z - x}{xz}\right)$, prove that $x^2 \dfrac{\partial u}{\partial x} + y^2 \dfrac{\partial u}{\partial y} + z^2 \dfrac{\partial u}{\partial z} = 0$.

9. If $u = x^3 F\left(\dfrac{y}{x}, \dfrac{z}{x}\right)$, prove that $x \dfrac{\partial u}{\partial x} + y \dfrac{\partial u}{\partial y} + z \dfrac{\partial u}{\partial z} = 3\,u$.

10. Let $f(x, y)$ be transformed into $g(u, v)$ by $x = u \cos \theta - v \sin \theta$, $y = u \sin \theta + v \cos \theta$, where θ is constant. Show that
$$\left(\frac{\partial g}{\partial u}\right)^2 + \left(\frac{\partial g}{\partial v}\right)^2 = \left(\frac{\partial f}{\partial x}\right)^2 + \left(\frac{\partial f}{\partial y}\right)^2$$
is an identity in u and v.

11. If $w = f(x, y)$ and $x = r \cosh \theta$, $y = r \sinh \theta$, find $\left(\dfrac{\partial w}{\partial r}\right)^2 - \dfrac{1}{r^2}\left(\dfrac{\partial w}{\partial \theta}\right)^2$ in

terms of $\dfrac{\partial w}{\partial x}$ and $\dfrac{\partial w}{\partial y}$.

12. Let $F(x, t) = f(x + 2t) + f(3x - 2t)$. Set $u = x + 2t$, $v = 3x - 2t$, and show by the chain rule that

$$F_1(x, t) = f'(u) + 3f'(v),$$
$$F_2(x, t) = 2f'(u) - 2f'(v).$$

Hence show that $F_1(0, 0) = 4f'(0)$, $F_2(0, 0) = 0$.

13. Let $z = yf(x^2 - y^2)$. Show that $y\dfrac{\partial z}{\partial x} + x\dfrac{\partial z}{\partial y} = \dfrac{xz}{y}$.

14. If $z = xy + xF\left(\dfrac{y}{x}\right)$, show that $x\dfrac{\partial z}{\partial x} + y\dfrac{\partial z}{\partial y} = xy + z$.

15. Suppose u is a function of r and $r = (x^2 + y^2 + z^2)^{1/2}$. Show that
$$\left(\dfrac{\partial u}{\partial x}\right)^2 + \left(\dfrac{\partial u}{\partial y}\right)^2 + \left(\dfrac{\partial u}{\partial z}\right)^2 = \left(\dfrac{du}{dr}\right)^2.$$

16. Consider the determinant

$$D = \begin{vmatrix} a_{11} & a_{12} & a_{13} \\ a_{21} & a_{22} & a_{23} \\ a_{31} & a_{32} & a_{33} \end{vmatrix} .$$

Let A_{ij} be the cofactor (appropriately signed minor) of a_{ij} in D. If the nine elements of D are regarded as independent variables, show that $\dfrac{\partial D}{\partial a_{ij}} = A_{ij}$.
Now suppose that the elements a_{ij} are differentiable functions of x, and let $a'_{ij} = \dfrac{d}{dx}(a_{ij})$. Show by the chain rule that

$$\dfrac{dD}{dx} = \begin{vmatrix} a'_{11} & a'_{12} & a'_{13} \\ a_{21} & a_{22} & a_{23} \\ a_{31} & a_{32} & a_{33} \end{vmatrix} + \begin{vmatrix} a_{11} & a_{12} & a_{13} \\ a'_{21} & a'_{22} & a'_{23} \\ a_{31} & a_{32} & a_{33} \end{vmatrix} + \begin{vmatrix} a_{11} & a_{12} & a_{13} \\ a_{21} & a_{22} & a_{23} \\ a'_{31} & a'_{32} & a'_{33} \end{vmatrix} .$$

17. Generalize the results of Exercise 16 for determinants of order n. State the final result as a verbal rule.

18. If

$$F(x, y, z) = \begin{vmatrix} f_1(x) & f_2(x) & f_3(x) \\ f_1(y) & f_2(y) & f_3(y) \\ f_1(z) & f_2(z) & f_3(z) \end{vmatrix} ,$$

calculate $\dfrac{\partial F}{\partial x}$, $\dfrac{\partial F}{\partial y}$, and $\dfrac{\partial F}{\partial z}$ by the results of Exercise 16.

19. Write out in detail and prove the version of Theorem III that corresponds to (6.5–1), (6.5–2), and (6.5–5).

20. In Example 2 let $G(s, t) = F(s^2 - t^2, t^2 - s^2)$. Formula (6.5–16) now becomes $tG_1(s, t) + sG_2(s, t) = 0$. Rewrite the chain-rule formulas in the solution of Example 2 without using the letters x, y, u.

21. From the situation expressed by (6.5–17) find $\left(\dfrac{\partial S}{\partial V}\right)_h$, $\left(\dfrac{\partial h}{\partial V}\right)_r$, $\left(\dfrac{\partial S}{\partial V}\right)_r$, $\left(\dfrac{\partial h}{\partial r}\right)_V$ and $\left(\dfrac{\partial r}{\partial S}\right)_h$.

22. (a) If $u = x^2 + y^2 + z^2$ and $z = xyv$, how many meanings are there for $\dfrac{\partial u}{\partial y}$?

(b) Find $\dfrac{\partial u}{\partial y}$ in each of its meanings, indicating the meaning in each case by proper subscripts.

23. (a) If $V = xyz$ and $S = xy + 2\,xz + 2\,yz$, find $\dfrac{\partial V}{\partial x}$ in each of its possible meanings, indicating the meaning in each case by proper subscripts. **(b)** Find

$$\left(\frac{\partial S}{\partial x}\right)_{y,\,V}, \quad \left(\frac{\partial S}{\partial V}\right)_{x,\,y} \quad \text{and} \quad \left(\frac{\partial V}{\partial S}\right)_{x,\,y}.$$

24. If $u = F(x, y)$ and $y = f(x)$, show that $\dfrac{du}{dx} = \dfrac{\partial u}{\partial x} + \dfrac{\partial u}{\partial y}\dfrac{dy}{dx}$. Explain carefully the difference between $\dfrac{\partial u}{\partial x}$ and $\dfrac{du}{dx}$.

25. If $G(x, y) = F(x, y, f(x, y))$, show that $G_1(x, y) = F_1(x, y, z) + F_3(x, y, z)\,f_1(x, y)$, where $z = f(x, y)$. Classify the variables and show that this is an instance of the chain rule. Write the corresponding formula for $G_2(x, y)$.

6.51   An application in fluid kinematics

A very good illustration of the occurrence of formulas such as (6.5–4) or (6.5–5) in physics is furnished in the very beginning of the study of the flow of a fluid.

To build up a mathematical model of an arbitrary fluid in motion, let us fix upon a certain portion of the fluid at a certain instant of time, say the fluid occupying a region R_0 at the time $t = 0$. Then, selecting an arbitrary particle of the fluid in R_0, we follow its motion as t increases. Each particle of the fluid will occupy a certain point of space at a certain time. At a general instant t the portion of the fluid which occupied R_0 at $t = 0$ will occupy a new region R; the particle which was at the point (x_0, y_0, z_0) of R_0 will have moved to a point (x, y, z) of R (see Fig. 40). We assume that the law of motion of the fluid is a definite

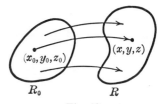

Fig. 40

(but perhaps very complicated) law, so that the co-ordinates (x, y, z) of the particle in R are determined by (i.e., are functions of) x_0, y_0, z_0, t, say

$$x = f(x_0, y_0, z_0, t),$$
(6.51–1)
$$y = g(x_0, y_0, z_0, t),$$
$$z = h(x_0, y_0, z_0, t).$$

These equations define the flow of the fluid for time subsequent to $t = 0$. In what follows we shall regard the functions f, g, h as known without specifying them in any particular manner.

The velocity of an individual particle at a given instant has components

$$\frac{\partial x}{\partial t}, \frac{\partial y}{\partial t}, \frac{\partial z}{\partial t}.$$

The velocity is a vector quantity; it is tangent to the path which is being followed by the particle under consideration.

Now let us consider the density of the fluid. If the mass of a substance is not distributed uniformly throughout its volume, it has variable density, and we must speak of the density *at a point*. Let us then consider the fluid in the region R at time t; let ρ be the density at the point x, y, z. Then ρ is a function of x, y, z. Conceivably ρ might be constant, but it need not be so. (The fluid might be the atmosphere, and R might be a region extending from sea level up to a height of 30,000 feet.) It is important to note that ρ does not depend merely on x, y, z. It is also a function of t, for the fluid which occupies a given region at time t need not have the same mass distribution as the fluid which occupies that region at some other time. Let us write

(6.51–2)
$$\rho = F(x, y, z, t).$$

If in this equation we keep t fixed and vary the point (x, y, z), we obtain the density of the fluid in R at time t. On the other hand, if we fix the point (x, y, z) and vary t, then we obtain the density at (x, y, z) of the fluid occupying R at different times.

There is still another important way of examining the variability of ρ. Imagine an observer moving along with a particle of the fluid. If he were capable of measuring ρ at his own position at any time, by what equation would he describe his observations? For him ρ is the composite function obtained by substituting x, y, z from (6.51–1) in (6.51–2). The result is that ρ is a function of x_0, y_0, z_0, t. Since the observer stays with the same particle always, x_0, y_0, z_0 are to be considered as constants, and ρ becomes a function of t alone, say

(6.51–3)
$$\rho = G(t).$$

Now let us ask the question: From the point of view of the observer moving with the particle, what is the rate of change of ρ? The answer is, of course, the value of the derivative $G'(t)$. Since G is a composite function we may express $G'(t)$ by the chain rule. We see in (6.51–2) that F depends on four variables, x, y, z, t. These are *first-class* variables. By equations (6.51–1) the variables x, y, z depend on other variables x_0, y_0, z_0, t. These are *second-class* variables. Note that t is both a first-class variable and a second-class variable. One way out of the perplexity which the student may at first experience in this situation may be found by an artifice of notation. Let us temporarily introduce a different letter, say θ, to stand for t in the set of first-class variables. We then write $\rho = F(x, y, z, \theta)$ and add the equation $\theta = t$ to the set (6.51–1). The chain rule then gives

$$\frac{dG}{dt} = \frac{\partial F}{\partial x}\frac{\partial f}{\partial t} + \frac{\partial F}{\partial y}\frac{\partial g}{\partial t} + \frac{\partial F}{\partial z}\frac{\partial h}{\partial t} + \frac{\partial F}{\partial \theta}\frac{\partial \theta}{\partial t}.$$

Having obtained the formula, we may forget the artifice, and write t again in place of θ. Note that $\dfrac{\partial \theta}{\partial t} = 1$. It is more usual, in the literature of this subject, to write the letters ρ, x, y, z instead of the functional symbols for these variables. When this is done we have

$$(6.51\text{–}4) \qquad \frac{d\rho}{dt} = \frac{\partial \rho}{\partial x}\frac{\partial x}{\partial t} + \frac{\partial \rho}{\partial y}\frac{\partial y}{\partial t} + \frac{\partial \rho}{\partial z}\frac{\partial z}{\partial t} + \frac{\partial \rho}{\partial t}.$$

The derivative of ρ with respect to t occurs here in two different forms, with different meanings.

EXERCISES

1. Let the fluid flow be the movement of the atmosphere. From the point of view of an observer standing on a street corner with the wind rushing by, what is the expression for the rate of change of density of air on the street corner? If the observer were to ride in the gondola of a freely-drifting balloon, what would be the expression for the rate of change of density of air at the gondola?

2. Suppose the atmosphere over a great plain is in a state such that at any given time ρ is a function only of altitude above the plain. Suppose that a free balloon is drifting along at a constant elevation of 500 feet. Explain why an observer carried by the balloon finds that $\dfrac{d\rho}{dt} = \dfrac{\partial \rho}{\partial t}$.

3. Suppose that, in (6.51–1), z is independent of t, and that
$$\rho = H(x^2 + y^2 + z^2, t).$$
Show that $\dfrac{d\rho}{dt} = \dfrac{\partial \rho}{\partial t}$ at all points on the z-axis.

6.52 Second derivatives by the chain rule

The purpose of this section is to illustrate by examples the use of the chain rule in dealing with second derivatives.

Example 1. If $F(x, y)$ is transformed into $G(r, \theta)$ by the equations $x = r \cos \theta$, $y = r \sin \theta$, find $\dfrac{\partial^2 G}{\partial r\, \partial \theta}$ in terms of derivatives of F with respect to x and y.

We start from the work already done in Example 1, § 6.5. From (6.5–13) we have

$$\frac{\partial G}{\partial \theta} = -\frac{\partial F}{\partial x} r \sin \theta + \frac{\partial F}{\partial y} r \cos \theta.$$

We now differentiate both sides of this equation with respect to r, bearing in mind that $\dfrac{\partial F}{\partial x}$ and $\dfrac{\partial F}{\partial y}$ depend on x and y, and are thus to be regarded as composite functions of r and θ. We have

(6.52–1)
$$\frac{\partial^2 G}{\partial r\, \partial \theta} = -\frac{\partial F}{\partial x} \frac{\partial}{\partial r} (r \sin \theta) - \frac{\partial}{\partial r}\left(\frac{\partial F}{\partial x}\right) r \sin \theta$$
$$+ \frac{\partial F}{\partial y} \frac{\partial}{\partial r} (r \cos \theta) + \frac{\partial}{\partial r}\left(\frac{\partial F}{\partial y}\right) r \cos \theta.$$

The problem now facing us is that of doing something further with expressions like $\dfrac{\partial}{\partial r}\left(\dfrac{\partial F}{\partial x}\right)$. Now $\dfrac{\partial F}{\partial x} = F_1(x, y)$. What we really have before us is the problem of finding $\dfrac{\partial H}{\partial r}$, where $H(r, \theta) = F_1(r \cos \theta, r \sin \theta)$. This is a problem exactly like that of finding $\dfrac{\partial G}{\partial r}$, except that we have F_1 in place of F, and H in place of G. Thus, just as in (6.5–13),

$$\frac{\partial H}{\partial r} = \frac{\partial F_1}{\partial x} \cos \theta + \frac{\partial F_1}{\partial y} \sin \theta.$$

This may be written

(6.52–2)
$$\frac{\partial}{\partial r}\left(\frac{\partial F}{\partial x}\right) = \frac{\partial^2 F}{\partial x^2} \cos \theta + \frac{\partial^2 F}{\partial y\, \partial x} \sin \theta.$$

In a similar way we find

(6.52–3)
$$\frac{\partial}{\partial r}\left(\frac{\partial F}{\partial y}\right) = \frac{\partial^2 F}{\partial x\, \partial y} \cos \theta + \frac{\partial^2 F}{\partial y^2} \sin \theta.$$

If we now use (6.52–2) and (6.52–3) in (6.52–1), we obtain

(6.52–4)
$$\frac{\partial^2 G}{\partial r\, \partial \theta} = -\frac{\partial F}{\partial x} \sin \theta - \left(\frac{\partial^2 F}{\partial x^2} \cos \theta + \frac{\partial^2 F}{\partial y\, \partial x} \sin \theta\right) r \sin \theta$$
$$+ \frac{\partial F}{\partial y} \cos \theta + \left(\frac{\partial^2 F}{\partial x\, \partial y} \cos \theta + \frac{\partial^2 F}{\partial y^2} \sin \theta\right) r \cos \theta.$$

In order to get this far we naturally assume that F and its partial derivatives F_1, F_2 are differentiable functions of x, y. Under this assumption it is true that $\dfrac{\partial^2 F}{\partial y\, \partial x} = \dfrac{\partial^2 F}{\partial x\, \partial y}$ (see Theorem IV, § 7.2). Therefore (6.52–4) may be written in a slightly more compact form. If we write $u = F(x, y) = G(r, \theta)$, (6.52–4) becomes

$$(6.52\text{–}5) \quad \frac{\partial^2 u}{\partial r\, \partial \theta} = - r \sin \theta \cos \theta \, \frac{\partial^2 u}{\partial x^2} + r(\cos^2 \theta - \sin^2 \theta) \frac{\partial^2 u}{\partial x\, \partial y}$$

$$+ r \sin \theta \cos \theta \, \frac{\partial^2 u}{\partial y^2} - \sin \theta \, \frac{\partial u}{\partial x} + \cos \theta \, \frac{\partial u}{\partial y} \, .$$

The work of finding formulas for $\dfrac{\partial^2 u}{\partial r^2}$ and $\dfrac{\partial^2 u}{\partial \theta^2}$ is similar.

Example 2. If $u = F(x, y)$ becomes $G(s, t)$ when we set

$$x = s + t, \, y = s - t,$$

find what $\dfrac{\partial^2 u}{\partial x^2} - \dfrac{\partial^2 u}{\partial y^2}$ becomes in terms of derivatives with respect to s and t.

For this purpose we express s and t in terms of x and y; we regard s, t as first-class variables and x, y as second-class variables. We have

$$s = \tfrac{1}{2}(x + y), \, t = \tfrac{1}{2}(x - y).$$

Then

$$(6.52\text{–}6) \qquad \frac{\partial u}{\partial x} = \frac{\partial u}{\partial s} \frac{\partial s}{\partial x} + \frac{\partial u}{\partial t} \frac{\partial t}{\partial x} = \frac{1}{2} \frac{\partial u}{\partial s} + \frac{1}{2} \frac{\partial u}{\partial t} \, .$$

Likewise,

$$(6.52\text{–}7) \qquad \frac{\partial u}{\partial y} = \frac{1}{2} \frac{\partial u}{\partial s} - \frac{1}{2} \frac{\partial u}{\partial t} \, .$$

Differentiating again,

$$\frac{\partial^2 u}{\partial x^2} = \frac{1}{2} \frac{\partial}{\partial x} \left(\frac{\partial u}{\partial s} \right) + \frac{1}{2} \frac{\partial}{\partial x} \left(\frac{\partial u}{\partial t} \right).$$

Now, by (6.52–6), with $\dfrac{\partial u}{\partial s}$ in place of u,

$$\frac{\partial}{\partial x} \left(\frac{\partial u}{\partial s} \right) = \frac{1}{2} \frac{\partial}{\partial s} \left(\frac{\partial u}{\partial s} \right) + \frac{1}{2} \frac{\partial}{\partial t} \left(\frac{\partial u}{\partial s} \right).$$

We find $\dfrac{\partial}{\partial x} \left(\dfrac{\partial u}{\partial t} \right)$ in the same way. Thus

$$(6.52\text{-}8) \quad \frac{\partial^2 u}{\partial x^2} = \frac{1}{4} \frac{\partial^2 u}{\partial s^2} + \frac{1}{4} \frac{\partial^2 u}{\partial t \, \partial s} + \frac{1}{4} \frac{\partial^2 u}{\partial s \, \partial t} + \frac{1}{4} \frac{\partial^2 u}{\partial t^2}.$$

We shall assume enough about u to insure that $\dfrac{\partial^2 u}{\partial t \, \partial s} = \dfrac{\partial^2 u}{\partial s \, \partial t}$. The student should show for himself that

$$(6.52\text{-}9) \quad \frac{\partial^2 u}{\partial y^2} = \frac{1}{4} \frac{\partial^2 u}{\partial s^2} - \frac{1}{2} \frac{\partial^2 u}{\partial t \, \partial s} + \frac{1}{4} \frac{\partial^2 u}{\partial t^2}.$$

Subtraction of (6.52–9) from (6.52–8) now gives the result

$$(6.52\text{-}10) \quad \frac{\partial^2 u}{\partial x^2} - \frac{\partial^2 u}{\partial y^2} = \frac{\partial^2 u}{\partial t \, \partial s}.$$

The basic point to be noted in using the chain rule in connection with second derivatives is this: Suppose u is a function of first-class variables x, y, $\cdots$ and of second-class variables s, t, $\cdots$. Then, if we have written down a chain-rule formula for one of the derivatives $\dfrac{\partial u}{\partial s}$, $\dfrac{\partial u}{\partial t}$, $\cdots$, this same formula is valid if we replace u throughout by any one of the derivatives $\dfrac{\partial u}{\partial x}$, $\dfrac{\partial u}{\partial y}$, $\cdots$. We are thus able to express symbols like $\dfrac{\partial}{\partial s} \left(\dfrac{\partial u}{\partial x} \right)$ entirely in terms of second derivatives of u with respect to the first-class variables x, y, $\cdots$.

EXERCISES

1. Find formulas comparable to (6.52–4) for $\dfrac{\partial^2 G}{\partial r^2}$ and $\dfrac{\partial^2 G}{\partial \theta^2}$ in the problem of Example 1. Hence show that

$$\frac{\partial^2 G}{\partial r^2} + \frac{1}{r} \frac{\partial G}{\partial r} + \frac{1}{r^2} \frac{\partial^2 G}{\partial \theta^2} = \frac{\partial^2 F}{\partial x^2} + \frac{\partial^2 F}{\partial y^2}.$$

2. If $u = F(x, y)$ becomes $G(r, \theta)$ when $x = r \cosh \theta$, $y = r \sinh \theta$, show that

$$\frac{\partial^2 u}{\partial r^2} + \frac{1}{r} \frac{\partial u}{\partial r} - \frac{1}{r^2} \frac{\partial^2 u}{\partial \theta^2} = \frac{\partial^2 u}{\partial x^2} - \frac{\partial^2 u}{\partial y^2}.$$

3. If $G(s, t) = F(e^s \cos t, e^s \sin t)$, show that $G_{11} + G_{22} = e^{2s}(F_{11} + F_{22})$, where G_{11} and G_{22} are evaluated at (s, t) and F_{11} and F_{22} are evaluated at $(e^s \cos t, e^s \sin t)$.

4. Give a complete proof of (6.52–9).

5. If $\xi = x + ct$, $\eta = x - ct$, and $u = F(x, t) = G(\xi, \eta)$, find $\dfrac{\partial^2 u}{\partial t^2} - c^2 \dfrac{\partial^2 u}{\partial x^2}$ in terms of derivatives with respect to ξ and η.

6. Show that $5 \dfrac{\partial^2 u}{\partial x^2} + 2 \dfrac{\partial^2 u}{\partial x \, \partial y} + 2 \dfrac{\partial^2 u}{\partial y^2}$ becomes $\dfrac{\partial^2 u}{\partial \xi^2} + \dfrac{\partial^2 u}{\partial \eta^2}$ if we set $\xi = \frac{1}{3}(x + y)$, $\eta = \frac{1}{3}(x - 2y)$.

7. If $u = \phi(x - ct) + \psi(x + ct)$, show that $\dfrac{\partial^2 u}{\partial t^2} = c^2 \dfrac{\partial^2 u}{\partial x^2}$.

8. (a) If $u = F(x^2 - 2xy)$, let $w = x^2 - 2xy$ and show that $\dfrac{\partial^2 u}{\partial x^2} = 2F'(w)$ $+ (2x - 2y)^2 F''(w)$.

(b) Find similar expressions for $\dfrac{\partial^2 u}{\partial x\, \partial y}$ and $\dfrac{\partial^2 u}{\partial y^2}$.

(c) Verify that $x\dfrac{\partial u}{\partial x} + (x - y)\dfrac{\partial u}{\partial y} = 0$.

9. (a) If $u = F(r)$ and $r = (x^2 + y^2 + z^2)^{1/2}$, show that $\dfrac{\partial^2 u}{\partial x^2} + \dfrac{\partial^2 u}{\partial y^2} + \dfrac{\partial^2 u}{\partial z^2}$ $= \dfrac{d^2 u}{dr^2} + \dfrac{2}{r}\dfrac{du}{dr}$. Hence find the form of $F(r)$ if this last expression is equal to zero when $r > 0$.

(b) Generalize the results of (a) for $u = F(r)$, $r = (x_1^2 + x_2^2 + \cdots + x_n^2)^{1/2}$.

10. If $V = \dfrac{1}{r} g\left(t - \dfrac{r}{c}\right)$, where c is constant and $r = (x^2 + y^2 + z^2)^{1/2}$, show that $\dfrac{\partial^2 V}{\partial x^2} + \dfrac{\partial^2 V}{\partial y^2} + \dfrac{\partial^2 V}{\partial z^2} = \dfrac{1}{c^2}\dfrac{\partial^2 V}{\partial t^2}$.

11. If $u = F[x + f(y)]$, show that $\dfrac{\partial u}{\partial x}\dfrac{\partial^2 u}{\partial x\, \partial y} = \dfrac{\partial u}{\partial y}\dfrac{\partial^2 u}{\partial x^2}$.

12. Show that $y\dfrac{\partial^2 u}{\partial x^2} + (x + y)\dfrac{\partial^2 u}{\partial x\, \partial y} + x\dfrac{\partial^2 u}{\partial y^2}$ becomes $-2t^2\dfrac{\partial^2 u}{\partial s\, \partial t} - 2t\dfrac{\partial u}{\partial s}$ when we set $s = y^2 - x^2$, $t = y - x$.

13. If $u = x^3 F(y/x, z/x)$, show that $x^2\dfrac{\partial^2 u}{\partial x^2} - y^2\dfrac{\partial^2 u}{\partial y^2} - z^2\dfrac{\partial^2 u}{\partial z^2} - 2yz\dfrac{\partial^2 u}{\partial y\, \partial z} + 4y\dfrac{\partial u}{\partial y} + 4z\dfrac{\partial u}{\partial z} = 6u.$

14. If $u = F(x, y) = G(s, t)$, where $s = xy$, $t = \dfrac{1}{y}$, find $\dfrac{\partial^2 u}{\partial x^2} + 2xy^2\dfrac{\partial u}{\partial x}$ $+ (y - y^3)\dfrac{\partial u}{\partial y} + x^2 y^2 u$ in terms of s, t, u, and derivatives of u with respect to s and t.

15. Prove that setting $x = e^s$, $y = e^t$ changes $x^2\dfrac{\partial^2 u}{\partial x^2} + y^2\dfrac{\partial^2 u}{\partial y^2} + x\dfrac{\partial u}{\partial x} + y\dfrac{\partial u}{\partial y}$ into $\dfrac{\partial^2 u}{\partial s^2} + \dfrac{\partial^2 u}{\partial t^2}$.

16. Suppose that $x = f(u, v)$, $y = g(u, v)$ transform $F(x, y)$ into $G(u, v)$. Suppose also that $\dfrac{\partial f}{\partial u} = \dfrac{\partial g}{\partial v}$ and $\dfrac{\partial f}{\partial v} = -\dfrac{\partial g}{\partial u}$.

Prove that $\dfrac{\partial^2 G}{\partial u^2} + \dfrac{\partial^2 G}{\partial v^2} = \left(\dfrac{\partial^2 F}{\partial x^2} + \dfrac{\partial^2 F}{\partial y^2}\right)\left[\left(\dfrac{\partial f}{\partial u}\right)^2 + \left(\dfrac{\partial f}{\partial v}\right)^2\right]$ is an identity in u and v.

17. If $u = x^n f(y/x) + x^{-n} g(y/x)$, show that

$$x^2 \frac{\partial^2 u}{\partial x^2} + 2 xy \frac{\partial^2 u}{\partial x \, \partial y} + y^2 \frac{\partial^2 u}{\partial y^2} + x \frac{\partial u}{\partial x} + y \frac{\partial u}{\partial y} = n^2 u.$$

18. (a) If $u = F(x, y) = G(s, t)$ and $x = f(s, t)$, $y = g(s, t)$, show that $G_{11} = F_{11} f_1^2 + F_{22} g_1^2 + 2 F_{12} f_1 g_1 + F_1 f_{11} + F_2 g_{11}$, where for convenience all the variables have been omitted from the functional symbols. (b) Write this same formula using the variable u and not using any of the functional symbols F, G, f, g. (c) Write analogous formulas for G_{12} and G_{22}.

19. Let
$$\xi_1 = a_{11} x_1 + a_{12} x_2$$
$$\xi_2 = a_{21} x_1 + a_{22} x_2$$

where the coefficients a_{11}, $\cdots$ are constants. With this change of variable, show

that $\quad a \dfrac{\partial^2 u}{\partial x_1{}^2} + 2 b \dfrac{\partial^2 u}{\partial x_1 \, \partial x_2} + c \dfrac{\partial^2 u}{\partial x_2{}^2} = \alpha \dfrac{\partial^2 u}{\partial \xi_1{}^2} + 2 \beta \dfrac{\partial^2 u}{\partial \xi_1 \, \partial \xi_2} + \gamma \dfrac{\partial^2 u}{\partial \xi_2{}^2},$

where the coefficients a, b, c, and α, β, γ are related in such a way that

$$\alpha\gamma - \beta^2 = (ac - b^2)(a_{11} a_{22} - a_{21} a_{12})^2.$$

20. If the change of variable (with constant coefficients)

$$\xi i = \sum_{j=1}^{n} c_{ij} x_j$$

changes $\quad \displaystyle\sum_{i,\,j=1}^{n} a_{ij} \frac{\partial^2 u}{\partial x_i \, \partial x_j} \quad$ into $\quad \displaystyle\sum_{k,\,l=1}^{n} b_{kl} \frac{\partial^2 u}{\partial \xi_k \, \partial \xi_l},$

show that $\qquad\qquad b_{kl} = \displaystyle\sum_{i,\,j=1}^{n} c_{ki} a_{ij} c_{lj}.$

6.53 Homogeneous functions. Euler's theorem

Consider the functions

$$x^2 + y^2, \quad \frac{xy}{x+y}, \quad x^2 y \log \frac{y}{x}.$$

Each of these functions has the interesting property that, if the variables x, y are replaced by tx, ty respectively, where t is a parameter, we obtain the original function multiplied by a power of t:

$$(tx)^2 + (ty)^2 = t^2(x^2 + y^2),$$

$$\frac{(tx)(ty)}{(tx) + (ty)} = t \frac{xy}{x+y},$$

$$(tx)^2(ty) \log \frac{ty}{tx} = t^3 \left(x^2 y \log \frac{y}{x} \right).$$

Functions of this type are called *homogeneous*. In general, we say that $F(x, y)$ is *homogeneous of degree n* if

(6.53–1) $\qquad\qquad F(tx, ty) = t^n F(x, y)$

for all values of x, y, and t for which $F(x, y)$ and $F(tx, ty)$ are defined. The degree n is a constant; it need not be an integer, and it may be negative. It may also be zero.

Example 1.   (a) $\dfrac{x}{x^2 + y^2}$ is homogeneous of degree $- 1$. (b) $x^{1/3} + xy^{-2/3}$ is homogeneous of degree $\frac{1}{3}$. (c) $\dfrac{x^2 - y^2}{x^2 + y^2}$ is homogeneous of degree 0.

If the fundamental relation (6.53–1) holds true only when t is restricted to positive (or nonnegative) values, we say that F is *positively homogeneous* of degree n. An example in which the limitation to nonnegative t is essential is furnished by the function $\sqrt{x^2 + y^2}$, which is positively homogeneous of degree 1. Since by definition the radical sign calls for the nonnegative square root of the radicand,

$$\sqrt{(tx)^2 + (ty)^2} = t\sqrt{x^2 + y^2}$$

holds only if $t \geqq 0$.

The definition of homogeneity is extended in an obvious way to functions of any number of variables.

The meaning of homogeneity can be interpreted geometrically. We refer back to § 5.4, where we discussed modes of representing a function. If (x, y) is a point distinct from the origin in the plane, the set of all points (tx, ty), as t varies, fills out the line determined by (x, y) and the origin (see Fig. 41). The fundamental relation (6.53–1) states that the value of the function F at the point (tx, ty) is t^n times the value of the function at the point (x, y). Thus, once we know the value of F at a point other than $(0, 0)$ on the line, we can compute its value at all other points on the line where it is defined. We must of course know the degree n. If the function is merely *positively* homogeneous, the requirement $t \geqq 0$ limits us to points (tx, ty) on the same side of the origin as (x, y) itself. It is clear that if F is positively homogeneous, and if we know its values at all points of the circle $x^2 + y^2 = 1$, we can compute its value at any other point where it is defined.

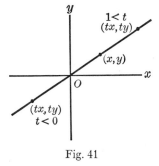

Fig. 41

Many simple or important functions arising in applied mathematics are homogeneous or positively homogeneous. For example, if F is such that its values are directly proportional to the nth power of the distance from the origin to (x, y), then F is positively homogeneous of degree n. In fact, we can write

$$F(x, y) = Ar^n, \; r = (x^2 + y^2)^{1/2}, \text{ whence } F(tx, ty) = t^n F(x, y) \text{ if } t \geqq 0.$$

One of the useful pieces of information about homogeneous functions is that furnished by a theorem named after the 18th century Swiss mathematician Leonard Euler.

EULER'S THEOREM (THEOREM IV).   *Let $F(x, y)$ be positively homogeneous of degree n. Then at any point where F is differentiable we have*

$$(6.53\text{-}2) \qquad x\,\frac{\partial F}{\partial x} + y\,\frac{\partial F}{\partial y} = nF(x, y).$$

PROOF.   Write $u = tx$, $v = ty$. Consider u, v as first-class variables, and t, x, y as second-class variables. The partial derivative of $F(u, v) = F(tx, ty)$ with respect to t is

$$F_1(u, v)\,\frac{\partial u}{\partial t} + F_2(u, v)\,\frac{\partial v}{\partial t}\,,$$

or $$xF_1(tx, ty) + yF_2(tx, ty),$$

provided that F is differentiable at (tx, ty). On the other hand, for positive t, $F(tx, ty) = t^n F(x, y)$, and $\dfrac{\partial}{\partial t}(t^n F(x, y)) = nt^{n-1}F(x, y)$. Thus

$$(6.53\text{-}3) \qquad xF_1(tx, ty) + yF_2(tx, ty) = nt^{n-1}F(x, y).$$

If we put $t = 1$ here we obtain the desired relation (6.53-2).

The theorem and the proof extend to functions of three or more independent variables. There is also a converse theorem; see Exercise 6.

Example 2.   If $F(x, y)$ is positively homogeneous of degree 2, and $u = r^m F(x, y)$, where $r^2 = x^2 + y^2$, show that

$$\frac{\partial^2 u}{\partial x^2} + \frac{\partial^2 u}{\partial y^2} = r^m\!\left(\frac{\partial^2 F}{\partial x^2} + \frac{\partial^2 F}{\partial y^2}\right) + m(m + 4)r^{m-2}F.$$

In working this problem it is convenient to observe that

$$2\,r\,\frac{\partial r}{\partial x} = 2\,x,\ \text{ or }\ \frac{\partial r}{\partial x} = \frac{x}{r}\,.$$

We have, therefore,

$$\frac{\partial u}{\partial x} = r^m\,\frac{\partial F}{\partial x} + mr^{m-1}\,\frac{\partial r}{\partial x}\,F = r^m\,\frac{\partial F}{\partial x} + mr^{m-2}xF.$$

Differentiating again,

$$\frac{\partial^2 u}{\partial x^2} = r^m\,\frac{\partial^2 F}{\partial x^2} + mr^{m-1}\,\frac{\partial r}{\partial x}\,\frac{\partial F}{\partial x} + mr^{m-2}x\,\frac{\partial F}{\partial x}$$

$$+ mr^{m-2}F + m(m - 2)r^{m-3}\,\frac{\partial r}{\partial x}\,xF$$

$$= r^m \frac{\partial^2 F}{\partial x^2} + 2\, mr^{m-2}x\, \frac{\partial F}{\partial x} + mr^{m-2}F + m(m-2)r^{m-4}x^2F.$$

Because of the symmetrical occurrence of x and y in r, we can write down at once

$$\frac{\partial^2 u}{\partial y^2} = r^m \frac{\partial^2 F}{\partial y^2} + 2\, mr^{m-2}y\, \frac{\partial F}{\partial y} + mr^{m-2}F + m(m-2)r^{m-4}y^2F.$$

If we now add, and take note of the relations

$$x\, \frac{\partial F}{\partial x} + y\, \frac{\partial F}{\partial y} = 2\,F, \quad x^2 + y^2 = r^2,$$

we find

$$\frac{\partial^2 u}{\partial x^2} + \frac{\partial^2 u}{\partial y^2} = r^m \left(\frac{\partial^2 F}{\partial x^2} + \frac{\partial^2 F}{\partial y^2} \right) + 4\, mr^{m-2}F + 2\, mr^{m-2}F$$
$$+ m(m-2)r^{m-2}F$$
$$= r^m \left(\frac{\partial^2 F}{\partial x^2} + \frac{\partial^2 F}{\partial y^2} \right) + m(m+4)r^{m-2}F.$$

This is the required result. Question for the student: Where was use made of the homogeneity of F?

EXERCISES

1. (a) If F is positively homogeneous of degree n, and if F, F_1, and F_2 are differentiable, the equation

$$x^2 \frac{\partial^2 F}{\partial x^2} + 2\, xy\, \frac{\partial^2 F}{\partial x\, \partial y} + y^2 \frac{\partial^2 F}{\partial y^2} = n(n-1)F(x, y)$$

is valid. Prove this by differentiating (6.53–3) partially with respect to t. **(b)** State and prove the corresponding result for third derivatives.

2. If F is positively homogeneous of degree n, and differentiable, its first partial derivatives are positively homogeneous of degree $n - 1$. Prove this.

3. Let $H(x, y)$ be positively homogeneous of degree p, and let $u = r^m H(x, y)$, where $r^2 = x^2 + y^2$. Show that

$$\Delta u = r^m \Delta H + m(2\,p + m)r^{m-2}H,$$

where, for any function $F(x, y)$, the notation ΔF means

$$\Delta F = \frac{\partial^2 F}{\partial x^2} + \frac{\partial^2 F}{\partial y^2}.$$

4. The equation $\Delta F = 0$ (see the explanation of notation at the end of Exercise 3) is called Laplace's equation in two dimensions. Here it is understood that (x, y) are rectangular co-ordinates of a point in the plane. If $H(x, y)$ is positively homogeneous of degree p, and $\Delta H = 0$, show that $\Delta(r^{-2p}H) = 0$ also. Use Exercise 3. Examples are furnished by taking $H(x, y)$ to be one of the functions x, y, $x^2 - y^2$, $2\, xy$, $3\, x^2y - y^3$, with appropriate values of p in each case.

5. (a) If x, y, z are rectangular co-ordinates in space, we write

$$\Delta F(x, y, z) = \frac{\partial^2 F}{\partial x^2} + \frac{\partial^2 F}{\partial y^2} + \frac{\partial^2 F}{\partial z^2}.$$

If $u = r^m H(x, y, z)$, where $r^2 = x^2 + y^2 + z^2$, and if H is positively homogeneous of degree p, work out the result analogous to that of Exercise 3. **(b)** What is the result for three dimensions comparable to that of Exercise 4? **(c)** Develop generalizations of the results of **(a)** and **(b)** for the case of n dimensions, taking $r^2 = x_1^2 + \cdots + x_n^2$.

6. Suppose that $F(x, y)$ is defined and differentiable in an open region R, and suppose that $x \, \partial F/\partial x + y \, \partial F/\partial y = nF(x, y)$ at each point of the region. Then, if (x, y) is in R, the relation $F(tx, ty) = t^n F(x, y)$ holds in any interval $t_0 < t < t_1$ (where $t_0 \geq 0$) provided that $t = 1$ is in this interval and provided that, for all such t, the points (tx, ty) are in R. To prove this converse of Euler's theorem, let $f(t) = F(tx, ty)$, where (x, y) is a fixed point of R. Use the hypothesis on F to prove that $tf'(t) = nf(t)$. From this, infer that $f(t)t^{-n}$ is a constant (depending on x, y). Then complete the proof.

6.6 Derivatives of implicit functions

In § 6.1 we dealt with the differentiation of implicit functions in a variety of particular situations. We did not attempt to deal with general cases in which the functions were merely indicated by some functional symbol. In practice it is necessary to have formulas to deal with implicit functions in terms of general functional notation.

A simple but typical case is that arising when z is defined as a function of x, y by an equation of the form

$$(6.6–1) \qquad\qquad F(x, y, z) = 0.$$

Suppose, for instance, that the equation is

$$x^2 + 2xz + z^2 - yz - 1 = 0,$$

so that in this case

$$(6.6–2) \qquad F(x, y, z) = x^2 + 2xz + z^2 - yz - 1.$$

Proceeding as in § 6.1, we have

$$2x + 2x \frac{\partial z}{\partial x} + 2z + 2z \frac{\partial z}{\partial x} - y \frac{\partial z}{\partial x} = 0,$$

$$2x \frac{\partial z}{\partial y} + 2z \frac{\partial z}{\partial y} - y \frac{\partial z}{\partial y} - z = 0,$$

$$\frac{\partial z}{\partial x} = -\frac{2x + 2z}{2x + 2z - y},$$

$$(6.6–3)$$

$$\frac{\partial z}{\partial y} = -\frac{-z}{2x + 2z - y}.$$

Now let us observe that if we regard x, y, z as independent in (6.6–2), then

$$\frac{\partial F}{\partial x} = 2x + 2z, \frac{\partial F}{\partial y} = -z, \frac{\partial F}{\partial z} = 2x + 2z - y.$$

Accordingly, equations (6.6–3) take on the form

$$(6.6\text{–}4) \qquad \frac{\partial z}{\partial x} = -\frac{\dfrac{\partial F}{\partial x}}{\dfrac{\partial F}{\partial z}}, \quad \frac{\partial z}{\partial y} = -\frac{\dfrac{\partial F}{\partial y}}{\dfrac{\partial F}{\partial z}}.$$

We are going to show that these formulas are general; that is, they hold when z is defined by (6.6–1), subject to certain general assumptions on the otherwise arbitrary function F.

We are not just now concerned with knowing what assumptions to make about F in order to guarantee that equation (6.6–1) does in fact define z as a function of x, y. Questions of this kind will be dealt with in Chapter VIII. We assume that z is a well-defined function of x, y, and that equation (6.6–1) is an identity in x, y when z is replaced by its functional value:

$$(6.6\text{–}5) \qquad\qquad z = f(x, y),$$
$$(6.6\text{–}6) \qquad\qquad F(x, y, f(x, y)) \equiv 0.$$

A simple instance of these last relations would be afforded by

$$z = \sqrt{1 - x^2 - y^2}, F(x, y, z) = x^2 + y^2 + z^2 - 1.$$

Another instance is

$$z = \log xy, F(x, y, z) = xy - e^z.$$

We also assume that F is a differentiable function of the *three independent* variables and that $f(x, y)$ has first partial derivatives. Under these conditions we shall prove formulas (6.6–4). Naturally we must assume that the denominator $\dfrac{\partial F}{\partial z}$ is not zero.

Under the foregoing hypotheses we look upon $G(x, y) = F(x, y, f(x, y))$ as a composite function of x, y. The first-class variables are x, y, z, and the second-class variables are x, y. The relations between the variables of the two classes are

$$x = x, y = y, z = f(x, y).$$

Therefore $\qquad \dfrac{\partial x}{\partial x} = 1, \dfrac{\partial x}{\partial y} = 0, \dfrac{\partial y}{\partial x} = 0, \dfrac{\partial y}{\partial y} = 1.$

The chain rule gives

$$\frac{\partial G}{\partial x} = \frac{\partial F}{\partial x}\frac{\partial x}{\partial x} + \frac{\partial F}{\partial y}\frac{\partial y}{\partial x} + \frac{\partial F}{\partial z}\frac{\partial z}{\partial x},$$

or
$$\frac{\partial G}{\partial x} = F_1 + F_3 \frac{\partial z}{\partial x}.$$

Likewise
$$\frac{\partial G}{\partial y} = F_2 + F_3 \frac{\partial z}{\partial y}.$$

Now by (6.6–6), $G(x, y) \equiv 0$, and so $\dfrac{\partial G}{\partial x} = \dfrac{\partial G}{\partial y} \equiv 0$. Therefore, assuming that $F_3 \neq 0$, we have

$$\frac{\partial z}{\partial x} = -\frac{F_1}{F_3}, \frac{\partial z}{\partial y} = -\frac{F_2}{F_3}.$$

These are the same equations as (6.6–4) except for the difference in notation. It is understood that the derivatives F_1, F_2, F_3 are evaluated with $z = f(x, y)$.

It is easily seen that the foregoing considerations may be generalized as follows. Suppose a function of n variables, say $u = f(x_1, \cdots, x_n)$ is determined implicitly as a solution of an equation $F(x_1, \cdots, x_n, u) = 0$. Then, under suitable assumptions of differentiability, we have

(6.6–7) $\quad \dfrac{\partial f}{\partial x_i} = - \dfrac{F_i(x_1, \cdots, x_n, f(x_1, \cdots, x_n))}{F_{n+1}(x_1, \cdots, x_n, f(x_1, \cdots, x_n))}, i = 1, \cdots, n.$

Example 1. Show that if $z = f(x, y)$ is a solution of the equation $F(x, y, z) = 0$, then the line normal to the surface $z = f(x, y)$ at a given point has direction ratios $F_1 : F_2 : F_3$, these partial derivatives being evaluated at the point in question.

It was shown in § 6.2 that the line normal to the surface has direction ratios $\dfrac{\partial z}{\partial x} : \dfrac{\partial z}{\partial y} : -1$. By (6.6–4) an equivalent set of ratios is

$$-\frac{F_1}{F_3} : -\frac{F_2}{F_3} : -1.$$

The ratios will not be altered if we multiply by $-F_3$. In this way we obtain the ratios $F_1 : F_2 : F_3$ for the direction of the normal line.

Next let us consider the case of two functions which arise as solutions of a pair of simultaneous equations. Suppose, for instance, that

(6.6–8) $\quad\quad u = f(x, y, z) \quad$ and $\quad v = g(x, y, z)$

are solutions of

(6.6–9) $\quad\quad F(x, y, z, u, v) = 0, \quad G(x, y, z, u, v) = 0.$

That is, suppose that

(6.6–10) $\quad\quad F(x, y, z, f(x, y, z), g(x, y, z)) \equiv 0$

is true for all values of x, y, z in some region, with a similar relation for G. We assume that the functions F, G, f, g are all differentiable. The problem which we pose is that of expressing the partial derivatives of f and g in terms of those of F and G. As usual, we write

$$F_1 = \frac{\partial F}{\partial x}, F_2 = \frac{\partial F}{\partial y}, \cdots, F_5 = \frac{\partial F}{\partial v}, \cdots.$$

The required formulas are then of the form

$$\frac{\partial u}{\partial x} = - \frac{\begin{vmatrix} F_1 & F_5 \\ G_1 & G_5 \end{vmatrix}}{\begin{vmatrix} F_4 & F_5 \\ G_4 & G_5 \end{vmatrix}}, \cdots, \frac{\partial u}{\partial z} = - \frac{\begin{vmatrix} F_3 & F_5 \\ G_3 & G_5 \end{vmatrix}}{\begin{vmatrix} F_4 & F_5 \\ G_4 & G_5 \end{vmatrix}},$$

(6.6–11)

$$\frac{\partial v}{\partial x} = - \frac{\begin{vmatrix} F_4 & F_1 \\ G_4 & G_1 \end{vmatrix}}{\begin{vmatrix} F_4 & F_5 \\ G_4 & G_5 \end{vmatrix}}, \cdots, \frac{\partial v}{\partial z} = - \frac{\begin{vmatrix} F_4 & F_3 \\ G_4 & G_3 \end{vmatrix}}{\begin{vmatrix} F_4 & F_5 \\ G_4 & G_5 \end{vmatrix}}.$$

This is on the assumption that the determinant appearing in the denominators is not equal to zero. The expressions $F_1, \cdots, G_5$ are evaluated with u and v given by (6.6–8).

We shall indicate briefly the manner of deriving formulas (6.6–11). We regard x, y, z, u, v as first-class variables, and x, y, z as second-class variables. Then, from the identity (6.6–10), we have (differentiating with respect to x)

$$F_1 + F_4 \frac{\partial u}{\partial x} + F_5 \frac{\partial v}{\partial x} = 0.$$

There is a similar equation involving G:

$$G_1 + G_4 \frac{\partial u}{\partial x} + G_5 \frac{\partial v}{\partial x} = 0.$$

These two linear equations are now solved for $\dfrac{\partial u}{\partial x}$ and $\dfrac{\partial v}{\partial x}$, with the results indicated in (6.6–11). The procedure is entirely similar for finding the derivatives of u and v with respect to y or z.

The student need not memorize the formulas (6.6–11), though it is useful to have ready reference to such formulas. It is the procedure for deriving the formulas which is important, and the student should be able to carry out the derivation himself.

Determinants such as those occurring in the formulas (6.6–11) are called *Jacobians*, after Carl Jacobi, a prominent German mathematician of the nineteenth century. There is a more compact notation which is often used for a Jacobian:

$$\frac{\partial(F, G)}{\partial(u, v)} = \begin{vmatrix} \dfrac{\partial F}{\partial u} & \dfrac{\partial F}{\partial v} \\ \dfrac{\partial G}{\partial u} & \dfrac{\partial G}{\partial v} \end{vmatrix}.$$

This is called a Jacobian of second order. The form of a general nth-order Jacobian involves n functions, each of n variables:

$$\frac{\partial(F_1, \cdots, F_n)}{\partial(u_1, \cdots, u_n)} = \begin{vmatrix} \dfrac{\partial F_1}{\partial u_1} & \dfrac{\partial F_1}{\partial u_2} & \cdots & \dfrac{\partial F_1}{\partial u_n} \\ \dfrac{\partial F_2}{\partial u_1} & \cdot & \cdots & \cdot \\ \cdot & \cdot & \cdots & \cdot \\ \cdot & \cdot & \cdots & \cdot \\ \dfrac{\partial F_n}{\partial u_1} & \cdot & \cdots & \dfrac{\partial F_n}{\partial u_n} \end{vmatrix}$$

In concluding this section we emphasize that here we are concerned with the *structure* of general formulas for the derivatives of implicit functions. Questions of existence and differentiability of implicit functions are taken up in Chapter VIII.

EXERCISES

1. If the equations

$$F(x, y, u, v, w) = 0, \quad G(x, y, u, v, w) = 0, \quad H(x, y, u, v, w) = 0$$

have solutions for u, v, and w as functions of x and y, show, taking for granted certain general conditions, that

$$\frac{\partial u}{\partial x} = - \frac{\dfrac{\partial(F, G, H)}{\partial(x, v, w)}}{\dfrac{\partial(F, G, H)}{\partial(u, v, w)}}.$$

Write five other similar formulas.

2. Suppose $y = f(x)$ is a solution of $G(x, y) = 0$, where $\dfrac{\partial G}{\partial y} \neq 0$, and let $g(x) = F(x, f(x))$. Show that $g'(x) = (F_1 G_2 - F_2 G_1)/G_2$, the right side being evaluated with $y = f(x)$.

3. If $z = f(x, y)$ is a solution of $F(x, y, z) = 0$ (with $F_3 \neq 0$), and if $H(x, y) = G(x, y, f(x, y))$, show that $\dfrac{\partial F}{\partial z} \dfrac{\partial H}{\partial y} = - \dfrac{\partial(F, G)}{\partial(y, z)}$. Write a similar formula involving $\dfrac{\partial H}{\partial x}$.

4. Suppose $u = f(x, y, z)$, $v = g(x, y, z)$ are solutions of $F(x, y, z, u, v) = 0$, $G(x, y, z, u, v) = 0$. Let $K(x, y, z) = H(x, y, z, f(x, y, z), g(x, y, z))$. Show that

$$\frac{\partial K}{\partial z} = \frac{\dfrac{\partial(F, G, H)}{\partial(z, u, v)}}{\dfrac{\partial(F, G)}{\partial(u, v)}}$$

under suitable conditions.

5. As an instance of (6.6–8) and (6.6–9) consider $u = x \sin xyz$, $v = y \cos xyz$ as solutions of $yu - xv \tan xyz = 0$, $y^2 u^2 + x^2 v^2 - x^2 y^2 = 0$. Verify all six of the formulas (6.6–11) in this case.

6. If the equation $F(x, y, z) = 0$ can be solved for each one of the variables in terms of the other two, show, taking for granted certain general conditions, that

$$\left(\frac{\partial x}{\partial y} \right)_z \left(\frac{\partial y}{\partial z} \right)_x \left(\frac{\partial z}{\partial x} \right)_y = -1.$$

7. Let $G(x, y, z, v) = 0$ have solutions $x = f(y, z, v)$, $y = g(x, z, v)$, $z = h(x, y, v)$, and by means of one of these equations at a time let $u = F(x, y, z)$ become a function of (y, z, v), (x, z, v), and (x, y, v) respectively. Show that, subject to certain conditions,

$$\left(\frac{\partial u}{\partial y} \right)_{x, v} - \left(\frac{\partial u}{\partial y} \right)_{z, v} = \left(\frac{\partial u}{\partial z} \right)_{x, y} \cdot \left(\frac{\partial z}{\partial y} \right)_{x, v} - \left(\frac{\partial u}{\partial x} \right)_{y, z} \cdot \left(\frac{\partial x}{\partial y} \right)_{z, v}.$$

An example is furnished by $xyv - z = 0$, $u = x^2 + y^2 + z^2$, and the student should check the meaning of the problem in terms of this special case if he feels an illustration to be necessary.

8. Suppose $u = f(x, y)$ is a solution of $F(x, y, u) = 0$, and that $y = g(x, z)$ is a solution of $G(x, y, z) = 0$. Let $H(x, z) = f(x, g(x, z))$. Show that $F_3 G_2 H_2 = F_2 G_3$ and $F_3 G_2 H_1 = F_2 G_1 - F_1 G_2$ are identities in x and z.

9. (a) Starting from (6.6–4), show that

$$\frac{\partial^2 z}{\partial x^2} = \frac{F_3^2 F_{11} - 2 F_1 F_3 F_{13} + F_1^2 F_{33}}{-F_3^3}.$$

(b) Derive analogous formulas for $\dfrac{\partial^2 z}{\partial y \, \partial x}$ and $\dfrac{\partial^2 z}{\partial y^2}$.

10. If $z = f(x, y)$ satisfies an equation of the form $z = F(ax + by + cz)$, where $a, b,$ and c are constants, show that $b \dfrac{\partial z}{\partial x} = a \dfrac{\partial z}{\partial y}$.

11. If $z = f(x, y)$ satisfies an equation of the form $F(x + y + z, x^2 + y^2 + z^2) = 0$, show that $(y - x) + (y - z) \dfrac{\partial z}{\partial x} + (z - x) \dfrac{\partial z}{\partial y} = 0$.

12. Suppose that the function $z = f(x, y)$ satisfies an equation of the form $F(ax + by + cz, Ax^2 + By^2 + Cz^2) = 0$, where a, b, c and A, B, C are constants. Show that $\dfrac{\partial z}{\partial x} = - \dfrac{aF_1 + 2 AxF_2}{cF_1 + 2 CzF_2}$.

13. If $z = \phi(x, y)$ satisfies the equation $F(f(x, y, z), g(x, y, z)) = 0$, show that

$$\frac{\partial z}{\partial y} = - \frac{F_1 f_2 + F_2 g_2}{F_1 f_3 + F_2 g_3} .$$

14. If $G_1(x_1, x_2, y)$, $G_2(x_1, x_2, y)$ and $f(x_1, x_2)$ are given, and if $g_i(x_1, x_2) = G_i(x_1, x_2, f(x_1, x_2))$ $(i = 1, 2)$, show that

$$\frac{\partial(g_1, g_2)}{\partial(x_1, x_2)} = \frac{\partial(G_1, G_2)}{\partial(x_1, x_2)} + \frac{\partial f}{\partial x_1} \frac{\partial(G_1, G_2)}{\partial(y, x_2)} + \frac{\partial f}{\partial x_2} \frac{\partial(G_1, G_2)}{\partial(x_1, y)} ,$$

with y replaced by $f(x_1, x_2)$ after the differentiations. This formula is used in the theory of first-order partial differential equations.

6.7   Extremal problems with constraints

Many interesting maximum or minimum problems arise in such a form that we are required to find an extremal value of a function, say $F(x, y, z)$, where the variables x, y, z are not independent of each other, but are restricted by some relation existing between them, this relation being expressed by an equation $G(x, y, z) = 0$.

Example 1.   Find the minimum value of

$$F(x, y, z) = (x - 3)^2 + (y - 2)^2 + (z - 1)^2$$

subject to the condition

$$G(x, y, z) = 2x - 3y - 4z - 25 = 0.$$

This problem has already been solved; see Example 1, § 6.3.

The equation $G(x, y, z) = 0$ is called a *constraint* on the variables x, y, z. It is immaterial whether the equation of constraint has the form $G(x, y, z) = 0$ or $G(x, y, z) = k$, where k is a specified constant, for the latter form of constraint can be written $G(x, y, z) - k = 0$.

An extremal problem with constraint may occur with any number of variables, and there may be more than one equation of constraint.

Example 2.   Find maximum and minimum values of $x^2 + y^2 + z^2$ subject to the two conditions

$$x + y - z = 0, \frac{x^2}{16} + y^2 + z^2 = 1.$$

In this problem there are two constraints on the three variables, so that there is actually only one independent variable.

Example 3.   Find the minimum value of $(x - u)^2 + (y - v)^2 + z^2$ subject to the condition $3x^2 + y^2 - 6x - 4y - 12z + 43 = 0$.

In this problem there are five variables and one constraint, which happens to involve only three of the five variables.

There are various methods for dealing with extremal problems with constraints. One method is to use the equation or equations of constraint to express certain of the variables in terms of the remaining variables. These latter variables are chosen as the independent variables, and the function whose extremal value is sought is then expressed in terms of the independent variables only. The solution is then carried out by standard methods. We shall call this general method the *method of direct elimination*. It is illustrated, in the case of the problem of Example 1 in the present section, by the solution given in Example 1, § 6.3.

A second method may conveniently be called the *method of implicit functions*. Suppose the problem is to find the point or points at which $F(x, y, z)$ has extreme values subject to the condition $G(x, y, z) = k$. Here we assume that F and G are given functions with continuous first partial derivatives, and that k is a given constant. Let us assume, for definiteness, that $\dfrac{\partial G}{\partial z} \neq 0$ and that the equation

$$(6.7\text{--}1) \qquad\qquad G(x, y, z) - k = 0$$

has a solution

$$(6.7\text{--}2) \qquad\qquad z = f(x, y).$$

We then seek to make the quantity

$$u = F(x, y, f(x, y))$$

a maximum or minimum. Accordingly we want to solve the equations

$$(6.7\text{--}3) \qquad\qquad \frac{\partial u}{\partial x} = 0, \ \frac{\partial u}{\partial y} = 0.$$

Now

$$(6.7\text{--}4) \qquad \frac{\partial u}{\partial x} = \frac{\partial F}{\partial x} + \frac{\partial F}{\partial z}\frac{\partial f}{\partial x}, \ \frac{\partial u}{\partial y} = \frac{\partial F}{\partial y} + \frac{\partial F}{\partial z}\frac{\partial f}{\partial y},$$

where z is replaced by $f(x, y)$ after the differentiations are performed.

We also have the identity

$$G(x, y, f(x, y)) - k = 0,$$

from which it follows by differentiation that

$$(6.7\text{--}5) \qquad \frac{\partial G}{\partial x} + \frac{\partial G}{\partial z}\frac{\partial f}{\partial x} = 0, \ \frac{\partial G}{\partial y} + \frac{\partial G}{\partial z}\frac{\partial f}{\partial y} = 0.$$

If we solve these equations for $\dfrac{\partial f}{\partial x}$ and $\dfrac{\partial f}{\partial y}$ and substitute in (6.7-4), we

obtain the equation $\dfrac{\partial u}{\partial x} = \dfrac{\dfrac{\partial F}{\partial x}\dfrac{\partial G}{\partial z} - \dfrac{\partial F}{\partial z}\dfrac{\partial G}{\partial x}}{\dfrac{\partial G}{\partial z}},$

and a similar equation for $\dfrac{\partial u}{\partial y}$. Equations (6.7–3) now take the form

(6.7–6) $\qquad \dfrac{\partial F}{\partial x}\dfrac{\partial G}{\partial z} - \dfrac{\partial F}{\partial z}\dfrac{\partial G}{\partial x} = 0, \quad \dfrac{\partial F}{\partial y}\dfrac{\partial G}{\partial z} - \dfrac{\partial F}{\partial z}\dfrac{\partial G}{\partial y} = 0,$

in which z is replaced by $f(x, y)$ after the differentiations are performed. Now the method of implicit functions for this extremal problem with constraint may be described as follows: We do not *actually* solve for z at the outset. Instead, we carry the work along and arrive at equations (6.7–6) as equations in *all three* variables. These two equations, together with the constraint (6.7–1), give us three equations which we solve as simultaneous equations in x, y, z. The required points of extreme value will be among the points found in this way. This general assertion is subject to some qualifications to rule out exceptional cases. We could, of course, think of y as a function of x and z, or of x as a function of y and z, the functional relation in each case being determined by the constraint. These alternatives would give us pairs of equations different from (6.7–6), but equivalent to them.

The implicit-function method was used in the solution of the problem of Example 3, § 6.3, with y regarded as a function of x and z. Our purpose just now is not to illustrate the method in particular cases, but to obtain the equations (6.7–6) in preparation for study of other aspects of extremal problems with constraints.

It is often useful to regard an extremal problem with constraint from a geometrical point of view. Consider, for instance, the problem of extremal value for the function $F(x, y, z)$ with the constraint $G(x, y, z) = k$ (a given constant). We shall suppose that the equation $G(x, y, z) = k$ defines a surface S possessing a tangent plane at each of its points with which we are concerned. We then consider the values of the function F at points of the surface S; we wish to find the points of S where F has maximum or minimum values. Now consider the family of surfaces

(6.7–7) $\qquad\qquad\qquad F(x, y, z) = C,$

where C is a parameter. Suppose that M is the absolute maximum value attained by F at points of S. Then, if $C > M$, the surface (6.7–7) and the surface S will have no points in common, whereas if $C = M$ the two surfaces will have at least one point (x_0, y_0, z_0) in common, and this will be a point where F attains its maximum value on S. We shall prove that the two surfaces $G(x, y, z) = k$ and $F(x, y, z) = M$ are tangent at this point (x_0, y_0, z_0). The direction ratios of the normals to the two surfaces are, respectively,

$$G_1 : G_2 : G_3 \quad \text{and} \quad F_1 : F_2 : F_3,$$

with the partial derivatives evaluated at (x_0, y_0, z_0) (see Example 1, § 6.6). The condition of tangency is therefore that these two sets of ratios be the same, i.e., that G_1, G_2, G_3 and F_1, F_2, F_3 be proportional. Now we saw earlier that equations (6.7–6) must hold at a point of extreme value, provided $G_3 \neq 0$. These equations may be written

$$F_1 G_3 - F_3 G_1 = 0, \quad F_2 G_3 - F_3 G_2 = 0,$$

or
$$F_1 = \frac{F_3}{G_3} G_1, \quad F_2 = \frac{F_3}{G_3} G_2,$$

whence it follows that F_1, F_2, F_3 and G_1, G_2, G_3 are proportional at (x_0, y_0, z_0). If it should happen that $G_3 = 0$ at the point, the same conclusion of proportionality can be reached by assuming that $G_2 \neq 0$ or $G_1 \neq 0$.

The argument and the conclusion are essentially the same in the case of a minimum rather than a maximum. Also, the surfaces will be tangent at a point of *relative* extreme value of F on the surface S. There may also be tangencies at points where F is neither a maximum nor a minimum. We sum up our findings in a formal statement.

THEOREM V. *Suppose the functions F and G have continuous first partial derivatives throughout a certain region of space. Let the equation $G(x, y, z) = k$ define a surface S, every point of which is in the interior of the region, and suppose that the three partial derivatives G_1, G_2, G_3 are never simultaneously zero at a point of S. Then a necessary condition for the values of $F(x, y, z)$ on S to attain an extreme value (either relative or absolute) at a point of S is that F_1, F_2, F_3 be proportional to G_1, G_2, G_3 at that point. If C is the value of F at the point, and if the constant of proportionality is not zero, the geometrical meaning of the proportionality is that the surface S and the surface $F(x, y, z) = C$ are tangent at the point in question.*

A fully detailed analytic proof of this theorem depends upon the theory of implicit functions, as developed in Chapter VIII. The geometric arguments which precede the statement of the theorem do not constitute a rigorous proof, but they give a good basis for understanding the theorem; with adequate knowledge of implicit-function theory, the argument given here can be developed into a complete proof.

Example 4. Consider the maximum and minimum values of $F(x, y, z)$ $= x^2 + y^2 + z^2$ on the surface of the ellipsoid

$$G(x, y, z) = \frac{x^2}{64} + \frac{y^2}{36} + \frac{z^2}{25} = 1.$$

Since $F(x, y, z)$ is the square of the distance from (x, y, z) to the origin, it is clear that we are looking for the points on the ellipsoid at maximum and minimum distances from the center of the ellipsoid. The maximum occurs at the ends of the longest principal axis, namely at ($\pm$ 8, 0, 0). The minimum occurs at the ends of the shortest principal axis, namely at (0, 0, $\pm$ 5). Consider the maximum point (8, 0, 0). The value of F at this point is 64, and the surface $F(x, y, z) = 64$ is a sphere. The sphere and the ellipsoid are tangent at (8, 0, 0), as asserted in the general theory. In this case the ratios $G_1 : G_2 : G_3$ and $F_1 : F_2 : F_3$ at (8, 0, 0) are $\frac{1}{4} : 0 : 0$ and 16 : 0 : 0, respectively.

This example brings out the fact that the tangency of the surfaces, or the proportionality of the two sets of ratios, is a necessary but not a sufficient condition for a maximum or minimum value of F; for the condition of proportionality exists at the points (0, $\pm$ 6, 0), which are the ends of the principal axis of intermediate length. But the value of F is neither a maximum nor a minimum at this point, as the student can readily see by considering the geometrical situation.

A similar geometrical interpretation can be given to the problem of extremal values for $F(x, y)$ subject to a constraint $G(x, y) = k$. Here we have a curve defined by the constraint, and a one-parameter family of curves $F(x, y) = C$. At a point of extremal value of F the curve $F(x, y) = C$ through the point will be tangent to the curve defined by the constraint. Suitable hypotheses must be made to rule out singular cases.

Example 5. Find the maximum value of $F(x, y) = xy$ subject to the constraint $x^2 + y^2 = 8$.

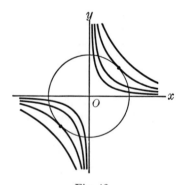

Fig. 42

Here we have a fixed circle; the curves $xy = C$ are hyperbolas. The hyperbola $xy = 4$ is tangent to the circle at the points (2, 2) and ($-$ 2, $-$ 2) (see Fig. 42). These are the points at which the maximum value is attained.

An advantage of the geometrical point of view is that we can often use it to conclude with certainty that F actually attains an absolute maximum or minimum subject to the given constraint. Suppose, for example, that the surface S forms a closed and bounded set of points in space. Then, since F is continuous at the points of S, we can infer that F actually attains an absolute maximum and an absolute minimum on S. This inference is based on a theorem analogous to Theorem III, § 3.2, and Theorem II, § 5.3. The general theorem covering this situation is Theorem IV, § 17.3.

Once we know that a maximum or minimum exists, we apply one of the methods which leads us to solve certain equations. If these equations have just a few solutions, as they often do in practice, it is an easy matter to tell which solutions give us the absolute maximum and which the absolute minimum.

6.8   Lagrange's method

In § 6.7 we discussed extremal problems with constraints, and described two methods for handling such problems. We referred to these methods as (1) the method of direct elimination, and (2) the method of implicit functions. In this section we shall explain and illustrate a method known as Lagrange's method; it was devised by the great 18th-century mathematician Joseph Louis Lagrange. Our explanations depend upon Theorem V and the discussions leading up to it in § 6.7.

In the case of the problem of extremal values of $F(x, y, z)$ subject to a constraint $G(x, y, z) = k$, Lagrange's method directs us to proceed as follows:

Form the function

$$(6.8–1) \qquad u = F(x, y, z) + \lambda G(x, y, z),$$

where λ is a constant, as yet undetermined in value. Treat x, y, z as independent variables, and write down the conditions

$$(6.8–2) \qquad \frac{\partial u}{\partial x} = 0, \; \frac{\partial u}{\partial y} = 0, \; \frac{\partial u}{\partial z} = 0.$$

Solve these three equations along with the equation of constraint

$$(6.8–3) \qquad G(x, y, z) = k$$

to find the values of the four quantities x, y, z, λ. More than one point (x, y, z) may be found in this way, but among the points so found will be the points of extremal values of F.

To understand the reason for the validity of Lagrange's method, observe that the equations (6.8–2) are precisely

$$(6.8–4) \qquad F_1 + \lambda G_1 = 0, \; F_2 + \lambda G_2 = 0, \; F_3 + \lambda G_3 = 0.$$

Here λ is a certain constant. These equations state, therefore, that at a point where they are satisfied, F_1, F_2, and F_3 are proportional to G_1, G_2, and G_3. But we know from Theorem V that such proportionality occurs at the points of the surface $G(x, y, z) = k$ where F has an extreme value. Thus the points of extreme value will be among those found by solving the four simultaneous equations (6.8–3) and (6.8–4). Thus Lagrange's method is justified in this type of problem.

The parameter λ occurring in Lagrange's method is called *Lagrange's multiplier*.

One of the great advantages of Lagrange's method over the method of implicit functions or the method of direct elimination is that it enables us to avoid making a choice of independent variables. This is sometimes very important; it permits the retention of symmetry in a problem where the variables enter symmetrically at the outset.

Example 1. Find the dimensions of the box of largest volume which can be fitted inside the ellipsoid

$$(6.8\text{–}5) \qquad \frac{x^2}{a^2} + \frac{y^2}{b^2} + \frac{z^2}{c^2} = 1,$$

assuming that each edge of the box is parallel to a co-ordinate axis.

Each of the eight corners of the box will lie on the ellipsoid. Let the corner in the first octant have co-ordinates (x, y, z); then the dimensions of the box are $2\,x$, $2\,y$, $2\,z$, and its volume is $V = 8\,xyz$. We wish to find the absolute maximum of V subject to the constraint (6.8–5). By the remarks at the end of § 6.7 we know that an absolute maximum exists. Following Lagrange's method, we set

$$u = 8\,xyz + \lambda \left(\frac{x^2}{a^2} + \frac{y^2}{b^2} + \frac{z^2}{c^2} \right).$$

The equations (6.8–2) in this case are

$$8\,yz + 2\,\lambda\,\frac{x}{a^2} = 0,$$

$$(6.8\text{–}6) \qquad 8\,zx + 2\,\lambda\,\frac{y}{b^2} = 0,$$

$$8\,xy + 2\,\lambda\,\frac{z}{c^2} = 0.$$

After dividing by 2, let us multiply these three equations by x, y, z respectively, and add. In view of equation (6.8–5) the result is

$$12\,xyz + \lambda = 0, \text{ or } \lambda = -\,12\,xyz.$$

We now put this result back into the first of the equations (6.8–6), and obtain, after a slight simplification,

$$yz(a^2 - 3\,x^2) = 0.$$

By symmetry we obtain the two further equations

$$zx(b^2 - 3 y^2) = 0, \ xy(c^2 - 3 z^2) = 0.$$

For maximum V it is clear that we want *positive* values of x, y, and z. The only possible solutions of (6.8–6) meeting these requirements are thus seen to be

$$x = \frac{a}{\sqrt{3}}, \ y = \frac{b}{\sqrt{3}}, \ z = \frac{c}{\sqrt{3}},$$

and

$$\lambda = - 12 \ xyz = - \frac{4}{\sqrt{3}} \ abc.$$

The box of maximum volume therefore has dimensions

$$\frac{2 \ a}{\sqrt{3}} \times \frac{2 \ b}{\sqrt{3}} \times \frac{2 \ c}{\sqrt{3}} .$$

Lagrange's method may be extended to the case of several constraints. The number of variables is immaterial. An undetermined multiplier is introduced corresponding to each constraint. We shall illustrate the extended method without formal proof.

Example 2.   Find the semiaxes of the ellipse in which the plane

(6.8–7) $$lx + my + nz = 0$$

cuts the ellipsoid

(6.8–8) $$\frac{x^2}{a^2} + \frac{y^2}{b^2} + \frac{z^2}{c^2} = 1.$$

Since the plane passes through the origin, it is clear that the ends of the semiaxes of the ellipse are at the points (x, y, z) where the function

$$F(x, y, z) = x^2 + y^2 + z^2$$

is a maximum or a minimum subject to the side-conditions (6.8–7) and (6.8–8). We set up the expression

$$u = x^2 + y^2 + z^2 + \lambda_1(lx + my + nz) + \lambda_2 \left(\frac{x^2}{a^2} + \frac{y^2}{b^2} + \frac{z^2}{c^2} \right),$$

using two Lagrange multipliers, λ_1, λ_2. The method of Lagrange leads us to write the equations $\dfrac{\partial u}{\partial x} = \dfrac{\partial u}{\partial y} = \dfrac{\partial u}{\partial z} = 0$, or

$$2 \ x + \lambda_1 l + 2 \ \lambda_2 \frac{x}{a^2} = 0,$$

(6.8–9) $$2 \ y + \lambda_1 m + 2 \ \lambda_2 \frac{y}{b^2} = 0,$$

$$2 \ z + \lambda_1 n + 2 \ \lambda_2 \frac{z}{c^2} = 0.$$

Theoretically we may solve these three equations together with the two equations of constraint to find the values of x, y, z, λ_1, λ_2. Actually, the solution presents great difficulties. However, since we are merely interested in the lengths of the semiaxes of the ellipse, it will suffice to solve for the value of $F(x, y, z)$. Let us write $p^2 = F(x, y, z)$, so that p is the length of the semimajor or semiminor axis according as F has a maximum or a minimum.

If we multiply equations (6.8–9) by x, y, z respectively, and add, we obtain

$$2(x^2 + y^2 + z^2) + \lambda_1 (lx + my + nz) + 2 \lambda_2 \left(\frac{x^2}{a^2} + \frac{y^2}{b^2} + \frac{z^2}{c^2} \right) = 0.$$

In view of (6.8–7) and (6.8–8), this becomes

$$2 p^2 + 2 \lambda_2 = 0, \text{ or } \lambda_2 = - p^2.$$

Returning to (6.8–9) with this result, we have

(6.8–10) $$2(a^2 - p^2)x + \lambda_1 a^2 l = 0,$$

and similar equations involving y and z. Except in special cases the situation will be such that x, y, z are all different from zero at the end of a semiaxis; and furthermore, p^2 will be different from a^2, b^2, and c^2. Thus in equation (6.8–10) we may assume $a^2 - p^2 \neq 0$ and $\lambda_1 \neq 0$ except in special cases. Thus

$$x = \lambda_1 \frac{a^2 l}{2(p^2 - a^2)} \, , \, y = \lambda_1 \frac{b^2 m}{2(p^2 - b^2)} \, , \, z = \lambda_1 \frac{c^2 n}{2(p^2 - c^2)} \, .$$

If we substitute these values in (6.8–7) and cancel the factor λ_1, we obtain the equation.

(6.8–11) $$\frac{a^2 l^2}{p^2 - a^2} + \frac{b^2 m^2}{p^2 - b^2} + \frac{c^2 n^2}{p^2 - c^2} = 0.$$

When cleared of fractions this equation is a quadratic in p^2 as an unknown. The two roots of the equation thus determine the lengths of the semiaxes of the ellipse. For some consideration of the exceptional special cases see Exercise 18.

The foregoing problem illustrates very clearly the merits of Lagrange's method in the preservation of symmetry.

EXERCISES

1. A rectangular box lies in the first octant, with one corner at the origin and the diagonally opposite corner on the plane $(x/a) + (y/b) + (z/c) = 1$ $(a, b, c > 0)$. Find the maximum possible volume of the box.

2. Apply Lagrange's method in finding the extreme values of $x^2 + y^2 + z^2$ sub-

ject to the constraint $(x^2/a^2) + (y^2/b^2) + (z^2/c^2) = 1$, where $a > b > c > 0$.

3. A triangle is such that the product of the sines of its angles is a maximum. Show that the triangle is equilateral.

4. Find the maximum value of $xyz/(a^3x + b^3y + c^3z)$ subject to the conditions $xyz = A^3$, x, y, $z > 0$ (a, b, c, A all > 0).

5. Find the minimum of $x + y + z$ subject to the conditions $(a/x) + (b/y) + (c/z) = 1$, x, y, $z > 0$ (a, b, c and x, y, z all > 0).

6. The perimeter of a triangle has a prescribed value $2s$. Determine the sides of the triangle so as to maximize the area.

7. Let $D = \begin{vmatrix} x & y \\ u & v \end{vmatrix}$. Find the maximum value of D^2 subject to the conditions $x^2 + y^2 = a^2$, $u^2 + v^2 = b^2$, where $a > 0$, $b > 0$. Solve the problem in two ways: (1) by Lagrange's method, and (2) by setting $x = a \cos \theta$, $y = a \sin \theta$, $u = b \cos \phi$, $v = b \sin \phi$, and using θ, ϕ as independent variables.

8. Find the minimum value of $x^3 + y^3 + z^3$ for positive x, y, and z, if it is required that $ax + by + cz = 1$, where a, b, c are positive constants.

9. Suppose a, b, c are positive constants. If x, y, z are positive and $ayz + bzx + cxy = 3abc$, show that $xyz \leq abc$.

10. Solve the following problems by Lagrange's method: (a) Example 1, § 6.3; (b) Example 3, § 6.3; (c) Example 2, § 6.7; (d) Example 3, § 6.7.

11. Let the lengths of the sides of a fixed triangle of area A be a, b, c. From an interior point O draw the perpendiculars to the sides of the triangle, and let their lengths be x, y, z corresponding to a, b, c. If now a parallelepiped is constructed with edges x, y, z and volume V, show that V is a maximum when the lines from O to the vertices of the triangle divide it into three equal areas. What is the maximum value of V?

12. For the situation described in Exercise 11, show that
$$(x^2 + y^2 + z^2)(a^2 + b^2 + c^2) \geq 4 A^2.$$

13. Find the minimum distance from $(0, 0, c)$ to the cone $z^2 = (x^2/a^2) + (y^2/b^2)$. Assume $c > 0$ and $0 < b < a$.

14. Given the ellipse $b^2x^2 + a^2y^2 = a^2b^2$, find the points (x, y) on the ellipse so that the line normal to the ellipse at (x, y) passes as far as possible from the origin. What is this greatest distance from the origin to a normal?

15. Find, by Lagrange's method, an extreme value of xyz subject to the conditions $(1/x) + (1/y) + (1/z) = c$, x, y, z all > 0 (c a positive constant). Is this value an absolute maximum, an absolute minimum, or neither?

16. A particle is to travel from A to P and thence to B, by a broken line as indicated in Fig. 43. Velocity from A to P is v_1, and from P to B is v_2. Show by

Lagrange's method that when the time of travel is least, $(\sin \theta_1)/(\sin \theta_2) = v_1/v_2$.

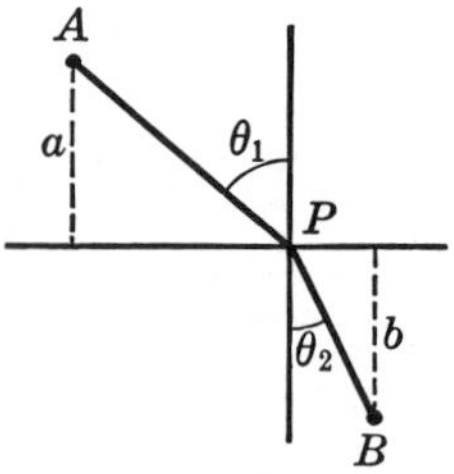

17. Let $p^2 = x^2 + y^2 + z^2$. Consider the extreme values of p^2, subject to the conditions $lx + my + nz = 0$, $(x^2 + y^2 + z^2)^2 = a^2x^2 + b^2y^2 + c^2z^2$, where $a > b > c > 0$. The minimum of p^2 is obviously zero. Show that, barring exceptional cases, the maximum is given by one of the roots of the equation

$$\frac{l^2}{p^2 - a^2} + \frac{m^2}{p^2 - b^2} + \frac{n^2}{p^2 - c^2} = 0.$$

Fig. 43

Observe from the second constraint that the maximum of p^2 is less than a^2.

18. This exercise is devoted to one of the exceptional cases which can arise in connection with the derivation of (6.8–11) in Example 2. Suppose $a > b > 0$ and $b = c$, so that the ellipsoid has circular cross sections in planes perpendicular to the x-axis. Suppose also that $l \neq 0$. Show that, in this case, the minimum p^2 is b^2 and that the maximum p^2 is the sole root of (6.8–11).

19. Find the maximum and minimum values of the squared distance from the origin to the first octant portion of the curve in which the plane $x + y + z = 12$ meets the surface $xyz = 54$.

20. Find the minimum of $x_1^2 + \cdots + x_n^2$ subject to the constraint $a_1x_1 + \cdots + a_nx_n = 1$, where $a_1^2 + \cdots + a_n^2 > 0$.

21. Find the maximum of $(\sum_{i=1}^{n} a_ix_i)^2$ subject to the condition $\sum_{i=1}^{n} x_i^2 = 1$. Assume that at least one of the a's $\neq 0$.

22. If $P(x, y)$ is a point on the ellipse $256\,x^2 + 81\,y^2 = 2304$ and $Q(u, v)$ is a point on the line $4\,x + 3\,y = 24$, find the minimum value of the distance PQ.

23. Find the minimum distance between the curves $x^2 + y^2 = 1$, $x^2y = 16$.

24. Let $y = f(x)$ be the equation of a smooth curve, and let $P(a, b)$ be a point not on the curve. Let D be the distance from P to a variable point Q of the curve. Prove by Lagrange's method that PQ is normal to the curve when D is a minimum or a maximum.

25. (a) Use the result of Exercise 24 to give a simple solution of Exercise 23. (b) Proceed as directed in (a) to solve Exercise 22.

26. If $P(x_0, y_0, z_0)$ is a fixed point outside the ellipsoid $(x^2/a^2) + (y^2/b^2) + (z^2/c^2) = 1$, and Q is on the ellipsoid, show that the line PQ is normal to the ellipsoid when the distance PQ is a minimum.

27. Let $G(x, y, z) = 0$ define a surface S with a tangent plane at each point, and let $P(a, b, c)$ be a point not on S. Show that, if Q is a point on S, the line PQ is normal to S whenever the distance PQ attains a relative extreme. From this result prove that, if A and B are variable points on two smooth nonintersecting surfaces, then, when A and B are located so as to make the distance between them an absolute minimum, the line AB is normal to both surfaces.

28. Show that the maximum value of $x^2y^2z^2$ subject to $x^2 + y^2 + z^2 = R^2$ is $(R^2/3)^3$. Deduce from this that

$$(x^2 y^2 z^2)^{1/3} \leq \frac{x^2 + y^2 + z^2}{3}$$

for all values of x, y, z. Generalize this result to n variables, and so obtain a proof of the inequality

$$(a_1 a_2 \cdots a_n)^{1/n} \leq \frac{a_1 + a_2 + \cdots + a_n}{n}$$

between the geometric mean and arithmetic mean of n positive numbers. What condition is both necessary and sufficient for equality?

29. In the case of two positive numbers, Exercise 28 gives

$$u^{1/2} v^{1/2} \leq \tfrac{1}{2} u + \tfrac{1}{2} v.$$

Derive from this the fact that

$$\left| \sum_{i=1}^{n} a_i b_i \right| \leq \left(\sum_{i=1}^{n} a_i^2 \right)^{1/2} \left(\sum_{i=1}^{n} b_i^2 \right)^{1/2}$$

if the a's and b's are any real numbers. This is known as *Cauchy's inequality.* Hint: Let

$$u = \frac{a_i^2}{A} \quad \text{and} \quad v = \frac{b_i^2}{B}, \qquad \text{where}$$

$$A = \sum_{i=1}^{n} a_i^2 \quad \text{and} \quad B = \sum_{i=1}^{n} b_i^2.$$

What condition is both necessary and sufficient for equality?

30. If $\alpha_1, \alpha_2, \cdots, \alpha_n$ are nonnegative numbers whose sum is 1, then $\alpha_1 x_1 + \alpha_2 x_2 + \cdots + \alpha_n x_n$ is called a *weighted arithmetic mean* of the x's. If the x's are positive, then one can define the corresponding *weighted* geometric mean to be $x_1^{\alpha_1} x_2^{\alpha_2}, \cdots, x_n^{\alpha_n}$. Show that if all the x's are positive, the weighted geometric mean is always less than or equal to the corresponding weighted arithmetic mean. Hint: Maximize $x_1^{\alpha_1} x_2^{\alpha_2}, \cdots, x_n^{\alpha_n}$ subject to the constraint $\alpha_1 x_1 + \alpha_2 x_2 + \cdots + \alpha_n x_n = C$. This will show that for all positive x's having a given weighted arithmetic mean C, the weighted geometric mean is less than or equal to C.

Notice that this result contains that of Exercise 28 as a special case.

31. If u, v, p, and q are positive and $(1/p) + (1/q) = 1$, then Exercise 30 gives

$$u^{1/q} v^{1/q} \leq \frac{1}{p} u + \frac{1}{q} v.$$

Deduce from this that under the above restrictions on p and q,

$$\left| \sum_{i=1}^{n} a_i b_i \right| \leq \left(\sum_{i=1}^{n} |a_i|^p \right)^{1/p} \left(\sum_{i=1}^{n} |b_i|^q \right)^{1/q},$$

where the a's and b's are any real numbers. This is known as *Hölder's inequality.* Notice that it is a generalization of Cauchy's inequality.

Hint: Let

$$u = \frac{|a_i|^p}{\sum\limits_{i=1}^{n} |a_i|^p} \quad \text{and} \quad v = \frac{|b_i|^q}{\sum\limits_{i=1}^{n} |b_i|^q}.$$

32. Use Hölder's inequality (Exercise 30) to deduce the inequality of Minkowski:

$$\left(\sum_{i=1}^{n} |x_i + y_i|^p\right)^{1/p} \leq \left(\sum_{i=1}^{n} |x_i|^p\right)^{1/p} + \left(\sum_{i=1}^{n} |y_i|^p\right)^{1/p}, \; p > 1.$$

Write $|x_i + y_i|^p = |x_i + y_i| \, |x_i + y_i|^{p/q} \leq |x_i| \, |x_i + y_i|^{p/q} + |y_i| \, |x_i + y_i|^{p/q}$, and then apply Hölder's inequality, once with $a_i = |x_i|$ and once with $a_i = |y_i|$. Observe that Minkowski's inequality obviously holds when $p = 1$.

6.9   Quadratic forms

A function of the type

$$(6.9\text{--}1) \qquad\qquad F(x, y) = Ax^2 + 2\,Bxy + Cy^2$$

is called a *homogeneous quadratic form* in x and y; the coefficients A, B, C may be any constants. Likewise, a function of the type

$$(6.9\text{--}2) \quad F(x, y, z) = Ax^2 + By^2 + Cz^2 + 2\,Dxy + 2\,Exz + 2\,Fyz$$

is called a homogeneous quadratic form in x, y, and z. The definition may be generalized for n variables, but we shall confine our attention to the cases of two and three variables. Quadratic forms are of much importance in algebra and geometry, and they arise frequently in a variety of contexts in applied mathematics.

For simplicity we shall devote nearly all our attention to the two-variable case. It will then be easy to understand the extension of our remarks to the three-variable case.

Consider the family of curves

$$(6.9\text{--}3) \qquad\qquad F(x, y) = k,$$

where k is a parameter. The nature of this family will depend upon the particular quadratic form. If they are ellipses, for example, they are all centered at the origin, and they all have the same axes of symmetry. The ellipses are all similar and similarly placed. The size of an individual ellipse is governed by the parameter k. For instance,

$$25\,x^2 + 36\,y^2 = k$$

is a family of ellipses, provided $k > 0$. The semiaxes of a particular ellipse

are $\qquad\qquad \dfrac{\sqrt{k}}{5} \quad$ and $\quad \dfrac{\sqrt{k}}{6}.$

There are three general possibilities as to the type of curve in the family. These possibilities are distinguished by the character of the *discriminant* of the quadratic form:

(6.9-4)
$$\Delta = \begin{vmatrix} A & B \\ B & C \end{vmatrix} = AC - B^2.$$

If $\Delta > 0$, the curves are ellipses; in this case $AC > 0$ and we may assume A and C are positive; we must then limit k to positive values, since otherwise the equation (6.9-3) will have no real locus. If $\Delta < 0$, the curves are hyperbolas except in the special case $k = 0$, when the locus is the pair of lines which are the asymptotes of all the hyperbolas in the family. If $\Delta = 0$, the family is an assemblage of parallel straight lines.

The facts stated in the foregoing paragraph can be established with the aid of a suitable rotation of the co-ordinate axes. First consider an arbitrary rotation of axes, with new co-ordinates x', y' related to the old co-ordinates x, y by equations

(6.9-5)
$$x = x' \cos \theta - y' \sin \theta,$$
$$y = x' \sin \theta + y' \cos \theta.$$

These equations transform the form $F(x, y)$ into a new form, say

$$G(x', y') = A'x'^2 + 2 B'x'y' + C'y'^2.$$

Now, it is a remarkable fact that the discriminant of $G(x', y')$ is the same as that of $F(x, y)$; in other words,

(6.9-6)
$$A'C' - B'^2 = AC - B^2.$$

Another very important fact is that, when $F(x, y)$ is given, it is always possible to choose the rotation of axes so that the $x'y'$ term is missing in $G(x', y')$, i.e., $B' = 0$. In this case the equation of the family (6.9-3) takes the form

(6.9-7)
$$A'x'^2 + C'y'^2 = k.$$

Now the discriminant is $\Delta = A'C'$, by (6.9-6), since $B' = 0$. Consequently $\Delta > 0$ means that A' and C' are of the same sign, so that the curves are ellipses; $\Delta < 0$ means that A' and C' are of opposite sign, so that the curves are hyperbolas; and $\Delta = 0$ means that $A' = 0$ or $C' = 0$, in either of which cases the curves are parallel lines.

It is of importance to know how to find the coefficients A', C' *without actually performing the rotation of axes*. The invariance of the discriminant provides us a method of accomplishing this. Consider the quadratic form

(6.9-8)
$$F(x, y) - \lambda(x^2 + y^2),$$

where λ is a parameter. Under a rotation of co-ordinate axes, this form is transformed into

(6.9-9)
$$G(x', y') - \lambda(x'^2 + y'^2),$$

because $x^2 + y^2 = x'^2 + y'^2$. Suppose that the rotation is such that $B' = 0$.

Then, the fact that the forms (6.9–8) and (6.9–9) have the same discriminant means that

(6.9–10)

$$\begin{vmatrix} A - \lambda & B \\ B & C - \lambda \end{vmatrix} = \begin{vmatrix} A' - \lambda & 0 \\ 0 & C' - \lambda \end{vmatrix} = (A' - \lambda)(C' - \lambda).$$

As a consequence, we see that $\lambda = A'$ and $\lambda = C'$ are the roots of the equation

(6.9–11)
$$\begin{vmatrix} A - \lambda & B \\ B & C - \lambda \end{vmatrix} = 0,$$

which is obtained by setting equal to zero the discriminant of the form $F(x, y) - \lambda(x^2 + y^2)$.

Example 1. Find the dimensions of the ellipse $73\ x^2 + 72\ xy + 52\ y^2 = 100$.

Here $A = 73$, $B = 36$, $C = 52$.

The equation (6.9–11) becomes

$$\lambda^2 - 125\ \lambda + 2500 = 0,$$

with roots $\lambda = 25$, $\lambda = 100$. This means, therefore, that the equation of the ellipse can be put in the form

$$25\ x'^2 + 100\ y'^2 = 100, \text{ or } \frac{x'^2}{4} + y'^2 = 1,$$

by a rotation of axes. The principal semiaxes of the ellipse are therefore 2 and 1.

In the foregoing discussion there are two crucial facts. The *first* of these is the invariance of the discriminant, as expressed in (6.9–6). This is a purely algebraic fact (see Exercise 13). The *second* important fact is the possibility of choosing a rotation so that $B' = 0$ in the new form $G(x', y')$. Let us now see how to prove that this choice is really possible.

We consider the values of the function $F(x, y)$ on the unit circle $x^2 + y^2 = 1$. Since F is continuous, there must be some point at which F attains its absolute maximum subject to the constraint $x^2 + y^2 = 1$. Let this point be (x_1, y_1). Now let the $x'y'$-axes be chosen in such a way that the point $x = x_1$, $y = y_1$ has co-ordinates $x' = 1$, $y' = 0$ (see Fig. 44). Then, if $F(x, y)$ becomes $G(x', y')$, it must be true that the absolute maximum of $G(x', y')$ subject to the constraint $x'^2 + y'^2 = 1$ occurs at the point $(1, 0)$ in the new co-ordinate system. Let us see what we obtain from

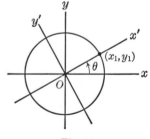

Fig. 44

this fact when we formulate and solve the extremal problem by the method of Lagrange. We write

$$u = A'x'^2 + 2 B'x'y' + C'y'^2 - \lambda(x'^2 + y'^2),$$

and set $\dfrac{\partial u}{\partial x'} = \dfrac{\partial u}{\partial y'} = 0$. We have written $-\lambda$ instead of λ so as to make our notation conform to that occurring earlier in the present section. The differentiation yields, after slight simplification, the equations

(6.9–12)
$$(A' - \lambda)x' + B'y' = 0,$$
$$B'x' + (C' - \lambda)y' = 0.$$

These equations must be satisfied when $x' = 1$, $y' = 0$, and for a certain value $\lambda = \lambda_1$. Therefore, we conclude that

$$A' - \lambda_1 = 0 \text{ and } B' = 0.$$

Thus $G(x', y') = \lambda_1 x'^2 + C'y'^2.$

We have now demonstrated that we can make $B' = 0$ by a suitable rotation of the axes. Observe that the determinant of the system of linear equations (6.9–12) is precisely the discriminant of the form $G(x', y') - \lambda(x'^2 + y'^2)$. This discriminant is a polynomial in λ, and, from the invariance of the discriminant, it follows that the roots of the polynomial are the same, no matter how the axes are rotated. Let the roots be denoted by λ_1, λ_2, with $\lambda_1 \geqq \lambda_2$. With $B' = 0$ we know that the roots are A' and C'. We have arranged matters so that $\lambda_1 = A'$ and $\lambda_2 = C'$. The quadratic form is then

$$G(x', y') = \lambda_1 x'^2 + \lambda_2 y'^2.$$

We repeat, for emphasis, that it is not necessary actually to perform the rotation of axes if we merely want to find λ_1, λ_2. They are the roots of the equation (6.9–11). The natural way of arriving at this equation is through applying Lagrange's method to the problem of extremizing $F(x, y)$ subject to the constraint $x^2 + y^2 = 1$, for the Lagrange method, applied to

$$Ax^2 + 2 Bxy + Cy^2 - \lambda(x^2 + y^2),$$

leads to the equations

$$(A - \lambda)x + By = 0,$$
$$Bx + (C - \lambda)y = 0.$$

These are homogeneous linear equations in x and y. We want a solution of them such that $x^2 + y^2 = 1$, and therefore such that x and y are not both zero. Such a solution will exist if and only if the equation (6.9–11) holds.

The generalization to the case of three variables is now a rather easy matter. The essential statement of the generalization is this: Given the quadratic form $F(x, y, z)$ as in (6.9–2), it is possible to make a rotation of the co-ordinate axes so that $F(x, y, z)$ becomes

(6.9–13) $$G(x', y', z') = \lambda_1 x'^2 + \lambda_2 y'^2 + \lambda_3 z'^2,$$

and the λ's are the roots of the equation

(6.9–14) $$\begin{vmatrix} A - \lambda & D & E \\ D & B - \lambda & F \\ E & F & C - \lambda \end{vmatrix} = 0.$$

A proof of this statement is indicated in Exercise 14. If one wishes, matters may be arranged so that $\lambda_1 \geq \lambda_2 \geq \lambda_3$.

Example 2. Reduce the quadratic form $xy + xz - yz$ to the form (6.9–13), and so identify the surface $xy + xz - yz = -1$.

The equation (6.9–14) takes the form

$$\begin{vmatrix} -\lambda & \frac{1}{2} & \frac{1}{2} \\ \frac{1}{2} & -\lambda & -\frac{1}{2} \\ \frac{1}{2} & -\frac{1}{2} & -\lambda \end{vmatrix} = 0$$

in this case. On expansion and simplification this becomes

$$-\tfrac{1}{4} + \tfrac{3}{4}\lambda - \lambda^3 = 0,$$

the roots of which are found to be $\frac{1}{2}, \frac{1}{2}, -1$. Thus, after a suitable rotation, our equation becomes

$$\tfrac{1}{2} x'^2 + \tfrac{1}{2} y'^2 - z'^2 = -1.$$

This defines a hyperboloid of two sheets with circular cross sections perpendicular to the z'-axis (for $|z'| > 1$).

A more symmetrical notation for quadratic forms is in some ways extremely desirable. If we write x_1, x_2, x_3 instead of x, y, z, a quadratic form in x_1, x_2, x_3 will have terms of all possible types $x_i x_j$, with i and j assuming the values 1, 2, 3. If we write a_{ij} for the coefficient of $x_i x_j$, the quadratic form will have the appearance

(6.9–15) $$\begin{aligned} F(x_1, x_2, x_3) = \quad & a_{11}x_1^2 + a_{12}x_1x_2 + a_{13}x_1x_3 \\ + \; & a_{21}x_2x_1 + a_{22}x_2^2 + a_{23}x_2x_3 \\ + \; & a_{31}x_3x_1 + a_{32}x_3x_2 + a_{33}x_3^2 \; . \end{aligned}$$

Since $x_1 x_2 = x_2 x_1$, we agree to make $a_{12} = a_{21} =$ half the total coefficient of $x_1 x_2$; similarly for a_{13} and a_{23}. The *discriminant* of the form is, by definition, the determinant

(6.9–16) $$\begin{vmatrix} a_{11} & a_{12} & a_{13} \\ a_{21} & a_{22} & a_{23} \\ a_{31} & a_{32} & a_{33} \end{vmatrix}.$$

The determinant appearing in (6.9–14) is seen to be the discriminant of the form $F(x, y, z) - \lambda(x^2 + y^2 + z^2)$.

EXERCISES

1. Find the dimensions of the ellipse $41\,x^2 - 24\,xy + 34\,y^2 = 25$.

2. Show that $F(x, y) = x^2 - 4\,xy - 2\,y^2 = 1$ is the equation of a hyperbola. Find a point on the unit circle at which $F(x, y)$ is a maximum. Then draw the xy-axes, the axes of symmetry of the hyperbola, and the hyperbola itself.

3. Find the maximum and minimum values of $F(x, y) = 9\,x^2 - 6\,xy + y^2$ on the circle $x^2 + y^2 = 1$. If λ_1 is the maximum value, show that $F(x, y) = \lambda_1$ is the equation of two lines, each tangent to the circle at a point where the maximum occurs.

4. Let $F(x, y, z) = y^2 + z^2 - \sqrt{2}\,xy + \sqrt{2}\,xz + 2\,yz$. Find the maximum and minimum values of this function on the surface of the unit sphere. What does (6.9–13) become in this case?

5. Reduce each of the following quadratic forms to the standard form (6.9–13):
(a) $y^2 + z^2 - \sqrt{2}\,xy - \sqrt{2}\,xz + 2\,yz$.
(b) $13\,x^2 + 13\,y^2 + 10\,z^2 + 8\,xy - 4\,xz - 4\,yz$.

6. (a) Put $F(x, y, z) = xy + yz + zx$ in the form (6.9–13).
(b) What is the maximum value of F on the unit sphere?
(c) At what points does it occur?

7. Determine the signs of λ_1, λ_2, λ_3 for each of the following quadratic forms, and so identify the type of each quadric surface. You need not find the actual value of the λ's.
(a) $x^2 + xy + yz = 1$.
(b) $yz + xz - xy = 1$.
(c) $x^2 + 2\,y^2 + 3\,z^2 - 2\,xy - 2\,yz = 2$.

8. Find maximum and minimum values of $17\,x^2 - 30\,xy + 17\,y^2$ when $5\,x^2 - 6\,xy + 5\,y^2 = 4$.

9. Find the minimum value of $x^2 + y^2 + xy$ subject to the condition $2\,x^2 + 6\,xy + 2\,y^2 = 9$.

10. Suppose that the locus of $Ax^2 + 2\,Bxy + Cy^2 = 1$ is an ellipse. Consider the problem of locating the maximum and minimum values of $x^2 + y^2$ on the ellipse. Apply Lagrange's method, starting with the expression $x^2 + y^2 - \lambda(Ax^2 + 2\,Bxy + Cy^2)$. Show that, in the resulting equations for locating the extreme values, λ must be a root of the equation

$$\begin{vmatrix} 1 - \lambda A & -B\lambda \\ -B\lambda & 1 - C\lambda \end{vmatrix} = 0,$$

and that the roots of this equation are the extreme values of $x^2 + y^2$. What is the relation between these roots and the semiaxes of the ellipse?

11. If x, y, λ are solutions of the simultaneous equations

$$(A - \lambda)x + By = 0,$$
$$Bx + (C - \lambda)y = 0, \qquad x^2 + y^2 = 1,$$

show that $Ax^2 + 2Bxy + Cy^2 = \lambda$. Hence show that, in the notation used in the text, λ_1 and λ_2 are respectively the maximum and minimum values of $F(x, y)$ when $x^2 + y^2 = 1$.

12. (a) Assume that $F(x, y) = k$ ($k > 0$) defines a family of ellipses. Let λ_1 and λ_2 be the roots of (6.9–11), with $\lambda_1 \geqq \lambda_2$. Show that the ellipse $F(x, y) = \lambda_1$ is externally tangent to the circle $x^2 + y^2 = 1$ at the ends of the minor axis of the ellipse, and that the ellipse $F(x, y) = \lambda_2$ is internally tangent to the circle at the ends of the major axis of the ellipse. Draw a figure showing these two ellipses, the circle, and the $x'y'$-axes. What can you infer from the figure about maximum and minimum values of $F(x, y)$ on the circle?

(b) Assume that $F(x, y) = k$ defines a family of hyperbolas. If λ_1 and λ_2 are the roots of (6.9–11) with $\lambda_1 \geqq \lambda_2$, explain why $\lambda_1 > 0$ and $\lambda_2 < 0$. Draw a figure showing the $x'y'$-axes, the unit circle, the hyperbolas $F(x, y) = \lambda_1$, $F(x, y) = \lambda_2$, and other members of the family.

13. Prove (6.9–6) by actually substituting (6.9–5) in (6.9–1) and computing A', B', C'. Also prove that $A' + C' = A + C$.

14. (a) Suppose that $F(x, y, z)$ in (6.9–2) becomes a new form $G(x', y', z')$ with coefficients A', B', $\cdots$, F' when we shift to new axes $x'y'z'$ obtained by a rotation from the xyz-system. Write the equations which correspond to (6.9–12) for the problem of making $G(x', y', z')$ a maximum on the unit sphere. If the new axes are chosen so that this maximum occurs at $x' = 1$, $y' = z' = 0$, show that $D' = E' = 0$, and that $G(x', y', z') = \lambda_1 x'^2 + B'y'^2 + C'z'^2 + 2F'y'z'$, where λ_1 is the maximum value of G on the unit sphere. Now explain how it is possible to choose a new set of axes x'', y'', z'', by a rotation from the $x'y'z'$-system, rotating about the x'-axis, in such a way that the form becomes $\lambda_1 x''^2 + \lambda_2 y''^2 + \lambda_3 z''^2$, where λ_2 and λ_3 are respectively the maximum and minimum values of G subject to the two constraints $x' = 0$, $y'^2 + z'^2 = 1$.

(b) It may be proved algebraically that the discriminant of a quadratic form $F(x, y, z)$ is equal to the discriminant of the new form $G(x', y', z')$ which is obtained from $F(x, y, z)$ by a rotation of axes. Use this fact to prove that the numbers λ_1, λ_2, λ_3 described in part (a) of this problem are the roots of the cubic equation (6.9–14).

MISCELLANEOUS EXERCISES

1. Choose a and b so that $\int_0^1 (\sqrt{x} - a - bx)^2 \, dx$ is as small as possible; $a + bx$ is then called a "least-square" approximation to $\sqrt{x}$ in the interval [0, 1].

2. If $\phi(u, v) = f(x, y)$, where $u = y^2 - x^2$, $v = y^2 + x^2$, show that

$$\frac{1}{4xy} \frac{\partial^2 f}{\partial x \, \partial y} = \frac{\partial^2 \phi}{\partial v^2} - \frac{\partial^2 \phi}{\partial u^2}.$$

3. (a) Show by a diagram the part of the xy-plane in which $f(x, y) = (a - x)(a - y)(x + y - a) \geqq 0$. Assume $a > 0$. (b) Find all points of the plane at which $f_1(x, y) = f_2(x, y) = 0$. (c) Which of these points yield relative maxima or minima of $f(x, y)$, and which do not? (d) Does f have any absolute extrema?

4. Let $S = 32\sqrt{3}\left(\dfrac{1}{x} + \dfrac{1}{y}\right) + xy(\sec\theta - \tfrac{1}{2}\tan\theta)$. Find the minimum value of S in the region $x > 0$, $y > 0$, $0 \leqq \theta < \pi/2$ of $xy\theta$-space. How do you know that the minimum does not occur when $\theta = 0$?

5. Find the point of occurrence of the maximum value of x^2yz^3 on the part of the plane $x + y + z = 24$ that lies in the first octant.

6. Find the maximum and minimum values of $x^2 + y^2 + z^2$ subject to the two conditions $(x^2/25) + (y^2/25) + (z^2/9) = 1$, $x + y + 2z = 0$.

7. Find the ratios y/x and x/z to make $x^2 + 2xy + x\sqrt{x^2 + z^2}$ a minimum when $3x^2y + x^2z$ has a prescribed positive value (x, y, z all > 0).

8. Find the maximum and minimum values of z subject to the conditions $x^2 + y^2 + z^2 = 25$, $(x - 2\sqrt{6})^2 + (y - 2\sqrt{3})^2 + (z - 6)^2 = 13$.

9. Suppose a, b, c, p given all positive, and $p < 1$. Find the maximum of $ax^p + by^p + cz^p$ subject to the conditions $x + y + z = 1$, $x, y, z > 0$.

10. Consider $S = 2(xy + yz + zx)$, subject to the conditions $ax + by + cz = 2A$, and $x, y, z > 0$, where a, b, c, A are fixed positive constants. Show that the maximum value of S is $8A^2[(a + b + c)^2 - 2(a^2 + b^2 + c^2)]^{-1}$.

11. Find the absolute maximum and minimum values of
$$f(x, y) = 3x^2 - 2(y + 1)x + 3y - 1$$
in the square $0 \leqq x \leqq 1$, $0 \leqq y \leqq 1$.

12. (a) Find the minimum distance from the point $(3, 4, 15)$ to the cone $4z^2 = x^2 + y^2$.

(b) Find the minimum distance from the point $(9, 12, -5)$ to the cone $4z^2 = x^2 + y^2$.

(c) Are there any relative extrema in addition to the absolute minimum in part (a)? in part (b)?

General Theorems of Partial Differentiation

7

7. Purpose of the chapter

This chapter is not primarily concerned with the technique of partial differentiation, but with statements and proofs of some of the important theorems about differentials and partial differentiation. We have separated the material of the chapter from that of Chapter VI for a number of reasons. In studying the subject of partial differentiation the student needs first of all to get acquainted with the new ideas which the subject presents to him. He needs to assimilate these ideas through the medium of examples and problems. He will want a reasoned development of the subject, but he will be more interested in mastery of technique and appreciation of some applications than in the details of the longer proofs, particularly as regards the proofs of theorems which he is quite willing to take for granted in the early stages. The theorems of §§ 7.1, 7.2, and 7.3 have been referred to in Chapter VI. The student needs to know these theorems, but he can very well go through Chapter VI without studying their proofs.

Sections 7.4, 7.5 are on a somewhat different footing. Every student who uses advanced calculus is likely to have need of Taylor's formula for a function of several variables. One meets references to the formula frequently in the literature of applied mathematics and in various branches of higher analysis. The law of the mean is merely a special case of Taylor's formula. We have put this material here rather than in Chapter VI because its applications are not so immediate.

The final section of the chapter deals with tests for maximum or minimum values of a function $F(x, y)$ in terms of the second partial derivatives. These tests are of interest and of theoretical importance. Corresponding tests for functions of more than two variables exist, but we have not included them in the book.

Within the limits of time of an ordinary year course in advanced calculus the instructor may wish to make only a selection from Chapter VII, with such a degree of emphasis on the proofs as he sees fit. The various sections are practically independent of each other, except that §§ 7.4 and 7.5 go together.

7.1 Sufficient conditions for differentiability

The concept of differentiability for a function of several variables was defined in § 6.4. To be differentiable at a given point a function must have first partial derivatives at that point. But this alone is not enough. We

212

may have a function $f(x, y)$ such that $f_1(0, 0)$ and $f_2(0, 0)$ exist, and yet such that f is not differentiable at $(0, 0)$; for an illustration see Exercise 7, § 6.4. The following theorem deals with sufficient conditions for differentiability:

THEOREM I.   *Suppose the function $f(x, y)$ is defined in some neighborhood of the point (a, b). Suppose one of the partial derivatives, say $\dfrac{\partial f}{\partial x}$, exists at each point of the neighborhood and is continuous at (a, b), while the other partial derivative is defined at least at the point (a, b). Then f is differentiable at (a, b).*

PROOF.   Let $z = f(x, y)$. We have to show that condition (6.4–7) is satisfied. In the notation used in connection with (6.4–7) we can write

(7.1–1)
$$\Delta z = f(a + \Delta x, b + \Delta y) - f(a, b + \Delta y) + f)a, b + \Delta y) - f(a, b).$$

We assume that $|\Delta x|$ and $|\Delta y|$ are less than δ, where δ is positive and so small that the square of side 2δ with center at (a, b) lies wholly in the region of definition of f and f_1. Considering $f(x, b + \Delta y)$ as a function of x alone, and applying the law of the mean, we have

(7.1–2)
$$f(a + \Delta x, b + \Delta y) - f(a, b + \Delta y) = \Delta x f_1(a + \theta \Delta x, b + \Delta y),$$

where $0 < \theta < 1$. By the continuity of f_1 at (a, b) we have

$$\lim_{(\Delta x, \Delta y) \to (0, 0)} f_1(a + \theta \Delta x, b + \Delta y) = f_1(a, b).$$

Also, by the definition of $f_2(a, b)$, we have

$$\lim_{\Delta y \to 0} \frac{f(a, b + \Delta y) - f(a, b)}{\Delta y} - f_2(a, b) = 0.$$

Consequently we may write

(7.1–3)
$$f_1(a + \theta \Delta x, b + \Delta y) - f_1(a, b) = \epsilon_1,$$
$$f(a, b + \Delta y) - f(a, b) - f_2(a, b)\Delta y = \epsilon_2 \Delta y,$$

where ϵ_1 and ϵ_2 are variable quantities which are zero when Δx and Δy are both zero, and which approach zero as Δx and Δy approach zero.

Turning now to (6.4–7), we see by the foregoing relations that

(7.1–4)
$$\frac{\Delta z - [f_1(a, b)\Delta x + f_2(a, b)\Delta y]}{|\Delta x| + |\Delta y|} = \frac{\epsilon_1 \Delta x + \epsilon_2 \Delta y}{|\Delta x| + |\Delta y|}.$$

Now
$$\left| \frac{\epsilon_1 \Delta x}{|\Delta x| + |\Delta y|} \right| = |\epsilon_1| \frac{|\Delta x|}{|\Delta x| + |\Delta y|} \leq |\epsilon_1|,$$

and similarly
$$\left| \frac{\epsilon_2 \, \Delta y}{|\Delta x| + |\Delta y|} \right| \leqq |\epsilon_2|.$$

Hence the left member of (7.1–4) does not exceed $|\epsilon_1| + |\epsilon_2|$. It therefore approaches zero as Δx and Δy approach zero. Condition (6.4–7) is thus fulfilled, and the proof is complete.

It will be observed that the conditions of the theorem are not symmetrical as regards x and y. We might equally well have assumed the mere existence of $f_1(a, b)$, while assuming the continuity of $f_2(x, y)$ at (a, b). In general, for a function of more than two variables, we assume the mere existence of one of the first partial derivatives, and the continuity of the other first partial derivatives. We may then conclude that the function is differentiable. The proof is similar to that of Theorem I. For most practical purposes the following statement is sufficient:

THEOREM II. *A function of several variables is differentiable at a point if the function and all its first partial derivatives are defined in some neighborhood of the point and if these derivatives are continuous at the point.*

EXERCISES

1. Let $f(x, y) = (x^4 + y^4)/(x^2 + y^2)$ if $x^2 + y^2 \neq 0$, and define $f(0, 0) = 0$. Show that f has first partial derivatives at all points, satisfying the inequalities $|f_1(x, y)| \leqq 6|x|$, $|f_2(x, y)| \leqq 6|y|$. Is f differentiable at $(0, 0)$?

2. Let $f(x, y) = (x^3 - y^3)/(x^2 + y^2)$ if $x^2 + y^2 \neq 0$, and define $f(0, 0) = 0$. Show that f has first partial derivatives at all points, but that these derivatives are discontinuous at $(0, 0)$. The function is *not* differentiable at $(0, 0)$. To prove this, show first that if it *were* differentiable, one could write
$$\frac{x^3 - y^3}{x^2 + y^2} = x - y + \epsilon(|x| + |y|),$$
where $\epsilon \to 0$ as $(x, y) \to (0, 0)$. Then show that this is impossible. SUGGESTION: Consider the situation when $y = -x$.

3. Let $f(x, y) = (x^2 + y^2)\sin \dfrac{1}{\sqrt{x^2 + y^2}}$ if $x^2 + y^2 \neq 0$, and define $f(0, 0) = 0$. Show that f has first partial derivatives at all points, but that these derivatives are discontinuous at $(0, 0)$. Show that $|f_1(x, y)| \leqq 2|x| + 1$. Prove that f is differentiable at $(0, 0)$.

This example shows that the hypotheses in Theorems I and II are sufficient, but not necessary, conditions for differentiability.

7.2 Changing the order of differentiation

We mentioned at the outset of Chapter VI that we ordinarily find the relation

(7.2–1)
$$\frac{\partial^2 f}{\partial y \, \partial x} = \frac{\partial^2 f}{\partial x \, \partial y}$$

to be valid for the functions $f(x, y)$ which we meet in everyday use of calculus. The relation (7.2–1) may be false in particular cases, however, and so it is well to know something of the conditions sufficient to guarantee its validity.

THEOREM III.   *Let the function $f(x, y)$ be defined in some neighborhood of the point (a, b). Let the partial derivatives f_1, f_2, f_{12} and f_{21} also be defined in this neighborhood, and suppose that f_{12} and f_{21} are continuous at (a, b). Then $f_{12}(a, b) = f_{21}(a, b)$. In other words, (7.2–1) holds at the point (a, b).*

PROOF.   We shall work entirely inside a square having its center at (a, b), and lying inside the neighborhood referred to in the theorem. Let h be a number different from zero such that the point $(a + h, b + h)$ is inside the square just referred to. Consider the expression

$$D = f(a + h, b + h) - f(a + h, b) - f(a, b + h) + f(a, b).$$

If we introduce the function

$$\phi(x) = f(x, b + h) - f(x, b),$$

we can express D in the form

(7.2–2) $$D = \phi(a + h) - \phi(a).$$

Now ϕ has the derivative

$$\phi'(x) = f_1(x, b + h) - f_1(x, b).$$

Hence ϕ is continuous, and we may apply the law of the mean to (7.2–2), with the result

(7.2–3) $$D = h\phi'(a + \theta_1 h) = h[f_1(a + \theta_1 h, b + h) - f_1(a + \theta_1 h, b)],$$

where $0 < \theta_1 < 1$. Next, let

$$g(y) = f_1(a + \theta_1 h, y).$$

The function g has the derivative

$$g'(y) = f_{12}(a + \theta_1 h, y).$$

Now we can write (7.2–3) in the form

$$D = h[g(b + h) - g(b)]$$

and apply the law of the mean. The result is

$$D = h^2 g'(b + \theta_2 h) = h^2 f_{12}(a + \theta_1 h, b + \theta_2 h),$$

where $0 < \theta_2 < 1$.

We might instead have started by expressing D in the form

$$D = \psi(b + h) - \psi(b),$$

where $\quad\quad \psi(y) = f(a + h, y) - f(a, y).$

This procedure would have led to an expression

$$D = h^2 f_{21}(a + \theta_4 h, b + \theta_3 h),$$

with $0 < \theta_3 < 1, 0 < \theta_4 < 1$. On comparing the two expressions for D we see that

(7.2–4) $\quad\quad f_{12}(a + \theta_1 h, b + \theta_2 h) = f_{21}(a + \theta_4 h, b + \theta_3 h).$

If we now make $h \to 0$, the points at which the derivatives in (7.2–4) are evaluated both approach (a, b). Hence, by the assumed continuity of f_{12} and f_{21}, we conclude that $f_{12}(a, b) = f_{21}(a, b)$. This completes the proof.

The conditions of Theorem III are not the only known sufficient conditions for the truth of (7.2–1). The theorem provides a useful working criterion, however. Another criterion is furnished by the following theorem:

THEOREM IV. *Let $f(x, y)$ and its first partial derivatives f_1, f_2 be defined in a neighborhood of the point (a, b), and suppose that f_1 and f_2 are differentiable at that point. Then $f_{12}(a, b) = f_{21}(a, b)$.*

This theorem requires more of the function f in some ways, and less in others, than Theorem III. We omit the proof, which begins very much like that of Theorem III.

We may use Theorem III or Theorem IV to prove that a mixed partial derivative of order higher than the second is independent of the order of performing the differentiations, provided we make appropriate assumptions about the continuity or differentiability. Thus, for example, suppose that $f(x, y)$ and all its partial derivatives of orders one, two, and three are continuous. (This is more than we actually need.) Then

$$\frac{\partial^3 f}{\partial y^2 \, \partial x} = \frac{\partial^3 f}{\partial y \, \partial x \, \partial y} = \frac{\partial^3 f}{\partial x \, \partial y^2}.$$

For $\quad\quad f_{122} = (f_{12})_2 = (f_{21})_2 = f_{212},$

and $\quad\quad f_{212} = (f_2)_{12} = (f_2)_{21} = f_{221}.$

By similar arguments we can deal with functions of more than two independent variables.

EXERCISES

1. Let $f(x, y) = xy\left(\dfrac{x^2 - y^2}{x^2 + y^2}\right)$ if $x^2 + y^2 \neq 0$, and define $f(0, 0) = 0$. Show

that $f_1(0, y) = -y$ and $f_2(x, 0) = x$ for all x and y. Then show that $f_{12}(0, 0) = -1$ and $f_{21}(0, 0) = 1$.

2. Define $f(x, y) = x^2 \tan^{-1} (y/x) - y^2 \tan^{-1} (x/y)$ if neither x nor y is zero, and $f(x, y) = 0$ if either $x = 0$ or $y = 0$ (or both). Show as in Exercise 1 that $f_{12}(0, 0) = -1, f_{21}(0, 0) = 1$.

3. Theorem III can be improved as follows:
Assume that f is defined in some neighborhood of the point (a, b), and that the partial derivatives f_1, f_2, f_{12} are also defined in this neighborhood. Suppose finally that f_{12} is continuous at (a, b). Then it is true that f_{21} is defined at (a, b) and that $f_{12}(a, b) = f_{21}(a, b)$.
(This theorem is due to H. A. Schwarz.)
Prove the theorem with the assistance of the following suggestions: Let h, k be small numbers different from zero. Let

$$\Delta = f(a + h, b + k) - f(a + h, b) - f(a, b + k) + f(a, b).$$

Show that there are numbers θ_1, θ_2 between 0 and 1, depending on h and k, such that

$$\Delta = hk f_{12}(a + \theta_1 h, b + \theta_2 k).$$

If $\epsilon > 0$, choose $\delta > 0$ so that $\left| \dfrac{\Delta}{hk} - f_{12}(a, b) \right| < \epsilon$ if $0 < |h| < \delta$ and $0 < |k| < \delta$. Why is this possible? Find the limit of $\dfrac{\Delta}{hk}$ with h fixed, as $k \to 0$, and so conclude that

$$\left| \frac{f_2(a + h, b) - f_2(a, b)}{h} - f_{12}(a, b) \right| \leq \epsilon$$

if $0 < |h| < \delta$. Now complete the proof of Schwarz's theorem.

4. To prove Theorem IV, start as in the proof of Theorem III, and obtain (7.2–3). Then, from the fact that f_1 is differentiable at (a, b), one can write

$$f_1(a + \theta_1 h, b + h) = f_1(a, b) + f_{11}(a, b)\theta_1 h + f_{12}(a, b)h + \epsilon_1 |h|,$$

where $\epsilon_1 \to 0$ as $h \to 0$. Explain why this is so.
Then go on to explain how to obtain the expression

$$D = h^2 f_{12}(a, b) + \epsilon |h| h,$$

where $\epsilon \to 0$ as $h \to 0$. Explain the derivation of the similar expression (where $\epsilon' \to 0$ as $h \to 0$)

$$D = h^2 f_{21}(a, b) + \epsilon' |h| h,$$

using the fact that f_2 is differentiable at (a, b). Now complete the proof of Theorem IV.

7.3   Differentials of composite functions

In § 6.4 we made the statement that a differentiable function of differentiable functions is differentiable. We now formulate this proposition in precise terms, and prove it. There may be any number of independent variables in each of the functions involved. For simplicity we deal with the case of two variables throughout.

THEOREM V. *Let $F(x, y)$ be defined in some neighborhood of the point (a, b), and let it be differentiable there. Let $f(s, t)$ and $g(s, t)$ be defined in some neighborhood of (s_0, t_0), and differentiable at that point. Assume further that*

$$f(s_0, t_0) = a, \; g(s_0, t_0) = b,$$

and consider the composite function

$$G(s, t) = F(f(s, t), g(s, t)).$$

Then G is differentiable at (s_0, t_0). Its differential may be written

$$(7.3\text{--}1) \qquad dG = \frac{\partial F}{\partial x} \, dx + \frac{\partial F}{\partial y} \, dy,$$

where

$$(7.3\text{--}2) \qquad dx = \frac{\partial f}{\partial s} \, ds + \frac{\partial f}{\partial t} \, dt,$$

$$dy = \frac{\partial g}{\partial s} \, ds + \frac{\partial g}{\partial t} \, dt.$$

It is to be understood that the partial derivatives of F are evaluated at (a, b), those of f and g at (s_0, t_0), and that ds, dt are independent variables.

PROOF. Let us write $u = G(s, t)$, $x = f(s, t)$, $y = g(s, t)$. For arbitrary Δs, Δt write $\Delta u = G(s_0 + \Delta s, t_0 + \Delta t) - G(s_0, t_0)$, with corresponding meanings for Δx, Δy. Note that $\Delta u = F(a + \Delta x, b + \Delta y) - F(a, b)$. Accordingly, the differentiability hypotheses mean that we may write

$$\Delta u = F_1 \, \Delta x + F_2 \, \Delta y + \epsilon(|\Delta x| + |\Delta y|),$$
$$(7.3\text{--}3) \qquad \Delta x = f_1 \, \Delta s + f_2 \, \Delta t + \delta(|\Delta s| + |\Delta t|),$$
$$\Delta y = g_1 \, \Delta s + g_2 \, \Delta t + \eta(|\Delta s| + |\Delta t|),$$

where the partial derivatives are evaluated as indicated in the statement of the theorem, and

$$\epsilon \to 0 \text{ as } \Delta x \text{ and } \Delta y \to 0,$$

while $\qquad \delta \text{ and } \eta \to 0 \text{ as } \Delta s \text{ and } \Delta t \to 0$

(compare with (6.4–8)). On combining formulas (7.3–3) we obtain

$$\Delta u = F_1(f_1 \, \Delta s + f_2 \, \Delta t) + F_1 \delta(|\Delta s| + |\Delta t|) + F_2(g_1 \, \Delta s + g_2 \, \Delta t)$$
$$+ F_2 \eta(|\Delta s| + |\Delta t|) + \epsilon(|\Delta x| + |\Delta y|).$$

This may be rewritten in the form

$$\Delta u = (F_1 f_1 + F_2 g_1)\Delta s + (F_1 f_2 + F_2 g_2)\Delta t$$

$$+ \left\{ F_1\delta + F_2\eta + \epsilon \frac{|\Delta x| + |\Delta y|}{|\Delta s| + |\Delta t|} \right\} (|\Delta s| + |\Delta t|).$$

If we can show that

(7.3–4)
$$\lim_{(\Delta s, \Delta t) \to (0, 0)} \left\{ F_1\delta + F_2\eta + \epsilon \frac{|\Delta x| + |\Delta y|}{|\Delta s| + |\Delta t|} \right\} = 0,$$

we shall have proved that G is differentiable, with the differential

$$dG = (F_1 f_1 + F_2 g_1)ds + (F_1 f_2 + F_2 g_2)dt.$$

This last form is equivalent to the combination of (7.3–1) and (7.3–2). Hence it remains only to prove (7.3–4).

Now F_1 and F_2 denote constants in (7.3–4), so that $F_1\delta + F_2\eta \to 0$. The functions f and g are continuous at (s_0, t_0) by Theorem II, § 6.4. Hence Δx and $\Delta y \to 0$, and therefore $\epsilon \to 0$ as Δs and $\Delta t \to 0$. Let M be a constant larger than the greatest of the numbers $|f_1|, |f_2|, |g_1|, |g_2|$. Then by (7.3–3) we see that

$$|\Delta x| \leqq (M + \delta)(|\Delta s| + |\Delta t|), \quad |\Delta y| \leqq (M + \eta)(|\Delta s| + |\Delta t|).$$

Hence
$$\epsilon \frac{|\Delta x| + |\Delta y|}{|\Delta s| + |\Delta t|} \leqq \epsilon(2 M + \delta + \eta).$$

This quantity approaches zero as Δs and Δt approach zero. The assertion (7.3–4) is therefore established.

7.4   The law of the mean

The student is already familiar with the law of the mean for functions of a single independent variable, in the form

(7.4–1) $$f(a + h) - f(a) = hf'(a + \theta h), 0 < \theta < 1$$

(see (1.2–4) and Theorem IV, § 1.2). It is useful to have a generalization of this result for functions of several independent variables. In seeking an appropriate form for such a generalization, we look upon (7.4–1) as furnishing a convenient formula for the difference between the values of the function f at two points of the x-axis, namely $x = a$ and $x = a + h$. This leads us, in the case of a function of two variables, to search for a means of expressing the difference

$$F(a + h, b + k) - F(a, b),$$

where the line segment joining the points (a, b), $(a + h, b + k)$ lies in the region of definition of the function F.

THEOREM VI. *Let F be defined in a region R of the xy-plane. Let L be the line segment with ends* (a, b), $(a + h, b + k)$. *We suppose that L lies in R and that all points of L except possibly the ends are interior points of R. Finally, we assume that F is continuous at each point of L and differentiable at each such point with the possible exception of the ends. Then, for a certain value of* θ, *such that* $0 < \theta < 1$, *we have*

$$(7.4\text{-}2)\quad F(a + h, b + k) - F(a, b) = hF_1(a + \theta h, b + \theta k)$$
$$+ kF_2(a + \theta h, b + \theta k).$$

PROOF. By introducing a parametric representation of the line segment L,

$$x = a + th, \, y = b + tk, \, 0 \leqq t \leqq 1,$$

we are able to regard the value of F along L as a function of the parameter t. Let us write

$$(7.4\text{-}3)\qquad\qquad f(t) = F(a + th, b + tk).$$

The derivative of this composite function is

$$(7.4\text{-}4)\quad f'(t) = hF_1(a + th, b + tk) + kF_2(a + th, b + tk).$$

Applying the ordinary law of the mean in the form (7.4–1), we have

$$(7.4\text{-}5)\qquad\qquad f(1) - f(0) = f'(\theta), \, 0 < \theta < 1.$$

On setting $t = 0, 1$ successively in (7.4–3), and $t = \theta$ in (7.4–4), we see that (7.4–5) becomes the formula (7.4–2). The proof is therefore complete.

Observe that the point $(a + \theta h, b + \theta k)$ is a point of the segment L somewhere between its ends (see Fig. 45).

It is clear from (7.4–2) that if the partial derivatives F_1 and F_2 have the value zero all along the line L, then F has the same value at both ends of L. This observation leads to the proof of a theorem which is analogous to Theorem V of § 1.2.

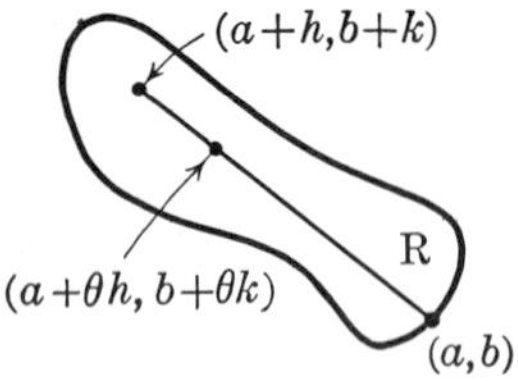

First we make a definition.

Fig. 45

DEFINITION. An open set S (in the plane or in space of three dimensions) is called *connected* if any two points of S can be joined by a path consisting of a finite number of line segments, the whole path lying entirely in S.

Later, in § 17.7, we shall define the term "connected set" for sets which are not necessarily open. That definition does not refer to joining points by segmental paths, but for open sets the definitions are equivalent.

Fig. 46

As an example of a nonconnected set, let S consist of all points of the plane for which $x^2 > 1$, that is, all points except those for which $-1 \leq x \leq 1$. Plainly S consists of two separated parts (see Fig. 46). Two points, one in each part, cannot be joined by a broken-line path lying entirely in S. This particular set S comes naturally to our attention if we study the function

$$f(x, y) = y + \sqrt{x^2 - 1}.$$

Now we come to the theorem.

THEOREM VII. *Let $F(x, y)$ be a function which is defined and differentiable throughout a connected open set S, and suppose that the first partial derivatives of F vanish at each point of S. Then $F(x, y)$ is constant in S.*

PROOF. Suppose A and B are any two points of S. Let them be joined by a path consisting of segments AP_1, P_1P_2, $\cdots$, $P_{n-1}P_n$, P_nB, all lying in S (see Fig. 47). By the comment just after the proof of Theorem VI we see that F has the same value at A as at P_1, the same value at P_1 as at P_2, and so on, so that F has the same value at B as at A. This means that F is constant in S.

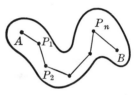

Fig. 47

Theorems VI and VII admit of immediate extension to functions of more than two independent variables. The extension of the law of the mean for three independent variables is

(7.4–6)
$$F(a + h, b + k, c + l) - F(a, b, c)$$
$$= hF_1(\bar{x}, \bar{y}, \bar{z}) + kF_2(\bar{x}, \bar{y}, \bar{z}) + lF_3(\bar{x}, \bar{y}, \bar{z}),$$

where $\bar{x} = a + \theta h, \bar{y} = b + \theta k, \bar{z} = c + \theta l.$

Formula (7.4–2) can also be written in the form

(7.4–7) $\quad F(x, y) - F(a, b) = F_1(X, Y)(x - a) + F_2(X, Y)(y - b),$

where (X, Y) is a certain point on the line joining (a, b) and (x, y).

EXERCISES

1. Let $F(x, y) = xy^2 - x^2y$. Find the appropriate value of θ in (7.4–2) if (a) $a = b = 0, h = 1, k = 2$; (b) $a = b = 0, h = 3, k = 2$; (c) $a = b = 1, h = 3, k = 2$.

2. Let $F(x, y)$ be the quadratic function $Ax^2 + 2\,Bxy + Cy^2$. Show that (7.4–7) holds with $X = \frac{1}{2}(x + a)$, $Y = \frac{1}{2}(y + b)$. What does this mean about the value of θ in (7.4–2)?

3. Taking $F(x, y) = \sin x \cos y$, prove that for some θ between 0 and 1 it is true that $\dfrac{3}{4} = \dfrac{\pi}{3} \cos \dfrac{\pi\theta}{3} \cos \dfrac{\pi\theta}{6} - \dfrac{\pi}{6} \sin \dfrac{\pi\theta}{3} \sin \dfrac{\pi\theta}{6}$.

4. (a) Write out formula (7.4–2) for $F(x, y) = \log(xe^{y^2})$, with $a = 1$, $b = 0$, $h = e - 1$, $k = 1$. (b) Write out (7.4–7) for this same function, with a, b, x, y arbitrary, except that $a > 0$, $x > 0$.

5. If $x \neq a$ in (7.4–7), show that $Y = b + \dfrac{y - b}{x - a}(X - a)$. Hence show that, under suitable conditions, $F(1, 0) - F(0, 1) = F_1(X, 1 - X) - F_2(X, 1 - X)$.

6. Let $F(x, y) = (1 - 2\,xy + x^2)^{-1/2}$. As a result of considering $F(1, 0) - F(0, 1)$, show that there is a number θ such that $0 < \theta < 1$ and

$$1 - \sqrt{2} = \sqrt{2}(1 - 3\,\theta)(1 - 2\,\theta + 3\,\theta^2)^{-3/2}.$$

7. Let $F(x, y, z) = xyz$. Find the appropriate value of θ in (7.4–6) if
(a) $a = b = c = 0$, $h = k = l = 1$; (b) $a = b = 0$, $c = 1$, $h = k = 1$, $l = -1$;
(c) $a = c = 0$, $b = 1$, $h = l = 1$, $k = 0$.

7.5   Taylor's formula and series

Just as we extended the ordinary law of the mean to functions of several variables, so we may extend the version of Taylor's formula given in § 4.3. The method is the same as that employed in the proof of Theorem VI, § 7.4. We write

$$(7.5-1) \qquad f(t) = F(a + th, b + tk)$$

and apply Taylor's formula to $f(t)$, using the two values $t = 0$, $t = 1$. From (4.3–7) with $a = 0$, $h = 1$ we have

$$(7.5-2) \quad f(1) = f(0) + f'(0) + \cdots + \frac{f^{(n)}(0)}{n!} + \frac{f^{(n+1)}(\theta)}{(n + 1)!}, 0 < \theta < 1.$$

The assumptions are that F and its partial derivatives of orders 1 to n inclusive are differentiable at all points along the line joining (a, b) and $(a + h, b + k)$. The main problem now is that of calculating the higher derivatives of f from (7.5–1). The first derivative is given by (7.4–4) in the previous section. Working from that formula, we see that

$$f''(t) = h[hF_{11} + kF_{12}] + k[hF_{21} + kF_{22}],$$

where all the partial derivatives on the right are evaluated at $(a + th, b + tk)$. Since $F_{12} = F_{21}$ (Theorem IV, § 7.2), we have

$$f''(t) = h^2 F_{11} + 2\,hkF_{12} + k^2 F_{22}.$$

This is sometimes written in the form

$$f''(t) = \left[\left(h\frac{\partial}{\partial x} + k\frac{\partial}{\partial y}\right)^2 F(x, y)\right]_{x = a + th,\, y = b + tk},$$

it being understood that

$$\left(h\frac{\partial}{\partial x} + k\frac{\partial}{\partial y}\right)^2 F(x, y) = h^2\frac{\partial^2 F}{\partial x^2} + 2\,hk\frac{\partial^2 F}{\partial x\,\partial y} + k^2\frac{\partial^2 F}{\partial y^2}.$$

The analogy with the pattern of the binomial expansion is now evident. We have $\quad f'''(t) = h^3 F_{111} + 3\,h^2 k F_{112} + 3\,hk^2 F_{122} + k^3 F_{222}$

$$= \left(h\frac{\partial}{\partial x} + k\frac{\partial}{\partial y}\right)^3 F(x, y),$$

with x and y set equal to $a + th$ and $b + tk$, respectively, in the partial derivatives. The general formula is

$$(7.5\text{-}3) \qquad f^{(n)}(t) = \left[\left(h\frac{\partial}{\partial x} + k\frac{\partial}{\partial y}\right)^n F(x, y)\right]_{x = a + th,\, y = b + tk}.$$

To get Taylor's formula for $F(x, y)$ we use (7.5-1) and (7.5-3) to substitute in (7.5-2). For $n = 1$ the result is

$$(7.5\text{-}4) \quad F(a + h, b + k) = F(a, b) + hF_1(a, b) + kF_2(a, b) + R_2,$$

where $\qquad R_2 = \dfrac{1}{2!}\left[\left(h\dfrac{\partial}{\partial x} + k\dfrac{\partial}{\partial y}\right)^2 F(x, y)\right]_{x = a + \theta h,\, y = b + \theta k},$

and $0 < \theta < 1$. For $n = 2$ the result is

$$(7.5\text{-}5) \quad F(a + h, b + k)$$
$$= F(a, b) + hF_1(a, b) + kF_2(a, b)$$
$$+ \frac{1}{2!}\,[h^2 F_{11}(a, b) + 2\,hk F_{12}(a, b) + k^2 F_{22}(a, b)] + R_3,$$

where $R_3 = \dfrac{1}{3!}\left[\left(h\dfrac{\partial}{\partial x} + k\dfrac{\partial}{\partial y}\right)^3 F(x, y)\right]_{x = a + \theta h,\, y = b + \theta k},$

and $0 < \theta < 1$. The extension to higher values of n is obvious. Observe that from $f^{(n)}(0)$ we get a homogeneous polynomial of degree n in h and k, the coefficient of $h^{n-p}k^p$ being $\dfrac{n!}{p!(n - p)!}\dfrac{\partial^n F}{\partial x^{n-p}\,\partial y^p}$, with the partial derivative evaluated at $x = a$, $y = b$.

Under certain conditions on the function F, the point (a, b), and the size of h and k, it may happen that $F(a + h, b + k)$ can be represented as the infinite series

(7.5-6) $F(a + h, b + k)$

$$= F(a, b) + \sum_{n=1}^{\infty} \frac{1}{n!} \left[\left(h \frac{\partial}{\partial x} + k \frac{\partial}{\partial y} \right)^n F(x, y) \right]_{x = a, y = b}.$$

This is the form of Taylor's series for a function of two variables. If we use only a specified number of terms of this series we get an approximate expression for $F(a + h, b + k)$. Frequently it is more convenient to write (x, y) in place of $(a + h, b + k)$. The typical term of the series then becomes a homogeneous polynomial in $(x - a)$ and $(y - b)$.

Example. Write Taylor's formula (7.5-4) with $n = 1$, and carry the series (7.5-6) through the term in $n = 2$, if $F(x, y) = 1/(xy)$, $a = 1$, $b = -1$.

From $F(x, y) = x^{-1}y^{-1}$ we readily find

$$\frac{\partial F}{\partial x} = - x^{-2}y^{-1}, \frac{\partial F}{\partial y} = - x^{-1}y^{-2},$$

$$\frac{\partial^2 F}{\partial x^2} = 2 x^{-3}y^{-1}, \frac{\partial^2 F}{\partial x \, \partial y} = x^{-2}y^{-2}, \frac{\partial^2 F}{\partial y^2} = 2 x^{-1}y^{-3}.$$

It is now easily seen that (7.5-4) becomes

$$\frac{1}{(1 + h)(- 1 + k)} = - 1 + h - k + R_2,$$

$$R_2 = \frac{h^2}{(1 + \theta h)^3(- 1 + \theta k)} + \frac{hk}{(1 + \theta h)^2(- 1 + \theta k)^2}$$

$$+ \frac{k^2}{(1 + \theta h)(- 1 + \theta k)^3}.$$

The series (7.5-6) begins

$$\frac{1}{(1 + h)(- 1 + k)} = - 1 + (h - k) + (- h^2 + hk - k^2) + \cdots.$$

Detailed verification should be supplied by the student as he reads this example.

If we write $x = 1 + h$, $y = - 1 + k$, the last formula becomes

$$\frac{1}{xy} = - 1 + [(x - 1) + (y + 1)]$$

$$+ [- (x - 1)^2 + (x - 1)(y + 1) - (y + 1)^2] + \cdots.$$

We shall not investigate the precise limitations on $(x - 1)$ and $(y + 1)$ which are necessary in this expansion.

EXERCISES

1. Write Taylor's formula (7.5-5) for $F(x, y) = \sin x \sin y$, using $a = 0$, $b = 0$, and $n = 2$.

2. Write Taylor's formula for $F(x, y) = \cos x \cos y$, using $a = 0$, $b = 0$, and $n = 2$.

3. Write Taylor's formula with $a = 3$, $b = 3$, $n = 3$ for $F(x, y) = x^3 + y^3 - 9xy + 27$.

4. Let $F(x, y) = \log(x + e^y)$. Expand according to Taylor's series in powers of $x - 1$ and y, going far enough to include all terms of degree 2 in these quantities.

5. Follow the instructions of Exercise 4 for $F(x, y) = \sin(e^y + x^2 - 2)$.

6. Write Taylor's series for $e^x \cos y$ in powers of x and y, going far enough to include all terms of third degree.

7. Write Taylor's series for $e^{-y^2 + 2xy}$ in powers of x and y, going far enough to include all terms of fourth degree.

8. Write Taylor's series for $x^2y + xy^2 + 1$ in powers of $x - 1$ and $y - 1$.

9. Write Taylor's series for $xy^3 - y^2 + y + 2$ in powers of $x - 1$ and $y - 2$.

10. (a) Find a linear function of x and y which is a good approximation for
$$F(x, y) = \tan^{-1}\left(\frac{x - y}{1 + xy}\right) \text{ when } x \text{ and } y \text{ are small.}$$

(b) Write the constant and linear terms in Taylor's series of $F(x, y)$ in powers of $x - 3$ and $y - \frac{1}{2}$.

11. Write out in full the expression
$$\frac{1}{4!}\left(h\frac{\partial}{\partial x} + k\frac{\partial}{\partial y}\right)^4 F(x, y).$$

What does the expression become (a) if $F(x, y) = x^4 - x^2y^2 + y^4$; (b) if $F(x, y) = \sin xy$ and if one sets $x = y = (\pi/2)^{1/2}$ after doing the differentiation?

12. (a) Carry on the work of the illustrative example in the text, showing that
$$\frac{\partial^n F}{\partial x^{n-p} \partial y^p} = (-1)^n (n - p)! \, p! \, \frac{1}{x^{n-p+1}y^{p+1}},$$

and that the polynomial of degree n in h and k in the Taylor's series is
$$(-1)^{n-1}[h^n - h^{n-1}k + h^{n-2}k^2 - \cdots + (-1)^n k^n].$$

(b) Assuming that $|h| < 1$ and $|k| < 1$, write
$$\frac{1}{(1 + h)(-1 + k)} = -(1 + h)^{-1}(1 - k)^{-1}$$
$$= -(1 - h + h^2 - h^3 + \cdots)(1 + k + k^2 + \cdots).$$

Then multiply the two series together term by term, and collect together the terms of like degree. Compare with the result found in **(a)**.

7.6   Sufficient conditions for a relative extreme

In § 6.3 we discussed relative maxima and minima for a function $f(x, y)$. In Theorem I of that section we reached the important conclusion that if

f attains a relative extreme value at an interior point of its region of defi-
nition, then necessarily the partial derivatives $\dfrac{\partial f}{\partial x}$ and $\dfrac{\partial f}{\partial y}$ vanish at the
point (provided these derivatives exist, of course). The conditions

(7.6–1) $\dfrac{\partial f}{\partial x} = 0, \dfrac{\partial f}{\partial y} = 0$

at a point do not in themselves guarantee a relative extreme, however.
In this section we wish to develop criteria which, taken together with
conditions (7.6–1), will guarantee a relative extreme, and enable us to
distinguish a relative maximum from a relative minimum.

It will be helpful if we begin by reviewing briefly the analogous considerations for a function of one variable. Suppose we have a function $y = f(x)$
defined on some interval having $x = a$ as an interior point. We suppose
f to be differentiable on the interval, and we assume that the second derivative exists at $x = a$.

THEOREM VIII.   *Under the conditions on f as just stated, suppose that*
   $f'(a) = 0$. Then
   (i) *If $f''(a) > 0$, f has a relative minimum at $x = a$;*
   (ii) *If $f''(a) < 0$, f has a relative maximum at $x = a$;*
   (iii) *If $f''(a) = 0$ no conclusion may be drawn; f may have a relative*
       maximum or minimum, or it may have neither.

PROOF.   Consider the value of f at any point $x = a + h$ near $x = a$.
We take $h \neq 0$, but it may be either positive or negative. By the law of
the mean,

(7.6–2) $f(a + h) - f(a) = hf'(a + \theta h), 0 < \theta < 1.$

Now, by the definition of $f''(a)$, we have

$$\lim_{\Delta x \to 0} \frac{f'(a + \Delta x) - f'(a)}{\Delta x} = f''(a).$$

Since $f'(a) = 0$ we can write this in the form

$$\frac{f'(a + \Delta x)}{\Delta x} = f''(a) + \epsilon,$$

where ϵ is a variable quantity such that $\epsilon \to 0$ as $\Delta x \to 0$. Consequently,
if we choose $\Delta x = \theta h$, we have

$$f'(a + \theta h) = (f''(a) + \epsilon)\theta h,$$

and (7.6–2) becomes

(7.6–3) $f(a + h) - f(a) = (f''(a) + \epsilon)\theta h^2.$

If we now suppose that $f''(a) \neq 0$, we see that, as soon as h is small enough to insure that $|\epsilon| < |f''(a)|$, the sign of the left member of (7.6-3) is the same as the sign of $f''(a)$. Consequently, if $f''(a) > 0$, we conclude that $f(a + h) > f(a)$ for all sufficiently small values of h different from zero. This means that f has a relative minimum at $x = a$. We have thus proved part (i) of the theorem; the same kind of argument is used for part (ii). If $f''(a) = 0$ no conclusion is reached, however, since we do not know the sign of ϵ in (7.6-3). The three examples $y = x^4$, $y = -x^4$, $y = x^3$, with $a = 0$, show that any of the possibilities mentioned in part (iii) may arise.

Let us now turn to the case of two independent variables. We have the following corresponding theorem:

THEOREM IX. *Suppose that $F(x, y)$ is defined and differentiable through-out a region R of which (a, b) is an interior point, and suppose that the first partial derivatives of F vanish at that point. Suppose further that the partial derivatives F_1 and F_2 are differentiable at (a, b). Let us write*

$$A = F_{11}(a, b), B = F_{12}(a, b), C = F_{22}(a, b).$$

Then
  (i) *If $B^2 - AC < 0$ and $A > 0$, F has a relative minimum at (a, b);*
  (ii) *If $B^2 - AC < 0$ and $A < 0$, F has a relative maximum at (a, b);*
  (iii) *If $B^2 - AC > 0$, F has neither a maximum nor a minimum at the point;*
  (iv) *If $B^2 - AC = 0$, no conclusion may be drawn and any of the behaviors of F described in parts (i)–(iii) may occur.*

PROOF. Note that part (iii) of Theorem IX has no counterpart in Theorem VIII. The method of proof of Theorem IX is similar to that of Theorem VIII. We consider the point $(a + h, b + k)$, where h and k are both small, but not both zero. By the law of the mean we have (with $0 < \theta < 1$)

$$F(a + h, b + k) - F(a, b) = hF_1(a + \theta h, b + \theta k) \\ + kF_2(a + \theta h, b + \theta k).$$

Since we assumed F_1 and F_2 differentiable, we may write (see (6.4–8))

$$F_1(a + \theta h, b + \theta k) - F_1(a, b) = \theta h F_{11}(a, b) + \theta k F_{12}(a, b) \\ + \epsilon_1(|\theta h| + |\theta k|),$$

where $\epsilon_1 \to 0$ as h and $k \to 0$. A similar expression may be written for F_2, with some quantity ϵ_2 in place of ϵ_1. Since $F_1(a, b) = F_2(a, b) = 0$, by hypothesis, we have

(7.6–4) $F(a + h, b + k) - F(a, b)$
$$= \theta[(Ah^2 + 2 Bhk + Ck^2) + (|h| + |k|)(\epsilon_1 h + \epsilon_2 k)].$$

This formula is the counterpart of (7.6–3). To get the information we desire from it, however, it is more convenient to express h and k in terms of polar co-ordinates with origin at (a, b). Let us write

$$h = r \cos \phi, \; k = r \sin \phi.$$

When these expressions are substituted in (7.6–4), a factor r^2 can be taken out, and we get

(7.6–5) $F(a + h, b + k) - F(a, b) = \theta r^2 [G(\phi) + \delta],$

where for abbreviation we have set

$$G(\phi) = A \cos^2 \phi + 2 B \sin \phi \cos \phi + C \sin^2 \phi,$$
$$\delta = (| \cos \phi | + | \sin \phi |)(\epsilon_1 \cos \phi + \epsilon_2 \sin \phi).$$

Here $0 < \theta < 1$, and $\delta \to 0$ as $r \to 0$. Since $G(\phi)$ is independent of r, it is clear that, if $G(\phi) \neq 0$, the sign of the left member of (7.6–5) will be the same as the sign of $G(\phi)$ when r is sufficiently small. Moreover, $G(\phi)$ is continuous, whence it follows that, if $G(\phi)$ is never zero, it always has the same sign, and, when r is sufficiently small, the sign of $G(\phi) + \delta$ will be the same as the sign of $G(\phi)$. Therefore, if $G(\phi)$ is always positive, F has a minimum at (a, b), while if $G(\phi)$ is always negative, F has a maximum at (a, b). On the other hand, if $G(\phi)$ is sometimes positive and sometimes negative, F has neither a maximum nor a minimum at (a, b). We shall show that the cases (i)–(iii) in the statement of Theorem IX lead to exactly these three types of behavior for $G(\phi)$.

Observe that

$$Ah^2 + 2 Bhk + Ck^2 = r^2 G(\phi).$$

This shows that the sign of $G(\phi)$ is the same as the sign of the quadratic function

$$f(h, k) = Ah^2 + 2 Bhk + Ck^2.$$

Now let us regard h, k as rectangular co-ordinates in a system with origin at the point $x = a, \; y = b$. Let h', k' denote rectangular co-ordinates in a system obtained from the hk-system by a rotation about the origin of the

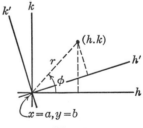

Fig. 48

system (see Fig. 48). As we saw in § 6.9, it is possible to choose the rotation in such a way that $f(h, k)$ becomes

$$(7.6\text{--}6) \qquad \lambda_1 h'^2 + \lambda_2 k'^2 = r^2 G(\phi),$$

where λ_1 and λ_2 are roots of the equation

$$(7.6\text{--}7) \qquad \begin{vmatrix} A - \lambda & B \\ B & C - \lambda \end{vmatrix} = \lambda^2 - (A + C)\lambda + AC - B^2 = 0.$$

Here $h^2 + k^2 = h'^2 + k'^2 = r^2$. Observe that the product of the roots of (7.6–7) is

$$(7.6\text{--}8) \qquad \lambda_1 \lambda_2 = AC - B^2,$$

and that the sum is

$$(7.6\text{--}9) \qquad \lambda_1 + \lambda_2 = A + C.$$

Everything now depends on the sign of the expression (7.6–6). It is clear that if λ_1 and λ_2 are both positive, $G(\phi)$ is positive, and we have the case of a minimum at (a, b), whereas if λ_1 and λ_2 are both negative, we have the case of a maximum at (a, b). Let us now consider cases (i) and (ii) of the theorem. The hypothesis $B^2 - AC < 0$ implies that A and C are of the same sign, and also, by (7.6–8), that λ_1 and λ_2 are of the same sign. Consequently, from (7.6–9) we see that $B^2 - AC < 0$ and $A > 0$ imply that λ_1 and λ_2 are positive, whereas $B^2 - AC < 0$ and $A < 0$ imply that λ_1 and λ_2 are negative. The conclusions in cases (i) and (ii) are therefore established by the foregoing arguments.

In case (iii), $B^2 - AC > 0$ implies that λ_1 and λ_2 are of opposite signs. Now we can choose ϕ so as to make $h' = 0$, $k' \neq 0$, $r^2 G(\phi) = \lambda_2 k'^2$, and we can also choose ϕ so as to make $h' \neq 0$, $k' = 0$, $r^2 G(\phi) = \lambda_1 h'^2$. Thus $G(\phi)$ can change sign, and so we have neither a maximum nor a minimum at (a, b).

Finally, if $B^2 - AC = 0$, at least one of the roots λ_1 and λ_2 is zero, by (7.6–8). No conclusion about a maximum or minimum can be drawn in this case. The reasons for this are clear from (7.6–6) and (7.6–5). As examples we cite the three functions

$$x^2, \; -y^2, \; x^4 - y^2.$$

Each of these functions comes under case (iv) of Theorem IX, with $a = b = 0$. The first has a minimum at $(0, 0)$, the second a maximum, while the third has neither.

A point at which all the first partial derivatives of a function are equal to zero is called a *critical point* of the function. If (a, b) is a critical point of $F(x, y)$, and if $B^2 - AC \neq 0$ at (a, b) (in the notation of Theorem IX), we call the critical point *nondegenerate*. The point is called a *saddle point* if $B^2 - AC > 0$.

It should be noted that, if we assume that F has partial derivatives of all orders, and that it can be represented by its Taylor's series (see § 7.5), then, when (a, b) is a critical point, the Taylor's series starts as follows:

$$F(a + h, b + k) = F(a, b) + \frac{1}{2!} [Ah^2 + 2\, Bhk + Ck^2] + \cdots.$$

Observe also that $AC - B^2$ is the discriminant of the quadratic form

$$Ah^2 + 2\, Bhk + Ck^2.$$

This gives us a clue to the proper method of defining nondegenerate critical points for functions of more than two variables. We illustrate briefly for the case of three variables. Suppose the origin is a critical point of the function $F(x_1, x_2, x_3)$, so that $F_1(0, 0, 0) = F_2(0, 0, 0) = F_3(0, 0, 0) = 0$. Let $a_{11} = F_{11}(0, 0, 0)$, $a_{12} = F_{12}(0, 0, 0)$, $a_{13} = F_{13}(0, 0, 0)$, and so on.

Then Taylor's series is

$$F(x_1, x_2, x_3) = F(0, 0, 0) + \frac{1}{2!} \sum_{i, j = 1}^{3} a_{ij}x_ix_j + \cdots.$$

The critical point is called nondegenerate in case the determinant

$$\begin{vmatrix} a_{11} & a_{12} & a_{13} \\ a_{21} & a_{22} & a_{23} \\ a_{31} & a_{32} & a_{33} \end{vmatrix}$$

is not zero. For nondegenerate critical points there is a generalization of Theorem IX. We consider the roots λ_1, λ_2, λ_3 of the equation

$$\begin{vmatrix} a_{11} - \lambda & a_{12} & a_{13} \\ a_{21} & a_{22} - \lambda & a_{23} \\ a_{31} & a_{32} & a_{33} - \lambda \end{vmatrix} = 0.$$

If these roots are all of the same sign, F has a *minimum* at the critical point if the roots are all *positive*, and a *maximum* there if they are all *negative*. But, if some roots are positive and some negative, there is neither a maximum nor a minimum at the critical point. These facts follow quickly from the discussion of quadratic forms in § 6.9. It is not actually necessary to assume the convergence of Taylor's series. All that is needed is that the first derivatives of F be differentiable at the critical point.

EXERCISES

1. Find all the critical points of each of the following functions. Test each critical point by Theorem IX, and state your conclusion.

(a) $y^2 + 3\, x^4 - 4\, x^3 - 12\, x^2 + 24$.
(b) $x^2 - 12\, y^2 + 4\, y^3 + 3\, y^4$.

(c) $x^4 + y^4 - 2x^2 + 4xy - 2y^2$.

(d) $x^2y^2 - 5x^2 - 8xy - 5y^2$.

(e) $2(x - y)^2 - x^4 - y^4$.

(f) $xy(12 - 3x - 4y)$.

(g) $x^3y^2(a - x - y)$, $a > 0$.

(h) $(1 - x)(1 - y)(x + y - 1)$.

(i) $x^2y(24 - x - y)^3$.

(j) $\dfrac{1}{xy} - \dfrac{4}{x^2y} - \dfrac{8}{xy^2}$.

(k) $\frac{5}{4}(x^2 + y^2) - 18x - 24y + 5\sqrt{x^2 + y^2} + 250$.

(l) $5(x^2 + y^2) - 24x - 32y - 60\sqrt{x^2 + y^2} + 1000$.

(m) $12x \sin y - 2x^2 \sin y + x^2 \sin y \cos y$.

2. If a and b are positive, show that $(a/x) + (b/y) + xy$ has a minimum at its only critical point. What is the situation if a and b are both negative? if they have opposite signs?

3. How many critical points has the function $(ax^2 + by^2)e^{-x^2-y^2}$ if $b > a > 0$? Discuss the nature of each of them.

4. Find the shortest distance from the point $(1, -1, 1)$ to the surface $z = xy$. Set up the squared distance as a function of x, y, find the critical points of the function, and test them by Theorem IX.

5. Discuss the problem of finding the shortest distance from the point $(0, 0, a)$ to the surface $z = xy$, where $a > 0$. Proceed as directed in Exercise 4. Separate the cases $0 < a \leq 1$ and $1 < a$.

6. If z is defined as a function of x, y by the equation $2x^2 + 2y^2 + z^2 + 8xz - z + 8 = 0$, find the points (x, y, z) at which z has a relative extreme, and test by second derivatives for a maximum or minimum in each case.

7. Proceed as directed in Exercise 6, starting from the equation

$$x^2 + 2y^2 + 3z^2 - 2xy - 2yz = 2.$$

8. Suppose that F is defined in the neighborhood of (a, b), and that $F_1(a, b) = 0$, $F_{11}(a, b) < 0$. Why is it impossible for F to have a relative minimum at (a, b)?

9. Locate the critical points of the function $xyz(x + y + z - 1)$. Show that there are six lines all of whose points are degenerate critical points, and one nondegenerate critical point. Is this a maximum or a minimum point? Can you answer this last question without second derivative tests?

10. Show that every critical point of the function $\dfrac{x^3 + y^3 + z^3}{xyz}$ is degenerate.

11. Locate the critical points of the function

$$F(x, y, z) = (ax^2 + by^2 + cz^2)e^{-x^2-y^2-z^2},$$

where $a > b > c > 0$. Show that there are two points of maximum value of F, one point of minimum value, and four critical points at which there is neither a maximum nor a minimum.

12. Study the function $F(x, y) = (y^2 - x^2)(y^2 - 2x^2)$. Show that there are four lines which divide the plane into eight regions, in each of which F has a constant sign. Discuss the critical points of the function. Are there any relative extrema?

13. Discuss the sign of the function $F(x, y) = (2x^2 - y)(x^2 - y)$ at various points of the plane, by appropriate consideration of the regions into which the plane is divided by the two parabolas $y = x^2$, $y = 2x^2$. Discuss the critical points of the function. Show that, along every straight line through the origin, the values of F reach a minimum at $(0, 0)$, but that F has neither a maximum nor a minimum at $(0, 0)$.

MISCELLANEOUS EXERCISES

1. Suppose $f(x, y)$ is differentiable at (a, b), with $A = f_1(a, b)$, $B = f_2(a, b)$. Let $F(r, \theta) = f(a + r \cos \theta, b + r \sin \theta)$. Then $F_1(0, \theta)$ exists and is equal to $A \cos \theta + B \sin \theta$, for every θ. Prove this directly from the definition of differentiability of f and the fact that $F_1(0, \theta) = \lim_{r \to 0} (1/r) \{F(r, \theta) - F(0, \theta)\}$.

2. If $F(x, y) = (1 - x)(1 - y)(x + y - 1)$, $a = b = \frac{2}{3}$, write Taylor's series for $F(a + h, b + k)$. What do you conclude about the sign of the difference $F(a + h, b + k) - F(a, b)$ when h and k are small?

3. Define $f(x, y) = (x^3 - y^3)/(x^2 + y^2)$ if $x^2 + y^2 \neq 0$, and $f(0, 0) = 0$. If we introduce cylindrical co-ordinates (r, θ, z) in the usual way, the surface $z = f(x, y)$ is represented by $z = r(\cos^3 \theta - \sin^3 \theta)$. Observe that the surface consists of a bundle of half-lines; the half-line corresponding to a fixed value of θ starts at the origin and passes through the cylinder $x^2 + y^2 = 1$ at a point for which $z = \cos^3 \theta - \sin^3 \theta$.

By plotting the curve $z = \cos^3 \theta - \sin^3 \theta$ with θ and z treated as plane rectangular co-ordinates, and then rolling up the plane to form a cylinder, one can visualize the surface. Do this. Does the surface have a tangent plane at the origin?

4. Suppose that f and ϕ are functions of a single variable, and that each function has continuous first and second derivatives. We shall suppose that $\phi(a) = c \neq 0$ and that $\phi'(a) \neq 0$. Let $F(x, y, z) = f(x) + f(y) + f(z)$, $G(x, y, z) = \phi(x)\phi(y)\phi(z)$. Consider the extremal problem for $F(x, y, z)$ subject to the constraint $G(x, y, z) = c^3$. Show that a possible solution of the problem occurs when $x = y = z = a$, and that the extreme will be a relative minimum if

$$f'(a) \left\{ \frac{\phi''(a)}{\phi'(a)} - \frac{\phi'(a)}{\phi(a)} \right\} < f''(a),$$

and a relative maximum if the inequality is reversed. As instances consider: first, $f(x) = x^2$, $\phi(x) = x$, $a > 0$; and second, $f(x) = e^{-x^2}$, $\phi(x) = x$.

Implicit-Function Theorems   **8**

8.   The nature of the problem of implicit functions

We have already acquired some familiarity with implicit functions, in §§ 6.1 and 6.6. Thus far what we have learned about implicit functions has been concerned almost entirely with techniques for differentiating such functions in concrete special cases (§ 6.1), or with general formulas for the derivatives of implicit functions in terms of functional notation (§ 6.6). In all this earlier work we have taken for granted the existence and differentiability of the implicit functions. In this chapter we shall inquire into these matters which have been taken for granted.

Consider then an equation in three variables,

$$(8-1) \qquad\qquad F(x, y, z) = 0.$$

In certain situations we say that an equation of this form has a solution

$$(8-2) \qquad\qquad z = f(x, y).$$

Our present purpose is to examine the following questions: *What does it mean to say that (8–2) is a solution of (8–1)? Under what conditions is a solution possible? What can be said about the differentiability of the function f in terms of what may be known about the function F?* The answers to these and related questions will occupy us in this chapter.

Questions of the same sort can be asked about other implicit-function situations. We may have $y = f(x)$ as a solution of $F(x, y) = 0$, or $z = f(u, v, w, x, y)$ as a solution of $F(u, v, w, x, y, z) = 0$. Or, there may be several functions which arise as solutions of a system of several equations. Under suitable conditions a set of r equations in $n + r$ variables may determine r of the variables as functions of the remaining n variables.

The nature of the problem is most easily understood, and the explanation of the theory is the simplest, in the case of an implicit function arising as a solution of a single equation in a small number of variables. We shall discuss the case of two variables first, and then the case of three variables. The discussion in this section is intended to provide motivation for, and understanding of, the formal statements of theorems in later sections.

Let F be a function of x and y, defined in a certain region of the xy-plane. Consider the equation

$$(8-3) \qquad\qquad F(x, y) = 0.$$

This equation expresses a condition which a point (x, y) may or may not satisfy. If there are some points which satisfy the condition, the set of all such points may be called the locus defined by (8–3). We know that, in many cases, the locus is some kind of curve. For instance, if $F(x, y) =$

$4x^2 + 9y^2 - 36$, the locus is an ellipse. Now let $f(x)$ be a function defined for a certain set of values of x. We say that $y = f(x)$ is a solution of (8–3) if all the points $(x, f(x))$ are part of the locus defined by (8–3), that is, if $F(x, f(x)) = 0$ for all the values of x which are involved in the definition of the function f. We assume without any further explicit mention that all functions which enter the discussion are single-valued.

In our work with implicit functions we do not attempt to get a solution of (8–3) in the form $y = f(x)$ such that it gives us *all* the locus defined by (8–3), for this may be impossible. Thus in the case of the ellipse $4x^2 + 9y^2 - 36 = 0$, part of the locus is given by the graph of $y = \frac{1}{3}(36 - 4x^2)^{1/2}$, and another part of the locus by the graph of $y = -\frac{1}{3}(36 - 4x^2)^{1/2}$. What we *do* attempt is to start with a particular point (x_0, y_0) of the locus defined by (8–3), and then to obtain a function $f(x)$, defined in some interval $x_0 - a < x < x_0 + a$, such that $y = f(x)$ is a solution of (8–3), and such that, *in a sufficiently restricted neighborhood of* (x_0, y_0), all the points for which $F(x, y) = 0$ are given by $y = f(x)$. This *localization* of the problem to a neighborhood of a particular point (x_0, y_0) is characteristic of all the treatment of implicit-function problems in this chapter.

Fig. 49 shows how, in the case of the equation $4x^2 + 9y^2 - 36 = 0$, localization of attention to a suitable neighborhood of one particular point of the locus leads to the determination of a solution $y = f(x)$ whose graph comprises all that part of the locus which is in the neighborhood. In Fig. 49 two such localizations are shown, the neighborhood being rectangular in each case. Observe that, if the center of one of the rectangular neighborhoods is a point (x_0, y_0) on the ellipse, the permissible size of the rectangle is governed by the consideration that every line parallel to the y-axis and passing through the interior of the rectangle shall intersect the ellipse exactly once inside the rectangle. It is always possible to choose the rectangle so as to satisfy this condition provided that the point

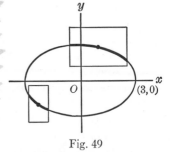

Fig. 49

(x_0, y_0) is not one of the points $(\pm 3, 0)$. These are the points where $\partial F/\partial y = 0$, with $F(x, y) = 4x^2 + 9y^2 - 36$. In our study of functions $y = f(x)$ defined implicitly as solutions of $F(x, y) = 0$ we shall always localize the problem within a neighborhood of a point at which $\partial F/\partial y \neq 0$. If we wanted to solve for x in terms of y instead of for y in terms of x, we would impose the requirement $\partial F/\partial x \neq 0$.

Let us turn now to a consideration of equations in three variables. Suppose that $F(x, y, z)$ is defined in a certain region in three-dimensional space. We consider the locus defined by the equation

(8–4) $$F(x, y, z) = 0,$$

and suppose that (x_0, y_0, z_0) is a point of this locus. We then confine our attention to points near (x_0, y_0, z_0), and ask the following question: Is it possible to find a rectangular box defined by certain inequalities

$$| x - x_0 | < a, | y - y_0 | < b, | z - z_0 | < c$$

such that every line parallel to the z-axis and passing through the interior of the box intersects the locus defined by (8–4) exactly once inside the box? If so, then to each pair (x, y) for which $| x - x_0 | < a$ and $| y - y_0 | < b$ there corresponds a unique z such that $| z - z_0 | < c$ and $F(x, y, z) = 0$. This defines z as a single-valued function of (x, y), say $z = f(x, y)$, and gives a solution of (8–4), that is, $F(x, y, f(x, y)) = 0$.

Under certain conditions it is possible to choose a box of the sort just described. We shall explain the plausibility of this statement from a geometrical point of view. Suppose that the locus defined by (8–4) is a surface, such as an ellipsoid, a hyperboloid, or a cone. Suppose that the surface is smooth, and that the tangent plane at the point (x_0, y_0, z_0) is not parallel to the z-axis. Then we expect the part of the surface near (x_0, y_0, z_0) to be almost like the nearby part of the tangent plane; it should then be represented by an equation $z = f(x, y)$, since each line parallel to the z-axis may be expected to intersect the surface exactly once in the vicinity of (x_0, y_0, z_0), provided that $x - x_0$ and $y - y_0$ are sufficiently small (see Fig. 50).

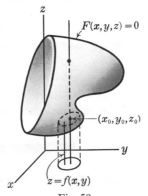

Fig. 50

The condition on F in order that the plane tangent to the surface at (x_0, y_0, z_0) shall not be parallel to the z-axis is $\dfrac{\partial F}{\partial z} \neq 0$ at the point, for the ratios $\dfrac{\partial F}{\partial x} : \dfrac{\partial F}{\partial y} : \dfrac{\partial F}{\partial z}$ define the direction of the normal to the surface; therefore $\dfrac{\partial F}{\partial z} = 0$ means that the normal is perpendicular to the z-axis.

Our discussions of the implicit-function problem for the two cases $F(x, y) = 0$ and $F(x, y, z) = 0$, with geometrical evidence for the solutions $y = f(x)$ and $z = f(x, y)$ in a localized form of the problem, suggest the kind of answers we may expect to some of the questions raised at the beginning of the present section. We do not yet have any real *proofs*, however. The proofs must come out of the analytical situation, for we do not really know the facts about geometry of the curves and surfaces except by an examination of the functions and equations.

8.1   The fundamental theorem

The following theorem is concerned with a precise statement bearing on the questions raised at the outset in § 8.

THEOREM I.　*Let $F(x, y, z)$ be a function defined in an open set S containing the point (x_0, y_0, z_0). Suppose that F has continuous first partial derivatives in S. Furthermore assume that*

$$F(x_0, y_0, z_0) = 0, \; F_3(x_0, y_0, z_0) \neq 0.$$

Under these conditions there exists a box-like region defined by certain inequalities

$$|x - x_0| < a, |y - y_0| < b, |z - z_0| < c,$$

lying in the region S and such that the following assertions are true:
　　Let R be the rectangular region $|x - x_0| < a, |y - y_0| < b$ in the xy-plane. Then

　　1. *For any (x, y) in R there is a unique z such that*

$$|z - z_0| < c \quad and \quad F(x, y, z) = 0.$$

Let us express this dependence of z on (x, y) by writing

(8.1–1)　　　　　　　　　　　　$z = f(x, y).$

　　2. *The function f is continuous in R.*
　　3. *The function f has continuous first partial derivatives given by*

$$f_1(x, y) = -\frac{F_1(x, y, z)}{F_3(x, y, z)}, f_2(x, y) = -\frac{F_2(x, y, z)}{F_3(x, y, z)},$$

where z is given by (8.1–1).

PROOF.　The first part of the proof is concerned with determining suitable values for the positive constants a, b, c which are mentioned in the theorem. Let A be a rectangular parallelepiped (box) with center at (x_0, y_0, z_0) such that the whole of A is entirely in the region S, and such that, moreover, $F_3(x, y, z)$ has everywhere in A the same sign which it has at (x_0, y_0, z_0). This choice of A is possible since S is open and F_3 is continuous (see Theorem III, § 5.3).

　　For definiteness let us assume $F_3 > 0$ in A. Consider the top and bottom faces of the box A. If we denote the height of the box by $2c$, these faces will lie in the planes $z = z_0 \pm c$. Since $F_3 > 0$, the value of F increases as we go upward along any line parallel to the z-axis. Since $F(x_0, y_0, z_0) = 0$, it follows that

$$F(x_0, y_0, z_0 + c) > 0 \quad and \quad F(x_0, y_0, z_0 - c) < 0.$$

Because of the continuity of F we see that F will be positive in a small rectangle with center at $(x_0, y_0, z_0 + c)$ in the plane $z = z_0 + c$, and negative in a small rectangle with center at $(x_0, y_0, z_0 - c)$ in the plane $z = z_0 - c$. Let us choose positive numbers a, b, so that these rectangles are determined by the inequalities

$$|x - x_0| < a, |y - y_0| < b.$$

We also take care to choose a and b so that the box B defined by the inequalities

$$|x - x_0| < a, |y - y_0| < b, |z - z_0| < c$$

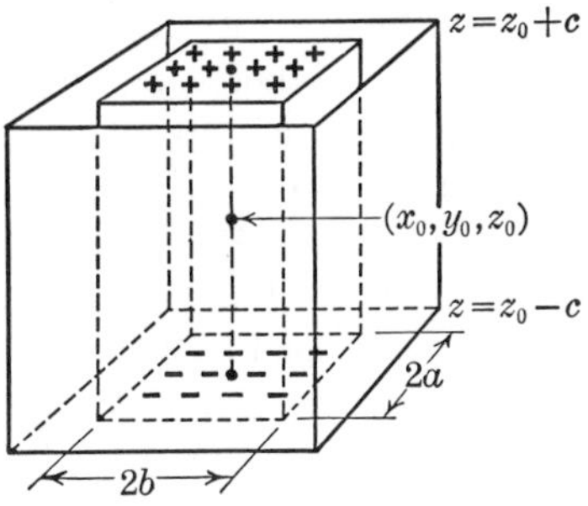

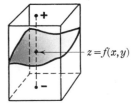

Fig. 51. *Boxes A and B*

is no larger than the box A (see Fig. 51).

Now consider the value of F along the segment in which a line parallel to the z-axis intersects the box B. As we go up along this segment the value of F increases. At the lower end of the segment, $F < 0$, while at the upper end $F > 0$. Hence there is *just one* point on the segment at which $F = 0$; for a given pair (x, y), the z-co-ordinate of this point is denoted by $z = f(x, y)$ (see Fig. 52). Assertion 1 of the theorem is now proved. Observe that thus far we have made no use of the partial derivatives F_1, F_2.

Fig. 52. *Box B*

Having obtained the function $f(x, y)$ by the foregoing argument, let us now prove that it is continuous. To prove continuity at (x_0, y_0) we must show that, given $\epsilon > 0$, there exists $\delta > 0$ such that

(8.1–2)
$$|f(x, y) - f(x_0, y_0)| < \epsilon \text{ when } |x - x_0| < \delta \text{ and } |y - y_0| < \delta.$$

We may assume that $\epsilon \leqq c$. Now $f(x_0, y_0) = z_0$; also,

$$F(x_0, y_0, z_0 + \epsilon) > 0, F(x_0, y_0, z_0 - \epsilon) < 0.$$

Hence, by the very argument used in proving assertion 1 of the theorem, we see that if we choose $\delta > 0$ so that

$$F(x, y, z_0 + \epsilon) > 0 \quad \text{and} \quad F(x, y, z_0 - \epsilon) < 0$$

when

(8.1–3) $|x - x_0| < \delta \quad \text{and} \quad |y - y_0| < \delta,$

then to each (x, y) satisfying (8.1–3) corresponds a unique z such that $|z - z_0| < \epsilon$ and $F(x, y, z) = 0$. This z must be equal to $f(x, y)$, however, by 1. Thus (8.1–2) holds, and the continuity at (x_0, y_0) is proved.

To prove continuity of f at any other point (x_1, y_1) of R, let $z_1 = f(x_1, y_1)$.

Observe that F satisfies at (x_1, y_1, z_1) the same hypotheses as are stated in the theorem relative to the point (x_0, y_0, z_0). Hence, by what has already been proved (as applied to this new situation), all the points (x, y, z) in the vicinity of (x_1, y_1, z_1) such that $F(x, y, z) = 0$ are furnished by a single-valued function (let us call it $g(x, y)$) which is continuous at (x_1, y_1). However, since all these points are in the box B, the uniqueness clause of conclusion 1 of the theorem assures us that $f(x, y) = g(x, y)$ when (x, y) is near (x_1, y_1). Hence f is continuous at (x_1, y_1).

It remains only to prove assertion 3 of the theorem. We shall deal with $\dfrac{\partial f}{\partial x}$; the treatment of $\dfrac{\partial f}{\partial y}$ is different only in the letters used. We employ the law of the mean (§ 7.4). Let (x, y) be a point of R, and let $z = f(x, y)$. We wish to show that

$$(8.1\text{--}4) \qquad \lim_{\Delta x \to 0} \frac{f(x + \Delta x, y) - f(x, y)}{\Delta x} = -\frac{F_1(x, y, z)}{F_3(x, y, z)} .$$

We assume that Δx is so small that $(x + \Delta x, y)$ is also in R, and write

$$\Delta z = f(x + \Delta x, y) - f(x, y).$$

Now, considering F as a function of x and z only, we have by the law of the mean,

$$(8.1\text{--}5) \qquad F(x + \Delta x, y, z + \Delta z) - F(x, y, z)$$
$$= \Delta x F_1(X, y, Z) + \Delta z F_3(X, y, Z),$$

where $\qquad X = x + \theta \, \Delta x, Z = z + \theta \, \Delta z, 0 < \theta < 1.$

The left member of (8.1–5) is zero, by the definition of the function f. Hence

$$(8.1\text{--}6) \qquad \frac{\Delta z}{\Delta x} = -\frac{F_1(X, y, Z)}{F_3(X, y, Z)} .$$

Now $\Delta z \to 0$ when $\Delta x \to 0$, by the continuity of f; therefore $X \to x$ and $Z \to z$. The truth of (8.1–4) is now seen to follow from (8.1–6), because F_1 and F_3 are continuous. The formula

$$f_1(x, y) = -\frac{F_1(x, y, f(x, y))}{F_3(x, y, f(x, y))}$$

has now been established. From it we see that f_1 is continuous (see Theorem IV, § 5.3). This completes the proof.

8.2   Generalization of the fundamental theorem

In the theorem of § 8.1 we dealt with the existence of a function $z = f(x, y)$ defined implicitly by an equation $F(x, y, z) = 0$. We chose to state the theorem for the case of three variables, but nothing essential in the theorem or its proof is really dependent upon the particular number three. The analytical details and the geometrical language presented in § 8.2 may all be modified easily to meet the situation of a different number of variables. We shall state the theorem formally in the general case ($n + 1$ variables). The proof will be omitted. The theorem tells us what we can be certain of, under appropriate conditions, in speaking of a function

$$z = f(x_1, \cdots, x_n)$$

defined implicitly by an equation of the form

$$F(x_1, x_2, \cdots, x_n, z) = 0.$$

We use geometrical language in speaking of the regions of definition of the above functions.

THEOREM II.   *Let $F(x_1, \cdots, x_n, z)$ be defined in an $(n + 1)$-dimensional neighborhood of the point $(a_1, \cdots, a_n, c)$. Suppose that F has continuous partial derivatives in this neighborhood, and furthermore, assume that*

$$F(a_1, \cdots, a_n, c) = 0, F_{n+1}(a_1, \cdots, a_n, c) \neq 0.$$

Under these conditions there exists a box-like region defined by certain inequalities

$$|x_1 - a_1| < A_1, \cdots, |x_n - a_n| < A_n, |z - c| < C,$$

lying in the above neighborhood, and such that the following assertions are true:

Let R be the n-dimensional region

$$|x_1 - a_1| < A_1, \cdots, |x_n - a_n| < A_n$$

in the space of the variables $x_1, \cdots, x_n$. Then

1. For any $(x_1, \cdots, x_n)$ in R there is a unique z such that

$$|z - c| < C \text{ and } F(x_1, \cdots, x_n, z) = 0.$$

Let us express this dependence of z on $(x_1, \cdots, x_n)$ by writing

$$z = f(x_1, \cdots, x_n).$$

2. The function f is continuous in R.
3. The function f has continuous first partial derivatives given by

$$\frac{\partial}{\partial x_i} f(x_1, \cdots, x_n) = -\frac{\dfrac{\partial}{\partial x_i} F(x_1, \cdots, x_n, z)}{\dfrac{\partial}{\partial z} F(x_1, \cdots, x_n, z)}, i = 1, \cdots, n,$$

where $z = f(x_1, \cdots, x_n)$.

EXERCISES

1. Is it true that the part of the locus defined by $x + y + z - \sin xyz = 0$ near the point $(0, 0, 0)$ can be represented in the form $z = f(x, y)$?

2. How can you be sure that the equation $e^z(x^2 + y^2 + z^2) - \sqrt{1 + z^2} + y = 0$ has a solution $z = f(x, y)$ which is continuous at $x = 1$, $y = 0$, with $f(1, 0) = 0$? Using the tangent plane as an approximation to the surface, calculate $f(1 + h, k)$ approximately when h and k are small.

3. Can the equation $(x^2 + y^2 + z^2)^{1/2} - \cos z = 0$ be solved uniquely for y in terms of x and z in the neighborhood of the point $(0, 1, 0)$? Can it be solved uniquely for z in terms of x and y in such a neighborhood?

4. Does there exist a function $f(x, y)$ continuous at $(1, -1)$, with $f(1, -1) = 0$, and such that $x^3 + y^3 + [f(x, y)]^3 = 3xyf(x, y)$ at all points of a neighborhood of $(1, -1)$?

5. Show that $z^3 + (x^2 + y^2)z + 1 = 0$ has a unique solution $z = f(x, y)$ defined for all x, y, and that f has continuous first partial derivatives everywhere.

6. Does the function $x(y - 1)\sqrt{z} + x^2z^3 + \sin x$ satisfy the hypotheses of Theorem I in a neighborhood of the point $(0, 0, 0)$?

7. For the purposes of this exercise let us make the following definition: A point (x_0, y_0, z_0) of the locus defined by $F(x, y, z) = 0$ is called a *regular point* of the locus if the part of the locus in a sufficiently small neighborhood of the point can be represented in at least one of the three forms $z = f(x, y)$, $x = g(y, z)$, $y = h(z, x)$, where f is defined and continuous for all values of (x, y) sufficiently close to (x_0, y_0), $f(x_0, y_0) = z_0$, and corresponding requirements are placed on the functions g, h.

(a) Now suppose that F has continuous first partial derivatives in a neighborhood of (x_0, y_0, z_0), that $F(x_0, y_0, z_0) = 0$, and that $\left(\dfrac{\partial F}{\partial x}\right)^2 + \left(\dfrac{\partial F}{\partial y}\right)^2 + \left(\dfrac{\partial F}{\partial z}\right)^2$ > 0 at the point. Prove that (x_0, y_0, z_0) is a regular point of the locus.

(b) Is the origin a regular point of the cone $x^2 + y^2 - z^2 = 0$?

(c) What is the locus defined by the equation $(x^2 + y^2 + z^2)^2 - a^2(x^2 + y^2 + z^2)$ $= 0$, $a \neq 0$? Are there any non-regular points of the locus?

(d) If a, b, c are positive, show that all points except $(0, 0, 0)$ of the locus defined by $(x^2 + y^2 + z^2)^2 - a^2x^2 - b^2y^2 - c^2z^2 = 0$ are regular.

8. In particular cases we may be able to arrive at information about $z = f(x, y)$ as a solution of $F(x, y, z) = 0$ by methods quite different from those used to prove Theorem I. As an example, consider the equation

$$ye^z + xz - x^2 - y^2 = 0.$$

Suppose that x and y are fixed, with $y \neq 0$. Consider w and z as variables, and draw the curve $w = e^z$ and the line

$$w = -\frac{x}{y}z + \frac{x^2 + y^2}{y}$$

on the same wz-co-ordinate axes. If the line intersects the curve, the z-co-ordinate of the point of intersection depends on the parameters x, y, say $z = f(x, y)$, and this gives a solution of $F(x, y, z) = 0$, where $F(x, y, z) = ye^z + xz - x^2 - y^2$.

(a) Plot the graphs of the curve and the line when $y \neq 0$ and $x/y > 0$, and show that there is a unique intersection.

(b) Show that if $x/y \leq 0$ there may be no intersection, one intersection (tangency), or two intersections. Show that, if x, y, z are such that the line is tangent to the curve, then $\dfrac{\partial F}{\partial z} = 0$.

9. (a) Give an example of an equation $F(x, y) = 0$, where F is continuous but the equation is not satisfied by any points (x, y).

(b) Give another example in which $F(x, y) = 0$ is satisfied by a certain point (x_0, y_0), but not by any other points near (x_0, y_0).

(c) Give an example in which the locus defined by $F(x, y, z) = 0$ is not a surface but a curve or a straight line.

10. Formulate an exactly worded theorem corresponding to Theorem I, but for the case of $y = f(x)$ defined as a solution of $F(x, y) = 0$, localized near a point (x_0, y_0). Give a detailed proof similar to the proof of Theorem I, and supply appropriate diagrams analogous to Figs. 51 and 52.

11. Is the locus of $y^2 + x^2 e^y = 0$ a curve?

12. What is the locus defined by

$$(e^{\sin x} - 1)^2 + (\sin y - 1)^2 = 0?$$

13. Suppose $F(x, y)$ has continuous first partial derivatives in a neighborhood of (a, b), that $F(a, b) = 0$, and that $\left(\dfrac{\partial F}{\partial x}\right)^2 + \left(\dfrac{\partial F}{\partial y}\right)^2 > 0$ at (a, b). Explain how you know that the part of the locus $F(x, y) = 0$ near (a, b) is a smooth curve without self-intersections.

14. If $F(x, y) = x^2 + y^2 - x^3$, find the solution $y = f(x)$ of $F(x, y) = 0$: (a) near the point $(5, 10)$; (b) near the point $(10, -30)$. (c) Near what points of the locus $F(x, y) = 0$ is there no solution $y = f(x)$ of the type considered in the discussion of Fig. 49?

15. If $F(x, y) = (y - x^2)^2 - x^5$, find the solution $y = f(x)$ of $F(x, y) = 0$: (a) near the point $(1, 0)$; (b) near the point $(1, 2)$. (c) What is the situation near the point $(0, 0)$?

16. What condition on a and b is sufficient to guarantee that the equation

$$yx^2 - y^3 + x^3 - a^2b - a^3 + b^3 = 0$$

has a solution $y = f(x)$ where f is continuous at $x = a$ and $f(a) = b$?

17. Can the part of the curve $y^n(x - y) - (x + y) = 0$ near the origin be represented in the form $y = f(x)$? Assume that $n \geq 1$. Draw a straight line segment which represents the curve approximately near the origin.

18. If $x_0 \neq 0$ and $x_0 \neq 1$, show that, if (x, y) is sufficiently near $(x_0, 0)$, the equation $\sin x^2 y - xy = 0$ is equivalent to $y = 0$.

19. (a) Find the unique solution $y = f(x)$ of $\tan y - xy = 0$ in the neighborhood of $(x_0, 0)$, if $x_0 \neq 1$.

(b) Show that, if (x_0, y_0) satisfies $\tan y - xy = 0$, and if $y_0 \neq 0$ and $\cos y_0 \neq 0$, the equation $\tan y - xy = 0$ defines uniquely a solution $y = f(x)$ such that f is continuous at x_0 and $y_0 = f(x_0)$.

20. Can the part of the curve $xe^y - y + 1 = 0$ near the point $(e^{-2}, 2)$ be represented in the form $y = f(x)$? Represent it in the form $x = g(y)$ and sketch the curve.

21. Make the following assumptions: (1) $F(x, y)$ is defined when $|x - x_0| < h$ and $|y - y_0| < k$; (2) $F(x_0, y_0) = 0$; (3) $F(x, y)$ is a continuous function of x for each y, and a continuous function of y for each x; (4) for each fixed x, $F(x, y)$ increases as y increases. Deduce that there exists a number c, $0 < c \leq h$, and a continuous function $f(x)$, defined when $|x - x_0| < c$, such that $|f(x) - y_0| < k$, and such that the graph of $y = f(x)$ comprises all points inside the rectangle $|x - x_0| < c$, $|y - y_0| < k$ at which $F(x, y) = 0$.

8.3 Simultaneous equations

In this section we shall discuss the problem of implicit functions as it arises in connection with simultaneous equations. We have already considered examples of the technique of finding the partial derivatives of functions defined implicitly by simultaneous equations (see §§ 6.1, 6.6). Our present concern is with *existence theorems* analogous to the fundamental theorems I, II given in §§ 8.1, 8.2.

Suppose we are given two equations in five variables. Experience with algebraic and trigonometric problems of this kind leads us to expect that, in certain cases at least, we may solve for two of the variables in terms of the remaining three. Thus, for instance, from equations

$$(8.3-1) \qquad F(x, y, z, u, v) = 0, \quad G(x, y, z, u, v) = 0,$$

we may be able to solve for u, v in terms of x, y, z:

$$(8.3-2) \qquad u = f(x, y, z), \quad v = g(x, y, z).$$

Or, again, if we have three equations in five variables, we may perhaps be able to express three of the variables in terms of the remaining two. The functional-notation expression of this situation would be that the equations

$$F(x, y, u, v, w) = 0, \quad G(x, y, u, v, w) = 0, \quad H(x, y, u, v, w) = 0$$

give rise to functions

$$u = f(x, y), \quad v = g(x, y), \quad w = h(x, y).$$

The general situation suggested by these cases is that in which we are given r equations in $n + r$ variables (n and r positive integers). In certain cases, at least, such a set of equations will define r of the variables as functions of the remaining n variables. In seeking to understand the implicit-function problem for r simultaneous equations, it is easiest to begin with the case $r = 2$. The total number of variables makes very little difference. For definiteness we consider the case of two equations in five variables, say in the form (8.3–1).

The simplest simultaneous-equation systems are the linear systems. Let us, as a preliminary special case, suppose that equations (8.3–1) are linear in u and v, but not necessarily linear in x, y, and z. Then they will have the form

$$A_1 u + B_1 v + C_1 = 0$$
$$\text{(8.3–3)} \qquad A_2 u + B_2 v + C_2 = 0,$$

where the coefficients A_1, B_1, $\cdots$, C_2 all depend on x, y, z. This system of equations can be solved for u, v in terms of x, y, z provided the determinant

$$D = \begin{vmatrix} A_1 & B_1 \\ A_2 & B_2 \end{vmatrix}$$

does not vanish. Now, regarding (8.3–3) as a particular instance of (8.3–1), we see that

$$F(x, y, z, u, v) = A_1 u + B_1 v + C_1,$$

and thus

$$\frac{\partial F}{\partial u} = A_1, \quad \frac{\partial F}{\partial v} = B_1.$$

Similar equations hold for G, and thus the determinant D is the same as the Jacobian determinant

$$\text{(8.3–4)} \qquad \frac{\partial(F, G)}{\partial(u, v)} = \begin{vmatrix} \dfrac{\partial F}{\partial u} & \dfrac{\partial F}{\partial v} \\ \dfrac{\partial G}{\partial u} & \dfrac{\partial G}{\partial v} \end{vmatrix}$$

It will be seen by referring back to § 6.6 that this same Jacobian arises in the denominators of the expressions for the partial derivatives of u and v as functions of x, y, z, assuming that such functions are defined by equations (8.3–1).

The foregoing considerations indicate that if we expect to solve equations (8.3–1) for u, v, we should make the assumption that the Jacobian (8.3–4) is different from zero. It must be kept in mind that in the general (nonlinear) case we are concerned not with the actual solution for u, v in the elementary sense of expressing u, v in terms of x, y, z by more or less simple formulas, but with the solution in the theoretical sense of knowing certainly that there exist functions (8.3–2) satisfying equations (8.3–1). A qualified guarantee of the existence of solutions in this theoretical sense is furnished by the following theorem:

THEOREM III. *Let S be a neighborhood of the point* $P_0 : (x_0, y_0, z_0, u_0, v_0)$ *in the 5-dimensional space of the co-ordinates* x, y, z, u, v. *Suppose that the functions* F, G *occurring in the system* (8.3–1) *are continuous and have continuous first partial derivatives in S. Also assume that both functions vanish at the point* P_0 *but that the Jacobian* (8.3–4) *does not vanish at the point. Under these conditions there exists a box-like region lying in S, defined by certain inequalities*

(8.3–5) $|x - x_0| < a, |y - y_0| < b, |z - z_0| < c,$

(8.3–6) $|u - u_0| < \alpha, |v - v_0| < \beta,$

such that the following assertions are true:

Let R be the region defined, in the 3-dimensional space of the co-ordinates x, y, z, *by the inequalities* (8.3–5). *Then*

1. *To any* (x, y, z) *in R there corresponds a unique pair of values* u, v *such that the inequalities* (8.3–6) *are satisfied and the functions* F, G *vanish (i.e., equations* (8.3–1) *are satisfied). This correspondence defines* u *and* v *as functions of* x, y, z, *say*

$$u = f(x, y, z), v = g(x, y, z).$$

2. *The functions* f, g *are continuous in R.*
3. *The functions* f, g *have continuous partial derivatives given by*

(8.3–7) $$\frac{\partial f}{\partial x} = -\frac{1}{J}\frac{\partial(F, G)}{\partial(x, v)}, \frac{\partial g}{\partial x} = -\frac{1}{J}\frac{\partial(F, G)}{\partial(u, x)}$$

and similar formulas with x *replaced by* y, *where*

(8.3–8) $$J = \frac{\partial(F, G)}{\partial(u, v)}.$$

PROOF. The proof rests very heavily on use of Theorem II, § 8.2. To begin with, we observe that the two partial derivatives $\dfrac{\partial F}{\partial v}$, $\dfrac{\partial G}{\partial v}$ cannot *both* vanish at the point P_0, for if they did, the Jacobian J would vanish there, contrary to the hypothesis. For definiteness assume that $\dfrac{\partial F}{\partial v}$ does not vanish at P_0. We are now able to apply Theorem II to the equation $F(x, y, z, u, v) = 0$, taking $n = 4$,

$$x_1 = x, x_2 = y, x_3 = z, x_4 = u, z = v.$$

As a result we obtain a function

$$v = \phi(x, y, z, u)$$

defined for (x, y, z, u) in a neighborhood of (x_0, y_0, z_0, u_0), and furnishing a solution of the equation $F(x, y, z, u, v) = 0$ for v. Next, we substitute in G, writing

$$H(x, y, z, u) = G(x, y, z, u, \phi(x, y, z, u)).$$

The equation $G(x, y, z, u, v) = 0$ is thus replaced by the equation $H(x, y, z, u) = 0$. We seek to solve this equation for u. As a condition for being able to solve, we need to know that $\dfrac{\partial H}{\partial u} \neq 0$. Now, by the rule for composite functions,

$$\frac{\partial H}{\partial u} = \frac{\partial G}{\partial u} + \frac{\partial G}{\partial v}\frac{\partial v}{\partial u}.$$

But we know from Theorem II that

$$\frac{\partial v}{\partial u} = \frac{\partial \phi}{\partial u} = -\frac{\dfrac{\partial F}{\partial u}}{\dfrac{\partial F}{\partial v}}.$$

Hence

$$\frac{\partial H}{\partial u} = \frac{\partial G}{\partial u} - \frac{\partial G}{\partial v}\frac{\dfrac{\partial F}{\partial u}}{\dfrac{\partial F}{\partial v}} = \frac{\dfrac{\partial G}{\partial u}\dfrac{\partial F}{\partial v} - \dfrac{\partial G}{\partial v}\dfrac{\partial F}{\partial u}}{\dfrac{\partial F}{\partial v}},$$

or

$$\frac{\partial H}{\partial u} = -\frac{J}{\dfrac{\partial F}{\partial v}}.$$

In a small neighborhood of the point P_0 neither J nor $\dfrac{\partial F}{\partial v}$ will vanish. We can then apply Theorem II to the equation $H(x, y, z, u) = 0$ to obtain a solution $u = f(x, y, z)$. Finally, substituting in ϕ we obtain v as a function of x, y, z:

$$v = g(x, y, z) = \phi(x, y, z, f(x, y, z)).$$

We shall omit the exact details of the limitation of the magnitudes of the differences $x - x_0$, $y - y_0$, ... in order to validate all the foregoing arguments. In applying Theorem II we are assured that the functions ϕ, f have continuous first partial derivatives. The function g, as a composite function, will then have continuous first partial derivatives also. The formulas (8.3–7) for the partial derivatives of f and g have already been obtained (see (6.6–11)). We can appeal to this earlier derivation now that we have proved the existence of f and g and the fact that they do possess continuous partial derivatives.

We shall not take space to state formally the analogue of Theorem III for systems of more than two equations. The nonvanishing of the appropriate Jacobian is the key condition. The proof of the general theorem for r equations may be made by mathematical induction on r. The proof for $r = 1$ is that of Theorem II; hence all that is necessary is to make the step from r to $r + 1$. This is not difficult, and may be patterned after the proof of Theorem III, which is the step from $r = 1$ to $r = 2$. Suggestions for this work are contained in Exercises 10, 11.

EXERCISES

1. Do there exist functions $f(x, y)$, $g(x, y)$, continuous in a neighborhood of $(0, 1)$, such that $f(0, 1) = 1$, $g(0, 1) = -1$, and such that

$$[f(x, y)]^3 + xg(x, y) - y = 0,$$
$$[g(x, y)]^3 + yf(x, y) - x = 0?$$

Explain your answer.

2. Suppose that the three equations

$$u^2 + v^2 + w^2 - x^2 = 0, \quad u^2 + v^2 - y^2 = 0, \quad u^2 + w^2 - z^2 = 0$$

are satisfied by a particular set of values $(x_0, y_0, z_0, u_0, v_0, w_0)$ of the variables. **(a)** What condition on this set of values is sufficient to insure that all "nearby" sets (x, y, z, u, v, w) satisfying the three equations are given by equations $u = f(x, y, z)$, $v = g(x, y, z)$, $w = h(x, y, z)$, where f, g, h are single-valued and continuous, with values u_0, v_0, w_0 respectively at (x_0, y_0, z_0)? **(b)** Solve the given equations explicitly for u^2, v^2, w^2, and from the solutions so found explain what happens if the sufficient condition in part **(a)** is not satisfied.

3. Suppose that $(x_0, y_0, z_0, u_0, v_0, w_0)$ satisfy the equations

$$u^2 + v^2 + w^2 = 1, \quad \frac{u^2}{x^2} + \frac{v^2}{y^2} + \frac{w^2}{z^2} = 1.$$

What are sufficient conditions which guarantee that all "nearby" sets satisfying these equations can be represented in the form

$$u = f(x, y, z, w), \quad v = g(x, y, z, w)?$$

4. Let (x_0, y_0, z_0, u_0) satisfy the equations

$$f(x) + f(y) + f(z) = F(u)$$
$$g(x) + g(y) + g(z) = G(u)$$
$$h(x) + h(y) + h(z) = H(u),$$

where all the functions involved have continuous derivatives. **(a)** State a sufficient condition for being able to solve for x, y, z in terms of u in the neighborhood of the given point. **(b)** What does the condition amount to in case $f(x) = x$, $g(x) = x^2$, $h(x) = x^3$?

5. The locus defined by the equations $x^2 + y^2 + z^2 - r^2 = 0$, $x + y + z - c = 0$ may be interpreted as the circle of intersection of a plane and a sphere. If (x_0, y_0, z_0) is a point of this locus, under what sufficient condition on x_0, y_0, z_0 may the part of the locus near (x_0, y_0, z_0) be represented in the form $y = f(x)$, $z = g(x)$?

6. Let (x_0, y_0, z_0) be a point of the locus defined by $z^2 + xy - a = 0$, $z^2 + x^2 - y^2 - b = 0$. (a) Under what sufficient conditions on x_0, y_0 may the part of the locus near (x_0, y_0, z_0) be represented in the form $x = f(z)$, $y = g(z)$? (b) What are sufficient conditions on x_0, y_0, z_0 for representing this part of the locus in the form $x = f(y)$, $z = g(y)$?

7. State carefully, in full detail, the analogue of Theorem III for the implicit-function problem arising from the equations

$$F(x, y, u, v, w) = 0, \quad G(x, y, u, v, w) = 0, \quad H(x, y, u, v, w) = 0,$$

where it is desired to find solutions

$$u = f(x, y), \quad v = g(x, y), \quad w = h(x, y)$$

in the neighborhood of $(x_0, y_0, u_0, v_0, w_0)$. For the proof assume that $\dfrac{\partial(F, G)}{\partial(v, w)} \neq 0$ and suppose that the equations $F = 0$, $G = 0$ are solved for v and w in terms of x, y, u. Let $K(x, y, u) = H(x, y, u, v, w)$ when v and w are expressed in terms of x, y, u. Show that

$$\frac{\partial K}{\partial u} = \frac{\partial(F, G, H)}{\partial(u, v, w)} \bigg/ \frac{\partial(F, G)}{\partial(v, w)}.$$

With these suggestions, write out a complete proof.

8. Suppose that $\quad F(x, y, u, v) = x^2 + y^2 - 2ux + 1$,
$\qquad\qquad\qquad\quad G(x, y, u, v) = x^2 + y^2 + 2vy - 1$.

(a) Interpret u and v as parameters, and plot the curves $F = 0$, $G = 0$ in the xy-plane, assuming $u^2 \geq 1$. (b) Now suppose x_0, y_0, u_0, v_0 satisfy the equations $F = 0$, $G = 0$, and that $u_0^2 > 1$. Explain geometrically why it is reasonable to expect that, if u and v differ but slightly from u_0 and v_0, respectively, the equations $F = 0$, $G = 0$ will determine a *unique* point (x, y), if this point is required to be sufficiently near (x_0, y_0). (c) Show that a set of values x_0, y_0, u_0, v_0 cannot satisfy the three equations $F = 0$, $G = 0$, $\dfrac{\partial(F, G)}{\partial(x, y)} = 0$ unless $u_0^2 = 1$. Use this result and an appropriate version of the implicit-function theorem for simultaneous equations to give an analytical explanation of the situation described in part (b).

9. Suppose that $(x_0, y_0, z_0, u_0, v_0, w_0)$ satisfy the equations

$$\frac{x^2}{a^2 + u} + \frac{y^2}{b^2 + u} + \frac{z^2}{c^2 + u} = 1,$$

$$\frac{x^2}{a^2 + v} + \frac{y^2}{b^2 + v} + \frac{z^2}{c^2 + v} = 1,$$

$$\frac{x^2}{a^2 + w} + \frac{y^2}{b^2 + w} + \frac{z^2}{c^2 + w} = 1,$$

where $0 < c < b < a$ and $-c^2 < u_0$, $-b^2 < v_0 < -c^2$, $-a^2 < w_0 < -b^2$. Show that the equations have a unique solution $x = f(u, v, w)$, $y = g(u, v, w)$, $z = h(u, v, w)$, where f, g, h are continuous at (u_0, v_0, w_0) and take on the values x_0, y_0, z_0 respectively there, provided that $x_0 y_0 z_0 \neq 0$. Prove conclusively that the appropriate Jacobian is not zero.

10. Assuming that $\dfrac{\partial(F_1, \cdots, F_{r-1})}{\partial(u_1, \cdots, u_{r-1})} \neq 0$,

let $u_i = \phi_i(x_1, \cdots, x_n, u_r)$ $(i = 1, \cdots, r - 1)$ be solutions of the first $r - 1$ equations in the system

$$F_1(x_1, \cdots, x_n, u_1, \cdots, u_r) = 0$$
$$\cdots \cdots \cdots \cdots \cdots \cdots \cdots$$
$$F_r(x_1, \cdots, x_n, u_1, \cdots, u_r) = 0$$

for $u_1, \cdots, u_{r-1}$ in terms of $x_1, \cdots, x_n, u_r$. Let

$$G(x_1, \cdots, x_n, u_r) = F_r(x_1, \cdots, x_n, \phi_1, \cdots, \phi_{r-1}, u_r),$$

where for convenience we have written merely ϕ_i instead of $\phi_i(x_1, \cdots, x_n, u_r)$. Show that

$$\frac{\partial G}{\partial u_r} \frac{\partial(F_1, \cdots, F_{r-1})}{\partial(u_1, \cdots, u_{r-1})} = \frac{\partial(F_1, \cdots, F_r)}{\partial(u_1, \cdots, u_r)}$$

is an identity in $x_1, \cdots, x_n, u_r$.

SUGGESTION: Differentiate the identities

$$F_1(x_1, \cdots, x_n, \phi_1, \cdots, \phi_{r-1}, u_r) = 0$$
$$\cdots \cdots \cdots \cdots \cdots \cdots \cdots \cdots \cdots$$
$$F_{r-1}(x_1, \cdots, x_n, \phi_1, \cdots, \phi_{r-1}, u_r) = 0$$
$$F_r(x_1, \cdots, x_n, \phi_1, \cdots, \phi_{r-1}, u_r) - G(x_1, \cdots, x_n, u_r) = 0$$

with respect to u_r and regard the resulting equations as a set of r linear equations in $\dfrac{\partial \phi_1}{\partial u_r}, \cdots, \dfrac{\partial \phi_{r-1}}{\partial u_r}, \dfrac{\partial G}{\partial u_r}$ as unknowns. Solving this system for $\dfrac{\partial G}{\partial u_r}$ by determinants gives the required result.

11. State the theorem analogous to Theorem III for functions $u_i = \phi_i(x_1, \cdots, x_n)$ defined by equations $F_i(x_1, \cdots, x_n, u_1, \cdots, u_r) = 0$, where $\dfrac{\partial(F_1, \cdots, F_r)}{\partial(u_1, \cdots, u_r)} \neq 0$. Use the results of Exercise 10 to prove the induction step from $r - 1$ to r, and so prove the theorem.

Transformations and Mappings   **9**

9. Introduction

In this chapter we shall be concerned with relations between two sets of variables. Suppose that u, v are the variables of one set, and that we write

(9–1) $$x = f(u, v), \qquad y = g(u, v),$$

where f and g are certain functions. Then x, y are the variables of a second set. We say that equations (9–1) define a *transformation*. A transformation is really a certain kind of function, for the transformation (9–1) is a law of correspondence whereby to certain pairs of numbers (u, v) there correspond certain pairs of numbers (x, y). The transformation is thus not a real-valued function of two real variables, but a function whose values are *pairs* of real numbers. We shall also consider transformations in which the number of variables in either set is three instead of two, for example,

(9–2) $$x = f(u, v, w), \qquad y = g(u, v, w), \qquad z = h(u, v, w).$$

In fact, the number of variables in either set might be any positive integer, and the numbers in the two sets might be different. For example, a set of parametric equations of a curve:

$$x = f(t), \qquad y = g(t), \qquad z = h(t),$$

may be regarded as a transformation in which the law of correspondence determines the triple (x, y, z) when t is given. Or, to give another example, the equations

$$x = \sin \phi \cos \theta, \qquad y = \sin \phi \sin \theta, \qquad z = \cos \phi$$

are parametric equations of the surface of the unit sphere $x^2 + y^2 + z^2 = 1$. These parametric equations define a transformation from the pair-variable (θ, ϕ) to the triple-variable (x, y, z).

In this chapter we confine our attention principally to transformations in which the number of variables in each set is the same. The most important cases in actual common applications are those in which the number in each case is either two or three, that is, transformations of one of the types (9–1), (9–2). Transformations such as those involved in rotation or rigid motion of axes come under these types. Also, the equations relating polar and rectangular co-ordinates in the plane, or spherical and rectangular co-ordinates in three-dimensional space, are of these types.

For brevity we shall say that a transformation such as (9–1) or (9–2) is *continuous* if the individual functions f, g, $\cdots$ which appear in these formulas are continuous functions of their arguments. A transformation will be called *differentiable* if these functions are differentiable, and *continuously differentiable* if the functions have continuous partial derivatives.

Consider the transformation (9–1). It may be possible to solve the equations defining the transformation for u and v in terms of x and y, so that we get equations of the form

$$(9–3) \qquad u = F(x, y), \qquad v = G(x, y).$$

In some cases the solution may be possible by explicit algebraic processes, or the functions F and G may be expressible in terms of familiar elementary functions. In other cases this may not be so, and the solution, if theoretically possible, must be understood to be possible in the same sense in which we talk about implicit functions defined as solutions of simultaneous equations (see § 8.3). In a later section we shall discuss the theoretical questions on this matter more carefully. For the moment let us assume that the transformation (9–1) is such that it sets up a *one-to-one correspondence* between certain pairs (u, v) and (x, y); that is, we assume that not only is the pair (x, y) determined when the pair (u, v) is given, but that no two different pairs (u, v) ever determine the same pair (x, y). In that case one of the pairs (x, y) determines uniquely the pair (u, v) which gives rise to it, and we may think of u and v as functions of x and y, as in (9–3). These latter equations determine a transformation from (x, y) to (u, v); we call it the *inverse* of the transformation (9–1).

The concept of the inverse of a transformation is an extension of the concept of an inverse function, as we meet it in studying functions of one variable in elementary calculus.

Example 1. The inverse of the transformation

$$x = 2u + 3v, \qquad y = u + 2v$$

is the transformation

$$u = 2x - 3y, \qquad v = -x + 2y.$$

Example 2. The transformation

$$(9–4) \qquad x = \frac{u^2 + v^2}{2}, \qquad y = \frac{u^2 - v^2}{2}$$

does not have a uniquely defined inverse, if all possible sets of values of u and v are considered. The equations can be solved for u^2 and v^2:

$$u^2 = x + y, \qquad v^2 = x - y.$$

If we consider the transformation (9–4) with the restriction that u and v are never to be assigned negative values, then the inverse transformation is

$$(9–5) \qquad u = \sqrt{x + y}, \qquad v = \sqrt{x - y}.$$

A different inverse transformation would be obtained if we were to place a different restriction on u and v.

We shall see in the following sections that the Jacobian determinant

$$(9\text{-}6) \qquad \frac{\partial(f, g)}{\partial(u, v)} = \begin{vmatrix} \dfrac{\partial f}{\partial u} & \dfrac{\partial f}{\partial v} \\ \dfrac{\partial g}{\partial u} & \dfrac{\partial g}{\partial v} \end{vmatrix}$$

plays an important part in studying the properties and uses of the transformation (9–1). The Jacobian (9–6) is called *the Jacobian of the transformation (9–1)*. It is common practice in mathematical literature to write x, y in place of f, g in (9–6), so that the Jacobian of the transformation is often denoted by $\dfrac{\partial(x, y)}{\partial(u, v)}$. The Jacobian of the inverse transformation is then denoted by $\dfrac{\partial(u, v)}{\partial(x, y)}$. These two Jacobians are reciprocals of each other; this will be shown in § 9.3 (Theorem II).

9.1 A theorem on inverse transformations

In this section we shall show the relation between the implicit-function theorem for simultaneous equations, as stated and proved in § 8.3, and the existence of an inverse of a given transformation.

THEOREM I. *Let $x = f(u, v)$, $y = g(u, v)$ define a continuously differentiable transformation for all pairs (u, v) in some neighborhood of a point (u_0, v_0). Let $x_0 = f(u_0, v_0)$, $y_0 = g(u_0, v_0)$, and suppose that the Jacobian $\dfrac{\partial(f, g)}{\partial(u, v)}$ is not zero at (u_0, v_0). Then there exist positive numbers a, b, α, β and functions $F(x, y)$, $G(x, y)$ defined when $| x - x_0 | < a$, $| y - y_0 | < b$, such that the following assertions are true:*

Let R be the rectangular region in the xy-plane defined by the inequalities $| x - x_0 | < a$, $| y - y_0 | < b$, and let S be the rectangular region in the uv-plane defined by the inequalities $| u - u_0 | < \alpha$, $| v - v_0 | < \beta$. Then

1. To any (x, y) in R corresponds a unique (u, v) in S such that $x = f(u, v)$, $y = g(u, v)$, and this unique pair is given by

$$(9.1\text{-}1) \qquad u = F(x, y), \qquad v = G(x, y).$$

2. The functions F, G are continuous and have continuous partial derivatives given by

$$(9.1\text{-}2) \qquad \frac{\partial F}{\partial x} = \frac{1}{J} \frac{\partial g}{\partial v}, \qquad \frac{\partial G}{\partial x} = -\frac{1}{J} \frac{\partial g}{\partial u},$$

$$\frac{\partial F}{\partial y} = -\frac{1}{J}\frac{\partial f}{\partial v}, \qquad \frac{\partial G}{\partial y} = \frac{1}{J}\frac{\partial f}{\partial u},$$

where

(9.1–3)
$$J = \frac{\partial(f, g)}{\partial(u, v)}.$$

In the formulas (9.1–2) *it is understood that* u, v *are to be expressed in terms of* x, y *by* (9.1–1).

PROOF.   This theorem is fundamentally just a special case of Theorem III, § 8.3, for the system of equations

(9.1–4) $f(u, v) - x = 0,$ $g(u, v) - y = 0;$

unlike the situation in the theorem referred to, there is no fifth variable z in our present equations. This is nothing essential, however. With the appropriate shifts in notation, (8.3–8) becomes (9.1–3), and the equations (8.3–7) become two of the equations in (9.1–2). There is nothing more to the proof. We just apply the earlier theorem to the equations (9.1–4).

It should be noted that the theorem does not guarantee anything about the size of the numbers a, b, α, β. However, they are chosen small enough to insure that the Jacobian (9.1–3) is not zero at any of the points (u, v) in the rectangle S corresponding to points (x, y) in the rectangle R. In general, in studying the inverse of a given transformation we stay away from points at which the Jacobian of the transformation is equal to zero.

Example.   The Jacobian of the transformation (9–4) is

$$\frac{\partial(x, y)}{\partial(u, v)} = -2\,uv,$$

and it is zero if either $u = 0$ or $v = 0$. By staying entirely in one quadrant of the uv-plane we can in this case avoid points where the Jacobian is zero. Suppose we choose the first quadrant, so that $u > 0$, $v > 0$. The inverse transformation is given by (9–5). Suppose we take $u_0 = 1$, $v_0 = 3$. Then $x_0 = 5$, $y_0 = -4$.

Let us see how the rectangles R and S must be chosen in this case. In the uv-plane we must keep away from the lines $u = 0$, $v = 0$. This means that, in the xy-plane, we must keep away from the lines $x + y = 0$, $x - y = 0$ (see (9–5) for realization of the truth of the last statement). Thus the rectangle R must lie above the line $x + y = 0$ and below the line $x - y = 0$, while the rectangle S must lie in the first quadrant (see Fig. 53 and Fig. 54).

These requirements mean that in this case the positive constants a, b, α, β must satisfy the inequalities $a + b < 1$, $\alpha < 1$, $\beta < 3$. Moreover, the size of R will depend on the size of S; if S is chosen with small dimensions, the dimensions of R will have to be small enough so that the inverse

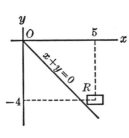

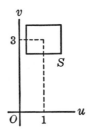

Fig. 53   Fig. 54

transformation will keep (u, v) in S when (x, y) is in R. Finally, we remind the student that Theorem I does not say that the inverse transformation is defined *only* for (x, y) in R; it says it is defined *at least* for (x, y) in R.

EXERCISE

In the equations $x = f(u, v)$, $y = g(u, v)$ regard u, v as first-class variables and x, y as second-class variables, the relation between the classes being that expressed by (9.1–1).

From this point of view the chain rule gives

$$1 = \frac{\partial f}{\partial u} \frac{\partial u}{\partial x} + \frac{\partial f}{\partial v} \frac{\partial v}{\partial x},$$

where $\dfrac{\partial u}{\partial x}$ means $\dfrac{\partial F}{\partial x}$ and $\dfrac{\partial v}{\partial x}$ means $\dfrac{\partial G}{\partial x}$. Carry on with the chain rule in this manner to obtain three other equations, and then solve them, thus obtaining equations (9.1–2).

9.2   Transformations of co-ordinates

The formulas connecting the rectangular co-ordinates (x, y) of a point and the rectangular co-ordinates (x', y') of the same point in a rotated system (see Fig. 55) are

(9.2–1)
$$x = x' \cos \phi - y' \sin \phi,$$
$$y = x' \sin \phi + y' \cos \phi.$$

Here (x', y') take the place of (u, v) in the general formulas (9–1). The transformation (9.2–1) has the inverse

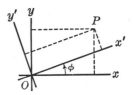

Fig. 55

(9.2–2)
$$x' = x \cos \phi + y \sin \phi,$$
$$y' = - x \sin \phi + y \cos \phi,$$

as the student should verify for himself. The student should also verify that

(9.2–3)
$$\frac{\partial(x, y)}{\partial(x', y')} = \frac{\partial(x', y')}{\partial(x, y)} = 1.$$

The familiar formulas connecting rectangular and polar co-ordinates in the plane are

(9.2–4) $x = r \cos \theta, \qquad y = r \sin \theta.$

The Jacobian of this transformation is

(9.2–5) $\dfrac{\partial(x, y)}{\partial(r, \theta)} = \begin{vmatrix} \cos \theta & -r \sin \theta \\ \sin \theta & r \cos \theta \end{vmatrix} = r.$

For convenience we shall assume that $r \geqq 0$ in our dealings with polar co-ordinates. If we attempt to solve for r, θ in terms of x, y, we find

(9.2–6)

$$r = (x^2 + y^2)^{1/2}, \qquad \sin \theta = \frac{y}{(x^2 + y^2)^{1/2}}, \qquad \cos \theta = \frac{x}{(x^2 + y^2)^{1/2}}.$$

To get this far we assume that $r \neq 0$, that is, that the Jacobian is not zero. Note that a unique formula for θ cannot be found, because θ may be changed by any multiple of 2π without affecting (9.2–6). This non-uniqueness does not violate the statement of Theorem I, § 9.1, however. (Why not?)

It is worth while to consider briefly the question of an explicit formula for θ in terms of x and y. From (9.2–6) one is tempted to write

$$\theta = \tan^{-1} \left(\frac{y}{x} \right).$$

There are many things wrong with this formula, however, if one is seeking the transformation inverse to (9.2–4). If we follow the usual custom of restricting the inverse tangent to its principal value, we limit ourselves to the range $-\pi/2 < \theta < \pi/2$. Any other convention about principal values leads to a similar difficulty. On the other hand, if we ignore principal values, and regard the inverse tangent as a multiple-valued function, we get into trouble, because the values we then get for θ will not all satisfy (9.2–6). Thus, if $x = -1$, $y = 1$, the values of θ such that $\tan \theta = -1$ are either in the second or fourth quadrant. The second-quadrant values satisfy (9.2–6), while the fourth-quadrant values do not. Still another difficulty is that $\tan^{-1}(y/x)$ is discontinuous at points on the y-axis, whereas the transformation (9.2–4) has a continuous inverse in the neighborhood of such a point (if the point is not the origin).

The inverse tangent (or the inverse cotangent) can be used if care is taken, of course. Thus, if $r_0 = \sqrt{2}$, $\theta_0 = \frac{3}{4}\pi$, $x_0 = -1$, $y_0 = 1$, the transformation inverse to (9.2–4) can be written

$$r = (x^2 + y^2)^{1/2}, \qquad \theta = \tan^{-1} \left(\frac{y}{x} \right) + \pi$$

as long as $x < 0$; here the principal value of the inverse tangent is to be used.

Let us now examine the nature of a rectangular or polar co-ordinate system from a point of view which will be fruitful in our systematic study of transformations. We consider the xy-plane with a fixed set of axes. A point (x_0, y_0) in the plane is located by the statement that it is at the intersection of the two lines $x = x_0$, $y = y_0$. A co-ordinate system is basically a method of locating points by setting up a correspondence between points and sets of numbers; the numbers corresponding to a point are called co-ordinates of the point. For an arbitrary rectangular co-ordinate system we need two one-parameter families of parallel lines, the lines of one family intersecting the lines of the other family at right angles. Let the equations of the two families be

$$(9.2\text{-}7) \qquad \begin{aligned} a_1 x + b_1 y &= u \\ a_2 x + b_2 y &= v, \end{aligned}$$

where the a's and b's are fixed, and u and v are parameters of the families. For the required perpendicularity we must have $b_1 : a_1 = -a_2 : b_2$, or

$$(9.2\text{-}8) \qquad a_1 a_2 + b_1 b_2 = 0.$$

If we now pick out values u_0, v_0, the two lines $u = u_0$, $v = v_0$ intersect at a unique point (x_0, y_0) (see Fig. 56). This point is determined by (u_0, v_0), and we may take these latter numbers as co-ordinates of the point in a new co-ordinate system. The xy-system and the uv-system are then related by the transformation (9.2-7). The uv-system will not in general be obtained from the xy-system by a rigid motion, for the units of distance along the u-axis and the v-axis will in general be different from each other and from the common unit of distance along the x and y axes.

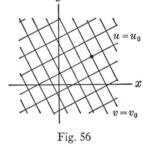

$u = u_0$

$v = v_0$

Fig. 56

If we drop the requirement (9.2-8), but insist that the two families in (9.2-7) not be identical, we get, in general, an *oblique* co-ordinate system.

The system of polar co-ordinates is also based on locating a point at an intersection, but in this case we have a family of concentric circles and a family of rays through the common center of the circles (see Fig. 57). The equation of the circles is

$$(9.2\text{-}9) \qquad x^2 + y^2 = r^2,$$

with parameter r. The family of rays is given by taking in each case the suitable half of the line

$$(9.2\text{-}10) \qquad y \cos \theta = x \sin \theta,$$

with θ as parameter. A point is located by specifying that it is at the intersection of a circle $r = r_0$ and a ray $\theta = \theta_0$. It will be observed that in general a point is at the intersection of just one circle and just one ray. The exceptional point is the origin. This is

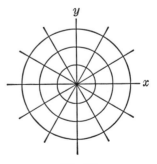

Fig. 57

precisely the point at which the Jacobian $\dfrac{\partial(x, y)}{\partial(r, \theta)} = 0$. It is called a *singular point* of the polar co-ordinate system.

In § 9.3 we shall study more general co-ordinate systems arising from the study of two one-parameter families of curves of a fairly arbitrary character. Co-ordinate systems of such a general type are called *curvilinear co-ordinate systems*.

9.3   Curvilinear co-ordinates

A system of curvilinear co-ordinates can be defined if we have two one-parameter families of curves subject to suitable conditions. The conditions may not be satisfied over the whole plane, but only in certain regions. Among the conditions which are normally required is the condition that each point in a region under consideration should lie on one and only one curve of each of the two families, and that these curves should not be tangent at the point. Before discussing the general theory it will be helpful to consider an example.

Example 1.   Suppose that the two families of curves are

(9.3–1) $y^2 = - u^2(x - u^2)$

(9.3–2) $y^2 = v^2(x + v^2)$,

where the parameters are u, v.

The curves of both families are parabolas with the x-axis as axis of symmetry. The parabolas (9.3–1) open to the left; the parabolas of the other family open to the right. The vertex of (9.3–1) is at $(u^2, 0)$, and its focus is a distance $\frac{1}{4} u^2$ from the vertex. The y-intercepts are $(0, \pm u^2)$. A number of the curves of both families are shown in Fig. 58.

The parameters u, v may be used as co-ordinates; that is, a point can be located by giving the values of u and v corresponding to the two parabolas which pass through the point. Actually, a u-parabola and a v-parabola intersect twice, so that a pair of values (u, v) does not determine a unique point. Conversely, a given point (x, y) does not determine u and v uniquely, because the equations of the parabolas are not affected by a change in the sign of u or v.

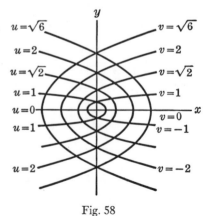

Fig. 58

Let us attempt to solve equations (9.3–1) and (9.3–2) for x and y. Subtracting (9.3–2) from (9.3–1), we find

$$0 = - x(u^2 + v^2) + u^4 - v^4,$$

or

$$x = \frac{u^4 - v^4}{u^2 + v^2} = u^2 - v^2,$$

provided u and v are not both zero. If this result is substituted in (9.3–2), we get

$$y^2 = u^2 v^2, \qquad \text{or} \qquad y = \pm\, uv.$$

There are thus two possible differentiable transformations,

(9.3–3) $$x = u^2 - v^2, \qquad y = uv,$$

and

(9.3–4) $$x = u^2 - v^2, \qquad y = -\, uv,$$

which may be used to determine a point (x, y) by giving values of u, v. Either transformation allows us to use u and v as curvilinear co-ordinates. Suppose, for example, that we use (9.3–3). There is some arbitrariness in the choice of signs for u and v, but the product uv must have the same sign as y. We may, for instance, use $u \geqq 0$ all the time; then we must use $v > 0$ when $y > 0$ and $v < 0$ when $y < 0$. Figure 58 has been labeled in accord with this choice. Other choices are possible, however. From (9.3–3) we find

$$\frac{\partial(x, y)}{\partial(u, v)} = 2(u^2 + v^2).$$

This is zero only when $u = v = 0$, which by (9.3–3) is equivalent to $x = y = 0$. The origin is called a *singular point* of the uv-curvilinear co-ordinate system. It is not possible to use u and v as co-ordinates, throughout a region having the origin as an interior point, in such a way as to have a one-to-one correspondence between (x, y) and (u, v) with the transformation from (u, v) to (x, y) and the inverse transformation from (x, y) to (u, v) both continuous throughout the region.

Now let us consider the general theory of curvilinear co-ordinates in the plane. Suppose we have given a continuously differentiable transformation

(9.3–5) $$x = f(u, v), \qquad y = g(u, v),$$

such that the Jacobian

$$J = \frac{\partial(f, g)}{\partial(u, v)}$$

is not equal to zero for a certain pair of values u_0, v_0. Denote the corresponding values of x and y by x_0, y_0. The following discussion will relate to pairs (x, y) sufficiently near (x_0, y_0) and pairs (u, v) sufficiently near (u_0, v_0) so that we can use the conclusions described in Theorem I, § 9.1. In particular, there is a continuously differentiable inverse transformation

$$(9.3\text{-}6) \qquad u = F(x, y), \qquad v = G(x, y).$$

Let us denote the Jacobian of the inverse transformation by j.

THEOREM II.   *The Jacobian of the inverse transformation* (9.3–6) *is the reciprocal of the Jacobian of the transformation* (9.3–5); *that is,*

$$(9.3\text{-}7) \qquad\qquad j = \frac{1}{J}.$$

PROOF.   From the formulas (9.1–2) we find at once that $j = J/J^2$, so that (9.3–7) holds. Observe that, as a consequence, $j \neq 0$.

The one-to-one correspondence between (x, y) and (u, v), as established by the transformation and its inverse, makes it possible for us to use (u, v) as co-ordinates for the point (x, y), since the point determines its co-ordinates by (9.3–6), and the co-ordinates determine the point by (9.3–5). It is desirable, however, to have a geometric interpretation of the uv-co-ordinate system. A geometric interpretation can be given in terms of two families of curves which form a mesh of quadrilaterals in somewhat the same way that the lines $x = $ constant, $y = $ constant, form a rectangular mesh. If we regard u as constant and let v vary, equations (9.3–5) are parametric equations of a curve, v being the parameter. We call such a curve a u-curve. Similarly, a v-curve is defined by (9.3–5) with v fixed and u as parameter. Alternatively, these curves may be thought of as defined by (9.3–6). With u constant, a u-curve is defined by $F(x, y) = u$. If we confine our attention to a small neighborhood of a point where $J \neq 0$, there will be exactly one u-curve and one v-curve through each point of the neighborhood, and these two curves will not be tangent at the point. We shall now show that the nontangency is a consequence of the fact that $J \neq 0$. The slope of a u-curve may be found from (9.3–5), with u constant and v the parameter. The slope is

$$\frac{dy}{dx} = \frac{\partial g}{\partial v} \bigg/ \frac{\partial f}{\partial v}.$$

Likewise, the slope of a v-curve is

$$\frac{dy}{dx} = \frac{\partial g}{\partial u} \bigg/ \frac{\partial f}{\partial u}.$$

These slopes are not equal, for their equality would imply that

$$\frac{\partial g}{\partial v} \frac{\partial f}{\partial u} = \frac{\partial g}{\partial u} \frac{\partial f}{\partial v}, \qquad \text{or} \qquad J = 0,$$

contrary to our assumption.

We can define curvilinear co-ordinates in three dimensions also. The basis for a set of such co-ordinates u, v, w is a transformation

(9.3–8) $\qquad x = f(u, v, w), \qquad y = g(u, v, w), \qquad z = h(u, v, w),$

with nonvanishing Jacobian, and the inverse transformation

(9.3–9) $\qquad u = F(x, y, z), \qquad v = G(x, y, z), \qquad w = H(x, y, z).$

The equations (9.3–9) define three one-parameter families of surfaces, called u-surfaces, v-surfaces, and w-surfaces respectively. A u-surface and a v-surface intersect in a curve, which we shall call a uv-curve. Similarly, there are vw-curves and wu-curves. The fact that the Jacobian of the transformation (9.3–8) is not zero implies that the three curves, one of each type, intersecting at a point, are pair-wise nontangent there.

The most familiar examples of curvilinear co-ordinates in three dimensions are cylindrical co-ordinates and spherical co-ordinates. A great variety of three-dimensional systems can be generated by starting with a two-dimensional system and rotating the plane about a line in the plane. The two families of curves in the plane generate surfaces in space. Half-planes through the axis of rotation form a third set of surfaces. Spherical co-ordinates are derived from plane polar co-ordinates in this way.

Example 2. Consider the transformation

(9.3–10) $\qquad x = uv \cos \theta, \qquad y = uv \sin \theta, \qquad z = u^2 - v^2,$

with Jacobian

(9.3–11) $\qquad\qquad \dfrac{\partial(x, y, z)}{\partial(u, v, \theta)} = 2\, uv(u^2 + v^2).$

Let us assume $u \geqq 0$, $v \geqq 0$, and write $r = (x^2 + y^2)^{1/2} = uv$. Then we see from (9.3–10) that

$$x = r \cos \theta, \qquad y = r \sin \theta.$$

The equations

(9.3–12) $\qquad\qquad r = uv, \qquad z = u^2 - v^2$

can be regarded as defining a set of plane curvilinear co-ordinates (u, v) in the rz-plane. With rotation about the z-axis, using r, θ as polar co-ordinates in the xy-plane, we obtain equations (9.3–10), which we can use to establish u, v, θ as curvilinear co-ordinates in space. The u-curves and v-curves in the rz-plane are parabolas (compare with equations (9.3–3) and (9.3–1), (9.3–2)), a few of which are shown in Fig. 59. In the three-dimensional system, the u-surfaces and v-surfaces are paraboloids of revolution about the z-axis, and the θ-surfaces are half-planes with the

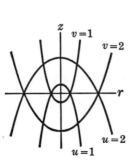

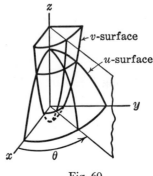

Fig. 59   Fig. 60

z-axis as edge (see Fig. 60). The transformation has a continuously differentiable inverse in the neighborhood of any point not on the z-axis, for
such points are obtained only when neither u nor v is zero, and the Jacobian
(9.3–11) is then not equal to zero. All points of the z-axis are singular
points of the $uv\theta$-coordinate system.

EXERCISES

1. (a) Sketch the u-curves and v-curves in the xy-plane if $x = u \cosh v$,
$y = v \sinh v$. (b) Does the uv-system of curvilinear co-ordinates have a singular
point? (c) To what part of the xy-plane does the system not apply?

2. Discuss the co-ordinates (u, v) related to rectangular co-ordinates (x, y)
by the equations $2x - y + u = 0$, $x - 2y + 2v = 0$. Draw and describe the
u-curves and the v-curves. Find the transformation and the inverse transformation
corresponding to (9.3–5) and (9.3–6) respectively. Verify Theorem II in this case.

3. Show that u, v can be used as curvilinear co-ordinates when $0 < x < \pi/2$
and $y > 0$ if $u = y/\tan x$, $v = y/\sin x$. Solve for x and y in terms of u and v. Draw
the curves $u = 1$, $v = 2$ in the xy-plane. Compute $\dfrac{\partial(u, v)}{\partial(x, y)}$ and $\dfrac{\partial(x, y)}{\partial(u, v)}$ and verify
that they are reciprocals.

4. Find the inverse of the transformation (9.3–3), assuming u and v positive.

5. Suppose $$x = \frac{2u}{u^2 + v^2}, \qquad y = \frac{-2v}{u^2 + v^2}.$$
(a) Show that

$$x^2 + y^2 = \frac{4}{u^2 + v^2},$$

and hence that

$$u = \frac{2x}{x^2 + y^2}, \qquad v = \frac{-2y}{x^2 + y^2}.$$

(b) Show that the u-curves are circles through the origin with centers on the
x-axis, and that the v-curves are circles tangent to the x-axis at the origin. Draw
several curves of each family. Locate the points with the following curvilinear
co-ordinates: $u = 1$, $v = -2$; $u = 0$, $v = 1$; $v = 0$, $u = -2$.

6. Let $x = \cos u \cosh v$, $v = \sin u \sinh v$, where $0 \leq u \leq \pi$ and v is arbitrary. (a) Describe the u-curves and the v-curves by name, and show how variation in the sizes of u and v affects the curves. Draw a number of curves of each family. (b) What are the singular points of the uv-curvilinear co-ordinate system? (c) Describe the points for which $u = 0$; for which $u = \pi/2$; for which $v = 0$.

7. Suppose that two families of curves in the xy-plane are defined by

$$\phi(x, y, u) = 0, \qquad \psi(x, y, v) = 0,$$

where ϕ and ψ are continuously differentiable functions such that

$$\frac{\partial(\phi, \psi)}{\partial(x, y)} \neq 0, \qquad \frac{\partial \phi}{\partial u} \frac{\partial \psi}{\partial v} \neq 0$$

for values of the variables under consideration. Show that these families of curves can be used to establish u and v as curvilinear co-ordinates. If the notations J, j are used as in Theorem II, show that

$$\frac{\partial(\phi, \psi)}{\partial(x, y)} = j \frac{\partial \phi}{\partial u} \frac{\partial \psi}{\partial v}.$$

8. (a) If $x = r \sin \phi \cos \theta$, $y = r \sin \phi \sin \theta$, $z = r \cos \phi$, find $\dfrac{\partial(x, y, z)}{\partial(r, \phi, \theta)}$.

(b) What is the geometric nature of an r-surface? a ϕ-surface? a θ-surface? a $\phi\theta$-curve? a θr-curve? an $r\phi$-curve?

(c) Solve for r, ϕ, θ in terms of x, y, z, assuming $r > 0$, $0 < \phi < \pi/2$, $0 < \theta < \pi/2$ for convenience. Compute $\dfrac{\partial(r, \phi, \theta)}{\partial(x, y, z)}$ and verify that it is the reciprocal of the Jacobian found in (a).

9. Discuss the three-dimensional system of curvilinear co-ordinates (u, v, θ), where $x = r \cos \theta$, $y = r \sin \theta$,

$$z = \frac{\sin v}{\cosh u + \cos v}, \qquad r = \frac{\sinh u}{\cosh u + \cos v},$$

and $r^2 = x^2 + y^2$. Begin by discussing and sketching the u-curves and v-curves in an rz-plane. The relevant equations to be obtained are

$$r^2 + z^2 - 2r \operatorname{ctnh} u + 1 = 0,$$
$$r^2 + z^2 + 2z \operatorname{ctn} v - 1 = 0.$$

Then rotate around the z-axis. Find the singular points of the co-ordinate system. The co-ordinates (u, v, θ) are known as *toroidal*, or *ring*, co-ordinates. Describe the u-surfaces and v-surfaces by name.

10. What is the analogue of Theorem II for a transformation $x = f(u)$ where there is just one variable in each set?

9.4 Mappings

Let us now study a transformation

(9.4-1) $\qquad\qquad x = f(u, v), \qquad y = g(u, v)$

and its inverse

(9.4–2) $u = F(x, y)$, $v = G(x, y)$

from a different point of view. Let us interpret (x, y) as rectangular co-ordinates in one plane, and (u, v) as rectangular co-ordinates in another plane. A pair of equations such as (9.4–1) can then be regarded as making the point (x, y) correspond to the point (u, v). We find it convenient to say that the point (u, v) is *mapped* into the point (x, y). Equations (9.4–1) are said to define a *mapping*, or a point transformation. If the functions f, g are defined in a certain region of the uv-plane, we say that this region is mapped into the xy-plane.

The concept of a mapping does not require that the transformation have an inverse. Nor is it necessary, for the mere concept of a mapping, to deal with continuous or differentiable transformations. However, we shall study continuously differentiable transformations for which the Jacobian is in general nonzero. The inverse transformation (9.4–2) then defines a mapping of a portion of the xy-plane into the uv-plane.

It is of interest, in studying mappings, to consider what becomes of a configuration of points in one plane, such as a curve or a region, when it is mapped into the other plane. The configuration in the second plane is called the *image* of the configuration in the first plane. When the mapping is continuous, with continuous inverse, the mapping process may be con-ceived intuitively as a deformation, with stretching and shrinking of varying amounts, but no tearing or puncturing. If the Jacobian of the mapping is zero at a certain point, however, there may fail to be an in-verse mapping. This can occur, for example, because several points in one plane are mapped into the same point in the other plane.

Example 1.   Consider the mapping

(9.4–3) $u = e^x \cos y$, $v = e^x \sin y$,

of the xy-plane into the uv-plane. We shall investigate the nature of the mapping by finding out what happens to lines $x = $ constant or $y = $ constant.
    We observe that

(9.4–4) $u^2 + v^2 = e^{2x}$, $v \cos y = u \sin y$.

Thus the image of any point on the line $x = x_0$ is a point on the circle in the uv-plane, with center at the origin and radius e^{x_0}. This circle is de-fined parametrically by equations (9.4–3) with $x = x_0$ and y as parameter. The point (u, v) goes counterclockwise once around the circle each time that y increases by 2π. The image of a point on the line $y = y_0$ lies on a straight line through the origin in the uv-plane, with slope $\tan y_0$. The image of the entire line $y = y_0$ is just one of the two rays composing the line $v \cos y = u \sin y$, however. For instance, if $y_0 = \pi/3$, the point (u, v) corresponding to (x, y_0) is given by

$$u = \tfrac{1}{2} e^x, \qquad v = \frac{\sqrt{3}}{2} e^x,$$

so that u and v are always positive. The mapping of the line $y = \pi/3$ onto the ray is indicated in Fig. 61; the part of the ray corresponding to $x < 0$ is shown by a dotted line.

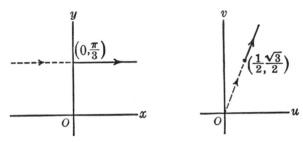

Fig. 61

If we now consider the strip in the xy-plane between the lines $y = 0$, $y = 2\pi$, we see that the image of the strip is the entire uv-plane with the exception of the origin. Line segments $x = x_0$ crossing the strip map into circles centered at the origin, of radius less than one if $x_0 < 0$, and of radius greater than one if $x_0 > 0$. Lines $y = y_0$ map into rays, the angle between the ray and the positive u-axis being y_0. The origin in the uv-plane is not obtained as the image of any point in the xy-plane, but $(u, v) \to (0, 0)$ as $x \to -\infty$. The nature of the mapping is suggested by Fig. 62a and Fig. 62b, in which certain corresponding areas are indicated by similar shading.

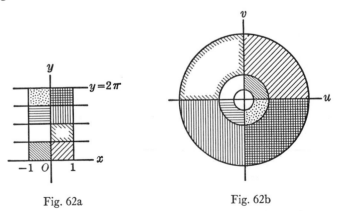

Fig. 62a   Fig. 62b

The equations (9.4–3) admit the period 2π for the variable y. Hence any other strip of width 2π parallel to the x-axis is mapped onto the uv-plane in a manner similar to that just described.

The Jacobian of the mapping is

$$\frac{\partial(u, v)}{\partial(x, y)} = \begin{vmatrix} e^x \cos y & - e^x \sin y \\ e^x \sin y & e^x \cos y \end{vmatrix} = e^{2x}.$$

It does not vanish for any value of x. The nonvanishing of the Jacobian means that if we fix our attention on any point (x_0, y_0) and its image (u_0, v_0), the mapping sets up a one-to-one correspondence between the points of a sufficiently small neighborhood of (x_0, y_0) and those of some neighborhood of (u_0, v_0). This is described by saying that the mapping is *locally* one-to-one, or one-to-one *in the small*. The xy-plane as a whole is not mapped in a one-to-one manner on the uv-plane, however, for many points in the xy-plane have the same image in the uv-plane (two such points are at least 2π units apart, however). Hence we say that the mapping is *not* one-to-one *in the large*.

Example 2.   Consider the mapping

$$(9.4–5) \qquad u = x^2 - y^2, \qquad v = 2xy.$$

In this example let us investigate the configurations in the xy-plane which map into lines $u = $ constant or $v = $ constant in the uv-plane. This is equivalent to regarding u and v as curvilinear co-ordinates and studying the u-curves and v-curves. We see readily that the u-curves are rectangular hyperbolas, with foci on the x-axis if $u > 0$, and on the y-axis

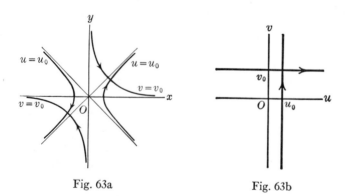

Fig. 63a   Fig. 63b

if $u < 0$. The v-curves are rectangular hyperbolas having the x and y axes as asymptotes. For $u = 0$ we have the pair of lines $y = \pm x$, and for $v = 0$ the pair of lines $x = 0$, $y = 0$.

The image of the hyperbola $u_0 = x^2 - y^2$ is the line $u = u_0$. Note, however, that each branch of the hyperbola maps onto the entire line. Likewise, each branch of the hyperbola $v_0 = 2xy$ maps onto the entire line $v = v_0$ (see Fig. 63a, Fig. 63b).

We now see that certain cells with curved sides in the xy-plane are mapped into rectangular cells in the uv-plane (see Fig. 64a, 64b).

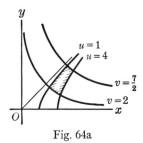

Fig. 64a

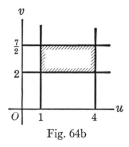

Fig. 64b

The Jacobian of the mapping (9.4–5) is

$$\frac{\partial(u, v)}{\partial(x, y)} = 4(x^2 + y^2).$$

Hence the mapping is locally one-to-one except in the neighborhood of the origin in the xy-plane.

The notion of a point transformation, or mapping, is not limited to two-dimensional problems. We may, for example, speak of mappings from xyz-space into uvw-space.

Observe that in the case of mappings from (x, y) to (u, v) which are locally one-to-one, we may interpret u and v as curvilinear co-ordinates in the xy-plane, or we may interpret x and y as curvilinear co-ordinates in the uv-plane. Similar remarks apply when there are more variables in each set.

The sign of the Jacobian has a significance which deserves mention. Consider a mapping with Jacobian not zero at (x_0, y_0), so that the mapping is, at least locally, one-to-one. If C is a small closed curve enclosing the point (x_0, y_0) in the xy-plane, the image of C in the uv-plane will be a small closed curve C' enclosing the point (u_0, v_0) which corresponds to (x_0, y_0). If the point (x, y) goes around C in the counterclockwise sense, the image point (u, v) will go around C'. But will (u, v) go counterclockwise also? The answer depends on the sign of the Jacobian of the mapping. If $J > 0$, (u, v) will go around C' in the same sense that (x, y) goes around C; but if $J < 0$, (u, v) will go in the sense opposite to that of (x, y). We shall not prove this fact just now. See Exercise 7 and § 15.32.

EXERCISES

1. Consider the mapping $x = au$, $y = bv$, where $a > 0$, $b > 0$. Find out what region in the uv-plane corresponds to the region in the xy-plane bounded by the ellipse $(x^2/a^2) + (y^2/b^2) = 1$.

2. Let R be the region in the xy-plane bounded by the lines $x - y = 0$, $x + y = 0$, $x - 2y = 2$. Find the region R' in the uv-plane onto which R is mapped by the equations $u = 2x - y$, $v = x - 2y$.

3. Find the image in the uv-plane of the triangular region bounded by $x = 0$, $y = 0$, $x + y = 1$, if $u = x + y$, $v = x - y$.

4. Find the image in the xy-plane of the rectangle in the uv-plane bounded by $u = 1$, $u = 4$, $v = 9$, $v = 16$, if the mapping is $x = \left(\dfrac{u + v}{2}\right)^{1/2}$, $y = \left(\dfrac{v - u}{2}\right)^{1/2}$.

5. Consider the mapping $u = \frac{1}{2}(x - y)$, $v = \sqrt{xy}$. Draw a number of the u-curves and v-curves in the xy-plane. What regions in the xy-plane map into the rectangle in the uv-plane bounded by $u = -\frac{3}{2}$, $u = \frac{3}{2}$, $v = 2$, $v = 4$?

6. Study the mapping $x = u - uv$, $y = uv$. Find the inverse transformation. Identify the u-curves and v-curves in the xy-plane, and the x-curves and y-curves in the uv-plane.

7. Let R be the region in the xy-plane bounded by the lines $y - x = 2$, $y - x = 6$, $y + x = 4$, $y + x = 8$. Let R' be the region in the uv-plane onto which R is mapped by the transformation $u = \frac{1}{2}(x + y)$, $v = \frac{1}{2}(x - y)$. Examine the direction in which a point goes around the boundary of R' when its correspondent goes around the boundary of R counterclockwise. Compare with the sign of the Jacobian and note the remark at the end of § 9.4.

8. A mapping from the xy-plane into the uv-plane is defined by

$$u = \frac{x}{1 - x - y}, \qquad v = \frac{y}{1 - x - y}.$$

(a) Find $\dfrac{\partial(u, v)}{\partial(x, y)}$. (b) Find the inverse transformation and $\dfrac{\partial(x, y)}{\partial(u, v)}$. (c) What are the u-curves and v-curves in the xy-plane? (d) What are the x-curves and y-curves in the uv-plane? (e) Find the region R in the xy-plane corresponding to the square in the uv-plane bounded by the lines $u = -\frac{1}{2}$, $u = -1$, $v = -1$, $v = -\frac{3}{2}$.

9. A mapping is defined by

$$u = \frac{x^2}{1 - x^2 - y^2}, \qquad v = \frac{y^2}{1 - x^2 - y^2}.$$

(a) What are the u-curves in the xy-plane if $u > 0$? if $-1 < u < 0$? if $u < -1$?

(b) Answer similar questions for the v-curves. (c) Find where $\dfrac{\partial(u, v)}{\partial(x, y)} = 0$.

(d) Draw the curves $u = \frac{1}{2}$, $u = 3$, $v = \frac{1}{3}$, $v = 1$ and mark four regions in the xy-plane which correspond to the rectangle bounded by the lines $u = \frac{1}{2}$, etc., in the uv-plane. (e) Follow directions similar to those in (d) for $u = -\frac{4}{3}$, $u = -2$, $v = -\frac{1}{2}$, $v = -\frac{1}{5}$.

9.5   Successive mappings

Suppose we have a transformation mapping (x, y) into (u, v), and a further transformation mapping (u, v) into (ξ, η). The effect of performing these two transformations in succession is to give a mapping from the xy-plane into the $\xi\eta$-plane.

Example 1.   Suppose

$$u = x + y, \qquad v = x - y,$$

and

$$\xi = uv, \qquad \eta = u + v.$$

Then

$$\xi = x^2 - y^2, \qquad \eta = 2\,x.$$

The single mapping which is produced by carrying out two successive transformations is called the *resultant*, or *product*, of the two transformations.

In the above example, we have

$$\frac{\partial(u, v)}{\partial(x, y)} = -2, \qquad \frac{\partial(\xi, \eta)}{\partial(u, v)} = v - u,$$

$$\frac{\partial(\xi, \eta)}{\partial(x, y)} = 4\,y,$$

as the student should verify for himself. Observe that

$$v - u = -2\,y.$$

Hence

$$\frac{\partial(\xi, \eta)}{\partial(u, v)} \frac{\partial(u, v)}{\partial(x, y)} = (v - u)(-2) = 4\,y = \frac{\partial(\xi, \eta)}{\partial(x, y)}.$$

This illustrates a general truth which we now state formally.

THEOREM III.   *Let T_1 denote a transformation from the xy-plane into the uv-plane, and let T_2 denote a transformation from the uv-plane into the $\xi\eta$-plane. Let the resultant transformation from the xy-plane into the $\xi\eta$-plane be denoted by T_3. Then the Jacobian of T_3 is the product of the Jacobians of T_1 and T_2, that is,*

$$(9.5\text{-}1) \qquad \frac{\partial(\xi, \eta)}{\partial(x, y)} = \frac{\partial(\xi, \eta)}{\partial(u, v)} \frac{\partial(u, v)}{\partial(x, y)}.$$

It is assumed that the transformations T_1 and T_2 are continuously differentiable. It is further assumed that the transformation T_2 is defined for points (u, v) obtained by application of the transformation T_1 to points (x, y) in some region R of the xy-plane.

PROOF OF THE THEOREM.   We regard ξ, η as functions of the first-class variables u, v, which are in turn functions of the second-class variables x, y. Thus

$$\frac{\partial \xi}{\partial x} = \frac{\partial \xi}{\partial u} \frac{\partial u}{\partial x} + \frac{\partial \xi}{\partial v} \frac{\partial v}{\partial x},$$

$$\frac{\partial \xi}{\partial y} = \frac{\partial \xi}{\partial u} \frac{\partial u}{\partial y} + \frac{\partial \xi}{\partial v} \frac{\partial v}{\partial y},$$

with similar equations for $\dfrac{\partial \eta}{\partial x}$, $\dfrac{\partial \eta}{\partial y}$. It is then a matter of straightforward multiplication to verify that

$$\frac{\partial \xi}{\partial x} \frac{\partial \eta}{\partial y} - \frac{\partial \xi}{\partial y} \frac{\partial \eta}{\partial x} = \left(\frac{\partial \xi}{\partial u} \frac{\partial \eta}{\partial v} - \frac{\partial \xi}{\partial v} \frac{\partial \eta}{\partial u} \right)\left(\frac{\partial u}{\partial x} \frac{\partial v}{\partial y} - \frac{\partial u}{\partial y} \frac{\partial v}{\partial x} \right).$$

This is exactly the relation (9.5–1). The result is recognized with less effort if one is familiar with the rule for writing the product of two determinants as another determinant. This is of especial importance in dealing with the generalization of the theorem for transformations in three or more variables.

EXERCISES

1. Use Theorem III to prove Theorem II (§ 9.3).

2. Verify the correctness of Theorem III when the transformations T_1 and T_2 are defined by $u = e^x \cos y$, $v = e^x \sin y$ and $\xi = u^2 + v^2$, $\eta = v/u$, respectively. Begin by finding the equations of the transformation T_3.

3. Let a_{ij} be the term in the ith row and jth column of a determinant of nth order whose value is A. Let b_{ij} and B refer in similar manner to a second determinant. Let C be the value of the determinant with elements c_{ij}, where $c_{ij} = \sum_{k=1}^{n} a_{ik}b_{kj}$. It is a theorem from algebra that $C = AB$. With this fact in mind, consider the generalization of Theorem III for transformations on n variables. Let T_1 be a transformation in which $u_1, \cdots, u_n$ are differentiable functions of $x_1, \cdots, x_n$, and let T_2 be a transformation in which $\xi_1, \cdots, \xi_n$ are differentiable functions of $u_1, \cdots, u_n$. Let T_3 be the resultant transformation, in which $\xi_1, \cdots, \xi_n$ are considered as functions of $x_1, \cdots, x_n$. Use the chain rule and the theorem on determinants to show that the Jacobian of T_3 is the product of the Jacobians of T_1 and T_2.

4. Consider the transformation (9.4–1) and its inverse (9.4–2), in the neighborhood of a point where $\dfrac{\partial F}{\partial x} \neq 0$. Then $u = F(x, y)$ has a solution $x = \phi(u, y)$. Let $G(\phi(u, y), y) = \Phi(u, y)$. Now consider an rs-plane. Show that the mapping (9.4–2) is the resultant of the mapping $r = F(x, y)$, $s = y$ from the xy-plane to the rs-plane, and the mapping $u = r$, $v = \Phi(r, s)$ from the rs-plane to the uv-plane. The Jacobians of these last two transformations are $\dfrac{\partial F}{\partial x}$ and $\dfrac{\partial \Phi}{\partial s}$ respectively.

Verify in the case of each of these mappings the correctness of the statement made at the end of § 9.4 about the significance of the sign of the Jacobian. This may be done by a direct inspection, since the mappings are of such simple types. Then use Theorem III to prove the correctness of the statement in question for the mapping from the xy-plane to the uv-plane. A similar procedure may be used when $\dfrac{\partial F}{\partial y} \neq 0$.

5. **(a)** Suppose $n > p$. Let $u_1, \cdots, u_n$ be differentiable functions of $x_1, \cdots, x_n$, with $u_{p+1}, \cdots, u_n$ the simple functions defined by $u_i = x_i$, $i = p + 1, \cdots, n$. Show that

$$\frac{\partial(u_1, \cdots, u_n)}{\partial(x_1, \cdots, x_n)} = \frac{\partial(u_1, \cdots, u_p)}{\partial(x_1, \cdots, x_p)}.$$

(b) If $u_1 = x_1, \cdots, u_p = x_p$, and the rest of the u's are differentiable functions of $x_1, \cdots, x_n$, show that

$$\frac{\partial(u_1, \cdots, u_n)}{\partial(x_1, \cdots, x_n)} = \frac{\partial(u_{p+1}, \cdots, u_n)}{\partial(x_{p+1}, \cdots, x_n)}.$$

6. Suppose $F_1, \cdots, F_n$ are continuously differentiable functions of $x_1, \cdots, x_n$, and that $\dfrac{\partial(F_1, \cdots, F_p)}{\partial(x_1, \cdots, x_p)} \neq 0$, where p is a fixed integer, $1 \leqq p < n$. Suppose the equations $u_1 = F_1(x_1, \cdots, x_n), \cdots, u_p = F_p(x_1, \cdots, x_n)$ are solved for $x_1, \cdots, x_p$ in terms of $u_1, \cdots, u_p$ and $x_{p+1}, \cdots, x_n$, and that these values of $x_1, \cdots, x_p$ are then substituted into $F_{p+1}, \cdots, F_n$, giving rise to functions $\psi_i(u_1, \cdots, u_p, x_{p+1}, \cdots, x_n)$, $i = p + 1, \cdots, n$. Show that

$$\frac{\partial(F_1, \cdots, F_n)}{\partial(x_1, \cdots, x_n)} = \frac{\partial(F_1, \cdots, F_p)}{\partial(x_1, \cdots, x_p)} \frac{\partial(\psi_{p+1}, \cdots, \psi_n)}{\partial(x_{p+1}, \cdots, x_n)},$$

where it is assumed that $u_1, \cdots, u_p$ are replaced by

$$F_1(x_1, \cdots, x_n), \cdots F_p(x_1, \cdots, x_n)$$

after calculating the derivatives in the last Jacobian.

SUGGESTIONS OF METHOD: Note first of all that $\psi_i(F_1, \cdots, F_p, x_{p+1}, \cdots, x_n) = F_i(x_1, \cdots, x_n)$, $i = p + 1, \cdots, n$. Let T_1 and T_2 be defined as follows:

T_1	T_2
$\xi_1 = F_1(x_1, \cdots, x_n)$	$u_1 = \xi_1$
$\cdots\cdots\cdots\cdots\cdots$	$\cdots\cdots$
$\xi_p = F_p(x_1, \cdots, x_n)$	$u_p = \xi_p$
$\xi_{p+1} = x_{p+1}$	$u_{p+1} = \psi_{p+1}(\xi_1, \cdots, \xi_n)$
$\cdots\cdots\cdots$	$\cdots\cdots\cdots\cdots\cdots$
$\xi_n = x_n$	$u_n = \psi_n(\xi_1, \cdots, \xi_n).$

Now apply Theorem III (for the case of n variables) and Exercise 5.

9.6   Identical vanishing of the Jacobian

In this section we shall inquire into the state of affairs when the Jacobian of a transformation is equal to zero throughout a region. We first investigate how such a thing may occur. It is convenient to deal with mappings of the xy-plane into the uv-plane. The reasoning can be extended to higher dimensions.

THEOREM IV. *Suppose that*

$$(9.6-1) \qquad u = F(x, y), \qquad v = G(x, y)$$

is a continuously differentiable mapping of a region R of the xy-plane into the uv-plane. Let R' be a region in the uv-plane which contains the image of R. Let $\Phi(u, v)$ be a function which is defined and continuous in R', such that there is no neighborhood of a point in R' throughout which $\Phi(u, v) = 0$. Finally, suppose that

$$(9.6-2) \qquad \Phi(F(x, y), G(x, y)) = 0$$

whenever (x, y) is a point of R. Then

$$(9.6-3) \qquad \frac{\partial(F, G)}{\partial(x, y)} = 0$$

at all points of R.

PROOF. Suppose the equation (9.6-3) fails to hold at some point (x_0, y_0) in R. Then the mapping (9.6-1) is locally one-to-one, and maps a neighborhood of (x_0, y_0) onto a neighborhood of the corresponding point (u_0, v_0). In view of this, (9.6-2) means that $\Phi(u, v) = 0$ throughout a neighborhood of (u_0, v_0). Since this violates the hypothesis, (9.6-3) must hold at every point of R.

Example 1. Theorem IV is illustrated by

$$F(x, y) = \sqrt{y} \sin x, \qquad G(x, y) = y \cos^2 x - y, \qquad \Phi(u, v) = u^2 + v,$$

with R the half-plane $y > 0$, R' the entire uv-plane.

When a pair of functions $F(x, y)$, $G(x, y)$ are such that the conditions stated in the hypotheses of Theorem IV are satisfied, F and G are said to be *functionally dependent* in the region R. The functional dependence is expressed by (9.6-2).

Now let us consider the converse question, of seeing what can be inferred about functional dependence of F and G if their Jacobian is zero throughout some region.

THEOREM V. *Suppose that F and G are continuously differentiable functions of x, y such that (9.6-3) holds in a neighborhood of the point (x_0, y_0). Suppose that either $\dfrac{\partial F}{\partial x} \neq 0$ or $\dfrac{\partial F}{\partial y} \neq 0$ at this point. Let $u_0 = F(x_0, y_0)$.*

Then there exists an interval of the u-axis centered at u_0, and a function $\phi(u)$ defined thereon, such that

(9.6–4) $$G(x, y) = \phi(F(x, y))$$

throughout a neighborhood of (x_0, y_0).

Before taking up the proof, we comment on the theorem. Observe that (9.6–4) is a particular relation of the form (9.6–2), with $\Phi(u, v) = \phi(u) - v$. If the hypothesis had been $\dfrac{\partial G}{\partial x} \neq 0$ or $\dfrac{\partial G}{\partial y} \neq 0$, the conclusion analogous to (9.6–4) would have been $F(x, y) = \psi(G(x, y))$. Theorem V is a "local" theorem, and not a strict converse of Theorem IV, for the theorem does not deal with a fixed region R, but with neighborhoods of a single point. The theorem does not say that the neighborhood in which (9.6–4) holds is the same as the neighborhood in which (9.6–3) is assumed to hold. It is possible to develop a theorem similar to Theorem V for regions of size fixed in advance; but the proof of such a theorem presents certain difficulties which we shall not attempt to consider.

PROOF OF THEOREM V.   Consider the equation $u = F(x, y)$. Suppose $\dfrac{\partial F}{\partial x} \neq 0$ at (x_0, y_0); then the implicit-function theorem guarantees the existence of a solution $x = f(u, y)$ giving all triples (x, y, u) near (x_0, y_0, u_0) for which $u = F(x, y)$. Moreover,

(9.6–5) $$\frac{\partial f}{\partial y} = - \frac{\dfrac{\partial F}{\partial y}}{\dfrac{\partial F}{\partial x}} .$$

All this applies when the differences $x - x_0$, $y - y_0$, $u - u_0$ are sufficiently small. Consider the function $G(x, y)$ as a function of u and y, with $x = f(u, y)$. The partial derivative with respect to y is

$$\frac{\partial G}{\partial x} \frac{\partial f}{\partial y} + \frac{\partial G}{\partial y} = \frac{\dfrac{\partial(F, G)}{\partial(x, y)}}{\dfrac{\partial F}{\partial x}} = 0$$

because of (9.6–5) and (9.6–3). Thus $G(f(u, y), y)$ is actually independent of y. Let us write $\phi(u) = G(f(u, y), y)$. Since $x = f(u, y)$ is equivalent to $u = F(x, y)$ for the values of the variables here in question, $\phi(u) = G(f(u, y), y)$ is equivalent to (9.6–4). If we assume $\dfrac{\partial F}{\partial y} \neq 0$ instead of $\dfrac{\partial F}{\partial x} \neq 0$, similar reasoning again leads to (9.6–4).

Example 2. The argument is illustrated by $F(x, y) = x^2 y^2$, $G(x, y) = - xy$, $x_0 = y_0 = 1$, $f(u, y) = \sqrt{u}/y$, $\phi(u) = - \sqrt{u}$. Observe that, with these same functions F, G, but with $x_0 = - 1$, $y_0 = 1$, we obtain $f(u, y) = - \sqrt{u}/y$, $\phi(u) = \sqrt{u}$.

Theorem V can be generalized to n functions of n variables. For $n = 3$ the hypotheses that

(9.6–6)
$$\frac{\partial(F, G, H)}{\partial(x, y, z)} = 0$$

in a neighborhood of (x_0, y_0, z_0), and that at least one of the Jacobians

$$\frac{\partial(F, G)}{\partial(x, y)}, \quad \frac{\partial(F, G)}{\partial(y, z)}, \quad \frac{\partial(F, G)}{\partial(z, x)}$$

is not zero at this point, leads to a conclusion of the form

(9.6–7) $$H(x, y, z) = \phi(F(x, y, z), G(x, y, z)),$$

where $\phi(u, v)$ is defined near $u_0 = F(x_0, y_0, z_0)$, $v_0 = G(x_0, y_0, z_0)$.

EXERCISES

1. Are the functions $x + y$, $x^2 + y^2$ functionally dependent in the neighborhood of any point?

2. Under what condition will the linear functions $ax + by$, $cx + dy$ be functionally dependent?

3. (a) Let $F(x, y, z) = x + y - z$, $G(x, y, z) = x - y + z$, $H(x, y, z) = x^2 + y^2 + z^2 - 2yz$. Verify that (9.6–6) holds in this case. Solve the equations $u = F(x, y, z)$, $v = G(x, y, z)$ for x and y in terms of u, v, z, and substitute the solutions for x and y in $H(x, y, z)$. Observe that the result is independent of z. From this work find the function $\phi(u, v)$ such that (9.6–7) holds.
(b) Show that this line of reasoning is applicable in the general case, provided
$$\frac{\partial(F, G)}{\partial(x, y)} \neq 0.$$

4. Show in each case that the functions are functionally dependent, and find the way in which the third function depends on the first two. Use the method of Exercise 3.
(a) $u = x + y + z$, $v = xy + yz + zx$, $w = x^2 + y^2 + z^2$.
(b) $u = x/(y - z)$, $v = y/(z - x)$, $w = z/(x - y)$.

MISCELLANEOUS EXERCISES

1. (a) Find the inverse of the transformation
$$u = \frac{x}{r^2}, v = \frac{y}{r^2}, w = \frac{z}{r^2}, \text{ where } r^2 = x^2 + y^2 + z^2.$$

(b) What are the u-surfaces?

(c) Calculate $\dfrac{\partial(u, v, w)}{\partial(x, y, z)}$.

2. If $t = r \cos \phi$, $x = r \sin \phi \cos \psi$, $y = r \sin \phi \sin \psi \cos \theta$, $z = r \sin \phi \sin \psi \sin \theta$, show that $t^2 + x^2 + y^2 + z^2 = r^2$ and find $\dfrac{\partial(t, x, y, z)}{\partial(r, \phi, \psi, \theta)}$.

This indicates how spherical co-ordinates may be introduced in four-dimensional space.

Vectors and Vector Fields   10

10.   Purpose of the chapter

In a sense, vectors are to the geometry of space of two or more dimensions what the real numbers are to the geometry of a straight line. The positive real numbers enable us to measure distances along the line, and the existence of negative numbers enables us to distinguish directions along the line. In space of two or more dimensions there are not just two directions, but infinitely many. A vector is a sort of generalized number. There are vectors of any specified direction and of any specified length. But, while it is appropriate to regard vectors as mathematical objects which are in some respects like real numbers, they do not have all the properties of real numbers. It is part of the purpose of this chapter to set forth the fundamental concepts about vectors and to sum up briefly the algebraic properties of vectors. We do not presume in the scope of one chapter to cover all of what is conventionally called vector analysis. The main purpose of the chapter is to provide a sufficient conceptual foundation for the use of vector language and notions about vector fields in certain later portions of the book, particularly Chapters XIV and XV. In the applications of calculus to the geometry of curves and surfaces, vectors play a decisive role. The integral theorems of Chapter XV are best understood and appreciated as theorems about vector fields. Nearly all the important applications of these theorems arise in contexts where vector fields are fundamental: in pure mathematics in connection with potential theory and various partial differential equation problems, and in mathematical physics wherever the formulation of physical concepts and laws draws upon the aforementioned branches of analysis.

10.1   Vectors in Euclidean space

We shall concern ourselves mainly with vectors in three dimensions. Parts (but not all) of the discussion of vectors can easily be adapted to the two-dimensional case, or can also be extended to space of four or more dimensions.

When we speak of Euclidean space of three dimensions we must at this point rely upon what the student has learned formally from studies of plane and solid geometry in school and from studies of analytic geometry in college, and also to a certain extent upon his intuitive or acquired notions of space, as built up from experience (e.g., observation of the shapes of objects such as books, houses, boxes; the study of maps; the physical concepts of space as learned from readings in astronomy, physics, etc.). This background of understanding of what Euclidean space is leaves something to be desired from the strictly mathematical point of view, for it does not make the foundations of the subject entirely explicit, nor does it indicate in what sense Euclidean geometry differs from the various geometries of non-Euclidean type.

We frequently find it convenient to deal with geometric problems with the help of a system of rectangular co-ordinates (x, y, z). It is important to understand, however, that geometrical matters are actually independent of the particular co-ordinate system which may be used in discussing them. The origin may be located where we please, and the orientation of the co-ordinate axes is subject to our arbitrary choice.

With these preliminary remarks let us now begin the discussion of vectors. Let O be an arbitrary point of space. If P is any point of space distinct from O, the directed line segment from O to P is called a vector. We denote it by $\overrightarrow{OP}$; the visual representation of the vector is the line segment OP with an arrowhead at P. Vectors will often be denoted by a single letter printed in boldface type. Thus, in Fig. 65, we may write $\mathbf{A}$ for the vector $\overrightarrow{OP}$. We also define a special vector $\mathbf{0}$ corresponding to the exceptional case in which P coincides with O. We call $\mathbf{0}$ the zero vector. Note that $\mathbf{0}$ is exceptional in two respects —it is the only vector whose length is zero, and it is the only vector in Euclidean space which has no direction.

Fig. 65

The set of all vectors as just defined is called the vector space with origin O. It is this vector space which is analogous to the real-number axis, as defined in § 2.6. In fact, we may consider the axis of reals to be a one-dimensional vector space. Just as there is a one-to-one correspondence between the set of all real numbers and the set of all points on a line (with the number 0 corresponding to an arbitrary point O of the line), so there is a one-to-one correspondence between the set of all vectors and the set of all points in space, with a special vector $\mathbf{0}$ corresponding to an arbitrarily selected point O of space (here space means Euclidean space of three dimensions).

In the above definition of vector, all vectors have the same initial point O. Clearly we could define another vector space with a different origin O'. These two spaces would look just alike, however, and one could be carried over into the other by a rigid motion in which the point O' is translated to the point O. Because of this congruence, we find it convenient to regard vectors belonging to different vector spaces as being equal if they coincide when the origins of the two spaces are brought together by a translation. This means, in effect, that two directed line segments are regarded as defining the same vector if they have the same length and the same direction.

Suppose that we introduce a system of rectangular co-ordinates with origin at O. Let $\mathbf{A}$ be the vector $\overrightarrow{OP}$, and let the co-ordinates of the point P be (A_1, A_2, A_3). Then we call A_1, A_2, A_3 the *components* of $\mathbf{A}$ in the given rectangular co-ordinate system (see Fig. 66). On occasion we find it convenient to write

$$\mathbf{A} = (A_1, A_2, A_3).$$

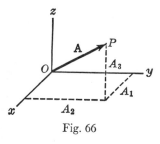

Fig. 66

It is to be observed, however, that the vector $\mathbf{A}$ itself is something independent of the co-ordinate system, whereas its components depend upon the co-ordinate system. A vector will have different components in different co-ordinate systems.

The length of a vector $\mathbf{A}$ is denoted by $||\,\mathbf{A}\,||$; this is sometimes called the norm of $\mathbf{A}$. Note that $||\,\mathbf{0}\,|| = |\,\mathbf{0}\,|$. Also observe that

(10.1–1) $$||\,\mathbf{A}\,|| = (A_1^2 + A_2^2 + A_3^2)^{1/2}.$$

10.2   The algebra of vectors

We now define the algebraic operations which are performed on vectors. The first of these operations is *addition*.

DEFINITION.   If $\mathbf{A}$ and $\mathbf{B}$ are vectors, we define the sum $\mathbf{C} = \mathbf{A} + \mathbf{B}$ as follows: Displace $\mathbf{B}$ without changing its direction until its initial point coincides with the terminal point of $\mathbf{A}$. The sum $\mathbf{A} + \mathbf{B}$ is then the vector from the initial point of $\mathbf{A}$ to the terminal point of the displaced vector $\mathbf{B}$ (see Fig. 67).

It is clear from Fig. 67 that the commutative law $\mathbf{A} + \mathbf{B} = \mathbf{B} + \mathbf{A}$ is satisfied. If either $\mathbf{A}$ or $\mathbf{B}$ is the vector $\mathbf{0}$, the foregoing definition must be understood to mean that

$$\mathbf{A} + \mathbf{0} = \mathbf{A}, \quad \mathbf{0} + \mathbf{B} = \mathbf{B},$$

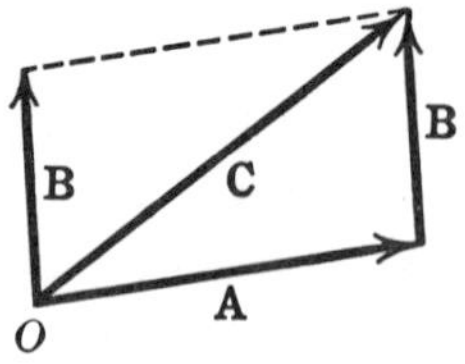

Fig. 67

for any vectors $\mathbf{A}, \mathbf{B}$.

The addition of vectors may be expressed in terms of components relative to a given rectangular co-ordinate system. If $\mathbf{A} = (A_1, A_2, A_3)$ and $\mathbf{B} = (B_1, B_2, B_3)$, the rule of addition is

(10.2–1) $$\mathbf{A} + \mathbf{B} = (A_1 + B_1, A_2 + B_2, A_3 + B_3);$$

that is, to find the components of the vector sum, we add the components of the separate vectors. Figure 68 illustrates the rule for the case of components in the y-direction.

The associative law of addition

$$(\mathbf{A} + \mathbf{B}) + \mathbf{C} = \mathbf{A} + (\mathbf{B} + \mathbf{C})$$

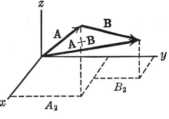

is also satisfied, for any three vectors. This may be seen geometrically, or from (10.2–1), by virtue of the known associativity of addition of real numbers.

Next we consider multiplication of vectors by real numbers.

Fig. 68

DEFINITION. If c is a positive real number and $\mathbf{A}$ is a vector other than $\mathbf{0}$, we define $c\mathbf{A}$ as the vector having the same direction as $\mathbf{A}$ and length c times the length of $\mathbf{A}$. If $c < 0$ we define $c\mathbf{A}$ as the vector of length $|\,c\,|$ times the length of $\mathbf{A}$ in the direction opposite to that of $\mathbf{A}$. If $c = 0$ or $\mathbf{A} = \mathbf{0}$, we define $c\mathbf{A} = \mathbf{0}$. For convenience we write $-1 \cdot \mathbf{A} = -\mathbf{A}$. Note that $1 \cdot \mathbf{A} = \mathbf{A}$.

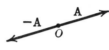

In terms of components the rule of multiplication by real numbers is

(10.2–2) $\qquad c\mathbf{A} = (cA_1,\ cA_2,\ cA_3).$

Fig. 69

The following algebraic laws are satisfied:

$$c(\mathbf{A} + \mathbf{B}) = c\mathbf{A} + c\mathbf{B},$$
$$(a + b)\mathbf{A} = a\mathbf{A} + b\mathbf{A},$$
$$a(b\mathbf{A}) = (ab)\mathbf{A}.$$

Subtraction of vectors is defined in terms of addition, and multiplication by -1:

$$\mathbf{A} - \mathbf{B} = \mathbf{A} + (-1 \cdot \mathbf{B}).$$

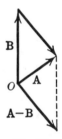

Observe that $\mathbf{A} - \mathbf{A} = \mathbf{0}$, and that the vector equation $\mathbf{X} + \mathbf{B} = \mathbf{A}$ has the unique solution

Fig. 70

$$\mathbf{X} = \mathbf{A} - \mathbf{B}.$$

The geometrical construction for subtraction should be noted (Fig. 70).

Thus far we have not defined multiplication of a vector by a vector. It is here that we first encounter a great difference between the algebra of vectors and the algebra of real numbers. As it turns out, there are two kinds of multiplication which are of importance. The first of these two kinds is the simpler to discuss. In this kind of multiplication the product of two vectors is a real number (not a third vector); we call this product the *dot product*. It is also called the *inner product*.

DEFINITION. Let $\mathbf{A}$ and $\mathbf{B}$ be two nonzero vectors, and let θ be the angle between them (take $0 \leqq \theta \leqq \pi$). The dot product of **A** and **B** is written **A · B**, and is defined as

(10.2-3) $\mathbf{A} \cdot \mathbf{B} = ||\mathbf{A}|| \, ||\mathbf{B}|| \cos \theta.$

In the event that either **A** or **B** is **0**, the angle θ is not well defined. Therefore, to complete the definition of dot products, we define

$$\mathbf{A} \cdot \mathbf{0} = \mathbf{0} \cdot \mathbf{A} = 0$$

for any **A**.

The dot product satisfies the three algebraic laws

$$\mathbf{A} \cdot \mathbf{B} = \mathbf{B} \cdot \mathbf{A},$$

(10.2-4) $(\mathbf{A} + \mathbf{B}) \cdot \mathbf{C} = \mathbf{A} \cdot \mathbf{C} + \mathbf{B} \cdot \mathbf{C},$

(10.2-5) $(c\mathbf{A}) \cdot \mathbf{B} = c(\mathbf{A} \cdot \mathbf{B}).$

The truth of the commutative law is immediately apparent from the symmetrical manner in which **A** and **B** enter in the definition of the dot product. To prove (10.2-4) and (10.2-5) we observe the following interpretation of the definition (10.2-3): **A · B** is the product of $||\mathbf{B}||$ and the projection of **A** on the directed line of the vector **B** (see Fig. 71). Now the projection of **A** + **B** on the directed line of **C** is the sum of the projections of **A** and **B** on that line. Hence (10.2-4) is true. A similar argument proves (10.2-5).

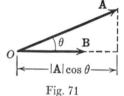

Fig. 71

If **B** = **A** in (10.2-3), then $\theta = 0$. Hence

(10.2-6) $\mathbf{A} \cdot \mathbf{A} = ||\mathbf{A}||^2.$

A fundamental fact about the product of two real numbers is that the product is zero if and only if one of the factors is zero. This property does not carry over to the dot product. We see from the definition that **A · B** can be zero either (a) because **A** or **B** is **0**, or (b) because **A** is perpendicular to **B**.

10.21 Orthogonal unit vectors

Let a fixed rectangular co-ordinate system be chosen with origin O. It is conventional, particularly in dealing with physical applications, to work with right-handed co-ordinate systems, and we shall ordinarily adhere to this convention. Now let **i**, **j**, **k** be vectors, each of unit length, in the directions of the positive x, y, and z axes, respectively (see Fig. 72). It is clear that if **A** is any vector, we can express it in the form

(10.21-1) $\mathbf{A} = A_1\mathbf{i} + A_2\mathbf{j} + A_3\mathbf{k},$

where A_1, A_2, A_3 are the components of $\mathbf{A}$ (see Fig. 73). We call $\mathbf{i}$, $\mathbf{j}$, $\mathbf{k}$ the *fundamental orthonormal triad* associated with this particular co-ordinate system. The word "orthonormal" is a combination of "orthogonal" and "normal." The vectors $\mathbf{i}$, $\mathbf{j}$, $\mathbf{k}$ form an *orthogonal set*; that is,

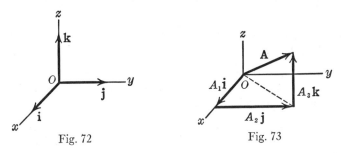

Fig. 72   Fig. 73

they are mutually perpendicular. A vector is said to be normalized, or normal, if it is of unit length. The orthonormal character of the triad $\mathbf{i}$, $\mathbf{j}$, $\mathbf{k}$ is expressed by the relations

(10.21–2)
$$\mathbf{i} \cdot \mathbf{i} = \mathbf{j} \cdot \mathbf{j} = \mathbf{k} \cdot \mathbf{k} = 1,$$
$$\mathbf{i} \cdot \mathbf{j} = \mathbf{j} \cdot \mathbf{k} = \mathbf{k} \cdot \mathbf{i} = 0.$$

Two vectors expressed in the form (10.21–1) are equal if and only if their corresponding components are equal. A vector equation is therefore equivalent to three ordinary equations.

We are now in a position to prove an important theorem about the expression of dot products in terms of components.

THEOREM I.   *In any given rectangular co-ordinate system the dot product of two vectors is expressed in terms of the components of the vectors by the formula*

(10.21–3) $\mathbf{A} \cdot \mathbf{B} = A_1B_1 + A_2B_2 + A_3B_3.$

PROOF. We express each vector in the form (10.21–1) and apply the distributive law repeatedly.

$$\mathbf{A} \cdot \mathbf{B} = (A_1\mathbf{i} + A_2\mathbf{j} + A_3\mathbf{k}) \cdot (B_1\mathbf{i} + B_2\mathbf{j} + B_3\mathbf{k}).$$

Because of (10.21–2) the result reduces to the simple form (10.21–3). The student should himself write out the nine terms of the product and see what becomes of them.

EXERCISES

1. Find $\mathbf{A} + \mathbf{B}$, $\mathbf{A} - \mathbf{B}$, and $\mathbf{A} \cdot \mathbf{B}$ in each case. Are $\mathbf{A}$ and $\mathbf{B}$ collinear, perpendicular, or neither?

(a) $\mathbf{A} = (1, 1, -1)$, $\mathbf{B} = (3, -2, -1)$.

(b) $A = (1, 4, 3)$, $B = (4, 2, -4)$.
(c) $A = (2, -1, 1)$, $B = (3, -4, -4)$.
(d) $A = (6, 4, -2)$, $B = (-9, -6, 3)$.

2. Find $||A||$ in each case.
(a) $A = 2i + 3j + 6k$. **(b)** $A = 4i + 2j - 4k$. **(c)** $A = 2i + j - 2k$.
(d) $A = 4i + 3j$.

3. If the angle between A and B is θ, where $A = (-2, -2, 1)$ and $B = (1, -2, 2)$, find $\cos \theta$.

4. Let A, B, C, D be four vectors in the vector space with origin O, and suppose that $B - A = C - D$. Show that the tips of these four vectors are consecutive vertices of a parallelogram.

5. Let A, B, C be noncoplanar vectors in the vector space with origin O. What is the vector $A + B + C$ in relation to the parallelepiped of which A, B, C are edges?

6. Let $A = (\cos\theta, \sin\theta, 0)$, $B = (\cos\phi, \sin\phi, 0)$, where $0 \leq \phi \leq \theta \leq 2\pi$. Draw a figure showing the positions of A and B in the xy-plane. Use Theorem I to obtain the trigonometric identity for $\cos(\theta - \phi)$.

7. If A and B are of unit length, and θ is the angle between them, express $||A - B||$ as a function of θ.

8. Let $C = A - B$ (see Fig. 70). Then $||C||^2 = (A - B) \cdot (A - B)$, by $(10.2-6)$. Use the distributive law to expand the dot product, and deduce the law of cosines from the result.

9. Let A be a vector of unit length, and let B be any vector. Let $C = (B \cdot A)A$, $D = B - C$. Prove that $D \cdot A = 0$. Draw a figure showing the relation of A, B, C, D. Work out the special cases **(a)** $A = i$, $B = 5i + 2j$; **(b)** $A = \frac{1}{5}(3i + 4j)$, $B = 5j$.

10. A set of vectors is called linearly independent if no one of the vectors of the set can be expressed as a finite sum of scalar multiples of the other members of the set. This is the same as saying that if $A_1, \cdots, A_n$ are members of the set, and $a_1A_1 + \cdots + a_nA_n = 0$, then $a_1 = a_2 = \cdots = a_n = 0$.

  (a) Show that a set of vectors is linearly independent if no vector is 0 and if the vectors are mutually perpendicular.

  (b) Let A_1, A_2, A_3 be linearly independent set of vectors. Define vectors B_i and C_i ($i = 1, 2, 3$) successively as follows:

$$B_1 = A_1, \quad C_1 = \frac{B_1}{||B_1||},$$

$$B_2 = A_2 - (A_2 \cdot C_1)C_1, \quad C_2 = \frac{B_2}{||B_2||},$$

$$B_3 = A_3 - (A_3 \cdot C_1)C_1 - (A_3 \cdot C_2)C_2, \quad C_3 = \frac{B_3}{||B_3||}.$$

Show that C_1, C_2, C_3 is an orthonormal set.

(c) Work out the details of part (b) for the special case $A_1 = 2\,i$, $A_2 = 3\,i + 4\,j$, $A_3 = i + 2\,j + 3\,k$.

10.22   Cross products

There is another kind of vector multiplication, in which the product of two vectors is another vector. We shall define this product presently. It is called the *cross product* (or, sometimes, the *vector product*), and is denoted by $A \times B$.

DEFINITION.   The cross product of the vector A by the vector B is defined as follows. If either A or B is 0 we define the product to be 0:

$$0 \times B = A \times 0 = 0.$$

Otherwise, let θ be the angle between A and B. We define $A \times B$ to be the vector of magnitude

$$||A \times B|| = ||A|| \; ||B|| \sin \theta$$

whose line is perpendicular to the plane of A and B, and whose direction is such that A, B, and $A \times B$ form a right-handed system (see Fig. 74). If the vectors A, B lie along the same line, they do not determine a plane. In this case $\sin \theta = 0$, however, and so $A \times B = 0$. Note that the magnitude of $A \times B$ is equal to the area of the parallelogram of which A, B are adjacent sides.

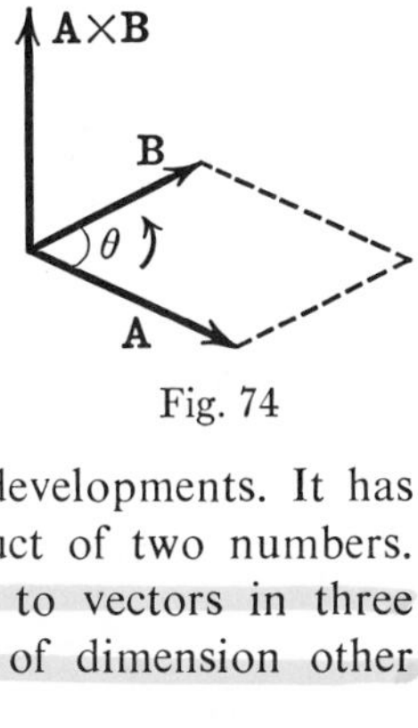

Fig. 74

The motivation for this definition, and its usefulness, will be better understood by the student after he has seen the occurrence of the cross product in physical applications and in later mathematical developments. It has only very limited analogies with the ordinary product of two numbers. Moreover, the cross product is something peculiar to vectors in three dimensions, having no analogue for vector spaces of dimension other than three.

The principal algebraic rules governing the cross product are

(10.22–1) $$A \times B = - (B \times A),$$

(10.22–2) $$(cA) \times B = c(A \times B),$$

(10.22–3) $$A \times (B + C) = (A \times B) + (A \times C).$$

Multiplication is not commutative, but anticommutative, as we see by (10.22–1). This law is apparent from the definition of the cross product, since B, A and $- (A \times B)$ form a right-handed system. The law (10.22–2) is also apparent from the definition of the cross product. The rule

(10.22–4) $$A \times (cB) = c(A \times B)$$

can be deduced from (10.22–1) and (10.22–2).

Now consider the proof of the distributive law (10.22–3). The law obviously holds if $A = 0$, so we consider the proof on the assumption that $A \neq 0$. If B is any vector, let B' denote the vector projection of B on the plane perpendicular to A through the origin (see Fig. 75). Clearly $||\,B'\,|| = ||\,B\,||\sin\theta$, and therefore $A \times B = A \times B'$. Now, projecting in this manner, we see that $B' + C'$ is the projection of $B + C$. Therefore, instead of proving (10.22–3), it is sufficient to prove

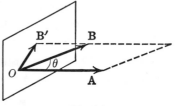

Fig. 75

(10.22–5) $$A \times (B' + C') = A \times B' + A \times C'.$$

The advantage here is that the vectors B', C' and $B' + C'$ either are 0 or are perpendicular to A.

Now let us consider a figure (Fig. 76) in which A is perpendicular to the plane of the page and is directed out toward the student as he reads the page. We may assume that neither B' nor C' is 0, and also that these vectors are not collinear, since in these cases (10.22–5) is certainly true, as we readily see (if B' and C' are collinear, C' is a multiple of B', and (10.22–5) then follows from (10.22–4)). Now, if V is any vector in the plane of the page, $A \times V$ will also lie in the plane of the page. It will make a 90° angle with V, and will have a length $||\,A\,||$ times that of V. Let us take V to be, successively, B', C', $B' + C'$. Then we see that the configuration of the vectors $A \times B'$, $A \times C'$, $A \times (B' + C')$ is similar to the configuration of B', C', $B' + C'$, but is

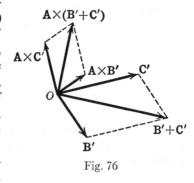

Fig. 76

$||\,A\,||$ times as large and is turned through an angle of 90° in the plane of the page (see Fig. 76). Necessarily, then, $A \times (B' + C')$ will be the diagonal of the parallelogram with adjacent sides $A \times B'$, $A \times C'$, because $B' + C'$ is the diagonal of the parallelogram with adjacent sides B', C'. Therefore (10.22–3) is true.

A second distributive law

$$(A + B) \times C = A \times C + B \times C$$

follows at once from (10.22–3) and (10.22–1).

Using the distributive law, we may obtain expressions for the components of $A \times B$ in terms of the components of the vectors A and B. We have

$$A \times B = (A_1 i + A_2 j + A_3 k) \times (B_1 i + B_2 j + B_3 k).$$

We expand this by repeated use of the distributive law, and use (10.22–2) and (10.22–3). Nine terms are obtained in this way. To simplify the result we observe the following multiplication table, which is easily worked out directly from the definitions.

$$
\begin{array}{lll}
i \times i = 0 & i \times j = k & i \times k = -j \\
j \times i = -k & j \times j = 0 & j \times k = i \\
k \times i = j & k \times j = -i & k \times k = 0
\end{array}
$$

As a consequence, we find

(10.22–6)
$$A \times B = (A_2 B_3 - A_3 B_2)i + (A_3 B_1 - A_1 B_3)j + (A_1 B_2 - A_2 B_1)k.$$

It is convenient to remember this formula by noting that if we write $C = A \times B$, then

$$C_1 = \begin{vmatrix} A_2 & A_3 \\ B_2 & B_3 \end{vmatrix}, \; C_2 = \begin{vmatrix} A_3 & A_1 \\ B_3 & B_1 \end{vmatrix}, \; C_3 = \begin{vmatrix} A_1 & A_2 \\ B_1 & B_2 \end{vmatrix}.$$

These two-row determinants are the cofactors of the elements of the first row of a three-row determinant having A_1, A_2, A_3, and B_1, B_2, B_3 as its second and third rows, respectively. Accordingly, as a memory device, we sometimes write

(10.22–7)
$$A \times B = \begin{vmatrix} i & j & k \\ A_1 & A_2 & A_3 \\ B_1 & B_2 & B_3 \end{vmatrix}.$$

EXERCISES

1. Find the indicated cross products.
(a) $(i + j + k) \times (i + 2j + 3k)$;
(b) $(2i - 3j - k) \times (2i - 5j + 3k)$;
(c) $(i - 2j - k) \times (i - 3j + 4k)$.

2. Find the area of the parallelogram of which the vectors $A = i - j + 2k$ and $B = 2i + 4j - k$ are adjacent sides.

3. (a) Let A, B, C be noncoplanar vectors from O with terminal points P, Q, R respectively. Explain why $\frac{1}{2}(B - A) \times (C - A)$ is a vector perpendicular to the plane of PQR, and of length equal to the area of the triangle PQR. (b) Find the area of the triangle formed by the points $(1, 1, -2)$, $(2, -1, 1)$, $(1, 3, -1)$.

4. Find $A \cdot (B \times C)$ and $B \cdot (A \times C)$ if

(a) $A = 2i - 3j + 5k$, $B = -i + 4j + 2k$, $C = 2i + 3j$;

(b) $A = 2i + 3j + k$, $B = i + 2j + 5k$, $C = -2i + 4j + 3k$.

5. (a) If $A = (A_1, A_2, A_3)$, etc., show that

$$A \cdot (B \times C) = \begin{vmatrix} A_1 & A_2 & A_3 \\ B_1 & B_2 & B_3 \\ C_1 & C_2 & C_3 \end{vmatrix}.$$

(b) How does it follow from (a) that $A \cdot (B \times C) = (A \times B) \cdot C$?

(c) What is the value of $A \cdot (B \times C)$ if any two of the vectors are equal?

(d) Explain why the numerical value of the determinant in (a) is equal to the volume of the parallelepiped having the vectors A, B, C as concurrent edges.

(e) If A, B, C are permuted in all possible ways in the product $A \cdot (B \times C)$, how many different values can be obtained?

(f) Find the values of $D \cdot (B - A)$ and $D \cdot (C - A)$, where

$$D = A \times B + B \times C + C \times A.$$

6. Let the pairs A, B and C, D each determine a plane. Write an equation involving dot and cross products expressing the condition that these two planes be perpendicular.

10.3   Rigid motions of the axes

By a *rigid motion* of the axes we mean a shift from a rectangular co-ordinate system xyz with origin O to another rectangular co-ordinate system $x'y'z'$ with origin O', both systems having the same unit of distance, and both systems having the same orientation (i.e., both being right-handed or both left-handed). Such a shift can be accomplished in two stages: by a translation to a new system with origin O' and axes parallel to the original axes, followed by a rotation about O'. The equations for a translation of the co-ordinate system are very simple, and need not concern us right now, since we regard the vector space with origin O' as not essentially different from the vector space with origin O. Accordingly, we confine our attention to a rotation of the co-ordinate axes about a fixed origin O. We deal exclusively with right-handed systems.

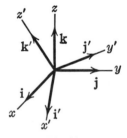

Fig. 77

Let i', j', k' be the fundamental orthonormal triad associated with the $x'y'z'$ system (see Fig. 77). We recall from analytic geometry the concept of the direction cosines of a directed line. Let the direction cosines of the x'-axis relative to the x-, y-, and z-axes respectively be l_1, m_1, n_1. By the definition of dot products we have

$$i' \cdot i = l_1, \qquad i' \cdot j = m_1, \qquad i' \cdot k = n_1.$$

Likewise $\qquad j' \cdot i = l_2, \qquad j' \cdot j = m_2, \qquad j' \cdot k = n_2$

are the direction cosines of the y'-axis relative to the axes of the xyz-system. With similar relations for the direction cosines of the z'-axis relative to the axes of the xyz-system, all the relations can be conveniently and compactly exhibited in a table, as follows.

(10.3–1)

	i	j	k
i'	l_1	m_1	n_1
j'	l_2	m_2	n_2
k'	l_3	m_3	n_3

Observe that the direction cosines of the x-axis relative to the axes of the $x'y'z'$-system are l_1, l_2, l_3, and so forth.

Now consider the vector $\overrightarrow{OP}$, where P is any point, with co-ordinates (x, y, z) and (x', y', z'), respectively, in the two systems. Then

(10.3–2)
$$\overrightarrow{OP} = x\mathbf{i} + y\mathbf{j} + z\mathbf{k}$$

and

(10.3–3)
$$\overrightarrow{OP} = x'\mathbf{i}' + y'\mathbf{j}' + z'\mathbf{k}'.$$

Now observe from (10.3–3) that $x' = \overrightarrow{OP} \cdot \mathbf{i}'$, $y' = \overrightarrow{OP} \cdot \mathbf{j}'$, and so on. If we form the dot product $\overrightarrow{OP} \cdot \mathbf{i}'$ from (10.3–2), we obtain

$$x' = \overrightarrow{OP} \cdot \mathbf{i}' = x\mathbf{i} \cdot \mathbf{i}' + y\mathbf{j} \cdot \mathbf{i}' + z\mathbf{k} \cdot \mathbf{i}',$$

or

$$x' = l_1 x + m_1 y + n_1 z.$$

By similar arguments we obtain the complete set of relations

(10.3–4)
$$\begin{aligned} x' &= l_1 x + m_1 y + n_1 z \\ y' &= l_2 x + m_2 y + n_2 z \\ z' &= l_3 x + m_3 y + n_3 z. \end{aligned}$$

These are the equations of transformation for the rotation of the co-ordinate system. The inverse transformation can be found in exactly the same way, starting from (10.3–3). The equations are

(10.3–5)
$$\begin{aligned} x &= l_1 x' + l_2 y' + l_3 z' \\ y &= m_1 x' + m_2 y' + m_3 z' \\ z &= n_1 x' + n_2 y' + n_3 z'. \end{aligned}$$

Consider now a vector $\mathbf{A}$ of the vector space with origin O. This vector will have components A_1, A_2, A_3 in the xyz-system, and A_1', A_2', A_3' in the $x'y'z'$-system. Since the components of $\mathbf{A}$ are merely the co-ordinates of the terminal point of A, we see that the two sets of components are related in exactly the same way that xyz and $x'y'z'$ are related, that is,

(10.3-6)
$$A'_1 = l_1 A_1 + m_1 A_2 + n_1 A_3$$
$$A'_2 = l_2 A_1 + m_2 A_2 + n_2 A_3$$
$$A'_3 = l_3 A_1 + m_3 A_2 + n_3 A_3,$$

with an inverse set of relations corresponding to (10.3-5). The two sets of relations are easily kept in mind by a table similar to (10.3-1):

(10.3-7)

	A_1	A_2	A_3
A'_1	l_1	m_1	n_1
A'_2	l_2	m_2	n_2
A'_3	l_3	m_3	n_3

It follows from these laws of transformation of components that if we know the components of a vector in one co-ordinate system, we can find its components in any system obtained by a rotation of the axes.

EXERCISES

1. Show that
$$i' = l_1 i + m_1 j + n_1 k, \quad i = l_1 i' + l_2 j' + l_3 k',$$
and obtain four other allied relations. Start from the fact that, if A is any vector, $A = (A \cdot i)i + (A \cdot j)j + (A \cdot k)k$.

2. What is the numerical value of $i \cdot (j \times k)$? Express this product in terms of i', j', k' by the results of Exercise 1, and deduce that
$$\begin{vmatrix} l_1 & l_2 & l_3 \\ m_1 & m_2 & m_3 \\ n_1 & n_2 & n_3 \end{vmatrix} = 1.$$
See Exercise 5a, § 10.22.

3. Observe that if one solves (10.3-5) for x' by Cramer's rule, and uses the result of Exercise 2, one finds
$$x' = (m_2 n_3 - m_3 n_2)x + (n_2 l_3 - n_3 l_2)y + (l_2 m_3 - l_3 m_2)z.$$
Comparing with (10.3-4), we surmise that
$$l_1 = \begin{vmatrix} m_2 & m_3 \\ n_2 & n_3 \end{vmatrix}, \quad m_1 = \begin{vmatrix} n_2 & n_3 \\ l_2 & l_3 \end{vmatrix}, \quad n_1 = \begin{vmatrix} l_2 & l_3 \\ m_2 & m_3 \end{vmatrix}.$$
Prove these results directly by considering the cross product $i' = j' \times k'$. What similar results are given as a consequence of the relation $i = j \times k$? Show that all the relations of this type are summarized in the statement: *Any element of the determinant appearing in Exercise 2 is equal to its own cofactor.*

4. Of what dot-product relation is the equation $l_1^2 + m_1^2 + n_1^2 = 1$ the expression? Answer the corresponding question for each of the equations
$$l_1 l_2 + m_1 m_2 + n_1 n_2 = 0, \ l_1 m_1 + l_2 m_2 + l_3 m_3 = 0.$$

5. Suppose the nine direction cosines in the table (10.3–1) are as follows:

$\dfrac{2}{7}$	$\dfrac{3}{7}$	$\dfrac{6}{7}$
$\dfrac{3}{7}$	$-\dfrac{6}{7}$	$\dfrac{2}{7}$
$\dfrac{6}{7}$	$\dfrac{2}{7}$	$-\dfrac{3}{7}$

(a) Find the components in the $x'y'z'$-system of the vector $14\,\mathbf{i} - 21\,\mathbf{j} + 7\,\mathbf{k}$.
(b) Find the components in the xyz-system of the vector $7\,\mathbf{i}' + 28\,\mathbf{j}' - 35\,\mathbf{k}'$.

6. Suppose the nine direction cosines in the table (10.3–1) are as follows:

$\dfrac{1}{\sqrt{6}}$	$\dfrac{-1}{\sqrt{2}}$	$\dfrac{1}{\sqrt{3}}$
$\dfrac{1}{\sqrt{6}}$	$\dfrac{1}{\sqrt{2}}$	$\dfrac{1}{\sqrt{3}}$
$\dfrac{-2}{\sqrt{6}}$	0	$\dfrac{1}{\sqrt{3}}$

(a) Find $\sqrt{2}(\mathbf{i} + \mathbf{j}) - \mathbf{k}$ in terms of $\mathbf{i}', \mathbf{j}', \mathbf{k}'$.
(b) Find $\mathbf{i}' - \mathbf{j}' + \sqrt{6}\,\mathbf{k}'$ in terms of $\mathbf{i}, \mathbf{j}, \mathbf{k}$.

10.4 Invariants

In dealing with a matter like the definition of the cross product, we have two courses open. One possibility is that of making the definition in geometric terms, without reference to any co-ordinate system. This is the course we followed in § 10.22. But there is another possibility. We might have said: Let us choose a fixed rectangular co-ordinate system xyz; then, given vectors $\mathbf{A} = (A_1, A_2, A_3)$, $\mathbf{B} = (B_1, B_2, B_3)$, let us form the vector $\mathbf{C}$ with components

(10.4–1)
$$C_1 = A_2 B_3 - A_3 B_2, \quad C_2 = A_3 B_1 - A_1 B_3, \quad C_3 = A_1 B_2 - A_2 B_1.$$

This vector we shall call $\mathbf{A} \times \mathbf{B}$. The three algebraic properties (10.22–1), (10.22–2), and (10.22–3) are immediate consequences of this definition. It is legitimate to define a vector in this way, but the procedure is not without its difficulties, for we have made the definition in terms of one particular co-ordinate system. From the definition (10.4–1) alone we have no right to assume that these same formulas hold in other co-ordinate systems obtained by rotation, that is, that

(10.4–2)

$$C'_1 = A'_2B'_3 - A'_3B'_2, \quad C'_2 = A'_3B'_1 - A'_1B'_3, \quad C'_3 = A'_1B'_2 - A'_2B'_1.$$

These formulas are in fact correct, but if we choose the second course for defining $\mathbf{A} \times \mathbf{B}$, we must *prove* that they are correct. We can give such a proof with the aid of the developments of § 10.3. We need, in particular, the results expressed in Exercise 3 of § 10.3; these results can all be derived with the definition of cross products in a fixed co-ordinate system xyz as given by (10.4–1).

We shall content ourselves with proving the first equation in (10.4–2); the others follow by symmetry. By the table (10.3–7) we have

$$C'_1 = l_1C_1 + m_1C_2 + n_1C_3,$$

or, using (10.4–1),

(10.4–3)

$$C'_1 = l_1(A_2B_3 - A_3B_2) + m_1(A_3B_1 - A_1B_3) + n_1(A_1B_2 - A_2B_1).$$

On the other hand, using the table (10.3–7) on $\mathbf{A}$ and $\mathbf{B}$, we have

$$\begin{aligned}
A'_2B'_3 &= (l_2A_1 + m_2A_2 + n_2A_3)(l_3B_1 + m_3B_2 + n_3B_3) \\
&= l_2l_3A_1B_1 + l_2m_3A_1B_2 + l_2n_3A_1B_3 \\
&\quad + m_2l_3A_2B_1 + m_2m_3A_2B_2 + m_2n_3A_2B_3 \\
&\quad + n_2l_3A_3B_1 + n_2m_3A_3B_2 + n_2n_3A_3B_3.
\end{aligned}$$

The product $A'_3B'_2$ yields a similar formula, the only difference being that the indices 2 and 3 are exchanged on l, m, n, but not on A and B. Subtracting and collecting terms, we find

$$\begin{aligned}
\text{(10.4–4)} \quad A'_2B'_3 - A'_3B'_2 &= (m_2n_3 - m_3n_2)(A_2B_3 - A_3B_2) \\
&\quad + (n_2l_3 - n_3l_2)(A_3B_1 - A_1B_3) \\
&\quad + (l_2m_3 - l_3m_2)(A_1B_2 - A_2B_1).
\end{aligned}$$

On comparing (10.4–3) with (10.4–4), and using the result of Exercise 3, § 10.3, we see that

$$C'_1 = A'_2B'_3 - A'_3B'_2.$$

This is what we set out to prove.

Once we have shown the truth of (10.4–2) for any system of co-ordinates obtained by rotating the original xyz-system, we have shown that the definition of $\mathbf{A} \times \mathbf{B}$ in (10.4–1), although *apparently* dependent upon the choice of one fixed co-ordinate system, is in fact independent of the particular co-ordinate system chosen, since for every $x'y'z'$-system the numbers C'_1, C'_2, C'_3 defined by (10.4–2) are components of one and the same vector.

The word *invariant* is used in mathematics (in the following way among other ways) to describe something which is apparently dependent upon a particular co-ordinate system, but which is actually unchanged when the co-ordinate system is changed in certain specified ways. As a simple example, consider a line in the xy-plane, with equation

$$ax + by + c = 0.$$

The perpendicular distance from the origin to this line is known to be

$$D = \frac{|c|}{\sqrt{a^2 + b^2}}.$$

If we keep the line fixed in the plane and rotate the co-ordinate axes, the coefficients a, b, c in the equation of the line will change, but D will not change, since the distance from the origin to the line is independent of the co-ordinate system. Therefore D is an invariant. It is called a numerical invariant, or a *scalar invariant*. Here the permitted changes of co-ordinates are those arising from rotations about the origin.

A vector is an invariant. If a vector is thought of as given by its components in a given co-ordinate system, the components actually depend on that co-ordinate system, and are not themselves numerical invariants. But as we pass from co-ordinate system to co-ordinate system, the components, changing in accord with the table (10.3–7), have in common a property which does not change, namely, that they are all components of the same vector.

We may know a thing to be an invariant because we can describe it in geometric terms, without reference to a co-ordinate system. But we can also demonstrate its invariance by investigating what happens under a specified change in the co-ordinate system; we must then show that the thing is unaltered by such a change. If direct geometric evidence of invariance is not available, we may be forced to fall back on the second procedure. Such a state of affairs arises when we come to deal with the concepts of *gradient*, *divergence*, and *curl*, later in this chapter. It was to prepare for these future situations that we took the trouble to prove, by examining the transformations of co-ordinates, that the cross product $\mathbf{A} \times \mathbf{B}$, as defined by (10.4–1), is a vector invariant with respect to rotations of the axes.

EXERCISES

1. By considering various dot products of the vectors $\mathbf{i}$, $\mathbf{j}$, $\mathbf{k}$, $\mathbf{i}'$, $\mathbf{j}'$, $\mathbf{k}'$, prove the relation

$$l_1^2 + l_2^2 + l_3^2 = 1$$

and two similar ones involving m's and n's. Also show that

$$m_1 n_1 + m_2 n_2 + m_3 n_3 = 0,$$

and prove two similar relations, one involving n's and l's, and one involving l's and m's. Then, using the results just described, give an algebraic proof that

$$A'_1B'_1 + A'_2B'_2 + A'_3B'_3 = A_1B_1 + A_2B_2 + A_3B_3,$$

thus verifying that $A \cdot B$ is a numerical invariant. The relevant properties of the direction cosines can be proved without using dot products. This leaves the way logically open for defining $A \cdot B$ by (10.21–3).

2. Show that, if the co-ordinate system is rotated 90° about the z-axis, so that the x'-axis is the same as the y-axis and the y'-axis coincides with the negative x-axis, the scheme (10.3–7) becomes

	A_1	A_2	A_3
A'_1	0	1	0
A'_2	-1	0	0
A'_3	0	0	1

Then show by an example that, if A and B are vectors, A_1B_3 is not a numerical invariant. Likewise show that the triple (A_1B_1, A_2B_2, A_3B_3) does not define a vector invariant, that is, that in general the vector having components (A_1B_1, A_2B_2, A_3B_3) in the xyz-system does not have components $(A'_1B'_1, A'_2B'_2, A'_3B'_3)$ in the $x'y'z'$-system. Consider, e.g., $A = i$, $B = i + j$.

3. Is $A_1 + A_2 + A_3$ a numerical invariant if $A = A_1i + A_2j + A_3k$? Justify your answer.

4. Let the nine direction cosines in the table (10.3–1) be specified as follows:

$\dfrac{1}{\sqrt{2}}$	$\dfrac{1}{\sqrt{2}}$	0
$\dfrac{1}{\sqrt{3}}$	$\dfrac{-1}{\sqrt{3}}$	$\dfrac{1}{\sqrt{3}}$
$\dfrac{1}{\sqrt{6}}$	$\dfrac{-1}{\sqrt{6}}$	$\dfrac{-2}{\sqrt{6}}$

Let $A = i + j$, $B = i - j + k$.

Calculate $A \times B$ in terms of i', j', k' in two ways: (a) by first expressing A and B in terms of i', j', k' and calculating $C = A \times B$ by formulas (10.4–2); (b) by calculating $C = A \times B$ from (10.4–1) and then transforming to the $x'y'z'$-system by (10.4–3), and the similar formulas for C'_2, C'_3. The purpose of this exercise is simply to give a concrete example illustrating how the cross product is invariant under rotation of axes.

10.5   Scalar point functions

The word "scalar" is used to contrast with the word "vector." It is customary, in any context where vectors and real numbers are both being discussed, to refer to real numbers as scalars. Thus, in this book, a scalar is a real number. The word may also be used as an adjective.

Let us recall the general meaning of the word "function." A function is a correspondence between two classes of objects; these two classes are called respectively the *domain of definition* of the function and the *range of values* of the function, or, more briefly, the *domain* and the *range* of the function. The function itself is the correspondence whereby to each object in the domain is assigned a corresponding object in the range. Let us now consider a case in which the domain is a class of points and the range is a class of real numbers. In such a case we shall call the function a *scalar point function*. If f denotes the function, P a variable point, and u the value of the function at P, we write $u = f(P)$.

Example 1.　At a fixed instant of time consider all the points in the earth's atmosphere, and let u be the temperature (in degrees centigrade) at the point P.

Example 2.　On the surface of a sphere of unit radius, let Q be a fixed point. If P is any point on the sphere, let u be the shortest great-circle distance from P to Q.

Example 3.　If C is a curve in a plane, and P is any point of C, let u be the radius of curvature of C at P, provided the curve is smooth enough to have a well-defined curvature.

The essential thing to be noted about the definition of a scalar point function is that it is independent of co-ordinate systems. We do not need co-ordinate systems to explain the concept of a scalar point function. And, if a co-ordinate system is used to locate various points, the function is essentially the correspondence between P and u, *not* a correspondence between the co-ordinates of P and u. There is, however, a link between a point function $f(P)$ and certain functions of the co-ordinates of P. This link is the concept of invariance under co-ordinate transformations. Suppose, for instance, that f is a given point function, defined throughout some region of Euclidean three-dimensional space. Choose a system of rectangular co-ordinates x, y, z, and let (x, y, z) be co-ordinates of P. Then the value of f at P will depend on (x, y, z), say $f(P) = F(x, y, z)$, so that the scalar point function f determines F as a real-valued function of the three variables x, y, z. Suppose another rectangular co-ordinate system x', y', z' is obtained from the xyz-system by a rotation of axes. When x, y, z are expressed in terms of x', y', z' by the equations of the form (10.3–5), $F(x, y, z)$ is transformed into some new function $G(x', y', z')$. But, since (x, y, z) and (x', y', z') refer to the same point P, we have $f(P) = G(x', y', z')$. The *scalar point function f is invariant under the transformation of co-ordinates.* We may also express $f(P)$ as a function of cylindrical or spherical co-ordinates.

Example 4. If O is a fixed point of space, consider the vector space with origin O. If P is an arbitrary point, denote the vector $\overrightarrow{OP}$ by $\mathbf{R}$. Let $\mathbf{A}$ be a fixed vector. Then

$$f(P) = \mathbf{A} \cdot \mathbf{R} \quad \text{and} \quad \phi(P) = ||\mathbf{R}||^2$$

are scalar point functions. If we introduce a fixed rectangular co-ordinate system with O as origin, and let $\mathbf{A} = (A_1, A_2, A_3)$, $\mathbf{R} = (x, y, z)$, then in this co-ordinate system we have

$$f(P) = A_1 x + A_2 y + A_3 z,$$
$$\phi(P) = (x^2 + y^2 + z^2).$$

It is possible to define directly the concepts of continuity and differentiability for scalar point functions, without reference to co-ordinate systems. Alternatively, however, the definitions may be made with reference to some arbitrarily chosen rectangular co-ordinate system. Thus, if $f(P) = F(x, y, z)$ in that system, we can say that f is continuous at P_0 if F is continuous at (x_0, y_0, z_0), with a similar definition for differentiability. Although these definitions are made with reference to a particular rectangular co-ordinate system, they are actually independent of the choice of that system. For example, if $F(x, y, z) = G(x', y', z')$, where the two systems are related by a rotation, and if F is differentiable, then G is differentiable also (by Theorem V, § 7.3).

10.51   Vector point functions

The concept of a vector point function is similar to the concept of a scalar point function in the matter of being independent of particular choices of co-ordinate systems. The difference is that the function values are vectors instead of scalars. The domain of definition of a vector point function is some set of points P. The range of values of the function is some set of vectors. Let f denote the function, and let $\mathbf{F}$ denote the vector corresponding to P. Then we write $\mathbf{F} = f(P)$ where $f(P)$ depends just on P itself and not on the co-ordinates we happen to be using.

Example 1. Let $\mathbf{R} = \overrightarrow{OP}$, and let $\mathbf{A}$ be a fixed vector. Then each of the expressions

$$\mathbf{R}, \quad (\mathbf{A} \cdot \mathbf{R})\mathbf{R}, \quad \mathbf{A} \times \mathbf{R}, \quad ||\mathbf{R}||^{-3}\mathbf{R}$$

defines a vector point function.

It is often convenient to introduce a co-ordinate system in order to deal with a vector point function. We shall be concerned mostly with rectangular co-ordinate systems. If we have an xyz-system with origin

O, let $\mathbf{i}$, $\mathbf{j}$, $\mathbf{k}$ be the fundamental orthonormal triad associated with the xyz-system. Suppose we have a function $\mathbf{F} = f(P)$; let the components of $\mathbf{F}$ be denoted by $F_1(x, y, z)$, $F_2(x, y, z)$, $F_3(x, y, z)$. Then

(10.51–1) $\mathbf{F} = F_1(x, y, z)\mathbf{i} + F_2(x, y, z)\mathbf{j} + F_3(x, y, z)\mathbf{k}.$

In another rectangular system $x'y'z'$, obtained from the xyz-system by rotation, $\mathbf{F}$ will have a different set of components, and with the triad $\mathbf{i}'$, $\mathbf{j}'$, $\mathbf{k}'$ for the new system, $\mathbf{F}$ will be expressed in the form

(10.51–2)
$$\mathbf{F} = F'_1(x', y', z')\mathbf{i}' + F'_2(x', y', z')\mathbf{j}' + F'_3(x', y', z')\mathbf{k}'.$$

The components F'_1, F'_2, F'_3 will be related to F_1, F_2, F_3 in the same way that A'_1, A'_2, A'_3 are related to A_1, A_2, A_3 in equations (10.3–6). Also, the two orthonormal triads are related in the manner indicated by table (10.3–1).

The value $\mathbf{F}$ of the vector point function is a vector invariant. Note, however, that an individual component of $\mathbf{F}$, such as $F_1(x, y, z)$, is not a scalar invariant, for in general $F_1(x, y, z) \neq F'_1(x', y', z')$.

If F_1, F_2, F_3 are any three functions of x, y, z, we may regard them as components of a vector point function $\mathbf{F}$ defined by (10.51–1), provided we understand that F_1, F_2, F_3 are the components of $\mathbf{F}$ in one fixed rectangular co-ordinate system, and that the components in any other rectangular system are obtained by transformations of the type (10.3–6).

A vector point function is often called a *vector field*, especially if the domain of definition is a region in two or three dimensions. This terminology springs from a certain mode of representing the vector function. Let the vector $\mathbf{F}$ be drawn in space with its initial point at the point P to which $\mathbf{F}$ corresponds. This allows one to portray the manner in which the magnitude and direction of $\mathbf{F}$ vary with the location of P. The geometrical configuration, whether actually drawn or merely conceived, is called a vector field (see Fig. 78).

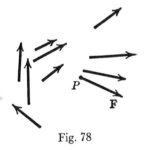

Fig. 78

Example 2. Consider the gravitational field of the earth. If O is the center of the earth, M is the mass of the earth, and r is the distance from O to an arbitrary point P outside the earth, then a particle of unit mass placed at P is attracted toward O by a force of magnitude

(10.51–3) $k \dfrac{M}{r^2},$

where k is a constant of proportionality. This force can be represented by a vector of magnitude given by (10.51–3), and of direction opposite to that of the vector $\mathbf{R} = \overrightarrow{OP}$. The magnitude of $\mathbf{R}$ is r. Therefore a vector of unit length in the direction opposite to that of $\mathbf{R}$ is

$$-\frac{1}{r}\,\mathbf{R}.$$

Thus the gravitational force $\mathbf{F}$ on the unit mass at P is

(10.51–4) $$\mathbf{F} = -k\,\frac{M}{r^3}\,\mathbf{R}.$$

A portrayal of this vector field is suggested by Fig. 79.

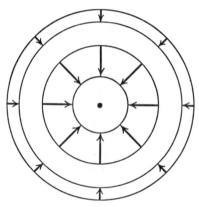

Fig. 79

EXERCISES

1. The equations

$$x = \frac{1}{\sqrt{2}}\,(x' - y'),$$

$$y = \tfrac{1}{2}(x' + y') + \frac{1}{\sqrt{2}}\,z',$$

$$z = -\tfrac{1}{2}(x' + y') + \frac{1}{\sqrt{2}}\,z'$$

define a rotation of axes.

(a) If $F(x, y, z) = 2\,x^2 - y^2 - z^2$ is the representation of a scalar point function in the xyz-system, find the representation $G(x', y', z')$ of the function in the $x'y'z'$-system.

(b) What is the table (10.3–7) for this rotation of axes? Find the xyz-representation $F(x, y, z)$ for the scalar point function for which $G(x', y', z') = x' + y' + \sqrt{2}\,z'$.

(c) Find the components of the vector field $\mathbf{F} = \sqrt{2}\,z\mathbf{i} + (y + z)\mathbf{j} + \sqrt{2}\,x\mathbf{k}$ in the $x'y'z'$-system.

(d) Express the vector field $\mathbf{F} = \sqrt{2}\,y'\mathbf{i}' + z'\mathbf{j}' + x'\mathbf{k}'$ in the xyz-system.

2. The equations

$$x' = \tfrac{1}{7}(2\,x + 3\,y + 6\,z),$$
$$y' = \tfrac{1}{7}(3\,x - 6\,y + 2\,z),$$
$$z' = \tfrac{1}{7}(6\,x + 2\,y - 3\,z)$$

define a rotation of axes.

(a) Express x, y, z in terms of x', y', and z'.

(b) Find the expression of $x'^2 + y'^2 - z'^2$ in the xyz-system.

(c) Express the vector field $-y\mathbf{i} + x\mathbf{j}$ in the $x'y'z'$-system.

(d) Do $x\mathbf{i} + 2\,y\mathbf{j} + 3\,z\mathbf{k}$ and $x'\mathbf{i}' + 2\,y'\mathbf{j}' + 3\,z'\mathbf{k}'$ represent the same vector field? Justify your answer.

10.6 The gradient of a scalar field

A scalar point function f may be thought of by imagining each point P at which f is defined as carrying a label with the value $f(P)$ of the function at that point (see the discussion of the second mode of representing functions in § 5.4). When a scalar function is represented in this way, it is often called a *scalar field*.

Let f be a scalar field defined throughout some region R, and suppose that f is differentiable in R. We are going to define the concept of the *gradient* of the field. The gradient of a scalar point function is a vector point function. It is convenient to use a rectangular co-ordinate system in the process of defining the gradient. However, we must take care to be certain that the definition gives us a result which is independent of the choice of the particular rectangular co-ordinate system.

Let a system of rectangular co-ordinates xyz, with origin O, be selected arbitrarily.

In this co-ordinate system let P have co-ordinates (x, y, z), and let $f(P) = F(x, y, z)$ be the representation of our point function. Form the vector field whose representation in the xyz-system is

$$(10.6\text{–}1) \qquad \frac{\partial F}{\partial x}\,\mathbf{i} + \frac{\partial F}{\partial y}\,\mathbf{j} + \frac{\partial F}{\partial z}\,\mathbf{k}.$$

This vector field, or vector point function, is what we shall call the gradient of the given scalar function f.

Presently we shall give an interpretation of the gradient which enables us to think of it apart from the co-ordinate system. But first let us show that the definition of the gradient yields the same vector field no matter what rectangular co-ordinate system is chosen. If a second co-ordinate system is selected, the two systems are related in such a way that one can be obtained from the other by either a translation or a rotation, or both. Let us consider the case where both systems have the same origin, and are related by a rotation of axes. This is the case of principal importance for our discussion of the gradient. The case where a translation may be involved is considered in Exercise 13.

Let the co-ordinates of the two systems be xyz and $x'y'z'$, related as in (10.3–4) and (10.3–5), and let the two representations of the scalar field be

$$f(P) = F(x, y, z) = G(x', y', z').$$

We wish to show that

$$(10.6\text{–}2) \quad \frac{\partial F}{\partial x} \mathbf{i} + \frac{\partial F}{\partial y} \mathbf{j} + \frac{\partial F}{\partial z} \mathbf{k} = \frac{\partial G}{\partial x'} \mathbf{i}' + \frac{\partial G}{\partial y'} \mathbf{j}' + \frac{\partial G}{\partial z'} \mathbf{k}'.$$

Once this is done, it will be clear that the definition of the gradient of f by expression (10.6–1) is invariant under a rotation of the axes.

To prove (10.6–2) we must show that the triple $\left(\dfrac{\partial F}{\partial x}, \dfrac{\partial F}{\partial y}, \dfrac{\partial F}{\partial z} \right)$ is related to the triple $\left(\dfrac{\partial G}{\partial x'}, \dfrac{\partial G}{\partial y'}, \dfrac{\partial G}{\partial z'} \right)$ just as the triple (A_1, A_2, A_3) is related to the triple (A_1', A_2', A_3') in (10.3–6) or (10.3–7). Now, by the chain rule,

$$\frac{\partial G}{\partial x'} = \frac{\partial F}{\partial x} \frac{\partial x}{\partial x'} + \frac{\partial F}{\partial y} \frac{\partial y}{\partial x'} + \frac{\partial F}{\partial z} \frac{\partial z}{\partial x'},$$

with similar equations for $\dfrac{\partial G}{\partial y'}$ and $\dfrac{\partial G}{\partial z'}$. From (10.3–5) we see that

$$\frac{\partial x}{\partial x'} = l_1, \quad \frac{\partial y}{\partial x'} = m_1, \quad \frac{\partial z}{\partial x'} = n_1.$$

Therefore
$$\frac{\partial G}{\partial x'} = l_1 \frac{\partial F}{\partial x} + m_1 \frac{\partial F}{\partial y} + n_1 \frac{\partial F}{\partial z}.$$

This is exactly like the first equation in (10.3–6). The other two relations of this type are established in the same way, and thus (10.6–2) is proved.

Frequently the value of a scalar function is denoted by a dependent variable, e.g., $u = f(P)$. In such cases the gradient is denoted by either of the symbols

$$\text{grad } u \quad \text{or} \quad \text{grad } f.$$

The symbolism ∇u or ∇f is also employed. Thus, in any rectangular co-ordinate system xyz, with corresponding orthonormal triad $\mathbf{i}, \mathbf{j}, \mathbf{k}$,

$$(10.6\text{–}3) \qquad \text{grad } u = \nabla u = \frac{\partial u}{\partial x} \mathbf{i} + \frac{\partial u}{\partial y} \mathbf{j} + \frac{\partial u}{\partial z} \mathbf{k}.$$

Example 1. If $\qquad u = 2x^2 + 3y^2 + 4z^2,$

$$\nabla u = 4x\mathbf{i} + 6y\mathbf{j} + 8z\mathbf{k}.$$

Let us now consider some properties of the gradient.

THEOREM II. *At any given point in the region where the differentiable scalar field $u = f(P)$ is defined let a direction be chosen. The rate of change* *of u (per unit distance) at the given point in the given direction is equal to the component of the gradient of u in that direction. If grad u $\neq$ 0 at the point, the direction of the gradient is that in which the rate of increase of u is greatest.*

PROOF. Let a rectangular co-ordinate system be chosen so that the given direction is that of the positive x-axis. The rate of change in the given direction is then $\partial u/\partial x$. But this is precisely the component of ∇u in the x-direction, by (10.6–3). The first assertion in the theorem is now justified. The correctness of the second assertion is an immediate consequence, since, of all components of a non-zero vector, the largest one is that in the direction of the vector itself.

The rate of change of a function at a given point in a given direction is called the *directional derivative.*

The importance of Theorem II is that it enables us to see the significance of the gradient without any reference to co-ordinate systems. A further grasp of the significance of the gradient is obtainable with the aid of the concept of level surfaces (see the latter part of § 5.4). Let us assume that the gradient is a continuous vector function, i.e., that $\dfrac{\partial u}{\partial x}$, $\dfrac{\partial u}{\partial y}$, $\dfrac{\partial u}{\partial z}$ are continuous functions. Let P_0 be a point at which these three partial derivatives are not all zero; this is the same as requiring that $\nabla u \neq 0$ at P_0. If u_0 is the value of u at P_0, it can be shown by the implicit-function theorem (§ 8.1) that, at least in some sufficiently small neighborhood of P_0, the points at which $u = u_0$ form a surface having a tangent plane whose normal varies continuously in direction as the point of tangency varies. This smooth surface is part of the level surface $u = u_0$.

THEOREM III.   *At any point where the gradient is not the vector* **0**, *the gradient is perpendicular to the level surface through the point.*

PROOF. Let the representation of the scalar field be $u = F(x, y, z)$. The equation of the level surface is $F(x, y, z) = u_0$. But we know (Example 1, § 6.6) that the direction of the normal to the surface is given by the ratios $\dfrac{\partial F}{\partial x} : \dfrac{\partial F}{\partial y} : \dfrac{\partial F}{\partial z}$. These partial derivatives are precisely the components of the gradient, and so the theorem is proved.

Example 2. Consider the scalar field $u = r^3$, where r is the distance OP (O fixed, P variable). Clearly u is constant on a surface if and only if r is constant on that surface. Thus the level surfaces are spheres with center O. Evidently u increases when r increases, the rate of increase being

$$\frac{du}{dr} = 3\, r^2.$$

It follows from Theorem III that ∇u is normal to the sphere at P. From Theorem II we see that the direction of ∇u is outward from the sphere at P, and that its length is $3\ r^2$. Now a unit vector in the direction of OP is

$$\frac{1}{r}\ \overrightarrow{OP}.$$

Therefore
$$\nabla u = 3\ r^2 \left(\frac{1}{r}\ \overrightarrow{OP} \right) = 3\ r\ \overrightarrow{OP}$$

in this case. This example illustrates how Theorems II and III sometimes furnish the means of calculating a gradient without resorting to rectangular co-ordinate systems. For a solution of this same problem in terms of rectangular co-ordinates see Exercise 5.

EXERCISES

1. Find ∇f in the case of each of the following functions.
(a) $f(P) = 2\ x^2 - 3\ xy + y^2 - 4\ xz + 6\ z^2$.
(b) $f(P) = x/(x^2 + y^2) + z/(y^2 + z^2)$.
(c) $f(P) = \log (x^2 + y^2) + z$.
(d) $f(P) = z/(x^2 + y^2 + z^2)^{3/2}$.
(e) $f(P) = e^{-x}(x^2 + y^2 + z^2)$.

2. In what direction from the point $(2, -1, 3)$ is the function $xy + yz + zx + xyz$ increasing most rapidly? Give direction cosines for the answer.

3. Find a vector indicating the direction and magnitude of most rapid *decrease* of the function $(x^2/16) + (y^2/25) - (z^2/9)$ at the point $(8, 25, -9)$. What is this rate of decrease?

4. (a) How fast is $4\ x^2 + 9\ y^2 + z^2$ changing at the point $(1, 1, 1)$ in a direction tangent to the ellipsoid $4\ x^2 + 9\ y^2 + z^2 = 14$? (b) What is the rate of change in the direction of the outward normal to the ellipsoid at the point?

5. Find ∇u by use of (10.6–3) if $u = (x^2 + y^2 + z^2)^{3/2}$. Compare with Example 2, noting that $r^2 = x^2 + y^2 + z^2$ and $\overrightarrow{OP} = x\mathbf{i} + y\mathbf{j} + z\mathbf{k}$.

6. Find ∇u if $u = x^2$, using Theorems II and III rather than (10.6–3). Check by using the latter formula.

7. Let $\mathbf{R}$ be the vector $\overrightarrow{OP}$ and r the distance OP. Find the gradient of each of the following functions by the method of Example 2. Express the answers in terms of $\mathbf{R}$ and r.
(a) $u = 1/r$;
(b) $u = r^2$;
(c) $u = 1/\sqrt{r}$;
(d) $u = \log (1/r)$;
(e) $u = r^n$;
(f) $u = e^{-r^2}$.

8. In the notation of Exercise 7 let $u = F(r)$, where F is a differentiable function of r. Show that $\nabla u = (1/r)F'(r)\mathbf{R}$: (a) by the method of Example 2; (b) by using

(10.6–3) with $r^2 = x^2 + y^2 + z^2$, $\mathbf{R} = x\mathbf{i} + y\mathbf{j} + z\mathbf{k}$. (c) Find the most general form for $F(r)$ if $\nabla u = -r^{-3}\mathbf{R}$.

9. Let r denote the distance from P to a fixed line in space. (a) If $f(P)$ is a function of r alone, what are the level surfaces of the scalar field? (b) If $f(P) = F(r)$, where F is differentiable, what can you say about the magnitude and direction of ∇f at P if P is not on the given line? (c) Let the fixed line be the z-axis. If P is the point (x, y, z), show that $\nabla f = [F'(r)/r](x\mathbf{i} + y\mathbf{j})$.

10. (a) If $u = f(P)$ and $v = g(P)$ are differentiable scalar fields, show that $\nabla(uv) = u\nabla v + v\nabla u$. (b) Use this result to find $\nabla(x/r^3)$.

11. Prove Theorem II along the following lines: Let $\mathbf{n}$ be a unit vector in a certain direction at the point $P_0(x_0, y_0, z_0)$. If the direction cosines of the direction are $\cos\alpha$, $\cos\beta$, $\cos\gamma$, then
$$\mathbf{n} = \cos\alpha\,\mathbf{i} + \cos\beta\,\mathbf{j} + \cos\gamma\,\mathbf{k}.$$
If s is distance measured positively along the line through P_0 in the direction of $\mathbf{n}$, with $s = 0$ at P_0, the line has parametric equations
$$x = x_0 + s\cos\alpha,\ y = y_0 + s\cos\beta,\ z = z_0 + s\cos\gamma.$$
Show that $\dfrac{du}{ds} = (\nabla u) \cdot \mathbf{n}$, and explain why this formula is equivalent to Theorem II.

12. If ϕ is colatitude and θ is longitude (both in radians) in a system of spherical co-ordinates related to rectangular co-ordinates by the equations $x = r\sin\phi\cos\theta,\, y = r\sin\phi\sin\theta,\, z = r\cos\phi$, what are (a) the level surfaces for $u = \phi$? (b) the level surfaces for $u = \theta$?
(c) Show that
$$\nabla\phi = \frac{\cos\theta\cos\phi}{r}\,\mathbf{i} + \frac{\sin\theta\cos\phi}{r}\,\mathbf{j} - \frac{\sin\phi}{r}\,\mathbf{k}.$$
This may be done either by using (10.6–3) or by careful use of Theorems II and III.
(d) Show that
$$\nabla\theta = -\frac{\sin\theta}{r\sin\phi}\,\mathbf{i} + \frac{\cos\theta}{r\sin\phi}\,\mathbf{j} = -\frac{y}{x^2 + y^2}\,\mathbf{i} + \frac{x}{x^2 + y^2}\,\mathbf{j}.$$
The remark in (c) applies here also.

13. Show that (10.6–2) holds if the xyz-system and the $x'y'z'$-system are related by a translation. In this case we have $\mathbf{i} = \mathbf{i}'$, $\mathbf{j} = \mathbf{j}'$, $\mathbf{k} = \mathbf{k}'$, since the vector space with origin O' is considered to be indistinguishable from the vector space with origin O (see § 10.1). Then discuss the case where both a translation and a rotation are needed to pass from one system to the other.

10.7   The divergence of a vector field

Let

(10.7–1) $\mathbf{F} = F_1(x, y, z)\mathbf{i} + F_2(x, y, z)\mathbf{j} + F_3(x, y, z)\mathbf{k}$

[handwritten margin note: subscripts here do not refer to partial derivatives here, but specify separate functions]

be a vector field with differentiable components. We remind the student that F_1, F_2, F_3 are the components of $\mathbf{F}$ in the xyz-co-ordinate system. In another rectangular co-ordinate system there will be another set of components. We also observe that the subscripts 1, 2, 3 do not refer to partial differentiation, as they did in Chapters VI and VII. The subscripts are simply labels on the separate functions.

The expression

$$(10.7-2) \qquad \frac{\partial F_1}{\partial x} + \frac{\partial F_2}{\partial y} + \frac{\partial F_3}{\partial z}$$

is of interest in relation to the vector field, for it has an important physical meaning in certain kinds of vector fields occurring in physical theories. It also occurs in a very important purely mathematical theorem which we shall study later (the divergence theorem, § 15.6). The importance of the expression (10.7-2) is closely connected with the fact that it is a *scalar invariant*, in the following sense: the expression (10.7-2) is unchanged in value if we make any rigid motion of the axes (see formula (10.7-4), further on). We shall deal with the case of a rotation of axes, leaving the simpler case of a translation for the student. Let the notation for the vector field in the rotated system be

$$\mathbf{F} = F'_1(x', y', z')\mathbf{i}' + F'_2(x', y', z')\mathbf{j}' + F'_3(x', y', z')\mathbf{k}'.$$

Then, by (10.3-6) we have

$$(10.7-3) \qquad F'_1 = l_1 F_1 + m_1 F_2 + n_1 F_3,$$

and other similar formulas. The xyz-system is related to the $x'y'z'$-system by equations (10.3-4) or (10.3-5). Now, by (10.7-3),

$$\frac{\partial F'_1}{\partial x'} = l_1 \frac{\partial F_1}{\partial x'} + m_1 \frac{\partial F_2}{\partial x'} + n_1 \frac{\partial F_3}{\partial x'}.$$

Also, by the chain rule,

$$\frac{\partial F_1}{\partial x'} = \frac{\partial F_1}{\partial x}\frac{\partial x}{\partial x'} + \frac{\partial F_1}{\partial y}\frac{\partial y}{\partial x'} + \frac{\partial F_1}{\partial z}\frac{\partial z}{\partial x'}$$

$$= l_1 \frac{\partial F_1}{\partial x} + m_1 \frac{\partial F_1}{\partial y} + n_1 \frac{\partial F_1}{\partial z},$$

with similar formulas for $\dfrac{\partial F_2}{\partial x'}$ and $\dfrac{\partial F_3}{\partial x'}$.

In the same way,

$$\frac{\partial F'_2}{\partial y'} = l_2 \frac{\partial F_1}{\partial y'} + m_2 \frac{\partial F_2}{\partial y'} + n_2 \frac{\partial F_3}{\partial y'},$$

$$\frac{\partial F_1}{\partial y'} = l_2 \frac{\partial F_1}{\partial x} + m_2 \frac{\partial F_1}{\partial y} + n_2 \frac{\partial F_1}{\partial z},$$

and so on. The student should himself write out all the formulas. From $\dfrac{\partial F'_1}{\partial x'}$ we obtain nine terms:

$$
\begin{aligned}
\frac{\partial F'_1}{\partial x'} = {}& l_1^2 \frac{\partial F_1}{\partial x} + l_1 m_1 \frac{\partial F_1}{\partial y} + l_1 n_1 \frac{\partial F_1}{\partial z} \\
&+ m_1 l_1 \frac{\partial F_2}{\partial x} + m_1^2 \frac{\partial F_2}{\partial y} + m_1 n_1 \frac{\partial F_2}{\partial z} \\
&+ n_1 l_1 \frac{\partial F_3}{\partial x} + n_1 m_1 \frac{\partial F_3}{\partial y} + n_1^2 \frac{\partial F_3}{\partial z} .
\end{aligned}
$$

The formulas for $\dfrac{\partial F'_2}{\partial y'}$ and $\dfrac{\partial F'_3}{\partial z'}$ are obtained by advancing the subscripts to 2 and 3 respectively on l, m, and n. Now

$$l_1^2 + l_2^2 + l_3^2 = 1,$$
$$l_1 m_1 + l_2 m_2 + l_3 m_3 = 0,$$

and so on (see Exercise 1, § 10.4). Hence it may be seen that

$$(10.7\text{–}4) \qquad \frac{\partial F'_1}{\partial x'} + \frac{\partial F'_2}{\partial y'} + \frac{\partial F'_3}{\partial z'} = \frac{\partial F_1}{\partial x} + \frac{\partial F_2}{\partial y} + \frac{\partial F_3}{\partial z} .$$

This proves the invariance of (10.7–2).

DEFINITION. The scalar function (10.7–2) is called the *divergence* of the vector field (10.7–1). It is denoted by div $\mathbf{F}$:

$$(10.7\text{–}5) \qquad \operatorname{div} \mathbf{F} = \frac{\partial F_1}{\partial x} + \frac{\partial F_2}{\partial y} + \frac{\partial F_3}{\partial z} .$$

Observe that div $\mathbf{F}$ is a scalar field associated with the vector field $\mathbf{F}$. We shall not try at this point to display the mathematical or physical importance of the concept of divergence. We remark merely that certain kinds of vector fields have the property that their divergence is everywhere zero. Such fields are called *solenoidal*. Among the most important solenoidal vector fields are those fields of force which are produced by the inverse square law of attraction or repulsion, e.g., gravitational or electrostatic fields. Also, if $\mathbf{F}$ denotes the velocity field of an incompressible fluid in a steady state of flow, the incompressibility is expressed by the fact that div $\mathbf{F} = 0$.

Example 1. The electrostatic field produced by a unit positive charge at O is

$$\mathbf{E} = \frac{1}{r^3}\, \overrightarrow{OP},$$

where $OP = r$. We shall show that the divergence of this field is zero wherever the field is defined (at all points except O). In a rectangular co-ordinate system let P have co-ordinates x, y, z. Then

$$r^2 = x^2 + y^2 + z^2, \ \overrightarrow{OP} = x\mathbf{i} + y\mathbf{j} + z\mathbf{k}.$$

Hence the components of $\mathbf{E}$ are

$$E_1 = \frac{x}{r^3}, \qquad E_2 = \frac{y}{r^3}, \qquad E_3 = \frac{z}{r^3}.$$

Now

$$\frac{\partial E_1}{\partial x} = \frac{r^3 - 3\,r^2 x\,\dfrac{\partial r}{\partial x}}{r^6}.$$

To find $\dfrac{\partial r}{\partial x}$ we have

$$2\,r\,\frac{\partial r}{\partial x} = 2\,x, \qquad \frac{\partial r}{\partial x} = \frac{x}{r}.$$

Hence

$$\frac{\partial E_1}{\partial x} = \frac{r^3 - 3\,rx^2}{r^6} = \frac{r^2 - 3\,x^2}{r^5}.$$

By symmetry,

$$\frac{\partial E_2}{\partial y} = \frac{r^2 - 3\,y^2}{r^5}, \qquad \frac{\partial E_3}{\partial z} = \frac{r^2 - 3\,z^2}{r^5};$$

therefore

$$\operatorname{div} \mathbf{E} = \frac{3\,r^2 - 3(x^2 + y^2 + z^2)}{r^5} = 0.$$

There is another common notation for the divergence; it is

(10.7–6) $$\operatorname{div} \mathbf{F} = \nabla \cdot \mathbf{F}.$$

The reason for this notation is one of formal appearance. The symbol ∇ was introduced in the notation for the gradient (see (10.6–3)). The symbol ∇ by itself is sometimes expressed as

(10.7–7) $$\nabla = \mathbf{i}\,\frac{\partial}{\partial x} + \mathbf{j}\,\frac{\partial}{\partial y} + \mathbf{k}\,\frac{\partial}{\partial z}.$$

In this form ∇ (which is read as "del") is called a *vector differential operator*. We say that the components of the operator ∇ in this particular co-ordinate system are

$$\frac{\partial}{\partial x}, \ \frac{\partial}{\partial y}, \ \frac{\partial}{\partial z}.$$

Recalling the formula

$$\mathbf{A} \cdot \mathbf{B} = A_1 B_1 + A_2 B_2 + A_3 B_3$$

for the dot product of two vectors, we see that appearances would lead us to write

$$\nabla \cdot \mathbf{F} = \frac{\partial}{\partial x} F_1 + \frac{\partial}{\partial y} F_2 + \frac{\partial}{\partial z} F_3.$$

The expression on the right here is in fact div $\mathbf{F}$, if the "product" of the symbols $\frac{\partial}{\partial x} F_1$ is understood to mean the derivative $\frac{\partial F_1}{\partial x}$ and so on. We thus have a justification of the notation (10.7–6).

Particular interest attaches to the divergence of the gradient of a scalar field. If the scalar field is $u = f(P)$ we see that

$$(10.7\text{–}8) \qquad \text{div (grad } u) = \frac{\partial^2 u}{\partial x^2} + \frac{\partial^2 u}{\partial y^2} + \frac{\partial^2 u}{\partial z^2}.$$

In the ∇ notation

$$\text{div (grad } u) = \nabla \cdot \nabla u.$$

It is customary to write $\nabla \cdot \nabla = \nabla^2$, so that

$$(10.7\text{–}9) \qquad \nabla^2 u = \frac{\partial^2 u}{\partial x^2} + \frac{\partial^2 u}{\partial y^2} + \frac{\partial^2 u}{\partial z^2}.$$

The left member of (10.7–9) is read as "del-squared of u," or "del-squared u." The equation

$$\frac{\partial^2 u}{\partial x^2} + \frac{\partial^2 u}{\partial y^2} + \frac{\partial^2 u}{\partial z^2} = 0$$

is of fundamental importance in many branches of applied mathematics. It is known as *Laplace's equation*, in honor of the researches of the famous French mathematician Pierre Simon de Laplace (1749–1827). Accordingly the expression $\nabla^2 u$ is often called the Laplacian of u.

Since the gradient and the divergence are both invariants with respect to rigid motions of the axes, it follows from (10.7–8) that $\nabla^2 u$ is a scalar invariant associated with the field u. In other words,

$$\frac{\partial^2 u}{\partial x'^2} + \frac{\partial^2 u}{\partial y'^2} + \frac{\partial^2 u}{\partial z'^2} = \frac{\partial^2 u}{\partial x^2} + \frac{\partial^2 u}{\partial y^2} + \frac{\partial^2 u}{\partial z^2},$$

if the $x'y'z'$-system is obtained from the xyz-system by a rigid motion.

Example 2. The electrostatic potential at P arising from a dipole of unit strength at O, oriented in the direction of the unit vector $\mathbf{n}$, is

$$V = \frac{\mathbf{n} \cdot \overrightarrow{OP}}{r^3} = \frac{\cos \theta}{r^2},$$

where $r = OP$ (see Fig. 80). The electrostatic field itself is $\mathbf{E} = -\nabla V$. Let us calculate $\nabla \cdot \mathbf{E} = -\nabla^2 V$. We choose a co-ordinate system with origin O so that $\mathbf{n}$ coincides with the direction of the positive z-axis. Then

Fig. 80

$$\mathbf{n} = \mathbf{k},\ \overrightarrow{OP} = x\mathbf{i} + y\mathbf{j} + z\mathbf{k},\ r^2 = x^2 + y^2 + z^2.$$

Hence $\mathbf{n} \cdot \overrightarrow{OP} = z$, and

$$V = \frac{z}{r^3}.$$

Then $\quad \dfrac{\partial V}{\partial x} = -3\,zr^{-4}\,\dfrac{\partial r}{\partial x},\quad 2\,r\,\dfrac{\partial r}{\partial x} = 2\,x,\quad \dfrac{\partial V}{\partial x} = -\dfrac{3\,xz}{r^5}.$

Likewise, by symmetry, $\quad \dfrac{\partial V}{\partial y} = \dfrac{-3\,yz}{r^5}.$

Next $\quad \dfrac{\partial V}{\partial z} = \dfrac{1}{r^3} - 3\,zr^{-4}\,\dfrac{\partial r}{\partial z} = \dfrac{1}{r^3} - \dfrac{3\,z^2}{r^5}.$

$$\frac{\partial^2 V}{\partial x^2} = -3\,z\left(\frac{1}{r^5} - 5\,xr^{-6}\,\frac{\partial r}{\partial x}\right) = -3\,z\left(\frac{1}{r^5} - \frac{5\,x^2}{r^7}\right),$$

and by symmetry we can write down a similar expression for $\dfrac{\partial^2 V}{\partial y^2}$. Finally,

$$\frac{\partial^2 V}{\partial z^2} = -3\,r^{-4}\,\frac{\partial r}{\partial z} - \frac{6\,z}{r^5} + 3\,z^2 5\,r^{-6}\,\frac{\partial r}{\partial z}$$

$$= -\frac{3\,z}{r^5} - \frac{6\,z}{r^5} + \frac{15\,z^3}{r^7}.$$

Collecting, we see that

$$\nabla^2 V = -3\,z\left(\frac{1}{r^5} - \frac{5\,x^2}{r^7}\right) - 3\,z\left(\frac{1}{r^5} - \frac{5\,y^2}{r^7}\right) - 3\,z\left(\frac{3}{r^5} - \frac{5\,z^2}{r^7}\right)$$

$$= -3\,z\left(\frac{5}{r^5} - \frac{5(x^2 + y^2 + z^2)}{r^7}\right) = 0.$$

EXERCISES

1. Show that, if u and $\mathbf{F}$ are differentiable scalar and vector fields, respectively, then

$$\nabla \cdot (u\mathbf{F}) = \nabla u \cdot \mathbf{F} + u\nabla \cdot F.$$

2. Show that, granted sufficient differentiability,

$$\nabla \cdot (u\,\nabla v) = u\,\nabla^2 v + \nabla u \cdot \nabla v.$$

3. (a) Let $\mathbf{F} = z\mathbf{i}$. If this vector field is interpreted to mean that $\mathbf{F}$ at (x, y, z)

is the velocity of the particle of a fluid at that point, describe in words the nature of the motion of the fluid. Find $\nabla \cdot \mathbf{F}$. What is happening to the volume of the part of the fluid which at a given instant occupies the cube with sides $x = 0$, $x = 1$, $y = 0$, $y = 1$, $z = 0$, $z = 1$?

(b) As in (a), interpret $\mathbf{F} = x\mathbf{i} + y\mathbf{j} + z\mathbf{k}$ as the velocity at (x, y, z) in a fluid flow. What is the nature of the motion? Find $\nabla \cdot \mathbf{F}$. Let V be the volume of the part of the fluid which at $t = 0$ occupies the sphere of radius r and center O. Find $\dfrac{dr}{dt}$ and $\dfrac{dV}{dt}$ in terms of r and show that $\dfrac{1}{V}\dfrac{dV}{dt} = \nabla \cdot \mathbf{F}$. This special situation gives some hint of the meaning of $\nabla \cdot \mathbf{F}$ in relation to expansion or contraction of volumes in an arbitrary fluid flow.

4. (a) In Exercise 8, § 10.6, it was shown that for a scalar field of the form $u = F(r)$, $\nabla u = \dfrac{1}{r} F'(r)\mathbf{R}$ where $r = OP$ and $\mathbf{R} = \overrightarrow{OP}$. Using this result and the formula in Exercise 1, show that

$$\nabla^2 u = F''(r) + \frac{2}{r} F'(r).$$

(b) Find $F(r)$ if $\nabla^2 u = 0$ when $r > 0$.

5. (a) Using the notation r and $\mathbf{R}$ as in Exercise 4, let $\mathbf{F} = \phi(r)\mathbf{R}$ and show that $\nabla \cdot \mathbf{F} = r\phi'(r) + 3\phi(r)$. (b) Find $\phi(r)$ if $\nabla \cdot \mathbf{F} = 0$ when $r > 0$. Show that in this case $\mathbf{F}$ can be interpreted as a force of attraction toward, or repulsion from, O according to the inverse square law.

6. (a) If $\mathbf{F} = \mathbf{A} \times \overrightarrow{OP}$, where $\mathbf{A}$ is any constant vector field, show that $\nabla \cdot \mathbf{F} = 0$. SUGGESTION: Choose axes so that $\mathbf{A}$ has the direction of the z-axis. This is not essential, but it simplifies the work.

(b) If a rigid solid body is rotating around the z-axis with angular speed ω, rotation in the xy-plane being in the counterclockwise sense as viewed by an observer looking down on the plane from the positive z-axis, the velocity of any point P of the body is $\mathbf{V} = \omega \mathbf{k} \times \overrightarrow{OP}$. Verify this. Write out the components of $\mathbf{V}$ and compute $\nabla \cdot \mathbf{V}$.

7. If $u = ax^2 + by^2 + cz^2$, find $\nabla^2 u$. If u is a general homogeneous polynomial of degree two in x, y, z, what is the condition on its coefficients necessary and sufficient to insure $\nabla^2 u = 0$?

8. Compare the demonstration that $\nabla \cdot \mathbf{F}$ is an invariant with the algebraic proof of the invariance of $\mathbf{A} \cdot \mathbf{B}$ (Exercise 1, § 10.4). Write out both demonstrations in detail and compare the steps.

10.8 The curl of a vector field

We have seen that div $\mathbf{F}$ is a scalar invariant which appears *formally* as the dot product $\nabla \cdot \mathbf{F}$ of the vector operator symbol ∇ and the vector $\mathbf{F}$. Let us now ask what we obtain formally from the cross product $\nabla \times \mathbf{F}$. By analogy with (10.22–7) we write

(10.8–1)
$$\nabla \times \mathbf{F} = \begin{vmatrix} \mathbf{i} & \mathbf{j} & \mathbf{k} \\ \dfrac{\partial}{\partial x} & \dfrac{\partial}{\partial y} & \dfrac{\partial}{\partial z} \\ F_1 & F_2 & F_3 \end{vmatrix}$$

In expanded form

(10.8–2)

$$\nabla \times \mathbf{F} = \left(\frac{\partial F_3}{\partial y} - \frac{\partial F_2}{\partial z} \right)\mathbf{i} + \left(\frac{\partial F_1}{\partial z} - \frac{\partial F_3}{\partial x} \right)\mathbf{j} + \left(\frac{\partial F_2}{\partial x} - \frac{\partial F_1}{\partial y} \right)\mathbf{k}.$$

The first question about this expression is: *Is it invariant with respect to rigid motions of the axes?* The student who has thoroughly understood the discussions of the invariance of the gradient and the divergence will appreciate the import of the present question; he will also be ready to carry through the proof of invariance of $\nabla \times \mathbf{F}$ on the basis of the experience he has acquired. In § 10.7 we have seen how to deal with the transformation of partial derivatives of the components of $\mathbf{F}$ under a rotation of the axes. We have to prove three relations, of which the relation

(10.8–3)

$$\frac{\partial F'_3}{\partial y'} - \frac{\partial F'_2}{\partial z'} = l_1 \left(\frac{\partial F_3}{\partial y} - \frac{\partial F_2}{\partial z} \right) + m_1 \left(\frac{\partial F_1}{\partial z} - \frac{\partial F_3}{\partial x} \right)$$
$$+ n_1 \left(\frac{\partial F_2}{\partial x} - \frac{\partial F_1}{\partial y} \right)$$

is typical. The details are almost identical with those of proving algebraically the invariance of the cross product $\mathbf{A} \times \mathbf{B}$ (see (10.4–3) and the following material). Here we need to establish relations such as

$$\frac{\partial F'_3}{\partial y'} = \left(l_2 \frac{\partial}{\partial x} + m_2 \frac{\partial}{\partial y} + n_2 \frac{\partial}{\partial z} \right)(l_3 F_1 + m_3 F_2 + n_3 F_3),$$

the expansion of the "product" on the right having an obvious meaning. We leave it for the student to complete the work himself.

The expression $\nabla \times \mathbf{F}$ defined in (10.8–2) is a vector invariant. Its importance is on the same level as that of the divergence, and for much the same reasons. We call $\nabla \times \mathbf{F}$ the *curl* of the field $\mathbf{F}$, and write

$$\text{curl } \mathbf{F} = \nabla \times \mathbf{F}.$$

The physical or geometrical significance of the curl cannot be explained in a few lines. Certain kinds of vector fields have the property that their curl is everywhere the vector $\mathbf{0}$. Such fields are called *irrotational*. This name comes from hydrodynamics, and is there used to describe the vector field of velocities of particles in certain types of fluid motion. The antithesis of irrotational fluid motion is *vortex* motion, of which a whirlpool affords the simplest example. In the physical applications to fields of force, the irrotational fields are the *conservative* fields, i.e., the fields in which a principle of conservation of energy applies.

Example 1. Consider the vector field $\mathbf{V} = \omega\mathbf{k} \times \overrightarrow{OP}$ representing the velocities of points P in a rigid body rotating about the z-axis (see Exercise 6(b), § 10.7). We have

$$\mathbf{V} = \begin{vmatrix} \mathbf{i} & \mathbf{j} & \mathbf{k} \\ 0 & 0 & \omega \\ x & y & z \end{vmatrix} = -\omega y\mathbf{i} + \omega x\mathbf{j}.$$

Here $V_1 = -\omega y$, $V_2 = \omega x$, $V_3 = 0$. Hence

$$\nabla \times \mathbf{V} = \begin{vmatrix} \mathbf{i} & \mathbf{j} & \mathbf{k} \\ \dfrac{\partial}{\partial x} & \dfrac{\partial}{\partial y} & \dfrac{\partial}{\partial z} \\ -\omega y & \omega x & 0 \end{vmatrix},$$

$$\nabla \times \mathbf{V} = 2\,\omega\mathbf{k}.$$

Thus the curl of the velocity field is a vector in the direction of the axis of rotation, having a magnitude equal to twice the angular speed of rotation.

Example 2. Let $\mathbf{F} = \dfrac{c}{r^3}\,\overrightarrow{OP}$, where c is a constant and r is the distance OP. This may be interpreted as the gravitational field of a mass concentrated at O (compare with Example 2, § 10.51). Show that

$$\nabla \times \mathbf{F} = \mathbf{0}.$$

The components of $\mathbf{F}$ are

$$F_1 = \frac{cx}{r^3}, \qquad F_2 = \frac{cy}{r^3}, \qquad F_3 = \frac{cz}{r^3},$$

and $r^2 = x^2 + y^2 + z^2$. Now

$$\frac{\partial F_2}{\partial x} = cy\left(-3\,r^{-4}\,\frac{\partial r}{\partial x}\right) = cy(-3\,r^{-4})\frac{x}{r} = -\frac{3\,cxy}{r^5}.$$

By a symmetrical calculation

$$\frac{\partial F_1}{\partial y} = -\frac{3\,cxy}{r^5}.$$

Hence the z-component of $\nabla \times \mathbf{F}$ is

$$\frac{\partial F_2}{\partial x} - \frac{\partial F_1}{\partial y} = 0.$$

The other components also vanish, by symmetry.

EXERCISES

1. If $\mathbf{F} = \nabla u$, show that $\nabla \times \mathbf{F} = 0$ at all points of the field. Assume that u is twice differentiable.

2. Show that $\nabla \cdot (\nabla \times \mathbf{F}) = 0$ in the case of any twice differentiable vector field.

3. Find $\nabla \times \mathbf{F}$ in each case.
(a) $\mathbf{F} = 2xz\mathbf{i} + 2yz^2\mathbf{j} + (x^2 + 2y^2z - 1)\mathbf{k}$.
(b) $\mathbf{F} = ax\mathbf{i} + by\mathbf{j} + cz\mathbf{k}$.
(c) $\mathbf{F} = y\mathbf{i} + z\mathbf{j} + x\mathbf{k}$.
(d) $\mathbf{F} = \dfrac{x\mathbf{i} + y\mathbf{j}}{x^2 + y^2}$.
(e) $\mathbf{F} = \dfrac{-y\mathbf{i} + x\mathbf{j}}{x^2 + y^2}$.
(f) $\mathbf{F} = -\dfrac{xz}{\sqrt{x^2 + y^2}}\,\mathbf{i} - \dfrac{yz}{\sqrt{x^2 + y^2}}\,\mathbf{j} + \sqrt{x^2 + y^2}\,\mathbf{k}$.

4. If u and $\mathbf{F}$ are differentiable scalar and vector fields, respectively, show that $\nabla \times (u\mathbf{F}) = u(\nabla \times \mathbf{F}) + \nabla u \times \mathbf{F}$.

5. If $\mathbf{E}$ and $\mathbf{F}$ are differentiable vector fields, show that
$$\nabla \cdot (\mathbf{E} \times \mathbf{F}) = \mathbf{F} \cdot (\nabla \times \mathbf{E}) - \mathbf{E} \cdot (\nabla \times \mathbf{F}).$$

6. If $\mathbf{F} = \mathbf{A} \times \overrightarrow{OP}$, where $\mathbf{A}$ is a constant vector field, show that $\nabla \times \mathbf{F} = 2\mathbf{A}$.

7. Let $\mathbf{R} = \overrightarrow{OP}$, $r = ||\,\mathbf{R}\,||$. (a) Find $\Delta \times \mathbf{R}$. (b) If $\mathbf{F} = \phi(r)\mathbf{R}$, where ϕ is differentiable, show that $\nabla \times \mathbf{F} = 0$.

8. If $\mathbf{A}$ is a constant vector field, and r, $\mathbf{R}$ have meanings as in Exercise 7, show that $\nabla \times (r^n\mathbf{A} \times \mathbf{R}) = (n + 2)r^n\mathbf{A} - nr^{n-2}(\mathbf{A} \cdot \mathbf{R})\mathbf{R}$.

9. Write out all the details of the proof of (10.8–3).

MISCELLANEOUS EXERCISES

1. Let $\mathbf{A}$ and $\mathbf{B}$ be constant, and let $\mathbf{R} = \overrightarrow{OP}$, $\mathbf{E} = \mathbf{R} - \mathbf{A}$, $\mathbf{F} = \mathbf{R} - \mathbf{B}$. Show that (a) $\nabla \cdot (\mathbf{E} \times \mathbf{F}) = 0$, (b) $\nabla \times (\mathbf{E} \times \mathbf{F}) = 2(\mathbf{B} - \mathbf{A})$, (c) $\nabla(\mathbf{E} \cdot \mathbf{F}) = \mathbf{E} + \mathbf{F}$.

2. Let $\mathbf{R} = \overrightarrow{OP}$, $r = ||\,\mathbf{R}\,||$, and let $\mathbf{A}$ be a constant vector.
(a) Find $\nabla \cdot (r^n\mathbf{A})$ and $\nabla \times (r^n\mathbf{A})$.
(b) Find $\nabla \cdot (r^n\mathbf{A} \times \mathbf{R})$.
(c) Show that $\nabla\left(\dfrac{\mathbf{A} \cdot \mathbf{R}}{r^3}\right) = -\nabla \times \left(\dfrac{\mathbf{A} \times \mathbf{R}}{r^3}\right)$.

3. With the notation of Exercise 2, and the additional assumption that $\mathbf{A}$ has unit length, show that
(a) $\nabla \cdot [(\mathbf{A} \cdot \mathbf{R})\mathbf{A}] = 1$,
(b) $\nabla \times [(\mathbf{A} \cdot \mathbf{R})\mathbf{A}] = 0$,
(c) $\nabla \cdot [(\mathbf{A} \times \mathbf{R}) \times \mathbf{A}] = 2$,
(d) $\nabla \times [(\mathbf{A} \times \mathbf{R}) \times \mathbf{A}] = 0$.

4. Find $\nabla \cdot (\nabla u \times \nabla v)$.

5. If $\mathbf{A}$ is a constant vector, show that $\nabla \cdot (\mathbf{A} \times \mathbf{F}) = -\mathbf{A} \cdot (\nabla \times \mathbf{F})$.

Linear Transformations 11

11. Introduction

The development of vector mathematics presented in the preceding chapter took place mostly during the second half of the nineteenth century. One of the leading contributors to the development was the American genius Josiah Williard Gibbs of Yale University. The first part of Chapter X, which deals with the ways vectors combine with other vectors and with scalars (real numbers), is called vector algebra. During the twentieth century, vector algebra expanded vastly into a subject called linear algebra, which has important applications in much of mathematics, especially advanced calculus.

This chapter will present enough linear algebra to enable us to use the subject to unify and extend the results obtained in the last several chapters. Prerequisite to full understanding of this chapter and the next is a little knowledge of simultaneous linear systems involving n equations in n unknowns. In particular, we shall use the following fact: A necessary and sufficient condition that such a system have a unique solution is that the determinant of the coefficient matrix be different from zero. In Chapter XII we occasionally use some of the most elementary rules for computing determinants.

A big step in the transition from vector algebra to linear algebra is the realization that functions from a vector space to a vector space are in themselves examples of vectors. The first part of this chapter will be devoted to explaining in detail what this means. We can begin with the once popular question, "What is a vector?" The common reply used to be that a vector is a quantity having both magnitude and direction. We shall soon see that neither magnitude nor direction is essential to vectors, and that most things having both (trains, for example) are not vectors. No satisfactory answer was arrived at until it was realized that one should not try to define a vector as an object having certain qualities, but rather as a member of a family of objects governed by certain rules. In particular, the important considerations are how to combine a vector with other vectors and with scalars through certain operations.

The problem is much like that of defining a checker. We might naturally begin by saying that it is a wooden or plastic disk, but such a beginning does not lead to anything satisfactory. If one were asked whether the top of a soft drink bottle is a checker, his first inclination would probably be to say no. Yet many games of checkers have been played with these objects. The important fact which emerges from all this is that there is no property intrinsic to a thing considered in isolation which can settle the question of whether it is or is not a checker. The answer depends entirely on whether it is a member of a set of objects which are moved according to certain well-defined rules.

Similarly, a mathematical object is a vector if and only if it is a member of a set whose members combine with each other and with real numbers according to certain well-defined rules—the rules of vector algebra. These rules have already been encountered either explicitly or implicitly in the early part of Chapter X. We shall collect and present them here as a set of abstract rules. The advantage of this is that it will enable us to apply them to sets of objects which we would be slow to recognize as vectors if we continued to associate vectors with physical or geometrical quantities. The rules actually apply to two sets—the set of real numbers R, whose elements we shall refer to as scalars, and the set $\mathcal{V}$ of vectors. In stating the rules we shall distinguish between the scalars and the vectors by using boldface type to refer to the latter.

On the set of vectors $\mathcal{V}$, there must be a binary operation, denoted by $+$ and called addition, which is commutative and associative, i.e., if $\mathbf{x}$ and $\mathbf{y}$ belong to $\mathcal{V}$, then their sum $\mathbf{x} + \mathbf{y}$ also belongs to $\mathcal{V}$ and

$$\mathbf{x} + \mathbf{y} = \mathbf{y} + \mathbf{x}$$

and

$$\mathbf{x} + (\mathbf{y} + \mathbf{z}) = (\mathbf{x} + \mathbf{y}) + \mathbf{z}$$

for all $\mathbf{x}$, $\mathbf{y}$, and $\mathbf{z}$ in $\mathcal{V}$.

There is a zero vector, denoted by $\mathbf{0}$, with the property that

$$\mathbf{x} + \mathbf{0} = \mathbf{x} \qquad \text{for all } \mathbf{x} \text{ in } \mathcal{V}.$$

The fact that this vector is denoted by the same symbol as the real number zero means that the student must be alert from now on to infer the correct meaning from the context in which the symbol occurs.

For each vector $\mathbf{a}$ in $\mathcal{V}$, there is a unique vector $\mathbf{x}$ such that $\mathbf{a} + \mathbf{x} = \mathbf{0}$. This vector $\mathbf{x}$ is denoted by $-\mathbf{a}$, of course.

There is an operation called multiplication of vectors by scalars, which is denoted merely by juxtaposition and which has the following properties.

$$c\mathbf{x} = \mathbf{x}c \text{ is in } \mathcal{V} \quad \text{for all } \mathbf{x} \text{ in } \mathcal{V} \text{ and all } c \text{ in } R.$$
$$c(\mathbf{x} + \mathbf{y}) = c\mathbf{x} + c\mathbf{y} \quad \text{for all } \mathbf{x} \text{ and } \mathbf{y} \text{ in } \mathcal{V} \text{ and all } c \text{ in } R.$$
$$(a + b)\mathbf{x} = a\mathbf{x} + b\mathbf{x} \quad \text{for all } a \text{ and } b \text{ in } R \text{ and all } \mathbf{x} \text{ in } \mathcal{V}.$$
$$(ab)\mathbf{x} = a(b\mathbf{x}) \quad \text{for all } a \text{ and } b \text{ in } R \text{ and all } \mathbf{x} \text{ in } \mathcal{V}.$$
$$1\mathbf{x} = \mathbf{x} \quad \text{for all } \mathbf{x} \text{ in } \mathcal{V}.$$
$$0\mathbf{x} = \mathbf{0} \quad \text{for all } \mathbf{x} \text{ in } \mathcal{V}.$$

Notice also that neither multiplication nor division of one vector by another is necessarily defined. In many applications of vectors it is convenient to define a dot product (frequently called a scalar product) like that in Chapter X. And in the study of three-dimensional space the vector product has been seen to be useful, but no concept of multiplying one vector by another is essential in defining vectors, and nowhere will we have occasion to define division for vectors.

We pause here to explain a useful abbreviation by the use of the symbol $\in$. If $\mathcal{S}$ is any set, the notation $a \in \mathcal{S}$ means "a is a member (or element) of $\mathcal{S}$." It can be variously read as "a belongs to $\mathcal{S}$," or "a is in $\mathcal{S}$." Other slight variations of the verbal rendering of the notation are useful. For instance, we may write one of the foregoing rules about vectors as follows: $(ab)\mathbf{x} = a(b\mathbf{x})$ for each $a,b \in R$ and each $x \in \mathcal{V}$. In this context $\in$ may be read as "in" or "belonging to."

The entire mathematical structure made up of $\mathcal{V}$, R, the two operations of vector addition and multiplication of a vector by a scalar, and the rules governing them is properly called a *vector space*. To avoid prolixity, however, this term is frequently applied just to the $\mathcal{V}$ when the rest of the structure is understood. The expressions *linear space* and *linear vector space* are frequently used as synonyms for vector space.

Example 1. Let $\mathcal{V}$ denote the set of all ordered n-tuples of real numbers and also define addition and scalar multiplication as follows: If $\mathbf{x} = (x_1, x_2, \cdots, x_n)$ and $\mathbf{y} = (y_1, y_2, \cdots y_n)$ then

$$\mathbf{x} + \mathbf{y} = (x_1 + y_1, x_2 + y_2, \cdots, x_n + y_n),$$

and if c is any scalar, $c\mathbf{x} = (cx_1, cx_2, \cdots, cx_n)$. It can easily be verified that $\mathcal{V}$ together with these operations constitutes a vector space. Remember that the zero vector is $(0, 0, \cdots, 0)$ and that it too is denoted by $\mathbf{0}$. This most important vector space is denoted by R^n. Much of advanced calculus is concerned with functions for which the domain lies in R^n and the range in R^m, where n and m are frequently the same.

Example 2. Here, $\mathcal{V}$ is the set of all real-valued continuous functions defined on $[0, 1]$. The reader already knows how to add two functions and how to multiply one by a scalar. He also knows that these operations, performed on continuous functions, always give continuous functions. We merely wish to point out that this long-familiar structure is a vector space, even though we have said nothing about "magnitude and direction" of the vectors.

Example 3. Let $\mathcal{V}$ stand for the collection of all 2×2 matrices. We add matrices simply by adding the elements in corresponding positions, and in order to multiply a matrix by a scalar, we multiply each element of the matrix by that scalar. Even though we have not associated any magnitude or direction with these matrices, they are nonetheless vectors.

11.1 Linear transformations

In Example 2 the vectors were functions—continuous functions from the unit interval [0, 1] to the real numbers. In this section, we introduce another very important kind of function space. The vectors in these spaces are the very simple functions called linear transformations. By a *linear transformation* we mean a function T from a vector space $\mathcal{U}$ to a vector space $\mathcal{S}$ having the following two properties:

(1) $T(\mathbf{x}) + T(\mathbf{y}) = T(\mathbf{x} + \mathbf{y})$ for all $\mathbf{x}$ and $\mathbf{y}$ in $\mathcal{U}$, and
(2) $T(c\mathbf{x}) = cT(\mathbf{x})$ for every scalar c, and every $\mathbf{x} \in \mathcal{U}$.

Notice that each of these two properties alone implies the important fact that *every linear transformation maps the zero vector of $\mathcal{U}$ into the zero vector of $\mathcal{S}$.*

The statement that f is a function from the set A to the set B is frequently abbreviated to $f : A \to B$. It means that f is defined at every point of A, and that for each a in A, $f(a) \in B$. In other words, the domain of f is all of A, but its range may be either all of B or just some proper subset. Therefore, when we say that T is a linear transformation, we shall imply that $T : \mathcal{U} \to \mathcal{S}$ for some pair of vector spaces $\mathcal{U}$ and $\mathcal{S}$, and that T satisfies Properties (1) and (2). The terms *linear function* and *linear map* are frequently used as synonyms for linear transformation.

In the important special case where $\mathcal{U} = \mathcal{S}$, the linear transformation is usually referred to as a *linear operator*. In other words, a linear operator is a linear transformation which maps a vector space into itself.

11.2 The vector space Hom(R^n, R^m)

For reasons connected with its role in abstract algebra, the set of all linear transformations from R^n to R^m is denoted by Hom(R^n, R^m). (Hom is an abbreviation for homomorphism.) We wish to prove that this set is a vector space. We define addition the same way we always have for functions; that is, if T_1 and T_2 belong to Hom(R^n, R^m), then their sum T is that function from R^n to R^m defined by

$$T(\mathbf{x}) = T_1(\mathbf{x}) + T_2(\mathbf{x}) \qquad \text{for all } \mathbf{x} \in R^n.$$

T is obviously then a well-defined function from R^n to R^m, but to show that it belongs to Hom(R^n, R^m) we must prove that it is linear. We therefore first verify Property (1).

$$T(\mathbf{x} + \mathbf{y}) = T_1(\mathbf{x} + \mathbf{y}) + T_2(\mathbf{x} + \mathbf{y}) = T_1(\mathbf{x}) + T_1(\mathbf{y}) + T_2(\mathbf{x}) + T_2(\mathbf{y})$$
$$= [T_1(\mathbf{x}) + T_2(\mathbf{x})] + [T_1(\mathbf{y}) + T_2(\mathbf{y})] = T(\mathbf{x}) + T(\mathbf{y}).$$

Property (2) is checked as follows:

$$T(c\mathbf{x}) = T_1(c\mathbf{x}) + T_2(c\mathbf{x}) = cT_1(\mathbf{x}) + cT_2(\mathbf{x}) = c[T_1(\mathbf{x}) + T_2(\mathbf{x})] = cT(\mathbf{x}).$$

This completes the proof that if T_1 and T_2 are any two elements of Hom(R^n, R^m), their sum must also belong to this set. It is even easier to prove that a scalar multiple of a linear transformation is a linear transformation. The other vector space axioms are obviously satisfied. The zero vector in this space Hom(R^n, R^m) is the linear transformation which maps every vector of R^n into the zero vector of R^m. The reader should verify that this function is linear.

11.3   Matrices

If A is a rectangular array of mn numbers arranged in m rows and n columns, and if $\mathbf{x} = (x_1, x_2, \cdots, x_n) \in R^n$, then we can multiply $\mathbf{x}$ by A in the following way:

$$Ax = \begin{pmatrix} a_{11} & a_{12} & \cdots & a_{1n} \\ a_{21} & a_{22} & \cdots & a_{2n} \\ \vdots & \vdots & \vdots & \vdots \\ a_{m1} & a_{m2} & \cdots & a_{mn} \end{pmatrix} \begin{pmatrix} x_1 \\ x_2 \\ \vdots \\ x_n \end{pmatrix} = \begin{pmatrix} a_{11}x_1 + a_{12}x_2 + \cdots + a_{1n}x_n \\ a_{21}x_1 + a_{22}x_2 + \cdots + a_{2n}x_n \\ \vdots & \vdots & \vdots & \vdots \\ a_{m1}x_1 + a_{m2}x_2 + \cdots + a_{mn}x_n \end{pmatrix}$$

$$= \begin{pmatrix} y_1 \\ y_2 \\ \vdots \\ y_m \end{pmatrix} = \mathbf{y}.$$

For each $\mathbf{x}$ in R^n, $\mathbf{y} \in R^m$, and thus,

$$\mathbf{y} = A\mathbf{x}$$

defines a function from R^n to R^m. It is easy to show (see Exercise 1) that this function is linear and therefore an element of Hom(R^n, R^m). In other words, every $m \times n$ matrix defines a linear transformation from R^n to R^m.

A less obvious fact of great importance is the converse of the preceding sentence. In other words, if T is any element of Hom(R^n, R^m), there is some $m \times n$ matrix, A, such that

$$T\mathbf{x} = A\mathbf{x}$$

for all $\mathbf{x} \in R^n$. A matrix A which is related to the transformation T in this way is said to represent T.

To prove that every linear transformation is represented by a matrix, it is convenient to consider the following special set of n vectors in R^n:

$$\begin{aligned} \mathbf{e}_1 &= (1, 0, 0, \ldots, 0) \\ \mathbf{e}_2 &= (0, 1, 0, \ldots, 0) \\ \mathbf{e}_3 &= (0, 0, 1, \ldots, 0) \\ &\vdots \\ \mathbf{e}_n &= (0, 0, \ldots, 0, 1). \end{aligned}$$

These vectors constitute what is called the *standard basis* in R^n. Every vector in the space can be expressed as a linear combination of these in an obvious way. Suppose $x = (x_1, x_2, \cdots , x_n)$. Then, by the rules for addition and multiplication by scalars in R^n, we see at once that

$$x = x_1e_1 + x_2e_2 + \cdots + x_ne_n.$$

There is, of course, a standard basis, for example, $u_1, u_2, \cdots , u_m$, in R^m. Since every vector in R^m is a linear combination of the u's,

(11.3–1)
$$
\begin{aligned}
Te_1 &= \alpha_{11}u_1 + \alpha_{12}u_2 + \cdots + \alpha_{1m}u_m \\
Te_2 &= \alpha_{12}u_1 + \alpha_{22}u_2 + \cdots + \alpha_{2m}u_m \\
&\ \vdots \qquad \vdots \qquad \vdots \qquad \vdots \qquad \vdots \\
Te_n &= \alpha_{n1}u_1 + \alpha_{n2}u_2 + \cdots + \alpha_{nm}u_m
\end{aligned}
$$

for some set of scalar coefficients, α_{ij}. These equations show what T does to the standard basis elements in R^n, and we shall see that this completely determines T. Let x be any vector in R^n, and suppose $x = (x_1, x_2, \cdots , x_n)$. Then $x = x_1e_1 + x_2e_2 + \cdots + x_ne_n$, and since T is linear,

$$Tx = T\sum_{i=1}^{n} x_ie_i = \sum_{i=1}^{n} x_iT(e_i) = \sum_{i=1}^{n} x_i \sum_{j=1}^{m} \alpha_{ij}u_j$$

$$= \sum_{j=1}^{m} \left(\sum_{i=1}^{n} \alpha_{ij}x_i \right) u_j = \sum_{j=1}^{m} y_ju_j,$$

which means that $y = (y_1, y_2, \cdots y_m)$, where

$$
\begin{aligned}
\alpha_{11}x_1 + \alpha_{21}x_2 + \cdots + \alpha_{n1}x_n &= y_1 \\
\alpha_{12}x_1 + \alpha_{22}x_2 + \cdots + \alpha_{n2}x_n &= y_2 \\
\vdots \qquad \vdots \qquad \vdots \qquad \vdots \qquad \vdots \\
\alpha_{1m}x_1 + \alpha_{2m}x_2 + \cdots + \alpha_{nm}x_n &= y_m.
\end{aligned}
$$

This can be written as $Ax = y$, where the elements a_{ij} of the matrix A are defined by $a_{ij} = \alpha_{ji}$. This can be expressed by saying that the $m \times n$ matrix, A, is the *transpose* of the $n \times m$ coefficient matrix of the system (11.3–1). The transpose of a matrix M is a matrix obtained by interchanging the rows and columns of M.

Observe that the zero element of $\text{Hom}(R^n, R^m)$ is represented by the $m \times n$ matrix in which all the entries are zero.

The matrix A which represents T was constructed with the help of what were called the standard bases in the domain and range spaces, R^n and R^m. There are other bases and if we had used them we would have obtained other matrix representations of T. This suggests that A might more properly be referred to as the standard representation of T. But since these are the only bases which we shall use, we shall sometimes speak only of "the" matrix representation of a linear transformation.

It is possible to summarize the results of this section up to this point by the following theorem, which is of basic importance, especially in the next few sections.

THEOREM I. *Each $m \times n$ matrix is the standard representation of a unique linear transformation from R^n to R^m, and conversely, every element of $\mathrm{Hom}(R^n, R^m)$ has a unique standard representation as an $m \times n$ matrix.*

If $T \in \mathrm{Hom}(R^n, R^m)$ and $L \in \mathrm{Hom}(R^m, R^p)$, then we can define a function $L \circ T$ from R^n to R^p as follows:

$$(L \circ T)(x) = L[T(x)] \qquad \text{for all } x \in R^n.$$

This is a composite function—the composite of L and T—and is easily proved to be linear (Exercise 2). Therefore, $L \circ T \in \mathrm{Hom}(R^n, R^p)$. It is important to be able to express the matrix which represents $L \circ T$ in terms of the matrices representing L and T. This can be done in the following straightforward way. Let A be the $m \times n$ matrix representing T and B, the $p \times m$ representation of L. Then,

$$(L \circ T)(x) = L[Tx] = B(Ax) = By,$$

where x is any vector of R^n and y is its image under T in R^m.

$$y_i = \sum_{j=1}^{n} a_{ij} x_j \qquad\qquad (i = 1, 2, \ldots, m)$$

$$(By)_k = \left(\sum_{i=1}^{m} b_{ki} y_i \right) \qquad (k = 1, 2, \ldots, p)$$

$$= \left(\sum_{i=1}^{m} b_{ki} \sum_{j=1}^{n} a_{ij} x_j \right) \qquad (k = 1, 2, \ldots, p)$$

$$= \sum_{j=1}^{n} \left(\sum_{i=1}^{m} b_{ki} a_{ij} \right) x_j \qquad (k = 1, 2, \ldots, p)$$

$$= \left(\sum_{j=1}^{n} c_{kj} x_j \right) \qquad\qquad (k = 1, 2, \ldots, p)$$

$$= Cx,$$

where C is the $p \times n$ matrix whose element in the position (k, j) is given by

$$c_{kj} = \sum_{i=1}^{m} b_{ki} a_{ij}.$$

The matrix C obtained in this way is called the *product* of B and A in that order; that is, $C = BA$. This definition of products shows that the number of columns in the first factor must equal the number of rows in the second, and that the product of a $p \times m$ matrix by an $m \times n$ matrix is a $p \times n$ matrix. Therefore, not only is matrix multiplication not commutative, the product is not even defined in both orders unless the two matrices are square and of the same size; even then the products in the two orders are not necessarily the same.

It should be remembered that the purpose in defining matrix multiplication as we have done is to make the product of two matrices represent the composition of the linear transformations represented by the two factors; that is, if A represents T, and B represents L, then the matrix BA represents the composite linear transformation $L \circ T$.

11.4   Some special cases

If $m = 1$, then $\text{Hom}(R^n, R)$ is the vector space consisting of all real-valued linear functions on R^n. The elements of this space are called *linear functionals* on R^n. From Theorem I we know that each linear functional on R^n is represented by a $1 \times n$ matrix. Such a matrix, consisting of 1 row and n columns, is just an ordered n-tuple of real numbers, and is therefore identifiable with some vector in R^n. In other words, if L is any linear functional on R^n, then there is some vector, $\mathbf{a}$, in R^n such that, for all $\mathbf{x}$,

$$L(\mathbf{x}) = \mathbf{a}\mathbf{x} = (a_1, a_2, \ldots, a_n) \begin{pmatrix} x_1 \\ x_2 \\ \vdots \\ x_n \end{pmatrix}.$$

We can regard this either as multiplying a vector by a $1 \times n$ matrix, or as multiplying a $1 \times n$ matrix by an $n \times 1$ matrix. Either way the result is the scalar $\sum_1^n a_i x_i$. This, of course, is the *scalar product* of $\mathbf{a}$ and $\mathbf{x}$, denoted by $\mathbf{a} \cdot \mathbf{x}$. (It is also called the *dot product*; see Theorem I in § 10.21 for the case $n = 3$.)

Therefore, for each linear functional L on R^n, there is a vector $\mathbf{a}$ in R^n which represents L in the sense that

$$L\mathbf{x} = \mathbf{a} \cdot \mathbf{x}.$$

It is trivial to prove that $\mathbf{a}$ is unique and that for each $\mathbf{a} \in R^n$ there is a unique linear functional, L, on R^n defined by $L\mathbf{x} = \mathbf{a}\mathbf{x}$. This means that $\text{Hom}(R^n, R)$ is practically the same as R^n.

In the special case $\text{Hom}(R, R^m)$, the linear transformations are, by Theorem I, represented by $m \times 1$ matrices, which are at once identified with vectors of R^m. In other words, if $T \in \text{Hom}(R, R^m)$, then there exists some $\mathbf{a} \in R^m$ such that, for every real number x,

$$Tx = \mathbf{a} \cdot x = \begin{pmatrix} a_1 \\ a_2 \\ \vdots \\ a_m \end{pmatrix} x = \begin{pmatrix} a_1 x \\ a_2 x \\ \vdots \\ a_m x \end{pmatrix}.$$

This says that every linear function from R to R^m is defined by taking scalar multiples of some fixed vector in R^m.

Finally, $\text{Hom}(R, R)$ is the set of all functions of the form $f(x) = ax$. These are all of the linear functions from the reals to the reals. This is not consistent with our high school algebra terminology, where elements such as $f(x) = ax + b$ are called linear functions. Such functions are properly called *affine*. They are obtained from linear functions simply by adding some constant. In the more general setting of functions from R^n to R^m, an affine function f would be defined by

$$f(x) = Tx + b,$$

where $T \in \text{Hom}(R^n, R^m)$ and $b \in R^m$.

11.5   Norms

We have seen that it is not necessary for a vector to have magnitude. In the familiar R^n spaces they do have magnitudes which are suggested by their geometrical representations. If $x = (x_1, x_2, x_3) \in R_3$, then we picture x as the line segment beginning at the origin and terminating at the point (x_1, x_2, x_3). (This is the representation used in § 10.1.) By the Pythagorean theorem, we know that the length of this line segment is $\sqrt{x_1^2 + x_2^2 + x_3^2}$, and is called the length of the vector, x. But if $x = (x_1, x_2, \ldots, x_n) \in R^n$, then we have no visual representation to go by and we are proceeding merely by analogy if we define the length of x to be $(\sum_1^n x_i^2)^{1/2}$. To decide whether analogy is a good guide in this matter, we must decide whether the function defined in this way, from R^n to the real numbers, has all the properties which we regard as essential for a length function.

There are only four properties which are regarded as absolutely necessary for length and they are the following:

(1) The length of every vector must be greater than or equal to zero.
(2) The length is zero for the zero vector and for no others.
(3) The length of any scalar multiple of a given vector must be the absolute value of the scalar times the length of the given vector.
(4) The length of the sum of two vectors must be less than or equal to the sum of their two lengths.

It is easily seen that the *square root of the sum of the squares* function, which is suggested by the geometry of R^2 and R^3, satisfies (1), (2), and (3). By using the Cauchy inequality (see Exercise 5), we can show that it also satisfies (4). We are therefore free to *define* the length of a vector x in R^n as $\sqrt{x_1^2 + \cdots + x_n^2}$, where $x = (x_1, x_2, \ldots, x_n)$.

Notice that when $n = 1$, we have the very special case where the real numbers themselves are considered as vectors. They do indeed form a somewhat trivial and artificial example of a vector space, with the above definition of a norm reducing to simply $\sqrt{x^2} = |x|$. In other words, the absolute value function is a norm for $R^1 = R$.

Since there are other functions from R^n to R satisfying the above four conditions, there are other ways we could define length. These other possibilities are useful in some situations, but for our purposes the definition which we have already introduced is sufficient for the R^n spaces.

It is frequently important to be able to associate a magnitude with a vector in some space other than R^n. In such spaces the word *norm* is used more often than *length*, although its defining properties are exactly those four which are considered essential for the length function. The norm of the vector x is usually denoted by $||x||$. For any vector space $\mathcal{V}$, to say that $||\ ||$ is a norm on $\mathcal{V}$ means that $||\ ||$ is a function from $\mathcal{V}$ to R such that

(1) $||x|| \geq 0$ for all $x \in \mathcal{V}$,
(2) $||x|| = 0$ if and only if x is the zero vector,
(3) $||cx|| = |c|\ ||x||$ for every real number c and all $x \in \mathcal{V}$,
(4) $||x + y|| \leq ||x|| + ||y||$ for all x and y in $\mathcal{V}$.

Property (4) is called the triangle inequality.

The difference between length and norm is not very sharp. The first tends to be reserved for vectors in the R^n spaces, whereas the second is the more general magnitude function and is applied freely to all vector spaces where there is some way of associating a magnitude with the vectors.

The four defining properties of a norm imply one additional property, which we shall find very useful in the next chapter, namely

(11.5–1) $$||x \pm y|| \geq \Big| ||x|| - ||y|| \Big|.$$

This is easily derived as follows:

$$x = y + (x - y).$$
$$\therefore ||x|| = ||y + (x - y)|| \leq ||y|| + ||x - y|| \qquad \text{by (4),}$$

and hence

$$||x - y|| \geq ||x|| - ||y||.$$

By starting with $y = x + (y - x)$, we obtain similarly

$$||x - y|| \geq ||y|| - ||x||.$$

Combining these two results we have

(11.5–2) $$||\,x - y\,|| \geq \Big| \,||\,x\,|| - ||\,y\,||\, \Big|.$$

Also,

$$||\,x + y\,|| = ||\,x - (-y)\,|| \geq \Big| \,||\,x\,|| - ||\,-y\,||\, \Big|$$

by (11.5–2), and therefore,

(11.5–3)
$$||\,x + y\,|| \geq \Big| \,||\,x\,|| - ||\,y\,||\, \Big|, \text{ since } ||\,-y\,|| = ||\,y\,||$$

by Property (3).
We have now completed the proof of (11.5–1).

Notice that if x is any nonzero vector in a space where we have a norm, then $||\,x\,|| > 0$. By Property (3),

$$\Big|\Big|\,\frac{x}{||\,x\,||}\,\Big|\Big| = \frac{1}{||\,x\,||}\,||\,x\,|| = 1.$$

Every vector whose norm is 1 is called a *unit vector*, so what we have shown is that when any nonzero vector is divided by its norm, the quotient is a unit vector. In any vector space with a norm, the *unit sphere* is defined to be the set of all unit vectors; that is, the equation of the unit sphere is $||\,x\,|| = 1$.

11.6 Metrics

One of the most important uses of a norm on a vector space is that it enables us to introduce a *metric*, which is a generalization of our notion of distance. In R^2 and R^3 we regard, as geometrically obvious, that the distance between any two elements (which we interchangeably call points and vectors) is the length of their difference. For example, if $x = (x_1, x_2)$ and $y = (y_1, y_2)$ are any two points in R^2, the distance between them is

$$\sqrt{(y_1 - x_1)^2 + (y_2 - x_2)^2},$$

which is the length of $y - x = (y_1 - x_1, y_2 - x_2)$. In any vector space the difference between any two vectors is also in the space, and if the space has a norm, we can define the distance between two vectors to be the norm of their difference. This gives us a function from ordered pairs of vectors to the real numbers.

One reason why the four properties which we used in defining a norm are correct is that they make this way of defining distance quite satisfactory.

If d is a function from ordered pairs of vectors to the real numbers, then, in order to satisfy our intuitive feelings about distance, it must obey the following four axioms:

D_1: $d(x, y) \geq 0$ for all x and y,
D_2: $d(x, y) = 0$ if and only if $x = y$,
D_3: $d(x, y) = d(y, x)$ for all x and y,
D_4: $d(x, y) + d(y, z) \geq d(x, z)$.

Axiom D_3 expresses the symmetry of distance, and Axiom D_4 is called the triangle inequality.

Suppose we consider a vector space on which there is defined a norm and suppose we introduce the function d defined by

$$(11.6\text{--}1) \qquad d(x, y) = ||\, x - y\, ||.$$

It is easy to show (see Exercise 11) that this function satisfies the above four axioms; it is therefore called a *metric*. A metric generalizes our idea of distance in the same sense that a norm generalizes our idea of length. Exercise 11 shows that every norm yields a metric by (11.6–1). Exercises 12 and 13 show that there can be metrics which do not come from any norm. This suggests that the relationship between norms and metrics might be interesting to explore. Since the exploration would be a digression from our main purpose, we shall not pursue it further. Our immediate goal is to define a norm on $\mathrm{Hom}(R^n, R^m)$, and to obtain a metric from it by (11.6–1). We shall deal with this in § 11.8. First, however, we must discuss the continuity of linear transformations.

11.7   Open sets and continuity

If E is a set and G is a set, then the set consisting of all objects which belong to E or to G, or to both, is called the *union* of E and G, which we denote by $E \cup G$. The set consisting of all members of both E and G is denoted by $E \cap G$ and is called the *intersection* of E and G. If $p(x)$ denotes some proposition involving x, then $\{x : p(x)\}$ denotes that set consisting of all x for which the proposition $p(x)$ is true. For example, $\{x : x \cdot a = 3\}$ is the set of all vectors whose inner product with the vector a is 3. Similarly, $\{x \in E : p(x)\}$ is that set of vectors belonging to the set E for which the proposition $p(x)$ is true.

For any vector space having a norm, we define the sphere of radius r centered at the vector a, denoted by $S(a, r)$, as follows:

$$S(a, r) = \{x : ||\, x - a\, || = r\}.$$

The *open ball*, $B(a, r)$, of radius r, centered at a, is given by

$$B(a, r) = \{x : ||\, x - a\, || < r\}.$$

The corresponding *closed ball*, $\overline{B(a, r)}$, is the union of these two sets; that is,

$$\overline{B(a, r)} = B(a, r) \cup S(a, r) = \{x : ||x - a|| \leq r\}.$$

Notice that in R^3, $S(a, r)$ is a spherical surface and $B(a, r)$ consists of the set of points lying inside this surface. $\overline{B(a, r)}$ is what might be called the solid sphere made up of the surface, together with the set of points lying inside. How would you describe $S(a, r)$, $B(a, r)$, and $\overline{B(a, r)}$ in R^2? in R? Notice that $S(a, r) \cap B(a, r)$ is the empty set.

If E is any subset of a normed vector space, we define an *interior point of* E to be a point which belongs to E and which is the center of some open ball contained in E. Notice that this is just a more general statement of our earlier definition in § 5.1. An *open set* can still be defined as one consisting entirely of interior points, and a *closed set* is still the complement of an open set.

By a *neighborhood* of a point, we shall simply mean an open set containing the point. This is a generalization of our earlier usage where the sets which we called neighborhoods were open intervals, open disks, or open rectangles. The really essential property of whatever we call a neighborhood of a point is that it be a set for which the point is an interior point. The definition which we shall use is the simplest way to get this property.

A subset E of a normed vector space is said to be *bounded* in case there is some number M such that $||x|| < M$ for all $x \in E$.

Let $\{x^k\}_{k=1}^{\infty}$ denote a sequence in a vector space $\mathcal{V}$ having a norm $||\ ||$. The superscript is not an exponent but is simply an index giving the order of the term in the sequence. As one would expect from the theory of sequences of real numbers (see § 1.62),

$$\lim_{k \to \infty} x^k = a$$

means that $a \in \mathcal{V}$ and if ϵ is any positive number, then there exists a positive integer N such that $||x^k - a|| < \epsilon$ whenever $N \leq k$. This is also expressed by saying that the sequence $\{x^k\}_{k=1}^{\infty}$ converges to a. In terms of our recently introduced notation, it can be expressed by saying that if ϵ is any positive number, then all but at most a finite number of the terms of the sequence lie in the ϵ-ball, $B(a, \epsilon)$, centered at a.

The definition of continuity still makes sense in all vector spaces having norms. Suppose that $\mathcal{X}$ and $\mathcal{Y}$ are normed vector spaces and f is a function from some subset D of $\mathcal{X}$ to $\mathcal{Y}$. To say that f is continuous at the point a of its domain D means that if ϵ is any positive number, then there exists some positive number δ such that

$$||f(a) - f(x)|| < \epsilon \text{ for all } x \in D \text{ such that } ||x - a|| < \delta.$$

In geometrical terms this says that if $B(f(a), \epsilon)$ is any open ball, centered at $f(a)$, in $\mathcal{Y}$, then there is some open ball $B(a, \delta)$, centered at a, in $\mathcal{X}$ such

that **f** maps the intersection $D \cap B(a, \delta)$ into $B(f(a), \epsilon)$. If **a** happens to be
an interior point of D, then we don't have to talk about the intersection of
D with $B(a, \delta)$—we can simply say that for some δ, **f** maps $B(a, \delta)$ into
$B(f(a), \epsilon)$. Notice that the number δ may depend on both **a** and ϵ. See Fig. 81
for a diagram which may help to picture the idea of continuity.

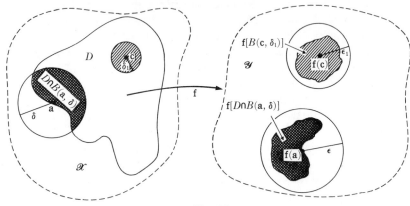

Fig. 81

Notice that since continuity is defined with the help of the norm, the
norm is automatically a continuous function. In fact, to say that the norm
is continuous at **a** means only that if ϵ is any positive number, there is
some positive number δ such that $\left| ||x|| - ||a|| \right| < \epsilon$ for all **x** such
that $||x - a|| < \delta$. By (11.5-2), $\left| ||x|| - ||a|| \right| \leq ||x - a||$, which
shows that we can take δ to be ϵ.

In § 5.3 we stated that the composition of continuous functions is continuous. We shall prove here a form of this statement which is appropriate
for our purposes.

THEOREM II. *Suppose that **f** is a function from some subset of $\mathcal{X}$ to $\mathcal{Y}$
and that **g** goes from some subset of $\mathcal{Y}$ to $\mathcal{Z}$, where $\mathcal{X}$, $\mathcal{Y}$, and $\mathcal{Z}$ are
normed vector spaces. Suppose further that **f** is defined at **a** and that **g** is
defined at $f(a) = b$. If **f** is continuous at **a** and **g** is continuous at **b**, then
the composition $g \circ f = \phi$ is continuous at **a**.*

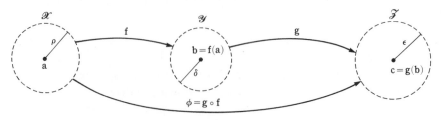

Fig. 82

PROOF. Let $g(b) = c$. Since g is continuous at b, if $B(c, \epsilon)$ is any open ball of radius $\epsilon > 0$, centered at c, there exists some $\delta > 0$ such that $g(y) \in B(c, \epsilon)$ for all y in $B(b, \delta)$ at which g is defined. See Fig. 82.

Since f is continuous at a, there is some $\rho > 0$ such that $f(x) \in B(b, \delta)$ for all $x \in B(a, \rho)$ at which f is defined. Therefore, for all those x in $B(a, \rho)$ at which ϕ is defined, $\phi(x) \in B(c, \epsilon)$.

Observe that the points of $B(a, \rho)$ at which ϕ is defined are those points x at which f is defined and for which $f(x)$ belongs to the domain of g.

THEOREM III. *If T is a linear transformation from R^n to R^m, then there exists some number M such that*

$$|| Tx || \leq M || x || \qquad \text{for all } x \in R^n.$$

PROOF. Let the standard matrix representation of T be the $m \times n$ matrix A. Then for each $x \in R^n$,

$$Tx = \begin{pmatrix} a_{11} & a_{12} & \cdots & a_{1m} \\ a_{21} & a_{22} & \cdots & a_{2m} \\ \vdots & \vdots & \vdots & \vdots \\ a_{m1} & a_{m2} & \cdots & a_{mn} \end{pmatrix} \begin{pmatrix} x_1 \\ x_2 \\ \vdots \\ x_n \end{pmatrix} = \begin{pmatrix} y_1 \\ y_2 \\ \vdots \\ y_m \end{pmatrix}.$$

$y = (y_1, y_2, \ldots, y_m)$ is that vector in R^m whose ith component y_i is given by

$$y_i = \sum_{j=1}^{n} a_{ij} x_j.$$

By the Cauchy inequality,

$$| y_i | \leq \left(\sum_{j=1}^{n} a_{ij}^2 \right)^{1/2} \left(\sum_{1}^{n} x_j^2 \right)^{1/2} = \left(\sum_{j=1}^{n} a_{ij}^2 \right)^{1/2} || x ||.$$

Now, let Q denote the maximum of the m numbers

$$\left(\sum_{j=1}^{n} a_{ij}^2 \right)^{1/2} \qquad (i = 1, 2, \cdots, m).$$

Then $| y_i | \leq Q || x ||$ for all i and $|| Ax || = || y || =$

$$\sqrt{y_1^2 + y_2^2 + \cdots + y_m^2} \leq \sqrt{mQ^2 || x ||^2} = \sqrt{m} Q || x ||.$$

This proves the theorem and shows that one possible choice for M is $\sqrt{m} Q$.

A transformation T for which there exists a number M such that $|| Tx || \leq M || x ||$ for all x is said to be *bounded*, and M is called a bound for the transformation. Later, we shall find it useful to have a bound M for a linear transformation in terms of the dominant element (or elements) of its matrix representation, A. Let

$$K = \max | a_{ij} |,$$

where $i \in (1, 2, \cdots , m)$ and $j \in (1, 2, \cdots , n)$. Then,

$$Q \leq \sqrt{nK^2} = K\sqrt{n}.$$

Therefore we have that

(11.7–1) $$|| Tx || \leq \sqrt{mn}\, K\, ||\, x\, ||$$

for all $x \in R^n$.

COROLLARY. *If T is a linear transformation from R^n to R^m, then T is continuous.*

PROOF. By the preceding theorem, T is bounded; that is, there is some number M such that $||\, Tx\, || \leq M\, ||\, x\, ||$ for all $x \in R^n$. Let y be any vector in R^n such that $||\, x - y\, || < \epsilon/M$. Then,

$$||\, Tx - Ty\, || = ||\, T(x - y)\, || \leq M\, ||\, x - y\, || < M\, \frac{\epsilon}{M} = \epsilon.$$

This says that for any open ball, centered at Tx, in R^m, there is some open ball, centered at x in R^n such that T maps all points of the second ball into the first. This is what is meant by continuity of T at x. Since x could be any point in R^n, we have proved that T is continuous everywhere.

One further fact is worth mentioning. Once the radius ϵ of the ball in R^m was chosen, we could see that a suitable radius for the ball centered at x in R^n was ϵ/M. This choice of radius does not depend on x—it is valid for every x. This implies more than merely the fact that T is continuous at every point, for T would be continuous even if the choice of suitable radius were dependent on x. What we have here is a special kind of continuity, known as *uniform continuity*.

11.8   A norm on Hom(R^n, R^m)

It is not at all obvious what is meant by the magnitude of a linear transformation. An approach which turns out to be quite useful in various connections is based on what the transformation T does to the lengths of vectors. Of course, T will usually have different effects on different vectors —lengthening some, shortening others, and leaving the lengths of others unchanged. To get around this complication, we might try defining the norm of T just in terms of its effect on that vector which it stretches the most; that is, we might consider the ratio $||\, Tx\, ||/||\, x\, ||$ and take the norm of T to be the maximum as x ranges over the space, excluding, of course, the zero vector where the ratio is not defined. This approach can be simplified by noticing that

$$\frac{||\, T(x)\, ||}{||\, x\, ||} = T\!\left(\frac{x}{||\, x\, ||}\right)$$

since T is linear and $x/\|x\|$ is some unit vector. This means that the set of values taken on by the ratio $\|Tx\|/\|x\|$ for all nonzero vectors is exactly the same as the set of values taken on by the function $\|T(u)\|$ when u is restricted to lie just on the unit sphere (the set of all vectors having norm 1).

Now, let T be any linear transformation from R^n to R^m. We want to show that

(11.8-1)
$$\|T\| = \max_{\|x\|=1} \|Tx\|$$

defines a norm on the vector space Hom(R^n, R^m). But first notice that $\|\ \ \|$ carries a different meaning in each of its three appearances in (11.8-1). Since $x = (x_1, x_2, \cdots, x_n) \in R^n$, $\|x\|$ involves the length function on R^n. $\|Tx\|$, on the other hand, refers to the length function on R^m, and on the left-hand side, $\|\ \ \|$ is a function from Hom(R^n, R^m) to the real numbers. It is general practice to use the same symbol, $\|\ \ \|$, for all norms and leave it to the reader to infer from the context which norm is meant.

Since both T and the norm are continuous functions, their composition, defined by $\|T(x)\|$, is a continuous function from R^n to R by Theorem II, § 11.7. The unit sphere in R^n is both closed and bounded (Exercise 15), so by Theorems I and II of § 5.3 the above function has a maximum value on this set. This shows that for each T,

$$\max_{\|x\|=1} \|T(x)\|$$

exists and therefore (11.8-1) does define a function from Hom(R^n, R^m) to the real numbers. It is obviously nonnegative, since the maximum of a nonnegative function is nonnegative. If T_0 is the zero element of Hom(R^n, R^m), then $T_0(x) = 0$ for all x; hence,

$$\|T_0\| = \max_{\|x\|=1} \|T_0(x)\| = 0.$$

If T is not the zero element, then there is some vector v in R^n such that $T(v) \neq 0$. Then $u = v/\|v\|$ is a unit vector such that

$$T(u) = T\left(\frac{v}{\|v\|}\right) = \frac{1}{\|v\|} T(v) \neq 0$$

$$\therefore \|T(u)\| > 0.$$

$$\therefore \|T\| = \max_{\|x\|=1} \|T(x)\| \geq \|T(u)\| > 0.$$

We have now proved that the function $\|\ \ \|$ defined by (11.8-1) satisfies Property (1) and (2) of § 11.5. The fact that it satisfies (3) is easy to prove and will be left as an exercise for the reader. To prove that this function satisfies the fourth property required of every norm is slightly more difficult.

What we must prove is that for all T_1 and T_2 belonging to $\text{Hom}(R^n, R^m)$,

$$\max_{||x||=1} \; ||T_1(x) + T_2(x)|| \leq \max_{||x||=1} \; ||T_1(x)||$$
$$+ \max_{||x||=1} \; ||T_2(x)||.$$

We know that for all $x \in R^n$,

$$||T_1(x) + T_2(x)|| \leq ||T_1(x)|| + ||T_2(x)||,$$

because $||\;\;||$ here is a norm on R^m. Therefore, the above inequality is certainly true for unit vectors. Hence, it follows that for all x on the unit sphere (i.e., for all x such that $||x|| = 1$),

$$||T_1(x) + T_2(x)|| \leq \max_{||x||=1} \; ||T_1(x)|| + \max_{||x||=1} \; ||T_2(x)||.$$

This says that for *all* unit vectors,

$$||T_1(x) + T_2(x)|| \leq ||T_1|| + ||T_2||.$$
$$\therefore \max_{||x||=1} \; ||T_1(x) + T_2(x)|| \leq ||T_1|| + ||T_2||.$$

But

$$\max_{||x||=1} \; ||T_1(x) + T_2(x)|| = ||T_1 + T_2||.$$
$$\therefore ||T_1 + T_2|| \leq ||T_1|| + ||T_2||.$$

This proves that the function defined by (11.8–1) is a norm on $\text{Hom}(R^n, R^m)$. An important property of this norm is that for all $x \in R^n$,

(11.8–2) $$||Tx|| \leq ||T|| \cdot ||x||$$

for each $T \in \text{Hom}(R^n, R^m)$. The proof of this fact (Exercise 14) is short. The reader should pause long enough now to distinguish clearly among the three meanings which the norm symbol has in (11.8–2).

The norm which we have defined for linear transformations is useful even though we have no way of evaluating it except in certain special cases. It will be sufficient for our purposes to have a practical bound for the norm, and we have already obtained one in the process of proving Theorem III of § 11.7. It was found in (11.7–1) that if $T \in \text{Hom}(R^n, R^m)$ and A is an $m \times n$ matrix representing T, then

$$||T|| \leq \sqrt{mn}\, K,$$

where $K = \max\limits_{\substack{1 \leq i \leq m \\ 1 \leq j \leq n}} |a_{ij}|$.

We get an important lower bound for $||T||$ by considering what T does to the unit vectors making up the standard basis in R^n. Recall that for each $r \in \{1,2, \cdots ,n\}$, $\mathbf{e}_r$ is that ordered n-tuple consisting of 1 in the rth place and 0's in the other $n - 1$ places.

$$\left|\left| T\mathbf{e}_r \right|\right| = \left|\left| A\mathbf{e}_r \right|\right| = \left|\left| (a_{1r},a_{2r}, \cdots ,a_{mr}) \right|\right| = \left(\sum_{i=1}^{m} a_{ir}^2 \right)^{1/2} = C_r.$$

C_r can be thought of as the length of the vector represented by the rth column in A. Let $C = \max C_r$ for $r \in \{1,2, \cdots ,n\}$. In other words, C is the length of the longest of the n column vectors in A. Since $\left|\left| Tu \right|\right| = C$ for at least one unit vector, $\left|\left| T \right|\right| \geq C$. From the definition of C_r we see that C_r is greater than or equal to $\max_i \left| a_{ir} \right|$. Hence, $C = \max_r C_r$ is greater than or equal to $\max_r(\max_i \left| a_{ir} \right|) = K$, and so $\left|\left| T \right|\right| \geq C \geq K$. Combining this with our previously obtained upper bound, we have

(11.8–3) $$K \leq \left|\left| T \right|\right| \leq \sqrt{mn} \, K.$$

The space $\mathrm{Hom}(R^n, R)$ is an interesting one from the point of view of computing norms. Recall that the elements of this space are called linear functionals and that each one of them is represented by a unique vector in R^n. Suppose that T is a linear functional on R^n and that $\mathbf{a} = (a_1,a_2, \cdots ,a_n)$ is its representing vector. The norm of $\mathbf{a}$, of course, is $\sqrt{a_1^2 + a_2^2 + \cdots + a_n^2}$. The norm of T is the maximum value of $\left|\left| Tx \right|\right|$ on the unit sphere, that is, on the set of vectors having unit norm in R^n.

$$\left|\left| T \right|\right| = \max_{||\mathbf{x}||=1} \left|\left| T\mathbf{x} \right|\right| = \max_{||\mathbf{x}||=1} \left| \mathbf{a} \cdot \mathbf{x} \right| =$$
$$\max_{||\mathbf{x}||=1} \left| a_1x_1 + \cdots + a_nx_n \right|.$$

By the Cauchy inequality,

$$\left| a_1x_1 + \cdots + a_nx_n \right| \leq \left(\sum_{1}^{n} a_i^2 \right)^{1/2} \left|\left| \mathbf{x} \right|\right| = \left|\left| \mathbf{a} \right|\right| \cdot \left|\left| \mathbf{x} \right|\right|.$$

So for all unit vectors, $\left|\left| Tx \right|\right| \leq \left|\left| \mathbf{a} \right|\right|$. Assuming that T is not the zero functional, $\mathbf{a}$ is not the zero vector and $\mathbf{u} = \mathbf{a}/\left|\left| \mathbf{a} \right|\right|$ is a unit vector. A simple calculation shows that $\left|\left| Tu \right|\right| = \left|\left| \mathbf{a} \right|\right|$. Therefore,

$$\max_{||\mathbf{x}||=1} \left|\left| Tx \right|\right| = \left|\left| \mathbf{a} \right|\right|, \text{ and hence, } \left|\left| T \right|\right| = \left|\left| \mathbf{a} \right|\right|.$$

This says that the norm of each nonzero linear functional on R^n is the same as the R^n norm of its representing vector. Since this statement about equality of norms is obviously also true for the zero functional, it is true in general.

11.9   End R^n

In the important special case where m and n are the same, $\mathrm{Hom}(R^n, R^n)$ is abbreviated to End R^n, and it denotes the vector space of all linear transformations from R^n to R^n. (End is an abbreviation for *endomorphism*.) Linear transformations from R^n to R^n are called *linear operators*. From Theorem I we know that they are represented by $n \times n$ matrices. An important subset of End R^n consists of the *invertible* operators. These are one-to-one functions and are also described as *nonsingular* or *reversible*. Saying that T is an invertible operator means that for each y in R^n, there is a unique x in R^n such that $Tx = y$. The unique x associated with each y in this way is denoted by $T^{-1}y$. This function T^{-1} is called the inverse of T. The linearity of T implies that of T^{-1} (Exercise 3); therefore, if T is an invertible member of End R^n, T^{-1} also belongs to End R^n.

One way of telling whether a linear operator is invertible is to look at the equation $Tx = y$ in matrix form. If A is a matrix representing T, we have $Ax = y$, or in scalar form,

$$\begin{pmatrix} a_{11}x_1 + a_{12}x_2 + \cdots + a_{1n}x_n \\ a_{21}x_1 + a_{22}x_2 + \cdots + a_{2n}x_n \\ \vdots \quad \vdots \quad \vdots \quad \vdots \\ a_{n1}x_1 + a_{n2}x_2 + \cdots + a_{nn}x_n \end{pmatrix} = \begin{pmatrix} y_1 \\ y_2 \\ \vdots \\ y_n \end{pmatrix}$$

We shall assume that the reader has had enough experience with simultaneous linear systems like this to know that: *For each y in R^n, a necessary and sufficient condition that there exist a unique solution x is that the determinant of the coefficient matrix be different from zero.* Such matrices are called *nonsingular* or *invertible*. So the invertible linear operators are those which are represented by invertible (or nonsingular) matrices.

Another important characterization of invertible operators is obtained by applying the italicized proposition of the preceding paragraph to the special case where all the y's in the above system are zero. In this case, the linear system is said to be *homogenous*. It is obvious at a glance that a homogenous system has the trivial solution $x_1 = x_2 = \cdots = x_n = 0$. If the determinant is different from zero, this trivial solution must be the only one—in other words, a nonsingular linear operator T maps only the zero vector into the zero vector. But if the determinant is zero, that is, if the operator is singular, then the trivial solution is not unique and this means that T maps some nonzero vector into the zero vector. *Hence, a necessary and sufficient condition that a linear operator be invertible is that it map only the zero vector into the zero vector.* It therefore follows that if a linear operator maps only the zero vector into the zero vector, then it must be invertible, and this is the property which we shall use in § 11.10 to prove that certain elements of End R^n are invertible.

Suppose that T is invertible. We have defined $T^{-1}y$ to be the unique solution of the equation $Tx = y$. Since there is a unique solution for all y, we have

$$T(T^{-1}y) = y \qquad \text{for all } y.$$

This means that the composition $T \circ T^{-1} = I$, where I is the identity transformation (i.e., I maps each vector into itself).

As in the case with all invertible functions, T^{-1} is also invertible, and in fact, $(T^{-1})^{-1} = T$. This means that for all y, the equation $T^{-1}x = y$ has the unique solution Ty (i.e., $T^{-1}(Ty) = y$ for all y). Another way to say this is that $T^{-1} \circ T = I$. This and the preceding paragraph together give us that

$$T \circ T^{-1} = T^{-1} \circ T = I.$$

In other words, every invertible linear operator commutes with its inverse and their composition is the identity operator. Conversely, it is easy to show (Exercise 20) that if two linear operators commute and their composition is the identity operator, then each is the inverse of the other. From this and the fact that composition is associative (Exercise 18), it is easy to prove (Exercise 19) that if T and L are invertible then their composition is invertible, and that

$$(T \circ L)^{-1} = L^{-1} \circ T^{-1}.$$

That is, the inverse of the composition of two invertible operators is the composition of their inverses *in the opposite order.*

Suppose that A and A^{-1} are the matrix representations of T and T^{-1}, respectively. Now, matrix multiplication was defined in such a way as to make the product of two matrices represent the composition, in the same order, of the operators which they represent; it follows, therefore, that

$$AA^{-1} = A^{-1}A = I.$$

Here we have used the same symbol, I, to represent both the identity operator and the so-called identity matrix which represents it. Thus, every nonsingular matrix commutes with its inverse, and the product is the identity matrix. From the fact that

$$Ix = x \qquad \text{for all } x,$$

it is easily seen that I is the $n \times n$ matrix with 1's along the main diagonal and 0's elsewhere.

If L and T are any two members of End R^n, then $L \circ T \in$ End R^n and by (11.8–2)

$$||(L \circ T)(x)|| = ||L(T(x))|| \leq ||L|| \cdot ||T(x)|| \qquad \text{for all } x.$$

Using (11.8–2) again, we get

$$||(L \circ T)(x)|| \leq ||L|| \; ||T|| \; ||x|| \qquad \text{for all } x \in R^n.$$

By the way we defined the norm of a linear transformation, this says that

(11.9–1) $$||L \circ T|| \leq ||L|| \; ||T||.$$

In words, the norm of the composition of two linear operators is less than or equal to the product of their norms. Suppose now that T happens to be invertible. Then $T \circ T^{-1} = I$, and since it is obvious that $||I|| = 1$, (11.9–1) gives

$$(11.9\text{–}2) \qquad ||T|| \ \ ||T^{-1}|| \geq 1,$$

or, the norm of the inverse of a linear operator is greater than or equal to the reciprocal of the norm of the operator.

11.10   The set of invertible operators

Since we have defined a norm on the vector space End R^n, we know what is meant by an open set in this space. It will be of importance to us in the next chapter to know that the set of all the invertible operators, which we shall denote by Ω, constitutes an open subset of End R^n. The following lemma is helpful in proving this:

LEMMA. *If* $T \in$ *End* R^n *and* $||T|| < 1$, *then* $I - T$ *is invertible, and*

$$||(I - T)^{-1}|| \leq \frac{1}{1 - ||T||} .$$

PROOF. Our plan is to show that if x is a vector of positive norm, then $(I - T)(x)$ also has positive norm. This will imply that $(I - T)$ maps nonzero vectors into nonzero vectors and is therefore invertible. So begin with the assumption that $||x|| > 0$. Then, with the aid of (11.5–2) we see that

$$||(I - T)x|| = ||Ix - Tx|| \geq \Big| ||Ix|| - ||Tx|| \Big| .$$

Also,

$$||Ix|| = ||x|| \qquad \text{and} \qquad ||Tx|| \leq ||T|| \ \ ||x||.$$

Therefore,

$$(11.10\text{–}1) \qquad ||(I - T)x|| \geq (1 - ||T||)||x||,$$

and since $||T|| < 1$, $||(I - T)x|| > 0$ for all x different from zero, and hence, $I - T$ is invertible.

Knowing that $I - T$ is invertible, we can write

$$y = (I - T)[(I - T)^{-1}y] \qquad \text{for all } y \text{ in } R^n.$$

Since (11.10–1) holds for all x in R^n, it holds in particular when $x = (I - T)^{-1}y$.

$$\therefore \ ||y|| = ||(I - T)[(I - T)^{-1}y]|| \geq (1 - ||T||)||(I - T)^{-1}y||,$$

and hence,

$$||(I - T)^{-1}\mathbf{y}|| \leq \frac{||\mathbf{y}||}{1 - ||T||} \qquad \text{for all } \mathbf{y} \text{ in } \mathbf{R}^n.$$

This implies that $||(I - T)^{-1}|| \leq 1/(1 - ||T||)$.

COROLLARY. *If* $||I - T|| < 1$, *then* T *is invertible.*

PROOF. By the lemma, $I - (I - T)$ is invertible, and $I - (I - T) = T$. This corollary shows that the identity transformation I is an interior point of Ω because the entire open ball of radius 1, centered at I, is contained in Ω.

THEOREM IV. *The set* Ω *of invertible linear operators is an open subset of End* $\mathbf{R}^n$.

PROOF. What we want to do is to show that every point of Ω is an interior point. In other words, we wish to show that if $T \in \Omega$, then there is some positive number ρ such that all operators within the distance ρ of T are invertible. We shall demonstrate how ρ may be chosen.

Since T is invertible, T^{-1} exists and we can write for all $L \in$ End $\mathbf{R}^n$,

$$L = T - (T - L) = T \circ [I - T^{-1} \circ (T - L)].$$

By (11.9-1), $||T^{-1} \circ (T - L)|| \leq ||T^{-1}|| \, ||T - L||$. Therefore, if we require that $||T - L|| < 1/||T^{-1}||$, then $||T^{-1} \circ (T - L)|| < 1$, and by the lemma, $I - T^{-1} \circ (T - L)$ is invertible. T is invertible by hypothesis, and since the composition of invertible operators is invertible,

$$L = T \circ [I - T^{-1} \circ (T - L)]$$

is invertible. In other words, the positive number ρ, which we are seeking, can be taken to be $1/||T^{-1}||$. Each invertible operator T is the center of an open ball of radius $1/||T^{-1}||$ which is wholly contained in Ω. This proves that every point of Ω is an interior point.

THEOREM V. *The function from* Ω *to* Ω *which associates with each operator its inverse is continuous.*

PROOF. As in the preceding theorem, the main thing is to get clearly in mind what it is we are trying to prove. The continuity asserted in this theorem is continuity at each point of Ω. Let T_0 denote any element of Ω. We know from Theorem IV that T_0 must be an interior point of Ω. So from the explanation of continuity in § 11.7 we see that our job is to prove that if ϵ is any positive number, there is some positive number δ (depending perhaps on both T_0 and ϵ) such that $||\, T^{-1} - T_0^{-1}\,|| < \epsilon$ for every operator satisfying the inequality $||\, T - T_0\,|| < \delta$.

From the outset we shall restrict attention to those operators T for which

$$(11.10\text{-}2) \qquad\qquad ||\, T - T_0\,|| < \frac{1}{||\, T_0^{-1}\,||} .$$

In the proof of Theorem IV it is shown that these are all invertible. Furthermore, $||\, T_0^{-1}\,(T_0 - T)\,|| \leq ||\, T_0^{-1}\,||\ ||\, T_0 - T\,|| < 1$, so by the Lemma at the beginning of this section, $I - T_0^{-1}(T_0 - T)$ is invertible and

$$(11.10\text{-}3) \quad ||\, [I - T_0^{-1}(T_0 - T)]^{-1}\,|| \leq \frac{1}{1 - ||\, T_0^{-1}(T_0 - T)\,||} .$$

It is easy to verify that

$$T^{-1} - T_0^{-1} = T_0^{-1}(T_0 - T)T^{-1},$$

from which we obtain the following results:

$$(11.10\text{-}4)$$
$$||\, T^{-1} - T_0^{-1}\,|| \leq ||\, T_0^{-1}\,||\ ||\, T_0 - T\,||\ ||\, T^{-1}\,||,\ \text{and}$$

$$(11.10\text{-}5) \qquad T^{-1} = [I - T_0^{-1}(T_0 - T)]^{-1}T_0^{-1}.$$

From (11.10-4) and (11.10-5) it follows that

$$||\, T^{-1} - T_0^{-1}\,|| \leq$$
$$||\, T_0^{-1}\,||\ ||\, T_0 - T\,||\ ||\, [I - T_0^{-1}(T_0 - T)]^{-1}\,||\ ||\, T_0^{-1}\,||.$$

Using (11.10-3) with this, we see that

$$(11.10\text{-}6) \qquad ||\, T^{-1} - T_0^{-1}\,|| \leq \frac{||\, T_0^{-1}\,||^2\,||\, T_0 - T\,||}{1 - ||\, T_0^{-1}(T_0 - T)\,||} .$$

We simplify this one step further by replacing $||\, T_0^{-1}(T_0 - T)\,||$ by $||\, T_0^{-1}\,||\,||\, T_0 - T\,||$ in the denominator on the right. This is permissible, since $||\, T_0^{-1}(T_0 - T)\,|| \leq ||\, T_0^{-1}\,||\,||\, T_0 - T\,|| < 1$, and the substitution of a larger quantity in this way makes the denominator smaller and the fraction larger. Thus

$$(11.10\text{-}7) \qquad ||\, T^{-1} - T_0^{-1}\,|| \leq \frac{||\, T_0^{-1}\,||^2\,||\, T_0 - T\,||}{1 - ||\, T_0^{-1}\,||\,||\, T_0 - T\,||} .$$

This result puts us in position to complete the proof, for it is clear that when $||\, T_0 - T\,||$ is made to approach zero, the expression on the right-hand side approaches zero. To be explicit however, we want to find a δ such that $||\, T - T_0\,|| < \delta$ will guarantee $||\, T^{-1} - T_0^{-1}\,|| < \epsilon$. An adequate choice of δ can be made as follows:

Let δ be the smaller of the numbers

$$\frac{1}{2\,||\, T_0^{-1}\,||}, \qquad \frac{\epsilon}{2\,||\, T_0^{-1}\,||^2} .$$

Then certainly (11.10–2) will be satisfied, so our argument as far as (11.10–7) is valid. But then also

$$|| T_0^{-1} ||^2 || T_0 - T || < \frac{\epsilon}{2}$$

in the numerator on the right in (11.10–7) and

$$|| T_0^{-1} || \ || T_0 - T || < \tfrac{1}{2},$$

so the denominator on the right in (11.10–7) is larger than $\tfrac{1}{2}$. Therefore,

$$|| T^{-1} - T_0^{-1} || < \frac{\epsilon/2}{1/2} = \epsilon.$$

This completes the proof that the function which maps T into T^{-1} is continuous. Is it also linear? Is Ω a subspace of End R^n?

EXERCISES

1. Let A be an $m \times n$ matrix. Show that the function $\mathbf{f} : R^n \to R^m$ defined by $\mathbf{f}(\mathbf{x}) = A\mathbf{x}$ is a linear transformation.

2. Given that $T \in \mathrm{Hom}(R^n, R^m)$ and $L \in \mathrm{Hom}(R^m, R^p)$, prove directly from the definition of linearity that $L \circ T \in \mathrm{Hom}(R^n, R^p)$.

3. Prove directly from the definition of a linear transformation that the inverse of an invertible linear operator is a linear operator.

4. A *permutation* of a set is a reversible function which maps that set onto itself. Let $\{k_i\}_{i=1}^n$ be a permutation of the first n positive integers, and let $\mathbf{f}$ denote that function from R^n to R^n which maps the vector $\{x_k\}_1^n$ into $\{x_{k_i}\}_{i=1}^n$. Prove that $\mathbf{f}$ is a linear operator on R^n and find its standard matrix representation, A. Find A^{-1}. Students who have had a course in modern algebra will recall that the set of all permutations on a set of n objects is called the *symmetric group of degree n*. The set of matrices which we have associated with this group is called a *representation* of the group. They also form a group, and it is easy to prove that the close relationship between the permutation group and the group of matrices is an *isomorphism*.

5. Using the Cauchy inequality (Exercise 29, § 6.8), prove that if $\mathbf{x}$ and $\mathbf{y}$ belong to R^n, then

$$|| \mathbf{x} + \mathbf{y} || \leq || \mathbf{x} || + || \mathbf{y} || .$$

What condition is both necessary and sufficient for the equality?

6. For each $\mathbf{x} \in R^n$, let

$$|| \mathbf{x} ||_p = \left(\sum_{i=1}^n | x_i |^p \right)^{1/p}.$$

Show that for each $p \geq 1$, $|| \ \ ||_p$ is a norm on R^n. HINT: Use Minkowski's inequality (Exercise 32, § 6.8).

7. Let $|| \ \ ||_1$ and $|| \ \ ||_2$ be two functions from R^n to R defined as follows:

$$|| \ x \ ||_1 = \sum_{i=1}^{n} | \ x_i \ |$$

$$|| \ x \ ||_2 = \max_{1 \leq i \leq n} | \ x_i \ | \ .$$

Show that $|| \ \ ||_1$ and $|| \ \ ||_2$ are norms on R^n. For the special case where $n = 2$, draw the unit sphere (circle) in these two norms on the same coordinate axes with the unit sphere in the Euclidean norm.

8. If $|| \ \ ||$ is the usual (Euclidean) norm on R^n and $|| \ \ ||_1$ is the first norm defined in Exercise 7, prove that there exist numbers α and β such that

$$\alpha \leq \frac{|| \ x \ ||_1}{|| \ x \ ||} \leq \beta \qquad \text{for all } x \neq 0.$$

α and/or β may depend on n but not on x.

9. Prove that if f and g are continuous, real-valued functions on $[a, b]$, then

$$\left| \int_{a}^{b} f(x)g(x)dx \right| \leq \left(\int_{a}^{b} f^2(x)dx \right)^{1/2} \left(\int_{a}^{b} g^2(x)dx \right)^{1/2}.$$

What condition is both necessary and sufficient for equality? HINT: Since

$$\int_{a}^{b} [\lambda f(x) + g(x)]^2 dx \geq 0 \qquad \text{for all } \lambda,$$

the quadratic equation in λ,

$$\lambda^2 \int_{a}^{b} f^2(x)dx + 2\lambda \int_{a}^{b} f(x)g(x)dx + \int_{a}^{b} g^2(x)dx = 0$$

cannot have two distinct real roots, and therefore the discriminant cannot be positive.

This inequality, which is reminiscent of the Cauchy inequality, is known as the Schwarz inequality. Notice that the Cauchy inequality can be proved by the method used here, starting from the fact that $\sum_{i=1}^{n} (\lambda a_i + b_i)^2 = 0$ can be written as a quadratic equation in λ which obviously cannot have distinct real roots. The similarity between these two equalities has led many authors to lump them together under the same name—the Cauchy-Schwarz inequality.

We have assumed that f, g, f^2, g^2, fg, and $(\lambda f + g)^2$ are all integrable. These things are proved in Chapter XVIII.

10. Let $\mathcal{C}$ denote the vector space of continuous functions on $[0, 1]$ (see Example 2, § 11) and let $|| \ \ ||_1$ and $|| \ \ ||_2$ be functions from $\mathcal{C}$ to R defined as follows:

$$|| f ||_1 = \max_{0 \leq x \leq 1} | f(x) | \ ,$$

$$|| f ||_2 = \left[\int_{0}^{1} f^2(x)dx \right]^{1/2}.$$

Prove that $||\ \ ||_1$ and $||\ \ ||_2$ are norms on $\mathcal{C}$. HINT: For $||\ \ ||_2$ use Exercise 9.

11. Prove that if $||\ \ ||$ is any norm on a vector space, $\mathcal{V}$, then the function d defined by

$$d(\mathbf{x}, \mathbf{y}) = ||\ \mathbf{x} - \mathbf{y}\ ||\qquad \text{for all } \mathbf{x} \text{ and } \mathbf{y} \text{ in } \mathcal{V}$$

is a metric.

12. Prove that if $\mathcal{V}$ is any vector space and d is the function defined by

$$
\begin{aligned}
d(\mathbf{x}, \mathbf{y}) &= 0 \qquad \text{if } \mathbf{x} = \mathbf{y}, \qquad \text{and}\\
d(\mathbf{x}, \mathbf{y}) &= 1 \qquad \text{otherwise,}
\end{aligned}
$$

then d is a metric. Prove that there cannot exist any norm, $||\ \ ||$, on $\mathcal{V}$ such that this particular metric is given by $||\ \mathbf{x} - \mathbf{y}\ ||$ for all $\mathbf{x}$ and $\mathbf{y}$ in $\mathcal{V}$.

13. Let ρ denote the usual distance function in the plane, $\mathbf{R}^2$, and define

$$d(\mathbf{x}, \mathbf{y}) = \frac{\rho(\mathbf{x}, \mathbf{y})}{1 + \rho(\mathbf{x}, \mathbf{y})}.$$

Show that d is also a metric on the plane. Show that it is not possible to define a norm on the plane such that d is expressible in terms of this norm by (11.6–1).

14. Prove (11.8–2).

15. Let $S(0, 1)$ denote the Euclidean unit sphere in $\mathbf{R}^n$, that is

$$S(0, 1) = \{\mathbf{x} \in \mathbf{R}^n : \sum_{i=1}^{n} x_i{}^2 = 1\}.$$

Prove that $S(0, 1)$ is closed.

16. Let $\{\mathbf{x}^k\}_{k=1}^{\infty}$ denote a sequence in $\mathbf{R}^n$, that is, $\mathbf{x}^k = (x_1{}^k, x_2{}^k, \cdots, x_n{}^k)$. Prove that a necessary and sufficient condition that

$$\lim_{k \to \infty} \mathbf{x}^k = \mathbf{y} = (y_1, y_2, \cdots, y_n)$$

is that $\lim_{k \to \infty} x_i{}^k = y_i$ for $i = 1, 2, \cdots, n$.

17. If $\mathbf{x}$ and $\mathbf{y}$ are any two points in $\mathbf{R}^n$, then the equation of the straight line determined by them can be written

$$\mathbf{z} = \mathbf{x} + t(\mathbf{y} - \mathbf{x}).$$

For a plane you may have thought of this as the parametric representation of the line, t being the parameter. In general, the right-hand side is a function from R to R^n. Notice that the function maps the interval $0 \leq t \leq 1$ into that linear segment between $\mathbf{x}$ and $\mathbf{y}$ inclusive. This means that we can define *the line segment determined by* $\mathbf{x}$ *and* $\mathbf{y}$ to be the set of points of the form $(1 - t)\mathbf{x} + t\mathbf{y}$ for $t \in [0, 1]$. Equivalently, we can define this segment to be the set of points

$$\lambda_1 \mathbf{x} + \lambda_2 \mathbf{y},$$

where $\lambda_1 \geq 0$, $\lambda_2 \geq 0$ and $\lambda_1 + \lambda_2 = 1$. To say that a subset E of a vector space is *convex* means that if $\mathbf{x}$ and $\mathbf{y}$ belong to E, then the line segment determined by $\mathbf{x}$ and $\mathbf{y}$ also lies in E. Prove that every ball—open or closed—is convex.

18. Suppose that $f : A \to B$; $g : B \to C$; and $h : C \to D$. Since composition of functions is a binary operation, we get a function going from A to D by forming either $h \circ (g \circ f)$ or $(h \circ g) \circ f$. Show that these two functions are the same. In other words show that although composition of functions is not necessarily commutative, it is always associative.

19. Using Exercise 18, show that if L and T are invertible linear operators, then $L \circ T$ is invertible, and $(L \circ T)^{-1} = T^{-1} \circ L^{-1}$. That is, the inverse of the composition of two invertible operators is the composition of their inverses *in the opposite order*.

20. Show that if T and L are two members of End R^n such that $T \circ L = L \circ T = I$. then they must be invertible and each is the inverse of the other.

21. Let A and B denote linear operators on R^n. Prove that if B is invertible and A commutes with B, then A commutes with B^{-1}.

22. Show that if $T \in$ End R^n and $||T|| < 1$, then

$$(I - T)^{-1} = I + T + T^2 + \cdots + T^k + T^{k+1}(I - T)^{-1}$$

for every positive integer k.

23. Show that the set of invertible operators is not a bounded subset of End R^n.

24. Show that the function which associates every invertible linear operator on R^n with its inverse is not uniformly continuous.

Differential Calculus of Functions from R^n to R^m **12**

12. Introduction

The purpose of this chapter is to unify much of our previous work by developing differential calculus in the general setting of functions from R^n to R^m. This will include the subjects of transformations introduced in Chapter IX, real-valued functions of several variables (i.e., functions from R^n to R), vector-valued functions of a single variable (i.e., functions from R to R^n), and even the one-variable functions of elementary calculus as special cases. A function from R^n to R^m is an ordered m-tuple of real-valued functions of n real variables, for example,

$$
\begin{aligned}
y_1 &= f^{(1)}(x_1, x_2, \cdots, x_n) \\
y_2 &= f^{(2)}(x_1, x_2, \cdots, x_n) \\
&\;\;\vdots \\
y_m &= f^{(m)}(x_1, x_2, \cdots, x_n).
\end{aligned}
$$

(12–1)

Here we have used superscripts to distinguish the *component functions* (or *coordinate functions*, as they are sometimes called), so that we can continue to use subscripts to denote partial derivatives. The vector $(x_1, x_2, \cdots, x_n)$ of R^n is mapped into the vector $(y_1, y_2, \cdots, y_m)$ of R^m. If we denote the first of these by $\mathbf{x}$ and the second by $\mathbf{y}$, and if we let $\mathbf{f}$ denote the ordered m-tuple of functions $(f^{(1)}, f^{(2)}, \cdots, f^{(m)})$, then we can abbreviate the notation of our function (12–1) to

$$\mathbf{y} = \mathbf{f}(\mathbf{x}).$$

Such functions sometimes go by the name of transformations, as in Chapter IX.

The special case where all of the $f^{(i)}$'s are linear has already been studied, in Chapter XI. In that case, (12–1) reduces to

$$
\begin{aligned}
y_1 &= a_{11}x_1 + a_{12}x_2 + \cdots + a_{1n}x_n \\
y_2 &= a_{21}x_1 + a_{22}x_2 + \cdots + a_{2n}x_n \\
&\;\;\vdots \qquad\;\; \vdots \qquad\;\; \vdots \qquad\;\; \vdots \qquad\;\; \vdots \\
y_m &= a_{m1}x_1 + a_{m2}x_2 + \cdots + a_{mn}x_n.
\end{aligned}
$$

(12–2)

This is frequently reduced to the matrix equation

$$\mathbf{y} = A\mathbf{x},$$

where A is the $m \times n$ coefficient matrix. It is important to think of (12.2) as a special case of (12.1), and to see that this special case consists of all the linear transformations from R^n to R^m. This is the vector space which we have already studied under the name $\mathrm{Hom}(R^n, R^m)$.

The relevance of linear algebra to differential calculus consists in the fact that it is possible to obtain very good local approximations to quite general functions, such as (12.1), by using linear functions, such as (12.2). We shall see that this enables us to deduce important information about the behavior of a nonlinear function near a point by studying the linear (or affine) functions of best approximation at that point. We begin by extending the idea of a differential to our more general setting.

12.1   The differential

Recall that in our earlier encounters with the differential, it turned out to be the best linear approximation to the increment in a function. In the simplest case, where f is a function from R to R,

$$\Delta f = f(x + h) - f(x)$$

was called the increment in f at x due to the increment h in the independent variable. When x is held fixed, Δf is a function of h. It is a nonlinear function in all except the trivial case where f is linear (or affine). The only linear functions of h are the constant multiples of h. If there is one such multiple of h, say ah, which approximates Δf so closely that

$$\lim_{h \to 0} \frac{\Delta f - ah}{h} = 0,$$

we say that ah is the differential, df of f at x. Then $df = ah$ and $a = f'(x)$.

In the case where f is a real-valued function of several real variables, it is to be regarded as a function from R^n to R, the several real variables being lumped together as a vector in R^n. It is not necessary that f be defined at all points of R^n—we shall require only that it be defined at least over some open subset of R^n. If $\mathbf{x}$ is an interior point of its domain, then the increment $\Delta f = f(\mathbf{x} + \mathbf{h}) - f(\mathbf{x})$ is defined for all sufficiently small $\mathbf{h}$ in R^n. For fixed $\mathbf{x}$, Δf is a function of $\mathbf{h}$. As we saw in § 11.4, the only linear functions from R^n to R are those of the scalar product form $\mathbf{a} \cdot \mathbf{h}$, where $\mathbf{a}$ is some fixed vector in R^n. If, for some $\mathbf{a}$, the linear function $\mathbf{a} \cdot \mathbf{h}$ approximates Δf so closely that

$$\lim_{\mathbf{h} \to 0} \frac{\Delta f - \mathbf{a} \cdot \mathbf{h}}{||\mathbf{h}||} = 0,$$

then we say that f is differentiable at $\mathbf{x}$, and $df = \mathbf{a} \cdot \mathbf{h}$ is the differential of f at $\mathbf{x}$.

These two special cases suggest how to define the differential of a function $\mathbf{f}$ from an open subset of R^n to R^m. We use boldface type to remind us that the values which the function takes on are vectors. We say that $\mathbf{f}$ is differentiable at the point **x** in its domain if there is a linear transformation, T, from R^n to R^m, such that

$$\lim_{\mathbf{h} \to 0} \frac{||\, \mathbf{f}(\mathbf{x} + \mathbf{h}) - \mathbf{f}(\mathbf{x}) - T\mathbf{h} \,||}{||\, \mathbf{h} \,||} = 0 \,.$$

This limit expresses the fact that $T\mathbf{h}$, which depends linearly on **h**, is a good approximation for $\mathbf{f}(\mathbf{x} + \mathbf{h}) - \mathbf{f}(\mathbf{x})$ when **h** is close to the zero vector in R^n. The differential of **f** at **x** is then $d\mathbf{f} = T\mathbf{h}$.

We must point out that there are two double meanings in the above equation. The zero on the left means the zero vector in R^n, whereas that on the right is the real number, zero. And the norm symbol, $||\ \ ||$, in the numerator means the norm on the space R^m, whereas the same symbol in the denominator refers to the R^n norm. Such double meanings are common in mathematics. The reader avoids confusion by careful attention to the context. It is also important to understand that in the denominator we must use the norm of **h** rather than **h** itself, because division by vectors is not defined.

It is not always easy to tell whether a particular function has a differential at a given point, but it is easy to prove directly from the above definition that if one does exist then it must be unique (see Exercise 2).

It is essential to make some remarks about the terminology and notation used in dealing with differentials. For a function **f** with domain in R^n and range in R^m, the differential at a point **x** of R^n is an element T of the vector space $\mathrm{Hom}(R^n, R^m)$. We have previously expressed this differential as

$$d\mathbf{f} = T\mathbf{h}.$$

This formula exhibits T acting on the independent variable **h**. There are occasions on which we wish to exhibit T without exhibiting **h**. Moreover, there are occasions on which we wish to exhibit the fact that T is related to **f** and that it depends on **x**. The most convenient way of doing this is to write

$$T = \mathbf{f}'(\mathbf{x}),$$

using the prime just as we do in the notation for the derivative. With this notation we then have

$$d\mathbf{f} = \mathbf{f}'(\mathbf{x})\mathbf{h}.$$

At various times hereafter we shall use each of the foregoing notations. Later, we shall be interested in cases in which **f** is such that T varies continuously as a function of **x**. In such a case we say that **f** is continuously differentiable.

The definition of the differential which we have just given emerged rather late in the history of calculus—not until this century, in fact. It quickly won wide acceptance because it has the properties which had long been regarded as essential in the differential of functions of one-variable calculus. For example, an immediate consequence of this definition is that differentiability implies continuity (Exercise 1). But the really clinching argument in its favor is that it implies so elegantly the chain rule, which tells us that differentiability is preserved under composition, and that the differential of the composition of two functions is simply the composition of their differentials.

THEOREM I (THE CHAIN RULE). *Suppose that* **f** *is a function from the open subset,* $\mathfrak{U}$, *of* R^n *to* R^m, *and that* **g** *is a function from some open subset of* R^m *to* R^p. *Suppose further that the domain of* **g** *contains the range of* **f**, *so that the composition* **g** $\circ$ **f** *is defined. If* **f** *has the differential* T *at* x_0, *and* **g** *has the differential* L *at* $\mathbf{f}(x_0)$, *then* **g** $\circ$ **f** *is differentiable at* x_0 *and its differential there is* $L \circ T$.

PROOF.   We know already that $L \circ T$ is a linear transformation from R^n to R^p. What we must prove is that

$$(12.1\text{–}1) \qquad \lim_{\mathbf{h} \to 0} \frac{||\,(\mathbf{g} \circ \mathbf{f})(x_0 + \mathbf{h}) - (\mathbf{g} \circ \mathbf{f})(x_0) - (L \circ T)[\mathbf{h}]\,||}{||\,\mathbf{h}\,||} = 0.$$

Let $\mathbf{f}(x_0 + \mathbf{h}) - \mathbf{f}(x_0) = \Delta \mathbf{f}$. Then

$$(\mathbf{g} \circ \mathbf{f})(x_0 + \mathbf{h}) = \mathbf{g}[\mathbf{f}(x_0 + \mathbf{h})] = \mathbf{g}[\mathbf{f}(x_0) + \Delta \mathbf{f}].$$

Since **g** has the differential L at $f(x_0)$,

$$\mathbf{g}[\mathbf{f}(x_0) + \Delta \mathbf{f}] = \mathbf{g}[\mathbf{f}(x_0)] + L[\Delta \mathbf{f}] + \epsilon(\Delta \mathbf{f})\,||\,\Delta \mathbf{f}\,||,$$

for some function ϵ from a neighborhood of the zero vector in R^m to R^p such that $\epsilon \to 0$ as $\Delta \mathbf{f} \to 0$. Therefore, if we put

$$\Delta(\mathbf{g} \circ \mathbf{f}) = (\mathbf{g} \circ \mathbf{f})(x_0 + \mathbf{h}) - (\mathbf{g} \circ \mathbf{f})(x_0),$$

then

$$\Delta(\mathbf{g} \circ \mathbf{f}) = L[\Delta \mathbf{f}] + ||\,\Delta \mathbf{f}\,||\,\epsilon.$$

Since **f** has the differential T at x_0,

$$\Delta \mathbf{f} = T[\mathbf{h}] + \eta(\mathbf{h})\,||\,\mathbf{h}\,||$$

for some η such that $\eta \to 0$ as $\mathbf{h} \to 0$.

$$\therefore \ \Delta(\mathbf{g} \circ \mathbf{f}) = L\{T[\mathbf{h}] + \eta\,||\,\mathbf{h}\,||\} + \epsilon\,||\,\Delta \mathbf{f}\,||,$$

or

$$\Delta(\mathbf{g} \circ \mathbf{f}) - (L \circ T)[\mathbf{h}] = ||\,\mathbf{h}\,||\,L[\eta] + \epsilon\,||\,\Delta \mathbf{f}\,||.$$

By (11.8–2),

$$||\Delta f|| \leq ||T|| \; ||h|| + ||\eta|| \; ||h||.$$

$$\therefore ||\Delta(g \circ f) - (L \circ T)[h]|| \leq ||h|| \; ||L|| \; ||\eta|| + ||\epsilon|| \; ||T|| \; ||h|| + ||\epsilon|| \; ||\eta|| \; ||h||,$$

or

$$\frac{||\Delta(g \circ f) - (L \circ T)[h]||}{||h||} \leq ||L|| \; ||\eta|| + ||\epsilon|| [||T|| + ||\eta||].$$

Since $\Delta f \to 0$ as $h \to 0$, $\epsilon(\Delta f) \to 0$ as $h \to 0$. Also recalling that $\eta(h) \to 0$ as $h \to 0$, we get

$$\lim_{h \to 0} \frac{||\Delta(g \circ f) - (L \circ T)[h]||}{||h||} = 0,$$

and this is (12.1–1.)

Notice that the *chain rule* expressed in the theorem just proved contains as special cases the chain rule of elementary calculus (Theorem II of Chapter I, given in § 1.11 and discussed in § 1.3) and the chain rule of Theorem V, Chapter VII, given in § 7.3. Here we see one of the advantages of using the ideas of linear algebra and vector spaces in calculus. There is just one general theorem which expresses the rule for finding the differential of the composition of two differentiable functions.

This thing which we have defined to be the differential is, beyond question, the main object of study in differential calculus. But, unfortunately, not everyone agrees at present as to what it should be called. Some very good authors call it the derivative and don't use the term *differential* at all. In our terminology we keep the word *derivative* with its old familiar meaning, which limits it to single-variable calculus. This is not a serious matter, but it does require that we apprise ourselves of the definition being used by the author or speaker from whom we are trying to learn.

12.2   The differential of functions from R to R^n

Consider a *continuous* function f with domain in R and range in R^n. It is a vector-valued function of a real variable. If t is the real variable and the components of f are $f_1, f_2, \cdots, f_n$, we can write

$$f(t) = (f_1(t), f_2(t), \cdots, f_n(t)),$$

so that f defines (and is defined by) the ordered n-tuple of real-valued functions.

It is easy to relate the continuity of f to the continuity of the component functions. In fact, $\mathbf{f}$ is continuous at t_0 if and only if each of the components $f_1, f_2, \cdots, f_n$ is continuous at t_0. The key to this is the fact that

$$||\, \mathbf{f}(t) - \mathbf{f}(t_0) \,|| = \left\{ \sum_{k=1}^{n} |f_k(t) - f_k(t_0)|^2 \right\}^{1/2}$$

Likewise, if we define the derivative of $\mathbf{f}$ at t_0 (when it exists) as the vector $\mathbf{f}'(t_0)$ such that

$$\lim_{t \to t_0} \left|\left|\, \frac{\mathbf{f}(t) - \mathbf{f}(t_0)}{t - t_0} - \mathbf{f}'(t_0) \,\right|\right| = 0,$$

we can show without great difficulty that the derivative $\mathbf{f}'(t_0)$ exists if and only if each of the component functions $f_1, f_2, \cdots, f_n$ has a derivative at t_0, in which case

$$\mathbf{f}'(t_0) = (f_1'(t_0), \cdots, f_n'(t_0)).$$

By analogy with the case $n = 1$, we would expect that $\mathbf{f}$ has a differential at t_0 if it has a derivative there, and furthermore, that the formula $d\mathbf{f} = \mathbf{f}'(t_0)dt$ ought to hold. It is easy to justify these expectations if we recall the fact emphasized in § 11.4, that every linear transformation L from R to R^n is representable in the form $L[h] = h\mathbf{v}$, where $\mathbf{v}$ is some fixed vector in R^n. The definition of the differential of $\mathbf{f}$ requires it to be expressible as $L[dt] = \mathbf{v}dt$, where $\mathbf{v}$ is such that

$$\lim_{dt \to 0} \frac{||\, \mathbf{f}(t + dt) - \mathbf{f}(t) - \mathbf{v}dt \,||}{|\, dt \,|} = 0.$$

But this can be rewritten as

$$\lim_{dt \to 0} \left|\left|\, \frac{\mathbf{f}(t + dt) - \mathbf{f}(t)}{dt} - \mathbf{v} \,\right|\right| = 0,$$

which shows that $\mathbf{f}$ must have a derivative at t and that $\mathbf{f}'(t) = \mathbf{v}$. The details of the argument are asked for in Exercise 4.

In order to prove an important result which we shall need later on, it is now necessary to consider integrals of continuous vector-valued functions defined on an interval of the real axis. Let $\mathbf{f}$ be such a function from the closed interval $[a, b]$ to R^n. In order to define the integral

$$(12.2\text{–}1) \qquad \int_a^b \mathbf{f}(t)\, dt,$$

we use the same procedure as that set forth in § 1.5 for defining the integral of a real-valued function. That is, we form an approximating sum

$$(12.2\text{–}2) \qquad \sum_{k=1}^{p} \mathbf{f}(t_k')\, \Delta t_k,$$

where $a = t_0 < t_1 < \cdots < t_p = b$, $t_k = t_k - t_{k-1}$, and $t_{k-1} \leq t_k' \leq t_k$. The approximating sum (12.2–2) is a vector whose components are similarly expressible as approximating sums. For example, the first component of the vector (12.2–2) is

$$(12.2–3) \qquad \sum_{k=1}^{p} f_1(t_k') \, \Delta t_k,$$

and this we immediately recognize as an approximating sum for the integral

$$(12.2–4) \qquad \int_a^b f_1(t) \, dt$$

of the first component function. By Exercise 6, the continuity of f implies that each of the component functions is continuous, and since continuous functions from closed bounded intervals to the real numbers are integrable, each of the component functions of f is integrable. Hence, the approximating sums in (12.2–2) converge to give

$$(12.2–5) \qquad \int_a^b f(t) \, dt = \left(\int_a^b f_1(t) \, dt, \cdots, \int_a^b f_n(t) \, dt \right).$$

That is, the integral of f is the vector whose components are the integrals of the components of f.

When f is continuous, the function whose value at t is $||f(t)||$ is also continuous. This is a consequence of the fact that

$$\left| \, ||f(t)|| - ||f(t_0)|| \, \right| \leq ||f(t) - f(t_0)||.$$

(The student should refer to (11.5–1)). So, now we know that $||f||$ is integrable, and we can proceed to compare the integrals of f and $||f||$.

THEOREM II.   *If f is a continuous function from $[a, b]$ to R^n, then*

$$(12.2–6) \qquad \left\| \int_a^b f(t) \, dt \right\| \leq \int_a^b ||f(t)|| \, dt.$$

PROOF.   The triangle inequality for norms (see (4) of § 11.5), can easily be seen to imply that if $u_1, u_2, \cdots, u_p$ are any vectors then

$$||u_1 + u_2 + \cdots + u_p|| \leq ||u_1|| + ||u_2|| + \cdots + ||u_p||.$$

Applying this result to the approximating sum (12.2–3) and observing that $\Delta t_k > 0$, we obtain the result

$$\left\| \sum_{k=1}^{p} f(t_k') \, \Delta t_k \right\| \leq \sum_{k=1}^{p} ||f(t_k')|| \, \Delta t_k.$$

If we now pass to the limit, the sums on the two sides in the foregoing inequality converge to integrals. Since the sum on the right converges to the integral of $||\mathbf{f}||$, while the expression on the left converges to the norm of the integral of $\mathbf{f}$, we see that the inequality (12.2–6) must be true.

Continuity of $\mathbf{f}$ is more than we need to prove (12.2–6). The same conclusion follows from the weaker hypothesis that $\mathbf{f}$ is merely integrable on $[a, b]$. However, the proof would require a more extensive knowledge of integration, and we will omit it, since the form of the theorem which we have given is adequate for our purpose.

Next, we apply Theorem II to obtain a theorem which we shall need in proving Theorem VIII of § 12.6.

THEOREM III.   *Suppose that $\mathbf{f}$ is a function from $[a, b]$ to R^n which has a derivative $\mathbf{f}'(t)$ at each point of $[a, b]$, and also suppose that the derivative $\mathbf{f}$ is a continuous function on $[a, b]$. Then there is some point ξ such that $a \leq \xi \leq b$ and*

$$(12.2\text{–}7) \qquad ||\mathbf{f}(b) - \mathbf{f}(a)|| \leq ||\mathbf{f}'(\xi)|| (b - a).$$

PROOF.   We consider the derivative function $\mathbf{f}'$ and its components $f_1', \cdots, f_n'$. We know by the fundamental theorem of calculus (Theorem VIII of § 1.53) that

$$(12.2\text{–}8) \qquad f_1(b) - f_1(a) = \int_a^b f_1'(t)\, dt;$$

there are exactly similar results for the other components. We infer that

$$(12.2\text{–}9) \qquad \mathbf{f}(b) - \mathbf{f}(a) = \int_a^b \mathbf{f}'(t)\, dt,$$

because the first component of the vector on the left side of the equality in (12.2–9) is the left member in (12.2–8), whereas the first component of the right member of (12.2–9) is the right member in (12.2–8), and likewise for the other components. From (12.2–9) and Theorem II we see that

$$(12.2\text{–}10) \qquad ||\mathbf{f}(b) - \mathbf{f}(a)|| \leq \int_a^b ||\mathbf{f}'(t)||\, dt.$$

Next, we apply the mean-value theorem for integrals (Theorem VI, § 1.51) to the right side of (12.2–10). Since $||\mathbf{f}'(t)||$ is a continuous function of t, there is some number ξ on the interval such that

$$(12.2\text{–}11) \qquad \int_a^b ||\mathbf{f}'(t)||\, dt = ||\mathbf{f}'(\xi)|| (b - a).$$

(This number ξ can actually be chosen so that $a < \xi < b$. See Exercise 4 in § 1.51). By combining (12.2–10) and (12.2–11) we obtain (12.2–7), which is the desired result.

12.3   The differential of functions from R^n to R

We have already considered the special case where $n = 2$ in § 6.4. We found there that if a function of two variables has a differential at (x_0, y_0), then it must be given by

$$df = f_1(x_0, y_0)h + f_2(x_0, y_0)k$$

for all (h, k) in R^2. The same argument used there shows that in general if the real-valued function f of n real variables has a differential at $\mathbf{a} = (a_1, a_2, \cdots, a_n)$, then

$$df = f_1(\mathbf{a})h_1 + f_2(\mathbf{a})h_2 + \cdots + f_n(\mathbf{a})h_n,$$

where $f_i(\mathbf{a})$ is $\partial f/\partial x_i$ evaluated at $\mathbf{a}$, and $\mathbf{h} = (h_1, h_2, \cdots, h_n) \in R^n$. This shows that df is the inner product of $\mathbf{h}$ with the vector $(f_1(\mathbf{a}), f_2(\mathbf{a}), \cdots, f_n(\mathbf{a}))$. This vector is the gradient of the function f evaluated at the point $\mathbf{a}$, and by a slight abuse of language, we shall represent it by $\mathbf{grad}\, f(\mathbf{a})$. The above expression for the differential then becomes

$$df = \mathbf{grad}\, f(\mathbf{a}) \cdot \mathbf{h}.$$

This merely tells us what the differential looks like if it exists. The following theorem, which has already been announced as Theorem II of Chapter VII, assures us that for most of the usual functions which we encounter, the differential does exist—at least for all but a few isolated exceptional points.

THEOREM IV.   *Let $\mathcal{U}$ denote an open subset of R^n and let f be a real-valued function defined in $\mathcal{U}$. Assume that all of the n first partial derivatives exist throughout $\mathcal{U}$, and that $\mathbf{a}$ is a point of $\mathcal{U}$ where they are continuous. Then f is differentiable at $\mathbf{a}$, and the differential of f at $\mathbf{a}$ is the linear function L from R^n to R which is defined by $df = L(\mathbf{h}) = \mathbf{grad}\, f(\mathbf{a}) \cdot \mathbf{h}$.*

PROOF.   Since $\mathcal{U}$ is open, there is some positive number, r, such that all points of R^n which are within a distance r of $\mathbf{a}$ must belong to $\mathcal{U}$. Let $\mathbf{h}$ be a vector of length less than r, and let Δf denote the increment in f at $\mathbf{a}$ due to the increment $\mathbf{h}$ in the independent variable. In order to study Δf, we shall represent it as the sum of a number of terms, to each of which we can apply the mean value theorem of elementary calculus. Suppose that $\mathbf{h} = (h_1, h_2, \cdots, h_n)$; for $k = 1, 2, \cdots, n$ let $\mathbf{h}^{(k)}$ be that vector which agrees with **h** in the first k components and is zero in the remaining $n - k$ components; also let $\mathbf{h}^{(0)}$ be the zero vector. Then it is easy to verify that

$$\Delta f = f(\mathbf{a} + \mathbf{h}) - f(\mathbf{a}) = \sum_{k=1}^{n} [f(\mathbf{a} + \mathbf{h}^{(k)}) - f(\mathbf{a} + \mathbf{h}^{(k-1)})].$$

Now $f(\mathbf{a} + \mathbf{h}^{(k)}) - f(\mathbf{a} + \mathbf{h}^{(k-1)})$ is the difference between the values of f at two points which differ only in that the kth coordinate of the first is $a_k + h_k$ and that of the second is a_k. For both of these points the other coordinates have the values $x_1 = a_1 + h_1$, $x_2 = a_2 + h_2$, $\cdots$, $x_{k-1} = a_{k-1} + h_{k-1}$, $x_{k+1} = a_{k+1}$, $\cdots$, $x_n = a_n$. In other words, we are holding all of the n real variables constant except the kth, and it varies from a_k to $a_k + h_k$. So we can apply the mean value theorem for functions of the one real variable, x_k, and obtain

$$f(\mathbf{a} + \mathbf{h}^{(k)}) - f(\mathbf{a} + \mathbf{h}^{(k-1)}) =$$
$$f_k(a_1 + h_1, \cdots, a_{k-1} + h_{k-1}, a_k + \theta h_k, a_{k+1}, \cdots, a_n)h_k$$

for some θ properly between 0 and 1. Here, $f_k = \partial f / \partial x_k$. Let us denote this point where the first partial derivative with respect to x_k is evaluated by ξ^k. Notice that

$$||\xi^k - \mathbf{a}|| \leq ||\mathbf{h}||$$

for $k = 1, 2, \cdots, n$. Then we have

(12.3–1)

$$f(\mathbf{a} + \mathbf{h}) - f(\mathbf{a}) = \sum_{k=1}^{n} f_k(\xi^k)h_k = \sum_{k=1}^{n} f_k(\mathbf{a})h_k + \sum_{k=1}^{n} [f_k(\xi^k) - f_k(\mathbf{a})]h_k.$$

We can write this in the form

$$f(\mathbf{a} + \mathbf{h}) - f(\mathbf{a}) - \operatorname{grad} f(\mathbf{a}) \cdot \mathbf{h} = \sum_{k=1}^{n} [f_k(\xi^k) - f_k(\mathbf{a})]h_k.$$

If we let $\delta_k = f_k(\xi^k) - f_k(\mathbf{a})$ and apply the Cauchy inequality, we see that

(12.3–2)

$$|f(\mathbf{a} + \mathbf{h}) - f(\mathbf{a}) - \operatorname{grad} f(\mathbf{a}) \cdot \mathbf{h}| \leq ||\mathbf{h}|| \left(\sum_{k=1}^{n} \delta_k^2 \right)^{1/2}.$$

Now let us define

$$\epsilon(\mathbf{h}) = \left(\sum_{k=1}^{n} \delta_k^2 \right)^{1/2}.$$

When we show in a moment that $\epsilon(\mathbf{h}) \to 0$ as $\mathbf{h} \to 0$, it will immediately follow from (12.3–2) that f is differentiable at $\mathbf{a}$, with differential equal to $\operatorname{grad} f(\mathbf{a}) \cdot \mathbf{h}$. We know by hypothesis that $f_k = \partial f / \partial x_k$ is continuous at $\mathbf{a}$.

Therefore, from the definition of δ_k and the fact that $||\xi^k - \mathbf{a}|| \leq ||\mathbf{h}||$ (noted above), we see that $\delta_k \to 0$ when $\mathbf{h} \to 0$; hence, $\epsilon(\mathbf{h}) \to 0$ also. This completes the proof.

In elementary calculus we study functions f which are real-valued functions of a single real variable. We can regard this as the special case $n = 1$ of either a function from R to R^n or a function from R^n to R. In either case, the vector which represents the differential of f at $x = a$ has but a single component, namely $f'(a)$.

12.4   The general case

Now let us consider a function $\mathbf{f}$ whose domain $\mathfrak{U}$ is an open subset in R^n and whose range is in R^m. We wish to examine the structure of this vector-valued function. We shall then discuss a method for determining when $\mathbf{f}$ is differentiable.

If $\mathbf{x}$ is a point in $\mathfrak{U}$, $\mathbf{f}(\mathbf{x})$ is a point in R^m, and we can represent it as a vector

(12.4–1) $$f(\mathbf{x}) = (f^{(1)}(\mathbf{x}), f^{(2)}(\mathbf{x}), \cdots, f^{(m)}(\mathbf{x})),$$

where each of the component functions $f^{(1)}, \cdots, f^{(m)}$ is a real-valued function on $\mathfrak{U}$. As we look at this representation of $\mathbf{f}$ it is quite natural to ask the following questions: If $\mathbf{f}$ is differentiable at a certain point $\mathbf{a}$, does that imply that each of the functions $f^{(1)}, \cdots, f^{(m)}$ is differentiable at $\mathbf{a}$? If so, how is the differential of $\mathbf{f}$ related to the differentials of the component functions? Is it true that if each of the component functions is differentiable at a point, the function $\mathbf{f}$ is itself differentiable at that point? The answers to these questions are furnished by the following theorem:

THEOREM V.   *Let $\mathbf{f}$ be a function from an open set $\mathfrak{U}$ in R^n, mapping $\mathfrak{U}$ into R^m. Let $\mathbf{f}$ have component functions $f^{(1)}, \cdots, f^{(m)}$ from $\mathfrak{U}$ into R, as shown in (12.4–1). Then $\mathbf{f}$ is differentiable at a point $\mathbf{a}$ of $\mathfrak{U}$ if and only if each of the component functions is differentiable at $\mathbf{a}$. When $\mathbf{f}$ is differentiable at $\mathbf{a}$, its differential there is the element of $\mathrm{Hom}(R^n, R^m)$ represented by the Jacobian matrix*

(12.4–2) $$\begin{pmatrix} \dfrac{\partial f^{(1)}}{\partial x_1} & \dfrac{\partial f^{(1)}}{\partial x_2} & \cdots & \dfrac{\partial f^{(1)}}{\partial x_n} \\[2mm] \dfrac{\partial f^{(2)}}{\partial x_1} & \dfrac{\partial f^{(2)}}{\partial x_2} & \cdots & \dfrac{\partial f^{(2)}}{\partial x_n} \\[2mm] \vdots & \vdots & \vdots & \vdots \\[2mm] \dfrac{\partial f^{(m)}}{\partial x_1} & \dfrac{\partial f^{(m)}}{\partial x_2} & \cdots & \dfrac{\partial f^{(m)}}{\partial x_n} \end{pmatrix}$$

in which the partial derivatives are evaluated at the point $\mathbf{a}$.

PROOF.   For the first half of the proof we assume that each of the component functions is differentiable at $\mathbf{a}$. We know that the differential of $f^{(k)}$ at $\mathbf{a}$ is $\mathbf{grad}\, f^{(k)}\, (\mathbf{a}) \cdot \mathbf{h}$, and that there is a real-valued function $\epsilon^{(k)}$, defined in a neighborhood of the origin in R^n and such that

(12.4–3) $f^{(k)}(\mathbf{a} + \mathbf{h}) - f^{(k)}(\mathbf{a}) - \mathbf{grad}\, f^{(k)}(\mathbf{a}) \cdot \mathbf{h} = \epsilon^{(k)}(\mathbf{h}) \,|\,|\,\mathbf{h}\,|\,|$,

while $\epsilon^{(k)}(\mathbf{h}) \to 0$ when $\mathbf{h} \to 0$. Let A denote the Jacobian matrix in (12.4–2), each partial derivative being evaluated at $\mathbf{a}$. Observe that the elements of the kth row of the Jacobian are the components of $\mathbf{grad}\, f^{(k)}(\mathbf{a})$. This means that the kth component of the vector $A\mathbf{h}$ is $\mathbf{grad}\, f^{(k)}(\mathbf{a}) \cdot \mathbf{h}$. Hence, if T is the element of $\mathrm{Hom}(R^n, R^m)$ represented by A, we see by (12.4–3) that

$$\mathbf{f}(\mathbf{a} + \mathbf{h}) - \mathbf{f}(\mathbf{a}) - T\mathbf{h}$$

is the vector whose kth component is $\epsilon^{(k)}(\mathbf{h}) \,|\,|\,\mathbf{h}\,|\,|$. By the way in which the norm in R^m is defined it then follows that

$$|\,|\,\mathbf{f}(\mathbf{a} + \mathbf{h}) - \mathbf{f}(\mathbf{a}) - T\mathbf{h}\,|\,| = [(\epsilon^{(1)})^2 + \cdots + (\epsilon^{(m)})^2]^{1/2} \,|\,|\,\mathbf{h}\,|\,| .$$

From this it follows directly that $\mathbf{f}$ is differentiable at $\mathbf{a}$, with differential $d\mathbf{f} = T\mathbf{h}$. This finishes the first part of the proof.

For the other part of the proof we start with the assumption that $\mathbf{f}$ is differentiable at $\mathbf{a}$. We know that the differential is of the form $d\mathbf{f} = T\mathbf{h}$, where T belongs to $\mathrm{Hom}(R^n, R^m)$, but we do not yet know how T is related to the component functions. Let T be represented by the matrix C and let $c_{k1}, c_{k2}, \cdots, c_{kn}$ be the elements of the kth row of C. In this situation we know that there is a function $\boldsymbol{\epsilon}$ from a neighborhood of the origin in R^n to R^m such that

(12.4–4) $\mathbf{f}(\mathbf{a} + \mathbf{h}) - \mathbf{f}(\mathbf{a}) - T\mathbf{h} = \boldsymbol{\epsilon}(\mathbf{h}) \,|\,|\,\mathbf{h}\,|\,|$

and $\boldsymbol{\epsilon}(\mathbf{h}) \to 0$ as $\mathbf{h} \to 0$. If $\epsilon^{(1)}, \cdots, \epsilon^{(2)}$ are the component functions of $\boldsymbol{\epsilon}$, we infer from (12.4–4) that

(12.4–5) $f^{(k)}(\mathbf{a} + \mathbf{h}) - f^{(k)}(\mathbf{a}) - \displaystyle\sum_{j=1}^{n} c_{kj} h_j = \epsilon^{(k)}(\mathbf{h}) \,|\,|\,\mathbf{h}\,|\,| .$

Now $|\,\epsilon^{(k)}(\mathbf{h})\,| \leq |\,|\,\boldsymbol{\epsilon}(\mathbf{h})\,|\,|$, and therefore, $\epsilon^{(k)}(\mathbf{h}) \to 0$ when $\mathbf{h} \to 0$. From this fact and (12.4–5) we now conclude that $f^{(k)}$ is differentiable at $\mathbf{a}$, and that its differential, as a linear function of $\mathbf{h}$, is given by the expression

$$\sum_{j=1}^{n} c_{kj} h_j .$$

But then we know from § 12.3 that the partial derivatives of $f^{(k)}$ exist at $\mathbf{a}$ and that c_{kj} is the value of $\partial f^{(k)}/\partial x_j$ evaluated at $\mathbf{a}$. In other words,

$$\mathbf{grad}\, f^{(k)}(\mathbf{a}) = (c_{k1}, \cdots, c_{kn}).$$

The rather formidable-looking matrix in (12.4–2) which represents the differential covers, as special cases, three kinds of differentials with which we are already familiar. For the single-variable functions of elementary calculus, $m = n = 1$ and (12.4–2) becomes the 1×1 matrix consisting of the number $f'(a)$ which represents the linear transformation T, defined by $T(h) = f'(a)h$, from R to R. For functions $\mathbf{f}$ from R to R^m, we get the $m \times 1$ matrix

$$\begin{pmatrix} f'_1(a) \\ f'_2(a) \\ \vdots \\ f'_m(a) \end{pmatrix} ,$$

where $\mathbf{f}(t) = (f_1(t), f_2(t), \cdots, f_m(t))$. Finally, in the case of a real-valued function of n variables; that is, $f : R^n \to R$, (12.4–2) becomes the $1 \times n$ matrix

$$(f_1(\mathbf{a}), f_2(\mathbf{a}), \cdots, f_n(\mathbf{a})), \quad \text{with } f_j = \frac{\partial f}{\partial x_j} ,$$

which we recognize immediately as just the gradient vector of f.

12.41 Directional derivatives

One important application of the chain rule is in the treatment of the directional derivative, which is of great importance in the study of functions from R^n to R. This concept has already been introduced and treated briefly in § 10.6. The main result there is found in Theorem II of that section. The proof there avoided the chain rule by introducing a special coordinate system. The following is an alternative approach.

We use the fact that each direction can be uniquely represented by a unit vector, $\bar{\mathbf{u}}$. Suppose that $f : R^n \to R$ and that $\mathbf{a}$ is an interior point of the domain of f; then, if t is a nonzero real number sufficiently near zero,

$$\frac{f(\mathbf{a} + t\mathbf{u}) - f(\mathbf{a})}{t}$$

is defined and is the average rate of change of f at $\mathbf{a}$ in the $\mathbf{u}$ direction. The denominator t is just the distance from $\mathbf{a}$ to $\mathbf{a} + t\mathbf{u}$ if t is positive, and is the negative of this distance if t is negative. If the limit of this average rate of change as $t \to 0$ exists, we call it the directional derivative of f at $\mathbf{a}$ in the direction $\mathbf{u}$, and denote it by $D_\mathbf{u}f(\mathbf{a})$; that is,

$$(12.41\text{–}1) \qquad D_\mathbf{u}f(\mathbf{a}) = \lim_{t \to 0} \frac{f(\mathbf{a} + t\mathbf{u}) - f(\mathbf{a})}{t} .$$

Notice that in the special case where $\mathbf{u}$ is the kth member $\mathbf{e}_k$ of the standard basis in R^n,

$$D_{\mathbf{e}_k}f(\mathbf{a}) = \frac{\partial f}{\partial x_k}\bigg|_{\mathbf{x} = \mathbf{a}} .$$

This shows that partial derivatives are special cases of directional derivatives, or conversely, the idea of a directional derivative generalizes the idea of a partial derivative.

Another important observation is that to define the directional derivative of a function f from R^n to R, we have introduced a new function, say F, from the real numbers to the real numbers, namely $F(t) = f(a + tu)$. We then defined in (12.41–1) the directional derivative of f at a in the direction u to be the ordinary derivative $F'(0)$ of F at 0. This makes it easy to prove the following theorem:

THEOREM VI.   *If a is a point at which the function $f : R^n$ to R is differentiable, then f has a directional derivative at a in every direction, u. Furthermore,*

$$D_u f(a) = \operatorname{grad} f(a) \cdot u .$$

PROOF.   As we have just pointed out, $D_u f(a) = F'(0)$, where F is the composition of the given function f with the affine function $g : R \to R^n$ defined by $g(t) = a + tu$. Our job is to find $F'(0)$, where $F(t) = (f \circ g)(t) = f[g(t)]$. Referring back to Theorem I we see that the hypotheses of that theorem are satisfied by our f and g here; therefore, we can write immediately

$$dF = L \circ T,$$

where T is the differential of g at 0 and L is the differential of f at $g(0) = a$. From Theorem I we know that L is represented by the $(1 \times n)$ matrix $(f_1(a), f_2(a), \cdots, f_n(a))$ and that T is represented by the $(n \times 1)$ matrix whose single column consists of the components of u. Matrix multiplication is defined in such a way that the matrix which represents the composition $L \circ T$ is the product of the matrices representing L and T. Thus, dF is represented by the matrix

$$(f_1(a), f_2(a), \cdots, f_n(a)) \begin{pmatrix} u_1 \\ u_2 \\ \vdots \\ u_n \end{pmatrix} = \sum_{i=1}^{n} f_i(a) u_i = \operatorname{grad} f(a) \cdot u.$$

This is just a (1×1) matrix, that is, a number. We should have expected this from the last paragraph of § 12.4, since the composite function F is merely a real-valued function of a single variable. We saw that the differentials of such functions are represented by their derivatives, and therefore,

$$F'(0) = D_u f(a) = \operatorname{grad} f(a) \cdot u.$$

12.5   Continuously differentiable functions

Suppose that f is a differentiable function from the open subset $\mathfrak{U}$ of R^n to R^m. This implies that f has a differential, which we shall call $T(\bar{x})$, at each point x of $\mathfrak{U}$. This function T from $\mathfrak{U}$ to $\text{Hom}(R^n,\ R^m)$ is determined by the function f from $\mathfrak{U}$ to R^m. We define the differentiable function f to be *continuously differentiable* in case the function T is continuous as a function of x. The student should write out what continuity means in this situation and observe that it involves the norm which we introduced on the vector space $\text{Hom}(R^n,\ R^m)$ in § 11.8.

In order to find a practical way of determining whether a given function is continuously differentiable, we introduce one more definition. The function f is said to belong to class $C^{(1)}$ in $\mathfrak{U}$ in case each of the mn first partial derivatives, $\partial f^{(i)}/\partial x_j$, is continuous throughout $\mathfrak{U}$. We are letting $f = (f^{(1)}, f^{(2)}, \cdots, f^{(m)})$, and we are saying that each of the functions $f_j^{(i)}$ is continuous from $\mathfrak{U}$ to R.

We now prove that being continuously differentiable and belonging to class $C^{(1)}$ are the same. For brevity, we indicate that a function f belongs to class $C^{(1)}$ by writing $f \in C^{(1)}$.

THEOREM VII. *A necessary and sufficient condition that f be continuously differentiable in $\mathfrak{U}$ is that f belong to class $C^{(1)}$ in $\mathfrak{U}$.*

PROOF OF SUFFICIENCY. Assume that $f \in C^{(1)}$ in $\mathfrak{U}$. By Theorem IV we know that each of the component functions of f is differentiable throughout $\mathfrak{U}$. Then Theorem V tells us that f is differentiable in $\mathfrak{U}$. We now know that the function T is at least defined for every point in $\mathfrak{U}$.

To show that T is continuous, let a denote any point in $\mathfrak{U}$. We wish to show that if ϵ is any positive number, then there must exist some positive number δ such that

$$||\ T(x) - T(a) < ||\ \epsilon \qquad \text{for all } x \text{ such that } ||\ x - a\ || < \delta.$$

Since both $T(x)$ and $T(a)$ are elements of the vector space $\text{Hom}(R^n,\ R^m)$, their difference is also in this space. By Theorem V we know that $T(a)$ is represented by the $m \times n$ matrix whose element in the ith row and jth column is $f_j^{(i)}(a)$. Therefore, $T(x) - T(a)$ is represented by the matrix whose element in the ith row and jth column is $f_j^{(i)}(x) - f_j^{(i)}(a)$. Since all mn of the functions $f_j^{(i)}$ are continuous in $\mathfrak{U}$, it is possible, for each $\epsilon > 0$, to find a $\delta > 0$ such that

$$\left| f_j^{(i)}(x) - f_j^{(i)}(\bar{a}) \right| < \frac{\epsilon}{\sqrt{mn}} \qquad \text{if} \qquad ||\ x - a\ || < \delta$$

for $i = 1, 2, \cdots, n$ and $j = 1, 2, \cdots, m$. Hence, by (11.8–3), the norm of the linear transformation $T(x) - T(a)$ is less than ϵ. In other words, given any $\epsilon > 0$, there exists some $\delta > 0$ (depending on both ϵ and a) such that

$$||\ T(x) - T(a)\ || < \epsilon \qquad \text{for all } x \text{ such that } ||\ x - a\ || < \delta.$$

PROOF OF NECESSITY. Assume that f is continuously differentiable, that is, that T is a continuous function of $\mathbf{x}$ throughout $\mathcal{U}$. Then if $\mathbf{a} \in \mathcal{U}$ and $\epsilon > 0$, there is a $\delta > 0$ such that

$$|| T(\mathbf{x}) - T(\mathbf{a}) || < \epsilon \qquad \text{if} \qquad || \mathbf{x} - \mathbf{a} || < \delta.$$

But by (11.8–3), every element of the $m \times n$ matrix representing $T(\mathbf{x}) - T(\mathbf{a})$ is less in absolute value than ϵ; that is,

$$|f_j^{(i)}(\mathbf{x}) - f_j^{(i)}(\mathbf{a})| < \epsilon \qquad \text{for all } \mathbf{x} \text{ such that } || \mathbf{x} - \mathbf{a} || < \delta,$$

and for all $i = 1,2, \cdots ,n$ and all $j = 1,2, \cdots ,m$. Since $\mathbf{a}$ could be any point in $\mathcal{U}$, this says that all of the n first partial derivatives of all m of the component functions of $\mathbf{f}$ are continuous at all points of $\mathcal{U}$. This means that $\mathbf{f} \in C^{(1)}$ in $\mathcal{U}$.

Notice that this theorem implies that a vector-valued function $\mathbf{f}$ is continuously differentiable if and only if each of its component functions is. Theorem V showed that differentiability of $\mathbf{f}$ at a point is equivalent to differentiability of each of the component functions at that point, and Exercise 6 showed that the continuity of $\mathbf{f}$ is equivalent to that of each of its component functions. The great practical significance of these results is that they enable us to investigate some important properties of vector-valued functions by studying the somewhat simpler component functions.

12.6   The fundamental inversion theorem

We are already familiar with problems associated with the solution of a system of n simultaneous linear equations in n unknowns. Such a system is of the form

(12.6–1)
$$\begin{aligned}
a_{11}x_1 + a_{12}x_2 + \cdots + a_{1n}x_n &= y_1 \\
a_{21}x_1 + a_{22}x_2 + \cdots + a_{2n}x_n &= y_2 \\
&\ \ \vdots \\
a_{n1}x_1 + a_{n2}x_2 + \cdots + a_{nn}x_n &= y_n ,
\end{aligned}$$

where the a's and the y's are given. We wish to find a set of x's which satisfy all the equations. With the help of vector and matrix notation, (12.6–1) is condensed to

$$A\mathbf{x} = \mathbf{y},$$

and the problem becomes one of finding the inverse of a linear operator. The $n \times n$ matrix A represents a linear operator on R^n, that is, a function which maps R^n into itself linearly; $\mathbf{y}$ is a given vector in R^n. The problem is to find a vector $\mathbf{x}$ in R^n which A maps into $\mathbf{y}$. A unique solution exists if and only if the determinant of A is different from zero. In this case, Cramer's rule gives, for each $\mathbf{y}$ in R^n, the unique vector $\mathbf{x}$ satisfying (12.6–1).

The rule which associates in this way the right x with each y is called the function which is inverse to the one represented by A. The inverse function is represented by the inverse matrix A^{-1}, and the equation $y = Ax$ is equivalent to $x = A^{-1}y$. Combining these two equations we get that $y = AA^{-1}y$ for all y in R^n and $x = A^{-1}Ax$ for all x in R^n. Therefore, $AA^{-1} = A^{-1}A = I$, the $n \times n$ identity matrix.

The main purpose of this chapter is to consider a generalization of the system (12.6–1), namely

$$(12.6\text{–}2) \quad \begin{aligned} f^{(1)}(x_1, x_2, \cdots, x_n) &= y_1 \\ f^{(2)}(x_1, x_2, \cdots, x_n) &= y_2 \\ &\vdots \\ f^{(n)}(x_1, x_2, \cdots, x_n) &= y_n, \end{aligned}$$

where $f^{(1)}, \cdots, f^{(n)}$ are given functions from R^n to R. Again, the problem is to solve for the x's in terms of the y's. Notice that (12.6–1) is just that special case of (12.6–2) in which all of the functions $f^{(1)}, f^{(2)}, \cdots, f^{(n)}$ are linear. (12.6–2) is usually abbreviated to

$$\mathbf{f}(\mathbf{x}) = \mathbf{y}.$$

Our goal is to find out when there is an inverse function $\mathbf{f}^{-1}$ which expresses $\mathbf{x}$ as a function of $\mathbf{y}$.

In Chapter IX, we made an approach to this problem with an inverse function theorem of § 9.1 derived from an implicit function theorem (Theorem III, § 8.3) derived in Chapter VIII. The proofs given there were highly appropriate for spaces of low dimension (say two or three), where all the steps could be written out in full detail. As was indicated in Chapters VIII and IX, the pattern of proof used there can also be extended to spaces of high dimension, but the details become so intricate that one loses sight of the main ideas. By using linear algebra we shall give a proof whose clarity is independent of the dimension of the space. Even more important, by using linear algebra we keep the basic ideas in the foreground and thereby deepen our understanding of the essence of differential calculus. The essential characteristic of our present method is that we deduce information about very general functions by approximating them with linear functions. Since good linear approximations can be expected to hold only in the neighborhood of some point, the information obtained is of a local nature, that is, valid near the point where the approximation is good.

To speak more precisely, let us assume that $\mathbf{f}$ is a function from the open subset $\mathcal{U}$ of R^n to R^n, and let $\mathbf{a}$ denote a point of $\mathcal{U}$ whose image under $\mathbf{f}$ is $\mathbf{b}$. Instead of asking whether the whole function $\mathbf{f}$ is invertible, we merely ask whether it is possible to find some neighborhood, $\mathcal{N}$, of $\mathbf{a}$ such that the restriction of $\mathbf{f}$ to $\mathcal{N}$ has an inverse. This may seem at first to have reduced the problem to insignificance, but actually the results which we shall obtain are still of far-reaching importance. We have emphasized the necessity of limiting attention to what may be only a small piece of f, only to remind us that the most we can hope for is a local result, that is, one valid in a neighborhood of some point. The main hypothesis of the theorem is the differentiability of f, but as we might expect from the simplest case—the case where $n = 1$—differentiability by itself is not enough (see Exercise 5). What we shall need is continuous differentiability, just as in the one-dimensional case.

The theorem which we are about to prove could well be called the most important theorem in differential calculus. It is the culmination of all that we have done in Chapters XI and XII. All of the material included in these two chapters was chosen primarily because of its importance in proving Theorem VIII and its corollary, Theorem IX. The proof is not short, involving as it does most of what we have covered so far, and we shall make it somewhat longer than is really necessary in an effort to make it more easily accessible. The notation will be varied at places where such variation seems to make the meaning clearer. In particular, we shall use $f'(x)$ to denote the differential of f at x when we wish to make it clear which point the differential is associated with.

THEOREM VIII. (FUNDAMENTAL INVERSE FUNCTION THE-OREM). *Let* f *be a function from some subset of* R^n *to* R^n, *and assume that* f *is continuously differentiable in some neighborhood of* $x = a$. *If the linear transformation* $f'(a)$ *is invertible, then there exist neighborhoods* $\mathcal{U}$ *and* $\mathcal{V}$ *of* a *and* $f(a)$, *respectively, such that the restriction of* f *to* $\mathcal{U}$ *is a one-to-one mapping of* $\mathcal{U}$ *onto* $\mathcal{V}$. *Moreover* f^{-1} *is continuously differentiable in* $\mathcal{V}$.

PROOF. Let $T_0 = f'(a)$ and write T for $f'(x)$ when that is convenient, it being understood that T depends on x. We know that $||\,T_0\,|| > 0$ and $||\,T_0^{-1}\,|| > 0$. We also know that, for any vector w in R^n, $w = T_0^{-1}T_0[w]$, and therefore,

(12.6–3)
$$||\,w\,|| \le ||\,T_0^{-1}\,||\,||\,T_0[w]\,|| \quad \text{or} \quad \frac{||\,w\,||}{||\,T_0^{-1}\,||} \le ||\,T_0[w]\,||.$$

This relation will be used later.

Since f is continuously differentiable, we can keep $||\,T - T_0\,||$ as small as we like by keeping $||\,x - a\,||$ sufficiently small. Recall that by Theorem IV of Chapter XI, T will be invertible if $||\,T - T_0\,|| < 1/||\,T_0^{-1}\,||$.

With the foregoing things in mind, we choose a positive number r so small that the closed ball $\overline{B(a, r)} = \{x : ||\,x - a\,|| \le r\}$ is contained in the neighborhood of the point **a** throughout which **f** is continuously differentiable, and also small enough that

$$(12.6\text{-}4) \quad ||\, T - T_0\, || < \frac{1}{2\, ||\, T_0^{-1}\, ||} \quad \text{for all } x \in \overline{B(a, r)}$$

We shall denote by A the open ball $B(a, r)$, which consists of the interior points of $\overline{B(a, r)}$, and we shall let S denote the sphere which is the surface of the ball; that is, $S = \{x : ||\, x - a\, || = r\}$.

Now let **u** and **v** be any pair of distinct points of $\overline{B(a, r)}$. Then $w = v - u \neq 0$. We are going to show that $f(v) \neq f(u)$. This is slightly more than we need in order to infer that when **f** is restricted to A, it defines a one-to-one mapping of A onto the image of A under **f**; we denote this image by $f(A)$. To proceed, we define as follows a function **F** from the closed unit interval $[0, 1]$ to R^n:.

$$F(t) = f(u + tw) - T_0[tw] = f(u + tw) - tT_0[w].$$

Obviously, $F(0) = f(u)$, and since $v = u + w$,

$$F(1) = f(v) - T_0[w].$$

By the chain rule,

$$F'(t) = f'(u + tw)[w] - T_0[w] = [f'(u + tw) - T_0][w].$$

This is some vector in R^n, which we would expect from the fact that **F** is a function from $[0, 1]$ to R^n. Since $\overline{B(a, r)}$ is convex, $u + tw \in \overline{B(a, r)}$ for all $t \in [0, 1]$ (Exercise 17, Chapter XI), we can let $f'(u + tw) = T$ in $(12.6\text{-}4)$ and arrive at

$$||\, F'(t)\, || \leq ||\, f'(u + tw) - T_0\, || \cdot ||\, w\, || < \frac{||\, w\, ||}{2\, ||\, T_0^{-1}\, ||} \leq \frac{||\, T_0[w]\, ||}{2}$$

for all $t \in [0, 1]$, using $(12.6\text{-}3)$ for the last step.

Since **F** satisfies the hypotheses of Theorem III, we know that there is some number $\xi \in [0, 1]$ such that $||\, F(1) - F(0)\, || \leq ||\, F'(\xi)\, ||$. Therefore,

$$(12.6\text{-}5) \quad ||\, F(1) - F(0)\, || = ||\, f(v) - f(u) - T_0[w]\, || < \frac{1}{2} ||\, T_0[w]\, ||.$$

Using $(11.5\text{-}1)$ we get that

$$||\, f(v) - f(u)\, || > \frac{1}{2} ||\, T_0[w]\, ||,$$

and by $(12.6\text{-}3)$ this gives

$$(12.6\text{-}6) \quad ||\, f(v) - f(u)\, || > \frac{||\, v - u\, ||}{2\, ||\, T_0^{-1}\, ||}$$

for all pairs of distinct points, $\mathbf{u}$ and $\mathbf{v}$, belonging to $\overline{B(\mathbf{a}, r)}$. This proves that $\mathbf{f}$ is one-to-one on $\mathbf{A}$.

The second part of the proof is devoted to showing that $\mathbf{f}(A)$ is open. We begin by showing that $\mathbf{f}(\mathbf{a})$ is an interior point of $\mathbf{f}(A)$, that is, that $\mathbf{f}(A)$ contains an open ball which has $\mathbf{f}(\mathbf{a})$ as its center. Notice that for every point $\mathbf{x}$ of the sphere S, $||\mathbf{x} - \mathbf{a}|| = r$, and, therefore by (12.6–6) $||\mathbf{f}(\mathbf{x}) - \mathbf{f}(\mathbf{a})|| > r/(2||T_0^{-1}||)$. This means that the image of the sphere S lies outside the sphere of radius $r/(2||T_0^{-1}||)$ centered at $\mathbf{f}(\mathbf{a})$. We shall show that the open sphere of half this radius centered at $\mathbf{f}(\mathbf{a})$ is a subset of $\mathbf{f}(A)$. This amounts to showing that for each $\mathbf{y}$ such that $||\mathbf{f}(\mathbf{a}) - \mathbf{y}|| < r/(4||T_0^{-1}||)$, there is an $\mathbf{x} \in A$ such that $\mathbf{f}(\mathbf{x}) = \mathbf{y}$ (see Fig. 83).

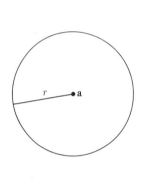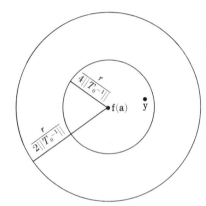

Fig. 83

Thus, we assume that $\mathbf{y}$ is at a distance less than $r/(4||T_0^{-1}||)$ from $\mathbf{f}(\mathbf{a})$ and define the real-valued function ϕ on $\overline{B(\mathbf{a}, r)}$ by

$$\phi(\mathbf{x}) = ||\mathbf{y} - \mathbf{f}(\mathbf{x})||.$$

In other words, $\phi(\mathbf{x})$ is the distance by which the image of $\mathbf{x}$ misses $\mathbf{y}$. Obviously ϕ is nonnegative. We shall proceed on the assumption that $\mathbf{y}$ does not belong to $\mathbf{f}(A)$ and show that this leads to a contradiction.

Since it has already been shown that $||\mathbf{f}(\mathbf{x}) - \mathbf{f}(\mathbf{a})|| > r/(2||T_0^{-1}||)$ for all $\mathbf{x} \in S$, whereas $||\mathbf{y} - \mathbf{f}(\mathbf{a})|| < r/(4||T_0^{-1}||)$, it is clear that ϕ cannot vanish on S. Therefore, since we are assuming that ϕ does not vanish at any point of A, then neither can it vanish at any point of the closed bounded set $\overline{B(\mathbf{a}, r)}$, which is composed of A together with S. Since $\mathbf{f}$ is continuous, the function ψ defined by $\psi(\mathbf{x}) = \mathbf{y} - \mathbf{f}(\mathbf{x})$ is a continuous function from $\overline{B(\mathbf{a}, r)}$ to R^n. By the remark in § 11.7 about the continuity of norms and by Theorem II of Chapter XI, which says that the composition of continuous functions is continuous, we see that ϕ is continuous. By Theorem II of Chapter V we know that there is some point of $\overline{B(\mathbf{a}, r)}$, say

u, where the value of ϕ is an absolute minimum. Then, knowing that ϕ does not vanish on S, we can infer that since ϕ does not vanish in A, $\phi(\mathbf{u}) > 0$.

Next, we want to show that **u** belongs to A rather than to S. This will follow if we can prove that there is at least one point of A where the value of ϕ is less than at any point of S. Referring to Fig. 83, it is easy to see that **a** itself is such a point. Since **y** lies inside the sphere of radius $r/(4\,||\,T_0^{-1}\,||\,)$ centered at $\mathbf{f}(\mathbf{a})$, it is clear that $\phi(\mathbf{a}) < r/(4\,||\,T_0^{-1}\,||\,)$, and since $\mathbf{f}(S)$ lies outside the sphere of radius $r/(2\,||\,T_0^{-1}\,||\,)$ centered at $\mathbf{f}(\mathbf{a})$, it is geometrically clear that $\phi(\mathbf{x}) > r/(4\,||\,T_0^{-1}\,||\,)$ when **x** is in S. It is also easy to prove this by using the properties of the norm and not relying on any geometrical intuition. The fact that $\phi(\mathbf{a}) < \phi(\mathbf{x})$ when **x** is in S does not mean, of course, that ϕ has its minimum value at **a**, but we can infer that if ϕ takes on its minimum at **u**, then **u** belongs to A and not to S. Knowing this, we can now arrive at a contradiction by showing that there must be many points **x** of A where $\phi(\mathbf{x}) < \phi(\mathbf{u})$.

Consider the nonzero vector $\mathbf{y} - \mathbf{f}(\mathbf{u})$. Since $T_0 = \mathbf{f}'(\mathbf{a})$ is nonsingular, there is a unique nonzero vector **z** in R^n such that $T_0[\mathbf{z}] = \mathbf{y} - \mathbf{f}(\mathbf{u})$. Since **u** belongs to the set A, and A is open, we know that there is an open ball centered at **u** and contained in A. Hence, for all values of t sufficiently near zero, $\mathbf{u} + t\mathbf{z}$ also belongs to A.

We now argue as follows, assuming that $0 < t < 1$ and that t is small enough to insure that $\mathbf{u} + t\mathbf{z}$ is in A:

$$\phi(\mathbf{u} + t\mathbf{z}) = ||\,\mathbf{f}(\mathbf{u} + t\mathbf{z}) - \mathbf{y}\,||$$
$$= ||\,\mathbf{f}(\mathbf{u} + t\mathbf{z}) - \mathbf{f}(\mathbf{u}) - T_0[t\mathbf{z}] + \mathbf{f}(\mathbf{u}) + T_0[t\mathbf{z}] - \mathbf{y}\,||$$
$$\leq ||\,f(\mathbf{u} + t\mathbf{z}) - \mathbf{f}(\mathbf{u}) - T_0[t\mathbf{z}]\,|| + ||\,f(\mathbf{u}) + T_0[t\mathbf{z}] - \mathbf{y}\,||\,.$$

Looking at the first term on the right, we see that we can apply (12.6–5) to it by letting $\mathbf{v} = \mathbf{u} + t\mathbf{z}$. The **w** in (12.6–5) is $\mathbf{w} = \mathbf{v} - \mathbf{u}$, which becomes $t\mathbf{z}$ in this case. Hence,

$$||\,\mathbf{f}(\mathbf{u} + t\mathbf{z}) - \mathbf{f}(\mathbf{u}) - T_0[t\mathbf{z}]\,|| < \frac{1}{2}\,||\,T_0[t\mathbf{z}]\,||\,.$$

Also, $$T_0[t\mathbf{z}] = tT_0[\mathbf{z}] = t[\mathbf{y} - f(\mathbf{u})];$$

thus $||\,\mathbf{f}(\mathbf{u} + t\mathbf{z}) - \mathbf{f}(\mathbf{u}) - T_0[t\mathbf{z}]\,|| < \dfrac{t}{2}\,||\,\mathbf{y} - \mathbf{f}(\mathbf{u})\,|| = \dfrac{t}{2}\,\phi(\mathbf{u}).$

Next,

$$\mathbf{f}(\mathbf{u}) + T_0[t\mathbf{z}] - \mathbf{y} = \mathbf{f}(\mathbf{u}) + t[\mathbf{y} - \mathbf{f}(\mathbf{u})] - \mathbf{y} = (1 - t)[\mathbf{f}(\mathbf{u}) - \mathbf{y}],$$

and so

$$||\,\mathbf{f}(\mathbf{u}) + T_0[t\mathbf{z}] - \mathbf{y}\,|| = (1 - t)\,||\,\mathbf{f}(\mathbf{u}) - \mathbf{y}\,|| = (1 - t)\phi(\mathbf{u}).$$

On combining these results we see that

$$\phi(\mathbf{u} + t\mathbf{z}) < \left(1 - \frac{t}{2}\right)\phi(\mathbf{u}) < \phi(\mathbf{u}).$$

Since $\phi(\mathbf{u})$ is the minimum value of ϕ, we have the desired contradiction.

What this contradiction proves is that $\mathbf{f}(\mathbf{a})$ is an interior point of $\mathbf{f}(A)$. To conclude that $\mathbf{f}(A)$ is open, we must show that every point of $\mathbf{f}(A)$ is an interior point. If $\mathbf{y}$ is any other point of $\mathbf{f}(A)$, then there is some unique point (call it $\mathbf{b}$) of A such that $\mathbf{f}(\mathbf{b}) = \mathbf{y}$. Now we know that $\mathbf{f}$ is continuously differentiable in a neighborhood of $\mathbf{b}$ and that $\mathbf{f}'(\mathbf{b})$ is invertible (because we chose r so small that $\mathbf{f}'(\mathbf{x})$ is invertible for every $\mathbf{x}$ in A). Hence, by the reasoning we have used, it follows that there is some open ball B with center at $\mathbf{b}$ such that B is contained in A and such that $\mathbf{f}(\mathbf{b})$ is an interior point of $\mathbf{f}(B)$, and therefore also an interior point of $\mathbf{f}(A)$.

We shall show that A and $\mathbf{f}(A)$ have all the properties of the sets $\mathfrak{U}$ and $\mathfrak{V}$, respectively, mentioned in the statement of the theorem. All that remains to be proved is that the inverse function $\mathbf{f}^{-1}$ from $\mathbf{f}(A)$ back to A is continuously differentiable throughout $\mathbf{f}(A)$. Our first task, then, is to show that $\mathbf{f}^{-1}$ is differentiable at each point of $\mathbf{f}(A)$. If we denote $\mathbf{f}^{-1}$ by $\mathbf{g}$ (for simplicity of notation), the differential of $\mathbf{g}$ (if it exists) at a point $\mathbf{y}$ of $\mathfrak{V}$ is an element $\mathbf{g}'(\mathbf{y})$ of End R^n. We shall finish by showing that not only does $\mathbf{g}'(\mathbf{y})$ exist at all points of $\mathbf{f}(A)$, but also that it is a continuous function of $\mathbf{y}$.

We begin by observing that the chain rule tells us what $\mathbf{g}'(\mathbf{y})$ has to be if it does exist. We apply the chain rule to the equation $\mathbf{f}[\mathbf{g}(\mathbf{y})] = \mathbf{y}$ for each $\mathbf{y}$ in $\mathbf{f}(A)$. This equation states that the composition $\mathbf{f} \circ \mathbf{g}$ is the identity function I on $\mathbf{f}(A)$. Since the identity function is itself a linear operator, it is its own differential (Exercise 6). Thus, if $\mathbf{y}$ is any point of $\mathbf{f}(A)$, the chain rule tells us that if $\mathbf{g}'(\mathbf{y})$ exists it must satisfy the equation

$$\mathbf{f}'[\mathbf{g}(\mathbf{y})] \circ \mathbf{g}'(\mathbf{y}) = I.$$

Since $\mathbf{f}'$ is invertible at each point of A, we can solve the above equation by multiplying both sides by the inverse of $\mathbf{f}'[\mathbf{g}(\mathbf{y})]$. This, of course, shows that if $\mathbf{g}'(\mathbf{y})$ exists, it must be $(\mathbf{f}'[\mathbf{g}(\mathbf{y})])^{-1}$. With this indication of what to expect, we can now undertake to prove that, in fact, $\mathbf{g}$ is differentiable at $\mathbf{y}$.

Let $\mathbf{y}$ and $\mathbf{k}$ be vectors, with $\mathbf{k} \neq 0$, such that $\mathbf{y}$ and $\mathbf{y} + \mathbf{k}$ are both in $\mathbf{f}(A)$. By the invertibility of $\mathbf{f}$, we know that there exist distinct unique vectors $\mathbf{x}$ and $\mathbf{x} + \mathbf{h}$ in A such that $\mathbf{f}(\mathbf{x}) = \mathbf{y}$ and $\mathbf{f}(\mathbf{x} + \mathbf{h}) = \mathbf{y} + \mathbf{k}$. Note that $\mathbf{h} \neq 0$, $\mathbf{x} = \mathbf{g}(\mathbf{y})$, and $\mathbf{x} + \mathbf{h} = \mathbf{g}(\mathbf{y} + \mathbf{k})$. Let us denote the differential of $\mathbf{f}$ at $\mathbf{g}(\mathbf{y})$ by T. From the preceding paragraph, we know that if $\mathbf{g}$ is differentiable at $\mathbf{y}$, its differential there must be T^{-1}. We shall show that the requirement for differentiability is satisfied, that is, that

(12.6–7) $$\lim_{\mathbf{k} \to 0} \frac{||\, \mathbf{g}(\mathbf{y} + \mathbf{k}) - \mathbf{g}(\mathbf{y}) - T^{-1}[\mathbf{k}] \,||}{||\, \mathbf{k} \,||} = 0.$$

We observe at the outset that

$$g(y + k) - g(y) - T^{-1}[k] = x + h - x - T^{-1}[f(x + h) - f(x)]$$
$$= T^{-1}T[h] - T^{-1}[f(x + h) - f(x)]$$
$$= T^{-1}\{T[h] - [f(x + h) - f(x)]\};$$

therefore

$$||\, g(y + k) - g(y) - T^{-1}[k]\, || \leq ||\, T^{-1}\, ||\; ||\, f(x + h) - f(x) - T[h]\, ||.$$

Since x and $x + h$ are in A and $h \neq 0$, it follows from (12.6–6) that

$$(12.6\text{–}8) \qquad ||\, k\, || = ||\, f(x + h) - f(x)\, || > \frac{||\, h\, ||}{2\, ||\, T_0^{-1}\, ||}$$

or

$$\frac{1}{||\, k\, ||} < \frac{2\, ||\, T_0^{-1}\, ||}{||\, h\, ||}.$$

Therefore,

$$\frac{||\, g(y + k) - g(y) - T^{-1}[k]\, ||}{||\, k\, ||}$$

$$\leq 2\, ||\, T_0^{-1}\, ||\; ||\, T^{-1}\, || \frac{||\, f(x + h) - f(x) - T[h]\, ||}{||\, h\, ||}.$$

This inequality holds for all nonzero k for which $||\, k\, ||$ is sufficiently small. When $||\, k\, || \to 0$, it follows from (12.6–8) that $||\, h\, || \to 0$. Since T is the differential of f at x, it then follows that

$$\frac{||\, f(x + h) - f(x) - T[h]\, ||}{||\, h\, ||} \to 0,$$

and thus we see from the foregoing inequality that condition (12.6–7) is satisfied.

The final step of the proof is that of showing that the differential of f^{-1} at y depends continuously on y. This comes out as a direct consequence of Theorem II of Chapter XI, which says that the composition of continuous functions is continuous. By Theorem V of Chapter XI we know that T^{-1} is a continuous function of T. We have assumed that T is continuous as a function of x (recall that $T = f'(x)$ and that f is assumed to be continuously differentiable). Finally, in the notation $x = f^{-1}(y)$, we know that x is continuous as a function of y. This is because we have just proved that f^{-1} is differentiable and we know from Exercise 1 that f^{-1} must therefore be continuous. Hence, by Theorem II of Chapter XI, T^{-1} is a continuous function of y; that is, f^{-1} is continuously differentiable.

We have now completed the proof of the inverse function theorem. The hypothesis that the differential of f at a be nonsingular is frequently expressed by saying that the determinant of the Jacobian matrix of f at a is not zero. Now observe that the Jacobian matrix of the linear system (12.6–1) is just the matrix of the coefficients. For each given **y**, a necessary and sufficient condition that (12.6–1) should determine **x** uniquely is that this coefficient matrix should have nonzero determinant. This is the theorem to which we referred in the first paragraph of Chapter XI as a prerequisite. A comparison between it and the inverse function theorem shows the latter to be a *partial* generalization of the former from linear to nonlinear functions.

There are three main pitfalls to be avoided in the application of the inverse function theorem. First is a tendency to forget the hypothesis of continuous differentiability. We usually remember to make sure that **f** is differentiable and that its differential at the point in which we are interested is invertible. It is easy to construct examples of functions from **R** to **R** which show that we also need continuous differentiability to guarantee invertibility (see Exercise 13). The way we test a function to see whether it is continuously differentiable is to take all the first partial derivatives and see if they are continuous. By Theorem IV, this is a necessary and sufficient condition for continuous differentiability.

Second, we must not think of the hypotheses of this theorem as necessary for invertibility—they are only sufficient. There are familiar examples of continuously differentiable functions which are invertible in some neighborhood of a point where the differential is not invertible. An elementary example is shown in Exercise 11.

Finally, it is easy to forget that the theorem gives only a local result which may be valid just for the restriction of the function to a very small neighborhood. And even people who have assimilated this fact are prone to think that if a function is locally invertible in some neighborhood of each point of its domain, then the whole function must be invertible. The falseness of this conclusion is brilliantly illuminated by Example 1 of § 9.4. That is one of the most instructive single examples to be found in this text, and anyone who studies it carefully will find his understanding greatly improved.

12.7   The implicit function theorem

An important introduction to this theorem is given in § 8 of Chapter VIII. The student is advised to reread it at this point. The introduction which we give here is somewhat similar but is intended to introduce the implicit function theorem as an extension to nonlinear analysis of a well-known theorem in linear algebra.

If one has three simultaneous homogeneous linear equations in five variables, for example,

(12.7–1)
$$a_1x_1 + a_2x_2 + \cdots + a_5x_5 = 0$$
$$b_1x_1 + b_2x_2 + \cdots + b_5x_5 = 0$$
$$c_1x_1 + c_2x_2 + \cdots + c_5x_5 = 0,$$

it makes sense to ask whether the system implies that some three of the unknowns, say x_3, x_4, and x_5 are functions of the other two. By transposing the terms involving the other two variables to the other side and using Cramer's rule, we see that a necessary and sufficient condition that x_3, x_4, and x_5 be implicitly determined uniquely as functions of x_1 and x_2 is that the determinant

$$\begin{vmatrix} a_3 & a_4 & a_5 \\ b_3 & b_4 & b_5 \\ c_3 & c_4 & c_5 \end{vmatrix}$$

be different from zero. In other words, by using the inverse function theorem for linear functions, we have obtained an implicit function theorem. We shall now obtain a partial extension to nonlinear functions, that is, to systems like

$$(12.7\text{–}2) \qquad \begin{aligned} f_1(x_1, x_2, x_3, x_4, x_5) &= 0 \\ f_2(x_1, x_2, x_3, x_4, x_5) &= 0 \\ f_3(x_1, x_2, x_3, x_4, x_5) &= 0 \end{aligned}$$

which reduce to (12.7–1) when the functions are linear. Notice that in the linear case, the determinant above, whose nonvanishing is so important, is the same as the Jacobian determinant

$$\begin{vmatrix} \dfrac{\partial f_1}{\partial x_3} & \dfrac{\partial f_1}{\partial x_4} & \dfrac{\partial f_1}{\partial x_5} \\[2ex] \dfrac{\partial f_2}{\partial x_3} & \dfrac{\partial f_2}{\partial x_4} & \dfrac{\partial f_2}{\partial x_5} \\[2ex] \dfrac{\partial f_3}{\partial x_3} & \dfrac{\partial f_3}{\partial x_4} & \dfrac{\partial f_3}{\partial x_5} \end{vmatrix}.$$

Expressed in this way, this same condition will also be of central importance in the nonlinear case.

The equations (12.7–2) can advantageously be written

$$(12.7\text{–}3) \qquad f_i(x_1, x_2, y_1, y_2, y_3) = 0 \qquad (i = 1, 2, 3).$$

The change in notation, writing y_1, y_2, y_3 instead of x_3, x_4, x_5, is to help us remember that we want to solve for the y's in terms of the x's. Further, the ordered pair (x_1, x_2) will be denoted by the vector $\mathbf{x}$ and we shall let $(y_1, y_2, y_3) = \mathbf{y}$. The three equations then become

$$f_i(\mathbf{x}, \mathbf{y}) = 0 \qquad (i = 1, 2, 3).$$

For each i, f_i is defined on ordered pairs of the form $(\mathbf{x}, \mathbf{y})$ where $\mathbf{x} \in \mathbf{R}^2$ and $\mathbf{y} \in \mathbf{R}^3$, and each of the three functions is real valued. Since an ordered triple of real-valued functions is commonly thought of as a vector-valued function taking its values in $\mathbf{R}^3$, the system of three equations can conveniently be thought of as

$$(12.7\text{–}4) \qquad \mathbf{f}(\mathbf{x}, \mathbf{y}) = 0.$$

We have to remember of course that $x \in R^2$ and $y \in R^3$, and that f takes its values in R^3.

If A and B are sets, then the set of all ordered pairs of the form (a, b), where $a \in A$ and $b \in B$ is called the *Cartesian product* of A and B and is denoted by $A \times B$. Ordered pairs (x, y) introduced in the preceding paragraph belong to $R^2 \times R^3$; thus we say that f is a function from some subset of $R^2 \times R^3$ to R^3. Since the points of $R^2 \times R^3$ are ordered pairs of the form (x, y), where x is an ordered pair of real numbers and y is an ordered triple of real numbers, (x, y) can be identified with an ordered quintuple of real numbers. Hence, it is common to identify $R^2 \times R^3$ with R^5.

In order to get clearly in mind what is meant by solving (12.7–4) for y as a function of x, consider the following very simple special cases.

$$f(x, y) = 3x + 2y - 5 = 0$$

This can be solved at a glance to get

$$y = \phi(x) = \frac{1}{2}(5 - 3x).$$

The essential feature of this function $\phi(x)$ is that if we substitute it for y in the equation $f(x, y) = 0$, we get an identity, namely

$$f(x, \phi(x)) = 3x + 2\phi(x) - 5 = 3x + (5 - 3x) - 5 \equiv 0.$$

In this special instance, the identity holds for all x. In more complicated cases, we may have to settle for identities which are valid only over some subset of the vector space to which x belongs. And in the more general case of the equation $f(x, y) = 0$, to solve for y in terms of x means to find a function $\phi(x)$ such that $f[x, \phi(x)] \equiv 0$, at least for all x belonging to some set in R^2.

Consider the set of all pairs (x, y) such that $f(x, y) = 0$. We call this the *solution set* for the given equation. To avoid dealing with a situation which is of no interest, we must assume that the equation does have solutions, that is, that the solution set is not empty. We are interested in knowing whether this set has the property that when (x_1, y_1) and (x_2, y_2) both belong and $x_1 = x_2$, then necessarily $y_1 = y_2$. If it does have this property, then y is determined as a function of x.

There are instances in which the solution set does not determine y uniquely as a function of x. For example, to take a case in which x and y are both in R, suppose $f(x, y) = x^2 + y^2 - 1$. Then $(0, 1)$ and $(0, -1)$ are both in the solution set, so that there are two values of y (instead of only one) corresponding to $x = 0$. But if we start with the pair $(0, 1)$ and confine attention to pairs (x, y) of the solution set for which x is close to 0 and y is close to 1, we find that this restricted portion of the solution set does define y uniquely as a function of x, the formula being $y = \sqrt{1 - x^2}$ (positive square root). This restriction of attention to all points (x, y) of the solution set close to a particular point (a, b) of the solution set is a standard feature of implicit function theorems.

Now for the implicit function theorem. We shall state and prove it for the case we have been discussing of three equations with five variables, but only trivial changes in notation are required to get a proof valid for m equations in n variables, where $m < n$.

THEOREM IX. (AN IMPLICIT FUNCTION THEOREM). *Suppose that $\mathcal{W}$ is an open subset of $R^5 = R^2 \times R^3$ and that $\mathbf{f}$ is a differentiable function of class $C^{(1)}$ from $\mathcal{W}$ to R^3. Assume further that there is a point $(\mathbf{a}, \mathbf{b})$ in $\mathcal{W}$ such that $\mathbf{f}(\mathbf{a}, \mathbf{b}) = 0$, and assume that the differential of $\mathbf{f}(\mathbf{a}, \mathbf{y})$, as a function of $\mathbf{y}$, is nonsingular at $\mathbf{y} = \mathbf{b}$. Then the equation $\mathbf{f}(\mathbf{x}, \mathbf{y}) = 0$ determines $\mathbf{y}$ uniquely as a $C^{(1)}$ function of $\mathbf{x}$ near $(\mathbf{a}, \mathbf{b})$. More precisely, there is some neighborhood E of $\mathbf{b}$ in R^3 and some neighborhood S of $\mathbf{a}$ in R^2 and a continuously differentiable function ϕ from S to R^3 such that $\phi(\mathbf{x}) \in E$ and $\mathbf{f}(\mathbf{x}, \phi(\mathbf{x})) = 0$ whenever $\mathbf{x}$ is in S. Furthermore, the only points $\mathbf{y}$ of E which satisfy $\mathbf{f}(\mathbf{x}, \mathbf{y}) = 0$ for some $\mathbf{x}$ in S are those for which $\mathbf{y} = \phi(\mathbf{x})$.*

PROOF.   Our main tool is the inverse function theorem for functions from R^n to R^n. Since our function $\mathbf{f}$ is from R^5 to R^3, the tool seems ill-suited to the task, but this disparity can easily be remedied. In the special case of the linear problem (12.7–1), we reduced it to a problem about functions from R^3 to R^3 by transposing the x_1 and the x_2 to the other side. In the nonlinear case (12.7–3), there is no way to get the x_1 and x_2 to the other side, so we artificially inflate the problem to one about functions from R^5 to R^5 in the following trivial way. We define functions $F_1, \cdots, F_5$ as follows:

$$F_1(x_1, x_2, y_1, y_2, y_3) = x_1$$
$$F_2(x_1, x_2, y_1, y_2, y_3) = x_2$$
$$F_3(x_1, x_2, y_1, y_2, y_3) = f_1(x_1, x_2, y_1, y_2, y_3)$$
$$F_4(x_1, x_2, y_1, y_2, y_3) = f_2(x_1, x_2, y_1, y_2, y_3)$$
$$F_5(x_1, x_2, y_1, y_2, y_3) = f_3(x_1, x_2, y_1, y_2, y_3)$$

Notice that this vector-valued function $\mathbf{F}$ can be defined more briefly as follows:

$$\mathbf{F}(\mathbf{x}, \mathbf{y}) = (\mathbf{x}, \mathbf{f}(\mathbf{x}, \mathbf{y})).$$

Since $\mathbf{f}$ is continuously differentiable, Theorem IV tells us that the same is true of **F**. By Theorem IV, the differential of **F** at (**a**, **b**) is represented by the Jacobian matrix

$$
\begin{pmatrix}
1 & 0 & 0 & 0 & 0 \\
0 & 1 & 0 & 0 & 0 \\
\dfrac{\partial f_1}{\partial x_1} & \dfrac{\partial f_1}{\partial x_2} & \dfrac{\partial f_1}{\partial y_1} & \dfrac{\partial f_1}{\partial y_2} & \dfrac{\partial f_1}{\partial y_3} \\
\dfrac{\partial f_2}{\partial x_1} & \dfrac{\partial f_2}{\partial x_2} & \dfrac{\partial f_2}{\partial y_1} & \dfrac{\partial f_2}{\partial y_2} & \dfrac{\partial f_2}{\partial y_3} \\
\dfrac{\partial f_3}{\partial x_1} & \dfrac{\partial f_3}{\partial x_2} & \dfrac{\partial f_3}{\partial y_1} & \dfrac{\partial f_3}{\partial y_2} & \dfrac{\partial f_3}{\partial y_3}
\end{pmatrix},
$$

where all the partial derivatives are evaluated at (**a**, **b**). We wish to show that the linear transformation represented by this matrix is nonsingular.

Notice that the 3×3 submatrix in the lower right-hand corner is nonsingular, because it represents the differential of **f**(**a**, **y**) as a function of **y** at **y** = **b**; in the upper left-hand corner we have the 2×2 identity matrix having determinant 1. So the determinant of the 5×5 matrix has the same value as that of the determinant of the 3×3 submatrix in the lower right-hand corner. In other words, the differential of **F** at (**a**, **b**) is nonsingular.

We observe that **F** maps (**a**, **b**) into the point (**a**, 0) of $R^2 \times R^3$. By applying the inverse function theorem (Theorem VIII) to **F**, we see that there must exist a neighborhood $\mathcal{U}$ of (**a**, **b**) in R^5 such that $\mathcal{U}$ is contained in $\mathcal{W}$ and such that **F** defines a one-to-one mapping of $\mathcal{U}$ onto a neighborhood $\mathcal{V}$ of (**a**, 0) in R^5. Moreover, the inverse mapping $\mathbf{F}^{-1}$ is of class $C^{(1)}$ on $\mathcal{V}$. It is easy to prove (see Exercise 15) that every neighborhood of (**a**, **b**) contains a neighborhood of this point which is a Cartesian product $D \times E$, where D is a neighborhood of **a** in R^2 and E is a neighborhood of **b** in R^3; we can assume that $\mathcal{U}$ itself is such a Cartesian product, and shall do this for convenience.

Let us write points of $\mathcal{V}$ in the form (**x**, **z**), where **x** is in R^2 and **z** belongs to R^3. To each (**x**, **z**) in $\mathcal{V}$, there corresponds a unique point $\mathbf{F}^{-1}(\mathbf{x}, \mathbf{z})$ = (**x**, **y**) in $\mathcal{U}$ such that $f(\mathbf{x}, \mathbf{y}) = \mathbf{z}$ and **F**(**x**, **y**) = (**x**, **z**). Thus, **y** is determined uniquely by **x** and **z**, and this defines **y** as a function of **x** and **z**, say **y** = **g**(**x** **z**). In particular, **g**(**a**, 0) = **b**. Then $\mathbf{F}^{-1}(\mathbf{x}, \mathbf{z})$ = (**x**, **g**(**x**, **z**)). Since $\mathbf{F}^{-1}$ is of class $C^{(1)}$, it is readily seen that **g** is also of class $C^{(1)}$.

Let S be the set of **x**'s such that (**x**, 0) is in $\mathcal{V}$. Clearly, S is contained in D, since (**x**, 0) comes from some point in $\mathcal{U} = D \times E$ by the mapping **F**. Observe that (**a**, 0) comes from (**a**, **b**). It is easy to prove (see Exercise 14) that S is an open subset of R^2. Let us define a function ϕ on S by the formula $\phi(\mathbf{x}) = \mathbf{g}(\mathbf{x}, 0)$. Observe that $\phi(\mathbf{x})$ is in R^3 and that (**x**, $\phi(\mathbf{x})$) is in $\mathcal{U}$ when **x** is in S. In particular, $\phi(\mathbf{a})$ = **b** and $\phi(\mathbf{x})$ is in E. Observe also that $\mathbf{F}^{-1}(\mathbf{x}, 0)$

$= (\mathbf{x}, \phi(\mathbf{x}))$. This means that $(\mathbf{x}, 0) = F(\mathbf{x}, \phi(\mathbf{x})) = (\mathbf{x}, f[\mathbf{x}, \phi(\mathbf{x})])$, and hence, that $f(\mathbf{x}, \phi(\mathbf{x})) = 0$. The function ϕ is of class $C^{(1)}$ on S, because $\phi(\mathbf{x}) = g(\mathbf{x}, 0)$ and we know that g is of class $C^{(1)}$.

Finally, to prove the last assertion in Theorem IX, assume that $\mathbf{y}$ is a point of E such that $f(\mathbf{x}, \mathbf{y}) = 0$ for some $\mathbf{x}$ in S. Then $F(\mathbf{x}, \mathbf{y}) = (\mathbf{x}, 0)$ is in $\mathcal{V}$, and hence, since the mapping of $\mathcal{U}$ onto $\mathcal{V}$ is one-to-one, we are assured that $\mathbf{y} = g(\mathbf{x}, 0) = \phi(\mathbf{x})$. This completes the proof.

In conclusion, we shall now state a more general form of the implicit function theorem. Except for minor changes in notation, the proof is the same as that for the special case just treated.

THEOREM X (THE IMPLICIT FUNCTION THEOREM). *Suppose that $\mathcal{W}$ is an open subset of R^{p+q} (which we shall identify with $R^p \times R^q$) and let f be a continuously differentiable function from $\mathcal{W}$ to R^q. Assume further that there is a point $(\mathbf{a}, \mathbf{b})$ in $\mathcal{W}$ such that $f(\mathbf{a}, \mathbf{b}) = 0$, and such that the differential of $f(\mathbf{a}, \mathbf{y})$ as a function of $\mathbf{y}$ is nonsingular at $\mathbf{y} = \mathbf{b}$. Then the equation $f(\mathbf{x}, \mathbf{y}) = 0$ determines $\mathbf{y}$ uniquely as a continuously differentiable function of $\mathbf{x}$ near $(\mathbf{a}, \mathbf{b})$. More precisely, there is some neighborhood E of $\mathbf{b}$ in R^q, some neighborhood S of $\mathbf{a}$ in R^p, and a continuously differentiable function ϕ from S to R^q such that $\phi(\mathbf{x}) \in E$ and $f(\mathbf{x}, \phi(\mathbf{x})) = 0$ whenever $\mathbf{x}$ is in S. Furthermore, the only points $\mathbf{y}$ of E which satisfy $f(\mathbf{x}, \mathbf{y}) = 0$ for some $\mathbf{x}$ in S are those for which $\mathbf{y} = \phi(\mathbf{x})$.*

12.8   The equivalence of norms in finite dimensional vector spaces

If one refers to the definitions of open set, continuity, and differentiability in § 11.7 and § 12.1, he will see that each of the three involves a norm. It was understood there that $||\ \ ||$ denoted the usual "square-root-of-the-sum-of-the-squares" norm which is commonly referred to as the *Euclidean norm*. From several of the exercises at the end of Chapter XI, however, we are aware of the fact that it is possible to define many other norms on R^n, and in terms of each of them, each of the three definitions mentioned above is still meaningful. But it is not at all obvious to what extent, if any, the concepts being defined depend on the norms being used.

Consider again, for example, the definition of continuity. The function f from the open subset $\mathcal{U}$ of R^n to R^m is said to be continuous at the point $\mathbf{a}$ of $\mathcal{U}$ if the following is true: for each $\epsilon > 0$, there exists a $\delta > 0$ such that $||f(\mathbf{x}) - f(\mathbf{a})|| < \epsilon$ if $||\mathbf{x} - \mathbf{a}|| < \delta$. It is usually assumed in this definition that the first norm is the Euclidean norm in R^m and the second is the Euclidean norm in R^n. Suppose that f happens to be continuous at $\mathbf{a}$ with this assumption about the norms. Would f still be continuous at $\mathbf{a}$ if we used different norms on R^n and R^m? This and the analogous questions about openness and differentiability are of profound importance in analysis, and we shall devote this entire section to their study. The term *analysis*, incidentally, does not have a very precise meaning; frequently, it is used to denote that branch of mathematics which studies all kinds of limiting processes in finite-dimensional vector spaces. When the spaces are infinite-dimensional, the term *functional analysis* is usually employed because many infinite dimensional spaces are spaces in which the vectors are functions, as in Example 2 of § 11.

As the title of this section suggests, our problems have a very gratifying solution in R^n, in that the three definitions are independent of the norm used. In particular, if the function in the preceding paragraph is continuous at **a** when the Euclidean norms are used, then it is continuous regardless of what norms are used in the definition of continuity. Before stating precisely what we mean by the equivalence of norms and proving that all norms on a finite dimensional vector space are equivalent, we shall consider a particular problem in R^2, where the simple geometry makes the essential ideas easy to grasp.

We have seen in Exercise 7 of Chapter XI that the function $|\ |\ \ |\ |_1$ from R^2 to R defined by

$$|\ |\ \mathbf{x}\ |\ |_1 = |\ x_1\ | + |\ x_2\ |$$

is a norm on R^2, and that the unit sphere $S_1(0, 1)$ in this norm is the square with its vertices at $(1, 0)$, $(0, 1)$, $(-1, 0)$, and $(0, -1)$. Throughout this section we shall always use $|\ |\ \ |\ |$ to denote the familiar Euclidean norm. $B(\mathbf{x}, r)$ will denote the open ball of radius r centered at $\mathbf{x}$ when the Euclidean norm is used, and $B_1(\mathbf{x}, r)$ denotes the corresponding ball for the norm $|\ |\ \ |\ |_1$.

$$B(\mathbf{x}, r) = \{\mathbf{y} : |\ |\ \mathbf{x} - \mathbf{y}\ |\ | < r\} = \{(y_1, y_2) : \sqrt{(x_1 - y_1)^2 + (x_2 - y_2)^2} < r\}$$
$$B_1(\mathbf{x}, r) = \{\mathbf{y} : |\ |\ \mathbf{x} - \mathbf{y}\ |\ |_1 < r\} = \{(y_1, y_2) : |\ x_1 - y_1\ | + |\ x_2 - y_2\ | < r\}.$$

From Fig. 84 it is clear that

(12.8–1) $$B\left(0, \frac{\sqrt{2}}{2}\, r\right) \subset B_1(0, r) \subset B(0, r),$$

where "$\subset$" means "is contained in" or "is a subset of."

It is more or less obvious that $B(\mathbf{x}, r)$ and $B_1(\mathbf{x}, r)$ are just the translates by the vector $\mathbf{x}$ of $B(0, r)$ and $B_1(0, r)$, respectively. The rest of this paragraph is only for those who wish to see a detailed justification of this statement. It should be sufficient to show that $B(\mathbf{x}, r)$ is what one gets by translating $B(0, r)$ by the vector $\mathbf{x}$. Therefore, suppose that $\mathbf{y} \in B(0, r)$. This means that $|\ |\ \mathbf{y}\ |\ | < r$. We wish to show that $\mathbf{y} + \mathbf{x} \in B(\mathbf{x}, r)$. But this merely means that $|\ |\ \mathbf{x} - (\mathbf{y} + \mathbf{x})\ |\ | < r$, which is equivalent to $|\ |\ \mathbf{y}\ |\ | < r$ and was given to start with. Thus, if $\mathbf{y} \in B(0, r)$ then $\mathbf{y} + \mathbf{x} \in B(\mathbf{x}, r)$.

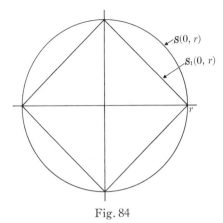

Fig. 84

Now conversely, assume that $\mathbf{y} \in B(\mathbf{x}, r)$, that is, $|| \mathbf{y} - \mathbf{x} || < r$. But this is exactly what it means to say that $\mathbf{y} - \mathbf{x} \in B(\mathbf{0}, r)$. Since $\mathbf{y} = \mathbf{x} + (\mathbf{y} - \mathbf{x})$, this means that if $\mathbf{y} \in B(\mathbf{x}, r)$, then there is some vector, namely $\mathbf{y} - \mathbf{x} \in B(\mathbf{0}, r)$, such that $\mathbf{y} = \mathbf{x} + (\mathbf{y} - \mathbf{x})$.

From the preceding paragraph and (12.8–1), we can now assert that for each $\mathbf{x} \in R^2$ and each positive number r,

$$(12.8\text{–}2) \qquad B\left(\mathbf{x}, \frac{\sqrt{2}}{2} r\right) \subset B_1(\mathbf{x}, r) \subset B(\mathbf{x}, r).$$

This shows that *every open ball in the $|| \;\; ||_1$ norm contains a concentric open ball in the Euclidean norm and vice versa.*

Now let E be any subset of R^2. We have said in § 11.7 that $\mathbf{x}$ is an interior point of E in case there is some positive number r such that $B(\mathbf{x}, r) \subset E$. By (12.8–2) we know that $B_1(\mathbf{x}, r) \subset B(\mathbf{x}, r)$, and therefore, $B_1(\mathbf{x}, r) \subset E$. Thus, if E contains an open ball in the $|| \;\; ||$ norm centered at $\mathbf{x}$ it also contains an open ball in the $|| \;\; ||_1$ norm centered at $\mathbf{x}$. And (12.8–2) also shows that the converse is true, that is, if E contains an open ball in the $|| \;\; ||_1$ norm centered at $\mathbf{x}$, then it also contains one in the $|| \;\; ||$ norm centered at the same point. This shows that when defining interior points of subsets of R^2, it makes no difference whether we use the Euclidean norm or the $|| \;\; ||_1$ norm. Since an open set is one consisting entirely of interior points, we have proved that we get the same open sets using one of these two norms as the other. What we wish to prove is that this is true not only in R^2 but in R^n, for every positive integer n.

The essential step in the argument which we have just completed was that every open ball in one norm contained a concentric open ball in the other. This was obvious from elementary plane geometry, but we can also prove it without the help of any figures as follows: Since

$$|| \mathbf{x} || = \sqrt{x_1{}^2 + x_2{}^2} \quad \text{and} \quad || \mathbf{x} ||_1 = | x_1 | + | x_2 | ,$$

it is obvious that

$$||\,x\,|| \le ||\,x\,||_1 \qquad \text{for all x.}$$

And by the Schwartz inequality,

$$||\,x\,||_1 \le \sqrt{2}\,\sqrt{x_1^2 + x_2^2} = \sqrt{2}\,||\,x\,||\,.$$

So we have that

$$||\,x\,|| \le ||\,x\,||_1 \le \sqrt{2}\,||\,x\,||\,,$$

from which we get

$$\frac{\sqrt{2}}{2} \le \frac{||\,x\,||}{||\,x\,||_1} \le 1 \qquad \text{for all } x \ne 0.$$

From these inequalities we can get (12.8–2) without the help of any pictures. Actually, it is just as easy to obtain the following more general result, which we shall find useful.

LEMMA I.   *If $||\ \ ||_2$ and $||\ \ ||_3$ are any two norms on a vector space, and if m and M are positive constants such that*

(12.8–3) $$0 < m < \frac{||\,x\,||_2}{||\,x\,||_3} \le M \qquad \text{for all } x \ne 0,$$

then

(12.8–4) $$B_2(x, mr) \subset B_3(x, r) \subset B_2(x, Mr),$$

where the subscripts on the B's of course indicate which norm is being used in the definition of the ball.

PROOF.   The proof of the first inclusion goes as follows:

$$y \in B_2(x, mr) \Leftrightarrow ||\,x - y\,||_2 < mr,$$

and by (12.8–3),

$$m\,||\,x - y\,||_3 \le ||\,x - y\,||_2.$$
$$y \in B_2(x, mr) \Rightarrow m\,||\,x - y\,||_3 < mr$$
$$\Rightarrow ||\,x - y\,||_3 < r$$
$$\Rightarrow y \in B_3(x, r).$$

This means that $B_2(x\ \ mr) \subset B_3(x, r)$. The second inclusion in (12.8–4) is proved similarly. In other words, condition (12.8–3) implies that each ball in one norm contains a concentric ball in the other. The simple reasoning which we used in comparing $||\ \ ||$ and $||\ \ ||_1$ in R^2 can be applied here to show that (12.8–3) implies that $||\ \ ||_2$ and $||\ \ ||_3$ determine the same open sets. This observation suggests the following concise definition of equivalence of norms:

DEFINITION. If $||\ ||_2$ and $||\ ||_3$ are any two norms on a vector space, then they are said to be *equivalent* in case there exist two numbers, m and M, such that

$$0 < m \le \frac{||\mathbf{x}||_2}{||\mathbf{x}||_3} \le M \qquad \text{for all } \mathbf{x} \ne 0,$$

or, equivalently in the R^n spaces,

(12.8–5) $$0 < m \le \frac{||\mathbf{x}||_2}{||\mathbf{x}||_3} \le M \qquad \text{for all } \mathbf{x} \text{ such that } ||\mathbf{x}|| = 1.$$

The fact that the above definitions are equivalent in R^n is seen by writing

$$\frac{||\mathbf{x}||_2}{||\mathbf{x}||_3} = \frac{\dfrac{1}{||\mathbf{x}||}\,||\mathbf{x}||_2}{\dfrac{1}{||\mathbf{x}||}\,||\mathbf{x}||_3} = \frac{\left|\left|\dfrac{\mathbf{x}}{||\mathbf{x}||}\right|\right|_2}{\left|\left|\dfrac{\mathbf{x}}{||\mathbf{x}||}\right|\right|_3} \qquad \text{if} \qquad \mathbf{x} \ne 0.$$

Since $\mathbf{x}/||\mathbf{x}||$ is a unit vector in the Euclidean norm, all the values which the ratio $||\mathbf{x}||_2/||\mathbf{x}||_3$ ever takes on are assumed on the Euclidean unit sphere, $\{\mathbf{x} : ||\mathbf{x}|| = 1\}$. Thus, (12.8–5) is sufficient to give that the two norms are equivalent, which in turn is enough to tell us that they determine the same open sets.

Having now defined equivalence of norms and pointed out one of its consequences, we shall proceed to show that, on finite dimensional vector spaces, all norms are equivalent. In the first part of this section we proved the equivalence of $||\ ||$ and $||\ ||_1$ on R^2. This very special result was obtained by relying largely on the simple geometry of the plane. In higher dimensional spaces the geometry is not so simple, and some norms are more complicated than the $||\ ||_1$ norm. But there is a property, common to all norms on R^n, which enables us to overcome these difficulties. Let $||\ ||_0$ denote any norm on R^n. Then this function from R^n to the non-negative real numbers has to be continuous when the usual definition of continuity is used, that is, when the norm used in the definition of continuity is the Euclidean norm. We prove this in the following lemma:

LEMMA II.   *Every norm on R^n is continuous with respect to the Euclidean norm.*

PROOF.   Suppose that $||\ ||_0$ is any norm on R^n, $||\ ||$ is the Euclidean norm, and $\{\mathbf{e}_k\}_{k=1}^n$ is the standard basis. For each $\mathbf{x} \in R^n$,

$$||\mathbf{x}||_0 = ||x_1\mathbf{e}_1 + x_2\mathbf{e}_2 + \cdots + x_n\mathbf{e}_n||_0$$
$$\le |x_1|\ ||\mathbf{e}_1||_0 + |x_2|\ ||\mathbf{e}_2||_0 + \cdots + |x_n|\ ||\mathbf{e}_n||_0$$
$$\le \left(\sum_1^n x_k^2\right)^{1/2} \left(\sum_1^n ||\mathbf{e}_k||_0^2\right)^{1/2} = B_0||\mathbf{x}||,$$

where $B_0 = \left(\sum_1^n || \mathbf{e}_k ||_0^2 \right)^{1/2}$.

In other words, for each norm $|| \quad ||_0$, there exists a number B_0 such that

$$|| \mathbf{x} ||_0 \leq B_0 || \mathbf{x} || \qquad \text{for all } \mathbf{x} \in R^n.$$

To see that $|| \quad ||_0$ is continuous at each point $\mathbf{x}$, let ϵ denote any positive number and observe that by (11.5–2) and the foregoing inequality,

$$\left| || \mathbf{x} ||_0 - || \mathbf{y} ||_0 \right| \leq || \mathbf{x} - \mathbf{y} ||_0 \leq B_0 || \mathbf{x} - \mathbf{y} || ,$$

and this can be made less than ϵ by requiring that $|| \mathbf{x} - \mathbf{y} ||$ be less than ϵ/B_0. Since this number ϵ/B_0 works for all $\mathbf{x}$, we have again an example of *uniform* continuity. (See Corollary of Theorem III, Chapter XI.)

We are now ready to prove our main result of this section.

THEOREM XI.   *If* $|| \quad ||_0$ *and* $|| \quad ||^*$ *are any two norms on* R^n, *then they are equivalent.*

PROOF.   Let $S(0, 1)$ denote the closed unit sphere in R^n. It is bounded by definition, and in Exercise 15 of Chapter XI we proved that it is closed. By Lemma II, both $|| \quad ||_0$ and $|| \quad ||^*$ are continuous with respect to the Euclidean norm. By Theorems I and II of Chapter V, we can infer that $|| \quad ||_0$ and $|| \quad ||^*$ have maximum and minimum values on $S(0, 1)$. Let M_0 and m_0 be the maximum and minimum values of $|| \quad ||_0$ on $S(0, 1)$, and let M^* and m^* be the corresponding extremes for $|| \quad ||^*$. Since a norm vanishes only on the zero vector, both m_0 and m^* are positive; therefore, if $\mathbf{x} \in S(0, 1)$, then

$$0 < \frac{m_0}{M^*} \leq \frac{|| \mathbf{x} ||_0}{|| \mathbf{x} ||^*} \leq \frac{M_0}{m^*} .$$

Letting $m_0/M^* = m$ and $M_0/m^* = M$, we have (12.8–5) and the proof is complete.

Even though we have proved the equivalence of norms on the R^n spaces, it may not be immediately clear that all norms on $\text{Hom}(R^n, R^m)$ are equivalent. The R^n spaces consist of ordered n-tuples of real numbers, whereas $\text{Hom}(R^n, R^m)$ consists of linear transformations. Recall, however, that each of these linear transformations has a unique matrix representation in terms of the standard bases. Each of these representing matrices is made up of $m \times n$ real numbers in m rows having n elements each. We can, therefore, get a one-to-one correspondence between $\text{Hom}(R^n, R^m)$ and $R^{m \times n}$ by associating each matrix with the $m \times n$-tuple obtained by stringing out the elements of the matrix in the natural order of reading, from left to right and top to bottom.

What we get in this way is more than merely a one-to-one correspondence—it is a one-to-one correspondence which is preserved by the vector space operations of addition and scalar multiplication. In other words, if T_1 and T_2 of $\text{Hom}(R^n, R^m)$ correspond to y_1 and y_2 of $R^{m \times n}$, then $T_1 + T_2$ corresponds to $y_1 + y_2$, and cT_1 corresponds to cy_1 for every scalar c. This special kind of one-to-one correspondence between vector spaces is called an *isomorphism*. It is important to us here because it gives us a one-to-one correspondence between the sets of all norms on the two spaces. If $|| \ \ ||_1$ is a norm on $\text{Hom}(R^n, R^m)$, then we can get a norm $|| \ \ ||_2$ on $R^{m \times n}$ by defining $|| \ y \ ||_2$ to be $|| \ T \ ||_1$, where T is that linear transformation which corresponds to y under the isomorphism described above. Similarly, for each norm on $R^{m \times n}$ one gets a norm on $\text{Hom}(R^n, R^m)$.

We have now shown that the isomorphism which we have introduced between $\text{Hom}(R^n, R^m)$ and $R^{n \times m}$ leads to a very natural one-to-one correspondence between the norms on the two spaces. From this we can easily show that the equivalence of norms on the first follows from the equivalence of those on the second. In doing this we shall use the definition of equivalence expressed in (12.8–3) instead of (12.8–5) because the Euclidean norm, $|| \ \ ||$, is not defined on $\text{Hom}(R^n, R^m)$.

Let $|| \ \ ||_0$ and $|| \ \ ||_*$ be any two norms on $\text{Hom}(R^n, R^m)$, and let $|| \ \ ||_0'$ and $|| \ \ ||_*'$ be the two corresponding norms on $R^{n \times m}$. For each T in the first space there is (under the isomorphism described above) a unique vector x in $R^{n \times m}$ and the values taken on by $|| \ \ ||_0$ and $|| \ \ ||_*$ at T are, respectively, those of $|| \ \ ||_0'$ and $|| \ \ ||_*'$ at x. Hence, the set of values taken on by $|| \ T \ ||_0/|| \ T \ ||_*$ as T varies over the nonzero members of $\text{Hom}(R^n, R^m)$ is exactly the same as the values taken on by $|| \ x \ ||_0'/|| \ x \ ||_*'$ as x varies over the nonzero vectors in $R^{n \times m}$. By the equivalence of $|| \ \ ||_0'$ and $|| \ \ ||_*'$, we know that there are numbers, m and M, such that

$$0 < m \le \frac{|| \ x \ ||_0'}{|| \ x \ ||_*'} \le M \qquad \text{for all } x \ne 0.$$

$$\therefore \ 0 < m \le \frac{|| \ T \ ||_0}{|| \ T \ ||_*} \le M \qquad \text{for all } T \ne 0.$$

Thus, $|| \ \ ||_0$ and $|| \ \ ||_*$ are equivalent.

This equivalence of norms which we have proved for the R^n spaces and the $\text{Hom}(R^n, R^m)$ spaces holds for all finite dimensional spaces. The main step in the proof is showing that every space of dimension k is isomorphic to R^k. Once this is done, the rest of the proof goes exactly as it did for R^k. (See Exercise 20.)

We now have shown that on a finite dimensional space all norms are equivalent, and that equivalent norms determine the same open sets. It follows from this that on finite dimensional spaces all limits turn out to be the same, regardless of the norm used. Suppose that $\mathbf{f}$ is a function from R^n to R^m. Then the sentence $\lim_{\mathbf{x} \to \mathbf{a}} \mathbf{f}(\mathbf{x}) = \mathbf{b}$ means that if $\epsilon > 0$, there exists a $\delta > 0$ such that $||\mathbf{f}(\mathbf{x}) - \mathbf{b}|| < \epsilon$ for all $\mathbf{x}$ satisfying $||\mathbf{x} - \mathbf{a}|| < \delta$. It is usually understood that the first norm is the Euclidean norm on R^m and that the second is the Euclidean norm on R^n. In more geometrical language, this can be put as follows: each open ball centered at $\mathbf{f}(\mathbf{a})$ in R^m contains the image of some open ball centered at $\mathbf{a}$ in R^n—where the balls are defined by the Euclidean norms. Since, by Lemma I, each open ball in one norm contains a concentric open ball in every other norm, one sees immediately that if $\lim_{\mathbf{x} \to \mathbf{a}} \mathbf{f}(\mathbf{x}) = \mathbf{b}$ for one particular choice of norms in R^n and R^m, then the same must be true for all choices of norms.

To say that $\mathbf{f}$ is continuous at $\mathbf{a}$ means the same as $\lim_{\mathbf{x} \to \mathbf{a}} \mathbf{f}(\mathbf{x}) = \mathbf{f}(\mathbf{b})$. So by the preceding paragraph, the continuity of $\mathbf{f}$ does not depend on the norm used. And similarly, differentiability is also independent of the choice of norm, since it is defined in terms of a certain limit existing and being equal to zero.

The norm which we actually used on $\text{Hom}(R^n, R^m)$ probably seemed somewhat artificial when we first introduced it in § 11.8. It is good to know that if one were to replace it with some other norm which might seem more natural to him, the set of invertible operators would still be open, the function in Theorem V of Chapter XI would still be continuous, and the inverse- and implicit-function theorems would still be true.

In contrast to the foregoing, however, if a space is not of finite dimension it is always possible to find two norms on the space which are not equivalent. This makes it possible for a sequence to converge in terms of one norm but fail to converge with respect to another—a possibility which can be realized in the space of a continuous function on the unit interval (see Example 2 of § 11). We shall refer here to this space as $\mathcal{C}$. It can be proven (see Exercise 10, Chapter XI) that one norm on this space can be defined by

$$||f||_1 = \max_{0 \le x \le 1} |f(x)|,$$

and that another norm on $\mathcal{C}$ can be defined as follows:

$$||f||_2 = \left(\int_0^1 f^2(x)\, dx \right)^{1/2}.$$

In the space $\mathcal{C}$, define

$$g_n(x) = 0 \quad \text{for } 2^{-n} \le x \le 1,$$
$$g_n(2^{-n-1}) = n^2 \quad \text{and} \quad g_n(0) = 0,$$

and let g^n have a straight-line graph on $[0, 2^{-n-1}]$ and on $[2^{-n-1}, 2^{-n}]$. If $\{f_n\}_1^\infty$ is that sequence in $\mathcal{C}$ defined by

$$f_n(x) = \sqrt{g_n(x)},$$

then $||f_n||_1 = n$ for all n, and

$$||f_n||_2 = n2^{-(n+1)/2} \qquad \text{for all } n.$$

From this it is easy to see that

$$\lim_{n \to \infty} f_n = 0 \qquad \text{(the zero function)}$$

in the norm $||\ \ ||_2$, but that in the $||\ \ ||_1$ norm, the sequence $\{f_n\}_1^\infty$ does not converge. If $||\ \ ||_1$ and $||\ \ ||_2$ were equivalent, then any sequence would either fail to converge in either norm, or else it would converge in both, and to the same limit. Hence, these two norms on $\mathcal{C}$ are not equivalent.

EXERCISES

1. Prove that if $\mathbf{f}$ is a function from some open subset of R^n to R^m, and if $\mathbf{f}$ is differentiable at some point $\mathbf{a}$ of its domain, then $\mathbf{f}$ must be continuous at $\mathbf{a}$.

2. Prove that if a function $\mathbf{f}$ from an open subset $\mathcal{U}$ of R^n to R^m is differentiable at a point $\mathbf{a}$ of its domain, then the differential is unique. HINT: The standard way to prove that there cannot be more than one of something is to assume that there are two and consider their difference. In this case the difference is easily shown to be a linear transformation whose norm has to be zero.

3. Prove that if $T \in \text{Hom}(R^n, R^m)$, then T is differentiable and that it is its own differential.

4. Let $\mathbf{f} = (f_1, f_2, \cdots, f_n)$ be a function from some linear interval $[a, b]$ to R^n. Prove that $\mathbf{f}$ is differentiable at t_0 in (a, b) if and only if each of the component functions is, in which case $d\mathbf{f} = \mathbf{v}\, dt$, where $\mathbf{v} = (f'_1(t_0), \cdots, f'_2(t_0))$.

5. Assume that $\mathbf{f}$ is a function from an open subset $\mathcal{U}$ of R^n to R^m, and that $\mathbf{a}$ is a point of $\mathcal{U}$ at which all mn of the first partial derivatives of the component functions of $\mathbf{f}$ exist. Prove that if T is that element of $\text{Hom}(R^n, R^m)$ whose standard representation is the Jacobian matrix of $\mathbf{f}$ at $\mathbf{a}$, then either $\mathbf{f}$ is differentiable at $\mathbf{a}$ with differential T, or else,

$$\lim_{\mathbf{h} \to 0} \frac{||\mathbf{f}(\mathbf{a} + \mathbf{h}) - \mathbf{f}(\mathbf{a}) - T(\mathbf{h})||}{||\mathbf{h}||}$$

does not exist.

6. If $\mathbf{f}$ is a function from some subset A of R^n to R^m, prove that $\mathbf{f}$ is continuous at the point $\mathbf{a}$ of A if and only if each of its component functions is continuous at $\mathbf{a}$.

7. (a) Show that the function f from R^2 to R defined as follows,

$$f(x, y) = \frac{xy}{x^2 + y^2} \qquad \text{if } (x, y) \neq (0, 0),$$

and
$$f(0, 0) = 0,$$

has partial derivatives with respect to both x and y at $(0, 0)$ but is not continuous there, and hence is certainly not differentiable at the origin.

(b) Let $g(x, y) = \dfrac{xy^2}{x^2 + y^4}$ if $(x, y) \neq (0, 0)$, and

$g(0, 0) = 0$.

Show that g has directional derivatives in all directions at the origin, but that it is not differentiable there. Is g continuous at the origin?

8. Define $F(x, y) = (2x^2 - y)(y - x^2)$ if $y \leq x^2$ or $2x^2 \leq y$, and define

$$F(x, y) = \frac{(2x^2 - y)(y - x^2)}{xy} \qquad \text{if} \qquad 0 < x^2 < y < 2x^2.$$

Show that **(a)** F is continuous at each point, including the point $(0, 0)$, and **(b)** F has a directional derivative in every direction at $(0, 0)$, but **(c)** F is not differentiable at $(0, 0)$.

9. Prove that $D_{\mathbf{u}}f(\mathbf{a}) = -D_{-\mathbf{u}}f(\mathbf{a})$.

(b) Assuming that $\mathbf{v} \neq -\mathbf{u}$, express $D_{\mathbf{u}}f(\mathbf{a}) + D_{\mathbf{v}}f(\mathbf{a})$ in terms involving just one directional derivative at $\mathbf{a}$.

10. Prove that if T is an invertible linear operator on R^n, and η is a function from some neighborhood of the origin in R^n to R^n, and $\lim_{\mathbf{x} \to 0} T\eta[(\mathbf{x})] = 0$, then

$$\lim_{\mathbf{x} \to 0} \eta(\mathbf{x}) = \mathbf{0}.$$

11. Observe that the function $f : R \to R$ defined by $f(x) = x^3$ is continuously differentiable everywhere and that the differential at 0 is singular (i.e., not invertible). The function f is nevertheless invertible—in every neighborhood of the origin, in fact. Construct functions from R^n to R^n exhibiting these properties.

12. Show that it is possible to prove the inverse function theorem in the special case where $n = 1$ without assuming the function to be continuously differentiable. It is sufficient that it merely be differentiable and that its derivative be different from zero at each point of an open interval (a, b). Show that one can then prove that the function has an inverse on (a, b), that its inverse function is differentiable at each point of its domain, and that the derivative of the inverse can be expressed very simply in terms of the derivative of the given function.

The authors do not know whether, in Theorem VIII, the hypothesis that $\mathbf{f}$ be continuously differentiable can be replaced by the hypothesis that $\mathbf{f}$ merely have an invertible differential throughout some neighborhood of $\mathbf{x} = \mathbf{a}$.

13. Consider the function $f : R \to R$ defined as follows:

$$f(x) = x \qquad \text{if } x \text{ is rational,}$$
$$f(x) = x^2 + x \qquad \text{if } x \text{ is irrational.}$$

Prove that f is not only continuous at 0 but differentiable there with an invertible differential. Then show that f is not invertible in any neighborhood of 0. The fact that f has a positive derivative at 0 does imply that there is some neighborhood of 0 in which $f(x) > f(0)$ if $x > 0$, and $f(x) < f(0)$ if $x < 0$, but there is no neighborhood of 0 throughout which f is either always increasing or always decreasing.

14. Suppose that $\mathcal{W}$ is an open subset of R^5 and that $(a_1, a_2, 0, 0, 0) \in \mathcal{W}$. Show that $\{(x_1, x_2): (x_1, x_2, 0, 0, 0) \in \mathcal{W}\}$ is an open subset of R^2.

15. Suppose that $\mathcal{W}$ is an open subset of $R^5 = R^2 \times R^3$ and that $(\mathbf{a}, \mathbf{b}) \in \mathcal{W}$ Prove that there are open sets $\mathcal{U}$ and $\mathcal{V}$ in R^2 and R^3 respectively such that $\mathbf{a} \in \mathcal{U}$, $\mathbf{b} \in \mathcal{V}$, and $\mathcal{U} \times \mathcal{V} \subset \mathcal{W}$. Notice that this argument can be extended to prove that every open set in $R^{p+q} = R^p \times R^q$ is the union of Cartesian products of open sets in R^p with open sets in R^q.

How can one distinguish geometrically between those subsets of the plane which are Cartesian products of subsets of the line and those which are not?

16. Write out in complete detail, with accompanying figures, a proof of Theorem X in the special case where $p = q = 1$. Draw a figure for the case which illustrates the possibility that the set S may consist of several disjoint sets (intervals).

17. Show that there cannot exist an invertible continuously differentiable function from an open subset of R^3 to R^2.

18. If $\{\mathbf{x}_k\}_{k=1}^{\infty}$ is a sequence in R^n, prove that the truth of the assertion

$$\lim_{k \to \infty} \mathbf{x}_k = \mathbf{a}$$

is independent of the norm used.

19. Prove that the two norms of Exercise 10, Chapter XI, are not equivalent. Use the definition following the proof of Lemma I.

20. Show that the proof that all norms on $\mathrm{Hom}(R^n, R^m)$ are equivalent can easily be modified to prove that if $\mathcal{V}$ is any vector space of finite dimension, then all norms on $\mathcal{V}$ are equivalent.

21. Show that the function in Theorem V of Chapter XI, which associates each invertible operator with its inverse, is differentiable. HINT: By Exercise 22 of Chapter XI one can write

$$(T_0 + H)^{-1} - T_0^{-1} = [T_0 (I + T_0^{-1}H)]^{-1} - T_0^{-1}$$
$$= (I + T_0^{-1}H)^{-1}T_0^{-1} - T_0^{-1} = [I - (-T_0^{-1}H)]^{-1}T_0^{-1} - T_0^{-1}$$
$$= \{I - T_0^{-1}H + (T_0^{-1}H)^2[I - (-T_0^{-1}H)]^{-1}\}T_0^{-1} - T_0^{-1}$$

assuming that $||H|| < 1/||T_0^{-1}||$.

Double and Triple Integrals   **13**

13. Preliminary remarks

This chapter is designed in such a way that it does not require of the student any previous knowledge of the subject of multiple integrals. Most students of elementary calculus will have had some acquaintance with double and triple integrals before coming to a course in advanced calculus. The extent of this acquaintance will vary considerably with the student, however, and it seems desirable here to start from the beginning.

Multiple integrals have important applications in geometry and the sciences. For these applications the *concept* of the integral (double or triple) is vital, quite apart from the important matter of knowing how to calculate the value of the integral. It is therefore very important for the student to pay attention to the *definitions* of double and triple integrals.

Naturally, the study of multiple integrals builds upon prior knowledge of ordinary definite integrals

$$\int_a^b f(x)dx.$$

Such integrals may be called *single* integrals, since they involve functions of a single independent variable, while multiple integrals involve functions of several independent variables.

In the study of integration, some of the salient matters to be considered are: the definition of the integral, the type of function which has an integral (i.e., which has an integral in the sense of the definition), properties of the integral, methods of finding the values of integrals, and applications. Of these matters, the question as to what types of functions are integrable is the most difficult. For most ordinary applications it is sufficient to deal with continuous functions. In Chapter XVIII we shall consider the theory of integration, paying special attention to questions of integrability; in particular, we shall prove in that later chapter that continuous functions are integrable. In the present chapter we shall avoid discussions of in tegrability, taking for granted the existence of the integrals which come to hand.

13.1 Motivations

Consider a thin plane sheet of metal covering a certain region R in the xy-plane. If the sheet is of uniform thickness and texture, the mass of any portion of the sheet will be directly proportional to the area of that portion. The constant of proportionality may be called the *areal density* of the sheet; it is the mass per unit area. If we denote this density by σ, the mass ΔM of an area ΔA of the sheet is

$$(13.1\text{–}1) \qquad \Delta M = \sigma \Delta A.$$

Let us now ask: How can we locate the center of mass (which is the same as the center of gravity) of the sheet? An intuitive attack on this question may be made as follows. Divide the sheet into a large number, say n, of small pieces. Denote the areas of these pieces by $\Delta A_1, \cdots, \Delta A_n$, and their masses by $\Delta M_1, \cdots, \Delta M_n$. If the maximum dimensions of the pieces are all sufficiently small, it is a reasonable approximation to regard each piece as a particle, all of the mass of the piece being thought of as concentrated at some point within the piece. We thus arrive at the picture of a system of particles, the mass ΔM_k being located at a point $P_k(x_k, y_k)$ in the kth piece.

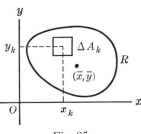

Fig. 85

The center of mass of this finite system of particles is at the point $(\bar{x}, \bar{y})$, where

$$(13.1\text{–}2) \qquad \begin{aligned} M\bar{x} &= x_1 \Delta M_1 + \cdots + x_n \Delta M_n, \\ M\bar{y} &= y_1 \Delta M_1 + \cdots + y_n \Delta M_n, \\ M &= \Delta M_1 + \cdots + \Delta M_n. \end{aligned}$$

It is plausible to suppose that, as the number of pieces is increased and the greatest dimension of all the pieces approaches zero, the point $(\bar{x}, \bar{y})$ located by formulas (13.1–2) will approach a limiting position which is the exact center of mass of the plate. If this limiting point has co-ordinates $(\bar{x}, \bar{y})$, we see from (13.1–1) and (13.1–2) that

$$(13.1\text{–}3) \qquad M\bar{x} = \lim (x_1\sigma \Delta A_1 + \cdots + x_n\sigma \Delta A_n),$$

with a similar formula for $\bar{y}$. The limit here is to be understood in much the same way as the limit defining a definite single integral (see § 1.63). In the present case the limit is a double integral, and we write

$$(13.1\text{–}4) \qquad M\bar{x} = \iint\limits_R x\sigma \, dA.$$

The expression on the right here is called the double integral of $x\sigma$ over the region R. There will be a similar formula for $\bar{y}$.

We shall see that double integrals can arise conceptually from many different geometrical or physical problems. For the present we select one further illustration, this time from geometry.

Consider a surface $z = f(x, y)$, where f is defined and continuous in a bounded closed region R of the xy-plane. Suppose that the values of f are everywhere positive, or perhaps zero, so that the surface never falls below the xy-plane. We now pose the problem: How can we calculate the volume which is under the surface and directly above the region R? This volume will be bounded laterally by a cylindrical surface composed of lines parallel to the z-axis erected at the point of the boundary of R.

The procedure for arriving at a formulation of the volume is very similar to the procedure used in expressing the area under a curve $y = f(x)$ from $x = a$ to $x = b$ as the definite integral $\int_a^b f(x)dx$. We divide the region R into a large number (say n) of small pieces (called subregions) of areas $\Delta A_1, \cdots, \Delta A_n$. This divides the volume under consideration into thin columns. Let the portion of the volume directly above the small area ΔA_k be denoted by ΔV_k, so that the total volume is

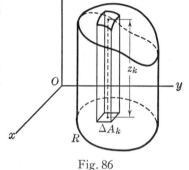

Fig. 86

$$V = \Delta V_1 + \cdots + \Delta V_n.$$

If $P_k(x_k, y_k)$ is any point in the kth subregion of R, let $z_k = f(x_k, y_k)$ be the distance from P_k up to the surface (see Fig. 86). If the dimensions of ΔA_k are sufficiently small, the expression

$$z_k \, \Delta A_k = f(x_k, y_k)\Delta A_k$$

is a good approximation to the volume ΔV_k. In fact, denoting by m_k and M_k the minimum and maximum values, respectively, of $f(x, y)$ in the kth subregion, we have

(13.1–5) $$m_k \, \Delta A_k \leqq \Delta V_k \leqq M_k \, \Delta A_k.$$

Since z_k lies between m_k and M_k, it is clear that ΔV_k and $z_k\Delta A_k$ do not differ by more than $(M_k - m_k)\Delta A_k$. It thus appears very plausible that the sum

(13.1–6) $$f(x_1, y_1)\Delta A_1 + \cdots + f(x_n, y_n)\Delta A_n$$

is a good approximation to V, and that we have exactly (using a summation symbol to abbreviate the expression (13.1–6)),

(13.1–7) $$V = \lim \sum_{k=1}^{n} f(x_k, y_k)\Delta A_k,$$

the limit being understood in the sense that the number n is increased indefinitely and the maximum dimension of the subregions $\Delta A_1, \cdots, \Delta A_n$ approaches zero. The choice of the point P_k in the kth subregion is arbitrary, and the exact shape of the subregions is immaterial. The limit on the right in (13.1–7) is equal to a double integral; the notation for the integral is

$$(13.1\text{–}8) \qquad\qquad V = \iint\limits_{R} f(x, y)\,dA.$$

If we compare the limits in (13.1–3) and (13.1–7), we see that they have the same form. In fact, the limit in (13.1–3) is the special case of that in (13.1–7) for which $f(x, y) = x\sigma$ (which happens to be independent of y). Limits of sums having the general form (13.1–6) occur in a variety of contexts, with widely different interpretations. The mathematical properties common to all such limits furnish us with a starting point for the general theory of double integrals.

13.2   Definition of a double integral

Let R be a closed, bounded region in the xy-plane, and let $f(x, y)$ be a function defined in R. In a very general theory of integration, we might seek to place no more restrictions on the function f and the region R than are absolutely necessary for the development of the theory. In the interests of simplicity, however, we shall make rather severe limitations on R, and we shall assume at the outset that the function f is continuous in R. Later it will be possible (and desirable) to broaden the treatment so that certain kinds of discontinuities of f are permitted.

The term "region" was defined in § 5.1. We are now concerned with closed, bounded regions. If R is such a region, it has an interior and a boundary. Since R is closed, the boundary is part of the region. The limitations we place on R are in the nature of assumptions about the character of the boundary. We have in mind, roughly speaking, that the boundary of R shall consist of a finite number of arcs of smooth curves joined together to form a closed curve, or possibly several (but a finite number of) such curves. A smooth curve is defined to be a curve with a continuously turning tangent. Circles, parabolas, and straight lines are among the simplest kinds of smooth curves. It is more difficult than one might suppose to be precise in describing the boundary of a region; we shall not attempt to express our assumptions more exactly than in the above statement. Hereafter in this chapter, in speaking of a region R in connection with a double integral, the foregoing assumptions will be taken for granted without explicit mention.

In defining a double integral, we start from approximating sums having the appearance of (13.1–6), but the subregions ΔA_k are chosen in a prescribed manner, and are not arbitrary in shape. Let two sets of lines be drawn, one set parallel to the x-axis, the other set parallel to the y-axis (see Fig. 87). The spacing of the lines need not be regular, but the spacing should be close enough so that the rectangles formed by the intersections of the two sets of lines are small in comparison with R. The network thus formed in the xy-plane is called a rectangular partition; one of the rectangles of the network is called a *cell*. Some of the cells will belong entirely to R; others will contain points which do not belong to R. For our purposes we discard

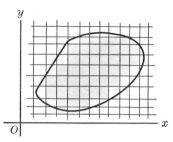

Fig. 87

these latter cells, retaining only those which do not go outside R (these are shaded in Fig. 87). Let the retained cells be numbered (in any order), and denote their areas by $\Delta A_1, \cdots, \Delta A_N$, N being the number of cells retained. It is convenient to refer to the kth cell as the cell ΔA_k. Now let P_k, with co-ordinates (x_k, y_k), be any point in the cell ΔA_k, and form the sum

$$(13.2\text{–}1) \quad f(x_1, y_1)\Delta A_1 + \cdots + f(x_N, y_N)\Delta A_N = \sum_{k=1}^{N} f(x_k, y_k)\Delta A_k.$$

The limit of this sum is defined as the double integral of the function f over the region R:

$$(13.2\text{–}2) \qquad \iint\limits_{R} f(x, y)\,dA = \lim \sum_{k=1}^{N} f(x_k, y_k)\Delta A_k.$$

The integral is the limit of the sum in the following sense: The integral is the real number to which we may approximate as closely as we please by the sums (13.2–1); to get any desired degree of approximation all that is necessary is to make the dimensions of all the relevant cells sufficiently small. The choice of P_k in ΔA_k is arbitrary, and the rectangular partition itself is arbitrary. In choosing partitions so that all the cells have very small dimensions, it is of course apparent that N will become very large.

The precise meaning of (13.2–2) is then as follows: If ϵ is any positive number, there is some corresponding positive number δ which depends upon ϵ (and also upon f and R) such that the inequality

$$(13.2\text{–}3) \qquad \left| \iint\limits_{R} f(x, y)\,dA - \sum_{k=1}^{N} f(x_k, y_k)\Delta A_k \right| < \epsilon$$

holds for all rectangular partitions in which the maximum cell dimensions are less than δ, and for all choices of the point P_k in the kth cell.

We take it for granted that the continuous function f is integrable, i.e., that the approximating sums (13.2–1) do actually converge to a limit in the sense just specified. This assumption is examined more closely in Chapter XVIII.

The student may already have speculated upon the fact that in forming the sum (13.2–1) we dealt only with those cells which belong *entirely* to R. We might have proceeded somewhat differently and retained not only those cells just mentioned, but also all those cells which touch the region R in any way (see Fig. 88). This procedure would give us more terms. The limiting value of the approximating sums would be the same as in (13.2–2), however. For, let ΔS denote the area of the additional cells, and let M be the maximum of $|f(x, y)|$ in R. Then the contribution of these additional cells to the approximating sum would not exceed $M\,\Delta S$. But ΔS approaches zero as the mesh of the partition is made finer and the maximum cell dimension approaches zero. Hence $M\,\Delta S \to 0$. In a fully detailed treatment of these matters the proof that $\Delta S \to 0$ turns out to depend on the nature of the boundary of R. The assumptions we made concerning this boundary are sufficient to insure that $\Delta S \to 0$.

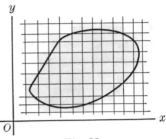

Fig. 88

The notation

(13.2–4) $$\iint\limits_{R} f(x, y)dx\, dy \text{ instead of } \iint\limits_{R} f(x, y)dA$$

is frequently used for the double integral. The letter which is used for area is of course immaterial. Thus, if we used S for the area of R, and $\Delta S_1, \cdots, \Delta S_N$ for the areas of the cells in the partition, we might denote the double integral by

$$\iint\limits_{R} f(x, y)dS.$$

13.21   Some properties of the double integral

A number of important properties of double integrals follow readily from the definition (13.2–2). Two of these properties are embodied in the formulas

(13.21–1) $$\iint\limits_{R} cf(x, y)dA = c \iint\limits_{R} f(x, y)dA,$$

(13.21–2)

$$\iint_R [f(x, y) + g(x, y)]dA = \iint_R f(x, y)dA + \iint_R g(x, y)dA.$$

In (13.21–1) c is a constant; *a constant factor may be taken outside the integral sign.* In (13.21–2) f and g are any two functions which are integrable in R; *the integral of a sum is the sum of the integrals.* These formulas are immediate consequences of the definition (13.2–2) and the fundamental theorems about limits (see § 1.6).

Another important property concerns the situation when the region R is composed of two regions R_1, R_2 which have no common points except for parts of their boundaries (see Fig. 89). The formula here is

(13.21–3) $$\iint_R f(x, y)dA = \iint_{R_1} f(x, y)dA + \iint_{R_2} f(x, y)dA.$$

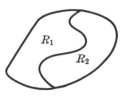

Fig. 89

The subregions R_1, R_2 are of course subject to the same assumptions as R as far as their boundaries are concerned.

13.22   Inequalities. The mean-value theorem

It is at once apparent from the definition of the double integral that

(13.22–1) $$\iint_R f(x, y)dA \geqq 0 \qquad \text{if } f(x, y) \geqq 0 \text{ in } R.$$

Hence, if $f(x, y) \geqq g(x, y)$ in R, we have

$$\iint_R f(x, y)dA \geqq \iint_R g(x, y)dA.$$

Now $$-|f(x, y)| \leqq f(x, y) \leqq |f(x, y)|,$$

and therefore

$$-\iint_R |f(x, y)|\, dA \leqq \iint_R f(x, y)dA \leqq \iint_R |f(x, y)|\, dA.$$

This result can be written

$$(13.22\text{–}2) \qquad \left| \iint_R f(x, y)dA \right| \leq \iint_R |f(x, y)| \, dA.$$

Let A be the area of R. Then, taking $f(x, y) \equiv 1$, we see that

$$\iint_R f(x, y)dA = \lim \sum_{k=1}^{n} \Delta A_k = A.$$

Hence, for any constant c,

$$(13.22\text{–}3) \qquad \iint_R c \, dA = cA.$$

Suppose now (returning to the case of an arbitrary continuous f) that m, M are numbers such that, in R,

$$m \leq f(x, y) \leq M.$$

Then $\qquad mA = \iint_R m \, dA \leq \iint_R f(x, y)dA \leq \iint_R M \, dA = MA,$

so that the integral of f has a value between mA and MA. Accordingly, we have the following theorem:

THEOREM I *(Mean-value theorem). If m and M are the minimum and maximum values of $f(x, y)$ in R, there is a number μ such that $m \leq \mu \leq M$ and*

$$(13.22\text{–}4) \qquad \iint_R f(x, y)dA = \mu A.$$

The number μ is called the *average* (or mean) *value* of f in R. We cannot as a rule find the value of μ unless we know the value of the integral, so that μ is in fact defined by (13.22–4). Nevertheless, even without exact knowledge of the value of μ, the fact that $m \leq \mu \leq M$ makes the formula (13.22–4) useful.

On occasion it is useful to know that there is some point $P(x, y)$ in R at which f takes on its average value μ. When f is continuous, there is always at least one such point P provided R is what is called a *connected* region, i.e., is not composed of two or more closed regions completely separated from each other. For a fuller discussion of what is meant by saying that a region is connected, the student is referred to § 17.7. This matter need not be considered any further at present, however.

13.23 A fundamental theorem

In our motivation of the definition of the double integral we used approximating sums which were obtained by decomposing the region R into subregions of arbitrary shape. In the definition (13.2–2), however, we restricted ourselves to subregions which are cells of a rectangular partition. It is important to know that we get the same limit of the approximating sums, no matter how the subdivision of R is made (so long as the pieces are sufficiently regular in shape that we can without ambiguity assign each of them an area). Assurance of this is given by the following theorem:

THEOREM II. *Let the double integral of the continuous function f over R be defined by* (13.2–2), *using rectangular partitions. Let R be divided in any manner into a finite number of subregions, of areas $\Delta A_1, \cdots, \Delta A_n$, the shapes being arbitrary except as qualified in the previous paragraph. Let a point $P_k(x_k, y_k)$ be chosen arbitrarily in ΔA_k. Then*

$$\iint_R f(x, y)dA = \lim \sum_{k=1}^{n} f(x_k, y_k)\Delta A_k.$$

Furthermore, it is not essential that the areas $\Delta A_1, \cdots, \Delta A_n$ completely fill out the area of R, provided that the amount of area omitted approaches zero in the limiting process.

This theorem appears to be intuitively evident from the geometrical interpretation of the double integral as a volume, as explained in § 13.1. A purely analytical proof may be given. In this proof the property (13.21–3) plays an important role. We forego the details.

Among other things, this theorem has the consequence that we are able to define the double integral of a continuous scalar point function over a region R; the integral is independent of co-ordinate systems, and is therefore a scalar invariant. Before reading the following brief remarks on this subject, the student will do well to read the first part of § 10.5.

Let R be a plane region of the type assumed in § 13.2, and let $f(P)$ be a continuous scalar point function defined in R. With an arbitrary choice of rectangular co-ordinates in the plane, let the representation of $f(P)$ be

$$f(P) = F(x, y),$$

P having co-ordinates (x, y). Consider the integral

(13.23–1) $$\iint_R F(x, y)dA,$$

as defined earlier in this chapter. If some other rectangular co-ordinate system is set up in the plane, denote the new co-ordinates of P by (x', y'), and the new representation of $f(P)$ by $\Phi(x', y')$. Then the integral

$$(13.23\text{–}2) \qquad\qquad \iint\limits_{R} \Phi(x', y')dA$$

has the same value as (13.23–1); for, the approximating sums converging to the integral (13.23–2), formed for a rectangular partition on the $x'y'$-coordinate system, will also converge to the integral (13.23–1), by virtue of Theorem II, since $F(x, y) = \Phi(x', y')$ when (x, y) and (x', y') refer to the same point. It follows that if we define the double integral of $f(P)$ over R by

$$(13.23\text{–}3) \qquad\qquad \iint\limits_{R} f(P)dA = \iint\limits_{R} F(x, y)dA,$$

then the integral is a scalar invariant.

13.3   Iterated integrals.   Centroids

We shall now learn how to calculate the value of a double integral by performing two successive single integrations. Our initial explanation of this method rests on the geometric interpretation of the double integral as a volume, as in the discussion which culminates in formula (13.1–8).

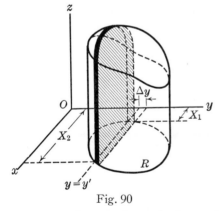

Fig. 90   Fig. 91

Our problem is to evaluate the integral

$$(13.3\text{–}1) \qquad\qquad \iint\limits_{R} f(x, y)dA,$$

where f is continuous in R. We shall for convenience assume that $f(x, y)$ is positive at each point of R. The integral (13.3–1) is equal to the volume under the surface $z = f(x, y)$ and directly above the region R. But we can calculate this volume in the following manner:

Let the region R be contained between the lines $y = a$, $y = b$ in the xy-plane (see Fig. 91), and suppose that an arbitrary line $y = y'$, where $a < y' < b$, intersects the boundary of R just twice, as shown in Fig. 91. The co-ordinates X_1, X_2 of these intersections will depend on y'.

Turning now to the three-dimensional picture (Fig. 90), consider the intersection of the plane $y = y'$ with the volume we are seeking. Let the area of this plane section of the volume be denoted by $S(y')$. This area may be expressed as an integral with respect to x, y' being held constant:

$$(13.3\text{--}2) \qquad S(y') = \int_{X_1}^{X_2} f(x, y')dx.$$

If Δy is a small positive number, the parallel plane $y = y' + \Delta y$ will also intersect the solid, and the volume of the slice between these two planes will be approximately $S(y')\Delta y$. Hence we should expect the total volume in question to be

$$(13.3\text{--}3) \qquad V = \int_a^b S(y)dy.$$

This is precisely the method which is followed in elementary calculus for finding the volumes of various kinds of solids, notably solids of revolution, pyramids, and so on. If we now drop the prime on y in (13.3–2), we see from (13.3–3) that the desired volume is

$$V = \int_a^b \left(\int_{X_1}^{X_2} f(x, y)dx \right)dy.$$

It is usual to write this expression in the form

$$(13.3\text{--}4) \qquad V = \int_a^b dy \int_{X_1}^{X_2} f(x, y)dx.$$

We have here what is called an *iterated integral*. Two successive integrations are indicated. First we integrate with respect to x, holding y constant in the integrand. The limits of integration X_1, X_2 generally depend on y, and are found by referring to Fig. 91 and taking account of the equations of the curves which form the boundary of R. The second integration is then performed with respect to y, between the limits $y = a$, $y = b$; these limits are the algebraically smallest and largest values, respectively, which y can assume in the region R.

We now have two expressions for the volume under the surface $z = f(x, y)$, one given by a double integral, and the other by an iterated integral. Therefore, we can state a theorem.

THEOREM III. *Let the region R have its extremes in the direction of the* *y-axis at y = a and y = b respectively (a < b). Let any line between these extremes and parallel to the x-axis cut the boundary of R in exactly two points, so that for a < y < b the boundary of R is formed by two curves*

$$x = X_1(y),\ x = X_2(y) \qquad (X_1 < X_2).$$

Then a double integral over R can be expressed as an iterated integral:

$$(13.3\text{--}5) \qquad \iint_R f(x, y)dA = \int_a^b dy \int_{X_1}^{X_2} f(x, y)dx.$$

It is of course also true that the double integral can be expressed as an iterated integral in which the first integration is with respect to y, provided the appropriate conditions on the boundary of R are fulfilled. See Fig. 92 and formula (13.3–6).

$$(13.3\text{--}6)$$

$$\iint_R f(x, y)dA = \int_c^d dx \int_{Y_1}^{Y_2} f(x, y)dy.$$

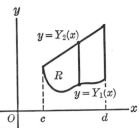

Fig. 92

We have been led to the discovery of Theorem III by an argument based on geometrical considerations which are highly plausible. The theorem itself is not dependent on the geometrical interpretation, however. Nor is the assumption that $f(x, y)$ is positive an essential one. The purely analytical proof of the theorem is discussed in Chapter XVIII, § 18.61.

Example 1. Let the region R lie in the first quadrant of the xy-plane, and be bounded by $y = 0$, $y^2 = x$, and $x + 2y = 3$. Let the surface be the plane $3x + 4y + 2z = 12$. Sketch the solid and find the volume for this particular instance of the foregoing discussion.

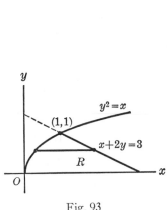

Fig. 93

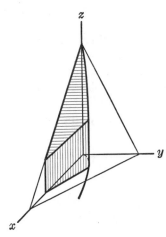

Fig. 94

The region R is shown in Fig. 93. The intersection of the parabola and line is at $(1, 1)$. The solid in question is shown in Fig. 94. The top surface is

$$z = \tfrac{1}{2}(12 - 3\,x - 4\,y).$$

Hence the volume is

$$V = \tfrac{1}{2} \iint\limits_{R} (12 - 3\,x - 4\,y)dA.$$

To express this as an iterated integral we read from Fig. 93 that a line parallel to the x-axis cuts the boundary of R at

$$x = y^2 \quad \text{and} \quad x = 3 - 2\,y.$$

Thus $X_1 = y^2$ and $X_2 = 3 - 2\,y$ in this case. The extreme values of y are $y = 0$, $y = 1$. Hence

$$V = \tfrac{1}{2} \int_0^1 dy \int_{y^2}^{3-2y} (12 - 3\,x - 4\,y)dx.$$

The first integration yields

$$\left[12\,x - \tfrac{3}{2}\,x^2 - 4\,xy \right]_{y^2}^{3-2y} = 36 - 24\,y - \tfrac{3}{2}(9 - 12\,y + 4\,y^2) - 12\,y + 8\,y^2$$
$$- (12\,y^2 - \tfrac{3}{2}\,y^4 - 4\,y^3)$$
$$= \tfrac{45}{2} - 18\,y - 10\,y^2 + 4\,y^3 + \tfrac{3}{2}\,y^4.$$

Hence

$$V = \tfrac{1}{2} \int_0^1 (\tfrac{45}{2} - 18\,y - 10\,y^2 + 4\,y^3 + \tfrac{3}{2}\,y^4)dy$$

$$= \tfrac{1}{2} \left[\tfrac{45}{2}\,y - 9\,y^2 - \tfrac{10}{3}\,y^3 + y^4 + \tfrac{3}{10}\,y^5 \right]_0^1$$

$$= \tfrac{1}{2}[\tfrac{45}{2} - 9 - \tfrac{10}{3} + 1 + \tfrac{3}{10}] = \tfrac{86}{15},$$

or

$$V = 5.733 \cdots.$$

Example 2. Express the double integral of Example 1 as an iterated integral in which the first integration is with respect to y.

When the integration is done in the order requested, we observe that the upper limit Y_2 of the formula (13.3–6) has two different expressions, according as $0 \leqq x \leqq 1$ or $1 \leqq x \leqq 3$ (see Fig. 95).

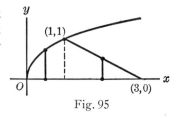

$$Y_2 = \sqrt{x}, \quad 0 \leqq x \leqq 1$$
$$Y_2 = \tfrac{1}{2}(3 - x), \quad 1 \leqq x \leqq 3.$$

Fig. 95

Under these conditions the iterated integral must be written as the sum of two iterated integrals:

$$\iint\limits_{R} f(x, y)dA = \int_0^1 dx \int_0^{\sqrt{x}} f(x, y)dy + \int_1^3 dx \int_0^{\frac{1}{2}(3-x)} f(x, y)dy.$$

In the present case $f(x, y) = \frac{1}{2}(12 - 3x - 4y)$.

We leave it for the student to complete the integrations and check with the result of Example 1.

The exercises following this section are intended to give the student practice in evaluating double integrals by means of iterated integrals. In these exercises the applications are limited to the calculation of volumes and the location of centers of gravity of thin sheets of constant areal density. The use of double integrals for this purpose was explained in § 13.1. If the thin sheet (often called a lamina) is specified as to shape and position by a region R in the xy-plane, and if the constant areal density is σ, the co-ordinates $(\bar{x}, \bar{y})$ of its center of gravity are given by

$$(13.3\text{-}7) \qquad M\bar{x} = \iint\limits_{R} x\sigma \, dA, \qquad M\bar{y} = \iint\limits_{R} y\sigma \, dA, \qquad \text{\textit{centroid of a lamina}}$$

M being the mass of the sheet. If A is the area of R, we have $M = \sigma A$. Dividing both sides of the formulas (13.3-7) by the constant factor σ, we have

$$(13.3\text{-}8) \qquad A\bar{x} = \iint\limits_{R} x \, dA, \qquad A\bar{y} = \iint\limits_{R} y \, dA.$$

It thus appears that $\bar{x}$ and $\bar{y}$ are independent of σ. The point $(\bar{x}, \bar{y})$ is thus a geometric characteristic of the region R, and is not affected by the material composing the sheet. This point is often called the *centroid* of R. It will be observed (compare Theorem I, § 13.22) that $\bar{x}$ is the mean value of x in R; a similar statement holds for $\bar{y}$.

EXERCISES

1. Compute the volume of each of the following regions by the use of a double integral. All literal constants a, b, c, etc. are assumed to be positive.

(a) The tetrahedron cut from the first octant by the plane $3x + 4y + 2z = 12$.

(b) The first octant section cut from the region inside the cylinder $x^2 + z^2 = a^2$ by the planes $z = 0$, $y = 0$, $x = y$.

(c) The hemisphere $x^2 + y^2 + z^2 \leq a^2$, $z \geq 0$.

(d) The region between the paraboloid $a^2z = H(a^2 - x^2 - y^2)$ and the xy-plane.

(e) The region bounded by the ellipsoid $(x^2/a^2) + (y^2/b^2) + (z^2/c^2) = 1$.

(f) The first octant portion of the region inside the cone $a^2y^2 = h^2(x^2 + z^2)$ and between $y = 0$ and $y = h$.

(g) The first octant region bounded by the co-ordinate planes and the cylinders $a^2y = b(a^2 - x^2)$, $a^2z = c(a^2 - x^2)$.

2. Find by double integrals the volumes of the tetrahedrons described:
(a) With plane faces $x = 0$, $z = 0$, $x + y = 5$, $8x - 12y + 15z = 0$;
(b) With vertices $(0, 0, 0)$, $(3, 0, 0)$, $(2, 1, 0)$, $(3, 0, 4)$;
(c) Cut from the first octant by the plane $(x/a) + (y/b) + (z/c) = 1$.

3. Interpret the iterated integral

$$\int_0^a dy \int_0^{\sqrt{a^2 - y^2}} \frac{2x + 4y}{3}\, dx$$

as the volume of a certain solid, and describe the solid geometrically. Calculate the volume.

4. Locate the centroids of the following plane regions, using double integrals. All literal constants are assumed to be positive.
(a) The triangle with vertices at $(0, 0)$, $(a, 0)$, (a, b).
(b) The triangle with vertices at $(0, 0)$, $(a, 0)$, (b, c), where $a > b$.
(c) The semicircular region $x^2 + y^2 \leq a^2$, $x \geq 0$.
(d) The region in the first quadrant and inside the ellipse $(x^2/a^2) + (y^2/b^2) = 1$.
(e) The first quadrant region bounded by $By^2 = H^2x$, $x = B$, $y = 0$.
(f) The region between the curve $x^3 = y^2$ and the line $x = 1$.
(g) The region bounded by $bx^2 = a^2y$ and the line $ay = bx$.

5. Locate the centroid of the region described by $x^n \leq y \leq 1$, $0 \leq x \leq 1$. What is the limiting position of the centroid as $n \to \infty$?

6. Solve Exercise 1(b) by use of a double integral over a suitable region in the xz-plane.

7. Interpret the integral

$$\int_{-a}^a dy \int_{a - \sqrt{a^2 - y^2}}^{a + \sqrt{a^2 - y^2}} x\, dx$$

as a double integral arising in the location of the centroid of a certain plane region, and hence write down the value of the integral without actually carrying out the integration.

8. Locate the centroids of each of the following plane regions:
(a) The region in the first quadrant between $x = 0$, $x = 1$, and between $y = x - x^2$, $y^2 = 2x$.
(b) The region bounded by the two parabolas $y = x^2 + x$, $y = 2x^2 - 2$.
(c) The region defined by $x^2 \leq y \leq 2 - x$, $0 \leq x \leq 1$.
(d) The region bounded by $y = 0$, $x + y = 2$, and the first quadrant part of $y = x^2$.

13.4   Use of polar co-ordinates

In suitable situations the evaluation of double integrals is greatly simplified by the use of polar co-ordinates. We shall explain the details of such use.

Suppose we wish to find the value of a double integral

$$(13.4\text{–}1) \qquad \iint\limits_{R} f(x, y)dA.$$

If we make the change to polar co-ordinates

$$x = r \cos \theta, \, y = r \sin \theta,$$

the function $f(x, y)$ becomes a function of r and θ, say

$$(13.4\text{–}2) \qquad f(x, y) = F(r, \theta).$$

We are going to appeal to Theorem II (§ 13.23). Instead of decomposing R by a rectangular partition, we make a subdivision based on a series of circles $r =$ constant and a series of rays $\theta =$ constant. These two series of curves are the two one-parameter families associated with the curvilinear co-ordinates r, θ (see § 9.2). We consider the cells of this subdivision which belong entirely to the region R, and number them consecutively in any order (see Fig. 96). Suppose that there are n cells and that ΔA_k is the area of the kth cell. According to Theorem II, referred to above,

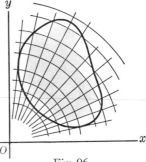

Fig. 96

$$(13.4\text{–}3) \qquad \iint\limits_{R} f(x, y)dA = \lim \sum_{k=1}^{n} f(x_k, y_k)\Delta A_k,$$

where (x_k, y_k) is a point which may be chosen arbitrarily in the kth cell. Let us agree to choose the point midway between the two circular arcs bounding the cell, and also midway between the two rays bounding the cell (see Fig. 97). If the radii of the circular arcs differ by Δr_k, and if the rays make an angle $\Delta\theta_k$ with each other, the area ΔA_k is easily computed in terms of $\Delta\theta_k$, Δr_k and the polar co-ordinates (r_k, θ_k) of the chosen point. The two circles are respectively

$$r = r_k - \tfrac{1}{2} \Delta r_k, \, r = r_k + \tfrac{1}{2} \Delta r_k.$$

Using the formula for the area of a sector of a circle, we have

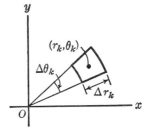

Fig. 97

$$\Delta A_k = \tfrac{1}{2}[(r_k + \tfrac{1}{2} \Delta r_k)^2 - (r_k - \tfrac{1}{2} \Delta r_k)^2]\Delta\theta_k,$$

or $\quad \Delta A_k = r_k \, \Delta r_k \, \Delta\theta_k.$

Hence in view of (13.4–2), (13.4–3) may be rewritten

$$(13.4\text{–}4) \qquad \iint_R f(x, y)dA = \lim \sum_{k=1}^{n} F(r_k, \theta_k)r_k \, \Delta r_k \, \Delta \theta_k.$$

Our next task is to work out a method of calculating the right member of (13.4–4). We do this with the aid of the concept of a point transformation, or mapping, from one plane to another, as developed in § 9.3. We regard (r, θ) as rectangular co-ordinates in one plane, and (x, y) as rectangular co-ordinates in another plane. The equations $x = r \cos \theta$, $y = r \sin \theta$ define a mapping from one plane to the other. Under this mapping, each of the cells of the subdivision of R shown in Fig. 96 corresponds to a *rectangular cell* in the $r\theta$-plane. The region R itself maps into a certain region in the $r\theta$-plane. Let us denote this region by T. Corresponding to the partition of R by the cells of the polar co-ordinate net, we have a rectangular partition of T. We may consider the function $F(r, \theta)$ as being defined in T, its value at (r, θ) being the same as the value of $f(x, y)$ at the corresponding point of R. If we number the cells in T in correspondence with their counterparts in R let ΔS_k be the area of the kth cell (see Fig. 98 and Fig. 99), then, in accordance with the notation of Fig. 97, the

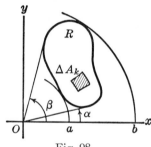

Fig. 98

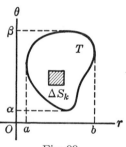

Fig. 99

dimensions of this kth cell are Δr_k by $\Delta \theta_k$, so that $\Delta S_k = \Delta r_k \, \Delta \theta_k$. Accordingly, we see that the limit on the right in (13.4–4) is precisely the limit which defines the double integral

$$\iint_T F(r, \theta)r \, dS$$

of the function $F(r \;\; \theta)r$ over the region T in the $r\theta$-plane. In the notation of (13.2–4) we may write

$$(13.4\text{–}5) \qquad \iint_R f(x, y)dx \, dy = \iint_T F(r, \theta)r \, dr \, d\theta.$$

Observe the factor r in the integrand on the right. This formula is the fundamental result we have been seeking. The last step is to express the integral on the right as an iterated integral with respect to r and θ. The limits of integration will of course depend on the particular region T and may be found by the method explained in § 13.3. In practice, however, these limits of integration are usually found directly by examining R. One must know the equations in polar co-ordinates for the curves forming the boundary of R. If we regard $f(x, y)$ and $F(r, \theta)$ as defining the same scalar point function in R, we may write

$$\iint_R f(x, y)dA \quad \text{and} \quad \iint_R F(r, \theta)dA$$

interchangeably. The use of polar co-ordinates in evaluating the double integral (13.4–1) is then summed up as follows:

$$(13.4\text{–}6) \qquad \iint_R F(r, \theta)dA = \int_\alpha^\beta d\theta \int_{R_1}^{R_2} F(r, \theta)r \, dr$$

$$= \int_a^b dr \int_{\Theta_1}^{\Theta_2} F(r, \theta)r \, d\theta.$$

In these iterated integrals α, β are the extreme values of θ, and a, b are the extreme values of r, in the region R. The inner limits R_1, R_2, Θ_1, Θ_2 are read off from the appropriate one of the two figures, as shown (Fig. 100a or Fig. 100b).

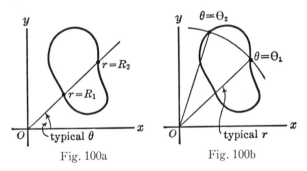

Fig. 100a   Fig. 100b

The use of polar co-ordinates may prove advantageous either by simplification of the integrand, or by simplification of the limits of integration in dealing with the iterated integrals. Experience and discernment are required to be able to judge whether or not to use polar co-ordinates. The student's first task is to practice the use of polar co-ordinates.

Example 1. Locate the centroid of the plane region R shown in Fig. 101 (above the x-axis and between the circles of radii a, b).

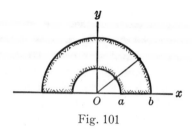

Fig. 101

The centroid is obviously on the y-axis, so $\bar{x} = 0$. The area A of R is $(\pi/2)(b^2 - a^2)$. Hence

$$\frac{\pi}{2}(b^2 - a^2)\bar{y} = \iint_R y \, dA.$$

The boundaries of R have very simple equations in polar co-ordinates. Therefore, we evaluate the double integral by an iterated integral in polar co-ordinates. Here

$$f(x, y) = y = r \sin \theta = F(r, \theta).$$

Also, $\alpha = 0$, $\beta = \pi$, $R_1 = a$, $R_2 = b$. Therefore, supplying the extra factor r in the integrand, we have

$$\iint_R y \, dA = \int_0^\pi d\theta \int_a^b r^2 \sin \theta \, dr$$

$$= \frac{b^3 - a^3}{3} \int_0^\pi \sin \theta \, d\theta = \tfrac{2}{3}(b^3 - a^3).$$

Then
$$\bar{y} = \frac{4}{3\pi} \frac{b^3 - a^3}{b^2 - a^2} = \frac{4}{3\pi} \frac{b^2 + ba + a^2}{b + a}.$$

For a semicircular region we put $a = 0$. In this case $\bar{y} = \dfrac{4}{3\pi} b$.

Example 2. Find the volume inside the cylinder $x^2 + (y - a)^2 = a^2$ and between the plane $z = 0$ and the paraboloid $4\,az = x^2 + y^2$.

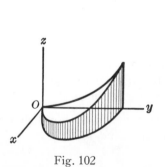

Fig. 102

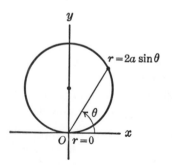

Fig. 103

The volume in question is given by

$$V = \iint_R \frac{1}{4a}(x^2 + y^2)dA,$$

where R is the region in the xy-plane bounded by the circle $x^2 + (y - a)^2$ $= a^2$. Half of the volume is shown in Fig. 102. Polar co-ordinates are convenient for this problem, because of both the form of the integrand and the equation of the boundary of R in polar co-ordinates. The polar equation of the boundary of R is $r = 2a\sin\theta$ (see Fig. 103). Hence we have

$$V = \frac{1}{4a}\iint_R r^2\, dA = \frac{1}{4a}\int_0^\pi d\theta \int_0^{2a\sin\theta} r^3\, dr.$$

$$V = a^3\int_0^\pi \sin^4\theta\, d\theta = 2a^3\int_0^{\pi/2} \sin^4\theta\, d\theta.$$

$$V = 2a^3 \frac{3\cdot 1}{4\cdot 2}\frac{\pi}{2} = \frac{3\pi}{8}a^3.$$

In evaluating the last integral we have used one of the standard tabulated formulas for the definite integrals

$$\int_0^{\pi/2} \sin^n\theta\, d\theta,\, n = 1, 2, \cdots.$$

These formulas are very convenient, and the student should acquaint himself with them.

EXERCISES

1. Calculate the volumes of the solids here described, using double integrals and polar co-ordinates.
   (a) Inside the cylinder $x^2 + y^2 = a^2$, between the planes $z = 0$, $z = x$, and in which $x \geq 0$.
   (b) Inside the sphere $x^2 + y^2 + z^2 = a^2$. Deal with the first octant only, and use symmetry.
   (c) Inside the cylinder $x^2 + y^2 = a^2$ and between $z = 0$ and $a^2 z = h(x^2 + y^2)$.
   (d) Between the cone $c^2(z - h)^2 = h^2(x^2 + y^2)$ and the plane $z = 0$.
   (e) Inside both the sphere $x^2 + y^2 + z^2 = 4a^2$ and the cylinder $x^2 + (y - a)^2$ $= a^2$.
   (f) Inside the cylinder $x^2 + y^2 = 2ax$ and between the plane $z = 0$ and the cone $z^2 = x^2 + y^2$.
   (g) Inside the prism bounded by the planes $y = x$, $y = 0$, $x = a/\sqrt{2}$, and between the plane $z = 0$ and the cone $az = h(x^2 + y^2)^{1/2}$.

2. Locate the centroids of the plane regions described as follows, using double integrals and polar co-ordinates:
   (a) In the first quadrant, between $x^2 + y^2 = 2ax$ and $y = 0$.

(b) Between $r = 2\,a\cos\theta$ and $r\cos\theta = a$, and on the side of the latter curve away from the origin.
(c) Inside the cardioid $r = a(1 + \sin\theta)$.
(d) Inside the first-quadrant loop of $r = a\sin 2\,\theta$.
(e) In the first quadrant, inside $r = 2\,a\cos\theta$ and outside $r = a$.
(f) Inside the loop of $r^2 = a^2\cos 2\,\theta$ which is bisected by the ray $\theta = 0$.

3. Find each of the two volumes into which the volume in Exercise 1(e) is divided by the cylinder $x^2 + y^2 = a^2$.

4. Find the volume inside the sphere $x^2 + (y - a)^2 + z^2 = a^2$ and between the planes $x = 0$, $y = x$.

5. Find the volume between the paraboloid $z = x^2 + y^2$ and the plane $z = x$.

13.5   Applications of double integrals

In § 13.1 we introduced the concept of a thin sheet of material substance. The concept of a distribution of matter without thickness is a very useful one. A plane region which carries such a mass distribution is called a lamina. A lamina is a mathematical idealization of a thin sheet, just as a particle is a mathematical idealization of a small, concentrated bit of matter. One may also speak of laminas which are curved surfaces, but here we shall deal only with plane laminas.

We wish to introduce the concept of a lamina of variable density. In the case of constant density, the density of a lamina is the ratio of mass to area:

$$(13.5-1) \qquad\qquad \sigma = \frac{M}{A}.$$

But we may imagine a lamina in which the mass is so distributed that various pieces of the lamina, although of equal area, will have different masses. For the general case, the density of a lamina is, by definition, an integrable function $\sigma(x, y)$ such that when it is integrated over any subregion ΔR of the lamina, it gives the mass of that portion:

$$(13.5-2) \qquad\qquad \Delta M = \iint\limits_{\Delta R} \sigma \, dA.$$

In particular, the total mass is

$$(13.5-3) \qquad\qquad M = \iint\limits_{R} \sigma \, dA.$$

We shall consider only the case of continuous densities. If the area of ΔR is ΔA we see from (13.5-2) by the mean-value theorem (§ 13.2) that

$$(13.5-4) \qquad \sigma' \leqq \frac{\Delta M}{\Delta A} \leqq \sigma'',$$

where σ' and σ'' are respectively the minimum and maximum values of the density in ΔR. Hence, if (x, y) is a fixed point of ΔR, and if we shrink the subregion ΔR so that its maximum diameter approaches zero, we see by the continuity of σ that

$$(13.5-5) \qquad \lim \frac{\Delta M}{\Delta A} = \sigma(x, y).$$

This relation replaces (13.5–1) when we deal with laminas of variable density.

Example 1. The density of a square lamina of side b varies in direct proportion to the distance from a particular corner of the square. Find the mass of the lamina.

Let us take the square as shown in Fig. 104, with the particular corner at the origin. We are given that

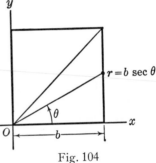

Fig. 104

$$\sigma = k\sqrt{x^2 + y^2},$$

k being a constant of proportionality. Hence,

$$M = k \iint_R \sqrt{x^2 + y^2} \, dx \, dy,$$

R being the square region. We meet difficulties if we proceed to the iterated integral in rectangular co-ordinates. It is better to use polar co-ordinates. The line $y = x$ divides the square into two triangles which are evidently of equal mass. The line $x = b$ has the polar equation $r = b \sec \theta$. Hence, if R_1 denotes the triangle below the line $y = x$,

$$M = 2k \iint_{R_1} r \, dA = 2k \int_0^{\pi/4} d\theta \int_0^{b \sec \theta} r^2 \, dr.$$

$$M = \frac{2 kb^3}{3} \int_0^{\pi/4} \sec^3 \theta \, d\theta$$

$$= \frac{2 kb^3}{3} \left[\tfrac{1}{2} \tan \theta \sec \theta + \tfrac{1}{2} \log \tan \left(\frac{\pi}{4} + \frac{\theta}{2} \right) \right]_0^{\pi/4}$$

$$M = \frac{kb^3}{3} \left[\sqrt{2} + \log \tan \frac{3\pi}{8} \right].$$

Many physical concepts are first formulated for systems of a finite number of discrete particles, and then extended to continuous distributions of matter by the use of integrals. The guiding principle is that of subdividing the continuous distribution into small parts. Each part is then replaced by a particle, which is obtained by concentrating all the mass of the part at some point within the part. The resulting system of particles is then regarded as an approximation to the continuous distribution, and physical attributes of the continuous mass are assumed to be obtained as limits of the corresponding physical attributes of the approximating system of particles. Where the physical attribute of the system of particles is expressed by a sum, that of the continuous mass will be expressed as the limit of a sum, and this limit will normally be a definite integral. Illustrations are furnished by such concepts as center of mass, moment of inertia, and gravitational or electrostatic attraction. In the case of a lamina of continuous density, when it is subdivided into small parts, of areas $\Delta A_1, \cdots, \Delta A_n$, the mass ΔM_k of the kth part will be expressed by

$$(13.5\text{--}6) \qquad \Delta M_k = \sigma(x_k, y_k)\Delta A_k,$$

where (x_k, y_k) is a suitably chosen point in the part (see (13.5–4) and the remarks at the end of § 13.22).

Using the general principle described in the foregoing paragraph we find that the center of mass (center of gravity) of a lamina is given by formulas (13.3–7). When σ is variable it cannot be taken from under the integral sign, of course. In this case we cannot use formulas (13.3–8), and M must be found by integration.

If L is a straight line in space, and if we have a system of particles of masses $m_1, \cdots, m_n$, the perpendicular distance from m_k to L being r_k, then the moment of inertia of the system about L as an axis is defined to be

$$I = m_1 r_1^2 + \cdots + m_n r_n^2.$$

To extend this definition to laminas in the xy-plane, let P_k denote the point (x_k, y_k) in (13.5–6), and let Q_k be the foot of the perpendicular drawn from P_k to L. Then for the approximating system of particles we have $r_k = P_k Q_k$, $m_k = \sigma(x_k, y_k)\Delta A_k$; therefore, the moment of inertia of the lamina about L is

$$I = \lim \sum_{k=1}^{n} (P_k Q_k)^2 \sigma(x_k, y_k)\Delta A_k.$$

This limit is clearly a double integral. If $D(x, y)$ denotes the perpendicular distance PQ from a typical point $P(x, y)$ of the lamina to the axis L, the double integral is

$$(13.5\text{--}7) \qquad I = \iint_R \sigma\, D^2\, dA.$$

In each particular problem D must be expressed as a function of the co-ordinates. For instance, if L is the y-axis, $D^2 = x^2$, while if L is the z-axis, $D^2 = x^2 + y^2$.

Moments of inertia are often expressed in the form $I = Mk^2$, M being the mass. The constant k is called the *radius of gyration*.

Example 2. Prove the following proposition: *Let a lamina occupy the region R, and let L be a line in the plane of the lamina. Let I denote the moment of inertia of the lamina about L, and let I_0 denote the moment of inertia about an axis L_0 parallel to L through the center of mass of the lamina. Then*

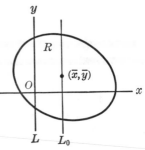

Fig. 105

$$(13.5\text{-}8) \qquad I = I_0 + Mh^2$$

where h is the distance between L and L_0.

It is convenient to locate our co-ordinate system so that L coincides with the y-axis. This is permissible, since the physical quantities are independent of the position of the co-ordinate axes. The distance from $P(x, y)$ to L is $D = |x|$. Hence

$$I = \iint_R \sigma x^2 \, dA.$$

The equation of L_0 is $x = \bar{x}$, and so

$$I_0 = \iint_R \sigma(x - \bar{x})^2 \, dA = \iint_R \sigma(x - h)^2 \, dA,$$

for $\bar{x} = h$.

Thus

$$I - I_0 = \iint_R \sigma[x^2 - (x - h)^2] dA.$$

Now $x^2 - (x - h)^2 = 2xh - h^2$. Therefore

$$I - I_0 = 2h \iint_R \sigma x \, dA - h^2 \iint_R \sigma \, dA = 2hM\bar{x} - h^2 M = Mh^2.$$

Thus (13.5-8) is proved.

If we set $\sigma = 1$ in (13.5-7), the resulting integral

$$(13.5\text{-}9) \qquad \iint_R D^2 \, dA$$

is called the *second moment* of the region R with respect to the axis L. This is a purely geometric characteristic of R in relation to L. Second moments are used in the theory of elasticity and strength of materials, particularly in the theory of the bending of beams. In the literature these second moments are often called moments of inertia. This is a misnomer, since no mass concept is involved. The physical dimensions of a moment of inertia are mass $\times$ (length)^2, while those of a second moment of a plane region are (length)^4. Second moments are also important in statistics and elsewhere. The integrals

$$\iint_R x \, dA, \quad \iint_R y \, dA$$

occurring in the formulas (13.3-8) for the centroid are, by contrast, called *first moments* (about the y-axis and x-axis, respectively).

EXERCISES

1. In each of the parts of this exercise a lamina of a certain shape is described, and the manner in which its density varies is defined. Find the mass and locate the center of mass of each lamina. Wherever it occurs in this exercise, k denotes a constant of proportionality.

(a) Triangular lamina with vertices at $(0, 0)$, $(a, 0)$, (a, b); $\sigma = kx$.

(b) The same lamina as in (a), but with $\sigma = ky$.

(c) The lamina occupying the region defined by $x^2 + y^2 \leq a^2$, $x \geq 0$, $y \geq 0$, with $\sigma = kx$.

(d) The lamina of (c), but with $\sigma = kxy$.

(e) The lamina of (c), but with $\sigma = k(x^2 + y^2)^{1/2}$.

(f) The lamina in the first quadrant, bounded by $bx^2 = a^2y$, $x = 0$, $y = b$, with $\sigma = kx$.

(g) The lamina of (f), but with $\sigma = k(b - y)$.

(h) The triangular lamina cut from the first quadrant by the line $x + y = a$, with σ directly proportional to the product of the distances from (x, y) to the sides of the triangle.

(i) The lamina in the first quadrant, bounded by $r = 2a \cos \theta$ and $\theta = 0$, with $\sigma = kr$.

(j) The lamina of (i), but with $\sigma = kr \sin 2\theta$.

2. For any distribution of mass, let I_x and I_y denote the moments of inertia of the distribution about the x-axis and the y-axis, respectively, and let J_0 denote the moment of inertia about the axis perpendicular to the xy-plane at the origin. Show that $J_0 = I_x + I_y$.

3. In each part of this exercise, a homogeneous lamina is described. Find I_x, I_y, and J_0 in each case (see Exercise 2).

(a) The circular lamina bounded by $x^2 + y^2 = a^2$.

(b) The annulus bounded by the two circles $x^2 + y^2 = r_i^2$ ($i = 1, 2, r_1 < r_2$).

(c) The rectangular lamina bounded by $x = \pm a$, $y = \pm b$.

(d) The triangular lamina bounded by $y = 0$, $x = a$, $ay = bx$.

(e) The elliptical lamina bounded by $b^2x^2 + a^2y^2 = a^2b^2$.

(f) The lamina bounded by $y^2 = 2 ax$ and $x = 2 a$.

(g) The lamina occupying the circular segment $x^2 + y^2 \leq b^2$, $x \geq b \cos \alpha$ where $0 < \alpha < \pi/2$.

(h) The lamina occupying the circular sector $0 \leq r \leq b$, $-\beta \leq \theta \leq \beta$, where $0 < \beta \leq \pi/2$.

4. For a lamina occupying a region R in the xy-plane, the double integral

$$U_{xy} = \iint_R \sigma xy \, dA$$

is called the *product of inertia* of the lamina with respect to the co-ordinate axes. Calculate this product of inertia for each of the following laminas, assuming constant density:

(a) The rectangular lamina bounded by $x = 0$, $x = a$, $y = 0$, $y = b$.

(b) The first-quadrant quarter of the circular lamina bounded by $x^2 + y^2 = a^2$.

(c) The triangular lamina with vertices at $(0, 0)$, $(a, 0)$, (a, b).

(d) The square lamina bounded by $x = -a$, $x = 2 a$, $y = -2 a$, $y = a$.

(e) The lamina composed of all except the third-quadrant portion of the region inside the circle $x^2 + y^2 = a^2$.

(f) The lamina bounded by $y^2 = 2 a(x + a)$ and $x = a$.

5. Products of inertia, as defined in Exercise 4, play an important role in the discussion of what happens to moments of inertia when the co-ordinate axes are rotated. Suppose the xy-system can be rotated into the $x'y'$-system by turning counterclockwise through an angle α. The equations relating the two systems are

$$x' = x \cos \alpha + y \sin \alpha,$$
$$y' = -x \sin \alpha + y \cos \alpha,$$

and a similar pair of equations giving x, y in terms of x', y'. For a given lamina, let $A = I_x$, $B = U_{xy}$, $C = I_y$ (using the notation defined in Exercises 2 and 4), and let A', B', C' denote the corresponding moments and products of inertia relative to the axes of the $x'y'$-system.

Show that

$$A' = A \cos^2 \alpha - 2 B \sin \alpha \cos \alpha + C \sin^2 \alpha,$$
$$A = A' \cos^2 \alpha + 2 B' \sin \alpha \cos \alpha + C' \sin^2 \alpha,$$
$$B' = B(\cos^2 \alpha - \sin^2 \alpha) + (A - C) \sin \alpha \cos \alpha,$$

and find three more formulas of this type. Hence prove that the curve

$$Ax^2 - 2 Bxy + Cy^2 = 1$$

has the equation

$$A'^2x'^2 - 2 B'x'y' + C'y'^2 = 1$$

when referred to the $x'y'$-axes.

This curve is called the *ellipse of inertia* for the given lamina, relative to the origin O. If the xy-axes coincide with the axes of symmetry of the ellipse, the xy-term in the equation must disappear, so that $B = 0$. In this position, the co-ordinate axes are called *principal axes of inertia* for the lamina relative to O. If $B \neq 0$, the position of the principal axes may be found by the method of choosing α so that $B' = 0$, i.e.,

$$B \cos 2\alpha + \tfrac{1}{2}(A - C) \sin 2\alpha = 0.$$

6. Show that, if one regards A' as a function of α, $\dfrac{dA'}{d\alpha} = -2 B'$; hence show that A' is either a maximum or a minimum when the $x'y'$-axes are principal axes of inertia.

7. Use the formula for A' in Exercise 5 to compute the following moments of inertia:
   (a) The lamina of Exercise 4 (a), about its diagonal.
   (b) The lamina of Exercise 4 (c), about the line $2\, ay = bx$.
   (c) The lamina of Exercise 4 (f), about the line $y = 2\, x$.

8. Find the principal axes of inertia for the following laminas:
   (a) The lamina of Exercise 4 (a), if $a = 2, b = 1$.
   (b) The lamina of Exercise 4 (b).
   (c) The lamina of Exercise 4 (d).
   (d) The lamina of Exercise 4 (e).
   (e) The lamina of Exercise 4 (f).

9. A lamina in the shape of the circle $x^2 + y^2 \leq a^2$ has density $\sigma = (x + y)^2$. Find its principal axes of inertia relative to its center, and the moments of inertia about these axes.

13.51   Potentials and force fields

In the theory of electrostatics, the concepts of charge and charge density are entirely analogous to the concepts of mass and mass density, with this exception: Charges may be either positive or negative, while we habitually think of masses as positive. A particle of electric charge e exerts an electrostatic force on another particle of charge e' according to the inverse-square law of Coulomb: The magnitude of the force is inversely proportional to the square of the distance between the charges, and directly proportional to the product of the charges. The force is directed along the line joining the charges, and like charges repel each other, while unlike charges attract. With proper choice of units (electrostatic units) the constant of proportionality may be taken as unity.

The vector form of Coulomb's law is as follows: Let e be at P, e' at P', and r be the distance PP'. Then the force exerted by e on e' is

$$(13.51\text{--}1) \qquad\qquad \mathbf{F} = \frac{ee'}{r^3}\, \overrightarrow{PP'}.$$

This should be compared with the analogous formula for gravitational attraction between two particles (see (10.51–4)).

Next we consider how to deal with the notion of electrostatic force produced by a continuous distribution of charge on a plane lamina. Consider a particle of unit positive charge at a fixed point Q, anywhere in space, but not on the lamina. Let σ be the charge density on the lamina, which we assume occupies a region R in the xy-plane. In the usual manner, we subdivide R and consider the force exerted on Q by the system of point charges which is obtained when we concentrate the charge Δe of each part ΔR of the lamina at a point P within the part. The contribution of this part to the total is a force

$$\Delta \mathbf{F} = \frac{\Delta e}{r^3} \overrightarrow{PQ},$$

where r is the distance PQ (see Fig. 106). All such vectors must be added, and then we must carry out the limiting process. Since Δe is approximately $\sigma \, \Delta A$, the total force exerted by the lamina is

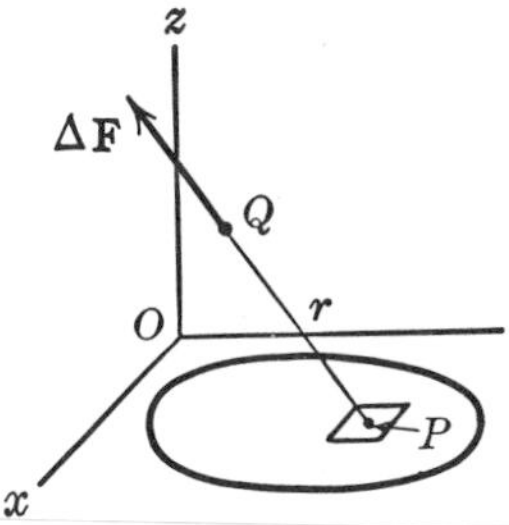

$(13.51\text{–}2)$ $\mathbf{F} = \iint\limits_R \dfrac{\sigma}{r^3} \overrightarrow{PQ} \, dA.$

Fig. 106

This is a *vector* double integral; i.e., the integrand is a vector function. We have not formally defined such integrals, but the work is entirely like that for scalar double integrals.

When it comes to actual computation of $\mathbf{F}$, we work with components of the vectors. It frequently occurs that because of symmetry we know in advance the direction of $\mathbf{F}$, and hence need only to deal with components of $\mathbf{F}$ in that one direction. If L is a directed line and if ψ is the angle which $\overrightarrow{PQ}$ makes with L, the component of $\mathbf{F}$ in the direction of L is

$(13.51\text{–}3)$ $F_L = \iint\limits_R \dfrac{\sigma}{r^2} \cos \psi \, dA.$ *force component in a given direction*

To evaluate, we must express the integrand in suitable co-ordinates and then pass to an iterated integral. *It must be kept in mind that the r in* (13.51–3) *is not necessarily the r of polar co-ordinates.*

The force exerted by a system of charges on a unit positive charge at Q is called the *field* at Q due to the system.

Example 1. Find the field due to a uniformly charged circular lamina of radius b, at a point Q a distance c from the center of the lamina along the perpendicular to it (Fig. 107).

The phrase "uniformly charged" means that the density is constant. We take the origin and the z-axis as shown in Fig. 107. Using polar co-ordinates in the xy-plane, we have the distance PQ given by

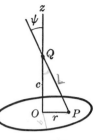

$$(PQ)^2 = r^2 + c^2.$$

By symmetry the force is evidently in the direction OQ, so that

$$\cos \psi = \frac{c}{\sqrt{r^2 + c^2}}.$$

Fig. 107

Hence, denoting the x, y, and z components of $\mathbf{F}$ by F_1, F_2, F_3, we have $F_1 = F_2 = 0$, and

$$F_3 = \iint_R \frac{\sigma c}{(r^2 + c^2)^{3/2}} \, dA = \sigma c \int_0^{2\pi} d\theta \int_0^b \frac{r \, dr}{(r^2 + c^2)^{3/2}},$$

$$F_3 = 2 \pi \sigma c \left(\frac{1}{c} - \frac{1}{\sqrt{b^2 + c^2}} \right) = 2 \pi \sigma \left(1 - \frac{c}{\sqrt{b^2 + c^2}} \right).$$

Observe that the force very near O is almost of amount $2 \pi \sigma$.

A systematic study of the theory of electrostatic fields is greatly simplified by introducing the concept of the *potential* of the field. The potential at a point Q, produced by a charge e at the point P, is defined to be

$$\frac{e}{r}, \qquad \text{where } r = PQ.$$

For the potential of several particles, the principle of superposition is used, and for continuous distributions of charge, the standard integral calculus procedure is employed. For a lamina on the region R, with charge density σ at P, the potential at Q is defined to be

(13.51–4) $$u(Q) = \iint_R \frac{\sigma}{PQ} \, dA.$$

The potential is a *scalar* point function. The electrostatic field is a *vector* point function. The relation between the two functions is shown in the fact that the gradient of the potential gives the negative of the field vector:

(13.51–5) $$\nabla_Q u(Q) = - \mathbf{F}.$$

The Q on the gradient symbol is to remind one that we must differentiate with respect to the co-ordinates of Q. If P is (x, y, z), and Q is (ξ, η, ζ), we have

(13.51–6) $$r^2 = (PQ)^2 = (\xi - x)^2 + (\eta - y)^2 + (\zeta - z)^2,$$

and

$$(13.51\text{–}7) \qquad u(Q) = \iint\limits_{R} \frac{\sigma(x, y)}{r} \, dx \, dy,$$

$$(13.51\text{–}8)$$

$$\mathbf{F} = \iint\limits_{R} \frac{\sigma(x, y)}{r^3} \left[(\xi - x)\mathbf{i} + (\eta - y)\mathbf{j} + (\zeta - z)\mathbf{k} \right] dx \, dy.$$

Formula (13.51–5) is then equivalent to

$$(13.51\text{–}9) \qquad \mathbf{F} \cdot \mathbf{i} = -\frac{\partial u}{\partial \xi} = \iint\limits_{R} \frac{\sigma(x, y)}{r^3} (\xi - x) dx \, dy,$$

and two similar formulas for the other components of $\mathbf{F}$.

It is only in very special instances that the potential can be computed in elementary form by integration. Usually the work leads to elliptic or other nonelementary integrals. Nevertheless, the study of the potential is very fruitful. Extensive consideration of the theory of potential functions is outside the scope of the present book.

Example 2. The lamina bounded by the lines $x = 0$, $x = a$, $y = 0$, $y = b$ in the xy-plane carries a charge of density $\sigma = xy$. Find the potential at the point $Q(0, 0, \zeta)$ on the z-axis.

The potential is

$$u = \iint\limits_{R} \frac{xy}{(x^2 + y^2 + \zeta^2)^{1/2}} \, dx \, dy = \int_0^b y \, dy \int_0^a \frac{x \, dx}{(x^2 + y^2 + \zeta^2)^{1/2}} \, .$$

The first integration gives

$$(a^2 + y^2 + \zeta^2)^{1/2} - (y^2 + \zeta^2)^{1/2},$$

so $\qquad u = \int_0^b [y(a^2 + y^2 + \zeta^2)^{1/2} - y(y^2 + \zeta^2)^{1/2}] dy,$

$$u = \tfrac{1}{3}(a^2 + b^2 + \zeta^2)^{3/2} - \tfrac{1}{3}(a^2 + \zeta^2)^{3/2} - \tfrac{1}{3}(b^2 + \zeta^2)^{3/2} + \tfrac{1}{3} \zeta^3.$$

EXERCISES

1. Find the potential at Q in Example 1, and verify that $F_3 = -\dfrac{\partial u}{\partial c}$ (assuming $c > 0$).

2. Find $F_3 = \mathbf{F} \cdot \mathbf{k}$ directly in Example 2, and then verify that $F_3 = -\dfrac{\partial u}{\partial \zeta}$ from the answer found in Example 2.

3. Find u and F_3 at the point Q in Example 1 if, instead of constant density, we have $\sigma = r$.

4. Find the potential at a corner of a uniformly charged square lamina of side b.

5. Find the potential at a point on the edge of a uniformly charged circular lamina of radius b. It is most convenient to take the point in question at the origin.

6. Find the potential at $Q(0, 0, b)$, where $b > 0$, due to a uniformly charged square lamina with corners at $(0, 0, 0)$, $(a, 0, 0)$, $(a, a, 0)$, $(0, a, 0)$. Set up the integral in polar co-ordinates, using the fact that the square can be divided by a diagonal so that each half contributes the same amount to the potential. The integral formula

$$\int \frac{\sqrt{a^2 + b^2 \cos^2 \theta}}{\cos \theta} \, d\theta = b \tan^{-1} \left(\frac{b \sin \theta}{\sqrt{a^2 + b^2 \cos^2 \theta}} \right)$$

$$+ \frac{a}{2} \log \frac{\sqrt{a^2 + b^2 \cos^2 \theta} + a \sin \theta}{\sqrt{a^2 + b^2 \cos^2 \theta} - a \sin \theta} + C$$

will be useful. The z-component of the field at Q may be computed from $F_3 = -\partial u / \partial b$ but it is perhaps easier to compute F_3 directly by integration.

7. It can be shown that $\partial u / \partial \xi$ can be computed from (13.51–7) by doing the differentiation under the integral sign, provided Q is a point not in the region R or on its boundary. Proceed from this to verify (13.51–9), using (13.51–6). Thus (13.51–5) is proved.

13.6   Triple integrals

We shall deal with the definition of a triple integral somewhat more briefly than we did with the definition of a double integral. We begin with a closed bounded region R in three dimensions, and let $f(x, y, z)$ be a function defined and continuous in R. As in § 13.2 we must make some assumptions about the character of the boundary of R. The precise nature of these assumptions need not be made explicit as long as we do not go carefully into questions of integrability. We shall for simplicity think of the boundary of R as consisting of a finite number of surfaces, each of which is smooth except possibly at certain isolated points (e.g., the vertex of a cone) or along certain curves (e.g., the edges of a cube or the rims of a solid right circular cylinder).

We take three sets of planes, parallel respectively to the x-, y-, and z-axes. The mesh of rectangular blocks which these planes form in space is called a rectangular partition. Those blocks, or cells, which belong entirely to R are numbered consecutively in any order. Let ΔV_k be the volume of the kth cell, and let its x-, y-, and z-dimensions be Δx_k, Δy_k, Δz_k respectively, so that $\Delta V_k = \Delta x_k \, \Delta y_k \, \Delta z_k$. Finally, let (x_k, y_k, z_k) be an arbitrarily selected point in the kth cell. Then we define the triple integral of the function f over

R by the following limit, as the maximum dimensions of all the cells approach zero:

$$(13.6\text{-}1) \qquad \iiint_R f(x, y, z)dV = \lim \sum_k f(x_k, y_k, z_k)\Delta V_k,$$

or, in another notation,

$$(13.6\text{-}2) \quad \iiint_R f(x, y, z)dx\, dy\, dz = \lim \sum_k f(x_k, y_k, z_k)\Delta x_k\, \Delta y_k\, \Delta z_k.$$

We take for granted that this limit exists and is independent of the particular method of forming the partitions and choosing the points (x_k, y_k, z_k).

The analogue of Theorem II, § 13.23, is true for triple integrals; that is, the integral is given by (13.6-1) when the subregions, instead of being rectangular blocks, are formed in any manner (so long as they are sufficiently regular in shape). They need not completely fill out the region R, provided that the amount of volume omitted approaches zero in the limit. These remarks are of importance for the understanding of what happens when we use cylindrical or spherical co-ordinates.

The properties of double integrals explained in § 13.21 extend at once to triple integrals. The same is true of the inequalities of § 13.22, and the mean-value theorem.

When it comes to devising an explanation of the evaluation of triple integrals by iterated integrals, we must proceed differently than in the case of double integrals, for no intuitive geometric procedure analogous to that of § 13.3 is available to us (a four-dimensional space would be required). There is a direct analytical method, however. This method could have been used for double integrals as well. We shall give a heuristic account of the method, thus making its plausibility clear. A fully rigorous account is rather long, and it seems advisable to leave the details for later study.

Let us first state the result. The letters x, y, z can be written in six possible orders. Corresponding to each such order there is an iterated integral evaluation of the triple integral, calling for three successive single integrations. The main problem of technique is that of learning how to write the limits of integration for the iterated integrals. The notation for an iterated integral is illustrated by

$$(13.6\text{-}3) \qquad \int_0^1 dy \int_0^{1-y} dx \int_{x+y}^1 (x^2 + y^2)dz.$$

The integrations in (13.6-3) are to be performed in the order z, x, y.

It will be enough to explain the transition from the triple integral to an iterated integral for one particular order of integration. Suppose this order is first with respect to z, then with respect to x, and finally with respect to y. Choosing a typical value of y, consider the cross section of R by a plane $y =$ constant, parallel to the xz-plane. We assume that R

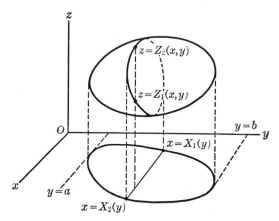

Fig. 108

is of such a shape that all the foregoing cross sections are plane regions of the type dealt with in our discussion of iterated integrals in two dimensions. As shown in Fig. 108, let the largest and smallest values of x in the cross section be respectively $X_1(y)$ and $X_2(y)$, and let $Z_1(x, y)$, $Z_2(x, y)$ be the values of z for which a typical line parallel to the z-axis in the cross section cuts the boundary of R. Finally, let $y = a$ and $y = b$ be the extreme values of y in the region R. Then

$$(13.6\text{-}4) \qquad \iiint_R f(x, y, z)dV = \int_a^b dy \int_{X_1}^{X_2} dx \int_{Z_1}^{Z_2} f(x, y, z)dz.$$

The formula (13.6-4) is the fundamental theorem about evaluating triple integrals by iterated integrals in rectangular co-ordinates. Before giving a heuristic justification of the formula we give an illustrative example.

Example.   Find the centroid of an octant of a solid sphere.

Let $x^2 + y^2 + z^2 = a^2$ be the equation of the surface of the sphere. We consider the first octant. Evidently $\bar{x} = \bar{y} = \bar{z}$, so we find $\bar{x}$ only. Analogous to (13.3-8) we have

$$V\bar{x} = \iiint_R x \, dV,$$

where V is the volume of R. In the present case $V = (\pi/6) \, a^3$. In the notation of (13.6-4) we see from Fig. 109 that

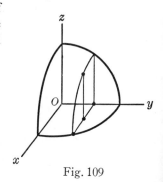

Fig. 109

$$Z_1 = 0, \; Z_2 = \sqrt{a^2 - x^2 - y^2}, \; X_1 = 0, \; X_2 = \sqrt{a^2 - y^2}.$$

Hence
$$\frac{\pi}{6} a^3 \bar{x} = \int_0^a dy \int_0^{\sqrt{a^2 - y^2}} dx \int_0^{\sqrt{a^2 - x^2 - y^2}} x \, dz$$

$$= \int_0^a dy \int_0^{\sqrt{a^2 - y^2}} x \sqrt{a^2 - x^2 - y^2} \, dx.$$

The x-integration yields

$$-\tfrac{1}{3}[a^2 - y^2 - x^2]^{3/2} \Big|_0^{\sqrt{a^2 - y^2}} = \tfrac{1}{3}(a^2 - y^2)^{3/2}.$$

Hence
$$\frac{\pi}{6} a^3 \bar{x} = \tfrac{1}{3} \int_0^a (a^2 - y^2)^{3/2} \, dy = \frac{a^4}{8} \frac{\pi}{2} \; ;$$

we omit the details of the last integration. Finally, then, $\bar{x} = \tfrac{3}{8} a$.

Now to explain (13.6–4). We go back to the definition (13.6–2). Let us single out all the cells of the partition which belong to R and lie in a particular column parallel to the z-axis (see Fig. 110). We may choose the points (x_k, y_k, z_k) so that the co-ordinates x_k, y_k are the same for all the points belonging to cells in the same vertical column. The values Δx_k and Δy_k will also be the same for all the cells in one column, and the area of the base of the column will be $\Delta x_k \Delta y_k$. Let us number the columns, say from 1 to N. Suppose ΔA_i is the area of the base of the ith column, and suppose the number of cells in the ith column is m_i. Let the points associated with these cells be $(x_i', y_i', z_{ij}), j = 1, \cdots, m_i$, and let their z-dimensions be Δz_{ij}. Then the sum in (13.6–2) can be written in the form

(13.6–5)
$$\sum_{i=1}^N \left(\sum_{j=1}^{m_i} f(x_i', y_i', z_{ij}) \, \Delta z_{ij} \right) \Delta A_i.$$

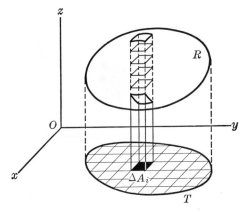

Fig. 110

The inside sum here is of the type occurring in the definition of a definite integral with respect to z. The interval of z-values that is being subdivided is approximately from the lower to the upper bounding surface of R, that is, from $Z_1(x'_i, y'_i)$ to $Z_2(x'_i, y'_i)$. Hence the inner sum is an approximation to

$$(13.6\text{--}6) \qquad \int_{Z_1(x'_i, y'_i)}^{Z_2(x'_i, y'_i)} f(x_i, y_i, z)dz.$$

For convenience let us write

$$(13.6\text{--}7) \qquad g(x, y) = \int_{Z_1(x, y)}^{Z_2(x, y)} f(x, y, z)dz.$$

Then the expression (13.6–6) is $g(x'_i, y'_i)$, and (13.6–5) is seen to be approximately equal to

$$(13.6\text{--}8) \qquad \sum_{i=1}^{N} g(x'_i, y'_i)\Delta A_i$$

if the cell dimensions in the z-direction are all sufficiently small. This sum, in turn, is of the type occurring in the definition of a double integral. If T is the plane region obtained by projecting the points of R perpendicularly on the xy-plane, the bases of the columns form a rectangular partition of T. When the dimensions of the cells of this partition are small enough, the sum (13.6–8) is very nearly equal to the double integral

$$\iint_T g(x, y)dA,$$

which in turn is equal to the iterated integral

$$(13.6\text{--}9) \qquad \int_a^b dy \int_{X_1}^{X_2} g(x, y)dx,$$

as we see from Fig. 108. We see, therefore, on combining (13.6–7) and (13.6–9), that the sum (13.6–5) is an approximation to the iterated integral

$$\int_a^b dy \int_{X_1}^{X_2} dx \int_{Z_1}^{Z_2} f(x, y, z)dz.$$

It may be shown in more detail that the approximation becomes better and better as we take the limit defining the triple integral, so that (13.6–4) is exactly true.

13.7   Applications of triple integrals

Triple integrals may be used to calculate the locations of centers of gravity, the masses of solids of variable density, moments of inertia, and other quantities of physical or geometrical significance. The fundamental principles of such applications are the same as those set forth in connection with double integrals (§ 13.5).

We shall use the Greek letter μ for volume density. The mass of a solid of variable density $\mu(x, y, z)$ occupying a region R is then

$$M = \iiint\limits_{R} \mu \, dV.$$

The center of gravity $(\bar{x}, \bar{y}, \bar{z})$ is found from the formula

$$M\bar{x} = \iiint\limits_{R} x\mu \, dV$$

and two other similar formulas. The moment of inertia about the z-axis is

$$I_z = \iiint\limits_{R} (x^2 + y^2)\mu \, dV.$$

The product of inertia relative to the planes $x = 0$ and $y = 0$ is

$$U_{xy} = \iiint\limits_{R} xy\mu \, dV.$$

Other moments of inertia I_x, I_y, and other products of inertia U_{yz}, U_{zx} are defined by analogous formulas.

Problems in gravitational attraction are mathematically almost identical with problems of electrostatic forces, since Newton's law and Coulomb's law are both inverse-square laws. There is a difference in sign, since two masses attract each other, whereas two positive charges repel. Newton's law for mass particles m and m' at P and P', a distance r apart, states that m exerts on m' a force

$$\mathbf{F} = k \, \frac{mm'}{r^3} \, \overrightarrow{P'P},$$

where k is a universal constant depending only on the units of mass, distance, and force. In theoretical work it is customary to choose units such that $k = 1$. We shall do this. The force of attraction on a unit mass at Q, produced by a solid of density μ occupying a region R, is

$$\mathbf{F} = \iiint\limits_{R} \frac{\mu}{r^3} \, \overrightarrow{QP} \, dV,$$

where $r = QP$, μ is evaluated at P, and integration is carried out with respect to the co-ordinates of P.

The concept of potential is useful in the theory of gravitational attraction. The potential at Q is defined to be

$$u(Q) = \iiint_R \frac{\mu}{r}\, dV.$$

The relation between the potential u and the gravitational field force $\mathbf{F}$ is expressed by the equation

$$\mathbf{F} = \nabla_Q u;$$

i.e., the field is the gradient of the potential. The situation is comparable to that in electrostatics (see § 13.51); there, however, the field is the *negative* of the gradient of the potential. The difference in sign arises from the difference in sign between Newton's and Coulomb's laws.

Example 1. The first octant portion of the solid inside the cylinder $x^2 + y^2 = a^2$ and between the planes $z = 0$, $z = h$ has density $\sigma = x$. Find its mass. We have

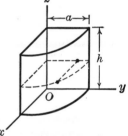

$$M = \iiint_R x\, dV = \int_0^h dz \int_0^a dy \int_0^{\sqrt{a^2 - y^2}} x\, dx;$$

$$M = \int_0^h dz \int_0^a \tfrac{1}{2}(a^2 - y^2)\,dy = \int_0^h \frac{a^3}{3}\, dz = \tfrac{1}{3}\, a^3 h.$$

The finding of the limits of integration is illustrated in Fig. 111.

Fig. 111

Example 2. Find the moment of inertia about the z-axis of the homogeneous tetrahedron bounded by the planes $z = x + y$, $x = 0$, $y = 0$, $z = 1$. The integral in this case is

$$I_z = \iiint_R \mu(x^2 + y^2)dV = \mu \int_0^1 dy \int_0^{1-y} dx \int_{x+y}^1 (x^2 + y^2)dz.$$

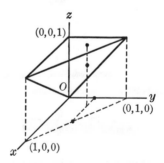

Fig. 112

The limits of integration are found by an examination of Fig. 112. Completion of the integration is left as an exercise for the student. The result is

$$I_z = \frac{\mu}{30} \cdot$$

Since the volume of the tetrahedron is $\frac{1}{6}$, the mass is $M = \mu/6$, whence $\mu = 6 M$ and $I_z = M/5$. This means that the radius of gyration of the tetrahedron about the z-axis is $k_z = 1/\sqrt{5}$.

EXERCISES

1. A homogeneous solid block is bounded by the planes $x = 0$, $x = a$, $y = 0$, $y = b$, $z = 0$, $z = c$. Show that $I_x = (M/3)(b^2 + c^2)$.

2. Locate the center of gravity of the tetrahedron in Example 2. Consider some of the other orders of integration, and select a convenient order for each part of the work.

3. (a) Locate the center of gravity of the solid tetrahedron cut from the first octant by the plane $(x/a) + (y/b) + (z/c) = 1$. (b) Find the moment of inertia of this tetrahedron about the y-axis.

4. Find $\bar{x}$ and $\bar{y}$ for the solid of Example 1.

5. The unit cube bounded by the planes $x = 0$, $x = 1$, $y = 0$, $y = 1$, $z = 0$, $z = 1$ has density $\mu = xz$. (a) Find the mass and locate the center of gravity of the cube. (b) Find the y-component of the attraction which the cube exerts on a unit mass at the origin.

6. A homogeneous solid is bounded by the plane $z = 0$ and the paraboloid $(x^2/a^2) + (y^2/b^2) + (z/c) = 1$. (a) Locate its center of gravity. (b) Find $I_x, I_y,$ and I_z for the solid.

7. Consider the homogeneous solid bounded by the ellipsoid
$$(x^2/a^2) + (y^2/b^2) + (z^2/c^2) = 1.$$
(a) Locate the center of gravity of the first octant portion of this solid.
(b) Calculate the moment of inertia of the entire solid about the x-axis.

8. The density of a cube of edge $2a$ is proportional to the square of the distance from its center, the density being unity at the center of each face. (a) Find the total mass and the average density. (b) Find I_z for this cube. (c) Find the ratio of the I_z in (b) to the value which I_z would have if the cube had the same total mass distributed uniformly throughout the volume.

9. If a solid has density at (x, y, z) equal to the distance from (x, y, z) to the origin, show that the potential at the origin is equal to the volume of the solid.

10. Suppose a solid has density at (x, y, z) equal to the cube of the distance from (x, y, z) to the origin. Show that the attraction of the solid on a unit mass at the origin is a force directed toward the centroid of the volume occupied by the

charge density = $\dfrac{\text{charge}}{\text{unit volume}}$

solid, and equal in magnitude to the product of the volume and the distance from the origin to the centroid.

11. Let I denote the moment of inertia of a solid about a certain axis, and let I_0 be the moment of inertia about a parallel axis through the center of gravity. If h is the distance between the two axes, prove that $I = I_0 + Mh^2$, where M is the total mass of the body.

12. Consider two rectangular co-ordinate systems with the same origin, the relations between the xyz-system and the $x'y'z'$-system being expressed in the notation of § 10.3. For a given body, show that

$$I_{x'} = I_x l_1^2 + I_y m_1^2 + I_z n_1^2 - 2 U_{yz} m_1 n_1 - 2 U_{zx} n_1 l_1 - 2 U_{xy} l_1 m_1.$$

This shows how products of inertia enter into consideration of the effect of rotation of axes upon moments of inertia.

The equation

$$I_x x^2 + I_y y^2 + I_z z^2 - 2 U_{yz} yz - 2 U_{zx} zx - 2 U_{xy} xy = 1$$

defines what is called the *ellipsoid of inertia* for the body relative to the origin O. A set of axes such that the products of inertia all vanish is called a set of *principal axes of inertia* for the body.

13.8   Cylindrical co-ordinates

If we use polar co-ordinates in a plane, and a rectangular co-ordinate along an axis perpendicular to the plane at the origin of the polar system, the combination is called a cylindrical co-ordinate system. Most commonly the polar co-ordinates are taken in the xy-plane (see Fig. 113), but there is no logical necessity for this choice. It is often convenient to evaluate a triple integral by an iterated integral in cylindrical co-ordinates. As we saw in § 13.6,

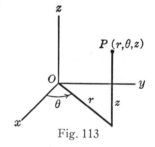

Fig. 113

$$(13.8\text{--}1) \qquad \iiint_R f(x, y, z)dV = \iint_T dA \int_{Z_1}^{Z_2} f(x, y, z)dz.$$

If we express the integrand in cylindrical co-ordinates, say $f(x, y, z) = F(r, \theta, z)$, the double integral in (13.8–1) may be evaluated as an iterated integral in polar co-ordinates. This leads to the result

$$(13.8\text{--}2) \qquad \iiint_R F(r, \theta, z)dV = \int_\alpha^\beta d\theta \int_{R_1}^{R_2} r \, dr \int_{Z_1}^{Z_2} F(r, \theta, z)dz.$$

Do not fail to observe the factor r which is introduced into the integrand of the iterated integral. The limits Z_1, Z_2 must be expressed in terms of

r and θ; the r and θ limits are found by inspection of the plane region T, the "shadow" of R on the xy-plane (see Fig. 110). The result (13.8–2) and others like it may also be obtained by an argument similar to that beginning after the Example in § 13.6. There are five other possible orders of integration. A systematic method for determining the limits of integration for any given order is illustrated in the following example:

Example.   Find the moment of inertia of a homogeneous right circular cone about its axis.

     Let the radius of the base be b, the altitude be h. The density μ is constant, so $M = \mu(\pi/3)b^2h$. We place the cone as shown in Fig. 114 (we draw only one-fourth the cone). Let the integration order be r, z, θ. We must first set up the triple integral:

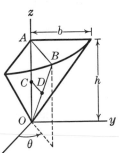

Fig. 114

$$ I = \iiint_R \mu(x^2 + y^2)dV = \mu \iiint_R r^2 \, dV. $$

Now picture a section of the region R made by holding the last integration variable (here θ) constant. In the present case this is the triangle OAB. Next assign the second integration variable z a typical value, and determine the range of freedom left to the first integration variable r. This process is indicated in Fig. 114 by the line CD. Since $OC = z$, the value of r at D is given by

$$ \frac{r}{z} = \frac{b}{h}. $$

The r limits of integration are therefore 0 and zb/h. Now let the line CD range in the z-direction as much as it may (from 0 to h); these are the z-limits of integration. Finally, let θ vary through all values necessary to have the θ-sections sweep out the entire region R. We see that the θ-limits of integration are 0 and 2π. Therefore (remembering the additional factor r),

$$ I = \mu \int_0^{2\pi} d\theta \int_0^h dz \int_0^{bz/h} r^3 \, dr = \frac{\mu\pi}{10} b^4h. $$

This may be written $I = \frac{3}{10} Mb^2$.

EXERCISES

    1. For the solid cone of the illustrative example find (a) I_y; (b) the location of the center of gravity of the first octant portion; (c) the attraction exerted on a unit mass at the origin; (d) the potential at a point $(0, 0, \zeta)$, where $\zeta < 0$.

2. Find the moment of inertia of a homogeneous solid sphere of radius a, about a diameter.

3. For a homogeneous solid right circular cylinder of height h and radius of base a, find the moments of inertia (**a**) about the axis of the cylinder; (**b**) about a line through the center of gravity of the cylinder, perpendicular to the axis of the cylinder; (**c**) the attraction exerted by the cylinder on a unit mass at the center of one end; (**d**) the potential at a point $(0, 0, \zeta)$, assuming the cylinder defined by $x^2 + y^2 \leqq a^2$, $0 \leqq z \leqq h$, and assuming $\zeta \geqq h$.

13.9   Spherical co-ordinates

To form a spherical co-ordinate system we start from an origin O and a fixed ray issuing from O. We shall take the ray as the positive z-axis; there is, however, no necessity for any one special relation between spherical and rectangular co-ordinates. The spherical co-ordinates are the distance $\rho = OP$ and the two angles θ, ϕ (see Fig. 115). The angle θ, sometimes called the *azimuth* of P, is the same as that used in plane polar co-ordinates. The angle ϕ is the *colatitude* of P. We always choose ϕ in the range $0 \leqq \phi \leqq \pi$. For most work ρ is taken nonnegative.

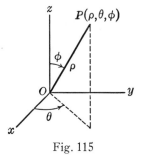

Fig. 115

The student should be aware that in some books the roles of θ and ϕ are reversed, so that θ denotes the colatitude. This is particularly true of European texts, and books on mathematical physics. With due caution the student should have no trouble accommodating himself to such differences in notation.

The standard formulas relating spherical to rectangular co-ordinates are

$$(13.9\text{--}1) \qquad x = \rho \sin \phi \cos \theta, \ y = \rho \sin \phi \sin \theta, \ z = \rho \cos \phi.$$

To explain the use of spherical co-ordinates in evaluating triple integrals, we employ a method like that of § 13.4. The curvilinear co-ordinates ρ, θ, ϕ give rise to a partition of R into cells which resemble cubes. A typical such cell is bounded by three pairs of surfaces, two each of the types $\rho = $ constant (spheres), $\theta = $ constant (half-planes through the z-axis), $\phi = $ constant (cones). Such a cell is generated by taking a shaded area as shown in Fig. 116 and turning it about the z-axis through an angle $\Delta\theta$. By formulas of elementary solid geometry the volume of this cell is found to be

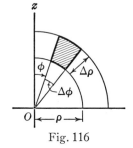

Fig. 116

$$(12.9\text{--}2) \qquad \Delta V = \tfrac{1}{3}[(\rho + \Delta\rho)^3 - \rho^3][\cos \phi - \cos (\phi + \Delta\phi)]\Delta\theta.$$

Now, by the law of the mean, we have

$$(\rho + \Delta\rho)^3 - \rho^3 = 3\,\rho'^2\,\Delta\rho,$$
$$\cos(\phi + \Delta\phi) - \cos\phi = -\sin\phi'\,\Delta\phi,$$

where ρ' is between ρ and $\rho + \Delta\rho$ and ϕ' is between ϕ and $\phi + \Delta\phi$.

Thus (13.9–2) becomes

$$\Delta V = \rho'^2 \sin\phi'\,\Delta\rho\,\Delta\theta\,\Delta\phi.$$

This may be expressed by saying that there is in each cell, say the kth one, a point $(\rho_k, \theta_k, \phi_k)$ such that the volume of the cell is

(13.9–3) $$\Delta V_k = \rho_k{}^2 \sin\phi_k\,\Delta\rho_k\,\Delta\theta_k\,\Delta\phi_k.$$

It is worth while observing that the expression $\rho^2 \sin\phi\,\Delta\rho\,\Delta\theta\,\Delta\phi$ is evidently a good approximation to ΔV in (13.9–2) when the dimensions of the cell are small, for three concurrent edges of the cell have the lengths

$$\Delta\rho,\ \rho\,\Delta\phi,\ \rho\sin\phi\,\Delta\theta.$$

Now consider a triple integral

$$\iiint\limits_R F(\rho,\theta,\phi)dV,$$

the integrand being expressed in terms of spherical co-ordinates. We express this integral as the limit of a sum

$$\sum_k F(\rho_k, \theta_k, \phi_k)\Delta V_k,$$

using the formula (13.9–3) for ΔV_k, so that

(13.9–4)
$$\iiint\limits_R F(\rho,\theta,\phi)dV = \lim \sum_k F(\rho_k,\theta_k,\phi_k)\rho_k{}^2 \sin\phi_k\,\Delta\rho_k\,\Delta\theta_k\,\Delta\phi_k.$$

By looking upon the equations (13.9–1) as defining a mapping between an xyz-space and a $\rho\theta\phi$-space, we interpret the limit on the right in (13.9–4) as defining a triple integral over the image of R in the $\rho\theta\phi$-space (compare with (13.4–4)), and so arrive at the iterated integral

(13.9–5) $$\iiint\limits_R F(\rho,\theta,\phi)dV = \int_\alpha^\beta d\theta \int_{\Phi_1}^{\Phi_2} d\phi \int_{R_1}^{R_2} F(\rho,\theta,\phi)\rho^2 \sin\phi\,d\rho,$$

or any one of five other iterated integrals. The essential thing to observe is the factor $\rho^2 \sin\phi$ on the right. The finding of the limits of integration follows the same principles illustrated in § 13.8 and earlier work.

Spherical co-ordinates seem to be particularly convenient for many gravitational-attraction problems.

Example. The smaller volume bounded by the sphere $x^2 + y^2 + z^2 = 4\,a^2$ and the plane $z = a$ is filled with a homogeneous solid. Find the gravitational field which this solid produces at the origin.

The density μ is constant; the force is in the z-direction, so $F_1 = F_2 = 0$. In the integral for F_3 we have $\psi = \phi$; hence

$$F_3 = \mu \iiint_R \frac{\cos \phi}{\rho^2} \, dV.$$

We integrate in this order: first ρ, then θ, and finally ϕ. Finding the limits is illustrated in Fig. 117 and Fig. 118. The equation of the plane $z = a$ is

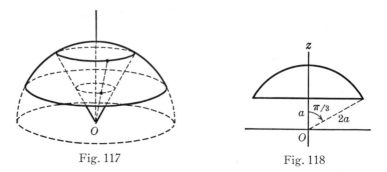

Fig. 117   Fig. 118

written as $\rho = a \sec \phi$; that of the sphere is $\rho = 2\,a$. Putting in the factor $\rho^2 \sin \phi$, we have

$$F_3 = \mu \int_0^{\pi/3} d\phi \int_0^{2\pi} d\theta \int_{a \sec \phi}^{2a} \sin \phi \cos \phi \, d\rho.$$

$$F_3 = 2\,\pi\mu \int_0^{\pi/3} (2\,a \sin \phi \cos \phi - a \sin \phi)d\phi$$

$$= 2\,\pi\mu a \left[\sin^2 \phi + \cos \phi \right]_0^{\pi/3} = \frac{\pi\mu a}{2}.$$

EXERCISES

1. Find the moment of inertia of a homogeneous spherical solid about a diameter, using spherical co-ordinates.

2. Locate the centroid of an octant of a solid sphere, using spherical co-ordinates.

3. Consider the homogeneous solid cone defined by $h^2(x^2 + y^2) \le b^2 z^2$,

$0 \leq z \leq h$. Find (a) the attraction it exerts on a unit mass at the origin; (b) the moment of inertia about the z-axis; (c) the moment of inertia about the y-axis.

4. Consider the homogeneous solid sphere whose surface is defined by $\rho = 2a \cos \phi$. It is divided into two parts by the plane $z = a$. (a) Find the attraction which each part exerts on a unit mass at the origin. (b) What is the ratio of the magnitudes of these attractions? (c) What is the combined attraction of both parts?

5. Suppose the sphere $x^2 + y^2 + (z - h)^2 = a^2$, where $0 < a \leq h$, is filled with matter of constant density μ. Show that its attraction on a unit mass particle at the origin is $\frac{4}{3}\pi\mu a^3/h^2$, and hence that the attraction is the same as though all the mass were concentrated at the center of the sphere. Begin by showing that a typical ray from O pierces the sphere at the two points given by $\rho = h \cos \phi \pm (h^2 \cos^2 \phi + a^2 - h^2)^{1/2}$, and that the range of ϕ values to be considered is $0 \leq \phi \leq \sin^{-1}(a/h)$.

6. In Exercise 5, suppose that $0 \leq h < a$, so that the origin is inside the sphere. In this case show that the attraction is $\frac{4}{3}\pi\mu h$. This shows that, if the sphere is divided up into a smaller concentric sphere of radius h, and a shell between the spheres of radii a and h, the net force of attraction due to the matter in the shell is zero, the total force being just that produced by the matter in the sphere of radius h, on whose surface O lies.

7. Show that the value of the integral

$$\int_0^\pi \frac{(c - \rho \cos \phi) \sin \phi}{(c^2 + \rho^2 - 2c\rho \cos \phi)^{3/2}} \, d\phi$$

is $2/c^2$ if $c > \rho \geq 0$, and is equal to 0 if $\rho > c \geq 0$. Also show that it is equal to $1/c^2$ if $\rho = c > 0$. Some of these results are useful in Exercises 8. If $c \neq \rho$ and $c\rho \neq 0$, the integral may be evaluated by making the substitution $t^2 = c^2 + \rho^2 - 2c\rho \cos \phi$, $t \, dt = c\rho \sin \phi \, d\phi$.

8. Consider a solid sphere of radius a, center at the origin, not necessarily of constant density, but with the mass distributed symmetrically about the center, so that μ depends only on ρ: $\mu = \mu(\rho)$. Show that the mass of the sphere is

$$M = 4\pi \int_0^a \rho^2 \mu(\rho) d\rho.$$

If a unit mass particle is h units from the center of the sphere, show that it is attracted with a force M/h^2 if $h \geq a$, and with a force

$$\frac{4\pi}{h^2} \int_0^h \rho^2 \mu(\rho) d\rho$$

if $0 < h < a$. If the density is constant, these results are the same as those obtained in a different way in Exercises 5, 6.

Curves and Surfaces 14

14. Introduction

Curves and surfaces are geometric entities with which the student is to some extent familiar. The simplest examples of these entities, such as the conic curves in the plane, and spheres, cylinders, cones, and other quadric surfaces in space, have been encountered repeatedly from analytic geometry through calculus. Geometrical interpretations of functions of one or two independent variables have led the student to think of curves and surfaces in quite general terms. In this chapter we propose to make a careful study of the means by which we render our intuitive notions about curves and surfaces amenable to precise mathematical treatment. This is done partly as an introduction to a branch of geometry—what is known as *differential geometry*—and partly as preparation for the following chapter on line and surface integrals.

A point to be emphasized is this: Our intuitive notions about curves and surfaces are all derived from relatively simple examples of these things. The general concepts of curves and surfaces are very inclusive, however, and in our studies we must remember that when we wish to *prove* something, we must appeal to the definitions and previously established theorems, not solely to our intuitions, which may present us with an oversimplified picture. Direct geometric visualization of the subjects of our discussion is, however, of great value, both for the suggestions we can derive and for the better understanding and retention of what we learn.

14.1 Representations of curves

Intuitively we think of a curve as a one-dimensional configuration, like the path of a moving particle, or as something we might obtain by bending and twisting a straight line. We shall define a curve by saying that it is an ordered configuration of points (x, y, z) given by three continuous functions of a parameter:

$$(14.1-1) \qquad x = f(t), y = g(t), z = h(t);$$

the range of the parameter is to be some interval (finite or infinite) of the real axis. We speak of (14.1–1) as a parametric representation of the curve. A curve may have more than one parametric representation. If we interpret t as time, (14.1–1) may be regarded as defining the path of a moving point. The point may pass through the same position in space several times; in this case the curve intersects itself. Evidently a curve in the above sense of the word is very general, and may not be very smooth.

Imagine, for instance, the track of a tiny particle in Brownian movement over a long period of time.

To avoid some of the complexities which may occur in dealing with curves in general, we introduce some restrictions. By an *arc* we mean a curve which does not intersect itself, which has two distinct ends, and which is represented in the form (14.1–1) with a finite range of the parameter t, say $a \leqq t \leqq b$, where $a < b$. A semicircle is an arc, but an entire circle is not. If a curve is defined by (14.1–1) with a finite range $a \leqq t \leqq b$ of the parameter, and if the points corresponding to $t = a$ and $t = b$ are coincident, we say that the curve is closed (it has no free ends). Such a closed curve without self-intersections is called a *simple closed curve*, or a closed *Jordan* curve (after a French mathematician). The prototype of an arc is the unit interval $0 \leqq x \leqq 1$, while the prototype of a simple closed curve is the unit circle $x^2 + y^2 = 1$. Continuous deformation (bending, twisting, stretching, shrinking) of an arc leaves it still an arc, provided no points are brought together which were originally distinct. The like can be said about closed Jordan curves. The boundary of a square is a closed Jordan curve. A parabola is neither an arc nor a closed curve. It may be regarded, however, as an infinite number of arcs joined end to end.

A curve is called *smooth* if two conditions are satisfied: (1) it does not intersect itself, and (2) it has a tangent line at each point, whose direction varies continuously as the point moves along the curve. The second of these conditions is satisfied if the functions f, g, h in (14.1–1) have continuous derivatives which do not all vanish together for any value of t. The direction of the tangent line is specified by the ratios

(14.1–2) $f'(t) : g'(t) : h'(t)$,

and we can find direction cosines by normalization (i.e., division by the square root of the sum of the squares of the three quantities in (14.1–2)).

The curves with which we normally deal either are smooth or each bounded portion of the curve is *sectionally smooth*, that is, composed of a finite number of smooth arcs joined end to end. Such a curve may have corners at the junction points. The periphery of a square is an example of a closed curve with corners.

Frequently we encounter curves as the intersections of two surfaces. A parametric representation of the form (14.1–1) may not immediately present itself, but may theoretically be derived from the analytical representations of the surfaces by implicit-function arguments.

14.2   Arc length

The length of a smooth arc from $t = a$ to $t = b$ is given by an integral:

$$(14.2-1) \qquad L = \int_a^b \left[\left(\frac{dx}{dt} \right)^2 + \left(\frac{dy}{dt} \right)^2 + \left(\frac{dz}{dt} \right)^2 \right]^{1/2} dt.$$

This formula, or the special form of it for plane curves, is familiar from elementary calculus. The derivation will now be given.

Consider any curve C, closed or not, without self-intersections, and defined by (14.1–1) with a finite parameter interval $a \leqq t \leqq b$. Consider a subdivision

$$a = t_0 < t_1 < t_2 < \cdots < t_n = b$$

of the interval (a, b) into n subintervals, and let P_k be the point of C corresponding to $t = t_k$. Joining the points $P_0, P_1, \cdots, P_n$ in order, we obtain a polygonal line inscribed on C. The length of the polygonal line is

(14.2–2) $$\overline{P_0 P_1} + \cdots + \overline{P_{n-1} P_n}.$$

Now consider what happens if we allow n to increase, and make the length of the longest of the segments $P_{k-1} P_k$ approach zero. If the sum (14.2–2) approaches a finite limit, we call this limit the length of the curve C, and we say that C is *rectifiable*. A curve may be rectifiable without being smooth or even sectionally smooth; we shall confine ourselves to showing that a smooth arc is rectifiable, with length given by (14.2–1). The length of a sectionally smooth curve is found by adding the lengths of its component arcs.

Consider a segment $P_{k-1} P_k$ of the polygonal line. Its length is

(14.2–3) $$[(\Delta x_k)^2 + (\Delta y_k)^2 + (\Delta z_k)^2]^{1/2},$$

where $$\Delta x_k = f(t_k) - f(t_{k-1}),$$

with similar formulas for Δy_k and Δz_k. By the law of the mean we find

$$\Delta x_k = f'(\alpha_k)\Delta t_k, \quad \Delta y_k = g'(\beta_k)\Delta t_k, \quad \Delta z_k = h'(\gamma_k)\Delta t_k,$$

where $\Delta t_k = t_k - t_{k-1}$ and the points $\alpha_k, \beta_k, \gamma_k$ are between t_{k-1} and t_k. Thus (14.2–2) becomes

(14.2–4) $$\sum_{k=1}^{n} [(f'(\alpha_k))^2 + (g'(\beta_k))^2 + (h'(\gamma_k))^2]^{1/2} \Delta t_k.$$

If the points $\alpha_k, \beta_k, \gamma_k$ were all the same, the limit of the sum (14.2–4) would be, by definition, the integral

$$\int_a^b [(f'(t))^2 + (g'(t))^2 + (h'(t))^2]^{1/2} \, dt,$$

which is the same as (14.2–1). Because the three points need not be the same, more argument is needed. The matter is covered, however, by a general theorem (Duhamel's principle) in the theory of integration, and we refer the student to Example 2, § 18.21, for the final discussion of the issue.

Let C be a smooth arc and let s be the length measured along C from $t = a$ to a variable point. It follows from (14.2–1), with the upper limit b replaced by a variable, that

(14.2–5) $$\frac{ds}{dt} = \left[\left(\frac{dx}{dt}\right)^2 + \left(\frac{dy}{dt}\right)^2 + \left(\frac{dz}{dt}\right)^2\right]^{1/2}$$

Hence

(14.2–6) $$ds^2 = dx^2 + dy^2 + dz^2.$$

The integral giving the arc length is frequently such that no evaluation of the integral can be made in terms of elementary functions. This is the case even with plane curves. Sometimes the arc length is expressible in terms of tabulated standard integrals, such as elliptic integrals.

Example. Consider the first octant portion of the curve of intersection of the sphere and cylinder

(14.2–7) $$x^2 + y^2 + z^2 = 4\,a^2, \qquad x^2 + (y - a)^2 = a^2,$$

as shown in Fig. 119.

It is convenient to use z as a parameter for this curve. If we eliminate x by subtracting the two equations in (14.2–7), we find

$$y = \frac{4\,a^2 - z^2}{2\,a}.$$

Substituting this result in the second of the equations (14.2–7), we find

$$x = \frac{z}{2\,a}\sqrt{4\,a^2 - z^2}.$$

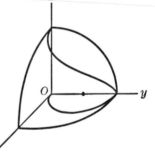

Fig. 119

As parametric equations of the curve, we have

$$x = \frac{z}{2\,a}\sqrt{4\,a^2 - z^2}, \qquad y = \frac{4\,a^2 - z^2}{2\,a}, \qquad z = z.$$

A direct calculation shows that

$$ds^2 = \frac{8\,a^2 - z^2}{4\,a^2 - z^2}\,dz^2.$$

The range of z is from 0 to 2 a, so the length of the first-octant portion of the curve is

(14.2–8) $$L = \int_0^{2\,a}\left(\frac{8\,a^2 - z^2}{4\,a^2 - z^2}\right)^{1/2} dz.$$

This integral is improper at the limit $z = 2\,a$, but it is convergent. It can be put in the form of a standard elliptic integral of the second kind. For further discussion of this problem see Exercise 8.

EXERCISES

The standard elliptic integral of the second kind is defined as

$$E(k, \phi) = \int_0^{\phi} \sqrt{1 - k^2 \sin^2 t}\ dt,$$

where $0 < k < 1$. If $\phi = \pi/2$, the integral is called *complete*. Values of this integral for various values of the parameters k, ϕ are given in many books of tables. In some of the exercises it is required that the arc length be expressed in the form of such a standard integral.

1. Consider the cylindrical helix $x = a \cos \theta$, $y = a \sin \theta$, $z = k\theta$. Show that the arc length from $\theta = \theta_0$ to $\theta = \theta_1$ is directly proportional to $\theta_1 - \theta_0$.

2. A point moves on a conical helix according to the formulas $x = t \cos \omega t$, $y = t \sin \omega t$, $z = t$, where t is the time. Find the speed of the point at $t = 0$ and at $t = 1$, and its average speed during this time interval.

3. Find the length of the curve $x = a(\theta - \sin \theta)$, $y = a(1 - \cos \theta)$, $z = 4\,a \sin(\theta/2)$, $0 \le \theta \le 2\,\pi$.

4. If $0 < b < 4\,a$, show that the length of the curve $x = a(\theta - \sin \theta)$, $y = a(1 - \cos \theta)$, $z = b \sin(\theta/2)$ from $\theta = 0$ to $\theta = 2\,\pi$ is

$$\int_0^{2\pi} \left[4\,a^2 - \left(4\,a^2 - \frac{b^2}{4} \right) \cos^2 \frac{\theta}{2} \right]^{1/2} d\theta$$

and that this is equal to $8\,aE\,(k,\,\pi/2)$, where $k^2 = 1 - (b^2/16\,a^2)$.

5. If $0 < 4\,a < b$, show that the arc length sought in Exercise 4 is $2\,bE\,(k,\,\pi/2)$, where $k^2 = 1 - (16\,a^2/b^2)$.

6. Show that the total circumference of an ellipse of major axis $2\,a$ and eccentricity e is $4\,aE\,(e,\,\pi/2)$. Start with the parametric form $x = a \sin t$, $y = b \cos t$, $b^2 = a^2(1 - e^2)$.

7. Consider the intersection of the ellipsoid $(x^2/a^2) + (y^2/b^2) + (z^2/c^2) = 1$ and the cylinder $x^2 + y^2 = b^2$, where $a > b$. Show that the first-octant portion of this curve has arc length

$$\frac{c\sqrt{a^2 - b^2}}{ak}\,E\left(k,\,\frac{\pi}{2} \right), \qquad \text{where } k^2 = \frac{c^2(a^2 - b^2)}{c^2(a^2 - b^2) + a^2 b^2}.$$

8. Show that the integral (14.2–8) is equal to $2\sqrt{2}\,aE(\sqrt{2}/2,\,\pi/2)$. This may be done either by making a suitable change of variable in (14.2–8) or by using the parametric representation of the original curve: $x = a \sin 2\,t$, $y = a + a \cos 2\,t$, $z = 2\,a \sin t$.

9. Find the length of the curve $x = a \cosh t$, $y = a \sinh t$, $z = at$, from $t = 0$ to $t = t_1$. This curve lies on the hyperbolic cylinder $x^2 - y^2 = a^2$.

10. A point starts at $(0, a, 0)$ and moves into the first octant along the curve $y = a \cosh (x/a)$, $z = a \sinh (x/a)$. Show that when it is at (x, y, z) it has traveled a distance $\sqrt{2}\, z$.

14.3   The tangent vector

Let C be a smooth arc. Let us choose a certain direction along the arc as the positive direction. At any point of C we shall define what we call the *tangent vector*. This is a vector of unit length along the tangent line in the positive direction. It is denoted by $\mathbf{T}$; the vector $\mathbf{T}$ is a vector point function (in the sense of § 10.51) defined along C. If the angles which $\mathbf{T}$ makes with the positive co-ordinate axes are α, β, γ, we can write

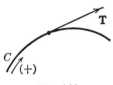

Fig. 120

$$\mathbf{T} = \cos \alpha \mathbf{i} + \cos \beta \mathbf{j} + \cos \gamma \mathbf{k}.$$

If s is arc length measured along C in the positive direction from some fixed point, the direction cosines are given by $\cos \alpha = \dfrac{dx}{ds}$, and so on, so that

(14.3–1)
$$\mathbf{T} = \frac{dx}{ds}\mathbf{i} + \frac{dy}{ds}\mathbf{j} + \frac{dz}{ds}\mathbf{k}.$$

If x, y, z are functions of t with continuous derivatives which are never all simultaneously zero,

(14.3–2)
$$\frac{ds}{dt} = \pm \left[\left(\frac{dx}{dt} \right)^2 + \left(\frac{dy}{dt} \right)^2 + \left(\frac{dz}{dt} \right)^2 \right]^{1/2},$$

with the plus sign chosen if s and t increase in the same sense along C; otherwise the minus sign is chosen. Since

$$\frac{dx}{dt} = \frac{dx}{ds}\frac{ds}{dt},$$

and so on, we see that $\mathbf{T}$ may be found from the formula

(14.3–3)
$$\mathbf{T} = \frac{1}{\left(\dfrac{ds}{dt} \right)} \left(\frac{dx}{dt}\mathbf{i} + \frac{dy}{dt}\mathbf{j} + \frac{dz}{dt}\mathbf{k} \right),$$

together with (14.3–2).

For many purposes it is convenient to discuss $\mathbf{T}$ entirely in vector notation, without the use of a co-ordinate system. If $P(x, y, z)$ is a variable point on C, we denote the vector $\overrightarrow{OP}$ by $\mathbf{R}$. In the co-ordinate notation

(14.3–4)
$$\mathbf{R} = x\mathbf{i} + y\mathbf{j} + z\mathbf{k}.$$

Equation (14.3–1) may be written

(14.3–5)
$$\frac{d\mathbf{R}}{ds} = \mathbf{T}.$$

This formula for the tangent vector can be derived directly by vector methods; we give the derivation, because it provides an instructive exercise in learning how to differentiate a vector function of a scalar variable.

We consider R as a function of s. Let the points P, P' on C correspond to s and $s + \Delta s$, respectively. Write

$$\overrightarrow{OP} = \mathbf{R},\ \overrightarrow{OP'} = \mathbf{R} + \Delta\mathbf{R},$$

so that
$$\Delta\mathbf{R} = \overrightarrow{PP'}.$$

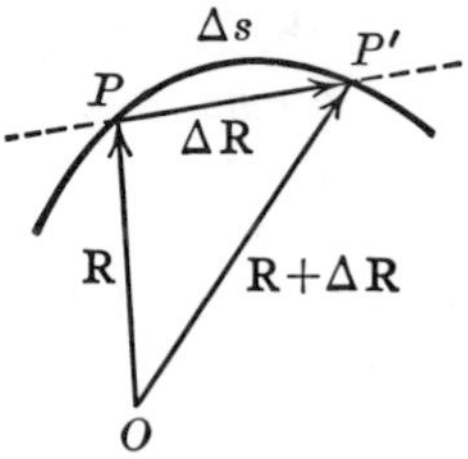

Fig. 121

The quotient $\Delta\mathbf{R}/\Delta s$ is a vector along the line of the chord PP'. Since the length of $\Delta\mathbf{R}$ is the length of the chord PP', we see that when P' approaches P the limit of the length of $\Delta\mathbf{R}/\Delta s$ is unity. Furthermore, the limiting direction of PP' is that of the tangent at P. Therefore

$$\frac{d\mathbf{R}}{ds} = \lim_{\Delta s \to 0} \frac{\Delta\mathbf{R}}{\Delta s} = \mathbf{T}.$$

Differentiation of $\mathbf{R}$ with respect to t gives

(14.3–6)
$$\frac{d\mathbf{R}}{dt} = \frac{d\mathbf{R}}{ds}\frac{ds}{dt} = \frac{ds}{dt}\mathbf{T}.$$

This is equivalent to formula (14.3–3). If t is the time variable, $\dfrac{d\mathbf{R}}{dt}$ is the *vector velocity* of the point P moving on C.

EXERCISES

1. Let $\mathbf{F}$ and $\mathbf{G}$ denote vector functions of a scalar variable t. Assuming that $\mathbf{F}$ and $\mathbf{G}$ are differentiable, prove the formulas

(a) $\dfrac{d}{dt}(\mathbf{F}\cdot\mathbf{G}) = \mathbf{F}\cdot\dfrac{d\mathbf{G}}{dt} + \dfrac{d\mathbf{F}}{dt}\cdot\mathbf{G}$,

(b) $\dfrac{d}{dt}(\mathbf{F}\times\mathbf{G}) = \mathbf{F}\times\dfrac{d\mathbf{G}}{dt} + \dfrac{d\mathbf{F}}{dt}\times\mathbf{G}$,

using the same method by which the rule for differentiating products is derived in elementary calculus.

2. If $\mathbf{F}$ in Exercise 1 is a vector of constant length, prove that $\mathbf{F}\cdot\dfrac{d\mathbf{F}}{dt} = 0$, and thus conclude that $\dfrac{d\mathbf{F}}{dt}$ is perpendicular to $\mathbf{F}$ unless $\dfrac{d\mathbf{F}}{dt} = 0$.

14.31   Principal normal. Curvature

In this section we continue with the notations used in § 14.3. We shall assume that x, y, z have second derivatives with respect to s. From (14.3–1) we have

$$(14.31\text{–}1) \qquad \frac{d\mathbf{T}}{ds} = \frac{d^2x}{ds^2}\,\mathbf{i} + \frac{d^2y}{ds^2}\,\mathbf{j} + \frac{d^2z}{ds^2}\,\mathbf{k}.$$

We shall prove that

$$(14.31\text{–}2) \qquad \mathbf{T} \cdot \frac{d\mathbf{T}}{ds} = 0.$$

This means that $\dfrac{d\mathbf{T}}{ds}$ is either $\mathbf{0}$ or perpendicular to $\mathbf{T}$. The demonstration may be given in several ways. From (14.2–6) we have

$$1 = \left(\frac{dx}{ds}\right)^2 + \left(\frac{dy}{ds}\right)^2 + \left(\frac{dz}{ds}\right)^2.$$

Therefore, differentiating with respect to s,

$$0 = 2\left[\frac{dx}{ds}\frac{d^2x}{ds^2} + \frac{dy}{ds}\frac{d^2y}{ds^2} + \frac{dz}{ds}\frac{d^2z}{ds^2}\right].$$

The right member of this formula is exactly $2\,\mathbf{T} \cdot \dfrac{d\mathbf{T}}{ds}$ as we see from (14.3–1) and (14.31–1) and the formula for the dot product in terms of components. Thus (14.31–2) is proved. Alternatively (14.31–2) is a special case of the result proved in Exercise 2, § 14.3.

When $\dfrac{d\mathbf{T}}{ds}$ is not $\mathbf{0}$, its direction is perpendicular to $\mathbf{T}$. A unit vector in the direction of $\dfrac{d\mathbf{T}}{ds}$ is called the *principal normal* to C at the point in question. The principal normal is denoted by $\mathbf{N}$, and the length of $\dfrac{d\mathbf{T}}{ds}$ is denoted by κ. Therefore

$$(14.31\text{–}3) \qquad \frac{d\mathbf{T}}{ds} = \kappa\mathbf{N}.$$

The scalar κ is called the *curvature* of C; it is given by

$$(14.31\text{–}4) \qquad \kappa = \left[\left(\frac{d^2x}{ds^2}\right)^2 + \left(\frac{d^2y}{ds^2}\right)^2 + \left(\frac{d^2z}{ds^2}\right)^2\right]^{1/2}.$$

This definition makes the curvature always either positive or zero. In the case of plane curves it is easy to see, by sketching a figure, that the principal normal lies in the plane of the curve and points toward the concave side of the curve. In the theory of plane curves, the principal normal is called simply the *normal*.

If $\dfrac{d\mathbf{T}}{ds} = \mathbf{0}$, our definition of $\mathbf{N}$ breaks down. This happens at all points of C if C is a straight line, and in this case there is no basis for calling any particular normal the principal one. Otherwise the vanishing of $\dfrac{d\mathbf{T}}{ds}$ is exceptional.

The principal normal and the curvature play an important role in connection with the acceleration of a particle moving in a curve.

We saw earlier that the vector velocity is

$$\mathbf{v} = \frac{d\mathbf{R}}{dt}.$$

The vector acceleration is therefore

$$\mathbf{a} = \frac{d\mathbf{v}}{dt} = \frac{d^2\mathbf{R}}{dt^2}.$$

Now, from (14.3–6), we see that

$$\frac{d^2\mathbf{R}}{dt^2} = \frac{ds}{dt}\frac{d\mathbf{T}}{dt} + \frac{d^2s}{dt^2}\mathbf{T}.$$

Also

$$\frac{d\mathbf{T}}{dt} = \frac{d\mathbf{T}}{ds}\frac{ds}{dt}.$$

Therefore, using (14.31–3), we see that

$$(14.31\text{–}5) \qquad \frac{d^2\mathbf{R}}{dt^2} = \frac{d^2s}{dt^2}\mathbf{T} + \kappa\left(\frac{ds}{dt}\right)^2\mathbf{N}.$$

The acceleration vector is made up of a component of magnitude $\dfrac{d^2s}{dt^2}$ along the tangent vector, and one of magnitude $\kappa\left(\dfrac{ds}{dt}\right)^2$ along the principal normal.

The reciprocal of the curvature is called the *radius of curvature* ρ:

$$(14.31\text{–}6) \qquad \rho = \frac{1}{\kappa}.$$

If we write $v = \left|\dfrac{ds}{dt}\right|$ for the speed of the particle moving along C, the component $\kappa\left(\dfrac{ds}{dt}\right)^2$ can be written $\dfrac{v^2}{\rho}$. For motion in a circular path this is the familiar form of the centripetal acceleration.

The tip of the vector $\rho\mathbf{N}$, drawn from the point P as initial point, is called the *center of curvature* of C corresponding to P.

The plane of $\mathbf{T}$ and $\mathbf{N}$ is called the *osculating plane* of C at P.

14.32 Binormal. Torsion

The vector

$$(14.32\text{–}1) \qquad\qquad \mathbf{B} = \mathbf{T} \times \mathbf{N}$$

is called the *binormal* vector of C at P. Observe that $\mathbf{T}$, $\mathbf{N}$, and $\mathbf{B}$ are mutually perpendicular unit vectors; moreover, they form a right-handed system, just as $\mathbf{i}$, $\mathbf{j}$, and $\mathbf{k}$ do. The binormal $\mathbf{B}$ is perpendicular to the osculating plane.

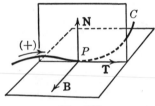

Fig. 122

There is a scalar quantity called *torsion* associated with C at each point. To define torsion we consider $\dfrac{d\mathbf{B}}{ds}$. From (14.32–1) we have (see Exercise 1(b), § 14.3)

$$\frac{d\mathbf{B}}{ds} = \mathbf{T} \times \frac{d\mathbf{N}}{ds} + \frac{d\mathbf{T}}{ds} \times \mathbf{N}.$$

We know by (14.31–3) that

$$\frac{d\mathbf{T}}{ds} \times \mathbf{N} = \kappa\mathbf{N} \times \mathbf{N} = \mathbf{0},$$

since the cross product of any vector with itself is zero. Therefore

$$(14.32\text{–}2) \qquad\qquad \frac{d\mathbf{B}}{ds} = \mathbf{T} \times \frac{d\mathbf{N}}{ds}.$$

Also, $\mathbf{B} \cdot \mathbf{B} = 1$, since $\mathbf{B}$ has unit length. Therefore, by Exercise 2, § 14.3,

$$(14.32\text{–}3) \qquad\qquad \mathbf{B} \cdot \frac{d\mathbf{B}}{ds} = 0.$$

Now, if $\dfrac{d\mathbf{B}}{ds} \neq 0$, the two preceding equations show that $\dfrac{d\mathbf{B}}{ds}$ is perpendicular both to $\mathbf{B}$ and to $\mathbf{T}$; it is therefore a multiple of $\mathbf{N}$, for $\mathbf{N}$ is also perpendicular to both $\mathbf{B}$ and $\mathbf{T}$. Thus we can write

(14.32–4)
$$\frac{d\mathbf{B}}{ds} = -\tau\mathbf{N},$$

where $-\tau$ is simply the proper multiple of $\mathbf{N}$ to give $\dfrac{d\mathbf{B}}{ds}$. The quantity τ thus defined is called the torsion of C at the point under consideration. If $\dfrac{d\mathbf{B}}{ds} = \mathbf{0}$, we define $\tau = 0$, so that (14.32–4) still holds.

For plane curves, $\mathbf{B}$ is constant, being a unit vector perpendicular to the plane. For such curves $\tau = 0$ at all points. A curve which does not lie in a single plane is called a *twisted* curve; its torsion measures, to some extent, the amount by which the curve is twisted.

For methods of finding $\mathbf{T}$, $\mathbf{N}$, $\mathbf{B}$, κ, and τ from the parametric equations of the curve, see Exercises 3, 4.

EXERCISES

1. If P and P' correspond to s and $s + \Delta s$, and if $\mathbf{T}$ and $\mathbf{T} + \Delta\mathbf{T}$ are the corresponding tangents to C, let $\Delta\theta$ be the angle between these tangents. Show that

$$\lim_{\Delta s \to 0} \frac{\Delta\theta}{\Delta s} = \kappa.$$

2. Prove that

$$\frac{d\mathbf{N}}{ds} = \tau\mathbf{B} - \kappa\mathbf{T}.$$

Formulas (14.31–3), (14.32–4), and the foregoing formula are called the Frenet formulas, or the Serret-Frenet formulas.

3. (a) Starting from (14.31–5), show that

$$\frac{d^3\mathbf{R}}{dt^3} = \left(\frac{d^3s}{dt^3} - \left(\frac{ds}{dt}\right)^3 \kappa^2\right)\mathbf{T} + \left(3\kappa\frac{ds}{dt}\frac{d^2s}{dt^2} + \left(\frac{ds}{dt}\right)^2\frac{d\kappa}{dt}\right)\mathbf{N} + \kappa\tau\left(\frac{ds}{dt}\right)^3\mathbf{B}.$$

Use the result in Exercise 2.

(b) Show that

$$\frac{d\mathbf{R}}{dt} \times \frac{d^2\mathbf{R}}{dt^2} = \kappa\left(\frac{ds}{dt}\right)^3\mathbf{B}, \qquad \left(\frac{d\mathbf{R}}{dt} \times \frac{d^2\mathbf{R}}{dt^2}\right) \cdot \left(\frac{d^3\mathbf{R}}{dt^3}\right) = \kappa^2\tau\left(\frac{ds}{dt}\right)^6.$$

(c) Show that

$$\kappa = \frac{|\mathbf{R}' \times \mathbf{R}''|}{|\mathbf{R}'|^3}, \qquad \tau = \frac{(\mathbf{R}' \times \mathbf{R}'') \cdot \mathbf{R}'''}{|\mathbf{R}' \times \mathbf{R}''|^2},$$

where primes denote differentiation with respect to t. In terms of x, y, z as functions of t, these formulas become

$$\kappa = \frac{[(y'z'' - y''z')^2 + (z'x'' - z''x')^2 + (x'y'' - x''y')^2]^{1/2}}{(x'^2 + y'^2 + z'^2)^{3/2}},$$

$$\tau = \frac{\begin{vmatrix} x' & y' & z' \\ x'' & y'' & z'' \\ x''' & y''' & z''' \end{vmatrix}}{(y'z'' - y''z')^2 + (z'x'' - z''x')^2 + (x'y'' - x''y')^2}.$$

4. To find **T**, **N**, and **B** from the parametric equations of the curve, observe first of all that **T** is a positive multiple of $\dfrac{d\mathbf{R}}{dt}$, provided $\dfrac{ds}{dt} > 0$, i.e., provided the positive sense along the curve is the same as that in which t increases. Since **T** is a unit vector, it follows that

$$\mathbf{T} = \frac{\mathbf{R'}}{|\mathbf{R'}|}.$$

Show with the help of Exercise 3(b) that

$$\mathbf{B} = \frac{\mathbf{R'} \times \mathbf{R''}}{|\mathbf{R'} \times \mathbf{R''}|}, \qquad \mathbf{N} = \frac{(\mathbf{R'} \times \mathbf{R''}) \times \mathbf{R'}}{|(\mathbf{R'} \times \mathbf{R''}) \times \mathbf{R'}|}.$$

If $\dfrac{ds}{dt} < 0$, the formulas for **T** and **B** require a minus sign on the right, but the formula for **N** is unchanged. If **T** and **B** have already been found, **N** is easily found from $\mathbf{N} = \mathbf{B} \times \mathbf{T}$.

5. (a) Find **T**, **N**, **B**, κ, and τ at $t = 0$ on the curve $x = t,\ y = t^2,\ z = t^3$.

(b) Find κ at $z = 1$ on the curve $x = \sqrt{z},\ y = 2\sqrt{z}$.

(c) Find κ and τ at $x = 3a$ on the curve $3ay = x^2,\ 2xz = a^2$.

(d) Find **T**, **N**, **B**, κ, and τ at $t = \pi$ on the curve $x = t - \sin t,\ y = 1 - \cos t,\ z = 4\sin(t/2)$.

(e) Find κ and τ in terms of t for the curve $x = \cos t,\ y = \sin t,\ z = e^t$.

(f) For $x = 3t - t^3,\ y = 3t^2,\ z = 3t + t^3$, show that $\kappa = \tau = \dfrac{1}{3(1 + t^2)^2}$.

(g) Find **T**, **N**, **B**, κ, and τ for the conical helix $x = 3t\cos t,\ y = 3t\sin t,\ z = 4t$, at $t = 0$.

6. Consider the helix $x = a\cos t,\ y = a\sin t,\ z = at\tan\alpha$. It lies on the cylinder $x^2 + y^2 = a^2$. It is a *screw curve*, with pitch angle α. As t increases, the point describes a right-handed screw path if $0 < \alpha < \pi/2$, and a left-handed screw path if $-\pi/2 < \alpha < 0$.

(a) Find the vectors **T**, **N**, **B** in terms of t. Show that **T** and **B** make constant angles with the z-axis, and that **N** is perpendicular to the z-axis and directed toward it.

(b) Show that $\qquad \kappa = \dfrac{\cos^2\alpha}{a}, \qquad \tau = \dfrac{\sin\alpha\cos\alpha}{a}$.

(c) Show that the center of curvature of the helix, corresponding to the point on the helix at the tip of **R**, is at $\mathbf{R_1} = \mathbf{R} + \rho\mathbf{N}$, and hence show that the center of curvature moves on a helix whose pitch angle is $(\pi/2) - \alpha$ and which lies on a coaxial cylinder of radius $a \tan^2 \alpha$. Show also that the center of curvature for this second helix is the original point on the original helix.

7. Show that, if a point moves so that its velocity and acceleration vectors always have unit length, $\kappa = 1$ at all points of the path.

8. For the curve $x = a \sin 2t$, $y = a + a \cos 2t$, $z = 2a \sin t$ (see Fig. 119) find $\mathbf{T}$, $\mathbf{N}$, $\mathbf{B}$, κ, and τ (a) at $t = \pi/2$; (b) at $t = \pi/4$; (c) at $t = 0$.

9. Suppose $|\mathbf{R}| = a$, so that the tip of $\mathbf{R}$ moves on a sphere of radius a with center at O. (a) Show that $\mathbf{R} = \alpha\mathbf{N} + \beta\mathbf{B}$, where α and β are related to ρ and τ by the equations $\alpha = -\rho$, $\beta\tau = -\dfrac{d\rho}{ds}$. (b) If $\tau \equiv 0$, show that the tip of $\mathbf{R}$ moves on a circle about the line of $\mathbf{B}$ as an axis. (c) If $\tau \neq 0$, show that

$$\rho^2 + \frac{1}{\tau^2}\left(\frac{d\rho}{ds}\right)^2 = a^2.$$

14.4   Surfaces

Speaking roughly, a surface is a configuration of points having a two-dimensional character; that is, a point moving on the surface, but otherwise unrestricted, has two degrees of freedom. This description of a surface is not a strict mathematical definition, however, and we need much greater precision in the work which is to follow.

Considerations about surfaces are of two kinds, which we may designate respectively as *local* and *in the large*. By a local property of a surface at a point we mean a property which can be completely described and analyzed by considering only the part of the surface in the immediate vicinity of the point. A property "in the large" is a characteristic of the surface as a whole, and cannot be defined merely in terms of the features of the surface near a single point. The property of *having a tangent plane at a given point* is a local property. A surface may have this property at some points but not at others. *Being a closed surface* is a property in the large (a spherical surface is closed, while a circular disk is not). There is clearly a difference in the large between the surface of a basketball and the surface of an inner tube. The mathematical name for the latter surface is *torus*. There is a classification of surfaces according to certain properties in the large and ignoring local properties. In this classification a spherical and an ellipsoidal surface are of the same type, while the torus is of a different type. Properties in the large are in many ways harder to study than local properties.

There are three common methods for the analytical representation of a surface. A locus in space defined by

(14.4–1) $z = f(x, y),$

where f is a single-valued continuous function defined on a region R of the xy-plane, is called a surface. In order to have the surface consist of one piece we assume that R is a connected region. A point (x, y, z) is called an interior point of the surface if the corresponding point (x, y) of R is interior to R. The points (x, y, z) corresponding to the boundary of R form the boundary or edge of the surface.

Often we have surfaces represented by equations of the form

$$(14.4\text{--}2) \qquad\qquad F(x, y, z) = 0.$$

If (x_0, y_0, z_0) is a point of such a surface, we can in many cases represent the portion of the surface near (x_0, y_0, z_0) in a form analogous to (14.4–1) by solving (14.4–2) for x, y, or z in terms of the two other variables.

For many purposes, especially in theoretical work, parametric representation of a surface is the most convenient. A locus defined by

$$(14.4\text{--}3) \qquad x = f(u, v), \qquad y = g(u, v), \qquad z = h(u, v),$$

where f, g, h are continuous functions defined on a connected region R of the uv-plane, is called a *parametric surface*. There may be other parametric representations of the same surface. This definition is so general that a parametric surface may not have any resemblance to what we would consider a surface to be, intuitively. It might, for example, intersect itself along infinitely many distinct curves in the vicinity of a single point. Hence, just as in dealing with curves in § 14.1, we find it necessary to introduce more restrictive definitions.

Suppose R is a closed rectangular region of the uv-plane: $a \leqq u \leqq b$, $c \leqq v \leqq d$, and suppose the equations (14.4–3) define a continuous one-to-one mapping of R onto a point set S in xyz-space. Then S is called a *simple surface element*. The one-to-one condition means that distinct points of R are not mapped into the same point of S. A simple surface element may be thought of as any configuration which may be obtained from a rectangular plane region by continuous deformation (bending, twisting, stretching, shrinking) without tearing and without bringing together any points which were originally distinct. In particular, a plane circular region is a simple surface element; so is a hemispherical surface.

If S is a simple surface element corresponding to a rectangular region R in the uv-plane, the points of S which correspond to the boundary of R form what is called the boundary of S. Other points of S are called interior points of the surface element.

The lateral surface of a right circular cylinder of finite length is not a simple surface element. It can, however, be regarded as composed of two simple surface elements joined together along the lines AB and CD (see Fig. 123). One of the elements is AA_1CDB_1B; the other is AA_2CDB_2B.

All surfaces which we shall subsequently consider may be thought of as being built up out of simple surface elements by matching together portions of the edges of the latter. This matching is always done in such a way that never more than two surface elements have an arc of edge in common. An infinite number of surface elements will be required to form a surface of infinite extent, such as a paraboloid or a complete cylinder, but, in considering surfaces in bounded portions of space, we shall assume that such

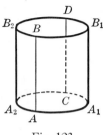

Fig. 123

surfaces can be formed from a finite number of simple surface elements. The exact manner of drawing the curves which divide a surface into elements is not unique; the curves can always be chosen so as to avoid the neighborhood of any particular point. Thus, in studying a local property of a surface, we may always assume that we are dealing with an interior point of a single surface element.

The boundary of a surface consists of the unmatched edges of its surface elements. If there are no unmatched edges, there is no boundary. A surface which has no boundary, and which lies in a bounded portion of space, is called a *closed surface*. For example, the surface of a sphere is closed, but the cylindrical surface in Fig. 123 is not closed.

A surface is called *smooth* at a point P if it has a tangent plane at each point in the neighborhood of P, and if the direction of the normal to this plane varies continuously from point to point. The whole surface is called smooth if it is smooth at every point.

The surface represented by (14.4–1) is smooth if f has continuous first partial derivatives. The direction of the normal in this case is specified by the ratios

$$(14.4\text{–}4) \qquad \frac{\partial z}{\partial x} : \frac{\partial z}{\partial y} : -1,$$

as we saw in § 6.2. In the representation (14.4–2) the surface is smooth at a point where the three partial derivatives $\dfrac{\partial F}{\partial x}, \dfrac{\partial F}{\partial y}, \dfrac{\partial F}{\partial z}$ do not all vanish, provided they are continuous in the neighborhood of the point. The normal direction is specified by the ratios

$$F_1 : F_2 : F_3$$

(see Example 1, § 6.6).

In the parametric case (14.4–3) let us suppose that the functions f, g, h have continuous first partial derivatives. We introduce the notations

$$(14.4\text{–}5) \qquad j_1 = \frac{\partial(y, z)}{\partial(u, v)}, \qquad j_2 = \frac{\partial(z, x)}{\partial(u, v)}, \qquad j_3 = \frac{\partial(x, y)}{\partial(u, v)}.$$

We assume that these three Jacobians do not vanish simultaneously. Under these conditions we can show that the surface is smooth, and that the normal direction is given by the ratios

$$(14.4\text{-}6) \qquad j_1 : j_2 : j_3.$$

For the proof, suppose $j_3 \neq 0$. Then, by the implicit-function theorem (Theorem III, § 8.3), the equations $x = f(u, v)$, $y = g(u, v)$ can be solved for u, v in terms of x, y. Substituting the solutions into $z = h(u, v)$ we obtain z as a function of x, y. This means that the portion of the surface near a point where $j_3 \neq 0$ can be represented in the form (14.4-1). By the rules for differentiating composite functions we shall have

$$(14.4\text{-}7) \qquad \frac{\partial z}{\partial x} = \frac{\partial z}{\partial u} \frac{\partial u}{\partial x} + \frac{\partial z}{\partial v} \frac{\partial v}{\partial x},$$

with a similar formula for $\dfrac{\partial z}{\partial y}$. If we regard u, v as dependent and x, y as independent in the equations $x = f(u, v)$, $y = g(u, v)$, we have

$$1 = \frac{\partial f}{\partial u} \frac{\partial u}{\partial x} + \frac{\partial f}{\partial v} \frac{\partial v}{\partial x},$$

$$0 = \frac{\partial g}{\partial u} \frac{\partial u}{\partial x} + \frac{\partial g}{\partial v} \frac{\partial v}{\partial x}.$$

Solving for $\dfrac{\partial u}{\partial x}$ and $\dfrac{\partial v}{\partial x}$, we find

$$\frac{\partial u}{\partial x} = \frac{\begin{vmatrix} 1 & f_2 \\ 0 & g_2 \end{vmatrix}}{\begin{vmatrix} f_1 & f_2 \\ g_1 & g_2 \end{vmatrix}}, \qquad \frac{\partial v}{\partial x} = \frac{\begin{vmatrix} f_1 & 1 \\ g_1 & 0 \end{vmatrix}}{\begin{vmatrix} f_1 & f_2 \\ g_1 & g_2 \end{vmatrix}},$$

or

$$\frac{\partial u}{\partial x} = \frac{1}{j_3} \frac{\partial y}{\partial v}, \qquad \frac{\partial v}{\partial x} = -\frac{1}{j_3} \frac{\partial y}{\partial u}.$$

The foregoing formulas are the same as some of those in (9.1-2), except for minor changes in notation. Hence, from (14.4-7)

$$\frac{\partial z}{\partial x} = \frac{1}{j_3} \left(\frac{\partial z}{\partial u} \frac{\partial y}{\partial v} - \frac{\partial z}{\partial v} \frac{\partial y}{\partial u} \right) = -\frac{j_1}{j_3}.$$

Similarly,

$$0 = \frac{\partial f}{\partial u} \frac{\partial u}{\partial y} + \frac{\partial f}{\partial v} \frac{\partial v}{\partial y},$$

$$1 = \frac{\partial g}{\partial u} \frac{\partial u}{\partial y} + \frac{\partial g}{\partial v} \frac{\partial v}{\partial y},$$

whence

$$\frac{\partial u}{\partial y} = -\frac{1}{j_3} \frac{\partial x}{\partial v}, \qquad \frac{\partial v}{\partial y} = \frac{1}{j_3} \frac{\partial x}{\partial u},$$

and
$$\frac{\partial z}{\partial y} = \frac{1}{j_3}\left(-\frac{\partial z}{\partial u}\frac{\partial x}{\partial v} + \frac{\partial z}{\partial v}\frac{\partial x}{\partial u}\right) = -\frac{j_2}{j_3}.$$

It follows that

$$\frac{\partial z}{\partial x} : \frac{\partial z}{\partial y} : -1 = -\frac{j_1}{j_3} : -\frac{j_2}{j_3} : -1.$$

Thus (14.4–4) and (14.4–6) define the same direction; this is what we set out to prove.

Once the direction of the normal is known, it is of course an easy matter to write out the equation of the tangent plane.

EXERCISES

1. Show that the parametric surface defined by $x = a \sin \phi \cos \theta$, $y = b \sin \phi \sin \theta$, $z = c \cos \phi$, $0 \leq \theta \leq 2\pi$, $0 \leq \phi \leq \pi$, is an ellipsoid, and that it is a sphere if $a = b = c$. The surface is not a simple surface element, however. Which part of the definition of a simple surface element is not satisfied in this case?

2. Explain how to divide the surface of a sphere into simple surface elements in several ways. In particular, if the sphere is $x^2 + y^2 + z^2 = 1$, describe a mode of division such that the points $(0, 0, \pm 1)$ are interior points of elements on which they lie. Describe a mode of division such that the points $(\pm 1, 0, 0)$ and $(0, \pm 1, 0)$ are interior points of the elements on which they lie.

3. For the case of each of the following parametric surfaces, obtain an equation of the surface in rectangular co-ordinates.
(a) $x = au \cos v$, $y = bu \sin v$, $z = u$ (elliptic cone).
(b) $x = u \cos v$, $y = u \sin v$, $z = ku^2$ (paraboloid of revolution).
(c) $x = a \sin u \cosh v$, $y = b \cos u \cosh v$, $z = c \sinh v$ (hyperboloid of one sheet).
(d) $x = r \cos \theta$, $y = r \sin \theta$, $z = (r^2/2) \sin 2\theta$ (hyperbolic paraboloid).
(e) $x = au \cos v$, $y = bu \sin v$, $z = u^2 \cos 2v$ (hyperbolic paraboloid).
(f) $x = a \cosh v$, $y = b \cosh v \cos u$, $z = c \cosh v \sin u$.

4. Show that the tangent plane to the ellipsoid $(x^2/a^2) + (y^2/b^2) + (z^2/c^2) = 1$ at (x_0, y_0, z_0) is $(x_0 x/a^2) + (y_0 y/b^2) + (z_0 z/c^2) = 1$.

5. Show that the direction of the normal to the surface in Exercise 3(e) is $-2bu \cos v : 2au \sin v : ab$.

6. Describe the parametric surface $x = a \cos u$, $y = a \sin u$, $z = v$, and find its equation in rectangular co-ordinates.

7. Describe the parameteric surface $x = 2u + v$, $y = u - v$, $z = 3u$, and find its equation in rectangular co-ordinates.

8. Show that the parametric surface $x = u + v$, $y = u - v$, $z = 4v^2$ is the parabolic cylinder $z = (x - y)^2$. Show that the tangent plane at the point corresponding to (u, v) is $4vx - 4vy - z = 4v^2$.

9. If the curve $y = f(x)$ in the xy-plane is revolved around the x-axis, show that the resulting surface can be represented parametrically in the form $x = u$, $y = f(u) \cos v$, $z = f(u) \sin v$. Assuming that $f'(u)$ is continuous and $f(u) > 0$, show that the direction of the normal is $f'(u) : -\cos v : -\sin v$.

10. (a) Consider the parametric surface $x = u + v$, $y = u^2 + v^2$, $z = u^3 + v^3$. Show that it is part of the surface $x^3 - 3xy + 2z = 0$, and, in fact, just the part for which $x^2 \leq 2y$ (u and v are required to be real, of course). **(b)** Find the direction of the normal to the surface at $(0, 8, 0)$, using (14.4–4), and also using (14.4–6). **(c)** Show that the direction of the normals is $6uv : -3(u + v) : 2$. Is this correct even when $u = v$? (Check with the result obtained from the rectangular equation when $u = v$, i.e., when $2y = x^2$.)

14.5   Curves on a surface

Consider a fixed surface. We are going to study local properties of curves on the surface, so we shall find it convenient to deal with a single simple surface element. We suppose equations (14.4–3) give a parametric representation of the element, and that the functions f, g, h have continuous first partial derivatives. If u and v are made to depend continuously on a parameter t, then x, y, z will become continuous functions of t, and we shall have a curve defined on the surface. If we suppose u and v are continuously differentiable functions of t, any arc of the curve will be rectifiable. We wish to obtain a formula for the differential of the arc-length in terms of u, v, du, and dv. No matter what the parameter is, we have the following formulas:

$$ds^2 = dx^2 + dy^2 + dz^2,$$

$$dx = \frac{\partial x}{\partial u} \, du + \frac{\partial x}{\partial v} \, dv,$$

and similar formulas for dy and dz.

Hence
$$ds^2 = \left(\frac{\partial x}{\partial u} \, du + \frac{\partial x}{\partial v} \, dv \right)^2 + \cdots .$$

On carrying out the details, we obtain the result

(14.5–1) $$ds^2 = E \, du^2 + 2 F \, du \, dv + G \, dv^2,$$

where $$E = \left(\frac{\partial x}{\partial u} \right)^2 + \left(\frac{\partial y}{\partial u} \right)^2 + \left(\frac{\partial z}{\partial u} \right)^2,$$

(14.5–2) $$F = \frac{\partial x}{\partial u} \frac{\partial x}{\partial v} + \frac{\partial y}{\partial u} \frac{\partial y}{\partial v} + \frac{\partial z}{\partial u} \frac{\partial z}{\partial v},$$

$$G = \left(\frac{\partial x}{\partial v} \right)^2 + \left(\frac{\partial y}{\partial v} \right)^2 + \left(\frac{\partial z}{\partial v} \right)^2.$$

We call attention to the structure of (14.5–1). It is a quadratic form in du and dv, with coefficients which are functions of u, v; these coefficients may be found directly from the parametric equations of the surface, as we see from (14.5–2). The quadratic form is called the *first fundamental form* of the surface. This form is of prime importance for what is called the metric differential geometry of the surface.

Example. The parametric equations

$$(14.5–3) \quad \begin{aligned} x &= (a + b \cos \phi) \cos \theta \\ y &= (a + b \cos \phi) \sin \theta \\ z &= b \sin \phi \end{aligned}$$

in which $0 < b < a$, and θ and ϕ have the ranges $0 \leq \theta \leq 2\pi, 0 \leq \phi \leq 2\pi$, define a torus. The part of the surface of the torus in the first octant is

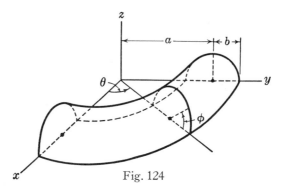

Fig. 124

shown in Fig. 124. This part is a simple surface element, corresponding to the ranges $0 \leq \theta \leq \pi/2, 0 \leq \phi \leq \pi$ of the parameters θ, ϕ. The entire torus is a closed surface, which may be divided into four simple surface elements, corresponding to the four squares in the $\theta\phi$-plane shown in Fig. 125.

Let us calculate the first fundamental form for the torus. We do this by direct calculation from (14.5–3), rather than by substitution in (14.5–1) and (14.5–2). Here θ and ϕ play the roles of u and v. We have

$$dx = -(a + b \cos \phi) \sin \theta \, d\theta - b \sin \phi \cos \theta \, d\phi,$$
$$dy = (a + b \cos \phi) \cos \theta \, d\theta - b \sin \phi \sin \theta \, d\phi,$$
$$dz = b \cos \phi \, d\phi.$$

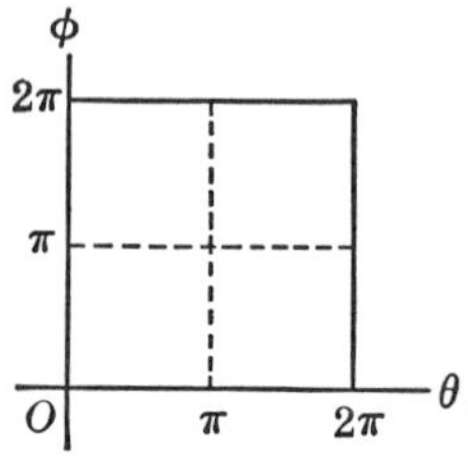

Hence, using the formula (14.2–6) for ds^2, we have

$$ds^2 = (a + b \cos \phi)^2(\sin^2 \theta + \cos^2 \theta)d\theta^2 \\ + b^2 \sin^2 \phi \, (\sin^2 \theta + \cos^2 \theta)d\phi^2 + b^2 \cos^2 \phi \, d\phi^2,$$

Fig. 125

(14.5–4) $$ds^2 = (a + b \cos \phi)^2 \, d\theta^2 + b^2 \, d\phi^2.$$

Observe that there is no term involving $d\theta \, d\phi$.

When a surface is defined in terms of parameters u, v, the curves $u = $ constant and $v = $ constant are called *co-ordinate curves*. Along a curve $v = $ constant (called a *v-curve*) we may consider u as the parameter of the curve. The direction of the tangent to this curve is given by the ratios

$$\frac{\partial x}{\partial u} : \frac{\partial y}{\partial u} : \frac{\partial z}{\partial u}.$$

Likewise the direction of the tangent to a *u*-curve ($u = $ constant) is

$$\frac{\partial x}{\partial v} : \frac{\partial y}{\partial v} : \frac{\partial z}{\partial v}.$$

If the two sets of curves intersect orthogonally at each point, we see from (14.5–2) that $F \equiv 0$. In this case, then, the first fundamental form becomes

$$ds^2 = E \, du^2 + G \, dv^2,$$

and no term in $du \, dv$ is present.

Along a *v*-curve $ds = \pm \sqrt{E} \, du$, and along a *u*-curve $ds = \pm \sqrt{G} \, dv$. Hence, if by direct geometric observation we can determine the values of $\dfrac{ds}{du}$ along a *v*-curve, we can find E without resorting to the work of calculating from (14.5–2). Similar remarks apply to G. In the case of a number of familiar surfaces, these short cuts are quite convenient.

These short cuts are available in the case of the torus of the illustrative Example. In that example the ϕ-curves are circles in horizontal planes, with centers on the z-axis. The radius of a ϕ-circle is $a + b \cos \phi$, and hence $ds = \pm (a + b \cos \phi) d\theta$ along the circle. The coefficient of $d\theta^2$ in the general formula for ds^2 is therefore $(a + b \cos \phi)^2$. Discussion of the θ-curves is left to the student. In this example the co-ordinate curves intersect each other at right angles. This accounts for the absence of the $d\theta \, d\phi$ term in the quadratic form.

From the directions of the tangents to the co-ordinate curves we may determine the direction of their common perpendicular, which is normal to the surface. Use of the usual scheme of three two-row determinants shows us that the direction of the normal is $j_1 : j_2 : j_3$. Thus we obtain the result (14.4–6) in a different way.

For certain purposes it is very convenient to use sets of variables with indices. Thus, we might write x_1, x_2, x_3 instead of x, y, z, and u_1, u_2 instead of u, v. If this is done the formulas for E, F, G can be written compactly in summation notation; for example,

$$F = \sum_{i=1}^{3} \frac{\partial x_i}{\partial u_1} \frac{\partial x_i}{\partial u_2}.$$

The formula for ds^2 can be written

$$ds^2 = \sum_{\alpha, \beta = 1}^{2} \sum_{i=1}^{3} \frac{\partial x_i}{\partial u_\alpha} \frac{\partial x_i}{\partial u_\beta} \, du_\alpha \, du_\beta.$$

In practice, the index notation is normally used when *tensor* methods are being employed. The conventions of tensor notation require the use of two kinds of indices, *upper* and *lower*, and for the present case the convention calls for the notations x^1, x^2, x^3, u^1, u^2. The upper indices are not exponents, but merely distinguishing marks. The coefficients in the formula for ds^2 are now denoted by g_{11}, g_{12}, g_{21}, g_{22}; g_{11} being the coefficient of $du^1 \, du^1$; g_{12} that of $du^1 \, du^2$; and so on. Thus

$$ds^2 = \sum_{\alpha, \beta = 1}^{2} g_{\alpha\beta} \, du^\alpha \, du^\beta,$$

where

$$g_{\alpha\beta} = \sum_{i=1}^{3} \frac{\partial x^i}{\partial u^\alpha} \frac{\partial x^i}{\partial u^\beta} \, .$$

Observe that $g_{12} = g_{21}$. In our previous notation, $g_{11} = E$, $g_{12} = F$, $g_{22} = G$. The four quantities g_{11}, g_{12}, g_{21}, g_{22} are the components of what is called the *fundamental metric tensor* of the surface. In all this discussion we have not defined the term *tensor*. To do so is outside the scope of this book. Suffice it to say that the concept of a tensor is an extension of the concept of a vector.

EXERCISES

1. Describe the θ-curves in the torus of Fig. 124, and, by direct inspection, find the formula for ds along a θ-curve.

2. Describe the θ-curves and ϕ-curves on the sphere $x = a \sin \phi \cos \theta$, $y = a \sin \phi \sin \theta$, $z = a \cos \phi$, and find the formula for ds along each curve, thus obtaining the first fundamental form $ds^2 = a^2(\sin^2 \phi \, d\theta^2 + d\phi^2)$.

3. Describe the u-curves and v-curves on the cylinder $x = u$, $y = a \sin v$, $z = a \cos v$, and find the general formula for ds^2.

4. Proceed as in Exercise 3 with the cone $x = u \sin v$, $y = mu$, $z = u \cos v$.

5. Calculate the first fundamental form for each of the following surfaces:
(a) $x = u \cos v$, $y = u \sin v$, $z = ku^2$.
(b) $x = a \sin u \cosh v$, $y = b \cos u \cosh v$, $z = c \sinh v$.
(c) $x = r \cos \theta$, $y = r \sin \theta$, $z = (r^2/2) \sin 2\theta$.
(d) $x = u + v$, $y = u - v$, $z = 4 v^2$.
(e) $x = u + v$, $y = u^2 + v^2$, $z = u^3 + v^3$.

6. (a) Let $\lambda = (\pi/2) - \phi$ be the latitude of a point on the sphere of Exercise 2. Let a curve on the sphere be given. If ω is the angle at which the curve crosses the circle of λ latitude, show that

$$\frac{d\lambda}{\cos \lambda \, d\theta} = \tan \omega$$

if a suitable convention is made about the sign of ω.

(b) If ω is constant, the curve is called a *rhumb line*, or a *loxodrome*. Find the equation in terms of λ and θ of the loxodrome through $\lambda = 0$, $\theta = 0$ with ω given $(0 < \omega < \pi/2)$. Show that, although the curve spirals infinitely often about both poles of the sphere, its length is finite and equal to $\pi a \csc \omega$.

14.6   Surface area

One of the first objectives in the study of surfaces with the aid of calculus is the derivation of formulas for calculating the area of a surface. There are various ways of getting at such formulas. We begin with a method which is frankly heuristic; that is, we formulate an argument which is plausible and which leads to a formula for the area as a double integral. This argument will not do, however, as the last word on the subject, for it is not based on an explicit definition of the area of a surface, but rather on our intuition of what ought to be true of such areas.

Consider a simple surface element, as given by parametric equations (14.4–3), with the parameters (u, v) representing a point varying over the rectangle $R: a \leqq u \leqq b, c \leqq v \leqq d$ in the uv-plane. Suppose that this rectangle is divided up into small cells by a rectangular partition after the fashion used in defining a double integral (§ 13.2). We shall suppose

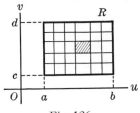

Fig. 126a

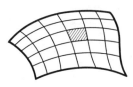

Fig. 126b

that the functions f, g, h entering in the parametric representation have continuous first partial derivatives in the closed region R. We shall use vector notation in what follows. We think of (14.4–3) as a vector mapping from the rectangle R into $\mathbf{R}^3$:

(14.6–1) $\mathbf{F}(u, v) = (f(u, v), g(u, v), h(u, v))$.

By Theorem VI of Chapter XII we know that $\mathbf{F}$ is continuously differentiable because we have assumed that the coordinate functions f, g, h have continuous first partial derivatives. Consider a typical cell in the interior of R and the corresponding cell of the surface element (Fig. 126a and Fig. 126b). We first seek to find an approximate expression for the area of this small cell of the surface element. The guiding idea is to think of the cell on the curved surface element as being approximately a plane parallelogram. The plane of this parallelogram is the plane tangent to the surface element at one corner of the cell under consideration. The parallelogram itself is the figure into which the cell in R is mapped by a certain affine transformation which we obtain by using the differential of $\mathbf{F}$. To explain this more clearly we must now introduce some additional notation.

In Fig. 127a we designate the lower left corner A of the typical cell in R by (u_i, v_j).

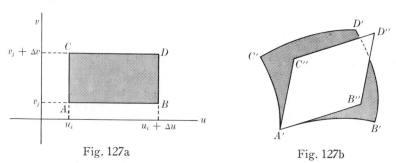

Fig. 127a   Fig. 127b

Denote the other corners by B, C, D as indicated, so that B is $(u_i + \Delta u, v_j)$ and C is $(u_i, v_j + \Delta v)$. In Fig. 127b is shown the cell $A'B'C'D'$ on the surface element, the image of the cell $ABCD$ under the mapping function $\mathbf{F}$. In particular, A' is the vector $\mathbf{F}(u_i, v_j)$. To locate a point in the cell $ABCD$, we write its coordinates as $(u_i + \delta u, v_j + \delta v)$, where δu and δv are increments in u and v, respectively (see Fig. 127a). Now consider the affine mapping which carries $(u_i + \delta u, v_j + \delta v)$ into that point (x, y, z) in R^3 which is the tip of the vector

$$(14.6\text{--}2) \qquad\qquad \mathbf{F}(u_i, v_j) + d\mathbf{F}(\delta u, \delta v).$$

Here, $d\mathbf{F}(\delta u, \delta v)$ is the differential of $\mathbf{F}$, calculated at the point A. We shall presently exhibit a formula for $d\mathbf{F}$ in (14.6–3). This affine mapping function (14.6–2) maps $ABCD$ into a parallelogram $A'B''C''D''$ (see Fig. 127b) in the plane tangent to the surface element at A'; furthermore, the line $A'B''$, which is the image of AB, is tangent to the curve $A'B'$ at A', and the line $A'C''$ is tangent to the curve $A'C'$ at A'. We shall calculate the area of the parallelogram $A'B''C''D''$ and use this as our approximation for the area of the cell $A'B'C'D'$ of the curved surface element.

Now we know by Theorem V of Chapter XII that the differential $d\mathbf{F}$ is the linear transformation from R^2 to R^3 represented by the Jacobian matrix

$$\left(\begin{matrix} \dfrac{\partial f}{\partial u} & \dfrac{\partial f}{\partial v} \\[2mm] \dfrac{\partial g}{\partial u} & \dfrac{\partial g}{\partial v} \\[2mm] \dfrac{\partial h}{\partial u} & \dfrac{\partial h}{\partial v} \end{matrix} \right),$$

where all the partial derivatives are evaluated at (u_i, v_j). In other words, with coordinate notation,

(14.6–3)

$$d\mathbf{F}(\delta u, \delta v) = \left(\frac{\partial f}{\partial y}\,\delta u + \frac{\partial f}{\partial v}\,\delta v,\ \frac{\partial g}{\partial u}\,\delta u + \frac{\partial g}{\partial v}\,\delta v,\ \frac{\partial h}{\partial u}\,\delta u + \frac{\partial h}{\partial v}\,\delta v \right).$$

We can now see that B'' is the tip of the vector $\mathbf{F}(u_i, v_j) + d\mathbf{F}(\Delta u, 0)$, so that $d\mathbf{F}(\Delta u, 0)$ is equal to the vector from A' to B''. Likewise, $d\mathbf{F}(0, \Delta v)$ is equal to the vector from A' to C''. Now, the area of the parallelogram $A'B''C''D''$ is equal to the magnitude of the cross product of the vector $A'B''$ and $A'C''$.

$$d\mathbf{F}(\Delta u, 0) \times d\mathbf{F}(0, \Delta v) = \begin{vmatrix} \mathbf{i} & \mathbf{j} & \mathbf{k} \\[2mm] \dfrac{\partial f}{\partial u}\,\Delta u & \dfrac{\partial g}{\partial u}\,\Delta u & \dfrac{\partial h}{\partial u}\,\Delta u \\[2mm] \dfrac{\partial f}{\partial v}\,\Delta v & \dfrac{\partial g}{\partial v}\,\Delta v & \dfrac{\partial h}{\partial v}\,\Delta v \end{vmatrix},$$

and therefore, the area of the parallelogram in question is

(14.6–4) $$\sqrt{j_1^2 + j_2^2 + j_3^2}\ \Delta u\, \Delta v,$$

where j_1, j_2, j_3 are the Jacobians defined in (14.4–5), evaluated at (u_i, v_j).

The next step in the process is to form the sum of all the terms of the form (14.6–4), corresponding to all of the cells into which R is divided in Fig. 126a. When we form this sum and then find its limit as the partition of R is refined, we obtain a double integral as the limit. This double integral is

(14.6–5) $$S = \iint\limits_R \sqrt{j_1^2 + j_2^2 + j_3^2}\ du\, dv.$$

We are led in this manner to define by this double integral the area S of the surface element into which R is mapped by (14.6–1).

This formula can be given another appearance, using the coefficients E, F, G from the quadratic form expressing ds on the surface in terms of

du and dv (see (14.5–2)). The alternative expression for the double integral is

(14.6–6)
$$S = \iint_R \sqrt{EG - F^2}\, du\, dv.$$

One way of passing from (14.6–5) to (14.6–6) is to make the straightforward calculation showing that

(14.6–7)
$$EG - F^2 = j_1^2 + j_2^2 + j_3^2.$$

In working problems it will sometimes be found to be easier to compute $j_1^2 + j_2^2 + j_3^2$ than $EG - F^2$ and vice versa.

Example 1. Compute the total area of the torus $x = (a + b \cos \phi) \cos \theta$, $y = (a + b \cos \phi) \sin \theta$, $z = b \sin \phi$, $0 < b < a$.

This torus was discussed in § 14.5 (see Fig. 124). The part in the first octant is a simple surface element corresponding to $0 \leq \theta \leq \pi/2$, $0 \leq \phi \leq \pi$, and the area of this part is one eighth of the total. From (14.5–4) we see that, if $u = \theta$ and $v = \phi$,

$$E = (a + b \cos \phi)^2, \qquad F = 0, \quad G = b^2.$$

Thus $\sqrt{EG - F^2} = b(a + b \cos \phi)$, and the total area is

$$S = 8 \int_0^\pi d\phi \int_0^{\pi/2} b(a + b \cos \phi)\,d\theta = 4\,\pi^2 ab.$$

This is in accord with the theorem of Pappus.

If we use (14.6–5) instead of (14.6–6) to calculate the first octant portion of the torus, we find that

$$j_1 = b(a + b \cos \phi)\cos \theta \cos \phi,$$
$$j_2 = b(a + b \cos \phi)\sin \theta \cos \phi,$$
$$j_3 = b(a + b \cos \phi)\sin \phi,$$

from which

$$j_1^2 + j_2^2 + j_3^2 = b^2(a + b \cos \phi)^2.$$

Thus, the identity (14.6–7) is verified in this particular case, and the integral for the area is calculated as before.

In the argument leading up to (14.6–5), it will be seen that the fact that the region R was a rectangle in the uv-plane was not essential. If R is any bounded closed region of the uv-plane of the type described in the discussion of double integrals in § 13.2, the discussion leading up to (14.6–4) applies to any cell in a rectangular partition of the type shown in Fig. 126a and (14.6–4) gives the area of the parallelogram which is the image of this cell under the affine mapping (14.6–2). Hence, the formulas (14.6–5) and (14.6–6) can be used to find the area of any portion of a smooth surface which is obtained by a one-to-one and continuously differentiable mapping from the region R in the uv-plane.

We now consider the special case of a surface defined by an equation $z = f(x, y)$ for all (x, y) belonging to some region R in the xy-plane. It will be assumed that f has continuous first partial derivatives in R. We can think of x and y as being the parameters u, v. This leads us to the following very special case of (14.4–3).

(14.6–8) $x = x, \qquad y = y, \qquad z = f(x, y).$

By simple calculations we see from (14.4–5) that

$$ j_1 = - \frac{\partial f}{\partial x}, \qquad j_2 = - \frac{\partial f}{\partial y}, \qquad j_3 = 1. $$

Consequently, the area of the portion of the surface corresponding to the plane region R is

(14.6–9) $S = \iint_R \left[1 + \left(\frac{\partial f}{\partial x} \right)^2 + \left(\frac{\partial f}{\partial y} \right)^2 \right]^{1/2} dx\, dy.$

Alternatively, if we wish to use (14.6–6), we can calculate as follows:

$$ dz = \frac{\partial z}{\partial x} dx + \frac{\partial z}{\partial y} dy, $$

$$ ds^2 = dx^2 + dy^2 + \left(\frac{\partial z}{\partial x} dx + \frac{\partial z}{\partial y} dy \right)^2 $$

$$ = \left[1 + \left(\frac{\partial z}{\partial x} \right)^2 \right] dx^2 + 2 \frac{\partial z}{\partial x} \frac{\partial z}{\partial y} dx\, dy + \left[1 + \left(\frac{\partial z}{\partial y} \right)^2 \right] dy^2. $$

Interpreting x as u and y as v, we have

$$ E = 1 + \left(\frac{\partial z}{\partial x} \right)^2, \qquad F = \frac{\partial z}{\partial x} \frac{\partial z}{\partial y}, \qquad G = 1 + \left(\frac{\partial z}{\partial y} \right)^2. $$

Hence the area of the surface is

(14.6–10) $S = \iint_R \left[1 + \left(\frac{\partial z}{\partial x} \right)^2 + \left(\frac{\partial z}{\partial y} \right)^2 \right]^{1/2} dx\, dy.$

Example 2.   Find the area of the upper half of the sphere $x^2 + y^2 + z^2 = a^2$ by using formula (14.6–10).

Here $z = \sqrt{a^2 - x^2 - y^2}, \qquad \dfrac{\partial z}{\partial x} = \dfrac{- x}{\sqrt{a^2 - x^2 - y^2}},$

with a similar formula for $\dfrac{\partial z}{\partial y}$. Thus the integrand in (14.6–10) becomes

$$\left[1 + \frac{x^2}{a^2 - x^2 - y^2} + \frac{y^2}{a^2 - x^2 - y^2} \right]^{1/2} = \frac{a}{\sqrt{a^2 - x^2 - y^2}} \,.$$

There is one difficulty. The hemisphere lies above the region R bounded by the circle $x^2 + y^2 = a^2$ in the xy-plane, and we see that the partial derivatives of z with respect to x and y are not defined on the boundary of this region. Moreover, the integral (14.6–10) in this case is

$$\iint\limits_R \frac{a}{\sqrt{a^2 - x^2 - y^2}} \, dx \, dy;$$

this is an improper integral, since the integrand becomes infinite at the boundary of R. We can get around the difficulty by considering the area above a smaller concentric circle of radius b, and then making $b \to a$ afterward. It is convenient to evaluate the double integral by an iterated integral in polar co-ordinates. The result is

$$S = \lim_{b \to a} \int_0^b dr \int_0^{2\pi} \frac{ar \, d\theta}{\sqrt{a^2 - r^2}} = 2\,\pi a \int_0^a \frac{r \, dr}{\sqrt{a^2 - r^2}} \,,$$

or $S = 2\,\pi a^2$. This is correct for the hemisphere.

We may put formula (14.6–10) in a different form. Let α, β, γ be the angles which the normal to the surface $z = f(x, y)$ makes with the positive z-axes, the positive direction of the normal being chosen so that γ is acute.

Now $\cos \alpha : \cos \beta : \cos \gamma = \dfrac{\partial z}{\partial x} : \dfrac{\partial z}{\partial y} : -1.$

Hence

(14.6–11) $\sec^2 \gamma = \left(\dfrac{\partial z}{\partial x} \right)^2 + \left(\dfrac{\partial z}{\partial y} \right)^2 + 1,$

and so from (14.6–10) we have

(14.6–12) $S = \iint\limits_R \sec \gamma \, dA.$

Let R be a connected region. Applying the mean-value theorem to formula (14.6–12), we conclude that there is a point P' on the surface such that, if γ' is the value of γ at P' and A is the area of R, then $S = A \sec \gamma'$. If a cylindrical surface parallel to the z-axis is constructed around the boundary of R, and if the plane tangent at P' to the original surface is constructed, then the area of this plane which is cut off within the cylinder is exactly $A \sec \gamma'$ (see Fig. 128). Hence, the area of the original surface

$z = f(x, y)$ is the same as the area cut from a certain one of its tangent planes by the cylinder parallel to the z-axis and intersecting the xy-plane in the boundary of the region R. If the region R is very small, it follows that $A \sec \gamma$ is a good approximation to S, no matter at which point P of the surface we evaluate γ. This remark is sometimes taken as the intuitive foundation for a derivation of formula (14.6–12), the procedure being to subdivide R into small parts, the area S being obtained as the limit of the sum of areas $\Delta A \sec \gamma$. It is this deriva- tion which is usually found in elementary calculus textbooks.

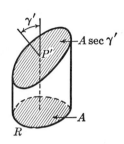

Fig. 128

Formulas (14.6–6) and (14.6–12) are the standard formulas of calculus for dealing with surface area. Where, however, is the *definition* of surface area? Are there surfaces which have area, and yet which are such that the area cannot be found by the integrals mentioned above, perhaps because of lack of sufficient smoothness? It is logically and aes- thetically desirable to have a definition of surface area which is directly geometric, and which does not put too many restrictions on the surface. A good definition ought not to depend upon the method of representing the surface analytically, and should not be limited to smooth surfaces. The demand for such a definition poses a very difficult problem, however. It may surprise the student to know that the problem has occupied the attention of many able mathematicians over the last fifty years, and that the end of research on the question is not yet in sight.

To present the concept of surface area to the student at the advanced calculus level, the most satisfactory logical approach seems to be the following: for a smooth simple surface element, with appropriate condi- tions on its parametric representation, formula (14.6–6) is to be taken as a *definition*; the discussion leading up to the formula is by way of motiva- tion. It can be shown that the area so defined is independent of the particular parametrization, and is therefore an intrinsic characteristic of the surface. This demonstration requires the theory of transformation of double in- tegrals, and is discussed in Chapter XV. In the particular case of surfaces $z = f(x, y)$, the formula (14.6–12) makes it clear that the area does not depend on the parametrization of the surface, but it must still be shown that the orientation of the z-axis is inessential, since the direction of this axis plays a role in the formula.

EXERCISES

1. Find the area of a sphere, using the parametric representation

$$x = a \sin \phi \cos \theta, \qquad y = a \sin \phi \sin \theta, \qquad z = a \cos \phi.$$

2. Find the area of the part of the cylinder $x^2 + z^2 = a^2$ inside the cylinder $y^2 = a(x + a)$.

3. Find the area of the part of the cone $x^2 + y^2 = z^2$ inside the cylinder $x^2 + y^2 = 2\,ax$.

4. Find the area of the part of the surface $z = xy$ inside the cylinder $x^2 + y^2 = a^2$.

5. Find the area of the surface element $x = au \cos v$, $y = bu \sin v$, $z = \frac{1}{2} u^2(a \cos^2 v + b \sin^2 v)$, $0 \leq u \leq 1$, $0 \leq v \leq 2\pi$. Identify the surface and the portion of it whose area is found.

6. A part of the surface $z^2 = 2\,xy$ can be parametrized by $x = u^2$, $y = v^2$, $z = \sqrt{2}\,uv$. (a) Find the area of the part of the surface above the rectangle $0 \leq x \leq a$, $0 \leq y \leq b$. (b) Find the area of the part of the surface above the region in the xy-plane between the xy-axes and the curve $x^{1/2} + y^{1/2} = 1$. Compare the solutions by (14.6–12) and (14.6–5).

7. Find the area defined by $x = r \cos \theta$, $y = r \sin \theta$, $z = \theta$, $0 \leq r \leq 1$, $0 \leq \theta \leq 2\pi$. Describe the surface.

8. Find the area of the spheroid $x = a \sin \phi \cos \theta$, $y = a \sin \phi \sin \theta$, $z = c \cos \phi$, distinguishing the cases $a \geq c > 0$, $c > a > 0$.

9. Find the total area of the part of the sphere $x^2 + y^2 + z^2 = 4\,a^2$ inside the cylinder $x^2 + y^2 = 2\,ay$, (a) using (14.6–12) and polar co-ordinates, (b) using the type of parametric representation occurring in Exercise 1. In (b) the main problem is to fine the proper region in the $\theta\phi$-plane to correspond to the first octant portion of the required area.

10. Find the total area of the part of the cylinder $x^2 + z^2 = a^2$ inside the cylinder $x^2 + y^2 = ax$.

11. Find the area of the portion of the surface $y^2 + z^2 = 4\,ax$ in the first octant, between $x = 0$ and $x = 3\,a$, and inside the cylinder $y^2 = ax$. Solve (a) by using (14.6–12), and (b) by using the parametric representation $x = r^2(4/a)$, $y = r \cos \theta$, $z = r \sin \theta$. Show that the latter method is equivalent to using the counterpart of (14.6–12) for projection on the yz-plane, and then introducing polar co-ordinates to do the integration.

12. Show that the area on the sphere $x^2 + y^2 + z^2 = c^2$ and inside the paraboloid $(x^2/a) + (y^2/b) = 2(z + c)$ is $4\,\pi c\sqrt{ab}$, provided that $0 < b \leq a \leq c$.

13. Prove the equality in (14.6–7).

14. On the surface $z = f(x, y)$ consider the locus of points where the angle γ is constant. Suppose the projection of this locus on the xy-plane is a closed curve bounding an area $A = \phi(\gamma)$. Show that it is plausible to think that the area of the part of the surface on which $\gamma_0 \leq \gamma \leq \gamma_1$ (where $0 \leq \gamma_0 \leq \gamma_1 \leq \pi/2$) is $\int_{\gamma_0}^{\gamma_1} \sec \gamma \phi'(\gamma)d\gamma$.

Check by applying to the hemisphere $z = \sqrt{a^2 - x^2 - y^2}$.

15. Find the area of the portion of the cylindrical surface $y^2 + z^2 = a^2$ which is in the first octant and inside the cylinder $(x - a)^2 + y^2 = a^2$. SUGGESTION: As one convenient possible parametrization use $u = \theta$, $v = x$, where $y = a \cos \theta$, $z = a \sin \theta$. From symmetry it may be seen that one half of the desired area comes from the part of the cylindrical surface corresponding to the part of the θx-plane defined by $a(1 - \sin \theta) \leq x \leq a$, $0 \leq \theta \leq \pi/2$.

14.7   Conformal maps

A parametric representation of a surface, with parameters u, v, is a mapping of a portion of the uv-plane onto the surface. This is a generalization of the notion of mapping considered in § 9.3.

Suppose Γ_1 and Γ_2 are two intersecting smooth curves in the uv-plane, and let their images on the surface be smooth curves C_1 and C_2. The mapping is called *conformal* if, for every such Γ_1, Γ_2, the curves C_1 and C_2 intersect at the same angle as Γ_1 and Γ_2 do. When the mapping is conformal, a small triangle in the uv-plane will map into a small curvilinear triangle on the surface, and since corresponding angles will be equal in the two figures, they will be similar in appearance. As a consequence, the mapping preserves the shapes of all figures, in a certain local sense. Of course it need not preserve the size, and distortion of size need not be uniform throughout.

If the mapping is conformal, this fact may be recognized by inspecting the formula for ds^2 on the surface.

THEOREM I.   *In order that the mapping be conformal, it is necessary and sufficient that $F \equiv 0$ and $E \equiv G$ in (14.5–2), so that the formula for ds^2 takes the form*

$$(14.7\text{–}1) \qquad\qquad ds^2 = E(du^2 + dv^2).$$

It is assumed that we are dealing with a simple surface element, that x, y, z have continuous first partial derivatives with respect to u, v, and that the three Jacobians j_1, j_2, j_3 do not all vanish at once.

Before undertaking to prove the theorem, let us consider the intersection of a line $v = $ constant and an arbitrary smooth curve Γ in the uv-plane. The image curves are a v-curve and the image C of Γ. Let σ be arc length along Γ in a specified direction, and let s denote arc length along C in the direction determined by the correspondence. Let the angles of intersection be α and θ respectively (see Fig. 129). We know from elementary calculus that

$$(14.7\text{–}2) \qquad\qquad \cos \alpha = \frac{du}{d\sigma}, \qquad d\sigma^2 = du^2 + dv^2.$$

We shall show that

(14.7–3)
$$\cos \theta = \frac{E\,du + F\,dv}{\sqrt{E}\,ds},$$

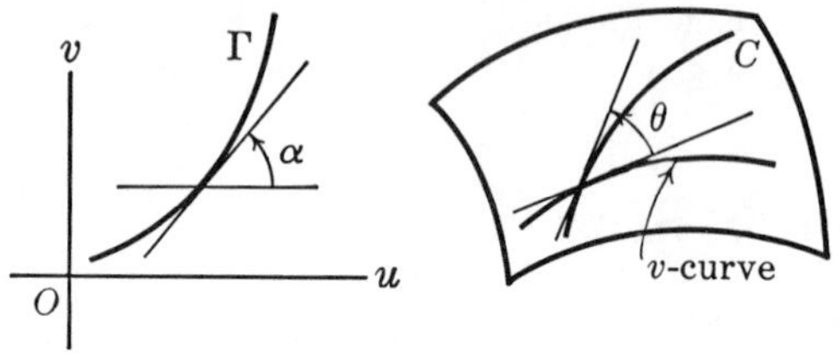

Fig. 129

where all the differentials are computed along C. To prove (14.7–3), we begin by considering the direction cosines of the tangents to the v-curve and the curve C. The tangent to the v-curve has direction cosines

$$\frac{1}{\sqrt{E}}\frac{\partial x}{\partial u}, \qquad \frac{1}{\sqrt{E}}\frac{\partial y}{\partial u}, \qquad \frac{1}{\sqrt{E}}\frac{\partial z}{\partial u}.$$

The factor $1/\sqrt{E}$ comes in to make the sum of the squares equal to unity, as we see by the first formula in (14.5–2). The direction cosines of the tangent to C are

$$\frac{dx}{ds}, \qquad \frac{dy}{ds}, \qquad \frac{dz}{ds}.$$

Therefore $\cos \theta = \dfrac{1}{\sqrt{E}} \left(\dfrac{\partial x}{\partial u}\dfrac{dx}{ds} + \dfrac{\partial y}{\partial u}\dfrac{dy}{ds} + \dfrac{\partial z}{\partial u}\dfrac{dz}{ds} \right).$

Now $\dfrac{\partial x}{\partial u}\,dx = \dfrac{\partial x}{\partial u}\left(\dfrac{\partial x}{\partial u}\,du + \dfrac{\partial x}{\partial v}\,dv \right),$

and there are similar expressions for $\dfrac{\partial y}{\partial u}\,dy$ and $\dfrac{\partial z}{\partial u}\,dz$. When these are collected together, and account is taken of the definitions of E and F (in (14.5–2)), we obtain precisely (14.7–3).

Now we shall prove Theorem I. The condition that the map be conformal is equivalent to the requirement that $\theta = \alpha$ for every possible choice of Γ. This condition is certainly necessary, by the very definition of conformality. It is also sufficient, for if $\theta_1 = \alpha_1$ and $\theta_2 = \alpha_2$ for two different curves Γ_1, Γ_2, then $\theta_2 - \theta_1 = \alpha_2 - \alpha_1$, and the angle between C_1 and C_2 is the same as the angle between Γ_1 and Γ_2. We therefore write the condition for a conformal map in the form

$$\frac{du}{d\sigma} = \frac{E\,du + F\,dv}{\sqrt{E}\,ds},$$

or

$$(14.7\text{-}4) \qquad \frac{du}{(du^2 + dv^2)^{1/2}} = \frac{E\,du + F\,dv}{\sqrt{E}(E\,du^2 + 2\,F\,du\,dv + G\,dv^2)^{1/2}}.$$

We see at once that this condition is satisfied if $F \equiv 0$ and $G \equiv E$. On the other hand, if (14.7-4) holds for all choices of Γ, and if we take Γ to be a line $u = $ constant, then $du = 0$ and dv is arbitrary, whence $F = 0$, by (14.7-4). If we now consider (14.7-4) once more for an arbitrary curve, and put $F = 0$, we have the condition

$$\frac{du}{(du^2 + dv^2)^{1/2}} = \frac{E\,du}{\sqrt{E}(E\,du^2 + G\,dv^2)^{1/2}},$$

which must hold for all curves Γ. This is equivalent to

$$\sqrt{E}(du^2 + dv^2)^{1/2} = (E\,du^2 + G\,dv^2)^{1/2},$$

which is equivalent to $E = G$, since du and dv are arbitrary. Thus we have shown that (14.7-4) holds for all choices of Γ if and only if $F \equiv 0$ and $E \equiv G$. This completes the proof of Theorem I.

The condition (14.7-1) can be written in the form

$$(14.7\text{-}5) \qquad ds = \sqrt{E}\,d\sigma,$$

where $d\sigma$ refers to arc length along a curve in the uv-plane, and ds refers to arc length along the image curve on the surface. Thus, in the mapping, the ratio $\dfrac{ds}{d\sigma}$ depends only on the particular point, and not on the direction in which one moves away from that point. The factor $\sqrt{E}$ is called the *scale factor* at the point.

The notion of conformality may be extended to the case of one surface mapped on another, where both surfaces may be curved. The following theorem may be proved.

THEOREM II.   *Let two smooth surfaces be in one-to-one correspondence. Let s_1 and s_2 denote the arc lengths along a pair of corresponding curves, one on each surface. The mapping is conformal if and only if the ratio of ds_2 to ds_1 at a point depends only on the point, and not on the particular curve through the point. If both surfaces are represented in terms of the same parameters u, v, this is equivalent to the requirement that $ds_2^2 = M\,ds_1^2$, where M is some function of u and v.*

EXERCISES

1. If the surface element lies in the xy-plane (i.e., if $z \equiv 0$ in the parametric equations), and if

$$\frac{\partial x}{\partial u} = \frac{\partial y}{\partial v}, \qquad \frac{\partial x}{\partial v} = -\frac{\partial y}{\partial u},$$

show that the map is conformal.

2. Is the map $x = a \cos u$, $y = a \sin u$, $z = av$ conformal? What is the surface?

3. Two surfaces are defined by

$$x_1 = r \cos \theta, \qquad y_1 = r \sin \theta, \qquad z_1 = r,$$
$$x_2 = \cos \theta, \qquad y_2 = \sin \theta, \qquad z_2 = \sqrt{2} \log r.$$

Identify the two surfaces and show that there is a conformal correspondence between them.

4. Show from (14.7–3) that

$$\tan \theta = \pm \frac{\sqrt{EG - F^2}\, dv}{E\, du + F\, dv}\,.$$

14.71   Cartography

In this section we shall discuss briefly some methods of mapping a spherical surface onto a plane. Mappings of this kind are useful in cartography, i.e., in the preparation of charts and maps for geographical, navigational, and other purposes.

A method of making a flat map of a portion of the earth's surface is often called a *map* projection, though usually the method does not involve a projection in the usual geometric sense of the word.

MERCATOR'S PROJECTION

In the Mercator method, a sphere is mapped onto a cylinder, and then the cylindrical surface is unrolled to give a flat map. Let us suppose for definiteness that the sphere is a scale model of the earth, and that the cylinder is tangent to the sphere at the equator. Let the origin be at the center of the sphere, with the positive z-axis through the north pole. A point $P_1(x_1, y_1, z_1)$ on the sphere may be located by giving its longitude $\theta(0 \leq \theta \leq 2\pi)$ and its latitude $\lambda\,(-\pi/2 \leq \lambda \leq \pi/2)$ (see Fig. 130).

If the radius of the sphere is b, then

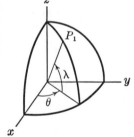

Fig. 130

$$\begin{aligned} x_1 &= b \cos \lambda \cos \theta, \\ (14.71\text{–}1) \qquad y_1 &= b \cos \lambda \sin \theta, \\ z_1 &= b \sin \lambda. \end{aligned}$$

The mapping of the sphere onto the cylinder is defined in the following way: Each point P_1 except the north and the south poles is mapped into a point $P(x, y, z)$ on the cylinder, according to the following specifications (see Fig. 131):

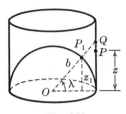

Fig. 131

(a) Let the line OP_1 be produced to meet the cylinder at Q, and draw the line parallel to the z-axis through the point Q.

(b) The point $P(x, y, z)$ corresponding to P_1 is located on this line, with $z = 0$ if $\lambda = 0$, $z > 0$ if $\lambda > 0$, and the value of z depends upon λ in such a way that the map is conformal.

It should be noted that P and Q are distinct except when $\lambda = 0$; for proof of this fact see Exercise 2.

The problem now is to find a formula from which to calculate z in terms of λ. The requirement of conformality enables us to find such a formula.

When the cylinder is unrolled onto a plane, let a uv-coordinate system be established, with the u-axis along the line which was wrapped around the equator and the v-axis along the line which was originally parallel to the z-axis and through the point $(b, 0, 0)$. The point $P(x, y, z)$ now becomes a point $P(u, v)$ with

$$(14.71\text{–}2) \qquad\qquad u = b\theta, \qquad v = z.$$

Arc length in the uv-plane is then given by

$$(14.71\text{–}3) \quad d\sigma^2 = du^2 + dv^2 = b^2 \, d\theta^2 + dz^2 = b^2 \, d\theta^2 + \left(\frac{dz}{d\lambda}\right)^2 d\lambda^2,$$

and arc length on the sphere is given by

$$(14.71\text{–}4) \qquad ds^2 = dx_1^2 + dy_1^2 + dz_1^2 = b^2 \cos^2 \lambda \, d\theta^2 + b^2 \, d\lambda^2.$$

Both arc-length formulas are now expressed in terms of the parameters θ, λ; the condition for conformality of the mapping is that ds^2 must be a multiple of $d\sigma^2$ (Theorem II, § 14.7). Therefore, we see that we must have

$$\frac{b^2}{b^2 \cos^2 \lambda} = \frac{\left(\dfrac{dz}{d\lambda}\right)^2}{b^2}, \quad \text{or} \quad \left(\frac{dz}{d\lambda}\right)^2 = b^2 \sec^2 \lambda.$$

This gives us a differential equation for z in terms of λ. Solving, with the requirement that $z = 0$ when $\lambda = 0$ and $z > 0$ when $\lambda > 0$, we find

$$dz = b \sec \lambda \, d\lambda,$$

$$(14.71\text{–}5) \qquad\qquad z = b \log \tan \left(\frac{\lambda}{2} + \frac{\pi}{2}\right),$$

or, alternatively,

$$(14.71\text{–}6) \qquad\qquad z = b \log (\sec \lambda + \tan \lambda).$$

Observe that $z \to + \infty$ as $\lambda \to \dfrac{\pi}{2}$. Also, $\dfrac{dz}{d\lambda} \to + \infty$ as $\lambda \to \dfrac{\pi}{2}$. Thus, on a Mercator chart, the lines representing parallels of latitude are spaced more and more widely apart as the latitude approaches 90°. There is, of course, extreme distortion of areas in the map of the polar regions.

STEREOGRAPHIC PROJECTION

Another method of mapping a sphere conformally on a plane may be described as follows: Let the sphere be placed tangent to the plane, with S the point of tangency and N the diametrically opposite point. From N draw a line intersecting the sphere at P and the plane at Q (see Fig. 132). The correspondence between P and Q defines what is called a *stereographic projection* of the sphere onto the plane. Although this type of projection has many mathematical uses, we limit our discussion to the proof that the map is conformal.

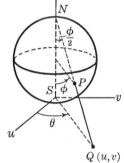

Fig. 132

Set up a system of rectangular co-ordinates u, v in the plane, and use the spherical co-ordinates θ, ϕ to locate P, as shown in Fig. 132. If the radius of the sphere is b, we see that

(14.71–7) $\qquad u = 2 b \tan \dfrac{\phi}{2} \cos \theta, \; v = 2 b \tan \dfrac{\phi}{2} \sin \theta$

because $\qquad\qquad SQ = 2 b \tan \dfrac{\phi}{2}.$

Therefore, by an easy calculation,

$$du^2 + dv^2 = 4 b^2 \tan^2 \frac{\phi}{2} \, d\theta^2 + b^2 \sec^4 \frac{\phi}{2} \, d\phi^2$$

$$= b^2 \sec^4 \frac{\phi}{2} (\sin^2 \phi \, d\theta^2 + d\phi^2).$$

Here we have used the identity

$$\sin \phi = 2 \sin \frac{\phi}{2} \cos \frac{\phi}{2}.$$

Now, if s refers to arc length along a curve on the sphere, and σ refers to arc length on the corresponding curve in the plane, we know that

$$ds^2 = b^2(\sin^2 \phi \, d\theta^2 + d\phi^2),$$

and so we see that

$$d\sigma^2 = du^2 + dv^2 = \sec^4 \frac{\phi}{2} \, ds^2.$$

This shows that the mapping is conformal.

A CONICAL MAP

There are various types of map projections which employ the procedure of mapping the sphere onto a cone, and then unrolling the conical surface onto a plane. One such method, which we shall now discuss, is called the *Lambert conformal projection with one standard parallel.*

Take a sphere of radius b and a cone with vertex V, tangent to the sphere along the parallel of latitude α (see Fig. 133). We make Q on the cone correspond to P on the sphere by requiring that Q lie on the line through V and the point in which OP produced meets the cone, that Q coincide with P along the parallel $\lambda = \alpha$, that P and Q be on the same side of the plane through this parallel if P is not on the parallel, and that the map be conformal.

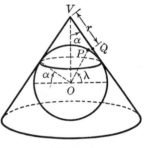

Let r be the distance VQ. The main problem is to find r as a function of λ. The work is much like that in our discussion of the Mercator projection, and the final result is

Fig. 133

$$(14.71\text{–}8) \qquad \log\left(\frac{r\tan\alpha}{b}\right) = \sin\alpha \log\left(\frac{\sec\alpha + \tan\alpha}{\sec\lambda + \tan\lambda}\right).$$

The details of the derivation are left as exercises.

EXERCISES

1. In the Mercator map of the sphere on the uv-plane, show that
$$ds^2 = \cos^2\lambda(du^2 + dv^2).$$

2. Show that, in the Mercator protection,
$$z = \frac{b}{2}\log\frac{b + z_1}{b - z_1},$$
and that the z-coordinate of Q is
$$h = \frac{bz_1}{\sqrt{b^2 - z_1^2}}.$$
From these results show that $h > z$ if $z_1 > 0$, so that Q is above P, as shown in Fig. 131.

3. If s_1 and s_2 are corresponding arc-lengths on the sphere and cone respectively in Fig. 133, and θ is the longitude angle, show that
$$ds_1^2 = b^2\cos^2\lambda \, d\theta^2 + b^2 \, d\lambda^2,$$
$$ds_2^2 = r^2\sin^2\alpha \, d\theta^2 + dr^2.$$
Go on to derive (14.71–8), using the conformality requirement. Show also that
$$ds_2^2 = \frac{r^2\sin^2\alpha}{b^2\cos^2\lambda} \, ds_1^2.$$

4. If $\phi = \dfrac{\pi}{2} - \lambda$ and $\beta = \dfrac{\pi}{2} - \alpha$, show that (14.71–8) becomes

$$r = b \tan \beta \left(\frac{\tan \dfrac{\phi}{2}}{\tan \dfrac{\beta}{2}} \right)^{\cos \beta}.$$

5. If $\lambda - \alpha$ is small in (14.71–8), show that the following *approximate* formulas for r hold:

$$r = b \operatorname{ctn} \alpha \, e^{(\alpha - \lambda) \tan \alpha},$$

$$r = b \operatorname{ctn} \alpha \left(1 - \frac{\lambda - \alpha}{\cos \alpha} \right)^{\sin \alpha}.$$

14.8 Envelopes of plane curves

The student may already have some acquaintance with the subject of envelopes, either from earlier studies in calculus, or from differential equations, where envelopes arise in connection with singular solutions. Much of the interest in envelopes stems from differential equations.

Let us begin with one-parameter families of curves in the xy-plane. If

$$(14.8\text{–}1) \qquad\qquad F(x, y, \alpha) = 0$$

is the equation of such a family, with α the parameter, it may be that there is a curve C in the plane with the following properties:

1. Each curve of the family is tangent to C.
2. C is tangent at each of its points to some curve of the family. If such a curve C exists, we call it the envelope of the family.

In certain situations we may be able to visualize the family and see that there is an envelope.

Example 1. The set of all straight lines at unit distance from the origin has the circle $x^2 + y^2 = 1$ as its envelope.

There are two problems which arise. The first is: If we know that the family (14.8–1) has an envelope, what analytical method is there for finding its equation? The second is: Is there any way of being certain, by examination of (14.8–1), whether or not there is an envelope? The solution of the first problem is much easier than that of the second.

If an envelope C exists, each point of it is characterized by a value of α, namely, that value which picks out the particular member of the family (14.8–1) which is tangent to C at the point. We may therefore regard C as given parametrically in terms of α:

$$(14.8\text{–}2) \qquad\qquad x = \phi(\alpha), \qquad y = \psi(\alpha).$$

We must then have the relation

(14.8–3) $$F[\phi(\alpha), \psi(\alpha), \alpha] \equiv 0$$

holding identically in α. We assume that the function F has continuous first partial derivatives and that ϕ and ψ have derivatives. We can therefore differentiate both sides of (14.8–3) with respect to α, obtaining

(14.8–4) $$F_1\phi'(\alpha) + F_2\psi'(\alpha) + F_3 \equiv 0,$$

where F_1 denotes $\partial F/\partial x$, etc., and we understand that in (14.8–4) these partial derivatives are evaluated with x, y given by (14.8–2). Now, the slope of a curve of the family (14.8–1) is

(14.8–5) $$\frac{dy}{dx} = -\frac{F_1}{F_2},$$

while the slope of the curve C defined by (14.8–2) is

(14.8–6) $$\frac{dy}{dx} = \frac{\psi'(\alpha)}{\phi'(\alpha)}.$$

At the point where C is tangent to a member of the family we then have

$$\frac{\psi'(\alpha)}{\phi'(\alpha)} = -\frac{F_1}{F_2}$$

or

(14.8–7) $$F_1\phi'(\alpha) + F_2\psi'(\alpha) = 0.$$

Combining this result with (14.8–4), we see that $F_3 = 0$ at the point in question. In other words,

(14.8–8) $$F_3[\phi(\alpha), \psi(\alpha), \alpha] \equiv 0$$

identically along C. The identities (14.8–3) and (14.8–8) together mean that x and y, as given by (14.8–2), are solutions of the simultaneous equations

(14.8–9) $$F(x, y, \alpha) = 0, \qquad F_3(x, y, \alpha) = 0,$$

where $F_3 = \dfrac{\partial F}{\partial \alpha}$. This result gives us an analytical method of finding the envelope. Either we solve (14.8–9) for x and y in terms of α, or we eliminate α between these two equations, obtaining an equation in x and y. The foregoing exposition does not guarantee that *all* of the locus thus found will constitute the envelope; it merely establishes that, if there is an envelope, it will be part (perhaps all) of the complete locus defined by (14.8–9).

Example 2. Find the envelope of the family of circles

$$(x - \alpha)^2 + y^2 = \frac{\alpha^2}{2}.$$

For a given α, the circle has center $(\alpha, 0)$ and radius $\dfrac{|\alpha|}{\sqrt{2}}$. It is clear from an inspection of Fig. 134 that there is an envelope, consisting of the lines $y = x$ and $y = -x$. In the present case equations (14.8–9) take the

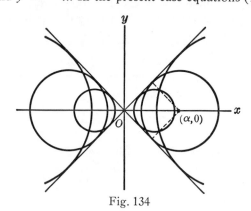

Fig. 134

form $\qquad (x - \alpha)^2 + y^2 - \dfrac{\alpha^2}{2} = 0,\ 2(x - \alpha)(-1) - \alpha = 0.$

If we solve for x and y, we find

$$x = \frac{\alpha}{2}, \qquad y = \pm \frac{\alpha}{2}.$$

Or, we may eliminate α, obtaining finally

$$y^2 - x^2 = 0.$$

In either case the locus found consists of the two lines $y = x,\ y = -x.$

Example 3. Consider the family $(y - \alpha)^2 = x^2(1 - x^2).$

These curves are all obtained from the one for which $\alpha = 0$ by translations parallel to the y-axis. It is apparent from Fig. 135 that the envelope consists of the two lines $x = 1,\ x = -1$. Let us apply the method of equations (14.8–9). We have to eliminate α between

$$(y - \alpha)^2 - x^2(1 - x^2) = 0$$

and $\qquad\qquad 2(y - \alpha)(-1) = 0.$

The result is $x^2(1 - x^2) = 0$, which yields as complete locus the *three* lines $x = 0,\ x = \pm 1$. The line $x = 0$ is clearly not part of the envelope. This shows that the locus defined by (14.8–9) may consist of something more than the envelope.

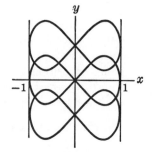

Fig. 135

Example 3 raises the following question: Suppose we find a curve C, defined by parametric equations $x = \phi(\alpha)$, $y = \psi(\alpha)$, such that equations (14.8–9) are satisfied, i.e.,

$$(14.8\text{–}10) \qquad F(\phi(\alpha), \psi(\alpha), \alpha) = 0, \qquad F_3(\phi(\alpha), \psi(\alpha), \alpha) = 0,$$

for a certain interval of values of α. Can it be determined *analytically* whether or not C is part of the envelope? We shall show that, if certain conditions are satisfied, C is certainly part of the envelope. These conditions are as follows:

At each point of C,

(a) $\dfrac{\partial F}{\partial x}$ and $\dfrac{\partial F}{\partial y}$ are not simultaneously zero,

(b) $\phi'(\alpha)$ and $\psi'(\alpha)$ are not simultaneously zero.

Now let C_α denote the curve of the original family (14.8–1) corresponding to a particular α. The first of the equations (14.8–10) tells us that C and C_α intersect at the point on C given by the particular α. Conditions (a) and (b) guarantee that both curves have tangents at this point. The only remaining problem is that of showing that C and C_α are tangent at the point. From (14.8–10) it follows that (14.8–4) holds, and hence also that (14.8–7) holds. There are now two cases to consider: case I, $\phi'(\alpha) \neq 0$ and case II, $\phi'(\alpha) = 0$. In case I we have $F_2 \neq 0$, for if $F_2 = 0$ it follows from (14.8–7) that $F_1 = 0$, which contradicts condition (a). Thus from (14.8–7) we have

$$- \frac{F_1}{F_2} = \frac{\psi'(\alpha)}{\phi'(\alpha)},$$

which means that C_α and C are tangent. In case II we must have $F_2 = 0$, for if $F_2 \neq 0$ we conclude from (14.8–7) that $\psi'(\alpha) = 0$, which contradicts condition (b). Thus, in case II, both C and C_α have tangents parallel to the y-axis, and the curves must be tangent. This shows that C is actually part of the envelope.

In Example 3 the line $x = 0$ has parametric equations $x = 0$, $y = \alpha$ (i.e., $\phi(\alpha) = 0$, $\psi(\alpha) = \alpha$). In this case the condition (a) is not satisfied.

Let us now turn to the second of the two problems which were mentioned immediately following Example 1. This is the problem of finding conditions on the function $F(x, y, \alpha)$ which are sufficient to guarantee the existence of an envelope of the family of curves. To solve the problem, we employ the theory of implicit functions, as developed in Chapter VIII. The objective is to show that, under suitable conditions, equations (14.8–9) define a curve C, and that at each point of C the conditions (a) and (b) are satisfied.

THEOREM III.   *Suppose that $F(x, y, \alpha)$ is continuous, with continuous first and second partial derivatives, in some neighborhood of the point*

(x_0, y_0, α_0) *in the three-dimensional xyα-space. At this point suppose that* F *and* $\dfrac{\partial F}{\partial x}$ *are both zero, but assume that at this point*

(14.8–11)
$$\begin{vmatrix} F_1 & F_2 \\ F_{31} & F_{32} \end{vmatrix} \neq 0$$

and

(14.8–12)
$$F_{33} \neq 0.$$

Then, with α *limited to some neighborhood of* $\alpha = \alpha_0$ *on the* α*-axis, equations* (14.8.9) *define a locus in the xy-plane which is a curve tangent at each of its points to the curve of the family* (14.8–1) *through that point.*

PROOF. The determinant in (14.8–11) is the Jacobian of F and $\dfrac{\partial F}{\partial \alpha}$ with respect to x and y. Hence, by Theorem III, § 8.3, we know that all triples of values (x, y, α) satisfying (14.8–9) and lying near (x_0, y_0, α_0) are furnished by equations of the form $x = \phi(\alpha)$, $y = \psi(\alpha)$, where ϕ and ψ have continuous derivatives. In view of the discussion following Example 3, it remains only to show that the conditions (a) and (b) are fulfilled. Now (14.8–11) and (14.8–12) continue to hold for values of (x, y, α) near (x_0, y_0, α_0), because of continuity. Condition (a) must be satisfied because of (14.8–11). Now the identity (14.8–8) holds, and therefore, differentiating with respect to α,

$$F_{31}\phi'(\alpha) + F_{32}\psi'(\alpha) + F_{33} \equiv 0,$$

the partial derivatives being evaluated with $x = \phi(\alpha)$, $y = \psi(\alpha)$. From this equation and (14.8–12) we see that ϕ' and ψ' are not both zero at once. Hence, condition (b) is fulfilled and the proof is complete.

The conditions of Theorem III, though sufficient, are not necessary, as is shown by the family $y = (x - \alpha)^3$ and its envelope $y = 0$.

EXERCISES

1. Find the envelopes of each of the following families. Describe the situation geometrically in each case.

(a) $y = \alpha x + (1/\alpha)$.

(b) $y^2 = 4\alpha(x - \alpha)$.

(c) $\alpha^3 x - \alpha y = 2$.

(d) $y^2 = \alpha^2(x - \alpha)$.

(e) $x + y \sin \alpha = a \cos \alpha$.

(f) $x^2 \sin \alpha + y^2 \cos \alpha = a^2$.

(g) $y = 2\alpha x + \alpha^4$.

(h) $y = \alpha x - \frac{1}{8}(1 + \alpha x^2)x^2$.

2. Find the envelope of all circles which are tangent to the y-axis and have their centers on the line $y = 2x$.

3. Find the envelope of all lines $(x/a) + (y/b) = 1$, (a) if $ab = c^2$ (constant); (b) if a and b are positive and $a + b = c$; (c) for which $a^2 + b^2 = c^2$.

4. Find the envelope of all ellipses $(x^2/a^2) + (y^2/b^2) = 1$, **(a)** for which the area is constant; **(b)** for which $a + b$ is constant.

5. Find the envelope of the family of circles having as diameters the chords of the ellipse $(x^2/a^2) + (y^2/b^2) = 1$ parallel to the y-axis.

6. As a circle of radius b rolls on a straight line, a fixed point P on the rim of the circle traces out a cycloid. Find the envelope of the diameter of the circle through the tracing point P.

7. A straight line moves so that the product of its perpendicular distances from two fixed points is constant. Show that the envelope is an ellipse having the two points as foci.

8. Find the envelope of all circles through the origin with the centers on the curve $xy = 1$.

9. A line through $(a, 0)$ cuts the y-axis at a variable point Q. Draw the line through Q perpendicular to the first one, and find the envelope of all such perpendiculars.

10. Let P be a variable point on the circle $x^2 + y^2 = a^2$, and let Q be the point $(-b, 0)$, where $b > a > 0$. Through P draw the line perpendicular to PQ, and find the envelope of all such perpendiculars.

11. Find the envelope of all circles through a fixed point of a given circle and having chords of the given circle as diameters. SUGGESTION: Use polar coordinates, taking $r = 2a \cos \theta$ for the given circle, and $r = 2a \cos \alpha \cos (\theta - \alpha)$ as the family (α the parameter). The envelope is the cardioid $r = a(1 + \cos \theta)$.

12. Does the family $(x - \alpha)^2 = y^3$ have an envelope?

13. Does the family $y^2 - x^2 - 2\alpha y + \alpha^2 = 0$ have an envelope? For what reason does Theorem III not apply in this case?

14. If a line meeting a curve be thought of as a ray of light and the curve a mirror, the incident and reflected rays make equal angles with the normal to the curve at the point of incidence. The envelope of the reflected rays is called a *caustic* curve.

(a) Show that if the incident rays are parallel to the x-axis and fall on the concave side of the semicircle $x^2 + y^2 = a^2$, $x \geq 0$, the caustic curve is defined by

$$4x = 3a \cos \theta - a \cos 3\theta, \quad -\frac{\pi}{2} \leq \theta \leq \frac{\pi}{2}.$$
$$4y = 3a \sin \theta - a \sin 3\theta,$$

This is half of an epicycloid of two cusps, produced by rolling a circle of radius $a/4$ on the outside of a circle of radius $a/2$.

(b) If the reflecting curve is a circle and the incident rays emanate from a point on the circle, show that the caustic curve is a cardioid.

15. Find the envelope of the normals to the parabola $y^2 = 4ax$.

16. Show that the envelope of the normals to the tractrix

$$y = -\sqrt{a^2 - x^2} + a \log \frac{a + \sqrt{a^2 - x^2}}{x}$$

is

$$y = a \log \frac{x + \sqrt{x^2 - a^2}}{a},$$

which is half of the catenary $x = a \cosh (y/a)$.

17. Show that the envelope of the normals to the curve $y = f(x)$ has parametric equations

$$x = \alpha - \frac{f'(\alpha)}{f''(\alpha)} [1 + (f'(\alpha))^2],$$

$$y = f(\alpha) + \frac{1}{f''(\alpha)} [1 + (f'(\alpha))^2].$$

The point (x, y) of the envelope is the center of curvature corresponding to the point $(\alpha, f(\alpha))$ on the original curve, so that the envelope of the normals is the *evolute*.

14.9   Envelopes of surfaces

The geometrical concept of the envelope of a family of surfaces is similar to the concept of the envelope of a family of plane curves.

Example 1.   Consider the family of spheres each of which has its center on the x-axis and is internally tangent to the paraboloid $y^2 + z^2 = 4x$.

This is clearly a one-parameter family; the parameter may be taken as the x-coordinate of the center of the sphere. Any one sphere is tangent to the paraboloid along a circle, and every point of the paraboloid is on such a circle. The paraboloid is what is called the envelope of the family of spheres.

In general, let

$$(14.9\text{–}1) \qquad\qquad F(x, y, z, \alpha) = 0$$

define a one-parameter family of surfaces, and suppose that there is a surface S (consisting perhaps of several separate parts) which is such that (1) each surface of the family is tangent to S along a curve, and (2) each point of S lies on such a curve. Under these conditions we shall call S the envelope of the family of surfaces.

The analytical procedure for finding envelopes of surfaces is almost identical with that used for finding envelopes of plane curves. We eliminate α between the two equations

$$(14.9\text{–}2) \qquad F(x, y, z, \alpha) = 0, \qquad F_4(x, y, z, \alpha) = 0,$$

where $F_4 = \dfrac{\partial F}{\partial \alpha}$. If this elimination yields an equation in x, y, z, and if there is an envelope, the envelope is part or all of the locus defined by the equation. Suitable restrictions must be placed on F, of course. A plausible but nonrigorous way of arriving at (14.9–2) may be explained as follows: Let C_α be the curve along which the envelope S is tangent to the curve (14.9–1), for a particular α. If two surfaces of the family, corresponding to parameter values α and $\alpha + \Delta\alpha$, intersect in a curve Γ, C_α is the limiting position of the curve Γ as $\Delta\alpha \to 0$. The student may be able to visualize this in Example 1. Now Γ is defined as the intersection of the surfaces $F(x, y, z, \alpha) = 0$ and $F(x, y, z, \alpha + \Delta\alpha) = 0$, and hence also of the surfaces

$$F(x, y, z, \alpha) = 0, \qquad \frac{F(x, y, z, \alpha + \Delta\alpha) - F(x, y, z, \alpha)}{\Delta\alpha} = 0.$$

Letting $\Delta\alpha \to 0$ here, we see that we should expect C_α to be defined by (14.9–2). Letting α vary, the set of all the C_α forms the envelope S. This argument is nonrigorous because, among other reasons, two nearby surfaces of the family need not intersect, even when there is an envelope.

A correct argument may be given as follows. Suppose part (or all) of the envelope S is represented parametrically in the form

$$(14.9\text{–}3) \qquad x = f(\alpha, \beta), \qquad y = g(\alpha, \beta), \qquad z = h(\alpha, \beta),$$

where β is a second parameter. Thus C_α is defined by (14.9–3) with α fixed and β variable. We assume that F, f, g, h all have continuous partial derivatives. Putting (14.9–3) into (14.9–1) we obtain an identity in α and β. Consequently, differentiating with respect to α,

$$(14.9\text{–}4) \qquad F_1 \frac{\partial f}{\partial \alpha} + F_2 \frac{\partial g}{\partial \alpha} + F_3 \frac{\partial h}{\partial \alpha} + F_4 \equiv 0$$

identically in α and β. But

$$(14.9\text{–}5) \qquad F_1 \frac{\partial f}{\partial \alpha} + F_2 \frac{\partial g}{\partial \alpha} + F_3 \frac{\partial h}{\partial \alpha} = 0,$$

since the direction of the normal to the surface of the family is $F_1 : F_2 : F_3$, and $\dfrac{\partial f}{\partial \alpha}, \dfrac{\partial g}{\partial \alpha}, \dfrac{\partial h}{\partial \alpha}$ are direction components of the tangent to a curve $\beta = $ constant on the envelope. From (14.9–4) and (14.9–5) we conclude that $F_4 = 0$. Thus the equations (14.9–2) must hold along C_α.

A guarantee that equations (14.9–2) will provide a portion of the envelope, locally, is that these equations hold at a point $(x_0, y_0, z_0, \alpha_0)$, and that at this point $F_4 \neq 0$ and at least one of the three partial derivatives F_1, F_2, F_3 is different from zero. Then we may solve $F_4(x, y, z, \alpha) = 0$ for α, say $\alpha = \phi(x, y, z)$. Let $G(x, y, z) = F[x, y, z, \phi(x, y, z)]$. Then, locally, (14.9–2) is equivalent to $G(x, y, z) = 0$. This is the equation of the envelope; for the direction of the normal to this surface is $G_1 : G_2 : G_3$, and at the point in question we have $G_1 = F_1$, $G_2 = F_2$, $G_3 = F_3$. To prove $G_1 = F_1$, for example, we note that $F_4 = 0$, and hence

$$G_1 = \frac{\partial G}{\partial x} = \frac{\partial F}{\partial x} + \frac{\partial F}{\partial \alpha}\frac{\partial \phi}{\partial x} = F_1 + F_4\frac{\partial \phi}{\partial x} = F_1.$$

Example 2. Find the envelope of the family of planes

$$(14.9\text{–}6) \qquad\qquad x \cos \theta + y \sin \theta + \frac{z}{\sqrt{3}} = 0.$$

Here the parameter is θ. Applying the method of (14.9–2), we have to eliminate θ between (14.9–6) and

$$(14.9\text{–}7) \qquad\qquad -x \sin \theta + y \cos \theta = 0.$$

From (14.9–7) we have $\tan \theta = (y/x)$, whence

$$\sin \theta = \frac{y}{\pm\sqrt{x^2 + y^2}}, \qquad \cos \theta = \frac{x}{\pm\sqrt{x^2 + y^2}}.$$

Combining with (14.9–6), we find

$$\frac{x^2 + y^2}{\pm\sqrt{x^2 + y^2}} + \frac{z}{\sqrt{3}} = 0.$$

In rationalized form this is

$$z^2 = 3(x^2 + y^2).$$

This is a right circular cone with axis along the z-axis, vertex at the origin, and semivertical angle 30°. Direct verification that the cone is the envelope may be had by studying the family of planes. They all pass through the origin. The normal to a plane has direction

$$\cos \theta : \sin \theta : \frac{1}{\sqrt{3}},$$

so that the acute angle γ between the normal and the positive z-axis is given by

$$\cos \gamma = \frac{\dfrac{1}{\sqrt{3}}}{\sqrt{\cos^2 \theta + \sin^2 \theta + \frac{1}{3}}} = \frac{1}{2};$$

therefore $\gamma = 60°$. The planes are evidently all tangent to the cone.

In the general case the curves C_α along which the envelope is tangent to members of the family are called *characteristic curves* of the family.

These curves play an important role in the study of partial differential equations of the first order.

In Example 2 the generators of the cone are the characteristic curves. There is also a concept of envelope for two-parameter families of surfaces.

Example 3. The family of all planes at unit distance from the origin is said to have the sphere $x^2 + y^2 + z^2 = 1$ as envelope. The planes clearly form a two-parameter family. Each plane is tangent to the sphere at one point, and the usual angular co-ordinates (θ, ϕ) of the point will serve as parameters to determine the particular plane.

In the case of two-parameter families, the envelope surface has a single member of the family tangent to it at each point. Suppose that

$$(14.9\text{–}8) \qquad\qquad F(x, y, z, \alpha, \beta) = 0$$

is the equation of the family. Under appropriate hypotheses, reasoning closely similar to that used earlier in this chapter shows that if there is an envelope it is part or all of the locus defined by the simultaneous equations

$$(14.9\text{–}9) \qquad\qquad F = 0, \qquad \frac{\partial F}{\partial \alpha} = 0, \qquad \frac{\partial F}{\partial \beta} = 0.$$

Example 4. Find the envelope of the two-parameter family of spheres

$$(14.9\text{–}10) \quad (x - a \cos \theta)^2 + (y - a \sin \theta)^2 + z^2 = \frac{a^2}{4} \text{ (parameters } a, \theta).$$

In addition to (14.9–10) we have the two equations obtained by differentiation:

$$- 2(x - a \cos \theta) \cos \theta - 2(y - a \sin \theta) \sin \theta = \frac{a}{2},$$

$$2(x - a \cos \theta)a \sin \theta - 2(y - a \sin \theta)a \cos \theta = 0.$$

On simplification, these become

$$x \cos \theta + y \sin \theta = \frac{3a}{4},$$

$$(14.9\text{–}11)$$

$$x \sin \theta - y \cos \theta = 0.$$

From the last equation, $\tan \theta = y/x$, and

$$\sin \theta = \frac{y}{\pm \sqrt{x^2 + y^2}}, \qquad \cos \theta = \frac{x}{\pm \sqrt{x^2 + y^2}}.$$

When these results are put in (12.9–11), we find

$$(14.9\text{–}12) \qquad\qquad a = \pm \tfrac{4}{3}\sqrt{x^2 + y^2}.$$

By expanding and using a trigonometric identity, we can write (14.9–10) in the form

$$x^2 + y^2 - 2\,ax\cos\theta - 2\,ay\sin\theta + z^2 + \tfrac{3}{4}\,a^2 = 0.$$

If we then use (14.9–12) and the first of equations (14.9–11), we obtain

$$x^2 + y^2 + z^2 = \frac{3\,a^2}{4} = \frac{3}{4}\cdot\frac{16}{9}\,(x^2 + y^2),$$

or
$$3\,z^2 = (x^2 + y^2).$$

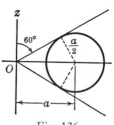

This is a right circular cone about the z-axis, with semivertical angle 60° and vertex at the origin. The fact that the cone is the envelope may be seen directly when we observe that the spheres (14.9–10) have their centers in the xy-plane, at distance a from the origin, and that the radius of such a sphere is $a/2$ (see Fig. 136).

Fig. 136

EXERCISES

1. Find the envelopes of the following families. Describe the situation geometrically in each case.
(a) $x^2 + y^2 + (z - \alpha)^2 = 1$.
(b) $x\cos\alpha + y\sin\alpha = p$ (p constant).
(c) $(x - 2\cos\alpha)^2 + (y - 2\sin\alpha)^2 + z^2 = 1$.
(d) $(x - \alpha)^2 + (y - 2\,\alpha)^2 + (z - 3\,\alpha)^2 = 16$.

2. Find the envelopes of the following families:
(a) $z = \alpha x + (1/\alpha)y + 1$.   (c) $z = \alpha x + 2\,\alpha y + 2\,\alpha^2$.
(b) $z = x\cos\alpha + y\sin\alpha$.   (d) $z = \alpha x + (1 - \alpha)y + \alpha - \alpha^2$.

3. Find the envelope of the family of all planes $x(a + t) + y(b + t) + z(c + t) = t^2$, where a, b, c are constants. What is the situation if $a = b = c$?

4. Find the envelope of the family of planes $(x/a) + (y/b) + (z/c) = 1$ (a) if a, b, c vary in such a way that $abc = p^3$ (a constant); (b) if a, b, c vary in such a way that $a^2 + b^2 + c^2 = p^2$ (a constant); (c) if a, b, c are positive variables whose sum remains constant.

5. Find the envelope of the two-parameter family $z = ax + by + ab$.

6. Consider all the planes tangent to the paraboloid $2\,z/c = (x^2/a^2) + (y^2/b^2)$ at the points where it is intersected by the plane $z = c$. Show that the envelope of these planes is the cone $(z + c)^2 = 2\,c^2[(x^2/a^2) + (y^2/b^2)]$.

7. Find the envelope of all the planes which are tangent to the ellipsoid $(x^2/a^2) + (y^2/b^2) + (z^2/c^2) = 1$ along the plane section $z = k$ ($0 < k < c$).

8. Consider all the planes tangent to an ellipsoid at the points in which the

ellipsoid is intersected by a plane through its center. Show that the envelope of these planes is an elliptic cylinder. SUGGESTION: Write the equation of the ellipsoid in the form $Ax^2 + By^2 + Cz^2 + 2\,Dyz + 2\,Ezx + 2\,Fxy = 1$ and take the plane of section to be $z = 0$. What can you say about the corresponding problem for a hyperboloid of one sheet?

9. Find the envelope of all spheres through the origin with centers on the parabola $y = 0$, $x^2 + 4\,az = 0$.

10. Find the envelope of all spheres through the origin with centers on the paraboloid $(x^2/a^2) + (y^2/b^2) = -(2\,z/c)$.

11. Find the envelope of all spheres through the origin with centers on the ellipsoid $(x^2/a^2) + (y^2/b^2) + (z^2/c^2) = 1$.

Line and Surface Integrals   15

15.   Introduction

In this chapter we consider some new concepts, the concepts of line integrals and surface integrals. These new kinds of integrals will be defined as limits of sums in the same general way that single and double integrals are defined. An ordinary single integral

$$\int_a^b f(x)dx$$

is an integral of a function which is defined along a line segment (an interval of a co-ordinate axis). There is a corresponding kind of integral for a function which is defined along a curve. Such an integral might well be called a *curvilinear* integral; the usual name is *line integral*, where *line* means, in general, a *curved line*. Likewise, the concept of a surface integral is a generalization of the concept of a double integral. A double integral

$$\iint_R f(x, y)dx\, dy$$

is an integral of a function which is defined on a region R in the xy-plane. A surface integral is an integral of a function which is defined on a surface. The double integral is a particular case, for the plane region R is a flat surface.

These new kinds of integrals have important applications to geometry and physics. They are also tools of great usefulness in analytical reasoning.

15.1   Point functions on curves and surfaces

In § 10.5 we introduced the concept of a scalar point function, the essential idea of which is that of considering the function values as associated with a point rather than with the co-ordinates of that point. A notation such as $f(P)$ signifies the value of the function f at the point P. Now a point function may be defined throughout some region of space, or merely on a curve or a surface. Thus, for instance, the curvature of a curve is a scalar point function defined only at points on the curve, while a unit vector normal to a smooth surface is a vector point function defined only on the surface.

We shall need the concept of continuity for point functions defined merely on curves or surfaces. Let $f(P)$ be defined on a curve C, and let P_0 be a fixed point of C. We say that f is continuous at P_0 if

$$\lim_{P \to P_0} f(P) = f(P_0),$$

it being understood, of course, that P can approach P_0 in any manner consistent with its being on C. The arithmetical form of this definition is as follows: f is continuous at P_0 if to each positive number ϵ corresponds some positive number δ (depending in general on P_0 and ϵ) such that $|f(P) - f(P_0)| < \epsilon$ whenever P is on C and at distance less than δ from P_0. The function is called continuous on C if it is continuous at each point of C. A similar definition is made for continuity on a surface.

A point function may, of course, be expressed as a function of co-ordinates when a co-ordinate system is introduced.

15.12   Line integrals

Let C be a curve in xyz-space. It may in particular be a plane curve, as for instance a curve in the xy-plane. There are two directions along a curve; we may arbitrarily choose to call one of these directions the positive direction, and the opposite direction the negative direction. When this choice has been made we say the curve is *oriented*, or given an orientation. If an arc is oriented, it has an initial point A and a terminal point B. An oriented simple closed curve has no initial or terminal point, but it is often convenient to select some point of the curve and regard it as both the initial and terminal point of the curve (see Fig. 137).

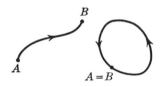

Fig. 137

In what follows we shall limit our discussion to curves which are formed by joining a finite number of arcs end to end. Such a curve may intersect itself a great deal. If it does not intersect itself at all, it is either a simple closed curve or else it may be regarded as a single arc.

Let C be an oriented curve with initial point A and terminal point B (A and B may coincide, as in Fig. 137). Let F denote a scalar point function which is defined and continuous along C. Let points P_0, P_1, $\cdots$, P_n be chosen in order along C, with $A = P_0$, $P_n = B$. Let Q_k be any point of C between P_{k-1} and P_k (see Fig. 138). In a given rectangular co-ordinate system let the co-ordinates of P_k be (x_k, y_k, z_k), and let $\Delta x_k = x_k - x_{k-1}$. Form the sum

(15.12–1) $$\sum_{k=1}^{n} F(Q_k)\Delta x_k;$$

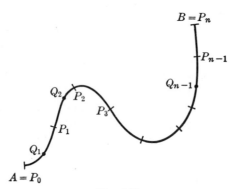

Fig. 138

if the sums of this sort have a limiting value as $n \to \infty$ and the greatest of the chord lengths $P_0P_1, \cdots, P_{n-1}P_n$ approaches zero, we denote the limit by

$$\int_C F(P)dx,$$

and call it the line integral of F with respect to x along C. If P is the point (x, y, z), and if $F(P)$ is denoted by $f(x, y, z)$, an alternative notation for the line integral is

$$\int_C f(x, y, z)dx.$$

Line integrals with respect to y or z are defined in the same way, with Δy_k or Δz_k replacing Δx_k in (15.12–1).

To compute the value of a line integral, we use some parametric representation of the curve C. Suppose the parametric equations of C are

$$x = \lambda(t), \quad y = \mu(t), \quad z = v(t), \quad a \leq t \leq b,$$

and suppose that x, y, and z have continuous derivatives with respect to t. We further suppose that the points A and B correspond to $t = a$ and $t = b$, respectively, and that (x, y, z) traces out C from A to B as t goes from a to b. Let the points P_k on C correspond to points t_k such that $a = t_0 < t_1 < \cdots < t_n = b$; let $\Delta t_k = t_k - t_{k-1}$, and let Q_k correspond to t_k', where $t_{k-1} \leq t_k' \leq t_k$. The sum (15.12–1) now takes the form

(15.12–2) $$\sum_{k=1}^{n} f(\lambda(t_k'), \mu(t_k'), v(t_k'))[\lambda(t_k) - \lambda(t_{k-1})].$$

By the law of the mean,

$$\lambda(t_k) - \lambda(t_{k-1}) = \lambda'(\tau_k)\Delta t_k,$$

where τ_k is some number between t_{k-1} and t_k.

Therefore $\qquad \int_C f(x, y, z)dx = \int_a^b f(\lambda(t), \mu(t), \nu(t))\lambda'(t)dt.$

In drawing this conclusion we use a standard theorem about definite integrals; this theorem appears as (18.21–4), § 18.21. It is a special case of Duhamel's principle. This argument shows that the limit defining the line integral exists provided C has a parametric representation in which x, y, z have continuous derivatives with respect to t.

The result may be written

$$(15.12\text{–}3) \qquad \int_C f(x, y, z)dx = \int_a^b f(x, y, z)\left(\frac{dx}{dt}\right)dt,$$

where the integral on the right is an ordinary definite integral of a function of t, whose integrand is found by expressing x, y, z in terms of t from the parametric equations of C.

Example 1. Find the values of

$$(a) \int_C (xy + y^2 - xyz)dx \qquad \text{and} \qquad (b) \int_C (x^2 - xy)dy$$

if C is the arc of the parabola $y = x^2$, $z = 0$ from $(-1, 1, 0)$ to $(2, 4, 0)$.

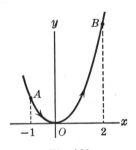

Fig. 139

Here we use x as the parameter. The integral (a) becomes

$$\int_C (xy + y^2 - xyz)dx = \int_{-1}^2 (x^3 + x^4)dx = \tfrac{15}{4} + \tfrac{33}{5} = \tfrac{207}{20}.$$

For (b) we have $dy/dx = 2\ x$, and so

$$\int_C (x^2 - xy)dy = \int_{-1}^{2} (x^2 - x^3)2x\ dx = \tfrac{15}{2} - \tfrac{66}{5} = -\tfrac{57}{10}.$$

It is not essential that the parameter be increasing as we go along the curve in the positive direction.

Example 2.   Consider the first quadrant arc C of the circle $x^2 + y^2 = 1$ in the xy-plane, oriented positively in the direction from $(0, 1)$ to $(1, 0)$ (see Fig. 140). With the parametric representation $x = \cos\theta$, $y = \sin\theta$, the initial point of C corresponds to $\theta = \pi/2$ and the terminal point to $\theta = 0$. In evaluating a line integral, the lower limit of integration will be $\theta = \pi/2$, and the upper limit $\theta = 0$. For instance, $dy = \cos\theta\ d\theta$, and so

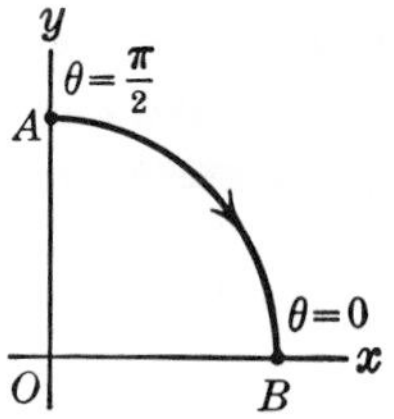

Fig. 140

$$\int_C x^2 y\ dy = \int_{\pi/2}^{0} \cos^2\theta \sin\theta \cos\theta\ d\theta = -\left. \frac{\cos^4\theta}{4}\right|_{\pi/2}^{0}$$

$$= -\tfrac{1}{4}.$$

The value of a line integral depends on the orientation which is assigned to the curve; if the orientation is reversed, the value of the integral is replaced by the negative of its former value. This is because the limits of integration are reversed in (15.12–3).

The value of a line integral does not depend upon the particular parametric representation of the curve which is used to calculate the value of the integral.

A sum of line integrals with respect to x, y, and z is often written with just one integral sign. Thus,

$$\int_C f(x, y, z)dx + g(x, y, z)dy + h(x, y, z)dz$$

means $$\int_C f(x, y, z)dx + \int_C g(x, y, z)dy + \int_C h(x, y, z)dz.$$

Example 3.   Compute the value of

(15.12–4) $$\int_C xz\ dx + x\ dy - yz\ dz$$

along the oriented curve shown in Fig. 141, consisting of a quarter circle in the xz-plane, and line segments in the xy-plane and yz-plane, respectively. Denote the three parts of C by C_1, C_2, C_3, respectively. On C_1 we choose x as parameter. Then $z = \sqrt{1 - x^2}$, $y = 0$, so $dy = 0$, and

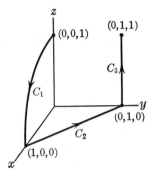

Fig. 141

$$\int_{C_1} xz\, dx + x\, dy - yz\, dz = \int_{C_1} xz\, dx + x \cdot 0 - 0 \cdot dz$$

$$= \int_0^1 x\sqrt{1 - x^2}\, dx = \tfrac{1}{3}.$$

For C_2 we use y as parameter; the equations are $x = 1 - y$, $z = 0$; so $dz = 0$, and

$$\int_{C_2} xz\, dx + x\, dy - yz\, dz = \int_{C_2} 0 \cdot dx + x\, dy - 0 \cdot 0$$

$$= \int_0^1 (1 - y)dy = \tfrac{1}{2}.$$

Finally, using z as parameter on C_3, we have $x = 0$, $y = 1$, $dx = 0$, $dy = 0$, and the integral over C_3 is just

$$\int_{C_3} - yz\, dz = \int_0^1 - z\, dz = -\tfrac{1}{2}.$$

Thus the line integral (15.12–4) has the value

$$\tfrac{1}{3} + \tfrac{1}{2} - \tfrac{1}{2} = \tfrac{1}{3}.$$

EXERCISES

1. Find the values of the following line integrals. All the curves are in the xy-plane

(a) $\int_C y^2\, dx - x\, dy$, along $y^2 = 4x$ from $(0, 0)$ to $(1, 2)$.

(b) $\int_C - y\, dx + x\, dy$, along $y^2 = 4x$ from $(4, 4)$ to $(0, 0)$.

(c) $\int_C (4x - y)dx$, along $y = 8x - 2x^2$ from $(4, 0)$ to $(0, 0)$.

(d) $\int_C (4x - y)dy$ along the curve of part (c).

(e) $\int_C x^2 \, dy$, along the curve $y = x^3 - 3x^2 + 2x$ from $(0, 0)$ to $(2, 0)$.

(f) $\int_C x \, dy - y \, dx$, along $2y = 3x + 2$ from $(2, 4)$ to $(4, 7)$.

(g) $\int_C (y - x)dx + x^2 y \, dy$, along $y^2 = x^3$ from $(1, -1)$ to $(1, 1)$.

(h) $\int_C (x^2 - y^2)dx + x \, dy$, along the first quadrant arc of $x^2 + y^2 = 4$, from $(0, 2)$ to $(2, 0)$.

2. Find the values of the following line integrals:

(a) $\int_C x^2 y^2 \, dx + xy^2 \, dy$, counterclockwise around the closed curve formed by parts of the line $x = 1$ and the parabola $y^2 = x$.

(b) $\int_C (x^2 - y^2)dx + 2xy \, dy$, counterclockwise around the square formed by the lines $x = 0$, $x = 2$, $y = 0$, $y = 2$.

(c) $\int_C \frac{xy^2}{x^2 + y^2} \, dy$, counterclockwise around the circle $x^2 + y^2 = a^2$.

(d) $\int_C x^2 y \, dx$, counterclockwise around the circle $x^2 + y^2 = a^2$.

(e) $\int_C -3y \, dx + 2x \, dy + 4z \, dz$, around the circle $x^2 + y^2 = 1$, $z = 1$, in the counterclockwise sense as viewed from $(0, 0, 2)$.

(f) $\int_C (x^2 - y^2)dx + x \, dy$, in the counterclockwise sense around the circle $x^2 + y^2 = 4$.

(g) $\int_C (x\sqrt{3} - y)dx + (y\sqrt{3} + x)dy$, counterclockwise around the circle $x^2 + y^2 = 1$.

(h) $\int_C -y^2 \, dx + x^2 \, dy$, counterclockwise around the closed curve formed by the upper half of the ellipse $(x^2/a^2) + (y^2/b^2) = 1$ and the x-axis from $x = -a$ to $x = a$.

3. Find the value of $\int_C \frac{-y}{x^2 + y^2} \, dx + \frac{x}{x^2 + y^2} \, dy$ in each of the following cases:

(a) If C is the counterclockwise arc of $x^2 + y^2 = 2$ from $(1, 1)$ to $(-\sqrt{2}, 0)$.
(b) If C is the line $x = 1$ from $(1, 0)$ to $(1, \sqrt{3})$.
(c) If C is the line $x + y = 1$ from $(0, 1)$ to $(1, 0)$.

4. Calculate $\int_C y \, dx - x \, dy + dz$, where C is the arc of the helix $x = a \sin t$, $y = a \cos t$, $z = t$, from $t = 0$ to $t = \pi/2$.

5. Calculate $\int_C \sqrt{y} \, dx + 2x \, dy + 3y \, dz$, where C is the arc of $x = t$, $y = t^2$, $z = t^3$ from $t = 1$ to $t = 2$.

6. Calculate $\int_C x \left(\frac{1 - y^2}{y^2 + z^2} \right)^{1/2} dx$ from $(0, 0, 1)$ to $(\sqrt{2}/2, \sqrt{2}/2, 0)$ along the

first-octant part of the curve of intersection of the plane $x = y$ and the cylinder $2y^2 + z^2 = 1$.

7. Calculate $\int_C (z/y)dx + (x^2 + y^2 + z^2)dz$ from (0, 1, 4) to (1, 0, 6) along the first-octant part of the intersection of $x^2 + y^2 = 1$ and $z = 2x + 4$.

8. Calculate $\int_C y\,dx - y(x - 1)dy + y^2z\,dz$ along the first-octant part of the curve $x^2 + y^2 + z^2 = 4$, $(x - 1)^2 + y^2 = 1$ from (2, 0, 0) to (0, 0, 2).

9. Calculate each of the integrals
(a) $\int_C z^2\,dx$, (b) $\int_C x^2\,dy$, (c) $\int_C y^2\,dz$ from (2, 0, 0) to $(0, 4/\sqrt{3}, 2/\sqrt{3})$ along the first-octant part of the ellipse defined by $x^2 + y^2 - z^2 = 4$, $2z = y$.

10. Consider the integral $\int_C (z + y)dx + (x + z)dy + (y + x)dz$, where C is

(a) The broken line joining (0, 0, 0), (1, 0, 0), (1, 1, 0), and (1, 1, 1) in that order.
(b) The straight line from (0, 0, 0) to (1, 1, 1).
(c) The broken line joining (0, 0, 0), (0, 0, 1), and (1, 1, 1) in that order.

Calculate the line integral in each case and show that the values are all equal.

11. Prove that the line integral in Exercise 10 has the same value for all curves C with initial point at (0, 0, 0) and terminal point at (1, 1, 1). HINT: If the curve is expressed in terms of a parameter t, consider $F'(t)$, where $F(t) = xy + yz + zx$ when x, y, z are expressed in terms of t.

12. Let C be the clockwise closed curve bounded by the lines $x = a$, $x = b$, the x-axis, and a curve $y = f(x)$, $a \leq x \leq b$, assuming that $a < b$ and that $f(x)$ is continuous and never negative. Using results from elementary calculus, show (a) that $\int_C y\,dx$ is the area enclosed by C; (b) that $\int_C \pi y^2\,dx$ is the volume generated when this area is revolved around the x-axis; (c) that $\int_C xy\,dx$ is the first moment of this area with respect to the y-axis; (d) that $\int_C \frac{1}{2} y^2\,dx$ is the first moment of the area with respect to the x-axis; (e) that $\int_C x^2y\,dx$ is the second moment of the area with respect to the y-axis. It can be shown later, after we have learned more about line integrals, that these same interpretations may be made for the foregoing line integrals if C is any sectionally smooth, simple closed curve in the xy-plane (except that in (b) we must require that the curve lie entirely on one side of the x-axis).

13. Using Fig. 142, explain why it appears correct to say that, if C is a simple closed curve oriented counterclockwise, $\int_C x\,dy$ is equal to the area enclosed by C.

14. Using Fig. 142 as a guide, set up a line integral with respect to y, giving the volume of the solid generated when the area enclosed by C is revolved around the x-axis. Assume, as in the figure, that the curve lies entirely above the x-axis.

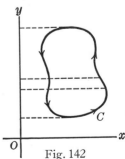

Fig. 142

15.13   Vector functions and line integrals.   Work

Consider a line integral of the form

$$(15.13-1) \qquad \int_C P \, dx + Q \, dy + R \, dz,$$

where P, Q, R are continuous functions defined along a certain oriented curve C. Such integrals often occur in connection with vector point functions, and we shall now indicate how the integral (15.13–1) can be expressed in a different notation by the use of vectors.

Let $\qquad\qquad \mathbf{F}(x, y, z) = P\mathbf{i} + Q\mathbf{j} + R\mathbf{k}$

be the vector function defined at each point of C in such a way that P, Q, R are its components in the xyz-coordinate system. Let s denote arc length along C, with $s = 0$ at the initial point of C and $s = l$ at the terminal point. We assume that C is smooth. Then the unit vector tangent to C in the positive direction at a given point is

$$\mathbf{T} = \frac{dx}{ds}\mathbf{i} + \frac{dy}{ds}\mathbf{j} + \frac{dz}{ds}\mathbf{k}.$$

Therefore, by the formula for dot products,

$$\mathbf{F} \cdot \mathbf{T} = P\frac{dx}{ds} + Q\frac{dy}{ds} + R\frac{dz}{ds}.$$

But, if we use s as the parameter along C, we know by (15.12–3) that

$$\int_C P \, dx = \int_0^l P\left(\frac{dx}{ds}\right) ds,$$

with similar formulas involving Q and R.

Therefore, we see that

$$(15.13-2) \qquad \int_C P \, dx + Q \, dy + R \, dz = \int_0^l (\mathbf{F} \cdot \mathbf{T}) ds.$$

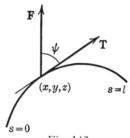

Fig. 143

Let ψ be the angle between the vectors **F** and **T**, with the understanding that $0 \leq \psi \leq \pi$; see Fig. 143. Since **T** has unit length and the length of **F** is

$$(P^2 + Q^2 + R^2)^{1/2},$$

we know that

$$\mathbf{F} \cdot \mathbf{T} = (P^2 + Q^2 + R^2)^{1/2} \cos \psi.$$

Therefore

$$(15.13\text{–}3) \quad \int_C P \, dx + Q \, dy + R \, dz = \int_0^l (P^2 + Q^2 + R^2)^{1/2} \cos \psi \, ds.$$

From this formula we can get a useful inequality concerning the value of the line integral. Let M be the maximum length of the vector **F**. Then

$$\left| (P^2 + Q^2 + R^2)^{1/2} \cos \psi \right| \leq M$$

at all points of C, and so

$$(15.13\text{–}4) \quad \left| \int_C P \, dx + Q \, dy + R \, dz \right| \leq Ml.$$

All these considerations apply to integrals of the form

$$\int_C P \, dx + Q \, dy$$

where P and Q are continuous functions defined along a curve C in the xy-plane. Here we can think of P and Q as components of a vector **F** lying in the xy-plane.

The line integral notation is often used for integrals with respect to s, if s is arc length in the positive direction along C. If g is any function of s, a commonly used notation is to write

$$\int_C g \, ds \quad \text{instead of} \quad \int_0^l g \, ds.$$

Example 1. Let C be the semicircle $(x - a)^2 + y^2 = a^2$, $y \geq 0$, from $(2a, 0)$ to $(0, 0)$. Let **F** be a vector of constant length $(|\mathbf{F}| = c)$ directed from (x, y) toward $(0, 0)$. Calculate

$$\int_C \mathbf{F} \cdot \mathbf{T} \, ds.$$

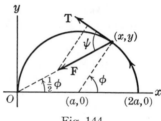

Fig. 144

In this case $\psi = (\pi/2) - (\phi/2)$ (see Fig. 144) and $\mathbf{F} \cdot \mathbf{T} = c \cos \psi = c \sin (\phi/2)$. Also, $s = a\phi$, so

$$\int_C \mathbf{F} \cdot \mathbf{T} \, ds = \int_0^\pi c \sin \frac{\phi}{2} \, a \, d\phi = 2 \, ac.$$

One of the important physical applications of line integrals is to the concept of work in analytic mechanics. The most elementary definition of work is that which is given when a constant force acts on a particle while the particle moves in a straight path along the line of action of the force. In this case the work is defined as the product of the magnitude of the force and the distance traversed. In elementary calculus this definition is generalized to cover the situation of a force of variable magnitude, the motion still being in a straight line. The work is then given by a definite integral.

A general definition of work can be made in terms of a line integral. Suppose a particle moves along a curve C, and that while so moving it is acted on by a force $\mathbf{F}$ which may vary both in magnitude and direction. The work done by this force is *defined* to be

$$(15.13\text{–}5) \qquad\qquad W = \int_C \mathbf{F} \cdot \mathbf{T} \, ds.$$

It is seen from Fig. 143 that this definition is in accord with physical intuition. Among other things we see that the component of $\mathbf{F}$ perpendicular to C contributes nothing to the work, which is all done by the tangential component. Also, (15.13–5) agrees with the more elementary definitions already referred to in the appropriate special cases.

If the x, y, and z components of $\mathbf{F}$ are F_1, F_2, F_3, respectively, the work can be expressed as the line integral

$$W = \int_C F_1 \, dx + F_2 \, dy + F_3 \, dz.$$

If time t is introduced as the parameter, we know that the vector velocity of the moving particle is

$$\mathbf{v} = \frac{ds}{dt} \, \mathbf{T}.$$

Thus $\mathbf{T} \, ds = \mathbf{v} \, dt, \, \mathbf{F} \cdot \mathbf{T} \, ds = \mathbf{F} \cdot \mathbf{v} \, dt,$

and the formula for work becomes

(15.13–6) $$W = \int_{t_0}^{t_1} \mathbf{F} \cdot \mathbf{v} \, dt,$$

where t_0 and t_1 are the initial and final values of t. This formula is convenient if the path of the particle is defined by giving x, y, z as functions of t. It should be noted, however, that the work does not depend on the particular law of motion, but only on the force and the path which is followed by the particle.

Example 2. A particle moves in the xy-plane according to the law

$$x = 64 \sqrt{3} \, t, \qquad y = 64 \, t - 16 \, t^2,$$

and is acted on by a force $\mathbf{F}$ which is directly proportional to the velocity in magnitude, but opposite in sense to the velocity. Find the work done by $\mathbf{F}$ from $t = 0$ to $t = 4$.

The components of velocity are

$$\frac{dx}{dt} = 64 \sqrt{3}, \qquad \frac{dy}{dt} = 64 - 32 \, t.$$

Hence the components of $\mathbf{F}$ are

$$F_1 = - \, 64 \sqrt{3} \, c, \qquad F_2 = - (64 - 32 \, t)c,$$

where c is a positive constant of proportionality. Hence

$$\mathbf{F} \cdot \mathbf{v} = - \, c \left\{ (64 \sqrt{3})^2 + (64 - 32 \, t)^2 \right\},$$

and the work is

$$W = \int_0^4 \mathbf{F} \cdot \mathbf{v} \, dt = - \, 1024 \, c \int_0^4 (16 - 4 \, t + t^2) dt,$$

$$W = - \, \frac{c}{3} \, (163{,}840).$$

EXERCISES

1. Find the value of the line integral in Example 1 if $\mathbf{F}$, instead of having constant length, always reaches just to the origin.

2. A point moves from $(0, 0)$ to $(2 \, a, 0)$ along the semicircle in Fig. 144. It is acted on by a force of constant magnitude 2, whose direction makes constant angles of $45°$ with both the positive co-ordinate axes. Find the work done by $\mathbf{F}$.

3. A particle of weight w descends from $(0, 2)$ to $(4, 0)$ along the parabola $8 \, y = (x - 4)^2$. It is acted on by gravity and also by a horizontal force of magnitude equal to the y-coordinate of the point, acting in the positive x-direction. Find the total work done by these two forces.

4. The cycloid $x = a(\theta - \sin \theta)$, $y = a(1 - \cos \theta)$ is generated by a point fixed in the circumference of a rolling circle. Let the point be acted on by a force of unit magnitude directed toward the center of the rolling circle. (a) Find the work done by the force as the particle moves from $\theta = 0$ to $\theta = \pi$. (b) How much of the work in (a) is done by the vertical component of the force?

   5. A weight is dragged along the x-axis from $x = 0$ to $x = 7$ (units in feet) by a string which passes over a pulley located at $(16, 12)$ in the xy-plane. If the tension in the string is constantly 10 pounds, find the work done by the pulling force.

15.2   Partial derivatives at the boundary of a region

   In this section we take up a few matters concerning the meaning and behavior of partial derivatives at the boundary of a region. The discussion is relevant to an exact understanding of some later parts of this chapter. The student may, if he wishes, go directly on to § 15.3, and read this section later.

   Let us recall the definition of a partial derivative. For the partial derivative of $f(x, y)$ with respect to x at (a, b) we define

$$f_1(a, b) = \lim_{x \to a} \frac{f(x, b) - f(a, b)}{x - a} \, .$$

Ordinarily we require that the limit be the same when $x \to a$ from the right as when $x \to a$ from the left. This presupposes that $f(x, y)$ is defined along the line $y = b$ for some distance on either side of $x = a$. If, however, $f(x, y)$ is not defined when $y = b$ and $x < a$, we require only that the limit exist as $x \to a$ from the right. The restriction to this kind of one-sided limit is typically necessary in considering the partial derivatives of f at a point on the boundary of the region in which the function is defined.

Example 1.   Let R be the region defined by $x^2 + y^2 \le 1$, and let $f(x, y) = (1 - x^2 - y^2)^{3/2}$. This function is not defined if $x^2 + y^2 > 1$. If (a, b) is a point on the boundary of R in the second quadrant, $f_1(a, b)$ must be understood as a limit in which $x \to a$ from the right, and $f_2(a, b)$ must be understood as a limit in which $y \to b$ from below. At most other boundary points the situation is similar. At $(0, 1)$, however, there is an even more unusual situation. Along the line $y = 1$, there is only the single point $x = 0$ which belongs to R. Therefore $f_1(0, 1)$ is not defined, because the quotient

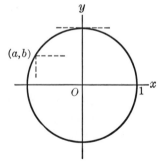

$$\frac{f(x, 1) - f(0, 1)}{x - 0}$$

Fig. 145

is not defined; *we cannot let $x \to 0$ either from the right or the left!* Similar considerations show that $f_1(0, -1)$ is undefined, and that $f_2(\pm 1, 0)$ are undefined.

Now let us look at the formula for $\dfrac{\partial f}{\partial x}$ when it is defined. The usual rules give

$$\frac{\partial f}{\partial x} = -3 x(1 - x^2 - y^2)^{1/2}.$$

Let us define $g(x, y) = -3 x(1 - x^2 - y^2)^{1/2}$

at each point of R. Then $f_1(x, y) = g(x, y)$ at all points where $f_1(x, y)$ is defined. But we observe that $g(x, y)$ is even defined at the points $(0, \pm 1)$, where $f_1(x, y)$ is *not* defined. Moreover, g is continuous in R, and $g(0, \pm 1) = 0$. Therefore, even though $f_1(0, 1)$ is not defined, it *is* true that

$$\lim_{(x, y) \to (0, 1)} f_1(x, y) = 0.$$

In situations like this example, it is customary to consider $f_1(0, 1)$ *as being defined* by the limiting value which $f_1(x, y)$ approaches as $(x, y) \to (0, 1)$. This is a conventional agreement which proves to be useful in practice. In the present case, this convention permits us to say that $\partial f/\partial x$ is continuous throughout the entire closed region R.

A general statement of this convention may be made as follows: If R is a closed region, and f is defined in R, we agree to say that $\dfrac{\partial f}{\partial x}$ is continuous in R if there is some function g which is defined and continuous in R and such that $\dfrac{\partial f}{\partial x} = g$ at all interior points of R. Similar conventions are made pertaining to $\dfrac{\partial f}{\partial y}$ and to partial derivatives of higher order. Also, similar conventions are made for functions $f(x, y, z)$ defined in a closed region of three-dimensional space.

The foregoing convention is useful when it comes to considering *normal derivatives* at the boundary of a region. Such considerations are quite important in the physical applications of line and surface integrals (in potential theory, for example).

Suppose a certain part of the boundary of R consists of a smooth arc. Let P_0 be a point of this arc, and let **n** be a vector of unit length perpendicular to the boundary at P_0, and pointing outward from R. We call **n** the *outer normal* at P_0 (see Fig. 146). Let α be the angle counterclockwise from the positive x-direction to the direction of **n**. If f is a function defined in R, with first partial derivatives which are

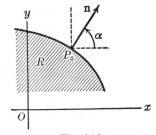

Fig. 146

continuous in R, the outer normal derivative of f at P_0 is defined to be

$$(15.2\text{–}1) \qquad \frac{\partial f}{\partial n} = \frac{\partial f}{\partial x} \cos \alpha + \frac{\partial f}{\partial y} \sin \alpha,$$

where $\dfrac{\partial f}{\partial x}$ and $\dfrac{\partial f}{\partial y}$ are evaluated at P_0.

Observe that $\dfrac{\partial f}{\partial n}$ is not, strictly speaking, the rate of change of f at P_0 in the direction of $\mathbf{n}$, because f is not defined outside of R, and therefore such a rate of change is meaningless. It *is* true, however, that $-\dfrac{\partial f}{\partial n}$ is the rate of change in the *inward* direction at P_0 (i.e., in the direction opposite to that of $\mathbf{n}$).

15.3 Green's theorem in the plane

The subject of the present section is a very important theorem relating to line integrals around closed curves in the plane. More precisely, the theorem exhibits an exact relation between a line integral taken around the curve (or curves) forming the boundary of a region and a certain double integral taken over the region. With the aid of the theorem we can transform certain double integrals into line integrals, and vice versa. Such transformations are of the highest usefulness in many mathematical arguments, as we shall subsequently see in some instances. Many other instances abound in the literature dealing with the partial differential equations of applied mathematics. A widespread usage sanctions the attachment to the theorem of the name of G. Green, an English mathematician of the early 19th century. There is also justification for calling it Gauss's theorem, after a great German mathematician of the same period.

THEOREM I. *Let R be a closed and bounded region of the xy-plane. Let the boundary of R consist of a finite number of simple closed curves which do not intersect each other, and each of which is sectionally smooth. Let $P(x, y)$ and $Q(x, y)$ be functions which are continuous and have continuous first partial derivatives in R. Let C denote the aggregate of curves forming the boundary of R, each oriented in such a way that the region is on the left as one advances along the curve in the positive direction. Then*

$$(15.3\text{–}1) \qquad \int_C P\,dx + Q\,dy = \iint_R \left(\frac{\partial Q}{\partial x} - \frac{\partial P}{\partial y} \right) dx\,dy.$$

DEFINITION. A region having the properties specified in Theorem I will be called a *regular* region.

There are certain difficulties in proving Green's theorem in the full generality of its statement. However, for regions of sufficiently simple shape the proof is quite easy. We shall begin by giving the proof for such easy cases, and then extending it somewhat. Then we shall proceed to illustrate the theorem in some special cases, and give some applications. We do not give a fully detailed proof of the theorem, but we give indications of such a proof. For further comments on the proof see § 15.31.

Before going further, we observe that the functions P and Q are independent of one another, and hence formula (15.3–1) may be broken down into two separate formulas, namely

$$(15.3\text{–}2) \qquad \int_C P \, dx = -\iint_R \frac{\partial P}{\partial y} \, dx \, dy,$$

$$(15.3\text{–}3) \qquad \int_C Q \, dy = \iint_R \frac{\partial Q}{\partial x} \, dx \, dy.$$

The proof of (15.3–1) is equivalent to proofs of (15.3–2) and (15.3–3) separately.

It should be mentioned that we assume the x and y axes have their usual relation to each other, i.e., that a counterclockwise rotation of 90° is needed to carry the positive x-axis into the position occupied by the positive y-axis. This arrangement is responsible for the minus sign in (15.3–2) and the lack of it in (15.3–3).

Let us assume that R has the simple form suggested by Fig. 147. That is, suppose that there is an interval $c \leqq x \leqq d$ such that for an x' outside the interval the line $x = x'$ does not intersect R, while for $c < x' < d$ the line $x = x'$ intersects R in an interval. The lines $x = c$, $x = d$ may intersect R either in an interval or a single point. The boundary of R then consists of a lower curve $y = Y_1(x)$, an upper curve $y = Y_2(x)$, and certain portions (either a segment or a point) of each of the lines $x = c$, d. The positive orientation of the boundary is shown in Fig. 147. Let C_1 and C_2 denote the oriented lower and upper curves, respectively.

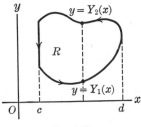

Fig. 147

DEFINITION. If a regular region R has the simple form just described, let us call it an *x-simple* region. Likewise we may define a *y-simple* region.

Now consider formula (15.3–2) for an x-simple region R. There are no contributions to the line integral from the portions of C on the lines $x = c$, $x = d$, since $dx = 0$ along a segment of either line. Hence

$$\int_C P \, dx = \int_{C_1} P \, dx + \int_{C_2} P \, dx.$$

But on C_1 we may take x as the parameter, and put $y = Y_1(x)$. Hence

$$\int_{C_1} P(x, y)dx = \int_c^d P(x, Y_1(x))dx.$$

Likewise,

$$\int_{C_2} P(x, y)dx = \int_d^c P(x, Y_2(x))dx = -\int_c^d P(x, Y_2(x))dx.$$

(Bear in mind the co-ordinates of the initial and terminal points in determining the limits of integration.) Thus

(15.3–4) $$\int_C P\, dx = -\int_c^d \{P(x, Y_2) - P(x, Y_1)\}dx.$$

Next consider the double integral, and use the iterated integral formula (13.3–6):

(15.3–5) $$\iint_R \frac{\partial P}{\partial y}\, dx\, dy = \int_c^d dx \int_{Y_1}^{Y_2} \frac{\partial P}{\partial y}\, dy.$$

The y integration may now be performed with x held constant. The result, by Theorem VIII, § 1.53, is

(15.3–6) $$\int_{Y_1}^{Y_2} \frac{\partial P}{\partial y}\, dy = P(x, Y_2) - P(x, Y_1).$$

On combining (15.3–6) with (15.3–5) and comparing with (15.3–4), we see the truth of (15.3–2) for an x-simple region R.

An entirely similar proof may be given for formula (15.3–3) if we assume that R is y-simple. The figure for this case would resemble Fig. 91 (page 385). Finally, if R is both x-simple and y-simple, we combine (15.3–2) and (15.3–3) to give (15.3–1). Green's theorem is thus easily proved for regions which are both x-simple and y-simple. In particular, a bounded region R is both x-simple and y-simple if its boundary consists of a single sectionally smooth convex curve. A rectangle is such a region.

There are x-simple regions which are not y-simple, and regions which are neither x-simple nor y-simple. On the other hand, many regions may be divided into a finite number of subregions, each of which is both x-simple and y-simple. For such a region it is easy to prove Green's theorem. For instance, suppose R is the region bounded between the circle and the large triangle in Fig. 148, with axes

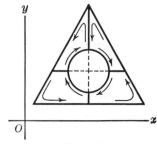

Fig. 148

as shown. This region is neither x-simple nor y-simple, but we can divide it into four subregions, each of which is both x-simple and y-simple. The formula of Green's theorem therefore holds for each of the subregions. If we add corresponding parts of the four formulas, the double integrals combine to give the correct double integral over the whole of R. Now consider the line integrals. In dividing R into parts, we introduced four interior connecting lines. Each of these lines occurs twice, but with opposite orientations in the two occurrences, since each line belongs to the boundary of two neighboring subregions. Hence, when all the line integrals are added, the contributions from these interior lines cancel out in pairs, leaving only the line integral around the total oriented boundary of R, that is, counterclockwise around the triangle and clockwise around the circle. In this way we obtain the proof of Green's theorem for this region. The idea of the proof may clearly be extended to any region which may be decomposed into a finite number of subregions which are both x-simple and y-simple. There are regions of the sort mentioned in Theorem I which cannot be thus decomposed; Green's theorem for such regions is discussed in § 15.31.

Example 1. Show that the area of a regular region R is given by any one of the formulas

$$(15.3\text{–}7) \quad A = -\int_C y \, dx, \quad A = \int_C x \, dy, \quad A = \tfrac{1}{2}\int_C - y \, dx + x \, dy,$$

where C is the positively oriented boundary of R.

We verify the third formula only, leaving the others to the student. Putting $P = -y$, $Q = x$ in Green's theorem, we get

$$\tfrac{1}{2}\int_C - y \, dx + x \, dy = \tfrac{1}{2}\iint_R (1 + 1) dx \, dy = A.$$

Example 2. Calculate $\int_C (x^2 - y^2) dx + 2 \, xy \, dy$, where C is the counterclockwise boundary of the square formed by the lines $x = 0$, $x = 2$, $y = 0$, $y = 2$.

Here we let R be the square region, and put $P = x^2 - y^2$, $Q = 2 \, xy$ in Green's theorem. The line integral is equal to

$$\iint_R \{2 \, y - (- 2 \, y)\} dx \, dy = 4 \iint_R y \, dx \, dy.$$

We could easily calculate the value of this double integral, but it is also possible to recognize its value by the following argument: If A is the area of R, and if $(\bar{x}, \bar{y})$ is the centroid of R, then, as we know,

$$\iint_R y \, dx \, dy = A\bar{y}.$$

In this case $A = 4$ and $\bar{y} = 1$. Hence

$$\int_C (x^2 - y^2)dx + 2 \, xy \, dy = 16.$$

Example 3.   Green's theorem can be used to prove the following theorem of Pappus: *If R is a regular region lying entirely on one side of the x-axis, and if R is revolved about the x-axis, the volume of the solid so generated is equal to $2\pi A\bar{y}$, where A and $\bar{y}$ have the same significance as in Example 2.*

For brevity, let us consider merely the case of an x-simple region, as pictured in Fig. 147. The reasoning can be extended to the most general regular region. The volume in question is, by elementary calculus,

$$V = \pi \int_c^d \{Y_2(x)\}^2 \, dx - \pi \int_c^d \{Y_1(x)\}^2 \, dx,$$

in the notation of Fig. 147. This is readily seen to be the same as the line integral

$$V = -\pi \int_C y^2 \, dx,$$

with C oriented counterclockwise. We now apply Green's theorem, with $P = -\pi y^2$, $Q = 0$. Thus

$$V = \iint_R 2\pi y \, dx \, dy = 2\pi A\bar{y}.$$

This is what we wanted to prove.

Example 4.   Use Green's theorem to deduce the integral formula

$$(15.3\text{–}8) \qquad \iint_R \left(\frac{\partial^2 u}{\partial x^2} + \frac{\partial^2 u}{\partial y^2} \right) dx \, dy = \int_C \frac{\partial u}{\partial n} \, ds,$$

where s refers to arc length along C and n refers to the outer normal to C.

Here it is assumed that R is a regular region and that u and its first and second partial derivatives are continuous in R.

To derive (15.3–8) we start by applying Green's theorem with

$$Q = \frac{\partial u}{\partial x}, \qquad P = -\frac{\partial u}{\partial y}.$$

Then we obtain

$$\iint_R \left(\frac{\partial^2 u}{\partial x^2} + \frac{\partial^2 u}{\partial y^2} \right) dx\, dy = \int_C - \frac{\partial u}{\partial y}\, dx + \frac{\partial u}{\partial x}\, dy.$$

To complete the derivation we have only to show that on C

$$(15.3\text{–}9) \qquad -\frac{\partial u}{\partial y}\frac{dx}{ds} + \frac{\partial u}{\partial x}\frac{dy}{ds} = \frac{\partial u}{\partial n}.$$

We know by (15.2–1) that

$$\frac{\partial u}{\partial n} = \frac{\partial u}{\partial x} \cos \alpha + \frac{\partial u}{\partial y} \sin \alpha.$$

We also know that

$$1(5.3\text{–}10) \qquad \frac{dx}{ds} = \cos \phi, \qquad \frac{dy}{ds} = \sin \phi,$$

where ϕ is the angle which the positive tangent to C makes with the posi-

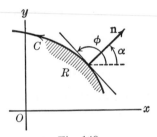

Fig. 149

tive x-axis. It is easy to show by sketching a figure that as one traverses a curve in such a way that the interior is always on the left, the outward normal lags behind the tangent by $\pi/2$ radians. It follows that $\alpha = \phi - (\pi/2)$ (see Fig. 149) and therefore

$$(15.3\text{–}11) \qquad \cos \alpha = \frac{dy}{ds}, \qquad \sin \alpha = -\frac{dx}{ds}.$$

The correctness of (15.3–9) is now apparent, and the proof of (15.3–8) is complete.

EXERCISES

1. Use Green's theorem to evaluate the following line integrals:
    (a) $\int_C 2\, xy\, dx - 3\, xy\, dy$, clockwise around the square bounded by $x = 3$, $x = 5$, $y = 1$, $y = 3$.

(b) $\int_C xy^2\,dx + 2\,x^2y\,dy$, counterclockwise around the ellipse $4\,x^2 + 9\,y^2 = 36$.

(c) $\int_C (x^2 + 2\,y^2)dy$, counterclockwise around the circle $(x - 2)^2 + y^2 = 1$.

(d) $\int_C e^x \sin y\,dx + e^x \cos y\,dy$, around the boundary of any regular region.

(e) $\int_C x^2y\,dx - y^2x\,dy$, counterclockwise around the region bounded by $y = \sqrt{a^2 - x^2}$ and $y = 0$. Use polar co-ordinates to evaluate the double integral.

(f) $\displaystyle\int_C \frac{-y\,dx + x\,dy}{x^2 + y^2}$, around the boundary of any regular region not containing the origin.

2. Calculate the line integrals of Exercise 2, § 15.12, parts **(a)**, **(d)**, **(f)**, **(g)**, and **(h)**, using Green's theorem.

3. Let C be any sectionally smooth simple closed curve in the xy-plane, oriented counterclockwise. Let R be the region bounded by C, and let R have area A and centroid $(\bar{x}, \bar{y})$. Show that

$$\frac{1}{2}\int_C x^2\,dy = A\bar{x}, \qquad \int_C xy\,dy = A\bar{y},$$

and, if R is a lamina of constant unit density, interpret

$$\int_C -x^2y\,dx \qquad \text{and} \qquad \int_C - x^2y\,dx + xy^2\,dy$$

as moments of inertia, specifying the axis of rotation in each case.

4. Let R be the region bounded by the rays $\theta = \alpha$, $\theta = \beta$ and the curve $r = f(\theta)$, as shown in Fig. 150. Use the third formula in (15.3–7) to show that the area of R is

$$A = \frac{1}{2}\int_\alpha^\beta \{f(\theta)\}^2\,d\theta,$$

by actually calculating the line integral, using r as parameter on the rays, and θ as parameter on the curved side of R. This gives a new derivation of a formula which is familiar from elementary calculus.

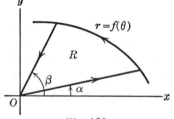

Fig. 150

5. Derive the formulas

$$\iint_R \left[u\,\Delta v + \frac{\partial u}{\partial x}\frac{\partial v}{\partial x} + \frac{\partial u}{\partial y}\frac{\partial v}{\partial y} \right] dx\,dy = \int_C u\,\frac{\partial v}{\partial n}\,ds,$$

$$\iint_R (u\,\Delta v - v\,\Delta u)dx\,dy = \int_C \left(u\,\frac{\partial v}{\partial n} - v\,\frac{\partial u}{\partial n} \right) ds,$$

where the notation

$$\Delta v = \frac{\partial^2 v}{\partial x^2} + \frac{\partial^2 v}{\partial y^2}$$

is employed. Model your work somewhat after that of Example 4.

6. If v satisfies the condition $\Delta v = 0$ (Laplace's equation) in R, show that

$$\int_C \frac{\partial v}{\partial n}\, ds = 0$$

and

$$\iint_R \left[\left(\frac{\partial v}{\partial x} \right)^2 + \left(\frac{\partial v}{\partial y} \right)^2 \right] dx\, dy = \int_C v\, \frac{\partial v}{\partial n}\, ds.$$

7. Green's theorem can be expressed in vector notation in two different ways. Consider the plane region R of the xy-plane as being part of three-dimensional space. If $F_1(x, y)$ and $F_2(x, y)$ are functions having continuous derivatives in R, let

$$\mathbf{F} = F_1(x, y)\mathbf{i} + F_2(x, y)\mathbf{j}$$

be a vector field defined in space. It has the special property that the z-component is zero and that the x- and y-components are independent of z. Show that

$$\text{(a)} \quad \iint_R (\nabla \cdot \mathbf{F})dx\, dy = \int_C \mathbf{F} \cdot \mathbf{n}\, ds$$

and

$$\text{(b)} \quad \iint_R (\nabla \times \mathbf{F}) \cdot \mathbf{k}\, dx\, dy = \int_C \mathbf{F} \cdot \mathbf{T}\, ds,$$

where $\nabla \cdot \mathbf{F}$ and $\nabla \times \mathbf{F}$ are as defined in §§ 10.7 and 10.8, and $\mathbf{n}$ and $\mathbf{T}$ are the unit outer normal and the unit positive tangent, respectively, on C. Formulas (15.3–10) and (15.3–11) will be found useful.

15.31 Comments on the proof of Green's theorem

In the statement of Green's theorem we required only that the region R be regular, whereas in the proof given in § 15.3 we assumed that R could be decomposed into a finite number of subregions, each both x-simple and y-simple. Now, not every regular region can be so divided, and so the proof of Theorem I, § 15.3, is not complete. An example of such a region is suggested by Fig. 151; here three sides of the region are formed by straight lines, but the fourth side (the top) is a smooth curve which oscillates more and more rapidly as it approaches the origin, and crosses the x-axis an infinite number of times in the interval shown. The equation $y = x^3 \sin(1/x)$ de-

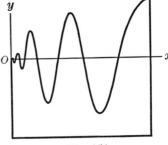

Fig. 151

fines such a curve. Such a region may be x-simple, but is not y-simple, and it cannot be divided into a finite number of y-simple subregions. It is complications of this sort at the boundary of R that make it somewhat difficult to give a complete proof of Theorem I.

One method for completing the proof may be described in outline as follows: Let R_1, R_2, $\cdots$, R_n, $\cdots$ be a sequence of regions, with oriented boundaries C_1, C_2, $\cdots$, having the following properties:

1. Each R_n is a regular region lying in R.
2. Each R_n can be divided into a finite number of subregions, each of which is both x-simple and y-simple.
3. As $n \to \infty$, R_n approaches R, and C_n approaches C in such a way

that

$$\iint_{R_n} \left(\frac{\partial Q}{\partial x} - \frac{\partial P}{\partial y} \right) dx \, dy \to \iint_{R} \left(\frac{\partial Q}{\partial x} - \frac{\partial P}{\partial y} \right) dx \, dy$$

and

$$\int_{C_n} P \, dx + Q \, dy \to \int_{C} P \, dx + Q \, dy.$$

Green's theorem holds for each R_n, by (1) and (2) and what has already been proved in § 15.3. It therefore holds for R, by (3). The crux of the problem now is the construction of the approximating regions R_n so that conditions (1), (2), (3) are satisfied. The details of proving such a construction possible are too intricate for consideration in this book.

There are also other approaches to the problem of a complete proof. One important fact is that any regular region may be divided into a finite number of subregions each of which is *either* x-simple or y-simple.

We conclude this section with the remark that, when a more advanced point of view is taken, and the integral concepts are suitably generalized, Green's theorem, that is, formula (15.3–1), remains true under hypotheses much less restrictive than those stated in Theorem I of § 15.3. The relaxation of restrictions applies not only to the character of the region R, but to the continuity requirements on the functions P, Q and their derivatives.

15.32   Transformation of double integrals

In this section we consider the effect on the double integral

(15.32–1)
$$\iint_{R} F(x, y) dx \, dy$$

when we make a change of variable by equations

(15.32–2) $x = f(u, v), \quad y = g(u, v).$

The discussion is based on the assumption that the student is familiar with Chapter IX, and with § 9.3 in particular. We assume that equations (15.32–2) establish a one-to-one mapping of a part of the uv-plane onto a part of the xy-plane. Let the inverse mapping be defined by the equations

(15.32–3) $$u = h(x, y), \quad v = k(x, y).$$

We suppose the functions f, g to be continuous, together with their first and second partial derivatives; we further assume that the Jacobian

$$J(u, v) = \frac{\partial(f, g)}{\partial(u, v)}$$

is always of the same sign—either always positive or always negative, and hence never zero. Let R be a regular region all of whose points are interior to the region of the xy-plane which is being mapped, and let R' be the corresponding region in the uv-plane. Then R' is also a regular region, as one may show with the aid of equations (15.32–3).

THEOREM II. *Let A be the area of R. Then, under the foregoing assumptions,*

(15.32–4) $$A = \iint\limits_{R'} |J(u, v)| \, du \, dv.$$

PROOF. We start off with the fact (see Example 1 of § 15.3) that

(15.32–5) $$A = \int_C x \, dy,$$

where C is the complete boundary of R, oriented in the usual positive sense, i.e., so that the region R is on the left as one advances along the curve in the positive direction. Let C' denote the boundary of R'. We orient C' by taking the positive sense along C' to be that which corresponds, under the mapping, to the positive sense along C; thus, as (x, y) moves along C in the positive sense, its image point (u, v) moves along C' in the positive sense. With this agreement we have

(15.32–6) $$\int_C x \, dy = \int_{C'} f(u, v) \left[\frac{\partial g}{\partial u} \, du + \frac{\partial g}{\partial v} \, dv \right],$$

since $$x = f(u, v) \quad \text{and} \quad dy = \frac{\partial g}{\partial u} \, du + \frac{\partial g}{\partial v} \, dv$$

hold for corresponding points of C and C'.

Next we apply Green's theorem in the uv-plane to the line integral in (15.32–6). Instead of $P \, dx + Q \, dy$ we have

$$f \frac{\partial g}{\partial u} \, du + f \frac{\partial g}{\partial v} \, dv.$$

Therefore, corresponding to

$$\frac{\partial Q}{\partial x} - \frac{\partial P}{\partial y} \quad \text{we have} \quad \frac{\partial}{\partial u}\left(f\frac{\partial g}{\partial v}\right) - \frac{\partial}{\partial v}\left(f\frac{\partial g}{\partial u}\right).$$

On carrying out the indicated differentiations, this latter expression is found to be precisely $J(u, v)$. Consequently,

$$(15.32\text{-}7) \qquad \int_{C'} f\frac{\partial g}{\partial u}\,du + f\frac{\partial g}{\partial v}\,dv = \pm \iint_{R'} J(u, v)du\,dv.$$

The choice of sign on the right is determined by the orientation of C'. If the orientation which we have given to C' coincides with the usual positive orientation of the boundary of R', the plus sign is correct; in the contrary case we must choose the minus sign. Combining (15.32–5), (15.32–6), and (15.32–7), we see that

$$A = \iint_{R'} \pm\, J(u, v)du\,dv.$$

Since A is positive and J is always of the same sign, it follows that the sign chosen in (15.32–7) must be the same as the sign of J. Whichever the sign, formula (15.32–4) is correct.

The last remarks enable us to justify the answer given to the question posed at the end of § 9.4. Suppose R is a circular region. The positive orientation of its circumference C is counterclockwise. The image of C will be a simple closed curve C', and R' will consist of the interior of C' and C' itself. Hence, the usual positive orientation of C' will also be counterclockwise. But the mapping of R onto R' induces a certain orientation of C'. *From the discussion in the foregoing paragraph we see that the induced orientation of C' is counterclockwise if and only if the Jacobian of the mapping is positive.*

If the regions R, R' are connected (if one is connected, so is the other) we can apply the law of the mean to (15.32–4), and obtain the following result: Let A' be the area of R'. *There is some point (u, v) in R' such that*

$$(15.32\text{-}8) \qquad\qquad A = |\,J(u, v)\,|\,A'.$$

The magnitude of the Jacobian is therefore a measure of the distortion of areas by the mapping. If $|\,J\,| = 1$ we say that the mapping is *equiareal*.

Let us now turn to the double integral (15.32–1).

THEOREM III. *If F is continuous in R and if the mapping meets the conditions stated prior to Theorem II, we have*

$$(15.32\text{-}9) \qquad \iint_{R} F(x, y)dx\,dy = \iint_{R'} F[f(u, v), g(u, v)]\,|\,J(u, v)\,|\,du\,dv.$$

PROOF. Let R be divided into a finite number of connected regular subregions $R_1, \cdots, R_n$ of areas $\Delta A_1, \cdots, \Delta A_n$. Let the corresponding regions in the uv-plane be $R_1', \cdots, R_n'$, of areas $\Delta A_1', \cdots, \Delta A_n'$. Choose a point (u_k, v_k) in R_k' such that (by (15.32–8))

$$\Delta A_k = |J(u_k, v_k)| \Delta A_k',$$

and let (x_k, y_k) be the corresponding point of R_k. For convenience write

$$G(u, v) = F[f(u, v), g(u, v)].$$

Then $\qquad \displaystyle\sum_{k=1}^{n} F(x_k, y_k)\Delta A_k = \sum_{k=1}^{n} G(u_k, v_k) |J(u_k, v_k)| \Delta A_k'.$

We now pass to the limit as $n \to \infty$ and the maximum dimensions of the subregions approach zero. By Theorem II, § 13.23, we obtain formula (15.32–9) as the result.

If we like we may interpret the variables u, v as curvilinear co-ordinates in the xy-plane; this point of view was discussed in § 9.3. If we consider a small cell in the mesh of co-ordinate curves, we see from (15.32–8) that the area of the cell with opposite vertices (u, v), $(u + \Delta u, v + \Delta v)$ is approximately $|J(u, v)| \Delta u \Delta v$. By expressing the uv-integral in (15.32–9) as an iterated integral, we obtain a formula for the evaluation of the double integral (15.32–1) by an iterated integral in curvilinear co-ordinates. For polar co-ordinates such a result is already known to us. If

$$x = r \cos \theta, \qquad y = r \sin \theta$$

we readily compute

$$\frac{\partial(x, y)}{\partial(r, \theta)} = r.$$

Thus, for the transformation to polar co-ordinates, the Jacobian is the familiar factor r.

EXERCISES

1. The transformation $x = u^2 - v^2$, $y = 2uv$ maps the rectangle $1 \leq u \leq 2$, $1 \leq v \leq 3$ of the uv-plane into a certain region R of the xy-plane. Make a sketch of R and find its area.

2. Consider the mapping $x = au + bv$, $y = -bu + av$, $(a, b$ constant$)$. If a certain region of unit area in the xy-plane corresponds to a region R' in the uv-plane, find the area of R'.

3. The equations $u = x^2y - y^3$, $v = 2xy^2$ map points near $(2, 1)$ in the xy-plane into points near $(3, 4)$ in the uv-plane. If R and R' are corresponding small regions containing these respective points, find the approximate ratio of the area of R to that of R'.

4. Calculate $\iint_R x\,dx\,dy$ and the area of R, and so find $\bar{x}$ for R, where R is the region bounded by the lines $x = 3\,y$, $2\,x + y = 0$, $x - 3\,y = 14$, $2\,x + y = 21$. Use the transformation $x = u + 3\,v$, $y = -2\,u + v$.

5. Use the transformation $x = au$, $y = bv$ to map the region R defined by $(x^2/a^2) + (y^2/b^2) \le 1$ onto the uv-plane. Evaluate the integral

$$\iint_R \left(\frac{x^2}{a^2} + \frac{y^2}{b^2} \right) dx\,dy$$

with the aid of this transformation and polar co-ordinates in the uv-plane.

6. Calculate $\iint_R \dfrac{dx\,dy}{x + y}$, where R is the region in the xy-plane bounded by the lines $x + y = 1$, $x + y = 4$, $y = 0$, $x = 0$. Use the transformation $x = u - uv$, $y = uv$.

7. Change the iterated integral $\displaystyle\int_0^1 dx \int_0^{1-x} e^{(y-x)/(y+x)}\,dy$ to a double integral and evaluate it with the aid of the transformation $u = x + y$, $v = x - y$.

8. Use the transformation $x = \sqrt{v - u}$, $y = u + v$ to evaluate the integral $\displaystyle\int_1^{\sqrt{2}} dx \int_{x^2}^{4-x^2} \dfrac{x\,dy}{x^2 + y}$.

9. Let R be the first-quadrant region bounded by the curves $x^2 - y^2 = 1$, $x^2 - y^2 = 4$, $x^2 + y^2 = 9$, $x^2 + y^2 = 16$. Calculate $\iint_R xy\,dx\,dy$ with the aid of the transformation $u = x^2 - y^2$, $v = x^2 + y^2$.

10. Find the area of the first-quadrant region bounded by $xy = 4$, $xy = 16$, $y = x + 3$, $y = x - 3$. Use the transformation $2\,u = x - y$, $v = \sqrt{xy}$.

11. Consider the mapping

$$u = \frac{2\,x}{x^2 + y^2}, \qquad v = \frac{2\,y}{x^2 + y^2}.$$

Draw the circles $u = \tfrac{1}{3}$, $u = \tfrac{1}{2}$, $v = \tfrac{1}{4}$, $v = 1$ in the xy-plane, and let R be the region bounded by them. Calculate the value of $\iint_R \dfrac{dx\,dy}{(x^2 + y^2)^2}$.

12. Find the region in the $r\theta$-plane corresponding to the region R in the xy-plane inside the circle $x^2 + y^2 = 2\,x$ and to the right of $x = 1$, assuming $x = r \cos\theta$, $y = r \sin\theta$, $r > 0$, $-\dfrac{\pi}{4} \le \theta \le \dfrac{\pi}{4}$. Transform $\iint_R \dfrac{x\,dx\,dy}{x^2 + y^2}$ to an integral in the $r\theta$-plane, and evaluate it.

13. Consider the families of curves $x^3 = uy$, $y^3 = vx$ in the xy-plane. Show that the first-quadrant area bounded by the curves $u = a^2$, $u = b^2$, $v = \alpha^2$, $v = \beta^2$ is $\frac{1}{2}(a - b)(\alpha - \beta)$, if $a > b > 0$, $\alpha > \beta > 0$.

14. Use the transformation $x = u + uv$, $y = v + uv$ to calculate the integral

$$\iint_R \frac{dx\, dy}{[(x - y)^2 + 2(x + y) + 1]^{1/2}} \,,$$ where R is the triangle with vertices at $(0, 0)$, $(2, 0)$, $(2, 2)$.

15.4   Exact differentials

An expression such as

$$(15.4–1) \qquad\qquad M(x, y)dx + N(x, y)dy$$

is called a *first-order differential form in two variables*. As examples we list

$$(y \sin x - 1)dx + \cos x \, dy,$$
$$(x^2 + y^2)dx - 2\,xy\,dy.$$

The functions M and N occurring in the form are assumed to be defined in some region of the plane. In practice certain continuity and differentiability restrictions will be imposed on the functions.

The purpose of this section is to study those particular kinds of differential forms which are the differentials of functions of two variables. The following forms are of this kind:

$$x\,dy + y\,dx = d(xy),$$

$$\frac{-y}{x^2 + y^2}\,dx + \frac{x}{x^2 + y^2}\,dy = d\tan^{-1}\left(\frac{y}{x}\right),$$

$$\frac{x}{x^2 + y^2}\,dx + \frac{y}{x^2 + y^2}\,dy = d\log(x^2 + y^2)^{1/2}.$$

A differential form is called an *exact differential* if it is the differential of some function u at all points of some region in the xy-plane. Since

$$du = \frac{\partial u}{\partial x}\,dx + \frac{\partial u}{\partial y}\,dy, \qquad\qquad du = M\,dx + N\,dy$$

this means that $M\,dx + N\,dy$ is exact if there is some differentiable function u such that

$$(15.4–2) \qquad\qquad M = \frac{\partial u}{\partial x}, \qquad N = \frac{\partial u}{\partial y}$$

at each point of some region.

There are many nonexact differential forms, as we shall see later.

In order to have enough precision for the statement of our theorems, we make the following formal definition:

DEFINITION.   The differential form $M\,dx + N\,dy$ is said to be exact *at the point* (a, b) if there is some single-valued differentiable function u defined in some neighborhood of (a, b), such that $du = M\,dx + N\,dy$ at all points of the neighborhood.

Observe that, in defining exactness of a form at a point, we actually place a condition on the form at all points of some neighborhood of the point, not merely at the point itself.

The fundamental problem relating to exact differentials may be stated in two parts:

(1) How can one tell by examining M and N whether or not the differential form is exact at a particular point?

(2) If the differential form is known to be exact, how can one actually find a function of which the form is the differential?

Both of these questions can be answered with the aid of line integrals, provided we assume that M and N have continuous first partial derivatives.

THEOREM IV.   *If M and N have continuous first partial derivatives at all points of some open rectangle, the differential form is exact at each point of the rectangle if and only if the condition*

(15.4–3) $$\frac{\partial M}{\partial y} = \frac{\partial N}{\partial x}$$

is satisfied throughout the rectangle. When this condition is satisfied, a function u such that $du = M\,dx + N\,dy$ is furnished by the line integral

(15.4–4) $$u(x, y) = \int_C M(s, t)\,ds + N(s, t)\,dt$$

along the path from (a, b) to (x, y) shown in Fig. 152, where (a, b) is the center of the rectangle.

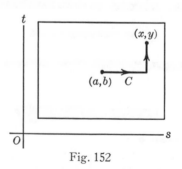

Fig. 152

REMARK. s and t are used in place of x and y as variables in the line integral, since (x, y) is a fixed point during the computation of the line integral.

PROOF OF THE THEOREM. The necessity of condition (15.4–3) is proved as follows: If the form is exact, there is some function u such that equations (15.4–2) hold. Then

$$\frac{\partial M}{\partial y} = \frac{\partial^2 u}{\partial y\, \partial x} \quad \text{and} \quad \frac{\partial N}{\partial x} = \frac{\partial^2 u}{\partial x\, \partial y}.$$

The second derivatives of u are continuous, since the first derivatives of M and N are continuous, by hypothesis. Therefore

$$\frac{\partial^2 u}{\partial y\, \partial x} = \frac{\partial^2 u}{\partial x\, \partial y},$$

and so the condition (15.4–3) is satisfied.

The sufficiency part of the proof consists in showing that if (15.4–3) holds and $u(x, y)$ is defined by (15.4–4), then $\dfrac{\partial u}{\partial x} = M$ and $\dfrac{\partial u}{\partial y} = N$. For this purpose we express the line integral in terms of ordinary integrals. Along the horizontal part of C, $t = b$ and s varies from a to x. Hence $dt = 0$, and the line integral over this part of C is equal to

$$\int_a^x M(s, b)\,ds.$$

On the vertical part of C, $s = x$ and t varies from b to y. Since s is constant during the integration, $ds = 0$, and the line integral over this part of C is equal to

$$\int_b^y N(x, t)\,dt.$$

Thus

$$(15.4\text{–}5) \qquad u(x, y) = \int_a^x M(s, b)\,ds + \int_b^y N(x, t)\,dt.$$

We may now think of x and y as variables. Doing this, we see that the first integral does not depend on y, while the second one involves y merely as a limit of integration. Therefore, by Theorem VII, § 1.52,

$$\frac{\partial u}{\partial y} = N(x, y).$$

All that remains to complete the proof is to show that $\dfrac{\partial u}{\partial x} = M$. To show this we shall prove that

(15.4–6) $u(x, y) = \int_a^x M(s, y)ds + \int_b^y N(a, t)dt.$

The required result will then follow when we differentiate with respect to x. The key to this part of the proof is the consideration of the line integral over the alterna-
tive path C_2 shown in Fig. 153. In this diagram
the former path C is denoted by C_1. The line
integral over C_2 is precisely the expression on
the right in (15.4–6). If R is the rectangle enclosed by C_1 and C_2, the counterclockwise boundary of R consists of C_1 and the reversal of C_2.
Therefore, by Green's theorem,

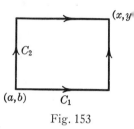

Fig. 153

$$\int_{C_1} M\, ds + N\, dt - \int_{C_2} M\, ds + N\, dt = \iint_R \left(\frac{\partial N}{\partial s} - \frac{\partial M}{\partial t} \right) ds\, dt.$$

But condition (15.4–3) in our present notation means that $\dfrac{\partial N}{\partial s} = \dfrac{\partial M}{\partial t}$.
Therefore, the line integrals over C_1 and C_2 are equal, and their common
value is $u(x, y)$. Actual calculation of the integral over C_2, using $s = a$,
$ds = 0$ on the first part, and $t = y$, $dt = 0$ on the second part, gives us
formula (15.4–6). The proof of Theorem IV is then complete.

The fact is that the line integral defining u can be taken over any path
from (a, b) to (x, y), so long as the path remains in the originally specified
rectangle. This sort of thing is discussed further in § 15.41.

Example 1.   Show that

$$(y - x^2)dx + (x + y^2)dy$$

is an exact differential in the whole xy-plane, and find a function of which
it is the differential.

Since $\dfrac{\partial}{\partial y} (y - x^2) = \dfrac{\partial}{\partial x} (x + y^2),$

we know that we have an exact differential. If (a, b) is an arbitrary point,
formula (15.4–5) becomes

$$u(x, y) = \int_a^x (b - s^2)ds + \int_b^y (x + t^2)dt,$$

or, after a simple calculation,

$$u = -\frac{x^3}{3} + xy + \frac{y^3}{3} + \frac{a^3}{3} - ab - \frac{b^3}{3}.$$

The point (a, b) is arbitrary, so the last three terms may be lumped into a single arbitrary constant C:

$$u = -\frac{x^3}{3} + xy + \frac{y^3}{3} + C.$$

In practice the function u is often found by a formal procedure based on (15.4–5), but not bringing in the point (a, b) explicitly. Let us write

$$u = \phi(x) + F(x, y),$$

where $\phi(x)$ and $F(x, y)$ represent the first and second terms, respectively, in (15.4–5). Since $\frac{\partial F}{\partial y} = N(x, y)$, we may think of F as an indefinite integral of N with respect to y, x being held constant. The condition $\frac{\partial u}{\partial x} = M$ means that

$$\phi'(x) + \frac{\partial F}{\partial x} = M.$$

Once F has been found we solve for $\phi(x)$ by integrating this last equation. The variable y is not involved in the equation, for

$$\frac{\partial}{\partial y}\left(\frac{\partial F}{\partial x} - M\right) = \frac{\partial^2 F}{\partial x \, \partial y} - \frac{\partial M}{\partial y} = \frac{\partial N}{\partial x} - \frac{\partial M}{\partial y} = 0.$$

Example 2. We illustrate the method on the previous example.

$$N = x + y^2, \quad F(x, y) = xy + \tfrac{1}{3} y^3;$$
$$u = \phi(x) + xy + \tfrac{1}{3} y^3,$$
$$\phi'(x) + y = M = y - x^2, \quad \text{or} \quad \phi'(x) = -x^2.$$

Thus
$$\phi(x) = -\tfrac{1}{3} x^3 + C,$$
$$u = -\tfrac{1}{3} x^3 + xy + \tfrac{1}{3} y^3 + C.$$

Once we have found a function u such that $du = M\,dx + N\,dy$, it is easy to evaluate the line integral of the differential form, as the next theorem shows.

THEOREM V. *Let $u = f(x, y)$ be a single-valued function with the differential $du = M\,dx + N\,dy$ at all points of some region. Let C be a curve lying in this region, and let C have initial point (x_0, y_0) and terminal point (x_1, y_1). We assume M and N continuous, and C sectionally smooth. Then*

(15.4–7) $$\int_C M\,dx + N\,dy = f(x_1, y_1) - f(x_0, y_0).$$

PROOF. We shall give the proof for a smooth arc. For a sectionally smooth curve the proof then follows by adding the results for constituent arcs. In terms of a parameter t, going from t_0 to t_1 as (x, y) goes from (x_0, y_0) to (x_1, y_1) along C,

$$\int_C M \, dx + N \, dy = \int_{t_0}^{t_1} \left(M \frac{dx}{dt} + N \frac{dy}{dt} \right) dt = \int_{t_0}^{t_1} \left(\frac{du}{dt} \right) dt$$

$$= u \Big|_{t=t_0}^{t=t_1} = f(x_1, y_1) - f(x_0, y_0).$$

Example 3. Evaluate the integral

$$\int_C \frac{x}{x^2 - y^2} \, dx + \frac{y}{y^2 - x^2} \, dy,$$

where C is a curve from $(1, 0)$ to $(5, 3)$ and lies between the lines $y = x$, $y = -x$.

We observe by inspection that

$$\frac{x \, dx}{x^2 - y^2} + \frac{y \, dy}{y^2 - x^2} = du,$$

where $u = \frac{1}{2} \log (x^2 - y^2)$. This function behaves properly as long as $x^2 > y^2$. Hence the integral in question is equal to

$$\tfrac{1}{2} \log (25 - 9) - \tfrac{1}{2} \log (1 - 0) = \log 4.$$

15.41   Line integrals independent of the path

Theorem V in the foregoing section shows that the line integral along C from (x_0, y_0) to (x_1, y_1) has the same value for all curves C starting at (x_0, y_0) and ending at (x_1, y_1), provided that C lies in the region where $u = f(x, y)$ is defined and $du = M \, dx + N \, dy$. Note, however, that we have emphasized the requirement that $f(x, y)$ be single-valued. The reason for insistence on this matter will be more apparent after we have considered the following example.

Example 1. Consider the line integral

(15.41–1) $$\int_C \frac{-y}{x^2 + y^2} \, dx + \frac{x}{x^2 + y^2} \, dy$$

along various curves from $(1, 0)$ to $(-1, 0)$.

In this case the differential form is exact at each point of the plane except $(0, 0)$, as the student should verify, using (15.4–3). It follows from Theorem IV that in any open rectangle which does not contain the origin there is some function defined whose differential is precisely the differential form appearing in (15.41–1). As a matter of fact, if r, θ are polar co-ordinates, it is easy to verify that

$$(15.41\text{–}2) \qquad\qquad d\theta = \frac{-y\,dx + x\,dy}{x^2 + y^2}.$$

We leave it for the student to verify this by calculating dx and dy in terms of dr and $d\theta$ from the equations $x = r\cos\theta$, $y = r\sin\theta$.

If we now attempt to find the value of the line integral (15.41–1) by applying Theorem V, we take $\theta = f(x, y)$. The question then arises: In what region of the plane may we regard θ as a single-valued differentiable function of x and y? Some standard procedure must be adopted so that a unique value of θ is assigned to each point. One possibility is to require $0 \leq \theta < 2\pi$. Then θ is a single-valued function defined at each point (x, y) except $(0, 0)$. The function is discontinuous (and hence not differentiable) at points on the positive x-axis, however, for the value of θ experiences a sudden jump as (x, y) crosses the positive x-axis. The value of θ might be standardized in a different way by requiring that $-\pi < \theta \leq \pi$; this would make θ discontinuous along the negative x-axis. There are infinitely many other ways of standardizing the definition of θ as a single-valued function, but there is no way in which θ can be defined so as to be single-valued and differentiable simultaneously at all points other than $(0, 0)$. Whatever method is used, θ must have at least one point of discontinuity on any closed curve which encircles the origin.

Now consider the line integral (15.41–1). If C goes from $(1, 0)$ to $(-1, 0)$ along a route which does not go through the origin or cross the negative y-axis, we may standardize θ by the requirement $-\pi/2 < \theta \leq 3\pi/2$. The value of the integral is then found by Theorem V; it is

$$\int_0^\pi d\theta = \pi.$$

For a path from $(1, 0)$ to $(-1, 0)$ in the lower half plane, we can define θ so that the discontinuities occur in the upper half plane, say by requiring $-3\pi/2 < \theta \leq \pi/2$. Then the value of the line integral is

$$\int_0^{-\pi} d\theta = -\pi.$$

Of course, the values of the line integrals for these two paths do not depend on the particular method which is chosen to standardize θ.

The foregoing example shows the following: If $M\,dx + M\,dy$ is a differential form such that $\dfrac{\partial M}{\partial y} = \dfrac{\partial N}{\partial x}$ at all points of a region D, it is not *necessarily true that M dx + N dy is the differential of a function which is single-valued and differentiable at all points of D.* This raises the question: Is there some kind of a condition which is sufficient to guarantee that if $\dfrac{\partial M}{\partial y} = \dfrac{\partial N}{\partial x}$ in D, then there exists a single-valued function $u = f(x, y)$ such that $du = M\,dx + N\,dy$ in D?

There is such a condition. To explain it we must introduce the concept of a simply connected region.

DEFINITION. A connected open set *D* is called a *simply connected* region if it has the property that whenever a simple closed curve *C* lies in *D*, all points inside *C* are also in *D*. If *D* is not simply connected it is called *multiply connected.*

The property of being simply connected is a property "in the large." The interior of a circle or rectangle is simply connected. The region between two concentric circles is multiply connected. So is the region consisting of the entire plane with the exception of one or any finite number of points.

Using this new concept, we can state the following theorem:

THEOREM VI. *Let D be a simply connected region. Let M and N have continuous first partial derivatives in D, such that*

(15.41–3)
$$\frac{\partial M}{\partial y} = \frac{\partial N}{\partial x}$$

at each point. Then there is a function $u = f(x, y)$, single-valued and differentiable in D, such that $du = M\,dx + N\,dy$.

PROOF. Let (x_0, y_0) and (x_1, y_1) be any points of D. Consider paths *C* from (x_0, y_0) to (x_1, y_1), restricted as follows: *C* consists of a finite number of straight line segments joined end to end, each segment being parallel

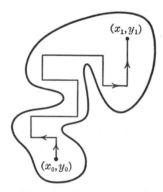

Fig. 154

to either the x or y axis (see Fig. 154). Further, C is to lie in D, and must not intersect itself. Such a curve C will be called an *elementary path*. Since D is connected, it can be shown that any two points of D can be joined by an elementary path in D. We omit the details of this.

Now, the idea of the proof is to show that

$$\int_C M\,dx + N\,dy$$

has the same value for any two elementary paths in D joining (x_0, y_0) to (x_1, y_1). Then, keeping (x_0, y_0) fixed, we define $f(x_1, y_1)$ as the value of the line integral. Since the value depends only on (x_1, y_1), and not on the particular elementary path, we get a single-valued function defined throughout D.

Suppose then that C_1 and C_2 are two such elementary paths in D from (x_0, y_0) to (x_1, y_1). They may coincide along certain line segments, but the noncoincident portions will have just a finite number of intersections, and these portions will form the boundaries of a finite number of regions. Each of these regions will lie entirely in D because of the assumption that D is simply-connected. The situation is like that suggested in Fig. 155, where these latter regions are shaded. If R is any one of these regions, the line integral around its complete boundary in the counterclockwise sense is zero, for the line integral is equal to

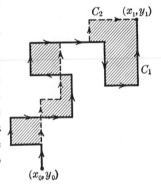

$$\iint_C \left(\frac{\partial N}{\partial x} - \frac{\partial M}{\partial y} \right) dx\,dy = 0$$

Fig. 155

by Green's theorem and the hypothesis (15.41–3). From this it is readily seen that the integrals over C_1 and C_2 are equal.

We now drop the subscripts on (x_1, y_1), and define $f(x, y)$ as the line integral from (x_0, y_0) to (x, y) over any elementary path in D. This function has partial derivatives

(15.41–4) $$\frac{\partial f}{\partial x} = M, \qquad \frac{\partial f}{\partial y} = N.$$

In fact, if we choose the path so that the last segment of C is horizontal, say from (a, y) to (x, y), then

$$f(x, y) = f(a, y) + \int_a^x M(s, y)\,ds,$$

and so the first equation in (15.41–4) must hold. The second equation also holds, by a similar argument in which the last segment of the path is parallel to the y-axis. This completes the proof of Theorem VI.

The student should make sure that he sees where the proof would break down if D were not simply connected.

The restriction to elementary paths is purely for the purposes of the proof of Theorem VI. Once the theorem is proved, we see that the line integral has the same value over any path from (x_0, y_0) to (x_1, y_1). The proof is by Theorem V, where in this case $f(x_0, y_0) = 0$.

A non–simply connected region can often be made simply connected by introducing certain barriers in the form of curves whose points are no longer considered as belonging to the region. Such curves are called *cuts*. For example, let D be the whole plane except for the origin. This is a non–simply connected region. But if we introduce the positive x-axis as a cut, the modified region (call it D_1) is simply connected. No curve enclosing the origin can lie wholly in D_1, and it is the fact that such curves do lie in D that makes D non–simply connected. As another example,

Fig. 156   Fig. 157

consider the region D shown in Fig. 156, lying inside the large curve and outside the two small ones. It is multiply connected, but it may be rendered into a simply connected region by making two cuts, as shown in Fig. 157. The student will readily see that the choice of cuts is not unique. The number of cuts needed in a given case is always the same, however, no matter how they are made.

EXERCISES

1. Test each form for exactness, and if it is exact, find a function of which the form is the differential. Practice the two methods illustrated in Examples 1 and 2, § 15.4. The point (a, b) may be taken as $(0, 0)$.

(a) $x \, dy + (y - 7)dx$.

(b) $(2y^2 - 3x)dx - 4xy \, dy$.

(c) $(2x + 5y - 7)dx + (5x - 8y + 3)dy$.

(d) $(2xy + x^2)dx + x^2 \, dy$.

(e) $xe^{xy} \sin y \, dx + (e^{xy} \cos y + y)dy$.

(f) $(xy \cos xy + \sin xy)dx + x^2 \cos xy \, dy$.

(g) $(4 x^3 + 10 xy^3 - 3 y^4)dx + (15 x^2y^2 - 12 xy^3 + 5 y^4)dy$.
(h) $(e^x \sin y - y)dx + (e^x \cos y - x - 2)dy$.

2. (a) Find a function u such that

$$du = \frac{1 + y^2}{x^3}\, dx - \frac{y + x^2y}{x^2}\, dy,$$

and describe the region or regions in which u is differentiable.

(b) Find the value of the line integral

$$\int_C \frac{1 + y^2}{x^3}\, dx - \frac{y + x^2y}{x^2}\, dy$$

from $(1, 0)$ to $(5, 2)$; from $(-3, 0)$ to $(-1, 4)$. In each case specify any essential limitations on the path C.

3. (a) Find a function u such that

$$du = \frac{dy}{\sqrt{y^2 - x^2}} - \frac{x\, dx}{y\sqrt{y^2 - x^2} + y^2 - x^2},$$

and describe the region or regions in which u is differentiable.

(b) Find the line integral of the differential form in **(a)** from $(3, 5)$ to $(5, 13)$, and specify any necessary limitations on the path.

4. Find a function of x alone, $w = \phi(x)$, which makes the differential form $w(x \sin y + y \cos y)dx + w(x \cos y - y \sin y)dy$ exact; then find the function of which it is the differential, if this function is equal to 0 at $(0, 0)$.

5. Let P_1 be $(1, 0)$, P_2 be $(-1, 0)$, and P be (x, y). Let θ_1 and θ_2 be the angles between the positive x-axis and P_1P and P_2P respectively. Let $u = \theta_1 + \theta_2$. Show that

$$du = -\left(\frac{y}{r_1^2} + \frac{y}{r_2^2}\right)dx + \left(\frac{x - 1}{r_1^2} + \frac{x + 1}{r_2^2}\right)dy,$$

where $r_1 = P_1P$, $r_2 = P_2P$. To make u a single-valued function it is necessary to make some definite agreement about the values of θ_1 and θ_2 at all points except P_1 and P_2.

(a) If it is agreed that $-\pi < \theta_1 \leq \pi$ and $0 \leq \theta_2 < 2\pi$, show that u is discontinuous if $y = 0$ and $x^2 > 1$. By making cuts along the lines of discontinuity, we get a simply connected region in which u is differentiable.

(b) If it is agreed that $0 \leq \theta_1 < 2\pi$ and $0 \leq \theta_2 < 2\pi$, where are the discontinuities of u?

(c) If $v = \theta_1 - \theta_2$ and the angles are chosen as in **(b)**, where is v discontinuous?

15.5   Further discussion of surface area

In § 14.6 we arrived at the formula

(15.5–1)
$$A = \iint_R \sqrt{EG - F^2}\, du\, dv$$

for the area of a surface represented parametrically, the parameters (u, v) ranging over a region R in the uv-plane. In the earlier discussion of this formula we pointed out that it is logically desirable to show that the number A found by the formula is really not dependent on the particular parametric representation of the surface which is used in calculating A. By using Theorem III, § 15.32, it is possible to give an analytic proof that A is independent of the parametrization.

Let another parametrization be given, in terms of parameters (r, t) ranging over a region R' in the rt-plane. Then, there is a one-to-one correspondence between the points of R' and those of R, since both of these regions are mapped in one-to-one fashion onto the same surface. Thus we may consider u and v as functions of r and t. There are now two formulas for ds^2 on the surface:

$$(15.5-2) \qquad ds^2 = E \, du^2 + 2 \, F \, du \, dv + G \, dv^2,$$

$$(15.5-3) \qquad ds^2 = E' \, dr^2 + 2 \, F' \, dr \, dt + G' \, dt^2.$$

The prime notation here has nothing to do with differentiation; we use it simply to indicate that E', F', G' are related to r and t in the same way that E, F, G are related to u and v.

What we wish to prove is that

$$(15.5-4) \qquad \iint\limits_{R} \sqrt{EG - F^2} \, du \, dv = \iint\limits_{R'} \sqrt{E'G' - F'^2} \, dr \, dt.$$

Now, regarding u and v as functions of r and t, we have

$$du = \frac{\partial u}{\partial r} \, dr + \frac{\partial u}{\partial t} \, dt,$$

and a similar formula for dv. We substitute these expressions for du and dv into (15.5-2), whereupon we get an expression for ds^2 in terms of dr and dt. If this expression is compared with (15.5-3), we see that

$$E' = E\left(\frac{\partial u}{\partial r}\right)^2 + 2 \, F \frac{\partial u}{\partial r} \frac{\partial v}{\partial r} + G\left(\frac{\partial v}{\partial r}\right)^2$$

$$(15.5-5)$$

$$F' = E \frac{\partial u}{\partial r} \frac{\partial u}{\partial t} + F\left(\frac{\partial u}{\partial r} \frac{\partial v}{\partial t} + \frac{\partial u}{\partial t} \frac{\partial v}{\partial r}\right) + G \frac{\partial v}{\partial r} \frac{\partial v}{\partial t},$$

$$G' = E\left(\frac{\partial u}{\partial t}\right)^2 + 2 \, F \frac{\partial u}{\partial t} \frac{\partial v}{\partial t} + G\left(\frac{\partial v}{\partial t}\right)^2.$$

Now let us consider the expression $E'G' - F'^2$. For convenience, we write

$$u_1 = \frac{\partial u}{\partial r}, \qquad u_2 = \frac{\partial u}{\partial t}, \qquad v_1 = \frac{\partial v}{\partial r}, \qquad v_2 = \frac{\partial v}{\partial t}.$$

Then, after carefully writing out the products involved in $E'G'$ and F'^2 from (15.5–5), we arrive at the final result

$$E'G' - F'^2 = (EG - F^2)(u_1 v_2 - u_2 v_1)^2,$$

or

$$(15.5\text{–}6) \qquad \sqrt{E'G' - F'^2} = \sqrt{EG - F^2} \left| \frac{\partial(u, v)}{\partial(r, t)} \right|.$$

Let us now transform the integral (15.5–1) by changing variables from (u, v) to (r, t). By Theorem III, § 15.32, the result is

$$\iint_R \sqrt{EG - F^2} \; du \; dv = \iint_{R'} \sqrt{EG - F^2} \left| \frac{\partial(u, v)}{\partial(r, t)} \right| dr \; dt.$$

This formula and (15.5–6) give us (15.5–4), and our proof is thus complete.

15.51   Surface integrals

A surface integral is a natural generalization of a double integral, and may be used for applications to such things as finding centers of gravity and moments of inertia of curved laminas, the potentials and components of force due to distributions of electrostatic charge on surfaces, and other quantities having physical or geometrical significance.

The direct intuitive formulation of the concept of a surface integral is as follows: Let S be a surface, and let $\phi(P)$ be a point function defined on S. Divide S into a number of surface elements $\Delta S_1, \ldots, \Delta S_n$, of areas $\Delta A_1, \ldots, \Delta A_n$. Let P_k be some point in the element ΔS_k. Then the surface integral of ϕ over S is defined as

$$(15.51\text{–}1) \qquad \iint_S \phi \; dA = \lim \sum_{k=1}^{n} \phi(P_k)\Delta A_k,$$

where the limit is taken as the maximum of the dimensions of the elements ΔS_k approaches zero.

In case the surface S is flat and lies in the xy-plane, this definition takes the same form as the expression of a double integral in Theorem II, § 13.23.

If $\phi(P)$ is expressed as a function $F(x, y, z)$, where (x, y, z) are the co-ordinates of P, the surface integral is denoted by

$$(15.51\text{–}2) \qquad \iint_S F(x, y, z)dA.$$

The conception of a surface integral is independent of all co-ordinate systems. It is also independent of the choice of parametric representation for the surface. However, to work with surface integrals, we must learn how to express surface integrals as ordinary double integrals. This is done with the aid of equations for the surface, whether in parametric form or otherwise.

Consider first the case in which S is defined by an equation $z = f(x, y)$, where (x, y) ranges over a region R in the xy-plane. In this case

(15.51–3)
$$\iint_S F(x, y, z)dA = \iint_R F(x, y, z) \sec \gamma \, dx \, dy,$$

where, in the integral on the right,
$$= \iint_R F(x, y, z) \frac{dx\,dy}{\cos \gamma}$$

$$z = f(x, y), \sec \gamma = \left[1 + \left(\frac{\partial z}{\partial x}\right)^2 + \left(\frac{\partial z}{\partial y}\right)^2\right]^{1/2};$$

γ is the acute angle between the normal to S at (x, y, z) and the positive z-axis. The derivation of (15.51–3) rests on the fact that if ΔS is a small piece of S which projects into a small rectangular cell of dimensions Δx by Δy in the region R, then the area ΔA of ΔS is given by

$$\Delta A = \sec \gamma \, \Delta x \, \Delta y,$$

where γ is evaluated at some point in ΔS (see the discussion following (14.6–12) in § 14.6).

Example 1. Find the moment of inertia of the hemispherical surface $z = \sqrt{a^2 - x^2 - y^2}$ about the x-axis, assuming the surface to be a homogeneous lamina of mass M.

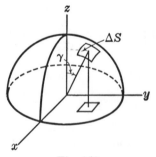

Fig. 158

The required moment of inertia is

$$I = \iint_S \sigma(y^2 + z^2)dA,$$

where σ is the constant density. Thus, by (15.51–3),

$$I = \iint_R \sigma(y^2 + z^2) \sec \gamma \; dx \; dy,$$

where R is the circular region $x^2 + y^2 \leq a^2$.

Now $$\sec \gamma = \frac{a}{z} = \frac{a}{\sqrt{a^2 - x^2 - y^2}}$$

and $y^2 + z^2 = a^2 - x^2$. Thus

$$I = \sigma a \iint_R \frac{a^2 - x^2}{\sqrt{a^2 - x^2 - y^2}} \; dx \; dy.$$

It is convenient to use polar co-ordinates to evaluate this integral. The iterated integral is

$$I = \sigma a \int_0^{2\pi} d\theta \int_0^a \frac{a^2 - r^2 \cos^2 \theta}{\sqrt{a^2 - r^2}} \; r \; dr.$$

We leave most of the calculations to the student.

$$I = \sigma a \int_0^{2\pi} \left(a^3 - \frac{2 a^3}{3} \cos^2 \theta \right) d\theta = \frac{4 \pi \sigma a^4}{3} \, .$$

Since $M = 2 \pi a^2 \sigma$, we have $I = \frac{2}{3} M a^2$.

If the surface is represented parametrically, with parameters (u, v), the appropriate expression for ΔA is

$$\Delta A = \sqrt{EG - F^2} \; \Delta u \; \Delta v$$

(see the discussion leading up to (14.6–6)), and, if $\phi(P)$ is expressed as some function $f(u, v)$, we have

$$(15.51\text{–}4) \qquad \iint_S f(u, v) dA = \iint_R f(u, v) \sqrt{EG - F^2} \; du \; dv,$$

where R is the region in the uv-plane which corresponds to the surface S. Here one should also remember that

$$EG - F^2 = j_1^2 + j_2^2 + j_3^2$$

(see (14.6–7)).

Example 2.   Find the electrostatic potential at $(0, 0, -a)$ of a uniformly distributed total charge e on the hemisphere in Example 1.

If σ is the constant density of charge, the potential in question is

$$u = \iint_S \frac{\sigma}{[x^2 + y^2 + (z + a)^2]^{1/2}} \; dA \, .$$

The student may wish to consult § 13.51 to refresh his memory on the concept of potential. See (13.51–4), in particular.

We use the parametric representation $x = a \sin \phi \cos \theta$, $y = a \sin \phi \sin \theta$, $z = a \cos \phi$. In this case, with $u = \phi$, $v = \theta$, it is readily found that

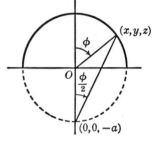

Fig. 159

$$ds^2 = a^2 \, d\phi^2 + a^2 \sin^2 \phi \, d\theta^2,$$

$$E = a^2, \quad F = 0, \quad G = a^2 \sin^2 \phi.$$

Also $[x^2 + y^2 + (z + a)^2]^{1/2} = 2\, a \cos \dfrac{\phi}{2} \cdot$

This result may be read directly from Fig. 159, or it may be worked out analytically.

The potential is thus

$$ u = \iint_S \frac{\sigma}{2\,a \cos \dfrac{\phi}{2}} \, dA = \sigma \int_0^{2\pi} d\theta \int_0^{\pi/2} \frac{a^2 \sin \phi}{2\,a \cos \dfrac{\phi}{2}} \, d\phi . $$

Since $\sin \phi = 2 \sin \dfrac{\phi}{2} \cos \dfrac{\phi}{2}$, we have

$$ u = 2 \, \pi \sigma a \int_0^{\pi/2} \sin \frac{\phi}{2} \, d\phi = 2 \, \pi \sigma a (2 - \sqrt{2}) . $$

Since $e = 2 \, \pi a^2 \sigma$, the result can be written

$$ u = \frac{(2 - \sqrt{2})e}{a} . $$

EXERCISES

1. Locate the centroid of the hemispherical surface $z = (a^2 - x^2 - y^2)^{1/2}$, using a surface integral.

2. Let S be the surface defined by $x^2 + y^2 + z^2 = 4$, $z \geq 1$.
(a) Find the value of $\iint_S (x^2 + y^2)z \, dA$, using (15.51–3).
(b) Find the value of $\iint_S [x^2 + y^2 + (z - 2)^2]dA$, using (15.51–4) and the parameters θ, ϕ as in Example 2.
(c) Find the value of $\iint_S (x^2 + y^2)dA$.

3. Calculate $\displaystyle \iint_S \frac{dA}{\sqrt{z - y + 1}}$ where S is defined by $2\,z = x^2 + 2\,y$, $0 \leq x \leq 1$, $0 \leq y \leq 1$.

4. Compute the value of $\iint\limits_{S} (x^2 + y^2 - 2 z^2)dA$ where S is the entire surface $x^2 + y^2 + z^2 = a^2$.

5. Compute the value of $\iint\limits_{S} \dfrac{dA}{\sqrt{2\,az - z^2}}$ where S is the part of the surface $x^2 + y^2 + (z - a)^2 = a^2$ which is inside the cylinder $x^2 + y^2 = ay$ and underneath the plane $z = a$.

6. Find the value of $\iint\limits_{S} xyz\, dA$, where S is the part of $x^2 + z^2 = 4$ in the first octant and between $y = 0$, $y = 1$.

7. Locate the centroid of the area on the surface of the sphere $x^2 + y^2 + z^2 = 4\,a^2$, inside the cylinder $x^2 + y^2 = 2\,ax$, and in the first octant.

8. If the surface in Exercise 7 is a homogeneous lamina, find its moment of inertia about the z-axis.

9. (a) If the cone $az = b\sqrt{x^2 + y^2}$ is parametrized by setting $x = r \cos \theta$, $y = r \sin \theta$, $z = br/a$, show that $EG - F^2 = (a^2 + b^2)r^2/a^2$.
   (b) If S is the part of the cone for which $0 \le z \le b$, and if S is a homogeneous lamina, show that its moment of inertia about the z-axis is $\frac{1}{2} Ma^2$, using a surface integral and (15.51–4). How does this compare with using (15.51–3) and polar co-ordinates in the xy-plane?

10. On the cylinder $x^2 + y^2 = 1$, $0 \le z \le 2$ use the θ and z of cylindrical co-ordinates as parameters, and calculate the value of the surface integral giving the electrostatic field strength produced at $(0, 0, 0)$ by a uniform density of charge on the cylinder.

11. Let S be a part of the surface of the sphere $x^2 + y^2 + z^2 = a^2$, and let it carry a uniformly distributed total mass M. Show that the gravitational field produced at the origin by this mass is $\mathbf{F} = (M/a^3)\mathbf{R}$, where $\mathbf{R}$ is the vector from 0 to the center of mass of S.

12. Suppose that S in Exercise 11 lies entirely on the hemisphere $z \ge 0$. Show that the z-component of the field at 0 is $\sigma A/a^2$, where σ is the density and A is the area of the projection of S on the xy-plane.

13. If the torus of Fig. 124 (page 438) carries a homogeneously distributed total mass M, show that its moment of inertia about the z-axis is $\frac{1}{2} M(2 a^2 + 3 b^2)$.

14. Let S be the sphere $x^2 + y^2 + z^2 = a^2$, and let it carry a uniform distribution of charge. Show that the field produced at $(0, 0, c)$ is of magnitude 0 if $|c| < a$, e/c^2 if $c > a$, and $\frac{1}{2} e/c^2$ if $c = a$. The substitution $t^2 = a^2 + c^2 - 2\,ac \cos \phi$ will be found useful in this exercise, as well as in Exercises 15, 16.

15. Solve the problem corresponding to Exercise 14 if the density of charge is $\sigma = \cos \phi$, where ϕ is the angle which OP makes with the positive z-axis.

16. Find the potential u at $(0, 0, c)$ of a uniform distribution of charge on the hemisphere $z = (a^2 - x^2 - y^2)^{1/2}$. Then calculate $- \dfrac{\partial u}{\partial c}$ to get the force.

15.6   The divergence theorem

The divergence theorem is a three-dimensional analogue of Green's theorem in the plane (§ 15.3). The main content of the theorem is the formula

(15.6–1)
$$\iiint\limits_{T} \left(\frac{\partial P}{\partial x} + \frac{\partial Q}{\partial y} + \frac{\partial R}{\partial z} \right) dV$$

$$= \iint\limits_{S} (P \cos \alpha + Q \cos \beta + R \cos \gamma) dA.$$

Here T is a region in three-dimensional space; S is the surface bounding T; P, Q, R are functions of x, y, z which are continuous and have continuous first partial derivatives in T; and $\cos \alpha$, $\cos \beta$, $\cos \gamma$ are the direction cosines of the line normal to S, directed *outward* from T.

The hypotheses covering T and S need to be made more precise. In the usual applications of formula (15.6–1) we may wish to take T to be one of the common solids, e.g., a cube, a sphere, or a right circular cylinder. Or, T might be the region contained between two concentric spheres, or the region which is left when a doughnut-shaped region is removed from the interior of a large ellipsoid. Thus, the surface S of T may actually consist of several detached pieces; in the last-mentioned example, S consists of the surface of the ellipsoid and the surface of the doughnut (a torus).

We saw in § 15.31 that there are not inconsiderable difficulties in the way of proving Green's theorem in the generality with which it is stated in § 15.3. There are difficulties of a corresponding kind, but even greater, in connection with the divergence theorem. It is not even an easy or brief matter to formulate reasonably general conditions on the region T such that (15.6–1) may be shown to hold true. However, if T is a sufficiently simple type of region, it is quite easy to prove formula (15.6–1). To describe the kind of region we have in mind let us begin with a description of the boundary. Let G be a regular region in the xy-plane. Let $Z_1(x, y)$ and $Z_2(x, y)$ be continuous functions defined in G and having continuous first partial derivatives there, and such that $Z_1 < Z_2$ at each interior point of G. Let S_1 and S_2 be the surfaces defined by $z_k = Z_k(x, y)$, $k = 1, 2$. Consider the cylindrical surface formed by erecting lines parallel to the z-axis at points of the boundary of G. Let S_3 be the portion of this cylindrical surface which is cut off

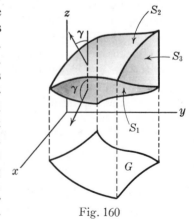

Fig. 160

between the surfaces S_1, S_2. It may happen that $Z_1 = Z_2$ on the boundary of G; in this event there is no surface S_3. Now let T be the region bounded above by S_2, below by S_1, and laterally by S_3. A region T formed in this manner will be called *xy-simple*. In a similar manner we may define what we mean by a *yz*-simple region or a *zx*-simple region. The region defined by $1 \leq x^2 + y^2 \leq 4, 0 \leq z \leq 1$ is *xy*-simple, but not *yz*-simple or *zx*-simple. The region defined by $x^2 + z^2 \leq y, 0 \leq y \leq 4$ is *zx*-simple.

We now prove the following lemma:

LEMMA. *Let T be an xy-simple region, and let S be the entire surface of T. Suppose that $F(x, y, z)$ and $\dfrac{\partial F}{\partial z}$ are continuous in T. Let γ be the angle between the positive z-axis and the outward drawn normal to S. Then*

$$(15.6\text{--}2) \qquad \iiint_T \frac{\partial F}{\partial z} \, dV = \iint_S F \cos \gamma \, dA.$$

PROOF. We begin by expressing the triple integral in (15.6–2) as an iterated integral in which the first integration is with respect to z. We have

$$\iiint_T \frac{\partial F}{\partial z} \, dV = \iint_G dx \, dy \int_{Z_1}^{Z_2} \frac{\partial F}{\partial z} \, dz \, .$$

In the z-integration, x and y are held constant. Hence

$$\int_{Z_1}^{Z_2} \frac{\partial F}{\partial z} \, dz = F(x, y, Z_2) - F(x, y, Z_1),$$

and the triple integral becomes

$$(15.6\text{--}3) \qquad \iiint_T \frac{\partial F}{\partial z} \, dV = \iint_G [F(x, y, Z_2) - F(x, y, Z_1)] dx \, dy.$$

Now consider the surface integral in (15.6–2). On the lateral surface S_3 of T we see that $\cos \gamma = 0$, for $\gamma = \pi/2$. Hence

$$\iint_S F \cos \gamma \, dA = \iint_{S_1} F \cos \gamma \, dA + \iint_{S_2} F \cos \gamma \, dA.$$

At a point on S_1 the outer normal extends downward, so that γ is obtuse and $\cos \gamma < 0$. On S_2, however, γ is acute (see Fig. 160). The surface integrals over S_1 and S_2 can be transformed into double integrals over G by formula (15.51–3). In the latter formula γ represents the *acute* angle between the positive z-axis and the *undirected* normal to the surface. Hence, since $\sec(\pi - \gamma) = -\sec\gamma$, we have

$$\iint_{S_1} F(x, y, z)\cos\gamma\, dA = \iint_G F(x, y, Z_1)\cos\gamma\sec(\pi - \gamma)dx\, dy$$

$$= -\iint_G F(x, y, Z_1)dx\, dy.$$

Likewise

$$\iint_{S_2} F(x, y, z)\cos\gamma\, dA = \iint_G F(x, y, Z_2)\cos\gamma\sec\gamma\, dx\, dy$$

$$= \iint_G F(x, y, Z_2)dx\, dy.$$

Thus we have shown that

$$\iint_S F\cos\gamma\, dA = \iint_G [F(x, y, Z_2) - F(x, y, Z_1)]dx\, dy.$$

On comparing this result with (15.6–3) we see that the lemma is proved.

Clearly, if T is yz-simple, we have a result analogous to (15.6–2) with $\dfrac{\partial F}{\partial x}$ instead of $\dfrac{\partial F}{\partial z}$, and $\cos\alpha$ instead of $\cos\gamma$; a corresponding result also holds for zx-simple regions. No additional proofs are needed, since the results differ from (15.6–2) in notation only, and the labeling of the axes is purely a matter of notation.

Now suppose that T is a region which is at once xy-simple, yz-simple, and zx-simple, and let S be its surface. Suppose that P, Q, R are functions which are continuous and have continuous first partial derivatives in T. At a point where S is smooth let $\mathbf{n}$ be a unit vector normal to S and extend outward from T, and let $\mathbf{n}$ make angles α, β, γ respectively with the positive x-, y-, and z-axes. Then by the lemma

$$\iiint_T \frac{\partial P}{\partial x}\, dV = \iint_S P\cos\alpha\, dA,$$

$$\iiint_T \frac{\partial Q}{\partial y}\, dV = \iint_S Q\cos\beta\, dA,$$

$$\iiint_T \frac{\partial R}{\partial z}\, dV = \iint_S R\cos\gamma\, dA.$$

Adding, we obtain the divergence theorem (15.6–1) for a region T of this restricted type.

Next we proceed to remove some of the restrictions on T. Let us call a region xyz-simple if it is at once xy-simple, yz-simple, and zx-simple. The region between two concentric spheres (say with centers at O) is not xyz-simple. But the three co-ordinate planes divide this region into eight parts, each of which is xyz-simple. In this subdivision process, certain additional surfaces are introduced as "interior partitions" in T. Each surface element of such an interior partition is on the boundary of two xyz-simple subregions. Let us say that a region T is xyz-standard if, by the introduction of a finite number of simple surface elements as interior partitions, we can divide T into a finite number of xyz simple subregions.

THEOREM VII. *Under the stated assumptions on P, Q, R, formula (15.6–1) holds when T is an xyz-standard region.*

PROOF.   Let $T_1, \cdots, T_n$ be the xyz-simple regions composing T, and let S_k be the entire surface of T_k. The surface S_k may consist partly of pieces of S and partly of interior partitions. By what we have already proved,

$$\iiint\limits_{T_k}\left(\frac{\partial P}{\partial x} + \frac{\partial Q}{\partial y} + \frac{\partial R}{\partial z}\right)dV = \iint\limits_{S_k}(P\cos\alpha + Q\cos\beta + R\cos\gamma)dA.$$

Here we have n formulas. We add them; the sum of the triple integrals is just the triple integral over T. Let us see what we get when we add the surface integrals. Suppose that T_j and T_k are adjacent regions, and let S_{jk} denote a surface element which is an interior partition between T_j and T_k. If $\mathbf{n}_j$ is the unit vector normal to S_{jk} outward from T_j, and $\mathbf{n}_k$ is the unit vector normal to S_{jk} outward from T_k, then $\mathbf{n}_j = -\mathbf{n}_k$, and so the direction cosines of $\mathbf{n}_j$ are the negatives of those of $\mathbf{n}_k$. The result is that, when the surface integrals over S_j and S_k are added, the integrals over S_{jk} cancel each other out. Thus, when all the surface integrals are added, the contributions from all the interior partitions cancel in pairs, and we are left with just the integral over the original boundary surface S. This shows that formula (15.6–1) holds true.

The theorem can be written in vector form. Indeed, it is from the vector form that the theorem derives its name.

THEOREM VIII.   (*The divergence theorem.*) *Let* $\mathbf{F}$ *be a vector point function which is defined and continuously differentiable in a bounded closed region T. Suppose that, for some choice of rectangular co-ordinate*

*system, T is an xyz-standard region. Let S be the surface of T, and $\mathbf{n}$
the unit outer normal vector to S. Then*

(15.6–4) $$\iiint_T \operatorname{div} \mathbf{F}\, dV = \iint_S \mathbf{F} \cdot \mathbf{n}\, dA.$$

This form of the theorem requires a few additional words of proof. Let
P, Q, and R be the x, y, and z components, respectively, of $\mathbf{F}$. Then

(15.6–5) $$\operatorname{div} \mathbf{F} = \frac{\partial P}{\partial x} + \frac{\partial Q}{\partial y} + \frac{\partial R}{\partial z},$$

(15.6–6) $$\mathbf{F} \cdot \mathbf{n} = P \cos \alpha + Q \cos \beta + R \cos \gamma,$$

so that (15.6–4) and (15.6–1) are equivalent. Now div $\mathbf{F}$ and $\mathbf{F} \cdot \mathbf{n}$ are
independent of the way in which the rectangular co-ordinate system is
chosen. Hence, if (15.6–4) is true for one choice of the co-ordinate system,
it is true for all choices. This shows us that T need not be an *xyz*-standard
region for *all* orientations of the axes; if it is an *xyz*-standard region for
some orientation, that is sufficient.

The restriction to *xyz*-standard regions is not absolutely necessary; it
is imposed in order to make the proof simple. Considerations similar to
those mentioned in § 15.31 will enable one to prove Theorem VIII for
certain regions which are never *xyz*-standard for any choice of axes. We
shall not discuss these generalizations, however.

Example 1. Let S be the surface of a region T for which the divergence
theorem is applicable. Let O be any fixed point in space, and let $P(x, y, z)$
be a variable point of S. Show that the volume of T is given by

(15.6–7) $$V = \tfrac{1}{3} \iint_S r \cos \psi\, dA,$$

where r is the distance OP and ψ is the angle between the directed line
OP and the outer normal to S at P.

To show this, take $P = x$, $Q = y$, $R = z$ in (15.6–5). Then from (15.6–4)
we have $\quad \iiint_T \operatorname{div} \mathbf{F}\, dV = \iiint_T \left(\frac{\partial P}{\partial x} + \frac{\partial Q}{\partial y} + \frac{\partial R}{\partial z}\right) dV = \iiint_T (1+1+1)\, dV$

$$\iiint_T 3\, dV = \iint_S r \cos \psi\, dA, \qquad = \iiint_T 3\, dV$$

which gives (15.6–7) at once.

The divergence theorem is useful in connection with the consideration
of solid angles. For an exposition of this matter we begin with a definition.
Let S be a surface element, and O a fixed point not on S. Assume that S
is not intersected more than once by any ray from O, and that no such
ray is tangent to S. As a point P varies over S, consider the point Q in

which the ray OP (extended, if necessary) intersects the unit sphere with center at O. These points Q fill out a certain portion of the surface of the unit sphere. The area ω of this portion is defined to be *the solid angle subtended by S at O.*

Example 2. Show that the solid angle is given by

$$(15.6\text{--}8) \qquad \omega = \iint_S \frac{\cos \psi}{r^2} \, dA = \iint_S \frac{\mathbf{n} \cdot \overrightarrow{OP}}{r^3} \, dA,$$

where r is the distance OP and $\mathbf{n}$ is the unit normal to S at P in such a direction that the angle ψ between $\mathbf{n}$ and $\overrightarrow{OP}$ is acute.

To derive (15.6–8) we use the divergence theorem as follows: Construct the sphere with radius a and center O. Choose a so small that S lies entirely outside this sphere. Draw all the rays joining O to S, and let Σ be the portion of the sphere pierced by these rays. Let T be the solid region formed by the bundle of rays cut off between Σ and S (see Fig. 161). For $\mathbf{F}$ we take the vector function

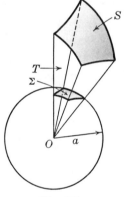

$\dfrac{\overrightarrow{OP}}{r^3}$ with components $\dfrac{x}{r^3}$, $\dfrac{y}{r^3}$, $\dfrac{z}{r^3}$. We apply the divergence theorem. Since $r^2 = x^2 + y^2 + z^2$, it is readily calculated that

$$\frac{\partial}{\partial x} \left(\frac{x}{r^3} \right) = \frac{r^2 - 3\, x^2}{r^5} \quad \text{and} \quad \operatorname{div} \mathbf{F} = 0.$$

Thus

$$\iiint_T (\operatorname{div} \mathbf{F}) \, dV = 0.$$

Fig. 161

Therefore, by (15.6–4), the integral of $\mathbf{F} \cdot \mathbf{n}$ over the entire surface of T is equal to zero. Here $\mathbf{n}$ refers to the outer normal. The surface of T consists of Σ, S, and the lateral surface formed by the rays joining O to the edge of S. Since $\mathbf{F}$ has the same direction as $\overrightarrow{OP}$, $\mathbf{F} \cdot \mathbf{n} = 0$ on this lateral portion, so that the surface integral over this portion is zero. Thus

$$(15.6\text{--}9) \qquad \iint_\Sigma \mathbf{F} \cdot \mathbf{n} \, dA + \iint_S \mathbf{F} \cdot \mathbf{n} \, dA = 0.$$

The integral over S in this equation is the same as the surface integral in (15.6–8). Thus we have to show that

$$(15.6\text{--}10) \qquad \iint_\Sigma \mathbf{F} \cdot \mathbf{n} \, dA = -\, \omega.$$

This is easy, however. On Σ the outer normal points away from T, which is toward O. But $\mathbf{F}$ points away from O, and has magnitude $1/r^2$. Thus, since $r = a$ on Σ, $\mathbf{F} \cdot \mathbf{n} = -1/a^2$, and the integral in (15.6–10) is

$$-\frac{1}{a^2} \iint_\Sigma dA = -\frac{1}{a^2} \text{ (area of } \Sigma\text{)}.$$

If $a = 1$, the area of Σ is ω, by the definition of the solid angle. But, in any case,

$$\frac{1}{a^2} \text{ (area of } \Sigma\text{)} = \omega,$$

since the area of Σ is proportional to a^2 when we vary a. This completes the proof of (15.6–10) and (15.6–8).

EXERCISES

1. Use (15.6–7) to show that the volume of any cone or pyramid is $\frac{1}{3} Bh$, where h is the altitude and B is the area of the base. Take O at the vertex.

2. If S is the hemispherical surface $\bar{z} = (a^2 - x^2 - y^2)^{1/2}$, show that the centroid of the solid hemisphere is given by

$$\bar{z} = \frac{3}{4 \pi a^3} \iint_S z^2 \cos \gamma \, dA,$$

where γ is the acute angle between the normal and the positive z-axis.

3. Use the divergence theorem to show that

$$\iint_S (x^2 \cos \alpha + y^2 \cos \beta + z^2 \cos \gamma)dA = \frac{8 \pi a^4}{3},$$

if S is the surface $x^2 + y^2 + z^2 = 2 az$ and α, β, γ are the direction angles of the outer normal.

4. What is the value of the integral in Exercise 3 if S is the surface of the cube bounded by $x = 0$, $x = a$, $y = 0$, $y = a$, $z = 0$, $z = a$?

5. Let $\mathbf{F} = r\overrightarrow{OP}$, where r is the distance OP. Use the divergence theorem to show that

$$\iint_S \mathbf{F} \cdot \mathbf{n} \, dA = 4 \iiint_T r \, dV.$$

By evaluating the surface integral, show that

$$\iiint_T (x^2 + y^2 + z^2)^{1/2} \, dV = \pi b^4$$

if T is the spherical region $x^2 + y^2 + z^2 \leq b^2$. Check by direct evaluation of the triple integral.

6. Let $\mathbf{F}$ be the vector function $(x^2 + y^2 + z^2)(x\mathbf{i} + y\mathbf{j} + z\mathbf{k})$. If T is the region $x^2 + y^2 + z^2 \leq b^2$, calculate $\iiint\limits_{T} (\nabla \cdot \mathbf{F})dV$ directly, and also by the divergence theorem. Observe that the value of the surface integral can be written down by inspection, since $\mathbf{F} \cdot \mathbf{n} = (x^2 + y^2 + z^2)^{3/2}$ on S.

7. Show that, in the notation of (15.6–1),

$$\iint\limits_{S} (x^2 + y^2)(x \cos \alpha + y \cos \beta)dA = 4 I_z,$$

where T is regarded as a homogeneous mass of unit density and I_z is its moment of inertia about the z-axis.

8. If $\mathbf{F} = x\mathbf{i} - y\mathbf{j} + z^2\mathbf{k}$, calculate $\iint\limits_{S} \mathbf{F} \cdot \mathbf{n}\, dA$, where S is the entire surface of any right circular cylinder of radius b with one base in the plane $z = 1$ and the other base in the plane $z = 3$.

9. Let S be the ellipsoidal surface

$$\frac{x^2}{a^2} + \frac{y^2}{b^2} + \frac{z^2}{c^2} = 1.$$

If (x, y, z) is a point on the ellipsoid, and D is the distance from the origin to the plane tangent to the ellipsoid at (x, y, z), show that $D^{-1} = \mathbf{F} \cdot \mathbf{n}$, where $\mathbf{n}$ is the unit outer normal to S at (x, y, z), and $\mathbf{F}$ is the vector with components x/a^2, y/b^2, z/c^2. Hence, show that

$$\iint\limits_{S} \frac{1}{D}\, dA = \tfrac{4}{3}\, \pi \left(\frac{bc}{a} + \frac{ca}{b} + \frac{ab}{c} \right).$$

10. Consider a limiting process in which a region T shrinks down onto a certain fixed point. If V is the volume of T, show that the divergence of a vector function at this point is given by

$$\operatorname{div} \mathbf{F} = \lim \frac{1}{V} \iint\limits_{S} \mathbf{F} \cdot \mathbf{n}\, dA.$$

This formula is sometimes used as an invariant definition of the divergence. It is useful for calculating div $\mathbf{F}$ in terms of various systems of curvilinear co-ordinates. See Exercise 11.

11. Let ρ, θ, ϕ be the usual spherical co-ordinates, and let F_ρ, F_θ, F_ϕ denote the components of a vector function $\mathbf{F}$ in the ρ, θ, ϕ directions respectively. Show that

$$\operatorname{div} \mathbf{F} = \frac{1}{\rho^2 \sin \phi} \left\{ \frac{\partial}{\partial \rho} (\rho^2 \sin \phi F_\rho) + \frac{\partial}{\partial \theta} (\rho F_\theta) + \frac{\partial}{\partial \phi} (\rho \sin \phi F_\phi) \right\}.$$

Obtain this by using the result in Exercise 10, using for T a region of the "volume element" type in spherical co-ordinates, with pairs of opposite faces corresponding to values ρ and $\rho + \Delta\rho$, θ and $\theta + \Delta\theta$, ϕ and $\phi + \Delta\phi$. Then let $\Delta\rho$, $\Delta\theta$, and $\Delta\phi$ approach zero. HINT: The volume of T is approximately $\rho^2 \sin \phi \, \Delta\rho \, \Delta\theta \, \Delta\phi$. The surface integral over the pair of faces corresponding to ρ and $\rho + \Delta\rho$ is approximately equal to

$$[F_\rho(\rho + \Delta\rho, \theta, \phi)(\rho + \Delta\rho)^2 \sin \phi - F_\rho(\rho, \theta, \phi)\rho^2 \sin \phi]\Delta\theta \, \Delta\phi,$$

and analogous results can be written down for the other pairs of faces.

This procedure can be adapted to any system of orthogonal curvilinear coordinates. The reasoning can be made quite precise with the aid of mean-value theorems.

15.61   Green's indentities

Among the important consequences of the divergence theorem are certain integral formulas known as Green's identities. These formulas play a significant role in connection with many of the partial differential equations of applied mathematics. The first identity is

(15.61–1)

$$\iiint_T \left[u \, \Delta v + \frac{\partial u}{\partial x} \frac{\partial v}{\partial x} + \frac{\partial u}{\partial y} \frac{\partial v}{\partial y} + \frac{\partial u}{\partial z} \frac{\partial v}{\partial z} \right] dV = \iint_S u \, \frac{\partial v}{\partial n} \, dA,$$

where we have used the notation

$$\Delta v = \frac{\partial^2 v}{\partial x^2} + \frac{\partial^2 v}{\partial y^2} + \frac{\partial^2 v}{\partial z^2}.$$

The expression Δv is called the Laplacian of v. In the surface integral, $\partial/\partial n$ refers to the directional derivative in the direction of $\mathbf{n}$, the outer normal to S. It is assumed that T is a region to which the divergence theorem applies, and that u and v, together with the first derivatives of u and the first and second derivatives of v, are continuous in T.

To prove the correctness of (15.61–1), choose

$$P = u \frac{\partial v}{\partial x}, \qquad Q = u \frac{\partial v}{\partial y}, \qquad R = u \frac{\partial v}{\partial z}$$

in the divergence theorem. Then

$$\frac{\partial P}{\partial x} + \frac{\partial Q}{\partial y} + \frac{\partial R}{\partial z}$$

is the expression under the triple integral in (15.61–1). On the other hand,

$$P \cos \alpha + Q \cos \beta + R \cos \gamma = u \frac{\partial v}{\partial n}$$

because $$\frac{\partial v}{\partial x} \cos \alpha + \frac{\partial v}{\partial y} \cos \beta + \frac{\partial v}{\partial z} \cos \gamma = \frac{\partial v}{\partial n}.$$

Thus (15.61–1) is just a particular case of the divergence theorem.

The identity can be written in a more compact form by using vector notation. If v is a scalar function, and ∇v is the gradient of v, we recall that the component of ∇v in any particular direction is the directional derivative of v in that direction. Hence, at a point on S

$$\nabla v \cdot \mathbf{n} = \frac{\partial v}{\partial n} \cdot$$

Likewise, we observe that the divergence of ∇v is the Laplacian of v:

$$\nabla \cdot (\nabla v) = \Delta v.$$

Since $\nabla \cdot (\nabla v)$ is conventionally written as $\nabla^2 v$, this latter notation is often used in place of Δv for the Laplacian of v. We may now write (15.61–1) in the form

$$(15.61\text{–}2) \qquad \iiint_T (u\,\nabla^2 v + \nabla u \cdot \nabla v)\,dV = \iint_S u\,\nabla v \cdot \mathbf{n}\,dA.$$

This is the particular case of (15.6–4) in which $\mathbf{F} = u\,\nabla v$.

We refer to (15.61–1) or (15.61–2) as *Green's first identity.*

Green's second identity is

$$(15.61\text{–}3) \qquad \iiint_T (u\,\Delta v - v\,\Delta u)\,dV = \iint_S \left(u\,\frac{\partial v}{\partial n} - v\,\frac{\partial u}{\partial n} \right) dA,$$

or, in a different notation,

$$(15.61\text{–}4) \qquad \iiint_T (u\,\nabla^2 v - v\,\nabla^2 u)\,dV = \iint_S (u\,\nabla v - v\,\nabla u) \cdot \mathbf{n}\,dA.$$

This is deduced from the first identity by exchanging u and v and subtracting. The second identity presumes that both u and v have continuous second derivatives in T.

EXERCISES

1. Show that

$$\iiint_T \nabla^2 u\,dV = \iint_S \frac{\partial u}{\partial n}\,dA.$$

2. If u is a function such that $\nabla^2 u = 0$ in T, show that

$$\iiint_T \left[\left(\frac{\partial u}{\partial x} \right)^2 + \left(\frac{\partial u}{\partial y} \right)^2 + \left(\frac{\partial u}{\partial z} \right)^2 \right] dV = \iint_S u\,\frac{\partial u}{\partial n}\,dA.$$

3. By putting $\mathbf{F} = \nabla u$ in Exercise 11, § 15.6, obtain the expression for the Laplacian in spherical co-ordinates:

$$\nabla^2 u = \frac{1}{\rho^2 \sin \phi} \left\{ \sin \phi\,\frac{\partial}{\partial \rho} \left(\rho^2\,\frac{\partial u}{\partial \rho} \right) + \frac{1}{\sin \phi}\,\frac{\partial^2 u}{\partial \theta^2} + \frac{\partial}{\partial \phi} \left(\sin \phi\,\frac{\partial u}{\partial \phi} \right) \right\}.$$

4. Show that the Laplacian can be expressed independently of all co-ordinate systems in the form

$$\nabla^2 u = \lim \frac{1}{V} \iint\limits_{S} \frac{\partial u}{\partial n}\, dA,$$

where the limit is taken as the surface S enclosing the volume V shrinks down on the point at which the Laplacian is evaluated.

15.62   Transformation of triple integrals

We shall now take up the problem for triple integrals which corresponds to the problem of § 15.32 for double integrals. We suppose that the two sets of variables (x, y, z), (u, v, w) are connected by certain equations

$$(15.62–1) \qquad x = f(u, v, w), \qquad y = g(u, v, w), \qquad z = h(u, v, w),$$

and that these equations establish a one-to-one mapping of a region in uvw-space onto a region in xyz-space. The functions f, g, h are assumed to be continuous, with continuous first and second partial derivatives. We write

$$J(u, v, w) = \frac{\partial(f, g, h)}{\partial(u, v, w)},$$

and assume that J has a constant sign. Let T be a closed and bounded region which is in the region of xyz-space which is being mapped, and let T' be the corresponding region of uvw-space. We assume that T and T' are regions to which the divergence theorem applies.

THEOREM IX.   *Let V be the volume of T. Then*

$$(15.62–2) \qquad V = \iiint\limits_{T'} |\, J(u, v, w)\,|\; du\; dv\; dw.$$

PROOF.   The proof is similar to that of Theorem II, § 15.32, but the calculations are somewhat more complicated. Let us begin by considering the surfaces S, S' of T and T' respectively. Suppose S_0 is a smooth simple surface element on S, and that S_0' is the corresponding element of S'. We assume that S_0 is represented parametrically in terms of certain parameters s, t. The correspondence set up by the mapping (15.62–1) gives us a representation of S_0' in terms of these same parameters. Let us write

$$j_1 = \frac{\partial(y, z)}{\partial(s, t)}, \qquad j_2 = \frac{\partial(z, x)}{\partial(s, t)}, \qquad j_3 = \frac{\partial(x, y)}{\partial(s, t)},$$

$$j_1' = \frac{\partial(v, w)}{\partial(s, t)}, \qquad j_2' = \frac{\partial(w, u)}{\partial(s, t)}, \qquad j_3' = \frac{\partial(u, v)}{\partial(s, t)}.$$

Also, let

$$D = \sqrt{j_1^2 + j_2^2 + j_3^2}, \quad D' = \sqrt{j_1'^2 + j_2'^2 + j_3'^2}.$$

We first show that

$$(15.62\text{–}3) \qquad j_3 = \frac{\partial(f, g)}{\partial(v, w)} j_1' + \frac{\partial(f, g)}{\partial(w, u)} j_2' + \frac{\partial(f, g)}{\partial(u, v)} j_3'.$$

There are similar formulas for j_1, j_2, but we shall not need them.

By (15.62–1) we see that

$$\frac{\partial x}{\partial s} = \frac{\partial f}{\partial u} \frac{\partial u}{\partial s} + \frac{\partial f}{\partial v} \frac{\partial v}{\partial s} + \frac{\partial f}{\partial w} \frac{\partial w}{\partial s},$$

with similar formulas for $\dfrac{\partial x}{\partial t}$, $\dfrac{\partial y}{\partial s}$, $\dfrac{\partial y}{\partial t}$.

In the interests of compactness in display, let us write

$$\frac{\partial f}{\partial u} = f_1, \qquad \frac{\partial f}{\partial v} = f_2, \qquad \frac{\partial u}{\partial s} = u_1, \qquad \frac{\partial u}{\partial t} = u_2,$$

and so on. Then we have

$$j_3 = \begin{vmatrix} \dfrac{\partial x}{\partial s} & \dfrac{\partial x}{\partial t} \\[2mm] \dfrac{\partial y}{\partial s} & \dfrac{\partial y}{\partial t} \end{vmatrix} = \begin{vmatrix} f_1 u_1 + f_2 v_1 + f_3 w_1 & f_1 u_2 + f_2 v_2 + f_3 w_2 \\[2mm] g_1 u_1 + g_2 v_1 + g_3 w_1 & g_1 u_2 + g_2 v_2 + g_3 w_2 \end{vmatrix}.$$

There are theorems on determinants by which the foregoing determinant may be expanded in the form

$$j_3 = \begin{vmatrix} f_2 & f_3 \\ g_2 & g_3 \end{vmatrix} \cdot \begin{vmatrix} v_1 & v_2 \\ w_1 & w_2 \end{vmatrix} + \begin{vmatrix} f_3 & f_1 \\ g_3 & g_1 \end{vmatrix} \cdot \begin{vmatrix} w_1 & w_2 \\ u_1 & u_2 \end{vmatrix} + \begin{vmatrix} f_1 & f_2 \\ g_1 & g_2 \end{vmatrix} \cdot \begin{vmatrix} u_1 & u_2 \\ v_1 & v_2 \end{vmatrix}.$$

This latter form is equivalent to (15.62–3).

It was shown in § 14.4 that the direction cosines of the normal to S_0 are given by

$$(15.62\text{–}4) \qquad \cos \alpha = \frac{j_1}{D}, \qquad \cos \beta = \frac{j_2}{D}, \qquad \cos \gamma = \frac{j_3}{D}.$$

In an arbitrary parametrization these may or may not refer to the outward direction of the normal, but since an exchange of the parameters s, t changes the sign of each of the Jacobians j_1, j_2, j_3, we may (and shall) assume that the parameters have been chosen so that (15.62–4) give the direction of the outer normal to S_0. The direction cosines of the outer normal to S_0' are

$$(15.62\text{–}5) \quad \cos \alpha' = \pm \frac{j_1'}{D'}, \qquad \cos \beta' = \pm \frac{j_2'}{D'}, \qquad \cos \gamma' = \pm \frac{j_3'}{D'},$$

where the sign may be $+$ or $-$, but is the same in all three equations. As we shall see, the sign depends upon the sign of $J(u, v, w)$.

Now, by the divergence theorem with $P = Q = 0$, $R = z$, we have

$$V = \iiint_T dx \, dy \, dz = \iint_S z \cos \gamma \, dA.$$

We show that

$$(15.62\text{–}6) \qquad \iint_S z \cos \gamma \, dA =$$

$$\pm \iint_{S'} h(u, v, w) \left[\frac{\partial(f, g)}{\partial(v, w)} \cos \alpha' + \frac{\partial(f, g)}{\partial(w, u)} \cos \beta' + \frac{\partial(f, g)}{\partial(u, v)} \cos \gamma' \right] dA,$$

where the choice of sign is the same as in (15.62–5). It is enough to carry out the demonstration for each pair of corresponding surface elements S_0, S_0'. By (15.51–4) and (15.62–4) we see that

$$\iint_{S_0} z \cos \gamma \, dA = \iint_R (z \cos \gamma) D \, ds \, dt = \iint_R z j_3 \, ds \, dt,$$

where R is the region in the st-plane corresponding to the element S_0. The last integral may be transformed into a surface integral over S_0' (again by (15.51–4)):

$$\iint_R z j_3 \, ds \, dt = \iint_{S_0'} \frac{z j_3}{D'} \, dA.$$

We now express the integrand in terms of u, v, w, using $z = h(u, v, w)$, (15.62–3) and (15.62–5). The result is the integrand on the right in (15.62–6), and thus the latter formula is established.

The next step is the application of the divergence theorem in the uvw-space, with

$$P = h \frac{\partial(f, g)}{\partial(v, w)}, \quad Q = h \frac{\partial(f, g)}{\partial(w, u)}, \quad R = h \frac{\partial(f, g)}{\partial(u, v)}.$$

A calculation shows that

$$\frac{\partial P}{\partial u} + \frac{\partial Q}{\partial v} + \frac{\partial R}{\partial w} = \frac{\partial(f, g, h)}{\partial(u, v, w)} = J(u, v, w);$$

there are certain second-derivative terms, but they cancel each other out. Thus, by (15.62–6) and the divergence theorem,

$$V = \pm \iiint_{T'} J(u, v, w) du \, dv \, dw.$$

This is equivalent to (15.62–2), for J is of constant sign, and V is of course positive.

It follows from the foregoing that the positive signs are to be chosen in (15.62–5) when $J > 0$, while the negative signs are to be chosen if $J < 0$.

With the aid of Theorem IX we can prove the following three-dimensional analogue of Theorem III, § 15.32:

THEOREM X. *If $F(x, y, z)$ is continuous in T and if the mapping meets the requirements stated prior to Theorem IX, we have*

(15.62–7)

$$\iiint\limits_{T} F(x, y, z)dx\, dy\, dz = \iiint\limits_{T'} F(x, y, z)\left|\frac{\partial(x, y, z)}{\partial(u, v, w)}\right| du\, dv\, dw,$$

where, on the right, $F(x, y, z)$ is to be expressed as a function of u, v, w by means of (15.62–1).

The proof is entirely analogous to that of Theorem III.

EXERCISES

1. Compute the Jacobian $J(u, v, w)$ for the transformation
$$x = \rho \sin \phi \cos \theta, \qquad y = \rho \sin \phi \sin \theta, \qquad z = \rho \cos \phi,$$
where $u = \rho$, $v = \phi$, $w = \theta$, and compare Theorem X for this case with the known results about evaluating triple integrals in terms of spherical co-ordinates.

2. Consider the transformation
$$x = u + 2v + 3w, \qquad y = 4u + 5v, \qquad z = 6u.$$
(a) If R is a region in xyz-space, and R' is the corresponding region in uvw-space, what is the volume of R if the volume of R' is 10?
(b) If the centroid of R is at (9, 45, 30), find the centroid of R'.

3. Use the transformation $x = au$, $y = bv$, $z = cw$ to calculate the triple integral $\iiint\limits_{R} (x^2 + y^2)dx\, dy\, dz$, where R is defined by $(x^2/a^2) + (y^2/b^2) + (z^2/c^2) \leq 1$.
The transformed integral may be evaluated by use of spherical co-ordinates in uvw-space.

4. Consider the transformation
$$x = \frac{v}{u} \cos w, \qquad y = \frac{v}{u} \sin w, \qquad z = v^2.$$
Let R be the region between the paraboloids $z = x^2 + y^2$, $z = 4(x^2 + y^2)$ and also between the planes $z = 1$, $z = 4$. Use the transformation to calculate the integral

$$\iiint\limits_{R} (x^2 + y^2)dx\, dy\, dz.$$

5. Consider the transformation

$$x = \frac{v}{1 + w}, \qquad y = \frac{vw}{1 + w}, \qquad z = u - v.$$

Show that the inverse transformation is

$$u = x + y + z, \qquad v = x + y, \qquad w = \frac{y}{x}.$$

Let R be the region in the first octant of xyz-space under the plane $x + y + z = 2$ and directly above the trapezoid in the xy-plane bounded by the lines $x + y = 1$, $x + y = 2$, $y = 0$, $y = x$. Show that the corresponding region R' in uvw-space is a prism with two faces perpendicular to the w-axis and three faces parallel to that axis. Then evaluate

$$\iiint_R \frac{xy + y^2}{x^3} \, dx \, dy \, dz$$

by transforming to an integral over R'.

6. Use the transformation

$$x = u(1 - v), \qquad y = uv(1 - w), \qquad z = uvw$$

to calculate

$$\iiint_R x \, dx \, dy \, dz \quad \text{and} \quad \iiint_R \frac{dx \, dy \, dz}{y + z}$$

where R is the tetrahedron cut from the first octant by the plane $x + y + z = 1$.

7. If curvilinear co-ordinates u, v, w are introduced into the first octant by the transformation

$$x = v \cos w, \qquad y = v \sin w, \qquad z = \sqrt{u - v^2},$$

describe the surfaces on which u, v, w respectively are constant. Describe the region R' in uvw-space which corresponds to the region R described as follows:

$$9 \leq x^2 + y^2 + z^2 \leq 16, \, 1 \leq x^2 + y^2 \leq 4, \, x \geq 0, \, y \geq 0, \, z \geq 0.$$

Calculate $\iiint_R z \, dx \, dy \, dz$ with the aid of the transformation.

8. Let curvilinear co-ordinates be introduced into the first octant by the transformation

$$x = v, \qquad y = w, \qquad z = (u - v^2 - w^2)^{1/2};$$

show that, for small values of Δu, Δv, Δw, the volume in the first octant bounded by the planes $x = v$, $x = v + \Delta v$, $y = w$, $y = w + \Delta w$, and the spheres $x^2 + y^2 + z^2 = u$, $x^2 + y^2 + z^2 = u + \Delta u$ is approximately

$$\frac{\Delta u \, \Delta v \, \Delta w}{2\sqrt{u - v^2 - w^2}}.$$

9. If R is the region defined by $x^2 + y^2 + z^2 \leq 1$, and $p^2 = a^2 + b^2 + c^2 > 0$, show that

$$\iiint_R \cos (ax + by + cz) dx \, dy \, dz = \frac{4 \pi}{p^3} (\sin p - p \cos p).$$

HINT: Make a rotation of the co-ordinate axes so that the plane $ax + by + cz = 0$ becomes the plane $z' = 0$. Then use cylindrical co-ordinates.

10. Suppose each point (x, y, z) moves according to the law

$$x = \xi(1 + t^2), \qquad y = \eta e^t, \qquad z = \zeta e^{2t},$$

so that the point is at (ξ, η, ζ) when $t = 0$. If the components of velocity are computed in terms of x, y, z we find

$$\frac{\partial x}{\partial t} = 2\, t\xi = \frac{2\, tx}{1 + t^2}, \qquad \frac{\partial y}{\partial t} = \eta e^t = y, \qquad \frac{\partial z}{\partial t} = 2\, \zeta e^{2t} = 2\, z,$$

so that the velocity is the vector

$$\mathbf{F} = \frac{2\, tx}{1 + t^2}\, \mathbf{i} + y\mathbf{j} + 2\, z\mathbf{k}.$$

Now let R_0 be any fixed region in $\xi\eta\zeta$-space, and let R be the region in xyz-space into which R_0 has been carried at time t. If V is the volume of R, show that

$$\frac{dV}{dt} = \iiint\limits_{R} (\nabla \cdot \mathbf{F})dx\, dy\, dz.$$

Actually, the validity of this equation does not depend on the particular equations of motion.

11. Show that the last formula in Exercise 10 holds for the case in which the equations of motion are

$$x = \frac{\xi}{1 + \xi t}, \qquad y = \eta + t, \qquad z = \zeta e^{-t}.$$

In this case the velocity vector is $\mathbf{F} = -x^2\mathbf{i} + \mathbf{j} - z\mathbf{k}$, and it is to be assumed that $\xi > 0$ in R_0.

15.7　Stokes's theorem

Stokes's theorem is the formula

$$(15.7\text{-}1) \qquad \iint\limits_{S}\left[\left(\frac{\partial R}{\partial y} - \frac{\partial Q}{\partial z}\right)\cos \alpha + \left(\frac{\partial P}{\partial z} - \frac{\partial R}{\partial x}\right)\cos \beta + \right.$$

$$\left. \left(\frac{\partial Q}{\partial x} - \frac{\partial P}{\partial y}\right)\cos \gamma\right] dA = \int_{C} P\, dx + Q\, dy + R\, dz.$$

Here S denotes a surface bounded by a curve C; P, Q, and R denote functions of x, y, z. The normal to S is constructed at each point, and a certain direction is chosen along each normal so that all these directions point away from S on the same side of S. This is called the positive side of S. The angles α, β, γ refer to the direction of the normal at (x, y, z) on S. The curve C is oriented in such a way that, if a person walks along C in the positive direction, standing on the positive side of S, the surface S is always on his left. If S is a simple surface element, this means that

the orientation of C appears as counterclockwise to a person standing on the positive side of the surface (see Fig. 162). A more exact description of the assumptions which we make about the surface S and the functions P, Q, R will be given later on.

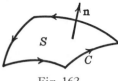

Fig. 162

For the particular case in which S is a region in the xy-plane, Stokes's theorem is identical with Green's theorem in the plane; for in that case, if we choose the positive side of S as that toward the positive z direction, we see that $\cos \alpha = 0$, $\cos \beta = 0$, $\cos \gamma = 1$. Therefore, since $dz = 0$ along C in this case, (15.7–1) becomes

$$\iint_S \left(\frac{\partial Q}{\partial x} - \frac{\partial P}{\partial y} \right) dA = \int_C P\, dx + Q\, dy,$$

and this is Green's theorem in the plane.

We shall now prove formula (15.7–1) for the case in which S is a simple surface element. If we pick out of (15.7–1) just those terms which involve P, we have the formula

(15.7–2) $$\iint_S \left(\frac{\partial P}{\partial z} \cos \beta - \frac{\partial P}{\partial y} \cos \gamma \right) dA = \int_C P\, dx .$$

There are two other similar formulas, one involving Q and one involving R. We shall prove (15.7–2). The other two formulas will then be established by cyclic permutation of the letters in the sets (P, Q, R), (x, y, z), (α, β, γ). Combination of the three formulas will then give us (15.7–1).

We shall assume that the surface element S is represented parametrically by a one-to-one mapping

$$x = f(u, v), \qquad y = g(u, v), \qquad z = h(u, v),$$

the parameters (u, v) ranging over a region G in the uv-plane. We assume that G is a regular region bounded by a simple closed curve Γ which is sectionally smooth. We also assume that the functions f, g, h have continuous first and second derivatives in G and that the Jacobians

$$j_1 = \frac{\partial(y, z)}{\partial(u, v)}, \qquad j_2 = \frac{\partial(z, x)}{\partial(u, v)}, \qquad j_3 = \frac{\partial(x, y)}{\partial(u, v)}$$

are never all zero at once. Let Γ be oriented counterclockwise in the uv-plane, and let C be oriented so that as (u, v) goes along Γ in the positive sense, (x, y, z) goes along C in the positive sense. Let

$$D = (j_1^2 + j_2^2 + j_3^2)^{1/2}.$$

Then for the direction of the normal toward the positive side of S we have

(15.7-3) $\cos \alpha = \pm \dfrac{j_1}{D}, \quad \cos \beta = \pm \dfrac{j_2}{D}, \quad \cos \gamma = \pm \dfrac{j_3}{D},$

where the same choice of sign is to be made in each equation. We shall show that the $+$ sign must be chosen. Select any point of S not on C. At this point j_1, j_2, and j_3 are not all zero, and we shall suppose for definiteness that $j_3 \neq 0$. Then near this point j_3 always has the same sign. Also, the normal near the point is never perpendicular to the z-axis and so $\cos \gamma$ is always of the same sign. The problem is to show that the sign of $\cos \gamma$ is the same as that of j_3. Let C_0 be a small closed curve on the portion of S we are considering. Let Γ_0 be the corresponding curve in the uv-plane, and let C_0' be the projection of C_0 in the xy-plane (see Fig. 163). Let the orientations of C_0 and Γ_0 agree with those of C and Γ respectively, and let the

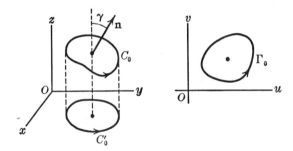

Fig. 163

orientation of C_0' be that which is naturally induced by the orientation of C_0. Now the equations $x = f(u, v)$, $y = g(u, v)$ may be regarded as defining a mapping from the uv-plane to the xy-plane. Under this mapping Γ_0 is mapped into C_0'. The Jacobian of the mapping is j_3, so, by the discussion in § 15.32, the orientation of C_0' will be counterclockwise (as viewed from the positive z-axis) if $j_3 > 0$, and clockwise if $j_3 < 0$. A little consideration of the relation between C_0' and C_0 then shows that $\cos \gamma > 0$ if $j_3 > 0$ and $\cos \gamma < 0$ if $j_3 < 0$. The case $j_3 > 0$ is illustrated in Fig. 163. This completes the demonstration that the $+$ signs are to be chosen in (15.7-3).

We are now ready to proceed with the proof of (15.7-2). In terms of the parameters u, v the surface integral becomes

$$\iint_G \left(\frac{\partial P}{\partial z} \cos \beta - \frac{\partial P}{\partial y} \cos \gamma \right) D \, du \, dv = \iint_G \left(\frac{\partial P}{\partial z} j_2 - \frac{\partial P}{\partial y} j_3 \right) du \, dv.$$

Here we have used (15.51–4) and (14.6–7). Next we show that

(15.7–4) $$\frac{\partial P}{\partial z} j_2 - \frac{\partial P}{\partial y} j_3 = \frac{\partial P}{\partial u} \frac{\partial x}{\partial v} - \frac{\partial P}{\partial v} \frac{\partial x}{\partial u}.$$

In fact, $$\frac{\partial P}{\partial u} = \frac{\partial P}{\partial x} \frac{\partial x}{\partial u} + \frac{\partial P}{\partial y} \frac{\partial y}{\partial u} + \frac{\partial P}{\partial z} \frac{\partial z}{\partial u},$$

with a similar formula for $\dfrac{\partial P}{\partial v}$. Therefore,

$$\frac{\partial P}{\partial u} \frac{\partial x}{\partial v} - \frac{\partial P}{\partial v} \frac{\partial x}{\partial u} = \frac{\partial P}{\partial y}\left[\frac{\partial y}{\partial u} \frac{\partial x}{\partial v} - \frac{\partial y}{\partial v} \frac{\partial x}{\partial u} \right] +$$
$$\frac{\partial P}{\partial z}\left[\frac{\partial z}{\partial u} \frac{\partial x}{\partial v} - \frac{\partial z}{\partial v} \frac{\partial x}{\partial u} \right],$$

and this is equivalent to (15.7–4).

The surface integral in (15.7–2) has now been reduced to the form

$$\iint\limits_{G} \left(\frac{\partial P}{\partial u} \frac{\partial x}{\partial v} - \frac{\partial P}{\partial v} \frac{\partial x}{\partial u} \right) du\, dv.$$

An easy calculation shows that this is the same as

(15.7–5) $$\iint\limits_{G} \left[\frac{\partial}{\partial u}\left(P \frac{\partial x}{\partial v} \right) - \frac{\partial}{\partial v}\left(P \frac{\partial x}{\partial u} \right) \right] du\, dv.$$

To this integral we now apply Green's theorem in the uv-plane. As a result, (15.7–5) is equal to the line integral

$$\int\limits_{\Gamma} P \frac{\partial x}{\partial u}\, du + P \frac{\partial x}{\partial v}\, dv.$$

But this is just $$\int\limits_{C} P\, dx,$$

and so we have completed the proof of (15.7–2) under the assumptions on S as stated earlier. We have assumed that P, Q, and R have continuous partial derivatives in some region containing S.

Stokes's theorem may be extended to more general surfaces by a process entirely similar to that employed in the proof of Green's theorem in the plane. The process is suggested by Fig. 164. It consists in dividing S into a finite number of simple surface elements by the construction of one or more "cuts," or interior dividing

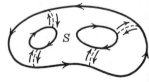

Fig. 164

lines. We assume that each element and its boundary takes its orientation from the overall orientation of S and C, and that each "cut" occurs as part of the boundary of just two surface elements, with opposite orientations in the two cases. If now we add the formulas of Stokes's theorem for the several elements, the contributions from the cuts to the line integrals cancel each other in pairs, and we obtain (15.7–1) as the final result.

Notice that it is taken for granted that there is an orientation of the boundary C of S, and a corresponding orientation of S itself, that is, a designation of a positive side of S. The assumption that such orientation of S is possible is a restriction placed on S, for there are surfaces which cannot be oriented, as we shall presently show.

The concept of orientation of a surface may be developed in the following way: We limit our discussion to surfaces which may be thought of as formed out of a finite number of simple surface elements in such a way that if two elements have an arc of common boundary, then no other element has any sub-arc of this arc as part of its boundary. Moreover, we require that if two elements have any points in common, the common points shall form a finite number of arcs on the boundary of each element. A simple surface element is oriented by assigning a positive direction to the simple closed curve forming its boundary. If S is a surface formed out of elements in the manner described, and if each element is given an orientation, we shall say that the elements are *coherently* oriented if the following two conditions are fulfilled:

(a) If Γ is an arc of common boundary between two elements S_1, S_2, then the orientation of Γ in S_1 is opposite to its orientation in S_2;

(b) If Γ is a simple closed curve forming part of the boundary of the whole surface S, then the orientations given to various arcs of Γ by the several elements of S shall all be consistent and given an orientation to Γ as a whole.

If the elements forming S can be coherently oriented, we shall say that S is *orientable*. Otherwise, S is said to be *nonorientable*. These definitions apply to both closed and nonclosed surfaces. When S is orientable we can designate a positive side of S by designating a positive side to each element in accord with the orientation of the boundary of the element, as explained near the beginning of this section. But if S is nonorientable we cannot designate a positive side of S, for there will be two adjacent elements of S for which the designations of the positive sides will be in conflict as we pass from one element to the other. A nonorientable surface is in fact *one-sided*, whereas an orientable surface is two-sided. The simplest nonorientable surface is the Moebius band, represented in Fig. 165. The student can easily make a

Fig. 165

model of this surface by cutting a long narrow strip of paper, and gluing the two ends together after giving one end a half twist. The Moebius band has a single simple closed curve C as boundary.

Stokes's theorem may be given a vector form, as follows:

THEOREM XI.   *Let S be an orientable surface as described above, formed from smooth surface elements with sectionally smooth boundaries. Let $\mathbf{F}$ be a vector field which is continuously differentiable in some open set containing S. Let $\mathbf{n}$ be the unit normal vector on the positive side of S. Then if S is not closed, and if C is the boundary of S,*

$$(15.7\text{--}6) \qquad \iint_S (\nabla \times \mathbf{F}) \cdot \mathbf{n}\, dA = \int_C \mathbf{F} \cdot \mathbf{T}\, ds,$$

where $\mathbf{T}$ is the unit vector tangent to C in the positive sense. If S is closed, the surface integral in (15.7–6) has the value zero.

The form (15.7–6) is equivalent to (15.7–1) with $\mathbf{F} = P\mathbf{i} + Q\mathbf{j} + R\mathbf{k}$, since, by (14.3–1),

$$\mathbf{T} = \frac{dx}{ds}\mathbf{i} + \frac{dy}{ds}\mathbf{j} + \frac{dz}{ds}\mathbf{k}.$$

If S is closed, the line integrals around the boundaries of the elements forming S all cancel out, because S has no boundary in this case. Hence, the surface integral over S must vanish.

EXERCISES

1. Let C be the curve of intersection of $x + y = 2b$ and $x^2 + y^2 + z^2 = 2b(x + y)$, oriented in the clockwise sense as viewed from the origin. Use Stokes's theorem to find the value of $\int_C y\, dx + z\, dy + x\, dz$.

2. Let S be the part of the surface $z = x^2 - y^2$ inside the cylinder $x^2 + y^2 = 1$, and let the positive side of S be such that γ is acute. Show that, if C is the boundary of S,

$$\iint_S (x \cos \alpha - y \cos \beta)\,dA = \int_C xy\, dz$$

is a special case of Stokes's theorem. Calculate the values of the line integral and the surface integral independently, without use of Stokes's theorem, and verify that they are equal. Observe that C may be parametrized by setting $x = \cos \theta$, $y = \sin \theta$.

3. Calculate the value of $\int_C y\, dx + z\, dy + x\, dz$ where C is the curve of intersection of $bz = xy$ and $x^2 + y^2 = a^2$, oriented counterclockwise around the cylinder as viewed from a point high upon the positive z-axis. Use Stokes's theorem and then convert to a double integral over a circle in the xy-plane.

4. Let C be the curve in which the cylinder $x^2 + y^2 = a^2$ is intersected by a plane parallel to the x-axis, and let C be oriented counterclockwise around the cylinder as viewed from a point high on the positive z-axis.

   (a) Show that $\int_C z(x^2 - 1)dy + y(x + 1)dz = 0$.

   (b) Show that $\int_C y(z - 1)dx + x(z + 1)dy = 2\pi a^2$.

   (c) If the plane is $y + z = a$, show that $\int_C z\, dx - x\, dz = 2\pi a^2$.

   (d) What is the value of the integral in part (c) if the plane is $z = 2(a + y)$?

5. Calculate $\int_C y^2\, dx + xy\, dy + zx\, dz$, where C is the curve of intersection of $x^2 + y^2 = 2\, ay$ and $y = z$.

6. Let S be the first octant portion of the surface of the sphere $x^2 + y^2 + z^2 = 4\, a^2$ which is inside the cylinder $x^2 + y^2 = 2\, ax$. Let the outer side of the sphere be the positive side of S. Calculate the integrals

$$\text{(a)} \int_C z\, dx - x\, dz, \quad \text{(b)} \int_C x\, dy - y\, dx, \quad \text{(c)} \int_C y\, dz - z\, dy,$$

where C is the boundary of S, oriented as in Stokes's theorem. Observe that, by suitable use of projection in evaluating the surface integrals which result from using Stokes's theorem, each line integral is equal to twice the area of an appropriate plane region.

7. If S is a simple surface element whose equation is $z = f(x, y)$, with (x, y) varying over a plane region R, let C be the boundary of S and Γ the boundary of R. Prove Stokes's theorem for S by converting the line integral over C to a line integral over Γ and the integral over S to an integral over R. Then use Green's theorem in the plane.

15.8   Exact differentials in three variables

An expression such as

(15.8–1) $P\, dx + Q\, dy + R\, dz$

is called a first-order differential form in three variables. Here P, Q, R denote functions of x, y, z. In this section we consider exact differentials and line integrals independent of the path.

DEFINITION.   The differential form (15.8–1) is said to be *exact at the point* (a, b, c) if there is some single-valued differentiable function $u = f(x, y, z)$ defined in some neighborhood of (a, b, c) such that

$$du = P\, dx + Q\, dy + R\, dz$$

at all points of the neighborhood.

The concept of an exact differential can be expressed in vector language, using the notion of the gradient of a scalar function. Let **F** be a vector field. It is called a *gradient field at a particular point* in case there is a scalar function u defined in some neighborhood of that point, with the property that **F** is the gradient of u in that neighborhood. In terms of the rectangular xyz-coordinate system, suppose that

$$(15.8\text{–}2) \qquad\qquad \mathbf{F} = P\mathbf{i} + Q\mathbf{j} + R\mathbf{k}.$$

Then $\nabla u = \mathbf{F}$ means that

$$\frac{\partial u}{\partial x} = P, \quad \frac{\partial u}{\partial y} = Q, \quad \frac{\partial u}{\partial z} = R,$$

or, equivalently, that

$$(15.8\text{–}3) \qquad\qquad du = P\,dx + Q\,dy + R\,dz.$$

Therefore, the differential form (15.8–1) *is exact at a point if and only if* **F** *is a gradient field at that point.*

We now state a theorem about line integrals independent of the path.

THEOREM XII.   *Suppose that P, Q, R are continuous in some region, and that in this region there is defined a single-valued differentiable function $u = f(x, y, z)$ such that* (15.8–3) *holds at all points of the region. Then, if C is any sectionally smooth curve in the region, with initial point (x_0, y_0, z_0) and terminal point (x_1, y_1, z_1), the equation*

$$(15.8\text{–}4) \quad \int_C P\,dx + Q\,dy + R\,dz = f(x_1, y_1, z_1) - f(x_0, y_0, z_0)$$

holds. Therefore the line integral is independent of the path from (x_0, y_0, z_0) to (x_1, y_1, z_1).

This theorem corresponds to Theorem V, § 15.4, and is proved in the same way.

Our next theorem corresponds to Theorem IV, § 15.4.

THEOREM XIII.   *Let D be an open region in the shape of a rectangular box, each face of which is parallel to a co-ordinate plane. Suppose P, Q, R are defined and have continuous first partial derivatives in D. Then, in order that there shall be a single-valued differentiable function $u = f(x, y, z)$ defined in D such that $du = P\,dx + Q\,dy + R\,dz$ at each point, it is necessary and sufficient that*

$$(15.8\text{–}5) \quad \frac{\partial R}{\partial y} - \frac{\partial Q}{\partial z} = 0, \quad \frac{\partial P}{\partial z} - \frac{\partial R}{\partial x} = 0, \quad \frac{\partial Q}{\partial x} - \frac{\partial P}{\partial y} = 0$$

at each point. Such a function can then be found by calculating the line integral of the differential form from an arbitrary fixed point (x_0, y_0, z_0) to the variable point (x, y, z) along any path lying in D.

PROOF. The conditions (15.8–5) are necessary, for if the function u exists we have

$$\frac{\partial u}{\partial z} = R, \quad \frac{\partial u}{\partial y} = Q, \quad \frac{\partial^2 u}{\partial y\, \partial z} = \frac{\partial^2 u}{\partial z\, \partial y},$$

and so $\dfrac{\partial R}{\partial y} = \dfrac{\partial Q}{\partial z}$; similar arguments show that all the equations (15.8–5) must be satisfied.

Let us now start with the assumption that equations (15.8–5) hold in D. Choose any fixed point (x_0, y_0, z_0) in D, and let (x_1, y_1, z_1) be any other point in D. Let C be the straight line segment from (x_0, y_0, z_0) to (x_1, y_1, z_1) and define

(15.8–6) $$f(x_1, y_1, z_1) = \int_C P\, dx + Q\, dy + R\, dz.$$

This gives us a definition for the value of the function at each point of D. It remains only to show that

(15.8–7) $$\frac{\partial f}{\partial x} = P, \quad \frac{\partial f}{\partial y} = Q, \quad \frac{\partial f}{\partial z} = R.$$

Now suppose that (a, b, c) is any third point of D, and let C_1, C_2 be the segments as shown in Fig. 166. We shall show that

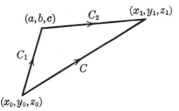

Fig. 166

(15.8–8) $$f(x_1, y_1, z_1) = f(a, b, c) + \int_{C_2} P\, dx + Q\, dy + R\, dz.$$

This is equivalent to showing that

$$\int_C = \int_{C_1} + \int_{C_2}$$

(with $P\ dx + Q\ dy + R\ dz$ understood as appearing under each integral sign). This in turn is equivalent to showing that

$$(15.8\text{–}9) \qquad\qquad \int_\Gamma P\ dx + Q\ dy + R\ dz = 0$$

where Γ is the oriented path all the way around the triangle in one direction. If S is the plane triangular region bounded by Γ, we see that (15.8–9) holds by virtue of (15.8–5) and Stokes's theorem (15.7–1).

Let us now consider x_1 as a variable, and y_1, z_1 as fixed, and let us take C_2 parallel to the x-axis, so that $b = y_1$, $c = z_1$. Then (15.8–8) becomes

$$f(x_1, y_1, z_1) = f(a, y_1, z_1) + \int_a^{x_1} P(x, y_1, z_1)dx.$$

From this equation it is clear that

$$\frac{\partial f(x_1, y_1, z_1)}{\partial x_1} = P(x_1, y_1, z_1)\,.$$

We have thus proved the first formula in (15.8–7). The other two are proved in the same way. This completes the proof of Theorem XIII. It will be observed that, for the proof we have given, the open region D need not be a rectangular box, provided it has the property that any two points of D can be joined by a line segment lying wholly in D. Such a region is called *convex*.

The substance of Theorems XII and XIII is capable of succinct statement in vector notation. Theorem XII says that

$$\int_C \mathbf{F} \cdot \mathbf{T}\ ds = f(x_1, y_1, z_1) - f(x_0, y_0, z_0)$$

provided $\mathbf{F} = \nabla f$ in the region where C lies. Theorem XIII says that $\mathbf{F}$ is the gradient of some function f in the region D if and only if $\nabla \times \mathbf{F} = \mathbf{0}$ in D. For, with $\mathbf{F}$ defined by (15.8–2), equations (15.8–5) mean precisely that $\nabla \times \mathbf{F} = \mathbf{0}$.

Example. Of what function is

$$(y^2 + 2\ z^2 x - 1)dx + 2\ yx\ dy + 2\ zx^2\ dz$$

the differential?

This differential form satisfies the conditions (15.8–5) at all points. We shall compute a function $f(x, y, z)$ by integrating from $(0, 0, 0)$ to (x, y, z) along a broken line as shown in Fig. 167. We use (r, s, t) as co-ordinates of a variable point along the path. Then

$$f(x, y, z) = \int_C (s^2 + 2\,t^2 r - 1)dr + 2\,sr\,ds + 2\,tr^2\,dt$$

$$= \int_0^x - dr + \int_0^y 2\,sx\,ds + \int_0^z 2\,tx^2\,dt,$$

$$f(x, y, z) = -x + y^2 x + z^2 x^2.$$

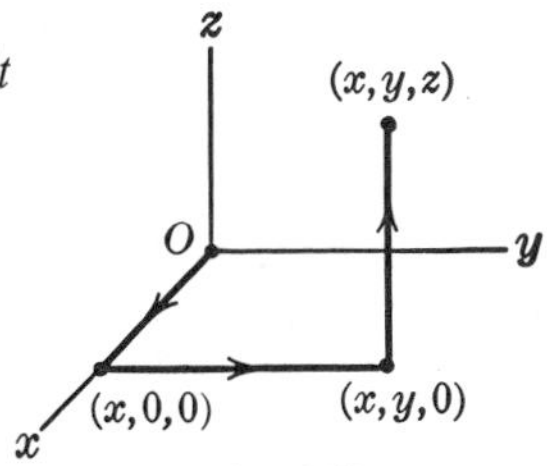

The differential of this function is the originally given differential form.

Fig. 167

Theorem XIII is not true for *all* open regions D. Consider, for example, the differential form

$$\frac{-y\,dx + x\,dy}{x^2 + y^2} + 2\,z\,dz,$$

where D is all of space except the z-axis. In this region the equations (15.8–5) are satisfied, and yet there is no single-valued function defined in D whose differential is equal to the foregoing differential form. The explanation lies in Example 1, § 15.41. The essential trouble is that, if a closed curve C in D encircles the z-axis, it is impossible to find a surface S in D, with C as a boundary, to which we can apply Stokes's theorem.

To get away from difficulties of this kind we need to develop a notion which for space of three dimensions is analogous to the notion of a simply connected region in two dimensions, as explained in § 15.41. We shall not attempt any systematic developments of this kind, however.

EXERCISES

1. In each of the following cases determine, by inspection or otherwise, a function whose differential is the differential form under the integral sign. Then use Theorem XII to evaluate the line integral over the specified path.

(a) $\int_C yz\,dx + zx\,dy + xy\,dz$, from $(1, 2, 3)$ to $(4, 5, 6)$;

(b) $\int_C 2\,xy^2 z\,dx + 2\,x^2 yz\,dy + x^2 y^2\,dz$, from $(0, 0, 0)$ to (a, b, c);

(c) $\int_C 2\,xyz^3\,dx + x^2 z^3\,dy + 3\,x^2 yz^2\,dz$, from $(1, 1, 1)$ to (p, q, r);

(d) $\int_C yz \cos (xz)dx + \sin (xz)dy + xy \cos (xz)dz$, from $(\pi^2/4, 1, 2/\pi)$ to $(\pi/2, -2, 3)$.

2. Let $u = x/r^2$ where $r^2 = x^2 + y^2 + z^2$, and let $\mathbf{F} = \nabla u$. Compute $\int_C \mathbf{F} \cdot \mathbf{T}\,ds$, where C is any curve from $(1, 0, 0)$ to (a, b, c) not passing through the origin.

3. Find a function whose differential is $2\,xy\,dx + (x^2 + \log z)dy + (y/z)dz$, by integrating along a path from $(0, 0, 1)$ to (x, y, z). Solve the problem in several ways: (a) Use a broken-line path from $(0, 0, 1)$ to $(x, 0, 1)$ to $(x, y, 1)$ to (x, y, z); (b) use a broken-line path from $(0, 0, 1)$ to $(0, 0, z)$ to $(0, y, z)$ to (x, y, z); (c) use the straight-line path from $(0, 0, 1)$ to (x, y, z).

4. Find a function whose differential is

$$\frac{2x}{z}\,dx + \frac{2y}{z}\,dy + \left(1 - \frac{x^2 + y^2}{z^2}\right)dz,$$

by integrating over some path from $(1, 1, 1)$ to (x, y, z).

5. Find a function whose differential is

$$e^{-xy}[(y - xy^2 + yz)dx + (x - x^2y + xz)dy - dz],$$

by integrating over some path from $(0, 0, 0)$ to (x, y, z).

6. Let C be the circle $x^2 + y^2 = a^2$, $z = 0$. If $P(x, y, z)$ is any point, let P' be the point $(x, y, 0)$, and let Q be the point in which the ray OP' (produced if necessary) intersects C. Let ϕ be the angle between the directions OQ and QP (see Fig. 168), with the convention that $0 < \phi < \pi$ if $z > 0$, $-\pi < \phi < 0$ if $z < 0$, and, if $z = 0$, that $\phi = 0$ if $x^2 + y^2 > a^2$ and $\phi = \pi$ if $x^2 + y^2 < a^2$. This leaves ϕ undefined at points of the circle C, but it is defined everywhere else.

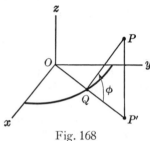

Fig. 168

(a) Show that, in general,

$$d\phi = \frac{-xz\,dx - yz\,dy}{\sqrt{x^2 + y^2}[(\sqrt{x^2 + y^2} - a)^2 + z^2]} + \frac{\sqrt{x^2 + y^2} - a}{(\sqrt{x^2 + y^2} - a)^2 + z^2}\,dz.$$

(b) At what points is ϕ discontinuous?

(c) At what points is ϕ continuous but not differentiable?

(d) If D is the region consisting of all of space except the z-axis and the circle C, show that the differential form in (a) satisfies conditions (15.8–5) in D.

(e) Describe a closed curve in D which is not the boundary of any surface lying in D and to which Stokes's theorem may be applied.

MISCELLANEOUS EXERCISES

1. If S is defined by $z = f(x, y)$, with (x, y) ranging over R, and if the positive side of S is chosen so that $\cos \gamma > 0$, show that

$$\iint_S (x \cos \alpha - y \cos \beta)dA = \iint_R \left(y \frac{\partial f}{\partial y} - x \frac{\partial f}{\partial x}\right)dx\,dy.$$

2. Find a function whose differential is

$$\left(e^x \cos y - \frac{3}{\sqrt{1 - x^2}}\right)dx - (e^x \sin y - 7 \sec^2 y)dy.$$

3. Suppose $a > 0$, $b > 0$. Let P be a point of intersection of $y^2 = -4\,a^2(x - a^2)$ and $y^2 = 4\,b^2(x + b^2)$, and let R be the region bounded by the first parabola, the x-axis, and the line OP. Show that $\displaystyle\iint_R \frac{dx\,dy}{\sqrt{x^2 + y^2}} = 2\,ab$ by using the transformation $x = u^2 - v^2$, $y = 2\,uv$.

4. Use (15.4–6) with $a = b = 0$ to find a function u such that

$$du = \frac{4\,xy\,dx + 2(1 + y^2 - x^2)dy}{(x^2 + y^2 - 1)^2 + 4\,y^2}.$$

The function so found has certain discontinuities. Where are they?

5. Find a region in the $r\theta$-plane which is mapped, by $x = r\cos\theta$, $y = r\sin\theta$, into the region R between $x^2 + y^2 = 1$, $x^2 + y^2 = 4$, and inside $x^2 + y^2 = 2\,x$.

Hence calculate $\displaystyle\iint_R \frac{x}{\sqrt{x^2 + y^2}}\,dx\,dy.$

6. Find a first-quadrant region in the uv-plane which maps into the region R defined by $1 \leq x^2 + y^2 \leq 4$, $y \geq 0$, if the mapping is $x = u^2 - v^2$, $y = 2\,uv$.

Calculate $\displaystyle\iint_R \frac{dx\,dy}{\sqrt{x^2 + y^2}}$ by transforming to the uv-plane, and check your result

by calculating the given integral in terms of polar co-ordinates.

7. Calculate $\displaystyle\iint_R \cos\left[\frac{\pi(2\,x - y)}{2(x - 2\,y)}\right]dx\,dy$, where R is the triangle bounded by

$x = y$, $x + y = 0$, and $x - 2\,y = 2$. Use the transformation $u = 2\,x - y$, $v = x - 2\,y$.

8. Let S be the surface of the elliptic cone $(x^2/a^2) + (y^2/b^2) = (z^2/c^2)$ between $z = 0$ and $z = c$. Show that the centroid is at $(0, 0, \frac{2}{3}\,c)$. Show that both the area and the first moment of S can be expressed in terms of the same complete elliptic integral of the second kind. For a hint as to a useful procedure, see Exercise 5, § 15.32.

9. Show that the potential at O due to a uniform charge on the surface of the torus $x = (a + b\cos\phi)\cos\theta$, $y = (a + b\cos\phi)\sin\theta$, $z = b\sin\phi$ is expressible in terms of the standard elliptic integrals

$$\int_0^{\pi/2} \sqrt{1 - k^2\sin^2\phi}\,d\phi, \qquad \int_0^{\pi/2} \frac{d\phi}{\sqrt{1 - k^2\sin^2\phi}},$$

where $\qquad\qquad k^2 = \dfrac{4\,ab}{(a + b)^2}.$

10. Find the area of the part of the surface $(x^2 + y^2 + z^2)^2 = x^2 - y^2$ in the first octant.

Point-Set Theory

16. Preliminary remarks

This chapter is a continuation of the study of limits and convergence. Chapter II may be regarded as part of the theory of point sets on a straight line. In Chapter V we began some study of point sets in a plane. Point-set theory is fundamental in the study of functions. Our primary aim in this chapter is to develop enough theory to deal with our later study of continuous functions, integration, and the theory of convergence of sequences and series. The principal theorems are the Bolzano-Weierstrass theorem, Cauchy's convergence condition, and the Heine-Borel theorem.

16.1 Finite and infinite sets

A *set* of points is defined by some condition which distinguishes the points which belong to the set from those which do not belong to it. The points which belong to a set are called *elements* of the set. In the study of limit concepts we are often obliged to deal with sets having an infinite number of elements. A set is called a *finite* set if there is some positive integer n such that the set has exactly n elements.

Sometimes we may define a set of points by a condition which is such that no points satisfy the condition. In this case we call the set empty, or *void*. The empty set is considered to be a finite set, the number of its elements being 0.

A set which is not finite is called *infinite*.

Example 1. Let S be the set of all points (x, y) in the xy-plane which are such that x and y are integers and $x^2 + y^2 < 100$. This is a finite set. Without determining exactly how many elements there are in S, we can easily see that the number does not exceed $(19)^2 = 361$.

Example 2. Let S be the set of all points (x, y) inside the circle $(x - 2)^2 + y^2 = 4$ and such that $y > x + 1$. This set is void, for the circle lies entirely below the line $y = x + 1$.

Perhaps the simplest infinite set is the set whose elements are the positive integer points $1, 2, 3, \cdots$ on the real number scale. Let us, for the time being, denote this set by the letter I.

DEFINITION. A point set S is called *denumerable*, or *denumerably infinite*, if there is a one-to-one correspondence between the elements of S and the elements of the set I of positive integers.

The existence of a one-to-one correspondence between the elements of S and those of I means that each point of S can be labeled with a unique positive integer n, with no two distinct points having the same label n, and that every positive integer is used as the label for some point. The elements of S can then be symbolized by $P_1, P_2, P_3, \cdots$.

The words *enumerable* and *countable* are synonyms for *denumerable*.

Infinite sets may be classified into those which are denumerable and those which are not. It can be shown that the set consisting of all the real numbers is nondenumerable, whereas the set consisting merely of the rational numbers is denumerable. The demonstrations of these facts may be based on Examples 3 and 4, below.

If a set is infinite, we can select out of it a subset which is denumerable. This simple but important fact is used in many arguments.

Example 3. The set S of all rational numbers x such that $0 < x < 1$ is denumerable.

To demonstrate the denumerability of S, let us make a display as follows:

$$\begin{array}{ccccc}
\frac{1}{2} & \frac{1}{3} & \frac{1}{4} & \frac{1}{5} & \frac{1}{6} \quad \cdots \\
\frac{2}{3} & \frac{2}{5} & \frac{2}{7} & \frac{2}{9} & \frac{2}{11} \quad \cdots \\
\frac{3}{4} & \frac{3}{5} & \frac{3}{7} & \frac{3}{8} & \frac{3}{10} \quad \cdots \\
& & \cdots & &
\end{array}$$

The numerators in the successive rows of the display are $1, 2, 3, \cdots$. The denominators in each row are increasing, but such that each fraction is proper and in its lowest terms. A little reflection shows us that each member of S appears just once in this array, and that each member of the array is an element of S. By labeling the elements $x_1, x_2, x_3, x_4, \cdots$ as shown in the following array, we see that S is denumerable.

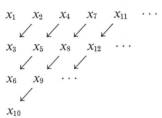

Example 4. Let S be the set of all nonterminating decimal fractions of the form $0.a_1a_2a_3 \cdots$, where $a_1, a_2, \cdots$ represent digits $0, 1, \cdots, 9$, and the case in which all the a's from some point onward are 0 is ruled out. This set is nondenumerable.

If S were denumerable, we could label *all* its members by a double subscript scheme as follows:

$$\text{1st element: } 0.a_{11}a_{21}a_{31} \cdots$$
$$\text{2d element: } 0.a_{12}a_{22}a_{32} \cdots$$
$$\text{3d element: } 0.a_{13}a_{23}a_{33} \cdots$$
$$\cdots$$

This leads to a contradiction, however, for there are elements of S not included in the foregoing list. In fact, consider the decimal $0.c_1c_2c_3 \cdots$, where the c's are chosen from $0, 1, \cdots, 9$ according to the rules: $c_1 \neq a_{11}$, $c_2 \neq a_{22}$, $c_3 \neq a_{33}$, $\cdots$, and each c_i is different from 0. This decimal belongs to S, but is different from each decimal in the proposed enumeration of the elements of S. Therefore, S is nondenumerable.

16.2   Point sets on a line

In dealing with point sets on a line we shall assume that a number scale has been established on the line. We shall usually think of it as the x-axis. Then each point is identified by its x-co-ordinate, and we shall refer to a point by referring to the corresponding real number.

DEFINITION.   By a *neighborhood* of a point x_0 we mean the set of all points x such that $x_0 - h < x < x_0 + h$, where h is some positive number. The neighborhood thus consists of all points between $x_0 - h$ and $x_0 + h$, not including these two points. For each choice of h we get a neighborhood of x_0.

This definition should be compared with the definition of a neighborhood in the discussion of point sets in the xy-plane; see § 5.1.

DEFINITION.   A point set S is called *open* if for each point x_0 of S there is some neighborhood of x_0 which belongs entirely to S.

Example 1.   If $a < b$, the set S of all x such that $a < x < b$ is an open set. It is called an *open interval*.

Example 2.   The set of all rational numbers r such that $0 < r < 1$ is not open. This is because any neighborhood of any point contains both rational and irrational numbers.

Example 3.   Let S be the set of all points on the x-axis except the points $1, \frac{1}{2}, \frac{1}{3}, \frac{1}{4}, \cdots$. This set is not open, for 0 is in the set, but no neighborhood of 0 lies wholly in the set.

DEFINITION.   In speaking of point sets on a line, if S is a set, the *complement* of S is defined as the set of all the points on the line which are not in S. The complement of S is denoted by $C(S)$.

For instance, if S is the set of Example 1, $C(S)$ consists of all x such that $x \leqq a$, together with all x such that $x \geqq b$.

DEFINITION.   A set is called *closed* if its complement is open.

Example 4.   If $a < b$, the set of all x such that $a \leqq x \leqq b$ is a closed set. It is called a *closed interval*.

It is very important to notice that, in our use of "open" and "closed" as adjectives describing sets, "*closed*" *is not the opposite of* "*open.*" If a set is not open, that does not mean that it is closed.

Example 5.   The set S of all x such that $0 < x \leqq 1$ is neither open nor closed. For, no neighborhood of 1 lies in S, which shows that S is not open. And no neighborhood of 0 lies in $C(S)$, which shows that $C(S)$ is not open, and hence that S is not closed.

The entire x-axis is an open set, by the definition of openness. The complement of this set is the void set. Also, the complement of the void set is the entire x-axis. It is convenient to agree, by convention, that the void set is open. This makes the entire x-axis a closed set. Thus, the entire x-axis is both open and closed, and the same is true of the void set. There are no other sets on the line which are both open and closed, however.

One of the most important concepts in this chapter is that of an *accumulation point* of a set.

DEFINITION.   Let S be a point set, and let y be a point which is not necessarily in S. We call y an accumulation point of S if in each neighborhood of y there is at least one point x which is in S and distinct from y.

In some books the term *limit point* is used instead of *accumulation point*.

A finite set can have no accumulation point, for if S is finite and y is any point, it is easily seen that if a small enough neighborhood of y is selected, the condition of the definition cannot be fulfilled. An infinite set may have no accumulation points, but it may also have one or several or infinitely many. There is no requirement that an accumulation point of S shall belong to S, but it may happen to belong to S in a particular case.

Example 6.   Let S be the set of all rational numbers. Every point on the number scale is an accumulation point of S.

This follows from the fact, demonstrated in § 2.5, that there is a rational number between any two distinct numbers. As a consequence, if x_0 is any number, every neighborhood of x_0 contains many rational numbers, so that x_0 is an accumulation point of S.

In the definition of y as a point of accumulation of S, the requirement of the definition actually makes it necessary for each neighborhood of y to contain infinitely many points of S. This is seen as follows: Let I_1 be some neighborhood of y, and let x_1 be a point of S in I_1, with $x_1 \neq y$. Now choose a neighborhood I_2 of y, small enough so that x_1 is not in I_2. There must be a point x_2 of S in I_2, with $x_2 \neq y$. We then repeat the argument, getting smaller and smaller neighborhoods $I_3, I_4, \cdots$, and points $x_3, x_4, \cdots$ all in S such that, for each k, x_k is in I_k but not in I_{k+1}, and $x_k \neq y$. The infinite sequence of distinct points belongs to the original neighborhood I_1.

When the full import of the foregoing paragraph has been realized by the student, he will not find it difficult to recognize points of accumulation in the situations which come to his attention.

There is an important relation between the concept of a closed set and the notion of accumulation point, as we see in the following theorem:

THEOREM I.   *A set is closed provided that each accumulation point of S is a member of S. Conversely, if S is closed, it contains each of its points of accumulation.*

PROOF.   Suppose every accumulation point of S belongs to S. If S were not closed, its complement would not be open. But, if $C(S)$ were not open, this would mean that $C(S)$ contains a point y which has no neighborhood lying entirely in $C(S)$. But if a neighborhood of y fails to lie wholly in $C(S)$, it must contain a point of S, and since y is in $C(S)$, the point of S cannot be the point y. Thus, if $C(S)$ were not open, it would contain a point y whose every neighborhood contains a point of S, and thus y would be a point of accumulation of S. This would contradict the supposition that every accumulation point of S belongs to S. Consequently, $C(S)$ must be open, and S closed. This proves the first assertion in Theorem I.

For the converse we suppose S is closed. If y is any accumulation point of S, we have to show that y is in S. If it were not, it would be in the complement $C(S)$, which is open. Then some neighborhood of y would also lie in $C(S)$. Such a neighborhood could not contain any points of S, and so y could not be an accumulation point of S. Thus we arrive at a contradiction unless we conclude that y is in S. This completes the proof.

EXERCISES

   1. For each of the following sets answer the question: Is the set open, closed, or neither?
   (a) All x such that $x < 1$.   (d) All rational numbers.
   (b) All x such that $x \geqq 0$.   (e) All irrational numbers.
   (c) All x such that either $x < 0$ or $x \geqq 1$.

   2. If S is the set of all x such that $0 \leqq x \leqq 1$, what points, if any, are points of accumulation of both S and $C(S)$?

   3. (a) If S is the set of all x such that $0 < x < 1$, are there any points of S which are points of accumulation of $C(S)$? (b) Are there any points of $C(S)$ which are points of accumulation of S?

   4. Prove that any finite set is closed.

   5. Prove that, if S is open, each of its points is a point of accumulation of S.

   6. For each of the following sets find the points of accumulation and specify whether or not the set is closed.
   (a) All numbers of the form $(n - 1)/n$, $n = 1, 2, 3, \cdots$.

(b) All numbers of the form $n/(n^2 + 1)$, $n = 0, 1, 2, 3, \cdots$.
(c) All numbers of the form $(-1)^n n/(n + 1)$, $n = 0, 1, 2, 3, \cdots$.

16.3   The Bolzano-Weierstrass theorem

A point set S on the x-axis is called *bounded* if there is some finite interval which contains all of S. In other words, S is bounded if there exist numbers a, b (with $a < b$) such that $a \leqq x \leqq b$ for every x in S.

THEOREM II.   *Suppose that S is a bounded, infinite set. Then there is at least one point of accumulation of S.*

This theorem is generally known as the Bolzano-Weierstrass theorem. Bernard Bolzano (1781–1848) of Prague was a pioneer in the rigorous study of point sets and other matters fundamental in analysis. Karl Weierstrass (1815-1897) was one of the great mathematicians of the nineteenth century.

PROOF.   For the proof we appeal to the theorem about nested intervals (Theorem VI, § 2.8). We assume that S lies in the closed interval $[a, b]$, and we denote this interval by I_1. Now divide I_1 into two equal parts by the midpoint $(a + b)/2$, and consider the two closed intervals $[a, (a + b)/2]$, $[(a + b)/2, b]$. Since every point of S lies in one or the other of these intervals, at least one of them must contain an infinite number of points of S. Let I_2 denote such a one of the two. We then bisect I_2 and obtain a new closed interval I_3. By repetition of the process we generate a nest of closed intervals $\{I_n\}$. By the manner of construction I_n contains infinitely many points of S, and the length of I_n is $(b - a)/2^{n-1}$. By the theorem referred to earlier in this paragraph, there is exactly one point common to all the intervals of the nest. This point, call it z, is an accumulation point of S, for any given neighborhood of z will contain I_n if n is sufficiently large. This is true because z is in I_n and the length of I_n approaches 0 as n increases. The neighborhood must therefore contain an infinite number of points of S.

We shall use the Bolzano-Weierstrass theorem as a basis for much of the later work in this chapter.

EXERCISES

1. Suppose S is a set having the number M as its least upper bound. If M is not a member of S, show that it is a point of accumulation of S. Give an example showing that, if M does belong to S, it need not be a point of accumulation of S. For the definition of least upper bounds see § 2.7.

2. For an alternative proof of Theorem II, let two classes L, R be defined as follows: A number x belongs to L if there are infinitely many elements of S to the

right of x on the scale; otherwise x belongs to R. Show that L and R form a cut (see § 2.4), and that the cut number is a point of accumulation of S.

16.31   Convergent sequences on a line

The notion of limit of a sequence has been discussed briefly in § 1.62. The student should review this earlier section at this time. Our present purpose is to develop some relations between the concept of a convergent sequence and the notion of a point of accumulation of a set.

If $\{x_n\}$ is a sequence of real numbers, we can consider the set consisting of all the distinct points on the number scale corresponding to the successive numbers x_1, x_2, x_3, $\cdots$. *It is to be emphasized that the set thus defined is not the same thing as the sequence.* A point set is just a collection of points; a sequence $\{x_n\}$ is a function defined on the positive integers, with x_n the value of the function corresponding to the particular positive integer n.

If $\{x_n\}$ is a sequence, the set of values x_1, x_2, x_3, $\cdots$ may be a finite set. This is true in the example

$$x_n = 1 - (-1)^n,$$

where the set of values consists of the two points 0, 2 If the set of values x_1, x_2, x_3, $\cdots$ is a finite set, the sequence will be convergent if and only if there is some integer N such that all the values x_n are the same when $n \geqq N$. This is easily seen by direct consideration of the definition of convergence (§ 1.62). The case in which the set of values is infinite is taken up in the theorem which follows:

THEOREM III.   *Suppose $\{x_n\}$ is a convergent sequence of real numbers, and suppose that S, the set of distinct points among the values x_1, x_2, x_3, $\cdots$, is an infinite set. Then S has just one point of accumulation, and this point is the limit of the sequence.*

PROOF.   Denote the limit by x. Consider any neighborhood of x, say the open interval from $x - \epsilon$ to $x + \epsilon$, where $\epsilon > 0$. By the definition of convergence there is some integer N such that x_n is in the above neighborhood if $n \geqq N$. Since S is infinite, there must be infinitely many distinct points of S represented among the values x_N, x_{N+1}, x_{N+2}, $\cdots$. These are all in the specified neighborhood of x, and so x must be an accumulation point of S. There can be no other accumulation point. For, if $y \neq x$, choose neighborhoods I_x and I_y of x and y, respectively, which are small enough so that no point belongs to both I_x and I_y. Then all but a finite number of points of S are in I_x, whereas f y were an accumulation point, an infinite number of points of S would have to be in I_y. This completes the proof of Theorem III.

The following theorem is useful:

THEOREM IV.　*Let S be any point set having an accumulation point y. Then there is a sequence $\{x_n\}$ such that $\lim_{n \to \infty} x_n = y$, and such that each x_n is an element of S and the values x_1, x_2, $\cdots$ are all distinct.*

Most of the argument needed to prove this theorem has already been given, in the second paragraph following Example 6, § 16.2. All that is necessary is to specify that the intervals I_1, I_2, I_3, $\cdots$ be chosen so that the length of I_n approaches 0 as $n \to \infty$. This will insure that $x_n \to y$.

We conclude this section with some considerations of *subsequences*. If $\{x_n\}$ is a sequence, a subsequence is another sequence $\{y_k\}$ formed by dropping out certain of the x_n's and retaining the rest in the order originally given. If the indices retained are n_1, n_2, n_3, $\cdots$, where $n_1 < n_2 < n_3 < \cdots$, then $y_1 = x_{n_1}$, $y_2 = x_{n_2}$, and so on. As particular examples of subsequences we cite:

$$x_2, x_4, x_6, x_8, \cdots \qquad (n_k = 2\,k),$$
$$x_1, x_2, x_6, x_{24}, \cdots \qquad (n_k = k!).$$

If $\{x_n\}$ is a sequence, convergent or not, we may consider a subsequence and inquire as to its convergence. A sequence may be divergent and yet contain a convergent subsequence. Indeed, it may contain many convergent subsequences, each with a different limit.

If a sequence is convergent, every subsequence formed from it is also convergent, and they all have the same limit, namely the limit of the original sequence.

There are many uses of the following theorem:

THEOREM V.　*Let $\{x_n\}$ be a bounded sequence. Then $\{x_n\}$ contains a convergent subsequence.*

PROOF.　Let S be the set of distinct points represented by the values x_1, x_2, $\cdots$. There are two cases to consider, according as S is finite or infinite. If S is finite, it has no accumulation point. But some point of S must be repeated an infinite number of times in the sequence, so that for some subsequence we must have $x_{n_1} = x_{n_2} = x_{n_3} = \cdots$. This subsequence is certainly convergent, the limit being the repeated value. If S is infinite, it must have at least one accumulation point, say y, by the Bolzano-Weierstrass theorem. But then, by Theorem IV, y is the limit of a sequence of distinct terms chosen from S. This sequence might not be a subsequence of $\{x_n\}$. For example, it might be x_{10}, x_5, x_{20}, x_{15}, x_{30}, x_{25}, $\cdots$, where the indices are not arranged in increasing order. But, by dropping out the terms whose indices are not in the natural order we get a subsequence (x_{10}, x_{20}, x_{30}, $\cdots$ in the foregoing example), and this subsequence will converge to y. This completes the proof.

EXERCISES

1. Suppose that $\{x_n\}$ is a sequence which is bounded and such that all the values x_1, x_2, x_3, $\cdots$ are distinct. Assume that the set of these values has just one point of accumulation, denoted by x. Prove that the sequence is convergent and that the limit is x.

2. Consider the sequence with terms 2, $\frac{1}{2}$, $\frac{4}{3}$, $\frac{1}{4}$, $\frac{6}{5}$, $\frac{1}{6}$, $\cdots$ where $x_n = \frac{1}{2}[1 - (-1)^n] + (1/n)$. Find two convergent subsequences with different limits.

3. Give an example of a sequence for which the set of values x_1, x_2, $\cdots$ is finite and there are three different convergent subsequences with distinct limits.

4. If all the terms of a sequence are distinct, and the set of values has just one point of accumulation, can the sequence fail to be convergent? Consider the case where $x_n = n[1 + (-1)^n] + (1/n)$.

16.4   Point sets in higher dimensions

In § 5.2 we defined neighborhoods for points in the xy-plane. Then we went on to define open sets, closed sets, and some other concepts. It was indicated that similar definitions can be made for point sets in three-dimensional space, starting from the definition of spherical neighborhoods.

If S is a point set in the plane, we define an accumulation point of S in exactly the same way as we did for point sets on the line: A point Q is called an accumulation point of S if each neighborhood of Q contains at least one point P which is in the set S and distinct from Q. This same definition is also used for point sets in three-dimensional space.

When we analyze and compare the fundamentals of point-set theory for point sets in 1, 2, or 3 dimensions, we observe the following things: (1) The verbal definition of an open set is the same in all three cases, but the word "neighborhood" has to be given the interpretation appropriate to the dimensionality. (2) The verbal definition of a closed set, namely, that S is closed if $C(S)$ is open, is the same in all three cases. (3) Theorem I of § 16.2, which states that a set is closed if and only if it contains all its points of accumulation, is valid for point sets in the plane or in space as well as for point sets on a line. The proof as given in § 16.2 does not depend upon the dimensionality.

It must be recognized that a point set on the x-axis may be open when it is considered as a point set on the line, but that it will not be open when considered as a point set in the xy-plane. This is because of the different meanings of the word "neighborhood" in the theories for one and two dimensions, respectively. Likewise, for example, the set of all points in the xy-plane for which $x^2 + y^2 < 1$ is open in the theory for two dimensions, but is not open when we regard it as a point set in the three-dimensional space of points (x, y, z).

We now turn to the Bolzano-Weierstrass theorem for higher dimensions. The statement is exactly the same as that already given in § 16.3 (Theorem II). Bounded sets in the plane or in space were defined in § 5.3. To prove the theorem for sets in the xy-plane we shall first discuss a generalization of Theorem VI, § 2.8.

Let R_1, R_2, $\cdots$, R_n, $\cdots$ be a sequence of closed rectangular regions, with sides parallel to the co-ordinate axes and such that R_2 is contained in R_1, R_3 is contained in R_2, and so on. Furthermore, let the dimensions of R_n approach 0 as $n \to \infty$. We shall call such a sequence of rectangles a *nest*. By an extension of the argument used in proving Theorem VI, § 2.8, it is easy to see that *if R_n is any nest of closed rectangles, there is one and only one point which is common to all the rectangles.*

Now let S be a bounded infinite point set in the xy-plane. Since S is bounded, it is contained in some sufficiently large closed rectangle with sides parallel to the axes, say R_1, defined by $a_1 \leqq x \leqq b_1$, $c_1 \leqq y \leqq d_1$. Now divide R_1 into four equal smaller rectangles, by the lines $x = (a_1 + b_1)/2$, $y = (c_1 + d_1)/2$. In at least one of the four closed smaller rectangles there must be an infinite number of points of S. Call such a one of the smaller rectangles R_2. The same procedure is now applied to R_2. By continuation we obtain a sequence R_1, R_2, $\cdots$, R_n, $\cdots$ forming a nest of closed rectangles, and, for each n, there are an infinite number of points of S in R_n. Let P be the unique point belonging to all the rectangles. This point P is easily seen to be an accumulation point of S.

This method of proof may be suitably extended to apply to point sets in three dimensions.

16.41 Convergent sequences in higher dimensions

The concept of a convergent sequence may be extended to the case of sequences whose terms are points in space of higher dimensions.

DEFINITION. Let P_n be a sequence of points. We say that the sequence is convergent to the limit P, and write $\lim_{n \to \infty} P_n = P$, if each neighborhood of P contains all but a finite number of the points P_1, P_2, P_3, $\cdots$.

For some purposes it is convenient to rephrase this definition in terms of distances between points. Let $d(P, Q)$ denote the distance between P and Q. This distance can be expressed in terms of the co-ordinates of P and Q, but just now it is more convenient to deal directly with the points and not with the co-ordinates. The meaning of $\lim_{n \to \infty} P_n = P$ is the following: If $\epsilon > 0$, there is some integer N such that $d(P_n, P) < \epsilon$ if $N \leqq n$.

From this definition it can easily be shown that, if $\{P_n\}$ is a sequence in the xy-plane, with co-ordinates (x_n, y_n), and if P is the point (x, y), then P_n converges to P if and only if x_n converges to x and y_n converges to y. A like situation obtains in three dimensions.

The relations between convergent sequences and points of accumulation are essentially the same in higher dimensions as in the case of points restricted to the x-axis. The Theorems III, IV, V of § 16.31 all remain true in higher dimensions. The proofs as given in § 16.31 remain valid with only a few minor modifications in notation to suit the higher dimensional cases.

EXERCISES

1. Let $\{P_n\}$ be the sequence $([1 - (-1)^n]/n, [1 + (-1)^n]/n)$. Is it convergent?

2. Let S be the set of points on the curve $y = \sin(1/x)$ $(x \neq 0)$ in the xy-plane. **(a)** Find a sequence of points P_n on the x-axis and in S such that P_n converges to $(0, 0)$. **(b)** Find a sequence of points P_n on the line $y = 1$ and in S such that P_n converges to $(0, 1)$. **(c)** What points must be adjoined to S in order to get a closed set?

3. Let S be the set of points (x, y) such that x is rational and $0 \leq x \leq 1$ while y is irrational and $0 < y < 1$. Is S open, closed, or neither? Describe the totality of points of accumulation of S.

16.5   Cauchy's convergence condition

The definition of convergence of a sequence is stated in a way which involves the limit of the sequence. Therefore one cannot use the definition to prove that a sequence is convergent unless one already knows what the limit is. This is often inconvenient, for it frequently happens that we want to prove that a sequence is convergent, even if we do not know precisely what the limit is. The following theorem provides a way out of the difficulty.

THEOREM VI.   *Let $\{P_n\}$ be a sequence of points. Let $d(P_m, P_n)$ denote the distance between P_m and P_n. A necessary and sufficient condition that the sequence be convergent is that $d(P_m, P_n)$ approach 0 as m and n become infinite. In other words, for each $\epsilon > 0$ there must be some N such that $d(P_m, P_n) < \epsilon$ whenever $N \leqq m$ and $N \leqq n$.*

We note that in the case of a sequence $\{x_n\}$ of points on the x-axis, the distance is

$$d(x_m, x_n) = |x_m - x_n|,$$

so that the condition becomes

$$| x_m - x_n | < \epsilon \quad \text{if} \quad N \leqq m \quad \text{and} \quad N \leqq n.$$

This condition on the sequence is called *Cauchy's condition* (see the reference to Cauchy in § 1.61).

PROOF THAT THE CONDITION IS NECESSARY. If $P_n \rightarrow P$ as $n \rightarrow \infty$, and $\epsilon > 0$, choose N so large that $d(P_n, P) < \epsilon/2$ if $N \leqq n$. This is possible, by the definition of convergence. If now $m \geqq N$ and $n \geqq N$, we have $d(P_m, P_n) \leqq d(P_m, P) + d(P, P_n) < (\epsilon/2) + (\epsilon/2) = \epsilon$, for the three points P_m, P_n, P lie either on a straight line or at the vertices of a triangle, and in either case the distance between one pair is not greater than the sum of the distances between the other two pairs. For points on a line we could write

$$| x_m - x_n | \leqq | x_m - x | + | x - x_n |,$$

which is an application of the inequality (2.2–9) with $a = x_m - x$, $b = x - x_n$.

PROOF THAT THE CONDITION IS SUFFICIENT. We now assume that Cauchy's condition is satisfied by the sequence $\{P_n\}$. First we shall prove that the sequence is bounded. Taking $\epsilon = 1$, Cauchy's condition assures us that there is some N such that $d(P_m, P_n) < 1$ if m and n are not less than N. In particular, $d(P_N, P_n) < 1$ if $n \geqq N$. This means that all points of the sequence with the possible exception of P_1, $\cdots$, P_{N-1} are less than unit distance from P_N. Certainly, then, there is some neighborhood with center at P_N large enough to contain *all* the points of the sequence, and the sequence must be bounded.

We now apply Theorem V, § 16.31 (the extension of this theorem to higher dimensions was discussed in § 16.41). The sequence $\{P_n\}$ contains a convergent subsequence P_{n_1}, P_{n_2}, $\cdots$. Let the limit of this subsequence be P. We shall complete the proof by showing that the original sequence is convergent to the limit P. Suppose $\epsilon > 0$. Choose N so that $d(P_m, P_n) < \epsilon/2$ if $m \geqq N$ and $n \geqq N$ (Cauchy's condition). Since the subsequence converges to P, we can choose one of the indices n_i so large that $d(P_{n_i}, P) < \epsilon/2$. We can also, at the same time, choose it larger than N. Then, if $n \geqq N$, we have

$$d(P_n, P) \leqq d(P_n, P_{n_i}) + d(P_{n_i}, P) < \frac{\epsilon}{2} + \frac{\epsilon}{2} = \epsilon,$$

by the way things have been arranged. Here we have used the same kind of inequality involving the distances between three pairs of points that we encountered in the necessity argument. The proof of Theorem VI is now complete.

Theorem VI is most often used as a tool for carrying on theoretical developments.

EXERCISE

Let a sequence be defined on the x-axis as follows: $x_1 = 1, x_2 = 2, x_3 = \frac{1}{2}[x_1 + x_2]$, and in general $x_{n+1} = \frac{1}{2}[x_{n-1} + x_n]$, $n = 2, 3, \cdots$. Show that $|x_m - x_n| \leq 1/2^{N-1}$ if $N \leq m$ and $N \leq n$, so that Cauchy's condition is fulfilled. SUGGESTION: Note that each term is midway between the two preceding terms.

16.6   The Heine-Borel theorem

In this section we are concerned with an important theorem of point-set theory which is widely known by the names of the two mathematicians Eduard Heine and Emile Borel. The central idea of this theorem probably originated in connection with the concept of *uniform continuity*. Heine defined this concept, and proved the very important theorem that if a function of one variable x is continuous at each point of a closed interval $a \leq x \leq b$, then it is uniformly continuous on that interval. We shall discuss the concept and the theorem in Chapter XVII, at which time we shall need to appeal to the Heine-Borel theorem. The student will doubtless find difficulty at first in appreciating the motivation of the Heine-Borel theorem; it may be best that he take up the study of the theorem when he needs it in reading parts of Chapter XVII. But since the theorem itself is a theorem of point-set theory, we put the exposition of it here in Chapter XVI.

Before stating the theorem we make a definition and consider some illustrations to motivate the theorem.

DEFINITION.   Let S be a point set, and suppose we have a collection of a certain number of open sets such that each point of S belongs to at least one of the open sets. Then we say that S is *covered* by the collection of open sets.

The number of open sets in the collection may be either finite or infinite.

Example 1.   Let S be the set of all points on the x-axis such that $0 < x \leq 1$. Let the collection of open sets be the sequence of open intervals

$$(\tfrac{1}{2}, \tfrac{3}{2}), (\tfrac{1}{4}, 1), (\tfrac{1}{8}, \tfrac{5}{8}), (\tfrac{1}{16}, \tfrac{3}{8}), \cdots,$$

the nth interval being $I_n = (1/2^n, (n + 2)/2^n)$. These intervals are shown schematically in Fig. 169. The set S is covered by this collection, for every x such that $0 < x \leq 1$ lies in I_n for some value of n.

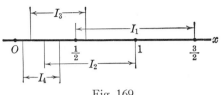

Fig. 169

Example 2. Let S be the set of points $0, \frac{1}{2}, \frac{1}{4}, \frac{1}{6}, \cdots, 1/(2n), \cdots$ on the x-axis. Let the collection of open sets be the open intervals

$$(-\tfrac{1}{10}, \tfrac{1}{10}), (\tfrac{1}{3}, 1), (\tfrac{1}{5}, \tfrac{1}{3}), (\tfrac{1}{7}, \tfrac{1}{5}), \cdots, \left(\frac{1}{2n+1}, \frac{1}{2n-1}\right), \cdots.$$

Note that S is covered by the collection of intervals.

Example 3. Let S be the set of all points on the x-axis such that $0 < x \le 2$. Consider the function $f(x) = 1/x$, and define a collection of open intervals as follows: Suppose $0 < \epsilon < \frac{1}{2}$. If x_0 is in S, denote by $I(x_0)$ the set of all x such that $x > 0$ and $|f(x) - f(x_0)| < \epsilon$, that is, the set of all $x > 0$ such that

$$\frac{1}{x_0} - \epsilon < \frac{1}{x} < \frac{1}{x_0} + \epsilon.$$

It is clear from Fig. 170 that $I(x_0)$ is the open interval from x_0' to x_0'' where x_0' and x_0'' are found by solving the equations

$$\frac{1}{x_0'} = \frac{1}{x_0} + \epsilon, \quad \frac{1}{x_0''} = \frac{1}{x_0} - \epsilon.$$

One easily finds

$$x_0' = \frac{x_0}{1 + \epsilon x_0}, \quad x_0'' = \frac{x_0}{1 - \epsilon x_0}.$$

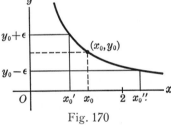

Fig. 170

Clearly $I(x_0)$ contains x_0. Consequently, as x_0 varies over S we get a collection of intervals $I(x_0)$ which covers S.

The Heine-Borel theorem is concerned with this question: If S is covered by a collection of open sets, under what conditions on S is it possible to choose a *finite* number of open sets from the collection in such a way that the new finite collection still covers S? Let us examine the preceding examples with this question in mind.

In Example 1 no finite number of the open intervals will suffice to cover S, for the first n intervals leave uncovered all the points x of S for which $0 < x \le 1/2^n$. Example 2 is different. The point O is covered by the special open interval $(-\tfrac{1}{10}, \tfrac{1}{10})$. This same interval also covers all but a finite number of the remaining points of S, namely, all points $1/(2 n)$ with $1/(2 n) < \frac{1}{10}$, or $n = 6, 7, 8, \cdots$. The remaining points are $\frac{1}{2}, \frac{1}{4}, \cdots, \frac{1}{10}$, and these are covered by the five intervals $(\frac{1}{3}, 1)$, $(\frac{1}{5}, \frac{1}{3})$, $\cdots$, $(\frac{1}{11}, \frac{1}{9})$. Hence six intervals suffice to cover S in Example 2. In Example 3 no finite selection of the intervals $I(x_0)$ will suffice to cover S, for, of any finite number of the $I(x_0)$, there will be one for which x_0' is furthest to the left, and this will leave all the x such that $0 < x \leq x_0'$ uncovered.

We now proceed directly to a formal statement and proof of the Heine-Borel theorem.

THEOREM VII.   *Let S be a bounded and closed point set, and let S be covered by a collection of open sets. Then a finite number of open sets may be chosen from the collection in such a way that S is covered by the new finite collection.*

All our illustrative examples were given for point sets on a line. However, the theorem holds equally for point sets on a line, in the plane, or in space. It is merely necessary to bear in mind the proper definitions of open and closed sets for each dimensionality.

PROOF OF THE THEOREM.   We shall give the proof for point sets on a line. But the method of proof may be immediately applied in higher dimensions.

The set S will lie on some closed interval $a \leq x \leq b$, since it is bounded. If S is a finite set there is no problem; we have only to choose one open set to cover each point of S, and this will be a finite collection. Hence let us assume that S is an infinite point set. We make the assumption that no finite number of the open sets of the given collection will suffice to cover S. From this assumption we shall deduce a contradiction, and thus the theorem will be proved.

We bisect the interval $[a, b]$ and consider the parts of S lying in each of the two closed subintervals. If each of these parts could be covered by a finite subcollection of the open sets, so could S itself. Therefore, for at least one of the subintervals the corresponding part of S cannot be covered by a finite number of the given open sets. We bisect this interval and proceed as before. Each time we bisect, an infinite number of points of S must belong in the subinterval which is retained, for otherwise this part of S could surely be covered by a finite number of the open sets. The repeated bisection process gives us a nest of closed intervals, and there is a unique point x_0 common to all the intervals of the next (Theorem VI, § 2.8). This point x_0 is clearly an accumulation point of S. It must therefore belong to S (Theorem I, § 16.2), because S is closed. Since x_0 is in S, there is some open set of the collection which contains x_0. This open set must therefore contain all except a finite number of the closed intervals of the nest which is shrinking down on x_0. This brings us to a contradiction, for each interval of the nest has the property that the part of S in it cannot be covered by a finite number of the given open sets. As explained earlier, our arrival at this contradiction completes the proof.

EXERCISE

Let S be the set $1 \leq x \leq 2$, and define the open intervals $I(x_0)$ as in Example 3. Show that S can be covered with a finite number N of these intervals, where N is the smallest integer exceeding $1/\epsilon$. SUGGESTION: Consider the intervals $I(x_0)$ for $x_0 = 1, 1 + \epsilon, 1 + 2\epsilon, \cdots$.

Fundamental Theorems on Continuous Functions

17

17. Purpose of the chapter

In Chapter III we proved several important theorems about continuous functions, confining ourselves always to the case of a single real variable. Analogues of some of these theorems for functions of two variables were considered in § 5.3, but no proofs were given. With the aid of the concepts and theorems developed in Chapter XVI we are now in a position to make a deeper study of continuity.

One important new concept will be introduced in the present chapter: the concept of uniform continuity. This is needed for the theory of integration, in Chapter XVIII.

17.1 Continuity and sequential limits

For the definition of continuity of a function of one variable, we refer the student to § 3. There is an alternative way of expressing continuity, using the notion of a convergent sequence. We express this as a theorem.

THEOREM I. *Suppose f is a function of x, defined on some interval I. Let x_0 be a point of I. Then f is continuous at x_0 if and only if $lim_{n \to \infty} f(x_n) = f(x_0)$ for every sequence $\{x_n\}$ which has terms that belong to I and which is such that $lim_{n \to \infty} x_n = x_0$.*

We shall not prove this theorem, because we are presently going to prove a more general theorem of which Theorem I is a special case.

In dealing with continuity it is quite worth while to arrange matters so that what we say applies just as well to functions of two or three variables as to functions of one variable. The terminology of point-set theory enables us to do this. In particular, the concept of continuity can be defined in terms of the concept of neighborhood, without explicit mention of ϵ and δ or use of inequalities.

DEFINITION. Let S be a point set, and let f be a function defined at each point of S. If P_0 is a certain point of S, the function is said to be continuous at P_0 provided that to each neighborhood V of the value $f(P_0)$ there corresponds a neighborhood U of the point P_0 such that the value $f(P)$ lies in the neighborhood V whenever the point P lies in the set S and also in the neighborhood U.

This definition is equivalent to the previously given definitions (in § 3 and § 5.3), though it is somewhat different in form from these earlier definitions. For example, if f is a function of one real variable, we may take the set S to be on the x-axis, and denote the point by x instead of P. The neighborhood V of $f(x_0)$ will be an interval defined by $|y - f(x_0)| < \epsilon$, or $f(x_0) - \epsilon < y < f(x_0) + \epsilon$, where ϵ is some positive number. The neighborhood U will be an interval defined by $|x - x_0| < \delta$, or $x_0 - \delta < x < x_0 + \delta$, where δ is some positive number. The statement that $f(x)$ is in V if x is in U is then the same as the statement that $|f(x) - f(x_0)| < \epsilon$ if $|x - x_0| < \delta$.

One thing which deserves to be emphasized is that the point set S on which f is defined is not subject to any restrictions. Thus, if we are talking about a function of two variables, and P has co-ordinates (x, y), the point set S can be any kind of point set in the plane. It does not need to be a region (as defined in § 5.1). For example, S could be the set of points on the parabola $y = x^2$, and $f(x, y)$ might be the radius of curvature at the point (x, y). Or, in three dimensions, S might be the surface of an ellipsoid, and $f(x, y, z)$ might be the distance from (x, y, z) to the origin.

We now come to the generalization of Theorem I.

THEOREM II.   *Suppose f is a function defined at each point of the set S. If P_0 is a particular point of S, the function is continuous at P_0 if and only if $\lim_{n \to \infty} f(P_n) = f(P_0)$ whenever $\{P_n\}$ is a sequence of points belonging to S such that $\lim_{n \to \infty} P_n = P_0$.*

PROOF.   Suppose f is continuous at P_0, and suppose P_n belongs to S and P_n converges to P_0. If V is any neighborhood of $f(P_0)$, we have to show that $f(P_n)$ is in V when n is sufficiently large. Now, since f is continuous at P_0, there is a neighborhood U of P_0 such that $f(P)$ is in V if P is in S and U. The fact that P_n converges to P_0 means that P_n is in U if n is sufficiently large. But then $f(P_n)$ must be in V, and one part of the proof is complete.

We now assume that $f(P_n)$ converges to $f(P_0)$ whenever $\{P_n\}$ is a sequence in S converging to P_0. We shall make the required proof by supposing that f is not continuous at P_0, and deducing a contradiction. The denial of continuity at P_0 may be phrased in this way: There is *some* neighborhood V of $f(P_0)$ such that, no matter what neighborhood U of P_0 is selected, some point P in U has the property that P is in S but $f(P)$ is not in V. Accordingly, let us choose a succession of neighborhoods U_1, $U_2, \cdots, U_n, \cdots$ closing down on P_0 (e.g. U_n consisting of all points at distance less than $1/n$ from P_0), and let P_n be a point in U_n with properties as described above. Then P_n must converge to P_0, but $f(P_n)$ does not converge to $f(P_0)$, because $f(P_n)$ is always outside the neighborhood V. This contradicts our initial assumption. Therefore the proof is complete.

17.2   The boundedness theorem

We give a generalization of Theorem II, § 3.1, and Theorem I, § 5.3.

THEOREM III.   *Let S be a bounded and closed point set, and let f be a function defined on S which is continuous at each point of S. Then the values of f are bounded, or, as we customarily say, f is bounded on S.*

The proof is rather simple. Suppose the values of f were not bounded. Then no finite interval of the real number scale contains all the values, and for each positive integer n there must be at least one point P_n in S such that $|f(P_n)| > n$. The sequence $\{P_n\}$ is bounded, and must therefore contain a convergent subsequence (see Theorem V, § 16.31, and the remarks about it in § 16.41). Denote the subsequence by $\{P_{n_i}\}$ and its limit by Q. Since S is closed, Q belongs to S. Then, since f is continuous at Q, $f(P_{n_i})$ converges to $f(Q)$, by Theorem II, § 17.1. This is in contradiction to $|f(P_{n_i})| > n_i$, which has the consequence that $|f(P_{n_i})| \to \infty$ as $i \to \infty$. To avoid the contradiction we are forced to conclude that f is bounded.

17.3   The extreme value theorem

We give a generalization of Theorem III, § 3.2, and Theorem II, § 5.3.

THEOREM IV.   *Let S be a non-empty bounded and closed set, and suppose f is a function defined on S and continuous at each point of S. Let m and M be the greatest lower bound and least upper bound, respectively, of the values of f on S. Then there is some point of S at which f has the value M, and there is also a point at which f has the value m.*

The existence of m and M is guaranteed by the theorem in § 17.2 and Theorem II, § 2.7.

We begin the proof by observing that since M is the least upper bound of the values $f(P)$, there must exist a sequence $\{P_n\}$ in S such that $f(P_n)$ converges to M. The sequence $\{P_n\}$ is bounded, and just as in the proof of Theorem III we obtain a subsequence $\{P_{n_i}\}$ converging to a point Q, with $f(P_{n_i})$ converging to $f(Q)$. But $f(P_{n_i})$ also converges to M, so $M = f(Q)$. This proves the theorem as regards M. The proof for m is essentially the same.

17.4   Uniform continuity

When we say that a function is continuous at a certain point, this describes a certain relation between the values of the function and the values of the independent variable near a particular value of the independent variable. The new concept of *uniform continuity*, which we are concerned with in this section, has to do with continuity of a function at many different points. The word "uniform" is used because, when we say that a function is uniformly continuous on a certain point set, this means that there is a certain quality about the continuity which is the same at all the points of the set. Before giving a formal definition of uniform continuity we shall have to see in what sense there can be recognizable differences in the continuity of a function at different points.

The type of difference we have in mind has to do with the amount by which the independent variable may be allowed to change if the value of the function is not permitted to change more than a specified amount.

Example 1. Consider the function $f(x) = x^2$. Suppose we start with some particular value x_0, and ask: How much may x differ from x_0 if $f(x)$ is required to differ from $f(x_0)$ by less than 2 units? The answer depends on the value of x_0. If $x_0 = 0$, the requirement on x is that $x^2 < 2$, or $|x| < \sqrt{2}$, so that x must differ from x_0 by less than $\sqrt{2}$. But if $x_0 = 8$ the requirement is that $|x^2 - 64| < 2$, or $62 < x^2 < 66$. Now $\sqrt{66} = 8.124 \cdots$, $\sqrt{62} = 7.874 \cdots$, so that the permissible difference between x and 8 is less than $\sqrt{66} - 8 = 0.124 \cdots$. The student will see at once from a graph of $y = x^2$ that the larger we make x_0, the smaller is the amount by which x may differ from x_0 if x^2 is to differ from x_0^2 by less than 2 units.

If instead of 2 units we specify ϵ units, where $\epsilon > 0$, the situation is essentially the same. If we want to find a number δ so that $|x^2 - x_0^2| < \epsilon$ whenever $|x - x_0| < \delta$ the value of δ will depend on x_0 as well as on ϵ, and with ϵ fixed, δ must be made smaller and smaller as x_0 gets larger. There is no single positive value of δ which is small enough to serve simultaneously for *all* values of x_0. This illustrates the *opposite* of uniform continuity.

DEFINITION. Let S be a point set on the x-axis, forming part or all of the set of values of x for which $f(x)$ is defined. Then f is said to be *uniformly continuous* on S provided that to each $\epsilon > 0$ there corresponds a $\delta > 0$ such that $|f(x) - f(x_0)| < \epsilon$ whenever x and x_0 are any points of S such that $|x - x_0| < \delta$.

Let us carry the discussion of Example 1 further in the light of this definition. The function $f(x) = x^2$ is *not* uniformly continuous on the set S of all x. In fact if S is any set which is not bounded, the function is not uniformly continuous on S, for if there is no limit to how large x_0 can be, no single δ can be found which meets the requirements of the definition. On the other hand, if S is a bounded set, the function *is* uniformly continuous on S. For instance, if S is the closed interval $0 \le x \le 8$, we can take $\delta = \epsilon/16$ in the definition of uniform continuity. To see this, observe that if x and x_0 belong to S,

$$| x^2 - x_0{}^2 | = | (x - x_0)(x + x_0) | \leq 16 | x - x_0 |$$

and so

$$| x^2 - x_0{}^2 | < \epsilon \quad \text{if} \quad | x - x_0 | < \frac{\epsilon}{16}.$$

Example 2. The function $f(x) = 1/x$ is not uniformly continuous on the set S defined by $0 < x \leq 1$, but it *is* uniformly continuous on the set S defined by $x \geq 1$. The first assertion follows from the fact that if $x_0 > 0$ and δ is chosen so that

$$\left| \frac{1}{x} - \frac{1}{x_0} \right| < \epsilon \quad \text{if} \quad | x - x_0 | < \delta,$$

the value of δ must approach 0 as $x_0 \to 0$. On the other hand, for the set S defined by $x \geq 1$ we can take $\delta = \epsilon$, because if x and x_0 belong to S and $| x - x_0 | < \epsilon$ we have

$$\left| \frac{1}{x} - \frac{1}{x_0} \right| = \frac{| x_0 - x |}{x x_0} \leq | x - x_0 | < \epsilon.$$

The essential theorem about uniform continuity will now be given.

THEOREM V. *Suppose S is a closed and bounded point set, and suppose the function f is defined and continuous at each point of S. Then f is uniformly continuous on S.*

PROOF. We make use of the Heine-Borel theorem. Suppose $\epsilon > 0$. If x' is any point of S, the definition of continuity assures us that there is some positive number h such that $| f(x) - f(x') | < \epsilon/2$ if x and x' belong to S and $| x - x' | < h$. The size of h will usually vary as x' is varied. Now consider the open interval $x' - (h/2) < x < x' + (h/2)$. When x' varies over S, the collection of all these open intervals covers the set S. By the Heine-Borel theorem (§ 16.6) a finite number of these intervals suffice to cover S. Let the centers of these intervals be denoted by $x_1, \cdots, x_n$ and let the corresponding values of h be $h_1, \cdots, h_n$. Choose δ as the smallest of the numbers $h_1/2, \cdots, h_n/2$. We shall show that this δ will serve as required in the definition of uniform continuity. Suppose x and x_0 belong to S and $| x - x_0 | < \delta$. Then x_0 belongs to one of the finite set of open intervals, say the one with end points $x_i \pm (h_i/2)$, so that $| x_0 - x_i | < h_i/2$. Now

$$| x - x_i | \leq | x - x_0 | + | x_0 - x_i | < \delta + \frac{h_i}{2}.$$

But $\delta \leq \dfrac{h_i}{2}$, and so $| x - x_i | < h_i$. The inequalities satisfied by $| x_0 - x_i |$ and $| x - x_i |$ guarantee that

$$| f(x_0) - f(x_i) | < \frac{\epsilon}{2} \quad \text{and} \quad | f(x) - f(x_i) | < \frac{\epsilon}{2} \cdot$$

Therefore $| f(x) - f(x_0) | \leq | f(x) - f(x_i) | + | f(x_i) - f(x_0) | < \epsilon.$
This completes the proof.

The definition of uniform continuity can be worded so as to apply to functions of more than one variable. It is merely necessary to write the condition involving ϵ and δ in the form "$| f(P) - f(P_0) | < \epsilon$ whenever P and P_0 are points of S such that $d(P, P_0) < \delta$." Here $d(P, P_0)$ is the distance between P and P_0. Theorem V remains valid for functions of several variables, and the proof as given above can be adapted easily to the new situation by using distances in place of absolute values.

17.5   Continuity of sums, products, and quotients

The process of addition may be considered as defining a function of two variables:

$$s(x, y) = x + y.$$

The same may be said of multiplication and division:

$$p(x, y) = xy, \quad q(x, y) = \frac{x}{y} \cdot$$

We use the letters s, p, q for these three functions because of the names sum, product, quotient. The following theorem states a fundamental fact about these functions:

THEOREM VI.   *The functions s and p are continuous at all points (x, y). The function q is continuous at all points where it is defined (i.e. at all points except those for which $y = 0$).*

The proof of this theorem is very similar to the proof of Theorem XIV, § 1.64, and on that account we omit the formal proof. Some suggestions for the student who wishes to work the details out for himself are given in the exercises following.

EXERCISES

1. Show that $| s(x, y) - s(x_0, y_0) | < \epsilon$ if (x, y) is in the square neighborhood of side ϵ with center at (x_0, y_0), and so certainly if (x, y) is in the circular neighborhood of radius $\epsilon/2$ with center at (x_0, y_0).

2. Let M be a number larger than $| x_0 |$ and $| y_0 |$. Let δ be the smaller of the

numbers $M - |y_0|$, $\epsilon/(2\,M)$. Show that $|p(x, y) - p(x_0, y_0)| < \epsilon$ if $|x - x_0| < \delta$ and $|y - y_0| < \delta$. Start from the fact that $xy - x_0 y_0 = (x - x_0)y + x_0(y - y_0)$ and so $|xy - x_0 y_0| \leq |x - x_0| |y| + |x_0| |y - y_0|$.

3. Study the proof of (1.64–4) and so show that $|q(x, y) - q(x_0, y_0)| < \epsilon$ if $|x - x_0| < \delta$ and $|y - y_0| < \delta$, where δ is the smaller of the numbers $\frac{1}{2}|y_0|$,

$$\frac{|y_0|^2 \epsilon}{2(|x_0| + |y_0|)} \cdot$$ Assume that $y_0 \neq 0$.

17.6   Persistence of sign

We shall now give a generalization of Theorem V, § 3.3.

THEOREM VII.   *Suppose f is continuous at a point P_0 of the set S on which it is defined, and suppose $f(P_0) \neq 0$. Then there is some neighborhood of P_0 such that at all points of S in this neighborhood the sign of $f(P)$ is the same as the sign of $f(P_0)$.*

The truth of this theorem depends merely on the fact that $f(P)$ is near $f(P_0)$ if P is near P_0. If $f(P_0) \neq 0$, all values of $f(P)$ sufficiently near $f(P_0)$ will have the same sign as $f(P_0)$.

17.7   The intermediate-value theorem

An intermediate-value theorem was stated as Theorem IV, § 3.3. We shall now consider how this theorem is to be generalized so as to apply to functions of more than one variable. The problem is of this nature: Suppose a function is defined on some point set, and is continuous at each point. If the function takes on the value 2 at one point and the value -3 at another point, does it also take on all values between -3 and 2? An example will show that such is not always true. For instance, let $f(x, y) = 1/(xy)$. Then $f(2, \frac{1}{4}) = 2$, $f(-\frac{1}{6}, 2) = -3$. But $f(x, y)$ never takes on the value 0, and 0 is between -3 and 2, in spite of the fact that f is continuous at each point of the set where it is defined. The explanation of the difficulty lies in the fact that the points $(2, \frac{1}{4})$ and $(-\frac{1}{6}, 2)$ are separated by the line $x = 0$, on which f is not defined.

If f is defined and continuous on a set S which is not separated into several disconnected parts, it can be proved that f has the property of taking on all values between every pair of distinct values. In order to make this assertion in an exact form, and prove it, we must first make some definitions.

DEFINITION.   Two point sets S_1 and S_2 are called *separated* if the following three conditions hold:

(1) Neither set is empty,

(2) No point belongs to both sets,

(3) Neither set has a point of accumulation belonging to the other set.

Example 1. The intervals $-1 < x < 0$ and $0 < x < 1$ are separated sets. The interior and exterior of the circle $x^2 + y^2 = 1$ are separated sets in the plane. The points on opposite sides of the plane $z = 0$ form two separated sets in space of three dimensions.

DEFINITION. A set S is called *connected* if it cannot be divided into two parts which are separated.

Example 2. An interval on the x-axis, with or without either end point, is a connected set. The first quadrant in the xy-plane ($x > 0$, $y > 0$) is a connected set. The surface of a sphere is a connected set.

If S is an open set which is connected, it may be shown that any two points of S can be joined by an arc which lies entirely in S. In fact, the arc may be taken to consist of a finite number of line segments joined end to end. In § 7.4 we *defined* an open set to be connected if it had this property. For open sets the two definitions are equivalent, but for sets which are not open the present definition is the proper one to use.

We now come to the general intermediate-value theorem.

THEOREM VIII. *Suppose S is a connected set and that f is a function which is continuous at each point of S. Suppose f takes on two different values C_1 and C_2 at points P_1 and P_2 in S. Then, for every number k between C_1 and C_2, there is some point of S at which f takes on the value k.*

PROOF. Let us suppose the notation is such that $C_1 < k < C_2$. We shall suppose that for some particular k there is no point P such that $f(P) = k$. We shall show that with this assumption we can divide S into two separated parts. This contradiction will complete the proof of the theorem.

Let S_1 be the set of those points of S for which $f(P) < k$, and S_2 the set for which $k < f(P)$. Observe that P_1 is in S_1, P_2 is in S_2, and each point of S belongs either to S_1 or to S_2. The sets S_1 and S_2 are separated. In fact, since conditions (1) and (2) in the definition of separated sets are clearly satisfied, we have only to verify the third condition. Now, it is impossible for a point of accumulation of S_2 to belong to S_1. For suppose Q were such a point. Then we could select a sequence $\{Q_n\}$ of points in S_2 such that Q_n converges to Q. Then $f(Q_n)$ converges to $f(Q)$, by continuity. But $f(Q_n) > k$, since Q_n is in S_2. Therefore the limit $f(Q)$ cannot be less than k. But $f(Q) < k$ since Q was assumed to be in S_1. This shows that no such point as Q can exist. The same argument shows that no point of accumulation of S_1 can belong to S_2. This proves that S_1 and S_2 are separated, and the proof is complete.

The Theory of Integration 18

18. The nature of the chapter

Heretofore, in considering integrals defined as limits of sums, we have taken it for granted that the limits do exist. The applications of integrals in calculating various geometrical and physical magnitudes are commonly of such a kind that we consider integrals of continuous functions, and it is plausible, from an intuitive standpoint, that the process of defining the integral of a continuous function does actually lead to a definite limit.

For certain purposes it becomes desirable and necessary to consider the integration of functions which may be discontinuous. Consequently it is necessary to lay down the fundamental definitions about integrals in such a way that the theory can be developed without depending on hypotheses about continuity. In this chapter we shall lay the groundwork of a general theory of integration. The form of this theory goes back to Bernhard Riemann (1826–1866), a German mathematician. The first purpose of this chapter is to set forth clearly the concept of the Riemann integral. A function which can be integrated according to Riemann's definition is called *integrable*. We show that continuous functions and certain kinds of discontinuous functions are integrable.

In certain ways the theory of multiple integrals is much more complicated than the theory of integrals of functions of one variable. Our discussion of multiple integrals is designed mainly to lead up to an analytical treatment of the relation between a multiple integral and its evaluation by iterated integrals. This is a topic on which we have not aimed to give the greatest possible generality, but a reasonably simple and comprehensible treatment of the kind of multiple integrals which ordinarily arise in calculus.

The chapter concludes with a very brief, conceptual, nontheoretical discussion of Stieltjes integrals.

18.1 The definition of integrability

Suppose that a, b are any real numbers such that $a < b$. We shall consider functions defined on the closed interval $[a, b]$. We shall not require the functions to be continuous, but we shall assume that each function is bounded.

Now let us subdivide the interval into any number of subintervals. This is done by inserting points between a and b. Thus, suppose

$$a = x_0 < x_1 < x_2 < \cdots < x_{n-1} < x_n = b.$$

Then we have n subintervals

$$[x_0, x_1], [x_1, x_2], \cdots, [x_{n-1}, x_n].$$

We shall refer to such a subdivision as a *partition* of $[a, b]$. The partition determined by these particular points will be called the partition $(x_0, x_1, \cdots, x_n)$.

In the closed subinterval $[x_{i-1}, x_i]$ let m_i and M_i denote respectively the greatest lower bound and least upper bound of the values $f(x)$. Then we form the two sums

(18.1–1)
$$s = m_1(x_1 - x_0) + m_2(x_2 - x_1) + \cdots + m_n(x_n - x_{n-1}),$$

(18.1–2)
$$S = M_1(x_1 - x_0) + M_2(x_2 - x_1) + \cdots + M_n(x_n - x_{n-1}).$$

We call s a *lower sum* and S an *upper sum*. Observe that these sums are dependent upon the particular function f and the particular partition.

Let us denote the greatest lower bound and least upper bound of $f(x)$ on all of $[a, b]$ by m, M respectively. Then $m \le m_i$ and $M_i \le M$ for $i = 1, 2, \cdots, n$. Of course it is also true that $m_i \le M_i$. We observe that

$$(x_1 - x_0) + (x_2 - x_1) + \cdots + (x_n - x_{n-1}) = b - a.$$

Therefore, it is readily seen that

(18.1–3) $$m(b - a) \le s \le S \le M(b - a).$$

The next step in our procedure is motivated by geometry. Suppose for a moment that the function f is continuous and that all its values are positive. The lower sum s then represents the area of a sum of rectangles all of which lie between the x-axis and the curve $y = f(x)$ (see Fig. 171). It then seems plausible to suppose that if we consider all possible partitions and the corresponding lower sums, the least upper bound of all these lower sums s will be exactly the area between the x-axis and the curve and between $x = a$ and $x = b$.

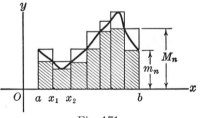

Fig. 171

It likewise seems plausible to suppose that this area is the greatest lower bound of all the upper sums S when all possible partitions are considered (again, see Fig. 171).

DEFINITION. As a result of the foregoing considerations, let us define

$I = $ least upper bound of all lower sums s,

$J = $ greatest lower bound of all upper sums S.

Then we shall say that the function f is *integrable* on $[a, b]$ if $I = J$, and in that case we shall call the common value of I and J the *definite integral* of f from $x = a$ to $x = b$, denote it by

$$\int_a^b f(x)dx.$$

This definition applies to each function which is defined and bounded on $[a, b]$. For each such function we get unique values for I and J, but it may happen that $I \neq J$. It is only when $I = J$ that the function is integrable.

Example 1.   Consider the function defined on $[0, 1]$ by setting $f(x) = 0$ if x is rational and $f(x) = 1$ if x is irrational. For this function and this interval $I = 0$, $J = 1$, so that the function is not integrable. To see the truth of this assertion, note that any subinterval contains both rational and irrational values of x, and therefore that $m_i = 0$, $M_i = 1$. Consequently, from the definitions of s and S we see that $s = 0$, $S = 1$, no matter how the partition is chosen. Accordingly, $I = 0$, $J = 1$.

The student will observe that the foregoing definition of $\int_a^b f(x)dx$ is not the same as the definition given in § 1.5. The definition which we are now considering makes no reference to the concept of a limit of approximating sums. Later on in this chapter we shall see that the two different forms of definition are equivalent. Our first concern, however, is to develop enough theory to furnish a practical means of telling when a function is integrable.

LEMMA I.   *Suppose we start with a certain partition, and then obtain a new partition by inserting some additional points. Let s, S refer to the lower and upper sums for the original partition, while s', S' are the sums for the new partition. Then $s \leq s'$ and $S' \leq S$.*

For simplicity we shall prove this on the supposition that just one new point is inserted. Suppose for definiteness that the new point ξ is between x_0 and x_1. Let

$$M_1' = \text{least upper bound of } f(x) \text{ for } x_0 \leq x \leq \xi,$$
$$M_1'' = \text{least upper bound of } f(x) \text{ for } \xi \leq x \leq x_1.$$

Then certainly $M_1' \leq M_1$ and $M_1'' \leq M_1$. Therefore

$$M_1'(\xi - x_0) + M_1''(x_1 - \xi) \leq M_1(\xi - x_0) + M_1(x_1 - \xi) = M_1(x_1 - x_0).$$

From this it follows that $S' \leq S$, because S' differs from S only in having $M_1'(\xi - x_0) + M_1''(x_1 - \xi)$ in place of $M_1(x_1 - x_0)$. The proof that $s \leq s'$ is similar. If more than one point is inserted we need only apply the argument several times.

LEMMA II. *For a given function and any two partitions, the lower sum for one partition is algebraically less than or equal to the upper sum for the other partition.*

PROOF. Denote the sums corresponding to the two partitions by s_1, S_1 and s_2, S_2. Now consider the third partition which is obtained by considering simultaneously all the points of subdivision of the first two partitions; denote the corresponding sums by s_3, S_3. We know by (18.1–3) that $s_3 \leq S_3$. Also, we know by Lemma I that $s_1 \leq s_3$ and $S_3 \leq S_2$. Therefore, combining the inequalities, we see that $s_1 \leq S_2$. This is what Lemma II asserts.

LEMMA III. *It is always true that $I \leq J$.*

This is an immediate consequence of Lemma II and the definitions of I and J.

We can now state an important criterion for integrability.

THEOREM I. *Suppose f is bounded on $[a, b]$, and suppose that corresponding to each positive ϵ there is a partition of $[a, b]$ such that the corresponding upper and lower sums satisfy the inequality $S - s < \epsilon$. Then f is integrable.*

PROOF. If the conditions as stated in the theorem are fulfilled, we have $S < s + \epsilon$. But $J \leq S$ and $s \leq I$, by the definitions of I and J. Therefore, combining inequalities, we see that $J \leq S < s + \epsilon \leq I + \epsilon$, or $J < I + \epsilon$. Since this conclusion is valid for every positive ϵ, we infer that $J \leq I$. But we also know that $I \leq J$ (Lemma III). Therefore $I = \cdot J$. This means that f is integrable, by definition.

Example 2. Suppose f is defined on $[0, 2]$ as follows:

$$f(x) = 1 \quad \text{if} \quad 0 \leq x < 1, \qquad f(x) = 2 \quad \text{if} \quad 1 \leq x \leq 2.$$

We can show by Theorem I that f is integrable.

For this purpose suppose a positive ϵ is assigned. Let h be a positive number smaller than 1 and also smaller than $\epsilon/2$. Consider the partition defined by

$$x_0 = 0, \qquad x_1 = 1 - h, \qquad x_2 = 1 + h, \qquad x_3 = 2.$$

It is readily seen from Fig. 172 that

$$m_1 = M_1 = 1, \qquad m_2 = 1, \qquad M_2 = 2, \qquad m_3 = M_3 = 2,$$

and therefore that

$$s = 1 \cdot (1 - h) + 1 \cdot 2h + 2(1 - h) = 3 - h,$$
$$S = 1 \cdot (1 - h) + 2 \cdot 2h + 2(1 - h) = 3 + h.$$

Consequently $S - s = 2h < \epsilon$. Therefore f is integrable, by Theorem I. In giving this argument we did not need to find the value of the integral, but it is not hard to see that

(18.1–4)
$$\int_0^2 f(x)dx = 3.$$

Note that the function is discontinuous at $x = 1$.

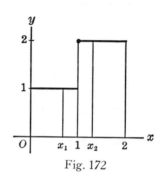

Fig. 172

The following theorem is the converse of Theorem I.

THEOREM II.   *If f is integrable on $[a, b]$, and if $\epsilon > 0$, there is a partition with upper and lower sums such that $S - s < \epsilon$.*

The proof is left as an exercise.

EXERCISES

1. Prove (18.1–4) in Example 2 by the following steps: First, $I \geqq 3$; next, $J \leqq 3$; and finally, $I = J = 3$. Explain each step fully.

2. Suppose f is defined as follows: $f(x) = 2$ if $0 \leqq x < 1$, $f(1) = 0$, $f(x) = -1$ if $1 < x < 2$, $f(2) = 3$, $f(x) = 0$ if $2 < x < 3$, $f(3) = 1$. **(a)** Prove that f is integrable, using an argument something like that in Example 2, but with six subintervals. **(b)** Find the value of $\int_0^3 f(x)dx$, using an argument like that of Exercise 1.

3. Suppose f is defined by the requirement that $f(x) = 2$ if x is a rational number of the form $p/2^q$, where p can take on all the values $0, \pm 1, \pm 2, \cdots$ and q can take on all the values $1, 2, \cdots$, and $f(x) = 1$ for all other values of x. Calculate I and J for this function on the interval $[0, 2]$, and thus prove that f is not integrable.

4. Prove Theorem II.

5. If f is integrable, so is the absolute-value function $\mid f(x) \mid$. Prove this by showing that if s, S refer to f, and s', S' refer to $\mid f \mid$, then $S' - s' \leqq S - s$. Then use Theorem I.

6. If f is integrable over the interval $[a, b]$, it is also integrable over any closed subinterval of $[a, b]$. Prove this, using Lemma I and Theorem I.

7. Suppose $a < b < c$ and that f is integrable over $[a, b]$ and also over $[b, c]$. Prove by Theorems I and II that f is integrable over $[a, c]$.

18.11   The integrability of continuous functions

Every continuous function is integrable. We state this in a formal theorem.

THEOREM III.   *If a function f is continuous at each point of $[a, b]$, it is integrable on that interval.*

PROOF.   The argument hinges on Theorem I and on the uniform continuity of the function. Suppose $\epsilon > 0$. Choose δ so that

$$(18.11\text{–}1) \qquad\qquad |f(x') - f(x'')| < \frac{\epsilon}{2(b - a)}$$

if x' and x'' are points of $[a, b]$ such that $|x' - x''| < \delta$. This may be done, since f is uniformly continuous (Theorem V, § 17.4). Now consider any partition $(x_0, x_1, \cdots, x_n)$ such that all the subintervals have length less than δ, and let s, S be the corresponding lower and upper sums. The boundedness of f is guaranteed by Theorem II, § 3.1. Now, in the interval $[x_{i-1}, x_i]$ we can choose a point x' so that $f(x')$ is as close as we like to M_i, and a point x'' so that $f(x'')$ is as close as we like to m_i. Since (18.11–1) holds, we conclude that

$$(18.11\text{–}2) \qquad\qquad M_i - m_i \leq \frac{\epsilon}{2(b - a)} \cdot$$

But then, by (18.1–1) and (18.1–2),

$$S - s = (M_1 - m_1)(x_1 - x_0) + \cdots + (M_n - m_n)(x_n - x_{n-1}),$$

and so

$$|S - s| \leq \frac{\epsilon}{2(b - a)} [(x_1 - x_0) + \cdots + (x_n - x_{n-1})] = \frac{\epsilon}{2} \cdot$$

Thus f is integrable, by Theorem I.

EXERCISE

If f is integrable on $[a, b]$, if $f(x) \geq 0$ for each x, and if there is at least one point ξ where f is continuous and $f(\xi) > 0$, prove that $\int_a^b f(x)dx > 0$.

18.12   Integrable functions with discontinuities

A function such that $f(x)$ never decreases as x increases can have points of discontinuity. Nevertheless, such a function is integrable over any closed interval on which it is defined.

THEOREM IV.   *Suppose $f(x)$ is defined when $a \leqq x \leqq b$, and that $f(x')$ $\leqq f(x'')$ if $x' < x''$. Then f is integrable on $[a, b]$.*

PROOF.   There are two cases to consider. If $f(a) = f(b)$, the hypothesis guarantees that $f(x)$ is constant on the interval, and in this case certainly the upper and lower sums are both equal to $f(a)(b - a)$, no matter what partition is chosen, and so the function is integrable. The other case is that in which $f(a) < f(b)$. If $\epsilon > 0$ we can choose a partition in which all the subintervals are so short that the conditions

$$x_i - x_{i-1} < \frac{\epsilon}{f(b) - f(a)} , \qquad i = 1, \cdots, n$$

are satisfied. Now, the special hypothesis on f assures that $m_i = f(x_{i-1})$ and $M_i = f(x_i)$. That is, as x goes from x_{i-1} to x_i, $f(x)$ increases from its smallest to its largest value in the subinterval. Consequently

$$S - s = [f(x_1) - f(x_0)](x_1 - x_0) + \cdots + [f(x_n) - f(x_{n-1})](x_n - x_{n-1}),$$

$$S - s < \{[f(x_1) - f(x_0)] + \cdots + [f(x_n) - f(x_{n-1})]\} \frac{\epsilon}{f(b) - f(a)} .$$

Because of cancellation of terms and the facts that $x_0 = a$, $x_n = b$, this last inequality becomes simply $S - s < \epsilon$. We then conclude that f is integrable, by Theorem I.

It is of course also true that f is integrable if $f(x)$ never increases as x increases. In this case $m_i = f(x_i)$ and $M_i = f(x_{i-1})$ on any subinterval $[x_{i-1}, x_i]$.

It is beyond the scope of this book to make an intensive study of the question as to exactly what kinds of discontinuous functions are integrable. The function of Example 1, § 18.1, is not integrable, but it is discontinuous at every point of the interval $[0, 1]$. For many practical purposes it is sufficient to know that if f is bounded and has only a finite number of points of discontinuity on $[a, b]$, then it is integrable. The discussion of the following example will illustrate a method by which the foregoing assertion may be proved.

Example 1.   Suppose f is defined on $[0, 2]$ by

$$f(x) = x \quad \text{if} \quad 0 \leqq x < 1, \qquad f(x) = x - 1 \quad \text{if} \quad 1 \leqq x \leqq 2.$$

This function is bounded on $[0, 2]$, and continuous except at $x = 1$. The graph is shown in Fig. 173. We shall use Theorem I to show that f is integrable. Suppose $\epsilon > 0$. We form a partition of $[0, 2]$ by taking $1 - (\epsilon/4)$ and $1 + (\epsilon/4)$ as two consecutive points in the partition. The remaining points are taken between 0 and $1 - (\epsilon/4)$ and $1 + (\epsilon/4)$ and 2, in a manner

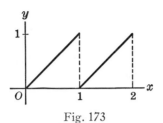

Fig. 173

to be specified presently. Observe that f is continuous on the intervals $[0, 1 - (\epsilon/4)]$, $[1 + (\epsilon/4), 2]$; the point of discontinuity has been enclosed in the subinterval $[1 - (\epsilon/4), 1 + (\epsilon/4)]$, whose length is $\epsilon/2$. Now consider the lower and upper sums s, S. Let s_1 represent the contributions coming to s from the subintervals of $[0, 1 - (\epsilon/4)]$, and let s_3 represent the contributions from the subintervals of $[1 + (\epsilon/4), 2]$. Let s_2 represent the contribution from the single subinterval $[1 - (\epsilon/4), 1 + (\epsilon/4)]$, so that $s = s_1 + s_2 + s_3$. With similar notations for upper sums we have $S = S_1 + S_2 + S_3$. Now, the least upper bound of $f(x)$ when $1 - (\epsilon/4) \leqq x \leqq 1 + (\epsilon/4)$ is 1, and the least value is 0. Therefore

$$ s_2 = 0\left(\frac{\epsilon}{2}\right) = 0, \quad S_2 = 1\left(\frac{\epsilon}{2}\right) = \frac{\epsilon}{2} . $$

Since f is continuous on $[0, 1 - (\epsilon/4)]$ and $[1 + (\epsilon/4), 2]$, we can choose the part of the partition in these intervals so that

$$ S_1 - s_1 < \frac{\epsilon}{4} \quad \text{and} \quad S_3 - s_3 < \frac{\epsilon}{4} . $$

This is by Theorems III and II. Then

$$ S - s = (S_1 - s_1) + (S_2 - s_2) + (S_3 - s_3) < \frac{\epsilon}{4} + \frac{\epsilon}{2} + \frac{\epsilon}{4} = \epsilon. $$

We then conclude by Theorem I that the function is integrable on $[0, 2]$.

The kind of argument employed in the foregoing discussion has a much more general application, and by means of it we can prove the following theorem.

THEOREM V. *If f is bounded on $[a, b]$, and if the points of $[a, b]$ at which f is discontinuous can be enclosed in a finite number of subintervals the sum of whose lengths can be made as small as we please, then f is integrable. In particular, f is integrable if it is bounded and has only a finite number of points of discontinuity.*

We omit the details of proof because of the similarity to the argument used in connection with Example 1.

Example 2. Suppose f is defined and bounded on [0, 1], and is continuous except at the points $\frac{1}{2}$, $\frac{1}{3}$, $\frac{1}{4}$, $\frac{1}{5}$, $\cdots$. Then it is integrable; for, no matter how small a positive ϵ we choose, we can enclose all the points of discontinuity in a finite number of subintervals of total length less than ϵ. To see that this is so, suppose $0 < \epsilon < 1$. As a first subinterval choose [0, $\epsilon/4$]. This interval will enclose all but a finite number of the points $\frac{1}{2}$, $\frac{1}{3}$, $\cdots$. If there are N points *not* so enclosed, we can enclose each of these remaining points in a subinterval of length $\epsilon/(2\,N)$. This gives us $N + 1$ subintervals in all, the sum of whose lengths is $3\,\epsilon/4$, and they enclose all the points at which f is discontinuous. Thus f is integrable, by Theorem V.

The condition for integrability in Theorem V is *sufficient*, but not *necessary*. The following theorem is true, but we shall not attempt to prove it.

THEOREM VI. *If f is bounded on [a, b], it is integrable if and only if the points at which f is discontinuous can be enclosed in a finite or denumerably infinite set of subintervals of total length as small as we please.*

18.2 The integral as a limit of sums

In this section we shall show the equivalence between the definition of the integral as given in § 18.1 and the definition which is used in elementary calculus (the one given in § 1.5).

It is desirable to introduce some special terminology and notation in connection with partitions. If [a, b] is a fixed closed interval, let us use symbols such as P, P_0, P', $\cdots$ for various partitions of the interval. If P is the partition determined by points $(x_0, x_1, \cdots, x_n)$, where $a = x_0$, $b = x_n$, the length of the longest subinterval in the partition is called the *mesh fineness* of P, and is denoted in symbols by $|\,P\,|$. Thus, by definition, $|\,P\,|$ is the maximum of the differences

$$x_1 - x_0, x_2 - x_1, \cdots, x_n - x_{n-1}.$$

The notion of the integral as a limit of sums is based on consideration of sums of the type

$$(18.2\text{-}1) \quad f(x_1')(x_1 - x_0) + f(x_2')(x_2 - x_1) + \cdots + f(x_n')(x_n - x_{n-1}).$$

Here $x_0, x_1, \cdots, x_n$ are the points of a partition P, and $x_1', \cdots, x_n'$ are any points chosen so that $x_{i-1} \leqq x_i' \leqq x_i$, $i = 1, \cdots, n$. The following theorem is fundamental in the theory of integration:

THEOREM VII. *Suppose f is bounded on the interval [a, b]. Then it is integrable if and only if the sums (18.2-1) approach a limit as the mesh fineness $|\,P\,|$ approaches 0. This limit is then the same as the integral defined in § 18.1.*

In order to prove Theorem VII it is best to begin by proving the following theorem, usually named after the French mathematician J. G. Darboux (1842–1917).

THEOREM VIII. *Suppose f is bounded on [a, b], and let s, S be the lower and upper sums corresponding to a partition P. Then s approaches I and S approaches J as $|P| \to 0$. This means that for any $\epsilon > 0$ there is some $\delta > 0$ such that*

$$|s - I| < \epsilon \quad and \quad |S - J| < \epsilon \quad if \quad |P| < \delta.$$

If we grant the truth of Darboux's theorem, it is rather easy to prove Theorem VII. Let us suppose that f is integrable. Now, if $x_{i-1} \leqq x_i' \leqq x_i$, we certainly have $m_i \leqq f(x_i') \leqq M_i$ and therefore

$$(18.2\text{–}2) \qquad s \leqq \sum_{i=1}^{n} f(x_i')(x_i - x_{i-1}) \leqq S.$$

As $|P| \to 0$, Darboux's theorem asserts that s and S approach I and J respectively. But $I = J = \int_a^b f(x)dx$, and so we see by (18.2–2) that the sums (18.2–1) must approach $\int_a^b f(x)dx$ as $|P| \to 0$. On the other hand, if we assume that the sums (18.2–1) approach some limit A, this means that all such sums lie between $A - \epsilon$ and $A + \epsilon$ if $|P|$ is sufficiently small. But, if we choose such a partition and keep it fixed, then by varying the choice of $x_1', \cdots, x_n'$ we can bring the sum (18.2–1) as close as we please to either s or S. Consequently we must have

$$A - \epsilon \leqq s \quad and \quad S \leqq A + \epsilon.$$

But then $S - s \leqq 2 \epsilon$. Since ϵ can be chosen as small as we please, we know by Theorem I that f is integrable. This concludes the proof of Theorem VII.

We still have to prove Theorem VIII. This proof is a bit intricate in detail. Let us first establish the following fact: If S is the upper sum corresponding to a partition P, and if S' is the new upper sum corresponding to a partition P' obtained from P by inserting a single additional point, then

$$(18.2\text{–}3) \qquad S - S' \leqq 2 C |P|,$$

where C is the least upper bound of $|f(x)|$ on $[a, b]$. To see this let us suppose for definiteness that the new point ξ is between x_0 and x_1, and use the notation as in the proof of Lemma I, § 18.1. Then

$$S - S' = M_1(x_1 - x_0) - M_1'(\xi - x_0) - M_1''(x_1 - \xi).$$

But since each of the M's is in absolute value less than or equal to C, we certainly have the inequality

$$S - S' \leqq C(x_1 - x_0) + C(\xi - x_0) + C(x_1 - \xi) = 2 C(x_1 - x_0),$$

and so (18.2–3) is true, for $| x_1 - x_0 | \leq | P |$. By an extension of this argument we see that

$$(18.2\text{–}4) \qquad\qquad S - S' \leq 2\,NC\,|\,P\,|$$

if P' is obtained from P by inserting N additional points, not more than one in any one of the original subintervals of P.

Now we are ready to prove Darboux's theorem. Suppose $\epsilon > 0$. Choose a fixed partition P_0 such that $S_0 < J + (\epsilon/2)$. This can be done, since J is the greatest lower bound of all possible upper sums. Suppose that h is the length of the *shortest* subinterval in P_0, and suppose the points forming P_0 are $\xi_0, \xi_1, \cdots, \xi_{N+1}$ (with $\xi_0 = a$, $\xi_{N+1} = b$). Choose δ so small that $\delta < h$ and $2\,NC\delta < \epsilon/2$. Consider any partition P such that $|\,P\,| < \delta$. Let P' be the partition formed by inserting the points $\xi_1, \cdots, \xi_N$ along with the points forming P. No more than one of the ξ's can fall into the same subinterval of P, so (18.2–4) will hold. Also, $S' \leq S_0$, by Lemma I. Thus

$$S - J = (S - S') + (S' - J) \leq (S - S') + (S_0 - J)$$

$$< 2\,NC\,|\,P\,| + \frac{\epsilon}{2} < \frac{\epsilon}{2} + \frac{\epsilon}{2} = \epsilon$$

by the way things have been arranged. We know, moreover, that $0 \leq S - J$. Therefore $|\,S - J\,| < \epsilon$ if $|\,P\,| < \delta$. This proves the part of Darboux's theorem referring to S and J. A similar argument can be given for s and I, but there is a device for deducing the case of s and I from the result already proved for S and J; see Exercise 3.

EXERCISES

1. Suppose f and g are integrable on $[a, b]$. Prove that the functions cf and $f + g$ are integrable on $[a, b]$, and that

$$\int_a^b cf(x)dx = c \int_a^b f(x)dx,$$

$$\int_a^b [f(x) + g(x)]dx = \int_a^b f(x)dx + \int_a^b g(x)dx.$$

Use Theorem VII and the basic theorems about limits, as stated in § 1.6.

2. If f is integrable on $[a, c]$ and $a < b < c$, prove that

$$\int_a^c f(x)dx = \int_a^b f(x)dx + \int_b^c f(x)dx.$$

Use Theorem VII and the result of Exercise 6, § 18.1.

3. Let us write $s(f)$, $I(f)$ to denote the dependence of s and I on the function f, and use similar notations for S, J. Show that $s(f) = -S(-f)$ and $I(f) = -J(-f)$. Now deduce the part of Theorem VIII dealing with s and I from the part dealing with S and J.

18.21   Duhamel's principle

It sometimes happens that we need to consider sums which resemble those occurring in the expression of an integral as a limit of sums but which do not have quite the right form to enable us to apply Theorem VII. It is useful to have a theorem which enables us to express the limit of certain sums as an integral, even though the situation is not one to which Theorem VII is applicable. We shall state such a theorem and discuss a few of its typical applications.

First let us introduce the concept of what we shall call a law of *partition weighting*. By this we mean a rule whereby to each partition P corresponds a set of numbers, one number being assigned as a "weight" to each subinterval of the partition. Thus, if P is the partition $(x_0, x_1, \cdots, x_n)$ of an interval $[a, b]$, the law of weighting will assign weights $\phi_1, \cdots, \phi_n$ to the respective subintervals $[x_0, x_1], \cdots, [x_{n-1}, x_n]$. As an example of a law of partition weighting, suppose f is a function defined on $[a, b]$, and let the weights $\phi_1, \cdots, \phi_n$ be the values $f(x_1'), \cdots, f(x_n')$, where x_i' is chosen in $[x_{i-1}, x_i]$ according to some kind of rule. In connection with a law of partition weighting we are going to consider sums of the form

$$\phi_1(x_1 - x_0) + \phi_2(x_2 - x_1) + \cdots + \phi_n(x_n - x_{n-1}).$$

We shall call these *weighted partition sums*. Our general purpose is to describe certain conditions under which these partition sums will approach an integral as a limit when the mesh fineness $|P|$ approaches 0. In the applications of this sort of thing, if we have a certain law of partition weighting, this law is likely to be of such a sort that it appears to be very nearly the same as a law in which the weights are of the type $f(x_1'), \cdots, f(x_n')$ derived from some function f. This will suggest that the limit of the partition sums is the integral $\int_a^b f(x)dx$. The problem is to justify this surmise by sound reasoning. The key lies in considering the size of the differences $\phi_1 - f(x_1'), \cdots, \phi_n - f(x_n')$.

THEOREM IX.   *Suppose we have a certain law of partition weighting, $\phi_1, \cdots, \phi_n$ corresponding to the subintervals of the partition $P: (x_0, x_1, \cdots, x_n)$. Suppose also that f is a function integrable on $[a, b]$, and suppose that there is some choice of points $x_1', \cdots, x_n'$ in each partition such that the maximum of the absolute values*

(18.21–1) $$|\phi_1 - f(x_1')|, \cdots, |\phi_n - f(x_n')|$$

approaches 0 as $|P| \to 0$. Then

(18.21–2) $$\lim_{|P| \to 0} \sum_{i=1}^{n} \phi_i(x_i - x_{i-1}) = \int_a^b f(x)dx.$$

PROOF. Suppose $\epsilon > 0$ and choose δ so small that the maximum of the absolute values in (18.21–1) is less than ϵ if $|P| < \delta$. Our hypothesis means that such a choice of δ is possible for each positive ϵ. Then $|P| < \delta$ implies

$$\left| \sum_{i=1}^{n} [\phi_i - f(x_i')](x_i - x_{i-1}) \right| \leq \sum_{i=1}^{n} |\phi_i - f(x_i')| (x_i - x_{i-1})$$

$$< \epsilon \sum_{i=1}^{n} (x_i - x_{i-1}) = \epsilon(b - a);$$

in other words,

$$(18.21\text{–}3) \qquad \lim_{|P| \to 0} \sum_{i=1}^{n} [\phi_i - f(x_i')](x_i - x_{i-1}) = 0.$$

But this is the same as the assertion

$$\lim_{|P| \to 0} \sum_{i=1}^{n} \phi_i(x_i - x_{i-1}) = \lim_{|P| \to 0} \sum_{i=1}^{n} f(x_i')(x_i - x_{i-1}),$$

and this is equivalent to (18.21–2), by Theorem VII.

We shall refer to Theorem IX as Duhamel's principle. Duhamel dealt with the problem of finding limits of sums of the same general type as those we have called partition sums associated with a law of partition weighting. In Duhamel's time and for long afterward in calculus textbooks the treatment of such matters was regularly couched in the phraseology of "infinitesimals." The treatment given here, avoiding the term "infinitesimal," is patterned after a formulation given by W. F. Osgood.

Example 1. Suppose that f and g are continuous functions defined on $[a, b]$. Let $F(x) = f(x)g(x)$. Consider a law of partition weighting which arises by choosing *two* points x_i', x_i'' in each subinterval of an arbitrary partition $(x_0, x_1, \cdots, x_n)$, and letting $\phi_i = f(x_i')g(x_i'')$. Then the limit of the corresponding partition sums is the integral of the function F. That is,

$$(18.21\text{–}4) \qquad \lim_{|P| \to 0} \sum_{i=1}^{n} f(x_i')g(x_i'')(x_i - x_{i-1}) = \int_{a}^{b} f(x)g(x)dx.$$

This is an application of Duhamel's principle. To see that it is a valid application we have to verify that the maximum of the quantities

$$(18.21\text{–}5) \qquad |f(x_i')g(x_i'') - F(x_i')| \qquad (i = 1, \cdots, n)$$

approaches 0 as $|P| \to 0$. The truth of (18.21–4) will then follow, for F is continuous, and therefore integrable. We make use of the fact that f is bounded and g is *uniformly* continuous. Let A be the maximum of

$| f(x) |$. The uniform continuity of g assures us that the maximum of the differences

$$| g(x_i'') - g(x_i') | \qquad (i = 1, \cdots, n)$$

approaches 0 as $| P | \to 0$, for $| x_i' - x_i'' | \leq | P |$. But, for the expression in (18.21–5) we can write

$$| f(x_i')[g(x_i'') - g(x_i')] | \leq A | g(x_i'') - g(x_i') |.$$

Since A is fixed, the assertion about (18.21–5) is seen to be true.

Formula (18.21–4) holds true under less restrictive assumptions on f and g; see Exercise 3.

Formula (18.21–4) can be extended in an obvious way to the case of products of three or more functions.

Example 2.　　Consider the derivation of the integral formula for arc length of a curve, as discussed in § 14.2. Let C be a smooth arc defined by

$$x = f(t), \quad y = g(t), \quad z = h(t), \quad a \leq t \leq b,$$

where f, g, h have continuous first derivatives. If $(t_0, t_1, \cdots, t_n)$ is a partition of $[a, b]$, the discussion in § 14.2 shows that the arc length of C is the limit

$$(18.21\text{–}6) \qquad \lim_{|P| \to 0} \sum_{i=1}^{n} \{(f'(\alpha_i))^2 + (g'(\beta_i))^2 + (h'(\gamma_i))^2\}^{1/2} \, \Delta t_i,$$

where $\Delta t_i = t_i - t_{i-1}$ and $\alpha_i, \beta_i, \gamma_i$ are certain points of the interval $[t_{i-1}, t_i]$. Now if we set

$$\phi_i = \{(f'(\alpha_i))^2 + (g'(\beta_i))^2 + (h'(\gamma_i))^2\}^{1/2},$$

we have a law of partition weighting. If it were true that $\alpha_i = \beta_i = \gamma_i$, the limit of the partition sums would be the integral

$$(18.21\text{–}7) \qquad \int_a^b \{(f'(t))^2 + (g'(t))^2 + (h'(t))^2\}^{1/2} \, dt.$$

To show that the limit in (18.21–6) is equal to the integral in (18.21–7) we compare ϕ_i with the corresponding expression in which β_i and γ_i are replaced by α_i, and show that Theorem IX is applicable (with the necessary changes in notation). To do this we make use of the fact that the function

$$\{(f'(t))^2 + (g'(u))^2 + (h'(v))^2\}^{1/2}$$

is uniformly continuous in the cubical region

$$a \leq t \leq b, \quad a \leq u \leq b, \quad a \leq v \leq b.$$

We omit the details.

EXERCISES

1. Suppose F is continuous on $[a, b]$ and that f is defined and has a continuous derivative on $[a, b]$. Using the standard notation relating to partitions, find the limit

$$\lim_{|P| \to 0} \sum_{i=1}^{n} F(x_i')[f(x_i) - f(x_{i-1})]$$

where x_i' is any point on the ith subinterval.

2. Suppose f and g are continuous on $[a, b]$, and $F(u, v)$ is a continuous function of u and v for all values of u and v obtained by setting $u = f(x)$, $v = g(y)$ and letting x, y vary over $[a, b]$. Use Theorem IX to show that

$$\lim_{|P| \to 0} \sum_{i=1}^{n} F(f(x_i'), g(x_i''))(x_i - x_{i-1}) = \int_a^b F(f(x), g(x))dx.$$

Explain carefully where uniform continuity comes into the argument.

3. Show that (18.21–4) is true if it is assumed that f is bounded and that $f(x)g(x)$ and $g(x)$ are integrable. For the proof it is enough to show that

$$\lim_{|P| \to 0} \sum_{i=1}^{n} f(x_i')[g(x_i'') - g(x_i')](x_i - x_{i-1}) = 0.$$

But, if $|f(x)| \leq A$, and if S and s are upper and lower sums for g, show that the expression following the limit symbol above is not larger than $A(S - s)$. Then use a certain theorem employing the fact that g is integrable.

18.3   Further discussion of integrals

In our theory of the Riemann integral we have been assuming that $a < b$. The cases of equal upper and lower limits and of lower limit larger than upper limit are handled exactly as in (1.52–2), (1.52–3). The way in which integrals over adjacent intervals are combined is indicated in Exercise 7, § 18.1, and Exercise 2, § 18.2. All these things are familiar from elementary calculus.

Inequalities for integrals are used a great deal. The basic fact is that if $f(x) \leq g(x)$ on $[a, b]$, then (assuming $a < b$)

$$(18.3\text{–}1) \qquad \int_a^b f(x)dx \leq \int_a^b g(x)dx.$$

In particular, since $f(x) \leq |f(x)|$ and $-|f(x)| \leq f(x)$, we have

$$\left| \int_a^b f(x)dx \right| \leq \int_a^b |f(x)| \, dx.$$

The inequality (18.3–1) is easily proved by analytical reasoning on either upper or lower sums. For the fact that $|f|$ is integrable if f is integrable, see Exercise 5, § 18.1.

Finally, in this miscellany of remarks, it should be stated that the product of two integrable functions is integrable. This result can be proved in various ways. It is an immediate consequence of the rather advanced Theorem VI. It can be proved by more elementary means, however, though the arguments are rather ingenious. We shall not give the details.

18.4   The integral as a function of the upper limit

In this section we consider

$$(18.4\text{--}1) \qquad F(x) = \int_a^x f(t)dt,$$

where $a \leq x \leq b$ and f is assumed to be integrable on $[a, b]$.

THEOREM X.   *The function F is continuous.*

This is very easily proved. Consider any two points x', x'', and suppose $x' < x''$. Then

$$F(x'') - F(x') = \int_a^{x''} f(t)dt - \int_a^{x'} f(t)dt = \int_{x'}^{x''} f(t)dt.$$

Let C be the least upper bound of $|f(t)|$ on $[a, b]$.

Then $\qquad |F(x'') - F(x')| \leq \int_{x'}^{x''} |f(t)| \, dt \leq C |x'' - x'|$.

The continuity of F is a direct consequence of this inequality.

In § 1.52 we considered the same formula (18.4–1), but on the assumption that f was continuous at each point of $[a, b]$. It was then proved (Theorem VII of § 1.52) that $F'(x) = f(x)$ at each point. In the present case we are not assuming the continu ty of f, but merely its integrability. With this less restrictive assumption we cannot conclude quite as much as before.

THEOREM XI.   *The formula $F'(x) = f(x)$ holds at each point where f is continuous.*

PROOF.   Let x_0 be a point at which f is continuous. If $\epsilon > 0$, choose δ so that

$$(18.4\text{--}2) \qquad |f(x) - f(x_0)| < \epsilon \quad \text{if} \quad |x - x_0| < \delta.$$

We wish to show that

$$(18.4\text{--}3) \qquad \lim_{h \to 0} \left\{ \frac{F(x_0 + h) - F(x_0)}{h} - f(x_0) \right\} - 0.$$

Now

$$F(x_0 + h) - F(x_0) = \int_a^{x_0+h} f(t)dt - \int_a^{x_0} f(t)dt = \int_{x_0}^{x_0+h} f(t)dt,$$

and

$$f(x_0) = \frac{1}{h} \int_{x_0}^{x_0+h} f(x_0)dt.$$

Therefore

$$(18.4\text{-}4) \qquad \frac{F(x_0 + h) - F(x_0)}{h} - f(x_0) = \frac{1}{h} \int_{x_0}^{x_0+h} [f(t) - f(x_0)]dt.$$

We now suppose that $0 < |h| < \delta$. Then the expression under the integral sign on the right side of (18.4–4) is in absolute value less than ϵ, by (18.4–2), since $|t - x_0| \leq |h| < \delta$. Consequently the whole right side of (18.4–4) is less than ϵ when $0 < |h| < \delta$. This proves (18.4–3) and thus proves the theorem.

EXERCISES

1. Use Theorem X to prove that the value of $\int_a^b f(x)dx$ is not changed if the function f is altered merely by changing the value of $f(b)$.

2. If f is integrable and $G(x) = \int_x^b f(t)dt$, show that G is continuous by finding the relation between F and G.

3. Prove the result corresponding to that in Exercise 1 if f is altered merely by changing the value of $f(a)$.

4. Suppose f is defined on $[0, 1]$ as follows: $f(x) = (-1)^n$ if $1/2^{n+1} < x \leq 1/2^n$, $n = 0, 1, 2, \cdots$, and $f(0) = 0$. Find the value of $\int_0^1 f(x)dx$ by using the result of Exercise 2.

5. If $F(x) = \int_0^x \sqrt{(1 - t^2)(2 - t^2)}dt$, sketch the graph of $y = F(x)$ when $-1 \leq x \leq 1$, using information derived from $F'(x)$ and $F''(x)$. The integral is an elliptic integral, and cannot be expressed in terms of elementary functions.

6. Show that $\dfrac{d}{dx} \displaystyle\int_x^b f(t)dt = -f(x)$ if f is continuous at x.

7. Let $F(x) = \int_{-1}^x |t| \, dt$. (a) Find $F'(x)$ by Theorem XI. (b) Find explicit formulas for $F(x)$ for the separate cases $x < 0$, $x > 0$. (c) Draw the graph of $y = F(x)$.

8. Let $f(x) = 1$ if $0 \leq x \leq 1$, $f(x) = -1$ if $-1 \leq x < 0$. Define $F(x)$ by (18.4–1) with $a = -1$. (a) Find separate explicit formulas for $F(x)$ if $-1 \leq x \leq 0$ and $0 \leq x \leq 1$. (b) Draw the graph of $y = F(x)$ and check the validity of Theorems X, XI. What about $F'(0)$?

18.41   The integral of a derivative

In this section we shall prove a generalization of Theorem VIII, § 1.53.

THEOREM XII. *Suppose f is integrable on* [a, b], *and suppose there is a function F which is continuous on* [a, b] *and such that, for each x on the open interval a* < x < b, *F has a derivative given by* $F'(x) = f(x)$. *Then*

$$(18.41-1) \qquad \int_a^b f(x)dx = F(b) - F(a).$$

PROOF. Consider any partition $(x_0, x_1, \cdots, x_n)$ of [a, b]. We apply the law of the mean (Theorem IV, § 1.2) to $F(x)$ on each of the subintervals $[x_{-1}, x_i]$, $i = 1, \cdots, n$. There is some point x_i' such that $x_{i-1} < x_i' < x_i$ and

$$F(x_i) - F(x_{i-1}) = (x_i - x_{i-1})F'(x_i').$$

Thus, noting that $F'(x_i') = f(x_i')$, we have

$$F(x_1) - F(x_0) = (x_1 - x_0)f(x_1')$$
$$F(x_2) - F(x_1) = (x_2 - x_1)f(x_2')$$
$$\vdots$$
$$F(x_n) - F(x_{n-1}) = (x_n - x_{n-1})f(x_n').$$

Adding, and recalling that $x_0 = a$, $x_n = b$, we see that

$$(18.41-2) \qquad F(b) - F(a) = \sum_{i=1}^{n} f(x_i')(x_i - x_{i-1}).$$

This kind of relation holds for each partition. Hence by Theorem VII we see that (18.41-1) is true.

The student will observe, on comparing the present theorem with Theorem VIII, § 1.53, that the conclusions in the two theorems are identical. In the earlier theorem we assumed that f was continuous, whereas here we have assumed merely that f is integrable.

Example.   Find the value of

$$\int_0^{2/\pi} \left(2 x \sin \frac{1}{x} - \cos \frac{1}{x}\right)dx.$$

Here

$$f(x) = 2 x \sin \frac{1}{x} - \cos \frac{1}{x}$$

when $x \neq 0$. We may define $f(0)$ in any manner we please, but f will be discontinuous at $x = 0$ in any case. However, f is continuous for other values of x, and bounded; therefore, it is integrable. To evaluate the integral we define

$$F(x) = x^2 \sin \frac{1}{x} \quad \text{if} \quad x \neq 0, \quad F(0) = 0.$$

The definition $F(0) = 0$ makes F continuous at $x = 0$. For $x \neq 0$ we readily verify that $F'(x) = f(x)$. We observe, incidentally, that $F'(0) = 0$; this may or may not be $f(0)$, depending on how we define $f(0)$. The conditions of Theorem XII are satisfied, and so

$$\int_0^{2/\pi} \left(2 x \sin \frac{1}{x} - \cos \frac{1}{x} \right) dx = F\left(\frac{2}{\pi} \right) - F(0) = \frac{4}{\pi^2}.$$

18.5   Integrals depending on a parameter

Our concern in this section is with integrals in which a parameter occurs under the integral sign or in the limits of integration, or in both places. The main objective is to learn about differentiation with respect to the parameter in such cases. We begin with the situation in which the parameter occurs solely in the integrand. Consider

$$(18.5\text{--}1) \qquad\qquad F(y) = \int_a^b f(x, y)dx.$$

We shall suppose that f is defined when (x, y) is a point of the rectangle in the xy-plane defined by $a \leq x \leq b, c \leq y \leq d$. We denote this rectangle by R.

It is important for us to know conditions on the function f that will guarantee the continuity of F. We do not attempt to give the most general (i.e. the least restrictive) conditions. For usefulness in practice the following theorem is convenient.

THEOREM XIII.   *If f is continuous at each point of R, then F is continuous for each y on the interval $[c, d]$.*

The proof depends upon the fact that f is uniformly continuous in R; see Exercise 9.

The next theorem deals with finding the derivative of $F(y)$.

THEOREM XIV.   *Suppose that $f(x, y)$ is an integrable function of x for each value of y, and that the partial derivative $\dfrac{\partial f(x, y)}{\partial y}$ exists and is a continuous function of x and y in the rectangle R. Then $F(y)$ has a derivative given by*

$$(18.5\text{-}2) \qquad\qquad F'(y) = \int_a^b \frac{\partial f(x, y)}{\partial y}\, dx.$$

The formula (18.5–2) is often called Leibniz's rule.

PROOF. To give the proof let us use the notation $f_2(x, y)$ for the partial derivative with respect to y. Formula (18.5–2) is equivalent to the statement

$$(18.5\text{-}3) \qquad \lim_{h \to 0} \left\{ \frac{F(y + h) - F(y)}{h} - \int_a^b f_2(x, y)dx \right\} = 0.$$

Now

$$F(y + h) - F(y) = \int_a^b [f(x, y + h) - f(x, y)]dx.$$

We apply the law of the mean to $f(x, y)$ as a function of y:

$$f(x, y + h) - f(x, y) = hf_2(x, y + \theta h).$$

Here θ is a number depending on x, y, h, and such that $0 < \theta < 1$. From this last formula we see that

$$\frac{F(y + h) - F(y)}{h} - \int_a^b f_2(x, y)dx$$

$$(18.5\text{-}4) \qquad = \int_a^b [f_2(x, y + \theta h) - f_2(x, y)]dx.$$

At this stage we make use of the fact that f_2 is uniformly continuous in R (by Theorem V, § 17.4). Suppose $\epsilon > 0$. Choose δ so that the values of f_2 at two different points of R differ by less than ϵ if the distance between the points is less than δ. Then certainly

$$| f_2(x, y + \theta h) - f_2(x, y) | < \epsilon$$

if $| h | < \delta$. Consequently, we see by (18.5–4) that

$$\left| \frac{F(y + h) - F(y)}{h} - \int_a^b f_2(x, y)dx \right| < \epsilon(b - a)$$

if $0 < | h | < \delta$. Since ϵ can be as small as we please, this proves (18.5–3).

Example 1. Find $F'(y)$ if

$$F(y) = \int_0^1 \log (x^2 + y^2)dx.$$

We can apply Theorem XIV with $0 \leq x \leq 1$ and y on any closed interval not containing $y = 0$. The result is

$$F(y) = \int_0^1 \frac{2y}{x^2 + y^2}\, dx.$$

The integration is readily performed, and we find

$$F'(y) = 2 \tan^{-1}\left(\frac{1}{y}\right).$$

If the parameter occurs merely in the limits of integration we can use Theorem XI and the standard chain rule for differentiating composite functions. We illustrate by an example.

Example 2. Suppose

$$F(y) = \int_{\sin y}^{e^y} \sqrt{1 + x^3} \, dx.$$

How is $F'(y)$ found in this case? We set

$$G(u, v) = \int_u^v \sqrt{1 + x^3} \, dx.$$

Then $G(u, v)$ becomes $F(y)$ if we put $u = \sin y$, $v = e^y$. Therefore

$$\frac{dF}{dy} = \frac{\partial G}{\partial u}\frac{du}{dy} + \frac{\partial G}{\partial v}\frac{dv}{dy}.$$

But, by Theorem XI,

$$\frac{\partial G}{\partial u} = -\sqrt{1 + u^3}, \qquad \frac{\partial G}{\partial v} = \sqrt{1 + v^3}.$$

(How does the minus sign get into the first formula?) Therefore

$$F'(y) = -\sqrt{1 + \sin^3 y} \cos y + \sqrt{1 + e^{3y}} \, e^y.$$

If the parameter occurs both under the integral sign and in the limits of integration, we use both Theorem XI and Theorem XIV and the chain rule.

Example 3. Suppose

$$F(y) = \int_0^{y^2} x^5(y - x)^7 \, dx.$$

Here we can define

$$G(u, v) = \int_0^u x^5(v - x)^7 \, dx$$

and put $u = y^2$, $v = y$. Then

$$\frac{\partial G}{\partial u} = u^5(v - u)^7, \qquad \frac{\partial G}{\partial v} = \int_0^u 7x^5(v - x)^6 \, dx.$$

Since $\dfrac{du}{dy} = 2y$, $\dfrac{dv}{dy} = 1$, the chain rule gives

$$F'(y) = y^{10}(y - y^2)^7 \cdot 2y + \int_0^{y^2} 7 x^5(y - x)^6 \, dx.$$

With a little practice the student will be able to handle problems of this kind without explicit introduction of auxiliary variables.

EXERCISES

1. If $F(u) = \int_0^4 \log (1 - u^2x^2)dx$, find $F'(u)$ by Leibniz's rule. What is the value of $F'(0)$? Show without performing any integrations that $F(u)$ has a maximum value at $u = 0$ and that the graph of $v = F(u)$ is concave downward in the uv-plane.

2. If $F(a) = \int_0^1 \dfrac{a \, dx}{\sqrt{1 - a^2x^2}}$, find $F'(a)$ by two different methods.

3. If $F(x) = \int_0^x e^{-x^2t^2} \, dt$, find $F'(x)$.

4. If $F(x) = \int_{x^2}^{\sin x} (x^2 - t^2)^n \, dt$, find $F'(x)$.

5. Suppose $\phi(y) = \int_0^y x^n(y - x)^m \, dx$, where m and n are positive integers. Calculate $\phi'(y)$, $\phi''(y)$, $\cdots$ without integration, and show that

$$\phi^{(m)}(y) = \dfrac{m!}{n + 1} y^{n+1}.$$

Then, observing that $\phi(0) = \phi'(0) = \cdots = \phi^{(m-1)}(0) = 0$, integrate $\phi^{(m)}(y)$ successively m times and so arrive at the formula

$$\phi(y) = \dfrac{m! \, n!}{(m + n + 1)!} y^{m+n+1}.$$

6. If $u = \int_{x - ct}^{x + ct} \phi(s)ds$, where ϕ has a continuous derivative, show that $\dfrac{\partial^2 u}{\partial t^2} = c^2 \dfrac{\partial^2 u}{\partial x^2}$.

7. If $u = \int_{1/x}^y dt \int_{1/t}^x f(s, t)ds$, show that $\dfrac{\partial^2 u}{\partial x \, \partial y} = f(x, y)$, assuming that f is continuous. Find the other second derivatives of u, assuming that f has continuous first partial derivatives.

8. Suppose $M(x, y)$ and $N(x, y)$ are continuous in a rectangle with center at (a, b), and that the partial derivatives $\dfrac{\partial M}{\partial y}$ and $\dfrac{\partial N}{\partial x}$ are continuous and equal in the rectangle. Show by Theorems XI, XII, and XIV that $\dfrac{\partial f}{\partial x} = M$ and $\dfrac{\partial f}{\partial y} = N$ if $f(x, y) = \int_a^x M(s, b)ds + \int_b^y N(x, t)dt$.

9. Prove Theorem XIII, using the uniform continuity of f in much the same way that the uniform continuity of f_2 was used in proving Theorem XIV.

18.6   Riemann double integrals

In this section we shall discuss the theory of Riemann double integrals

$$(18.6-1) \qquad \iint_R f(x, y)\,dA$$

in a manner paralleling the theory for single integrals, as developed in the first part of this chapter. We shall be much briefer in the theory of double integrals, and many proofs will be omitted.

It is simplest to begin with the case in which R is a rectangle with sides parallel to the co-ordinate axes. Suppose these sides are $x = a$, $x = b$, $y = c$, $y = d$, where $a < b$ and $c < d$. If we form a partition of $[a, b]$ by points $(x_0, x_1, \cdots, x_m)$, and a partition of $[c, d]$ by points $(y_0, y_1, \cdots, y_n)$, the lines $x = x_i$ and $y = y_j$ form a rectangular partition of the rectangle R into rectangular cells. The numbers m, n need not be equal; the total number of cells in the partition of R is $N = mn$.

Now suppose that f is a function which is defined and bounded in R. If the cells are numbered in any order, let $m_1, \cdots, m_N$ be the greatest lower bounds of $f(x, y)$ in the various cells, and let $M_1, \cdots, M_N$ denote the corresponding least upper bounds of $f(x, y)$. If the areas of the cells are $\Delta A_1, \cdots, \Delta A_N$, we form the *lower sum*

$$s = m_1\,\Delta A_1 + \cdots + m_N\,\Delta A_N$$

and the *upper sum*

$$S = M_1\,\Delta A_1 + \cdots + M_N\,\Delta A_N.$$

Then we denote the least upper bound of all possible lower sums by I, and the greatest lower bound of all possible upper sums by J.

DEFINITION.   If $I = J$, we say that f is integrable over R, and we define the double integral (18.6–1) as the common value of I and J.

Starting from this definition we can obtain analogues of the lemmas and theorems in § 18.1. It can be proved that if f is continuous in the closed rectangle R, it is integrable. The proof, using uniform continuity, is much like the proof of Theorem III, § 18.11.

A function may have certain points of discontinuity and yet be integrable. For example, if f is continuous in R except at certain points which lie on a finite number of smooth curves, it can be shown that f is integrable. An instance is furnished by the function defined by $f(x, y) = 1$ if $x^2 + y^2 \leqq 1$, $f(x, y) = 0$ if $x^2 + y^2 > 1$ and $-2 \leqq x \leqq 2$, $-2 \leqq y \leqq 2$. This function is discontinuous at the points of the circle $x^2 + y^2 = 1$, but it is integrable over the square in which it is defined.

It will be very useful in some later work to be able to deal with integrable functions having some points of discontinuity. On that account we shall discuss such matters a little further here. As an aid in this discussion we introduce the concept of the *outer content* of a set of points in the *xy*-plane.

DEFINITION. Let T be a point set in the rectangle R. Consider any rectangular partition of R. Select all those cells of the partition which contain points of T, and let A denote the sum of the areas of all these cells. If we consider all possible partitions, the greatest lower bound of the values of A is defined as the outer content of T.

It will be seen that if we define a function f such that $f(x, y) = 1$ if (x, y) is a point of T, and $f(x, y) = 0$ elsewhere in R, then A is the upper sum S for this function, and consequently the number J is the outer content of T.

It is also quite evident that the outer content of T depends on T alone, not on the choice of the rectangle R containing T.

THEOREM XV.   *Let f be any function which is bounded in R, and let T be the set of points in R at which f is discontinuous. Then f is integrable provided the outer content of T is 0.*

This theorem is analogous to Theorem V, § 18.12. We omit the proof. The condition of zero outer content is sufficient, but not necessary, for f to be integrable.

It is easy to see that a straight line segment has zero outer content. Also, any smooth arc has zero outer content (see Exercise 5). The boundaries of regions of familiar shapes, such as squares, circles, ellipses, all have zero outer content.

Next we turn to the definition of double integrals over regions which are not rectangles with sides parallel to the axes. We confine our attention to bounded regions with boundaries of outer content zero. For convenience we shall refer to regions of this type as *Riemann regions*.

Suppose G is a Riemann region. Let R be a rectangle containing G and having its sides parallel to the axes. Suppose f is a function which is defined in G. We shall define a new function throughout R by setting $g(x, y) = f(x, y)$ at points of G and $g(x, y) = 0$ at points of R not in G. We then say that f is integrable over G if g is integrable over R, and in that case we define

$$(18.6\text{--}2) \qquad \iint\limits_{G} f(x, y)\,dA = \iint\limits_{R} g(x, y)\,dA.$$

This procedure makes everything depend on the theory of integrals over rectangles. Alternative procedures are possible. We could define upper and lower sums directly for f and the region G. In that case it becomes necessary to decide what to do with cells of a partition in case the cells contain points in G as well as points not in G. The fact that the boundary of G has zero outer content makes it immaterial whether such cells are ignored or not.

Darboux's theorem (Theorem VIII, § 18.2) can be generalized to the case of double integrals, and as a result we get the important fact that a double integral can be expressed as a limit of sums (the analogue of Theorem VII). Thus we make connection with the earlier definition of double integrals in § 13.2.

EXERCISES

1. Suppose $f(x, y) = \tan^{-1}\left(\dfrac{1}{x - y}\right)$ if $x \neq y$, and define $f(x, y) = 0$ when $x = y$. Is f integrable over the square $0 \leq x \leq 1, 0 \leq y \leq 1$?

2. Suppose $f(x, y) = \sin(1/xy)$ if $xy \neq 0$, and define $f(x, y) = 1$ if x or $y = 0$. Is f integrable over the square $-1 \leq x \leq 1, -1 \leq y \leq 1$?

3. Suppose $f(x, y) = 0$ if $x + y$ is rational, and $f(x, y) = 1$ if $x + y$ is irrational. Is f integrable over the square $0 \leq x \leq 1, 0 \leq y \leq 1$?

4. Suppose $f(x)$ is a continuous function defined when $a \leq x \leq b$. Let T be the set of points on the graph of $y = f(x)$ in the xy-plane. Is it possible for T to have positive outer content? Justify your answer.

5. Let C be a smooth arc with parametric equations $x = f(t)$, $y = g(t)$, $a \leq t \leq b$. Divide $[a, b]$ into n equal parts by points $t_0, \cdots, t_n$; let $\delta = (b - a)/n$ and let (x_i, y_i) be the point of C corresponding to t_i. Choose M so that $|f'(t)|$ and $|g'(t)|$ do not exceed M on $[a, b]$. Use the law of the mean to show that the points of C for $t_{i-1} \leq t \leq t_i$ satisfy the inequalities $|x - x_i| \leq M\delta$, $|y - y_i| \leq M\delta$, and hence that C is covered by n squares of total area $4 M^2(b - a)\delta$. What do you conclude about the outer content of C?

18.61   Double integrals and iterated integrals

In § 13.3 we gave an account of the procedure for expressing a double integral as an iterated integral. The procedure is summed up in the two formulas (13.3–5), (13.3–6), and a formal statement about the first of these formulas was made in Theorem III, § 13.3. However, we did not actually prove the theorem; we merely made an argument for its plausibility, by interpreting both the double integral and the iterated integral as expressions for a certain volume. The student will do well to read § 13.3 as far as Example 1 before proceeding further with the present section.

We now wish to give a strictly analytical proof of the relation between double integrals and iterated integrals. This proof is necessary if we are to have a firm logical justification for the procedures used in evaluating double integrals. The proof is, moreover, very instructive for the prospective student of more advanced mathematics, for it is fairly representative of a type of proof which is encountered in a variety of different forms in higher analysis. The basic principle is that of showing that certain limit processes of rather complicated nature can be replaced by two successive limit processes of simpler nature.

We begin with the case of a double integral over a rectangular region.

THEOREM XVI.   *Let R be the rectangle $a \leq y \leq b$, $c \leq x \leq d$, where $a < b$, $c < d$. Suppose that $f(x, y)$ is defined and integrable over R. For each x suppose that $f(x, y)$ is integrable with respect to y over $[a, b]$, and let the function*

$$(18.61\text{--}1) \qquad \phi(x) = \int_a^b f(x, y)dy$$

be integrable over $[c, d]$. Then

$$(18.61\text{--}2) \qquad \iint_R f(x, y)dA = \int_c^d dx \int_a^b f(x, y)dy.$$

PROOF.   Let $(x_0, \cdots, x_m)$ be a partition of $[c, d]$ into m equal parts and $(y_0, \cdots, y_n)$ a partition of $[a, b]$ into n equal parts. We then obtain a partition of R into mn cells, each of the same area. Let $\Delta x_i = x_i - x_{i-1}$, $\Delta y_j = y_j - y_{j-1}$. Now, if we add up all the mn terms $f(x_i, y_j)\Delta x_i \Delta y_j$ where $1 \leq i \leq m$ and $1 \leq j \leq n$, we get an approximation to the value of the double integral of $f(x, y)$ over R, and we can make this approximation as close as we please by taking m and n sufficiently large. That is, if $\epsilon > 0$, there is some integer q depending on ϵ such that

$$(18.61\text{--}3) \qquad \left| \sum_{i=1}^m \sum_{j=1}^n f(x_i, y_j)\Delta x_i \Delta y_j - \iint_R f(x, y)dA \right| < \epsilon$$

if $q \leq m$ and $q \leq n$.

Next, since $f(x_i, y)$ is an integrable function of y, we know by Theorem VII, § 18.2, that for each i

$$\lim_{n \to \infty} \sum_{j=1}^n f(x_i, y_j)\Delta y_j = \int_a^b f(x_i, y)dy.$$

It then follows from (18.61–3) that

$$(18.61\text{--}4) \qquad \left| \sum_{i=1}^m \left(\int_a^b f(x_i, y)dy \right)\Delta x_i - \iint_R f(x, y)dA \right| \leq \epsilon$$

if $q \leq m$. Using the definition of $\phi(x)$ [see (18.61–1)], we can rewrite (18.61–4) as

$$\left| \sum_{i=1}^{m} \phi(x_i)\Delta x_i - \iint_R f(x, y)dA \right| \leq \epsilon.$$

This means that

$$\lim_{m \to \infty} \sum_{i=1}^{m} \phi(x_i)\Delta x_i = \iint_R f(x, y)dA.$$

But since ϕ is integrable we know that this last limit is $\int_c^d \phi(x)dx$.

Therefore

$$\int_c^d \phi(x)dx = \iint_R f(x, y)dA.$$

This is the same as (18.61–2).

There is, naturally, a corresponding theorem in which the roles of x and y are reversed.

If $f(x, y)$ is continuous in the rectangle, all the assumptions in Theorem XVI are satisfied, as we see by using Theorem III, § 18.11, and Theorem XIII, § 18.5.

Let us now consider Theorem III, § 13.3, in either of the forms (13.3–5), (13.3–6). It is assumed that f is continuous in the region under consideration (see Fig. 91 and Fig. 92). The boundaries of these regions have zero outer content, for they are made up of graphs of continuous functions, and it is easy to show that the graph of a continuous function, e.g. $x = X_1(y)$, $a \leq y \leq b$, has zero outer content (the proof depends on uniform continuity). Now let the region be enclosed in the smallest possible rectangle with sides parallel to the axes, and extend the definition of f by setting $f(x, y) = 0$ in the part of the rectangle outside the region R. This makes f integrable over the rectangle, and the conditions for applying Theorem XVI are satisfied. Thus the double integral over the rectangle is equal to the iterated integral in each of the two possible orders. But, since $f = 0$ outside of the region R, the integrations need not be carried beyond the boundary of R. Thus, for example, the integral with respect to x across the width of the rectangle becomes simply

$$\int_{X_1}^{X_2} f(x, y)dx.$$

In this way we see that (13.3–5) and (13.3–6) are both true.

18.7   Triple integrals

The theory of Riemann triple integrals may be developed in a manner closely analogous to the theory of double integrals. The ideas are not fundamentally different from those in § 18.6. In dealing with integrals over regions in three dimensions, it is assumed that the regions have boundaries of outer content zero. Regions of familiar shapes, e.g. spheres, cones, and cylinders, are of the required type. Likewise, the reduction of triple integrals to iterated integrals of lower order is handled in much the same way as the corresponding problem for double integrals in § 18.61. We omit the details.

18.8   Improper integrals

In the Riemann theory of integration the functions are assumed to be bounded, and the intervals or regions of integration are assumed to be bounded. Under these conditions, if the function is integrable, the integral is said to be a *proper* integral. There are some extensions of the definitions of integrals. When the Riemann theory is taken as basic, as it is in this book, any integral whose definition does not come within the framework of the Riemann theory, but which is defined by a limiting process depending on the Riemann theory, will be called an *improper* integral. An integral may be improper because it is an integral over an unbounded interval or region; or it may be improper because it is the integral of an unbounded function.

Example 1. The integral $\int_0^1 x^{-1/2} \, dx$ is improper because the function $f(x) = x^{-1/2}$ is not bounded on $[0, 1]$. However, the integral $\int_c^1 x^{-1/2} \, dx$ is proper for each c if $0 < c < 1$, and we define

$$\int_0^1 x^{-1/2} \, dx = \lim_{c \to 0^+} \int_c^1 x^{-1/2} \, dx.$$

The limit exists and has the value 2, as is easily verified.

Example 2. $\int_1^\infty \dfrac{dx}{x^\alpha}$ is improper, because the interval of integration is not finite. We define

$$\int_1^\infty \frac{dx}{x^\alpha} = \lim_{b \to \infty} \int_1^b \frac{dx}{x^\alpha}$$

if the limit exists, which is if and only if $\alpha > 1$, as may be verified by the student. The value in that case is $(\alpha - 1)^{-1}$.

These two examples are instances of the definition of improper integrals as limits of proper integrals. Similar procedures can be used to define improper double and triple integrals. For a more thorough discussion of improper integrals see Chapter XXII.

18.9   Stieltjes integrals

The purpose of this section is to give a brief elementary introduction to the subject of Stieltjes integrals. These integrals are named after a Dutch mathematician, T. J. Stieltjes (1856–1894). They have long been used by mathematicians as a tool in theoretical investigations. The current tendency is in the direction of a wider usage, and the student of pure or applied mathematics will sometimes encounter the Stieltjes integral in his reading before he has had a chance to learn much about the integral. The common practice is to discuss the theory of the Stieltjes integral in graduate courses in the theory of functions of a real variable, along with (or after) the study of functions of bounded variation. Our intent here is to break the ice much earlier, particularly for the student who may never take the advanced courses just mentioned. Our discussion will necessarily deal only with the rudiments, and will stress definitions, basic properties, and illustrations. Proofs are omitted. In large part they are very similar to proofs in the theory of Riemann integrals.

The Stieltjes integral involves two functions f and g, each defined on a closed interval $[a, b]$. It is denoted by

$$(18.9\text{--}1) \qquad \int_a^b f(x)dg(x).$$

In the special case in which g is the simple function $g(x) = x$, the Stieltjes integral (18.9–1) becomes the Riemann integral

$$\int_a^b f(x)dx.$$

To define the Stieltjes integral (18.9–1) we start with a partition P of $[a, b]$: $(x_0, x_1, \cdots, x_n)$ and a set of points $x_1', \cdots, x_n'$, one in each subinterval of the partition. We then define

$$(18.9\text{--}2) \qquad \int_a^b f(x)dg(x) = \lim_{|P| \to 0} \sum_{i=1}^n f(x_i')[g(x_i) - g(x_{i-1})],$$

provided the sums converge to a unique limit as $|P| \to 0$. Here $|P|$ denotes the mesh fineness of the partition P, as defined in § 18.2.

Example 1.   Find the value of the integral if $f(x) = 1$ and $g(x) = x^2$. The sum in this case is

$$(x_1^2 - x_0^2) + (x_2^2 - x_1^2) + \cdots + (x_n^2 - x_{n-1}^2) = x_n^2 - x_0^2 = b^2 - a^2.$$

Therefore $$\int_a^b f(x)dg(x) = b^2 - a^2$$

in this case. This argument works just as well for any $g(x)$ when $f(x) = 1$ for all x. Thus

(18.9–3) $$\int_a^b 1 \cdot dg(x) = g(b) - g(a),$$

no matter what kind of a function g is. In particular, g need not be continuous in (18.9–3).

If g is constant on $[a, b]$, all the differences $g(x_i) - g(x_{i-1})$ are zero, and so we see that

(18.9–4) $$\int_a^b f(x)dg(x) = 0$$

when g is constant on $[a, b]$. This is true no matter what kind of function f is.

One of the important practical uses of Stieltjes integrals involves the case in which g is a discontinuous function which has a finite number of discontinuities at which it jumps suddenly in value, but remains constant in value in the open intervals between the points of discontinuity. The simplest case is that in which the only discontinuities are at the end-points a and b.

Example 2. Suppose $g(x) = c$ if $a < x < b$, and let $g(a)$ and $g(b)$ have any values whatever (see Fig. 174) In this case we can show that

(18.9–5) $$\int_a^b f(x)dg(x) = f(a)[c - g(a)] + f(b)[g(b) - c],$$

where f is any continuous function.

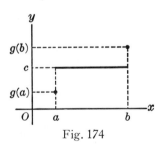

Fig. 174

To see this, let $(x_0, x_1, \cdots, x_n)$ be any partition of $[a, b]$. Then the differences $g(x_i) - g(x_{i-1})$ will all be zero as long as both points x_{i-1}, x_i are in the open interval, since $g(x) = c$ there. Hence in (18.9–2) only the terms for $i = 1$, $i = n$ remain, and

$$\int_a^b f(x)dg(x) = \lim_{|P| \to 0} \{ f(x_1')[g(x_1) - g(x_0)] + f(x_n')[g(x_n) - g(x_{n-1})] \}.$$

But $x_0 = a$, $x_n = b$, $g(x_1) = g(x_{n-1}) = c$. Also, $x_1' \to a$ and $x_n' \to b$ when $|P| \to 0$. Therefore, because of the continuity of f, $f(x_1') \to f(a)$, $f(x_n') \to f(b)$. When these results are assembled, we see that (18.9–5) holds.

Observe that the differences $c - g(a)$, $g(b) - c$ are the amounts by which the value $g(x)$ jumps at the points of discontinuity as x moves to the right.

If there are more points of discontinuity it is easy to extend the formula (18.9–5).

Example 3. Suppose $[a, b]$ is divided into N parts by the partition $(a_0, a_1, \cdots, a_N)$, and suppose $g(x)$ has the value c_i in the interior of the ith subinterval (see Fig. 175). The values of $g(x)$ at the points $a_0, \cdots, a_N$

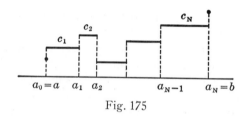

Fig. 175

can be quite independent of the values $c_1, \cdots, c_N$. If f is continuous on $[a, b]$ we can show that

$$(18.9–6) \quad \int_a^b f(x)dg(x) = f(a)[c_1 - g(a)] + f(a_1)[c_2 - c_1] + \cdots$$
$$+ f(a_{N-1})[c_N - c_{N-1}] + f(b)[g(b) - c_N].$$

Observe that the values of g at $a_1, \cdots, a_{N-1}$ (the *interior* points of discontinuity) do not enter into the formula.

The proof of (18.9–6) can be given with the aid of the general formula

$$(18.9–7)$$

$$\int_a^b f(x)dg(x) = \int_a^{a_1} f(x)dg(x) + \int_{a_1}^{a_2} f(x)dg(x) + \cdots + \int_{a_{N-1}}^b f(x)dg(x).$$

This last formula is valid for any functions f, g such that all the integrals exist.

As a concrete illustration of the kind of thing arising in Example 3, consider the moment of inertia of a number of mass particles distributed

Fig. 176

along the x-axis. Let the masses $m_1, \cdots, m_n$ be located as shown in Fig. 176, with $x_1 < x_2 < \cdots < x_n$. The moment of inertia of this mass system about the y-axis is

$$(18.9\text{–}8) \qquad\qquad I = m_1 x_1^2 + \cdots + m_n x_n^2.$$

We shall show how (18.9–8) can be written as a Stieltjes integral. Take any closed interval $[a, b]$ containing all the points $x_1, \cdots, x_n$. For any x

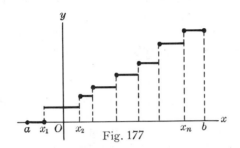

Fig. 177

in $[a, b]$ define $g(x) = 0$ if $a \leq x \leq x_1$, and, if $x_1 < x$, define $g(x)$ as the sum of all the masses m_i for which $a \leq x_i \leq x$. The graph of $g(x)$ appears in Fig. 177.

The exact definition of $g(x)$ is

$$
\begin{aligned}
g(x) &= 0 && \text{if} \quad a \leq x \leq x_1 \\
g(x) &= m_1 && \text{if} \quad x_1 < x < x_2 \\
g(x) &= m_1 + m_2 && \text{if} \quad x_2 \leq x < x_3 \\
&\kern-1em\cdots\cdots\cdots\cdots\cdots\cdots\cdots\cdots\cdots\cdots\cdots\cdots\cdots \\
g(x) &= m_1 + m_2 + \cdots + m_n && \text{if} \quad x_n \leq x \leq b.
\end{aligned}
$$

In this case there is no jump at b, and there is a jump at a if and only if $a = x_1$. Formula (18.9–6) now becomes

$$\int_a^b f(x)dg(x) = f(x_1)m_1 + \cdots + f(x_n)m_n.$$

In particular, if $f(x) = x^2$ we see that

$$\int_a^b x^2 dg(x) = m_1 x_1^2 + \cdots + m_n x_n^2 = I.$$

This is the desired expression of (18.9–8) as a Stieltjes integral.

It is important to know some of the general conditions under which it is certain that the limit (18.9–2) defining the Stieltjes integral will exist. As we have seen, there are interesting and important cases where the integral exists when g has certain discontinuities. The most important conditions sufficient to guarantee the existence of the integral are that f be continuous and that g be either (1) a nondecreasing function, (2) a nonincreasing function, or (3) the sum of a nondecreasing function and a nonincreasing function. A function g is called *nondecreasing* if its values never decrease as x increases; it is called *nonincreasing* if $g(x)$ never increases as x increases. A function g of the third type in the foregoing classification is said to be of *bounded variation* on $[a, b]$. There is an alternative way of defining such functions, based on the notion of computing the sum of the total increase and the total decrease of the values of the function as x goes from a to b. We do not have enough space to discuss in detail the interesting and important properties of functions of bounded variation.

It can be shown that if $\int_a^b f(x)dg(x)$ exists, then $\int_a^b g(x)df(x)$ also exists, and

$$(18.9\text{--}9) \qquad \int_a^b f(x)dg(x) = f(b)g(b) - f(a)g(a) - \int_a^b g(x)df(x).$$

This is called the formula of integration by parts.

Still another important fact about Stieltjes integrals is that if g has a continuous derivative and if f is integrable in the Riemann sense, then

$$(18.9\text{--}10) \qquad \int_a^b f(x)dg(x) = \int_a^b f(x)g'(x)dx,$$

where the integral on the right is a Riemann integral.

Example 4. Consider a linear mass distribution on the interval $a \leq x \leq b$. Suppose the density ρ at the point x is a continuous function of x, denoted by $\rho(x)$. Then the mass on the interval $[a, x]$ is

$$m(x) = \int_a^x \rho(t)dt.$$

We know that $m'(x) = \rho(x)$; therefore (18.9–10) applies with $m(x)$ in place of $g(x)$. In particular, the total mass is

$$M = \int_a^b 1 \cdot dm(x) = \int_a^b \rho(x)dx$$

and the abscissa $\bar{x}$ of the center of mass is given by

$$M\bar{x} = \int_a^b x \, dm(x) = \int_a^b x\rho(x)dx.$$

The function $m(x)$ in Example 4 is called a *cumulative mass distribution function*. This same name applies to the function $g(x)$ which is discussed in connection with Fig. 177, but in that case there is no continuous density function. One of the great uses of the Stieltjes integrals is in the unification of ideas and formulas about mass distributions, whether discrete or continuous.

The notion of a distribution function occurs also in the theory of probability, with applications in statistics, and Stieltjes integrals play an important part in the general formulation of such concepts as *first moment, variance, and mathematical expectation.*

EXERCISES

1. Suppose $g(x) = 1$ if $0 \leq x < 1$ and $g(x) = 4$ if $1 < x \leq 2$. Calculate the values of $\int_0^2 x^k \, dg(x)$ for $k = 0, 1, 2$.

2. Suppose $g(x) = x$ if $0 < x < 1$, $g(x) = 2 - x$ if $1 < x < 2$, $g(0) = 1$, $g(1) = 0$, $g(2) = 2$. Calculate $\int_0^1 x \, dg(x)$, $\int_1^2 x \, dg(x)$, $\int_0^2 x \, dg(x)$, and check by (18.9–7).

3. Suppose $g(x) = 5 - n$ if $n - 1 < x < n$, for $n = -1, 0, 1, 2, 3$, and $g(n) = n$. Compute $\int_{-2}^3 x^k \, dg(x)$ for $k = 0, 1, 2$.

4. If $g(x) = e^{-x}$, find (a) $\int_0^1 x \, dg(x)$, (b) $\int_0^1 g(x)dg(x)$.

5. If $g(x) = \tan^{-1} x$ find (a) $\int_{-\sqrt{3}}^1 x \, dg(x)$ and (b) $\int_{-1}^{\sqrt{3}} g(x)dg(x)$.

6. Let $g(x) = [x]$ (the greatest integer n such that $n \leq x$).
(a) Find $\int_0^a x \, dg(x)$ for the successive values $a = \frac{1}{2}, 1, \frac{3}{2}, 2$.
(b) Find $\int_0^3 g(x) \, d(\sqrt{1 + x^3})$.

7. A uniform rod 10 feet long weighs 5 pounds and carries three concentrated weights of 1 pound each fastened at each end and at the point $x = 5$ (the rod extending from $x = 0$ to $x = 10$). Define $w(x) = $ total weight on $[0, x]$ if $x > 0$, $w(0) = 0$. (a) Draw the graph of $w(x)$. (b) Calculate $\int_0^{10} x \, dw(x)$ and $\int_0^{10} (x - 5)^2 \, dw(x)$.

8. Let $\phi(x)$ be continuous and positive on $[a, b]$. Let $A(x)$ be the area between $y = \phi(x)$ and $y = 0$ from a to x.
(a) Express the total area under the curve, and the first and second moments of this area with respect to the y-axis, as Stieltjes integrals.
(b) Express the y-co-ordinate of the centroid of the area by a formula involving a Stieltjes integral with $dA(x)$.

9. If the area in Exercise 8 is revolved around the x-axis, let $V(x)$ be the volume generated by $A(x)$. Express the co-ordinate $\bar{x}$ for the centroid of the total volume by a formula involving a Stieltjes integral.

10. Use (18.9–5) and (18.9–7) to prove (18.9–6).

Infinite Series 19

19. Definitions and notation

Perhaps the simplest of all infinite series is the geometric series

(19–1) $$1 + x + x^2 + x^3 + \cdots + x^n + \cdots.$$

Let us briefly summarize the main facts about this series. We start from the algebraic identity

$$1 - x^{n+1} = (1 - x)(1 + x + x^2 + \cdots + x^n), \quad n \geq 1$$

and rewrite it in the form

(19–2) $$\frac{1}{1 - x} = 1 + x + \cdots + x^n + \frac{x^{n+1}}{1 - x}, \quad x \neq 1.$$

If $-1 < x < 1$ we see that $\lim\limits_{n \to \infty} x^{n+1} = 0$; therefore *remainder*

$$\lim_{n \to \infty} (1 + x + \cdots + x^n) = \frac{1}{1 - x}.$$

We write this last result in the form

(19–3) $$\frac{1}{1 - x} = 1 + x + x^2 + \cdots + x^n + \cdots, \quad -1 < x < 1.$$

Here we have an infinite-series representation of the function $(1 - x)^{-1}$, valid subject to certain limitations on x.

Let us now make some formal definitions. Suppose that $u_0, u_1, u_2, \cdots, u_n, \cdots$ is an infinite sequence of numbers. The expression

(19–4) $$u_0 + u_1 + u_2 + \cdots + u_n + \cdots$$

is called an *infinite series*, and the numbers $u_0, u_1, \cdots$ are called the *terms* of the series. The series is in a formal sense just a certain collection of mathematical symbols arranged in a certain way. Now we consider the sequence of numbers $s_0, s_1, s_2, \cdots, s_n, \cdots$ formed as follows:

$$s_0 = u_0$$
$$s_1 = u_0 + u_1$$
$$s_2 = u_0 + u_1 + u_2$$
$$\vdots$$
$$s_n = u_0 + u_1 + \cdots + u_n.$$

If the sequence s_n has a limit as $n \to \infty$ we say that the series (19–4) is *convergent*; if the limit of s_n is s we say that s is the *sum* (or value) of the series, and we write

$$s = u_0 + u_1 + u_2 + \cdots + u_n + \cdots.$$

If the series is not convergent we say that it is *divergent*, and we do not assign it any sum. The numbers $s_0, s_1, \cdots, s_n, \cdots$ are called the *partial sums* of the series.

We have made the definitions in the case where the terms are numerical constants. If the u's are functions of a variable x, they assume definite numerical values when x is given a fixed value, and then all the foregoing definitions apply. In such a case the partial sums s_n will also be functions of x.

Let us return to the series (19–1). We have seen that it is convergent if $-1 < x < 1$, the sum then being given by (19–3). For example,

$$\frac{3}{2} = 1 + \frac{1}{3} + \frac{1}{3^2} + \frac{1}{3^3} + \cdots + \frac{1}{3^n} + \cdots .$$

But if $x \leq -1$ or $1 \leq x$ the series is divergent. If $x \geq 1$ this is clear at once, for then $s_n \geq n + 1$, and so $s_n \to +\infty$ as $n \to \infty$. If $x = -1$ the series becomes

$$1 - 1 + 1 - 1 + \cdots ,$$

with $s_0 = 1, s_1 = 0, s_2 = 1, s_3 = 0, \cdots$. This sequence $\{s_n\}$ is not convergent, and so the series is divergent. How do the partial sums behave when $x < -1$?

From a given series (19–4) we may form new series by omitting a certain finite number of terms at the beginning, e.g.,

$$u_2 + u_3 + u_4 + \cdots ,$$
$$u_{100} + u_{101} + \cdots .$$

All such series will be convergent if the original series is convergent, and divergent if the original series is divergent (why?). Likewise we may form a new series by multiplying each term of the original series by the same constant c:

$$cu_0 + cu_1 + \cdots + cu_n + \cdots .$$

If $c \neq 0$ this series will be convergent if and only if the original series is convergent (why?). If the original series had sum s, the new series has sum cs.

We frequently use the summation symbol $\sum$ in dealing with series:

(19–5) $$\sum_{n=1}^{N} u_n = u_1 + u_2 + \cdots + u_N,$$

(19–6) $$\sum_{i=m}^{n} u_i = u_m + u_{m+1} + \cdots + u_n.$$

Observe that the index n in (19–5) and the index i in (19–6) are *dummy indices*; this means that the expression is not altered if the index letter is changed:

$$\sum_{n=1}^{N} u_n = \sum_{k=1}^{N} u_k.$$

A sum such as (19–6) is in certain respects analogous to an integral $\int_a^b f(x)dx$, in which x is a dummy variable. The u_i of (19–6) corresponds to $f(x)$; the limits of summation $i = m$ and $i = n$ correspond to the limits $x = a$ and $x = b$ on the integral.

The infinite series (19–4) is now denoted by

(19–7) $$\sum_{n=0}^{\infty} u_n,$$

whether it is convergent or not. Often, for convenience in printing, we write $\sum u_n$ in place of (19–7), suppressing the limits 0 and ∞.

Finally, we note that u_n is not necessarily the nth term of the series. The index notation merely expresses the fact that u_n is a function of n. *It is very important to observe that if the series* (19–7) *is convergent, then* $u_n \to 0$ *as* $n \to \infty$. For $u_n = s_n - s_{n-1}$. If the series is convergent, s_n and $s_{n-1} \to s$ as $n \to \infty$, and so $\lim u_n = \lim s_n - \lim s_{n-1} = 0$. But although $u_n \to 0$ is a *necessary* condition for convergence, it is by no means *sufficient*. A series may diverge, even though $u_n \to 0$ (see Example 1, § 19.2).

EXERCISES

1. For what values of x is each of the following series convergent? Express the sum of the series as a simple function of x in each case.

(a) $a + ax + ax^2 + \cdots + ax^n + \cdots$, $a \neq 0$;

(b) $8x + 8x^3 + 8x^5 + \cdots + 8x^{2n+1} \cdots$;

(c) $cx^2 + cx^4 + cx^6 + \cdots + cx^{2n} + \cdots$;

(d) $\dfrac{10}{x} + \dfrac{10}{x^2} + \dfrac{10}{x^3} + \cdots$;

(e) $1 + \dfrac{1}{1+x} + \dfrac{1}{(1+x)^2} + \cdots + \dfrac{1}{(1+x)^n} + \cdots$;

(f) $2 + 2\left(\dfrac{1-x}{1+x}\right) + 2\left(\dfrac{1-x}{1+x}\right)^2 + \cdots$;

(g) $e^x + e^{2x} + e^{3x} + \cdots + e^{nx} + \cdots$;

(h) $\log x + (\log x)^2 + \cdots + (\log x)^n + \cdots$.

2. Explain why the series

$$\frac{1+1}{1+2} + \frac{1+2}{1+4} + \cdots + \frac{1+n}{1+2n} + \cdots$$

is divergent.

3. Prove that the series

$$\frac{2}{\sqrt{1000 + 4}} + \frac{3}{\sqrt{1000 + 9}} + \cdots + \frac{n}{\sqrt{1000 + n^2}} + \cdots$$

is divergent.

4. Discuss the behavior of the partial sums s_n in the case of each of the following series:

(a) $1 - 2 + 3 - 4 + 5 - \cdots$;

(b) $1 + \frac{1}{2} - 1 + \frac{1}{4} - 1 + \frac{1}{8} - 1 + \cdots$;

(c) $(1 - \frac{1}{2}) + (\frac{1}{2} - \frac{1}{3}) + (\frac{1}{3} - \frac{1}{4}) + \cdots$;

(d) $(\sqrt{2} - \sqrt{1}) + (\sqrt{3} - \sqrt{2}) + (\sqrt{4} - \sqrt{3}) + \cdots$.

Which, if any, of the series are convergent? One can tell that two of the series are divergent without investigating the partial sums s_n. Which are these series, and how does one know by quick inspection that they are divergent?

5. Is the series

$$\frac{1^1}{(101)!} + \frac{2^2}{(102)!} + \cdots + \frac{n^n}{(100 + n)!} + \cdots$$

convergent or divergent? Justify your answer.

19.1 Taylor's series

In order to become familiar with a number of interesting examples of infinite series, let us examine how we are led to study the representation of functions by infinite series. The starting point is Taylor's formula with remainder (Chapter IV, especially § 4.3). For instance, in Examples 3 and 4 of § 4.3 we saw that if $x > -1$,

$$\log (1 + x) = x - \tfrac{1}{2} x^2 + \tfrac{1}{3} x^3 - \cdots + (- 1)^{n-1} \frac{1}{n} x^n + R_{n+1},$$

where

$$| R_{n+1} | \leq \frac{| x |^{n+1}}{n + 1} \quad \text{if} \quad 0 \leq x \leq 1$$

and

$$| R_{n+1} | \leq \frac{| x |^{n+1}}{1 + x} \quad \text{if} \quad - 1 < x \leq 0.$$

These inequalities show that $R_{n+1} \to 0$ as $n \to \infty$ when x is limited as indicated. Therefore

$$\log (1 + x) = \lim_{n \to \infty} \left[x - \tfrac{1}{2} x^2 + \tfrac{1}{3} x^3 - \cdots + (- 1)^{n-1} \frac{1}{n} x^n \right]$$

if $- 1 < x \leq 1$. According to the definitions in § 19 this result may be written in the form

(19.1–1)

$$\log (1 + x) = x - \tfrac{1}{2} x^2 + \tfrac{1}{3} x^3 - \cdots + (- 1)^{n-1} \frac{1}{n} x^n + \cdots.$$

As a special instance, let $x = 1$; then

(19.1–2) $\log 2 = 1 - \frac{1}{2} + \frac{1}{3} - \frac{1}{4} + \cdots + (-1)^{n-1} \dfrac{1}{n} + \cdots$.

Another series representation of $\log 2$ may be obtained by putting $x = -\frac{1}{2}$ in (19.1–1). We find

$$\log \tfrac{1}{2} = -\log 2 = -\tfrac{1}{2} - \tfrac{1}{2}(-\tfrac{1}{2})^2 + \tfrac{1}{3}(-\tfrac{1}{2})^3 - \cdots,$$

(19.1–3)

$$\log 2 = \frac{1}{2} + \frac{1}{2} \cdot \frac{1}{2^2} + \frac{1}{3} \cdot \frac{1}{2^3} + \cdots + \frac{1}{n} \cdot \frac{1}{2^n} + \cdots.$$

Now suppose that f is a function which has derivatives of all orders on an open interval containing the point $x = a$. According to Taylor's formula (4.3–8) we have, for any x of this interval,

$$f(x) = f(a) + f'(a)(x - a) + \cdots + \frac{f^{(n)}(a)}{n!} (x - a)^n + R_{n+1},$$

where the remainder R_{n+1} can be expressed in a variety of different forms. If we can show, for a particular x, that $\lim R_{n+1} = 0$, then it follows that $f(x)$ can be represented by an infinite series

(19.1–4) $$f(x) = \sum_{n=0}^{\infty} \frac{f^{(n)}(a)}{n!} (x - a)^n$$

(with the usual convention that $n! = 1$ if $n = 0$). This is called *Taylor's series expansion of $f(x)$ about the point $x = a$* (also sometimes called the expansion of $f(x)$ in powers of $x - a$). Observe that we are *not* asserting that (19.1–4) is always true. It *will* be true if $\lim_{n \to \infty} R_{n+1} = 0$, but this is a matter to be investigated for each particular function and each particular x.

Example 1. The function e^x can be represented by Taylor's series for all values of x, no matter how the point $x = a$ is chosen.

In order to prove what has just been stated, let us first choose $a = 0$. Then, since $f^{(n)}(x) = e^x$ for all orders n, Taylor's formula with remainder is

$$e^x = 1 + x + \frac{x^2}{2!} + \cdots + \frac{x^n}{n!} + R_{n+1}.$$

Lagrange's form of the remainder is

$$R_{n+1} = \frac{e^X}{(n+1)!} x^{n+1},$$

with X between 0 and x. Thus certainly $-|x| < X < |x|$, and $0 < e^X < e^{|x|}$; therefore

(19.1–5) $$|R_{n+1}| \leq e^{|x|} \frac{|x|^{n+1}}{(n+1)!}.$$

We shall prove that $\lim_{n \to \infty} R_{n+1} = 0$. From (19.1–5) we see that it will be sufficient to prove that $\lim_{n \to \infty} \dfrac{|x|^n}{n!} = 0$. Now choose an integer N such that $N \geq 2\,|x|$. Then, if $n > N$,

$$\frac{|x|^n}{n!} = \frac{|x|^N}{N!} \frac{|x|^{n-N}}{(N+1)(N+2)\cdots n}$$

$$\leq \frac{|x|^N}{N!} \left(\frac{N}{N+1}\right)\left(\frac{N}{N+2}\right)\cdots\left(\frac{N}{n}\right)\frac{1}{2^{n-N}},$$

$$\frac{|x|^n}{n!} \leq \frac{|x|^N}{N!} \left(\frac{1}{2}\right)^{n-N}.$$

Keeping N fixed and letting $n \to \infty$, we see that $(\frac{1}{2})^{n-N} \to 0$, and so the desired result is attained. Therefore the series representation

$$(19.1-6) \qquad e^x = 1 + x + \frac{x^2}{2!} + \cdots + \frac{x^n}{n!} + \cdots$$

is valid for all values of x.

Now consider any point $x = a$. Taylor's formula for e^x now takes the form

$$e^x = e^a + e^a(x - a) + \cdots + e^a \frac{(x-a)^n}{n!} + e^X \frac{(x-a)^{n+1}}{(n+1)!},$$

with X between a and x. A very slight modification of the foregoing argument shows that the remainder approaches zero as $n \to \infty$, so that the representation

$$(19.1-7) \qquad e^x = e^a + e^a(x - a) + \cdots + e^a \frac{(x-a)^n}{n!} + \cdots$$

is valid for all values of x.

Example 2. The functions $\sin x$ and $\cos x$ can be represented by Taylor's series for all values of x, no matter how the point $x = a$ is chosen. The series take particularly simple forms when $a = 0$, namely

$$(19.1-8) \qquad \sin x = x - \frac{x^3}{3!} + \frac{x^5}{5!} - \frac{x^7}{7!} + \cdots$$

$$(19.1-9) \qquad \cos x = 1 - \frac{x^2}{2!} + \frac{x^4}{4!} - \frac{x^6}{6!} + \cdots.$$

The details are left to the student (see Exercise 1).

In general the problem of finding a Taylor's series representation for a given function, and proving that the representation is valid, can be a very difficult task, especially if we approach the problem directly as in the foregoing examples. It may prove to be a very complicated matter to calculate the derivatives $f^{(n)}(x)$ for higher values of n, and one cannot always expect to find a manageable general formula for the coefficient $\dfrac{f^{(n)}(a)}{n!}$. Moreover, if one cannot get a reasonably simple formula for $f^{(n+1)}(x)$, there is very little chance of proving that $R_{n+1} \to 0$ as $n \to \infty$ by means of the standard formulas for the remainder (Lagrange, Cauchy, or the integral form (4.2–8)). There are, however, more advanced methods for attacking the problem of finding out whether a function can be represented by Taylor's series. The most important of these methods belongs to the theory of functions of a complex variable, and is beyond the scope of this book.

In certain particular cases which are interesting and important there are special devices which enable us to find series representations without the necessity of calculating a general formula for $f^{(n)}(x)$. One such device is considered in § 19.11. Other devices are considered in Chapter XXI.

EXERCISES

1. Prove the validity of the series expansions (19.1–8) and (19.1–9) for $\sin x$ and $\cos x$.

2. Prove the validity of the expansion

$$(1 - x)^{-1/2} = 1 + \tfrac{1}{2} x + \frac{1 \cdot 3}{2 \cdot 4} x^2 + \cdots + \frac{1 \cdot 3 \cdots (2\,n - 1)}{2 \cdot 4 \cdots 2\,n} x^n + \cdots$$

when $-1 < x < 1$. Suggestion: See Exercise 7, § 4.3. For the case $0 < x < 1$ it is convenient to use the fact that $na^n \to 0$ as $n \to \infty$ if $|a| < 1$. Prove this fact separately.

19.11 A series for the inverse tangent

In the formula (19–2) let us put $x = -t^2$. Then

(19.11–1)
$$\frac{1}{1 + t^2} = 1 - t^2 + t^4 - \cdots + (-1)^n t^{2n} + (-1)^{n+1} \frac{t^{2n+2}}{1 + t^2}.$$

Integrating both sides of this algebraic identity, we have, for any x,

$$\int_0^x \frac{dt}{1 + t^2} = \tan^{-1} x = x - \frac{x^3}{3} + \frac{x^5}{5} - \cdots + (-1)^n \frac{x^{2n+1}}{2\,n + 1}$$

$$+ (-1)^{n+1} \int_0^x \frac{t^{2n+2}}{1 + t^2}\, dt.$$

We apply Theorem V of § 4.4 to the integral on the right:

$$\int_0^x \frac{t^{2n+2}}{1 + t^2}\, dt = \frac{1}{1 + X^2} \int_0^x t^{2n+2}\, dt = \frac{1}{1 + X^2} \frac{x^{2n+3}}{2\,n + 3},$$

where X is between 0 and x. Now if $|x| \leq 1$,

$$\lim_{n \to \infty} \frac{x^{2n+3}}{2n+3} = 0,$$

and therefore we conclude that

(19.11–2)
$$\tan^{-1} x = x - \frac{x^3}{3} + \frac{x^5}{5} - \cdots + (-1)^n \frac{x^{2n+1}}{2n+1} + \cdots.$$

This series representation is valid if $-1 \leq x \leq 1$. In particular, if $x = 1$, we find the interesting result

(19.11–3)
$$\frac{\pi}{4} = 1 - \tfrac{1}{3} + \tfrac{1}{5} - \tfrac{1}{7} + \tfrac{1}{9} - \cdots.$$

The series (19.11–2) is actually the Taylor's series expansion of $\tan^{-1} x$ in powers of x, even though we did not find the terms of the series by calculating the successive derivatives of $\tan^{-1} x$. This latter procedure would lead to great complications, as the student may find if he attempts it. But it can be shown in the general theory of power series, that if a function can be represented by a series of powers of $x - a$, then the series is actually the Taylor's series expansion of the function about the point $x = a$.

EXERCISES

1. Show that $\pi = 16 \tan^{-1} \tfrac{1}{5} - 4 \tan^{-1} \tfrac{1}{239}$. Start by setting $\phi = \tan^{-1} \tfrac{1}{5}$ and computing $\tan 2\phi$ and $\tan 4\phi$ by repeated use of the formula for the tangent of twice an angle. Then compute $\tan \left(4\phi - \dfrac{\pi}{4} \right)$. If the inverse tangents in this formula for π are computed by the series (19.11–2), a value of π accurate to a number of decimal places may be found without excessive labor. As a sample, let the student obtain the approximation $\pi = 3.141593$, of which the first five decimal places are accurate and the sixth is correct as a rounded-off figure.

2. Show that $2 \tan^{-1} \tfrac{1}{10} - \tan^{-1} \tfrac{1}{5} = \tan^{-1} \tfrac{1}{515}$. Combine this with the result of Exercise 1 to obtain the formula

$$\frac{\pi}{4} = 8 \tan^{-1} \frac{1}{10} - 4 \tan^{-1} \frac{1}{515} - \tan^{-1} \frac{1}{239}.$$

Try this formula along with (19.11–2) in the computation of π.

19.2   Series of nonnegative terms

For a deeper study of the general problem of representation of functions by infinite series it is essential to have a certain fund of general knowledge about infinite series as such. We turn then to a different problem. Suppose we have before us a series, no matter how obtained. What can we do to

determine whether or not the series is convergent? We begin by considering the special case in which all the terms of the series are nonnegative.
Special though this case is, what we learn about it will have important
applications to the general problem.

THEOREM I. *Suppose that $u_n \geq 0$ for every n. Then the series $\sum u_n$ is convergent if and only if the sequence $\{s_n\}$ of partial sums is bounded.*

PROOF. By definition $s_n = u_0 + u_1 + \cdots + u_n$. Since $u_n \geq 0$, it is
clear that $0 \leq s_n \leq s_{n+1}$. If the partial sums are bounded, the sequence
$\{s_n\}$ has a limit, by Theorem III, § 2.7. On the other hand, if the sequence
is not bounded, then $s_n \rightarrow + \infty$ as $n \rightarrow \infty$. This completes the proof of
the theorem.

Example 1. The series

$$(19.2\text{--}1) \qquad 1 + \tfrac{1}{2} + \tfrac{1}{3} + \cdots + \frac{1}{n} + \cdots$$

is divergent. It is called the harmonic series.

We prove the divergence by showing that the partial sums are not
bounded. Let

$$s_n = 1 + \tfrac{1}{2} + \cdots + \frac{1}{n}.$$

Then

$$s_{2n} = s_n + \frac{1}{n+1} + \frac{1}{n+2} + \cdots + \frac{1}{2n} > s_n + \frac{1}{2},$$

since

$$\frac{1}{n+1} + \frac{1}{n+2} + \cdots + \frac{1}{2n} > n \cdot \frac{1}{2n} = \frac{1}{2}.$$

With $s_{2n} > s_n + \tfrac{1}{2}$ for every n, it is plainly impossible for $\{s_n\}$ to be bounded.
We have

$$s_1 = 1, \; s_2 = \tfrac{3}{2}, \; s_4 > s_2 + \tfrac{1}{2} = 2, \; s_8 > s_4 + \tfrac{1}{2} > \tfrac{5}{2},$$

and in general

$$s_{2^n} > \frac{n+2}{2}.$$

For a series with negative as well as positive terms, convergence of the
series is *not* guaranteed by boundedness of the partial sums. For instance,
the series

$$1 - 1 + 1 - 1 + 1 - \cdots$$

is not convergent, yet for its partial sums we have $s_n = 1$ or $s_n = 0$, depending on the oddness or evenness of n. Thus these partial sums are bounded.

THEOREM II. *Let $\sum a_n$ and $\sum b_n$ be two series of nonnegative terms, and suppose that, for all values of n after some fixed index N, it is true that $a_n \leqq b_n$. Then if the series $\sum b_n$ is convergent, so is $\sum a_n$, and if the series $\sum a_n$ is divergent, so is $\sum b_n$.*

PROOF. In discussing convergence or divergence we may drop the terms with index less than N. Then, for any $n > N$,

$$a_N + a_{N+1} + \cdots + a_n \leqq b_N + b_{N+1} + \cdots + b_n.$$

The proof of the theorem is an immediate consequence of this inequality and Theorem I.

Example 2. The series

$$(19.2\text{--}2) \qquad 1 + \frac{1}{2^2} + \frac{1}{3^3} + \cdots + \frac{1}{n^n} + \cdots$$

is convergent. For $\dfrac{1}{n^n} < \dfrac{1}{n!}$ if $n > 1$, and the series

$$(19.2\text{--}3) \qquad 1 + \frac{1}{2!} + \frac{1}{3!} + \cdots + \frac{1}{n!} + \cdots$$

is known to be convergent (see the series (19.1–6) for e^x with $x = 1$; see also Example 5, § 1.62).

THEOREM III. *Let $\sum a_n$ and $\sum b_n$ be two series of positive terms, and suppose that a_n/b_n approaches a nonzero limit as $n \to \infty$. Then either both series are convergent, or both are divergent.*

PROOF. Suppose that $a_n/b_n \to c$, and $c \neq 0$. Then, for all sufficiently large values of n we shall have $\frac{1}{2} c < a_n/b_n < \frac{3}{2} c$, whence

$$b_n < \left(\frac{2}{c}\right) a_n \quad \text{and} \quad a_n < \left(\frac{3}{2} c\right) b_n.$$

Since the convergence or divergence of a series is not affected by multiplying each of its terms by the same nonzero constant, the foregoing inequalities together with Theorem II are sufficient to prove Theorem III.

Example 3. The series

$$\frac{2^2}{3!} + \frac{3^2}{4!} + \cdots + \frac{(n+1)^2}{(n+2)!} + \cdots$$

is convergent. We prove this by using the convergent series (19.2–3) and applying Theorem III:

$$\frac{(n+1)^2}{(n+2)!} \bigg/ \frac{1}{n!} = \frac{(n+1)^2}{(n+1)(n+2)} = \frac{n+1}{n+2} \to 1.$$

Example 4. The series

$$1 + \frac{1}{3} + \frac{1}{5} + \cdots + \frac{1}{2n-1} + \cdots$$

is divergent. We prove this by using the series (19.2–1) and applying Theorem III.

$$\frac{1}{2n-1} \bigg/ \frac{1}{n} = \frac{n}{2n-1} \to \frac{1}{2}.$$

EXERCISES

1. Test the following series for convergence or divergence by Theorem II.

(a) $\displaystyle\sum_{n=1}^{\infty} \frac{1}{\sqrt{n}}$.

(b) $\displaystyle\sum_{n=2}^{\infty} \frac{1}{\log n}$.

(c) $1 + \dfrac{1}{1 \cdot 3} + \dfrac{1}{1 \cdot 3 \cdot 5} + \cdots + \dfrac{1}{1 \cdot 3 \cdots (2n-1)} + \cdots$.

(d) $\dfrac{1}{2 \cdot 2} + \dfrac{2}{3 \cdot 2^2} + \cdots + \dfrac{n}{(n+1)2^n} + \cdots$.

2. Prove that $(n+1)/n! < 8/2^n$ if $n \geq 1$. What do you conclude about the following series?

$$\frac{2}{1!} + \frac{3}{2!} + \cdots + \frac{n+1}{n!} + \cdots .$$

3. Compare the series

$$\frac{4 \cdot 2}{3!} + \frac{5 \cdot 3}{4!} + \cdots + \frac{(n+3)(n+1)}{(n+2)!} + \cdots$$

with a suitable multiple of the series (19.2–3), and use Theorem II to establish convergence. Also prove the convergence of the series by using the series (19.2–3) and Theorem III.

4. Compare the series

$$\frac{3}{5} + \frac{3 \cdot 5}{5 \cdot 10} + \cdots + \frac{3 \cdot 5 \cdot 7 \cdots (2n+1)}{5 \cdot 10 \cdot 15 \cdots 5n} + \cdots$$

with a suitable multiple of the series

$$1 + \frac{1}{2} + \cdots + \frac{1}{2^{n-1}} + \cdots ,$$

and use Theorem II to establish convergence.

5. Test the following series for convergence or divergence by Theorem III.

(a) $\displaystyle\sum_{n=1}^{\infty} \frac{n+1}{(n+2)n!}$.

(c) $\displaystyle\sum_{n=1}^{\infty} \frac{n^2}{n^3+1}$.

(b) $\displaystyle\sum_{n=1}^{\infty} \frac{2n+1}{n^2+n}$.

(d) $\displaystyle\sum_{n=1}^{\infty} \frac{\sqrt{n+1}}{n^{n+\frac{1}{2}}}$.

(e) $\displaystyle\sum_{n=1}^{\infty} \frac{(n+1)(n+2)}{n^2 \cdot 2^n}$.

(f) $\displaystyle\sum_{n=1}^{\infty} n^{(1-n)/n}$ (see Exercise 17, § 1.62). $\lim_{n\to\infty} n^{1/n} = 1$

(g) $\displaystyle\sum_{n=1}^{\infty} \frac{(n+1)^n}{n^{2n}}$ (see (1.62-5)). $e = \lim_{n\to\infty} \left(1+\frac{1}{n}\right)^n$

6. Prove that, if $\sum a_n$ and $\sum b_n$ are series of positive terms with $\sum b_n$ convergent and $a_n/b_n \to 0$, then $\sum a_n$ is convergent. State and prove a comparison theorem in which part of the hypothesis is that $a_n/b_n \to \infty$.

7. Let $\sum a_n$ be a convergent series of positive terms, and let $\sum b_n$ be a series such that each b_n is *some* term of the series $\sum a_n$. Prove that the series $\sum b_n$ is convergent.

8. Show that the series

$$\frac{1}{1 \cdot 2} + \frac{1}{2 \cdot 3} + \cdots + \frac{1}{n(n+1)} \cdots$$

is convergent by noting that

$$\frac{1}{n(n+1)} = \frac{1}{n} - \frac{1}{n+1}$$

and finding a simple expression for the sum of the first n terms. What can you infer about the series

$$\frac{1}{1^2} + \frac{1}{2^2} + \frac{1}{3^2} + \frac{1}{4^2} + \cdots$$

and

$$\frac{1}{1 \cdot 2} + \frac{1}{3 \cdot 4} + \frac{1}{5 \cdot 6} + \cdots ?$$

9. Prove that the series

$$\frac{1}{1} \log \frac{2^2}{1 \cdot 3} + \frac{1}{2} \log \frac{3^2}{2 \cdot 4} + \cdots + \frac{1}{n} \log \frac{(n+1)^2}{n(n+2)} + \cdots$$

is convergent by showing directly that the sum of the first n terms is less than

$\log \dfrac{2(n+1)}{n+2} < \log 2$.

19.21   The integral test

For this section the student will need to be familiar with a few of the most elementary things about improper integrals of the type

$$\int_a^{\infty} f(x)dx = \lim_{b \to +\infty} \int_a^b f(x)dx.$$

The integral is called *convergent* if the limit exists, and *divergent* if the limit does not exist. We refer the student to § 18.8 and § 22.

The idea of the integral test is to relate the convergence or divergence of a series

$$u_1 + u_2 + u_3 + \cdots + u_n + \cdots$$

to the convergence or divergence of a certain improper integral $\int_a^\infty f(x)dx$. A relationship of this sort can be established for certain kinds of series. For the series

$$\frac{1}{1^2} + \frac{1}{2^2} + \frac{1}{3^2} + \cdots + \frac{1}{n^2} + \cdots$$

the appropriate integral to consider is $\int_1^\infty \frac{dx}{x^2}$. In general one wants a function $f(x)$ such that $f(n) = u_n$.

THEOREM IV. *Let $f(x)$ be a function which is positive, continuous, and nonincreasing as x increases for all values of $x \geq N$, where N is some fixed positive integer. Let the terms of an infinite series be given by $u_n = f(n)$ when $n \geq N$. Then the series $\sum u_n$ converges or diverges according as the improper integral $\int_N^\infty f(x)dx$ is convergent or divergent.*

PROOF.   If $m \leq x \leq m + 1$, we have $f(m + 1) \leq f(x) \leq f(m)$, and therefore

$$u_{m+1} = f(m + 1) \leq \int_m^{m+1} f(x)dx \leq f(m) = u_m.$$

We set m successively equal to $N, N + 1, \cdots, n$ and add. The result is

$$(19.21\text{--}1) \quad u_{N+1} + \cdots + u_{n+1} \leq \int_N^{n+1} f(x)dx \leq u_N + \cdots + u_n.$$

Suppose now that the integral $\int_N^\infty f(x)dx$ is convergent. Then from the first inequality in (19.21–1) it follows that

$$u_{N+1} + \cdots + u_{n+1} \leq \int_N^\infty f(x)dx.$$

This shows that the partial sums of the series $\sum_{n=N+1}^\infty u_n$ are bounded, and hence that this series is convergent (by I, § 19.2). The series $\sum_{n=1}^\infty u_n$ is then convergent also. On the other hand, suppose the integral $\int_N^\infty f(x)dx$ is divergent. Since $f(x) > 0$ this can only happen if $\int_N^{n+1} f(x)dx \to + \infty$ as $n \to \infty$. It then follows from the second inequality in (19.21–1) that $u_N + \cdots + u_n \to + \infty$, and hence that the series is divergent. This completes the proof.

Example.   The series

(19.21–2) $$\frac{1}{1^p} + \frac{1}{2^p} + \frac{1}{3^p} + \cdots + \frac{1}{n^p} + \cdots$$

is convergent if $p > 1$ and divergent if $p \leq 1$.

These results are established by considering the integral $\int_1^\infty \dfrac{dx}{x^p}$ and applying Theorem IV. If $p > 1$ we have

$$\int_1^b \frac{dx}{x^p} = \frac{x^{-p+1}}{-p+1} \Big|_1^b = \frac{1}{1-p}\left(\frac{1}{b^{p-1}} - 1\right),$$

$$\lim_{b \to \infty} \int_1^b \frac{dx}{x^p} = \int_1^\infty \frac{dx}{x^p} = \frac{1}{p-1}.$$

We leave it for the student to verify that the integral diverges if $p \leq 1$. If $p = 1$ this gives us a new proof that the harmonic series is divergent (see Example 1, § 19.2).

The proof of Theorem IV gives us some estimates of the sum of the series if it is convergent. From (19.21–1) we have, on letting $n \to \infty$,

(19.21–3) $$\sum_{n=N+1}^\infty u_n \leq \int_N^\infty f(x)dx \leq \sum_{n=N}^\infty u_n.$$

Actually neither inequality can be an equality. That is, (19.21–3) remains true if the sign $\leq$ is replaced by $<$ at both places where it occurs. We leave it for the student to supply the argument in justification of this assertion.

EXERCISES

1. Test the following series for convergence or divergence by Theorem IV. For the convergent series give an upper bound for the sum of the series with the aid of (19.21–3).

(a) $\displaystyle\sum_{n=1}^\infty \frac{1}{\sqrt{n+1} - 1}$.

(e) $\displaystyle\sum_{n=2}^\infty \frac{1}{n(\log n)^2}$.

(b) $\displaystyle\sum_{n=2}^\infty \frac{2n}{n^4 - 4}$.

(f) $\displaystyle\sum_{n=4}^\infty \frac{1}{n \log n[\log(\log n)]^2}$.

(c) $\displaystyle\sum_{n=1}^\infty \frac{1}{(2n-1)(2n)}$.

(g) $\displaystyle\sum_{n=2}^\infty \frac{1}{(\log n)^{10}}$.

(d) $\displaystyle\sum_{n=2}^\infty \frac{1}{n \log n}$.

(h) $\displaystyle\sum_{n=2}^\infty \frac{(\log n)^2}{n^3}$.

2. Show that the series

$$\frac{1}{2(\log 2)^p} + \frac{1}{3(\log 3)^p} + \cdots + \frac{1}{n(\log n)^p} + \cdots$$

is convergent if $p > 1$, divergent if $p \leq 1$.

3. For what values of p does the series $\sum\limits_{n=4}^{\infty} \dfrac{1}{n \log n[\log(\log n)]^p}$ converge?

4. Show that the series $\sum\limits_{n=2}^{\infty} \dfrac{1}{(\log n)^p}$ diverges for all values of p.

5. For what values of p and q is the series $\sum\limits_{n=2}^{\infty} \dfrac{(\log n)^q}{n^p}$ convergent?

6. Let $C_n = 1 + \frac{1}{2} + \cdots + (1/n) - \log n$. Put $N = 1$, $f(x) = 1/x$ in (19.21–1), with $n - 1$ in place of n, and show that $0 < 1/n \leqq C_n$. Also show that $C_n - C_{n+1} > 0$. The sequence C_n is therefore convergent, since it is decreasing and bounded below. The number $C = \lim_{n \to \infty} C_n$ is known as Euler's constant. Its value is approximately 0.577.

Show by a similar argument that

$$\lim_{n \to \infty} \left[1 + \frac{1}{\sqrt{2}} + \cdots + \frac{1}{\sqrt{n}} - 2(\sqrt{n} - 1) \right] \text{ exists.}$$

19.22   Ratio tests

In this section we deal with series all of whose terms are positive. In many cases it proves to be useful to consider the ratio u_{n+1}/u_n of two successive terms of the series $\sum u_n$.

THEOREM V.   *Let $\sum u_n$ and $\sum v_n$ be two series of positive terms, and suppose that the inequality*

$$(19.22\text{–}1) \qquad \frac{v_{n+1}}{v_n} \leqq \frac{u_{n+1}}{u_n}$$

holds for all values of n. Then if $\sum u_n$ is convergent, so is $\sum v_n$.
   The same conclusion can of course be drawn if (19.22–1) holds merely for all sufficiently large values of n.

PROOF.   We have

$$\frac{v_2}{v_1} \leqq \frac{u_2}{u_1}, \qquad \frac{v_3}{v_2} \leqq \frac{u_3}{u_2}, \qquad \cdots, \qquad \frac{v_n}{v_{n-1}} \leqq \frac{u_n}{u_{n-1}}.$$

Hence

$$v_n = v_1 \frac{v_2}{v_1} \cdot \frac{v_3}{v_2} \cdots \frac{v_n}{v_{n-1}} \leqq v_1 \frac{u_2}{u_1} \cdot \frac{u_3}{u_2} \cdots \frac{u_n}{u_{n-1}} = \frac{v_1}{u_1} u_n,$$

so that $v_n \leqq C u_n$, where C is the positive constant v_1/u_1. The convergence of $\sum v_n$ now follows by an application of Theorem II (§ 19.2).

One of the most useful applications of Theorem V is obtained by choosing for $\sum u_n$ the geometric series $1 + r + r^2 + \cdots + r^n + \cdots$, which for positive r is convergent if $r < 1$. For this choice of u_n we have $u_{n+1}/u_n = r$. The following theorem is an immediate corollary of Theorem V.

THEOREM VI. *The series $\sum v_n$ of positive terms is convergent if there is a positive number $r < 1$ such that $v_{n+1}/v_n \leqq r$ for all sufficiently large values of n.*

The ratio v_{n+1}/v_n need not approach a limit as $n \to \infty$, but it does so in many cases arising in practice.

THEOREM VII. *Let $\sum v_n$ be a series of positive terms, and suppose that the ratio v_{n+1}/v_n approaches a limit t as $n \to \infty$. Then the series is convergent if $0 \leqq t < 1$ and divergent if $t > 1$. If $v_{n+1}/v_n \to + \infty$ we write $t = + \infty$; in this case the series diverges.*

PROOF. If $0 \leqq t < 1$, choose r so that $t < r < 1$. Then v_{n+1}/v_n is close to t when n is large, and so $v_{n+1}/v_n < r$ for all sufficiently large values of n. The convergence of the series is then assured by Theorem VI. On the other hand, if $t > 1$, then $v_{n+1} > v_n$ if n is large enough, and in this case the terms of the series cannot approach zero as $n \to \infty$. Hence the series cannot be convergent (see the final paragraph of § 19).

No conclusion can be drawn if $t = 1$, for this case can occur both with convergent and with divergent series. For instance, in the case of the series (19.21–2) it turns out that $t = 1$ for all values of p.

Example. The series $\sum\limits_{n=1}^{\infty} \dfrac{n^2}{2^n}$ is convergent.

Here we have $v_n = \dfrac{n^2}{2^n}$, $\dfrac{v_{n+1}}{v_n} = \dfrac{(n+1)^2}{2^{n+1}} \dfrac{2^n}{n^2} = \dfrac{1}{2}\left(\dfrac{n+1}{n}\right)^2$,

and $\lim\limits_{n \to \infty} \dfrac{v_{n+1}}{v_n} = \dfrac{1}{2}$.

Further uses of the theorems of this section will be developed in § 19.4.

EXERCISES

1. Test the following series for convergence or divergence by Theorem VII.

(a) $\sum\limits_{n=1}^{\infty} \dfrac{1 \cdot 3 \cdots (2n-1)}{2 \cdot 5 \cdots (3n-1)}$.

(b) $\sum\limits_{n=1}^{\infty} \dfrac{(n!)^2 2^n}{(2n+2)!}$.

(c) $\sum\limits_{n=1}^{\infty} \dfrac{n!}{2^n}$.

(d) $\sum\limits_{n=1}^{\infty} n(\tfrac{3}{4})^n$.

(e) $\sum\limits_{n=1}^{\infty} \dfrac{n^n}{n!}$.

(f) $\sum\limits_{n=1}^{\infty} \dfrac{2 \cdot 4 \cdots 2n}{4 \cdot 7 \cdots (3n+1)}$.

(g) $\sum\limits_{n=2}^{\infty} \dfrac{n^{n-1}}{3^{n-1}(n-1)!}$.

2. If $0 < r < 1$ and p is any positive integer, show that the series

$$\sum_{n=p+1}^{\infty} n(n-1) \cdots (n-p)r^n$$

is convergent.

3. Suppose $\sum u_n$ and $\sum v_n$ are series of positive terms satisfying (19.22–1). Prove that, if $\sum v_n$ is divergent, so is $\sum u_n$. Is this theorem equivalent to Theorem V?

4. Suppose $u_n > 0$ and $\dfrac{u_{n+1}}{u_n} \leqq 1 - \dfrac{2}{n} + \dfrac{1}{n^2}$. Show that $\sum u_n$ is convergent.

5. Suppose $u_n > 0$ and $\dfrac{u_{n+1}}{u_n} \geqq 1 - \dfrac{1}{n}$. Show that $\sum u_n$ is divergent.

19.3 Absolute and conditional convergence

Theorems I–VII all deal with series whose terms are nonnegative. Thus far we have developed no general methods for discussing the convergence of a given series if its terms are unrestricted as to sign. On the other hand, in § 19.1 and § 19.11 we have incidentally met some examples of convergent series with both positive and negative terms, e.g.

$$(19.3\text{--}1) \quad \log 2 = 1 - \frac{1}{2} + \frac{1}{3} - \frac{1}{4} + \cdots + (-1)^{n-1}\frac{1}{n} + \cdots,$$

$$(19.3\text{--}2) \quad \frac{1}{e} = 1 - 1 + \frac{1}{2!} - \frac{1}{3!} + \cdots + (-1)^n \frac{1}{n!} + \cdots.$$

Now there is an important difference between the two series (19.3–1) and (19.3–2). Each of the series is convergent as it stands. But let us consider the series which are obtained if we make the alteration of changing the signs of all the negative terms. We then have

$$(19.3\text{--}3) \qquad 1 + \frac{1}{2} + \frac{1}{3} + \cdots + \frac{1}{n} + \cdots$$

$$(19.3\text{--}4) \qquad 1 + 1 + \frac{1}{2!} + \cdots + \frac{1}{n!} + \cdots.$$

The change from (19.3–1) to (19.3–3) has given us a divergent series (the harmonic series (19.2–1)), whereas in the change from (19.3–2) to (19.3–4) we still have a convergent series (its value is e; see (19.1–6)).

If a series of positive terms is to be convergent, the terms with large index n must be so small that even the sum of arbitrarily many of them is small. But with a series of terms in which infinitely many are positive and infinitely many are negative, the series may be convergent because of a partial cancelling out effect, e.g. a negative term offsetting the cumulative effect of one or more positive terms. This kind of process may operate to produce a convergent series even though the series would not converge if all the terms were replaced by their absolute values. This situation is illustrated by the convergent series (19.3–1) and its divergent counterpart (19.3–3). With this background of explanation we now make a definition.

DEFINITION. Let $\sum u_n$ be a given series. Consider the series $\sum |u_n|$ obtained by putting $|u_n|$ in place of u_n. We say that $\sum u_n$ is *absolutely convergent* provided that the series $\sum |u_n|$ is convergent.

The careful student will observe at once that the definition of *absolute convergence* of $\sum u_n$ does not in itself make any statement about the mere *convergence* of $\sum u_n$. It is in fact true, however, that if $\sum |u_n|$ is convergent, then so is $\sum u_n$. Before proving this we must take up a general criterion for convergence.

THEOREM VIII. *A series $\sum u_n$ is convergent if and only if to each $\epsilon > 0$ there corresponds some integer N (depending on ϵ and the particular series) such that*

(19.3–5) $|u_{m+1} + u_{m+2} + \cdots + u_n| < \epsilon$

for all integers m, n such that $N \leq m < n$.

PROOF. This is a direct consequence of Cauchy's convergence condition (Theorem VI, § 16.5). We write

$$s_n = u_1 + u_2 + \cdots + u_n .$$

Then

$$s_n - s_m = u_{m+1} + \cdots + u_n .$$

The condition (19.3–5) is now seen to be equivalent to $|s_n - s_m| < \epsilon$ and the theorem is merely the statement of Cauchy's condition for the sequence $\{s_n\}$.

Theorem VIII is not very useful for making direct tests to find whether a series is convergent. But it is of fundamental importance in the general theory of infinite series, and we shall use it in proving various things about series. The first such application deals with absolute convergence.

THEOREM IX. *If the series $\sum u_n$ is absolutely convergent, it is convergent.*

PROOF. We are assuming that $\sum |u_n|$ is convergent. By Theorem VIII, therefore, we know that to any $\epsilon > 0$ there corresponds some N such that

$$|u_{m+1}| + |u_{m+2}| + \cdots + |u_n| < \epsilon$$

whenever $N \leq m < n$. Now

$$| u_{m+1} + u_{m+2} + \cdots + u_n | \leq | u_{m+1} | + \cdots + | u_n |.$$

Hence $| u_{m+1} + \cdots + u_n | < \epsilon$ whenever $N \leq m < n$. This is precisely condition (19.3–5), and so we conclude that the series $\sum u_n$ is convergent.

DEFINITION. A series is called *conditionally* convergent if it is convergent, but not absolutely convergent.

The series (19.3–1) is conditionally convergent.

We shall consider some of the differences between absolutely convergent and conditionally convergent series. In the series $\sum u_n$ let us denote by $a_1, a_2, a_3, \cdots$ the positive terms, taken in the order of their occurrence; let the negative terms be denoted by $-b_1, -b_2, -b_3, \cdots$. Thus, in the series

$$1 - \tfrac{1}{2} + \tfrac{1}{3} - \tfrac{1}{4} + \cdots$$

we have

$$a_1 = 1, \quad a_2 = \tfrac{1}{3}, \quad a_3 = \tfrac{1}{5}, \quad \cdots,$$
$$b_1 = \tfrac{1}{2}, \quad b_2 = \tfrac{1}{4}, \quad b_3 = \tfrac{1}{6}, \quad \cdots.$$

We now consider the two series $\sum a_n$ and $\sum b_n$, each of which consists entirely of positive terms.

THEOREM X. *If the series $\sum u_n$ is absolutely convergent, then each of the series $\sum a_n, \sum b_n$ is convergent, and $\sum u_n = \sum a_n - \sum b_n$. But if the series $\sum u_n$ is conditionally convergent, then each of the series $\sum a_n, \sum b_n$ is divergent.*

The theorem is illustrated for the absolutely convergent case by the series (19.3–2), which is the difference of the two convergent series

$$\left(1 + \frac{1}{2!} + \frac{1}{4!} + \cdots \right) - \left(1 + \frac{1}{3!} + \frac{1}{5!} + \cdots \right).$$

In the case of the series (19.3–1) each of the constituent series

$$1 + \tfrac{1}{3} + \tfrac{1}{5} + \cdots, \quad \tfrac{1}{2} + \tfrac{1}{4} + \tfrac{1}{6} + \cdots$$

is divergent.

PROOF OF THE THEOREM. Suppose that $\sum u_n$ is absolutely convergent, with $M = \sum | u_n |$. Then

$$| u_1 | + | u_2 | + \cdots + | u_n | \leq M,$$

no matter how large n is. Now consider a partial sum of the series $\sum a_n$, say $a_1 + \cdots + a_m$. Since each a_i is a positive term somewhere in the series $\sum u_n$, the terms $a_1, \cdots, a_m$ all occur in the sum $| u_1 | + \cdots + | u_n |$ if n is sufficiently large. But then we see that

$$a_1 + a_2 + \cdots + a_m \leq M.$$

It follows by Theorem I (§ 19.2) that the series $\sum a_n$ is convergent. In the same way we see that $b_1 + \cdots + b_m \leqq M$, since each b_i is a term somewhere in the series $\sum |u_n|$. Thus the series $\sum b_n$ is convergent.

Now suppose that, in the sum $u_1 + \cdots + u_n$, the number of positive terms is p_n and the number of negative terms is q_n. Then

(19.3–6) $u_1 + \cdots + u_n = (a_1 + \cdots + a_{p_n}) - (b_1 + \cdots + b_{q_n})$.

In the case of absolute convergence we let $n \to \infty$ and obtain the result

$$\sum_{n=1}^{\infty} u_n = \sum_{n=1}^{\infty} a_n - \sum_{n=1}^{\infty} b_n.$$

It may happen that there are only a finite number of a's or a finite number of b's, or possibly none of one kind or the other. In these cases the series is of course absolutely convergent if it is convergent at all, since its terms from some point onward are all of one sign. Let us then consider the case in which there are infinitely many terms of each sign, so that p_n and $q_n \to \infty$ as $n \to \infty$. Let

$$s_n = u_1 + \cdots + u_n, \; A_{p_n} = a_1 + \cdots + a_{p_n}, \; B_{q_n} = b_1 + \cdots + b_{q_n},$$

so that (19.3–6) becomes $s_n = A_{p_n} - B_{q_n}$. We also have

(19.3–7)
$$|u_1| + \cdots + |u_n| = (a_1 + \cdots + a_{p_n}) + (b_1 + \cdots + b_{q_n})$$
$$= A_{p_n} + B_{q_n}.$$

Now suppose that the series $\sum u_n$ is convergent, and consider the series $\sum a_n, \sum b_n$. If either of these latter series is convergent, so is the other, by virtue of the relation $s_n = A_{p_n} - B_{q_n}$. For instance, if $\sum a_n$ is convergent, B_{q_n} approaches a limit, since s_n and A_{p_n} each approach limits, and $B_{q_n} = A_{p_n} - s_n$. But to say that $\lim_{n \to \infty} B_{q_n}$ exists is equivalent to saying that $\sum b_n$ is convergent, since $\{B_{q_n}\}$ is a nondecreasing sequence, and is therefore convergent if and only if the partial sums of the series $\sum b_n$ are bounded. But if both the series $\sum a_n, \sum b_n$ are convergent, the series $\sum |u_n|$ is convergent, by (19.3–7). Thus we see that, if $\sum u_n$ is conditionally convergent, both of the series $\sum a_n, \sum b_n$ must be divergent. This completes the proof of Theorem X.

EXERCISES

1. If $\sum c_n$ is a convergent series of positive terms, the series $\sum c_n x^n$ is absolutely convergent when $|x| \leqq 1$. Prove this.

2. Show that, if $0 \leqq r < 1$, the series $\sum r^n \sin n\theta$ and $\sum r^n \cos n\theta$ are absolutely convergent for all values of θ.

3. Which of the following series are absolutely convergent?

(a) $1 - \dfrac{1}{3^2} + \dfrac{1}{5^2} - \dfrac{1}{7^2} + \cdots$.

(b) $1 - \dfrac{1}{2} + \dfrac{1}{3!} - \dfrac{1}{4} + \dfrac{1}{5!} - \cdots$.

(c) $\dfrac{1}{\log 2} - \dfrac{1}{2 \log 3} + \dfrac{1}{3 \log 4} - \cdots + (-1)^n \dfrac{1}{(n-1) \log n} + \cdots$.

(d) $\displaystyle\sum_{n=1}^{\infty} \dfrac{\sin n\theta}{n^2}$.

4. Consider the series $\sum u_n$, where $u_n = \dfrac{1}{n} \log \dfrac{(n+1)^2}{n(n+2)}$. Show that, if

$1 \le m < n$, $u_{m+1} + \cdots + u_n < \dfrac{1}{m+1} \log \dfrac{(m+2)(n+1)}{(m+1)(n+2)} < \dfrac{1}{m+1} \log 2$.

Since $u_n > 0$, this shows that condition (19.3–5) is satisfied when N is chosen so that $\dfrac{\log 2}{N+1} \le \epsilon$. Therefore, by Theorem VIII, the series is convergent. Show by a similar argument that $\sum a_n \log b_n$ is convergent if the following conditions are satisfied: (a) $a_n \ge a_{n+1}$ and $a_n \to 0$ as $n \to \infty$; (b) $b_n > 1$; (c) there is a constant M such that $b_1 \cdot b_2 \cdots b_n \le M$ for all n.

19.31   Rearrangement of terms

Suppose that we have two convergent series with values s and t respectively:

$$s = u_1 + u_2 + \cdots + u_n + \cdots,$$
$$t = v_1 + v_2 + \cdots + v_n + \cdots.$$

We may combine these series by adding corresponding terms, and the result will be a new convergent series whose value is $s + t$:

$$s + t = w_1 + w_2 + \cdots + w_n + \cdots,$$

where $w_1 = u_1 + v_1$, $w_2 = u_2 + v_2$, $\cdots$, $w_n = u_n + v_n$, $\cdots$. This assertion is easily proved as a direct application of the rule that the limit of a sum is the sum of the limits, for

$$(w_1 + \cdots + w_n) = (u_1 + \cdots + u_n) + (v_1 + \cdots + v_n).$$

Now let us consider the question: Suppose the series $\sum u_n$ is convergent; what can we say about a series which is obtained by using the same terms u_n, but *in a different order*? It may surprise the student to learn that (a) if the new series is convergent, it does not necessarily have the same sum as the original series, and (b) the new series may be divergent.

Example 1. By rearranging the terms in the series (19.3–1), whose sum is log 2, we can obtain the following result:

(19.31–1)
$$\tfrac{3}{2}\log 2 = 1 + \tfrac{1}{3} - \tfrac{1}{2} + \tfrac{1}{5} + \tfrac{1}{7} - \tfrac{1}{4} + \tfrac{1}{9} + \tfrac{1}{11} - \tfrac{1}{6} + \cdots.$$

The scheme in the rearrangement is to take two positive terms and one negative term, then two more positive terms and another negative term, and so on. To prove (19.31–1) we note that

(19.31–2)
$$\log 2 = 1 - \tfrac{1}{2} + \tfrac{1}{3} - \tfrac{1}{4} + \tfrac{1}{5} - \tfrac{1}{6} + \tfrac{1}{7} - \cdots,$$
$$\tfrac{1}{2}\log 2 = \tfrac{1}{2} - \tfrac{1}{4} + \tfrac{1}{6} - \tfrac{1}{8} + \tfrac{1}{10} - \tfrac{1}{12} + \tfrac{1}{14} - \cdots.$$

In this last series we may insert zero terms without affecting the value. Thus

(19.31–3)
$$\tfrac{1}{2}\log 2 = 0 + \tfrac{1}{2} + 0 - \tfrac{1}{4} + 0 + \tfrac{1}{6} + 0 - \cdots.$$

On adding (19.31–2) and (19.31–3) term by term we get

$$\tfrac{3}{2}\log 2 = 1 + 0 + \tfrac{1}{3} - \tfrac{1}{2} + \tfrac{1}{5} + 0 + \tfrac{1}{7} - \tfrac{1}{4} + \cdots.$$

This is the same as (19.31–1) when we drop out the zero terms.

Example 2. In this example we shall not be so explicit. The series (19.31–2) is conditionally convergent, so that the series composed of its positive terms is divergent. This means that by taking k large enough we can make

$$1 + \frac{1}{3} + \frac{1}{5} + \cdots + \frac{1}{2k-1}$$

as large as we please. We now rearrange the series (19.31–2) according to the following plan: First take just enough of the positive terms to obtain a sum greater than 2. Then take the first negative term, $-\tfrac{1}{2}$. This leaves us with more than $\tfrac{3}{2}$. Now take just enough more positive terms to increase the total sum beyond 4, and then take one more negative term, $-\tfrac{1}{4}$. Continuing in this way, we build up partial sums such that when the term $-1/(2n)$ is taken, the total exceeds $2n - (1/n)$. The series so formed must diverge.

What we have just seen in Examples 1 and 2 is typical of conditionally convergent series. In fact, it may be proved that the terms in a conditionally convergent series may be rearranged so as to produce a series which has any desired sum, or such that the partial sums tend to $+\infty$ or to $-\infty$. By contrast, no such thing can happen with absolutely convergent series.

THEOREM XI.　*Let the series $\sum u_n$ be absolutely convergent, with sum s. Let $\sum v_n$ be any series obtained by a rearrangement of the terms of $\sum u_n$ (i.e. every v_i is some u_j and every u_k is some v_l). Then $\sum v_n$ is convergent, with sum s.*

PROOF. First let us prove the assertion on the assumption that all the u's (and hence all the v's) are nonnegative. Since $s = \sum u_n$ and since each v_i is some u_j, it is clear that the partial sums of the series $\sum v_n$ cannot exceed s. Thus the series $\sum v_n$ must be convergent and its sum s' must satisfy the inequality $s' \leqq s$. Reversing the role of $\sum u_n$ and $\sum v_n$, we see that $s \leqq s'$. Therefore $s' = s$. This completes the proof for the case of series of nonnegative terms. Now, in the general case of an absolutely convergent series, we have

$$\sum u_n = \sum a_n - \sum b_n$$

in the notation of Theorem X (§ 19.3). In the rearranged series $\sum v_n$, the separation into positive and negative terms yields

$$\sum v_n = \sum a_n' - \sum b_n',$$

where $\sum a_n'$ is a rearrangement of $\sum a_n$ and $\sum b_n'$ is a rearrangement of $\sum b_n$. By what has just been proved for series of positive terms, we have

$$\sum a_n' = \sum a_n, \quad \sum b_n' = \sum b_n.$$

Hence the series $\sum v_n$ is convergent, with the same sum as $\sum u_n$.

EXERCISES

1. Show that $\log 2 = 1 + \frac{1}{3} + \frac{1}{5} - \frac{1}{2} - \frac{1}{4} - \frac{1}{6} + + + - - - \cdots$.

2. Show that $\log 2 - 1 = \frac{1}{3} - \frac{1}{2} + \frac{1}{5} - \frac{1}{4} + \frac{1}{7} - \frac{1}{6} + \cdots$.
Here the positive terms have odd denominators, and the negative terms have even integers as denominators. The terms alternate in sign.

3. Show that $\frac{1}{2} \log 2 = 1 - \frac{1}{2} - \frac{1}{4} + \frac{1}{3} - \frac{1}{6} - \frac{1}{8} + \frac{1}{5} - - + \cdots$.

4. Let $\sum u_n$ be a convergent series, and let $\sum v_n$ be a rearrangement of it. In the rearrangement, suppose that no term of the original series is moved more than N places from its original position, where N is a fixed number. Show that the new series is convergent and has the same value as the old one.

19.32 Alternating series

The simplest type of series having both positive and negative terms is the type in which the successive terms alternate in sign. Many commonly occurring series are of this type, and have the additional property that the magnitude of the terms steadily decreases toward zero as n increases. Concerning such series we have a theorem of practical importance as a test for convergence.

THEOREM XII. *Suppose the terms of the series $\sum u_n$ are alternately positive and negative, that $|u_{n+1}| \leqq |u_n|$ for all n, and that $u_n \to 0$ as $n \to \infty$. Then the series is convergent.*

PROOF. For convenience let us write the series in the form

$$c_1 - c_2 + c_3 - c_4 + \cdots + c_{2n-1} - c_{2n} + \cdots,$$

where $c_n > 0$. There are two kinds of partial sums, depending on whether the sum ends with a positive or a negative term:

$$s_{2n-1} = c_1 - c_2 + \cdots + c_{2n-1}$$
$$s_{2n} = c_1 - c_2 + \cdots - c_{2n}.$$

If we consider the sequences $\{s_{2n-1}\}$ and $\{s_{2n}\}$ separately, we observe that

$$s_1 \geqq s_3 \geqq s_5 \geqq \cdots$$
$$s_2 \leqq s_4 \leqq s_6 \leqq \cdots.$$

These inequalities depend on the fact that $c_{n+1} \leqq c_n$. For example,

$$s_7 = s_5 - c_6 + c_7 = s_5 - (c_6 - c_7) \leqq s_5$$

since $c_6 - c_7 \geqq 0$, and

$$s_8 = s_6 + c_7 - c_8 \geqq s_6$$

since $c_7 - c_8 \geqq 0$. Moreover, $c_1 - c_2 \leqq s_n \leqq c_1$ for all values of n (see Fig. 178). Therefore each of the sequences $\{s_{2n-1}\}$, $\{s_{2n}\}$ is convergent,

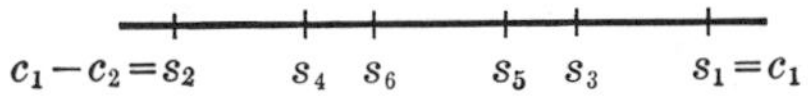

$$c_1 - c_2 = s_2 \qquad s_4 \ \ s_6 \qquad s_5 \ \ s_3 \qquad s_1 = c_1$$

Fig. 178

being monotonic and bounded. But $s_{2n-1} - s_{2n} = c_{2n}$, and since $c_{2n} \to 0$, it follows that both sequences have the same limit. But then the sequence $\{s_n\}$, where n runs through *all* positive integers, is convergent, its limit being the common limit of $s_1, s_3, s_5, \cdots$ and $s_2, s_4, s_6, \cdots$. The theorem is therefore proved.

It is worth while noting that in a series of the kind just described, the partial sum $u_1 + \cdots + u_n$ differs from the sum of the series by not more than $|u_{n+1}|$, i.e. by not more than the first term not included in the partial sum.

Example 1. The series

$$\frac{1}{\sqrt{1}} - \frac{1}{\sqrt{2}} + \frac{1}{\sqrt{3}} - \cdots + (-1)^{n-1} \frac{1}{\sqrt{n}} + \cdots$$

is convergent, since the conditions of Theorem XII are satisfied.

Example 2. Compute e^{-1} with an error not exceeding 0.005 by using (19.3–2):

$$e^{-1} = 1 - 1 + \frac{1}{2!} - \frac{1}{3!} + \frac{1}{4!} - \cdots.$$

It will be sufficient to stop with the term involving $\dfrac{1}{n!}$ if n is chosen so that

$\dfrac{1}{(n + 1)!} < 0.005$. Thus we use the approximation

$$e^{-1} = 1 - 1 + \frac{1}{2!} - \frac{1}{3!} + \frac{1}{4!} - \frac{1}{5!},$$

the next term in the series being

$$\frac{1}{6!} = \frac{1}{720} = 0.0014 \text{ (approximately)}.$$

This gives $e^{-1} = \dfrac{44}{120} = \dfrac{11}{30} = 0.3666 \cdots,$

with an error less than 0.0014. Our work actually shows that

$$0.3666 < e^{-1} < 0.3681.$$

EXERCISES

1. Show by Theorem XII that each of the following series is convergent. Prove carefully that each condition of the theorem is fulfilled.

(a) $\dfrac{1}{2^2} - \dfrac{2}{3^2} + \dfrac{3}{4^2} - \cdots + (-1)^{n+1} \dfrac{n}{(n + 1)^2} + \cdots.$

(b) $\log \frac{3}{2} - \log \frac{4}{3} + \log \frac{5}{4} - \cdots.$

(c) $\dfrac{1}{1 + 1} - \dfrac{\sqrt{2}}{1 + 2} + \dfrac{\sqrt{3}}{1 + 3} - \cdots.$

(d) $\dfrac{1}{2 \cdot 4} - \dfrac{1 \cdot 3}{2 \cdot 4 \cdot 6} + \cdots + (-1)^{n+1} \dfrac{1 \cdot 3 \cdots (2n - 1)}{2 \cdot 4 \cdots 2n \cdot (2n + 2)} + \cdots.$

(e) $\displaystyle\sum_{n=1}^{\infty} (-1)^{n+1} \dfrac{3 \cdot 6 \cdots 3n}{1 \cdot 4 \cdots (3n - 2)} \cdot \dfrac{1}{n^2}.$

2. Test the following series for convergence or divergence.

(a) $\dfrac{\log 2}{\sqrt{2}} - \dfrac{\log 3}{\sqrt{3}} + \dfrac{\log 4}{\sqrt{4}} - \cdots.$

(b) $\dfrac{1}{2} - \left(\dfrac{2}{3}\right)^2 + \left(\dfrac{3}{4}\right)^3 - \cdots + (-1)^{n+1} \left(\dfrac{n}{n + 1}\right)^n + \cdots.$

(c) $(1 - \log 2) - (1 - 2 \log \frac{3}{2}) + (1 - 3 \log \frac{4}{3}) - \cdots.$

(d) $\frac{1}{3} - \frac{2}{3} + \frac{1}{6} - \frac{2}{6} + \frac{1}{9} - \frac{2}{9} + \cdots.$

3. Show that the conditions of Theorem XII are fulfilled by the series

$\displaystyle\sum_{n=N}^{\infty} (-1)^n \dfrac{(\log n)^p}{n^q}$ for any fixed positive q and any p, provided the initial value $n = N$ is large enough (how large will depend on p and q).

4. The series

$$1 - \frac{1}{\sqrt{2}} + \frac{1}{\sqrt{3}} - \frac{1}{\sqrt{4}} + \frac{1}{\sqrt{5}} - \cdots$$

is convergent, by Theorem XII. Why is it conditionally convergent? Show that the rearrangement

$$1 + \frac{1}{\sqrt{3}} - \frac{1}{\sqrt{2}} + \frac{1}{\sqrt{5}} + \frac{1}{\sqrt{7}} - \frac{1}{\sqrt{4}} + + - \cdots$$

diverges to $+\infty$. SUGGESTION. Compare the sum S_{3n} of the first 3 n terms of the rearranged series with the sum s_{2n} of the first 2 n terms of the original series, and show that

$$S_{3n} - s_{2n} > \frac{n}{\sqrt{4n-1}}.$$

Since s_{2n} approaches a limit as $n \to \infty$, it follows that $S_{3n} \to +\infty$. How does one then show that $S_n \to +\infty$?

19.4 Tests for absolute convergence

In a very large number of important cases the most convenient test for absolute convergence is the following one, which is known as d'Alembert's ratio test (after J. le R. d'Alembert, 1717?–1783).

THEOREM XIII. *Let $\sum u_n$ be a series with all its terms different from zero. Then the series is absolutely convergent if there is a positive number $r < 1$ such that $\left| \frac{u_{n+1}}{u_n} \right| < r$ for all sufficiently large values of n. In particular, this condition is satisfied if the limit*

$$t = \lim_{n \to \infty} \left| \frac{u_{n+1}}{u_n} \right|$$

exists and $t < 1$. But if $t > 1$ the series is divergent.

PROOF. The assertions regarding absolute convergence are direct applications of Theorems VI and VII (§ 19.22) to the series $\sum |u_n|$. If $t > 1$ the general term u_n cannot approach zero, and the series $\sum u_n$ cannot be convergent.

The theorem makes no assertion about what happens if $t = 1$. The series may then converge absolutely, or conditionally, or it may diverge.

Theorem XIII is very useful in dealing with power series, as the following example shows.

Example 1. The series

$$(19.4\text{–}1) \qquad 1 - \tfrac{1}{2} x + \frac{1 \cdot 3}{2 \cdot 4} x^2 - \frac{1 \cdot 3 \cdot 5}{2 \cdot 4 \cdot 6} x^3 + \cdots$$

$$+ (-1)^n \frac{1 \cdot 3 \cdots (2n-1)}{2 \cdot 4 \cdots 2n} x^n + \cdots$$

is absolutely convergent if $|x| < 1$, and divergent if $|x| > 1$.

Here we have (calling the first term u_0)

$$u_n = (-1)^n \frac{1 \cdot 3 \cdots (2n-1)}{2 \cdot 4 \cdots 2n} x^n,$$

$$u_{n+1} = (-1)^{n+1} \frac{1 \cdot 3 \cdots (2n+1)}{2 \cdot 4 \cdots (2n+2)} x^{n+1},$$

$$\left| \frac{u_{n+1}}{u_n} \right| = \frac{2n+1}{2n+2} |x|, \qquad \lim_{n \to \infty} \left| \frac{u_{n+1}}{u_n} \right| = |x|.$$

The assertions about (19.4–1) are now seen to follow by application of Theorem XIII (with $t = |x|$). For the present we avoid a discussion of the behavior of the series (19.4–1) when $|x| = 1$ (but see Example 3).

Example 2. If m is any number except 0 or a positive integer, the series

(19.4–2)

$$1 + \frac{m}{1!} x + \frac{m(m-1)}{2!} x^2 + \cdots + \frac{m(m-1) \cdots (m-n+1)}{n!} x^n + \cdots$$

is absolutely convergent if $|x| < 1$ and divergent if $|x| > 1$.

The ratio $\frac{u_{n+1}}{u_n}$ is readily found to be

$$\frac{m(m-1) \cdots (m-n)}{m(m-1) \cdots (m-n+1)} \frac{n!}{(n+1)!} \frac{x^{n+1}}{x^n} = \frac{m-n}{n+1} x,$$

so that

$$\lim_{n \to \infty} \left| \frac{u_{n+1}}{u_n} \right| = \lim_{n \to \infty} \left| \frac{m-n}{n+1} \right| |x| = |x|.$$

Application of Theorem XIII gives the result as asserted.

For cases in which $\lim_{n \to \infty} \left| \frac{u_{n+1}}{u_n} \right| = 1$ there is a convenient test which is sometimes effective. It was first established by L. J. Raabe in 1832, and is known as Raabe's test. The essential idea of the test is to use Theorem V (§ 19.22), taking one of the series in that theorem to be the series

$$\frac{1}{1^p} + \frac{1}{2^p} + \cdots + \frac{1}{n^p} + \cdots,$$

which is known to converge if $p > 1$ and diverge if $p \leq 1$. Let $a_n = n^{-p}$. Then

$$\frac{a_{n+1}}{a_n} = \frac{n^p}{(n+1)^p} = \left(\frac{n+1}{n} \right)^{-p} = \left(1 + \frac{1}{n} \right)^{-p}.$$

Hence, by Theorem V, we have the following criterion. *If $p > 1$ and if*

$$(19.4\text{-}3) \qquad\qquad \left| \frac{u_{n+1}}{u_n} \right| \leqq \left(1 + \frac{1}{n} \right)^{-p}$$

for all sufficiently large values of n, the series $\sum u_n$ *is absolutely convergent.*
This criterion in itself is not very useful, however, for it is not easy in
practice to tell whether the inequality (19.4-3) is satisfied. To improve
matters let us proceed as follows: We set $f(x) = (1 + x)^{-p}$ and expand
$f(x)$ in powers of x by Taylor's formula with remainder. We use (4.3-7)
with $a = 0$, $h = x$, $n = 1$. The result is

$$(1 + x)^{-p} = 1 - px + \frac{x^2}{2} \frac{p(p + 1)}{(1 + \theta x)^{p+2}}, \quad 0 < \theta < 1.$$

This is valid if $x > -1$. We put $x = 1/n$. Thus

$$\left(1 + \frac{1}{n} \right)^{-p} = 1 - \frac{p}{n} + \frac{A_n}{n^2}, \qquad A_n = \frac{p(p + 1)}{2\left(1 + \dfrac{\theta}{n} \right)^{p+2}}.$$

The exact form of A_n is unimportant. The only essential thing is that
A_n is bounded as $n \to \infty$. Now (19.4-3) can be put in the form

$$\left| \frac{u_{n+1}}{u_n} \right| \leqq 1 - \frac{p}{n} + \frac{A_n}{n^2} \,;$$

we rewrite this as

$$(19.4\text{-}4) \qquad\qquad p - \frac{A_n}{n} \leqq n \left\{ 1 - \left| \frac{u_{n+1}}{u_n} \right| \right\}.$$

Now when $p > 1$ and n is very large, the expression $p - (A_n/n)$ is greater
than 1. The form of the inequality now suggests that we consider the limit

$$(19.4\text{-}5) \qquad\qquad t = \lim_{n \to \infty} n \left\{ 1 - \left| \frac{u_{n+1}}{u_n} \right| \right\},$$

provided this limit exists.

THEOREM XIV. *(Raabe's test.) Suppose that the limit* (19.4-5) *exists
(either as a finite limit or as* $+ \infty$ *or* $- \infty$*). Then the series* $\sum u_n$ *is absolutely convergent if* $t > 1$, *but not if* $t < 1$.

PROOF. Suppose $t > 1$, and choose p so that $1 < p < t$ (here t may
be $+ \infty$). Then certainly we shall have

$$p < n \left\{ 1 - \left| \frac{u_{n+1}}{u_n} \right| \right\}$$

for all sufficiently large values of n. Consequently the inequality (19.4-4)
must hold, since $A_n > 0$. This is equivalent to (19.4-3), and therefore the

series $\sum u_n$ is absolutely convergent. To show that the series does not converge absolutely if $t < 1$ it suffices to show that if we take p such that $t < p < 1$, then

$$\left(1 + \frac{1}{n}\right)^{-p} \leqq \left|\frac{u_{n+1}}{u_n}\right|$$

for all sufficiently large values of n. For, by Theorem V, the convergence of $\sum |u_n|$ would imply the convergence of $\sum n^{-p}$, and the latter series is divergent. We omit the details of the demonstration, which are very similar to the details of the first part of the proof, with all the inequalities reversed.

Raabe's test is indecisive if $t = 1$ (see Exercise 9).

Example 3.   Consider the series of Example 1 when $x = \pm 1$. In that case

$$\left|\frac{u_{n+1}}{u_n}\right| = \frac{2n+1}{2n+2}, \qquad n\left\{1 - \left|\frac{u_{n+1}}{u_n}\right|\right\} = \frac{n}{2n+2},$$

so that the limit t in (19.4–5) is $\frac{1}{2}$. Therefore, by Theorem XIV, the series (19.4–1) is not absolutely convergent when $|x| = 1$. This shows that it actually diverges when $x = -1$, for the terms are all positive in that case. When $x = 1$, however, the terms of the series are alternating in sign, and we can show that the series is conditionally convergent. The series is

$$1 - \frac{1}{2} + \frac{1 \cdot 3}{2 \cdot 4} - \frac{1 \cdot 3 \cdot 5}{2 \cdot 4 \cdot 6} + \cdots + (-1)^n \frac{1 \cdot 3 \cdots (2n-1)}{2 \cdot 4 \cdots 2n} + \cdots.$$

It is clear that the terms steadily decrease in magnitude as $n \to \infty$, so that by Theorem XII all we need to show is that

$$(19.4\text{–}6) \qquad \lim_{n \to \infty} \frac{1 \cdot 3 \cdots (2n-1)}{2 \cdot 4 \cdots 2n} = 0.$$

Let

$$c_n = \frac{1 \cdot 3 \cdots (2n-1)}{2 \cdot 4 \cdots 2n}.$$

Then

$$c_n < \frac{2 \cdot 4 \cdots 2n}{3 \cdot 5 \cdots (2n+1)} = \frac{1}{c_n}\frac{1}{2n+1},$$

and therefore

$$c_n^2 < \frac{1}{2n+1}, \qquad c_n < \frac{1}{\sqrt{2n+1}};$$

the truth of (19.4–6) is now evident.

There is a somewhat more powerful test than Raabe's test. It is essentially a test for series of positive terms, and therefore may be used to test for absolute convergence by applying the test to $\sum |u_n|$. The test reads as follows:

THEOREM XV.   *Suppose that* $\left|\dfrac{u_{n+1}}{u_n}\right|$ *can be expressed in the form*

(19.4–7)
$$\left|\frac{u_{n+1}}{u_n}\right| = 1 - \frac{p}{n} + \frac{A_n}{n^q},$$

where $q > 1$ and the sequence $\{A_n\}$ is bounded. Then the series $\sum u_n$ is absolutely convergent if $p > 1$, and not absolutely convergent (either divergent or conditionally convergent) if $p \leq 1$.

The test in Theorem XV is due to Karl Friedrich Gauss (1777–1855), one of the greatest mathematicians in history. We leave the proof to the student, with certain guiding suggestions; see Exercise 10.

Example 4.   Consider the series (19.4–2) when $|x| = 1$. As before, assume that m is not zero or a positive integer. Then

$$\left|\frac{u_{n+1}}{u_n}\right| = \left|\frac{m-n}{n+1}\right|.$$

For sufficiently large values of n we shall have $|m - n| = n - m$, and hence

$$\left|\frac{u_{n+1}}{u_n}\right| = \frac{n-m}{n+1} = \frac{1 - \dfrac{m}{n}}{1 + \dfrac{1}{n}} = 1 - \frac{m+1}{n} + \frac{m+1}{n^2\left(1 + \dfrac{1}{n}\right)},$$

so that (19.4–7) holds with $p = m + 1$, $q = 2$, $A_n = \dfrac{m+1}{1 + (1/n)}$. It follows that the series (19.4–2) is absolutely convergent when $|x| = 1$ if and only if $m + 1 > 1$, i.e. if and only if $m > 0$.

There is a test due to Cauchy which, although not as easy to apply in many cases as d'Alembert's ratio test (Theorem XIII), is more powerful (i.e. has wider theoretical applicability). It has important applications in the theory of power series.

THEOREM XVI.   (*Cauchy's root test.*) *The series $\sum u_n$ is absolutely convergent if there is a positive number $r < 1$ such that $|u_n|^{1/n} \leq r$ for all sufficiently large values of n. This condition will be satisfied if $\lim_{n \to \infty} |u_n|^{1/n}$ exists and is less than unity. But, if $|u_n|^{1/n} \geq 1$ for an infinite number of values of n, the series $\sum u_n$ is divergent.*

PROOF.   In the first case we have $|u_n| \leq r^n$, and the absolute convergence follows by comparison with the convergent geometric series $\sum r^n$. In the second case u_n cannot approach zero as $n \to \infty$ (since $|u_n| \geq 1$ for infinitely many values of n). Hence the series is divergent.

Example 5.   The series

$$\sum_{n=0}^{\infty}\left(1 + \sin\frac{n\pi}{2}\right)^n \frac{x^n}{2^n}$$

is absolutely convergent if $|x| < 1$, and divergent if $|x| \geqq 1$. For,

$$|u_n|^{1/n} = \left(1 + \sin\frac{n\pi}{2}\right)\frac{|x|}{2} \leqq |x|$$

for all n, and $\quad |u_n|^{1/n} = |x|$

if $n = 1, 5, 9, 13, \cdots$. The statements about the series now follow immediately from Theorem XVI.

Cauchy's test is indecisive if $|u_n|^{1/n} \to 1$, as we see by applying it to the series $\sum n^{-p}$ with $p = 1$ and 2, successively.

EXERCISES

1. Apply d'Alembert's ratio test to each series and state explicitly the conclusions which you draw from use of the test. In parts (a)–(e) make a separate investigation and decide whether the series converges or diverges for each of the values of x for which the ratio test is indecisive.

(a) $\displaystyle\sum_{n=1}^{\infty} \frac{x^n}{n^2}$.

(b) $\displaystyle\sum_{n=1}^{\infty} \frac{n+1}{n^2(n+2)}\, x^n$.

(c) $\displaystyle\sum_{n=1}^{\infty} \frac{n!}{x^n}$.

(d) $\displaystyle\sum_{n=1}^{\infty} \frac{3 \cdot 5 \cdots (2n+1)}{5 \cdot 10 \cdots 5n}\, x^n$.

(e) $\displaystyle\sum_{n=1}^{\infty} \frac{n!}{n^n}\, x^n$.

(f) $\displaystyle\sum_{n=1}^{\infty} \frac{n!}{3 \cdot 5 \cdots (2n+1)}\, x^n$.

(g) $\displaystyle\sum_{n=1}^{\infty} \frac{n^p}{2^{n-1}}\, x^n$.

(h) $\displaystyle\sum_{n=1}^{\infty} \frac{(2n)^n}{(n+1)^{n+1}}\, x^n$.

2. Apply Cauchy's root test to each series and state explicitly the conclusions which you draw from use of the test.

(a) $\displaystyle\sum_{n=1}^{\infty} \left(\frac{x}{n}\right)^n$.

(b) $\displaystyle\sum_{n=1}^{\infty} n^p x^n, \, p > 0$.

(c) $\displaystyle\sum_{n=2}^{\infty} \frac{1}{(\log n)^n}$.

(d) $\displaystyle\sum_{n=2}^{\infty} \frac{1}{(\log n)^{\log n}}$.

(e) $\frac{1}{3} + \frac{1}{2} + (\frac{1}{3})^2 + (\frac{1}{2})^2 + (\frac{1}{3})^3 + (\frac{1}{2})^3 + \cdots$.

(f) $\frac{1}{2}x + \frac{1}{2} \cdot \frac{4}{3}x^2 + (\frac{1}{2})^2 \cdot \frac{4}{3}x^3 + (\frac{1}{2})^2(\frac{4}{3})^2 x^4 + (\frac{1}{2})^3(\frac{4}{3})^2 x^5 + (\frac{1}{2})^3(\frac{4}{3})^3 x^6 + \cdots$.

(g) $\displaystyle\sum_{n=1}^{\infty} \frac{[1 + (1/n)]^{2n}}{e^n}$.

(h) $\displaystyle\sum_{n=0}^{\infty} \left(3 + 2\cos\frac{n\pi}{2}\right)^n \left(\frac{x}{2}\right)^n$.

3. Apply Raabe's test to each series and state explicitly the conclusions which you draw from use of the test. In (e) and (f) use the formula for $(1+x)^{-p}$ developed for the proof of Raabe's test.

(a) $\displaystyle\sum_{n=1}^{\infty} \frac{2 \cdot 4 \cdots 2n}{3 \cdot 5 \cdots (2n+1)}$.

(b) $\displaystyle\sum_{n=1}^{\infty} \frac{2 \cdot 4 \cdots 2n}{5 \cdot 7 \cdots (2n+3)}$.

(c) $\displaystyle\sum_{n=1}^{\infty} \frac{5 \cdot 6 \cdots (4+n)}{8 \cdot 9 \cdots (7+n)}\, (n+1)$.

(d) $\displaystyle\sum_{n=1}^{\infty} \frac{4 \cdot 5 \cdots (3+n)}{8 \cdot 9 \cdots (7+n)} \cdot \frac{5 \cdot 6 \cdots (4+n)}{n!}$.

(e) $\displaystyle\sum_{n=1}^{\infty}\left[\frac{1\cdot 3\cdots(2n-1)}{2\cdot 4\cdots 2n}\right]^{p}.$ (f) $\displaystyle\sum_{n=1}^{\infty}\left[\frac{2\cdot 4\cdots 2n}{5\cdot 7\cdots(2n+3)}\right]^{p}.$

(g) $\displaystyle\sum_{n=1}^{\infty}\frac{1\cdot 3\cdots(2n-1)}{2\cdot 4\cdots(2n)}\cdot\frac{3\cdot 7\cdots(4n-1)}{5\cdot 9\cdots(4n+1)}.$

4. Test each series in Exercise 3 by Gauss's test. Observe that Gauss's test is more effective than Raabe's test in parts (e), (f), (g).

5. Let $0 < a < b < 1$. Show by Cauchy's root test that the series

$$a + b + a^2 + b^2 + a^3 + b^3 + \cdots$$

is convergent. Attempt to test the series by d'Alembert's ratio test and state clearly the outcome of your attempt.

6. Consider the series

$$a + ab + a^2b + a^2b^2 + a^3b^2 + \cdots,$$

in which each term is derived from its predecessor by multiplying alternately by b and by a. Show that, if $0 < a < 1$ and $0 < b < 1$, the series may be proved convergent by d'Alembert's ratio test. Assuming a and b to be positive, prove that the series is convergent if $ab < 1$ and divergent if $ab \geqq 1$.

7. Assume that none of the numbers a, b, c is a negative integer or zero. Prove that the series

$$\frac{ab}{c} + \frac{a(a+1)b(b+1)}{2!\,c(c+1)} + \frac{a(a+1)(a+2)b(b+1)(b+2)}{3!\,c(c+1)(c+2)} + \cdots$$

is absolutely convergent if $c > a + b$ and divergent if $c \leqq a + b$. Why is conditional convergence not possible?

8. (a) Suppose $a_n > 0$ and $t = \displaystyle\lim_{n \to \infty} \frac{\log(1/a_n)}{\log n}$. Show that the series $\sum a_n$ is convergent if $t > 1$ and divergent if $t < 1$. SUGGESTION: Show that $t > 1$ implies $a_n < n^{-p}$ if $1 < p < t$ and n is sufficiently large.

(b) Apply the test of part **(a)** to each of the series

$$\sum_{n=1}^{\infty}\left(1 - \frac{1}{\sqrt{n}}\right)^{n}, \quad \sum_{n=2}^{\infty}\frac{1}{(\log n)^{\log n}}, \quad \sum_{n=2}^{\infty}\frac{1}{(\log n)^{\log(\log n)}}.$$

9. For each of the series

$$\sum_{n=1}^{\infty}\left[\frac{1\cdot 3\cdots(2n-1)}{2\cdot 4\cdots 2n}\right]^{2}, \quad \sum_{n=2}^{\infty}\frac{1}{n(\log n)^{2}}$$

show that $t = 1$ in Raabe's test. Show by other methods that the first series is divergent and the second is convergent. This shows that Raabe's test is indecisive when $t = 1$.

10. Prove Theorem XV with the aid of the following suggestions. If $p \neq 1$, the conclusions are easily drawn with the aid of Raabe's test. Write out the argument explicitly. For the case $p = 1$, let

$$v_n = \frac{1}{(n-1)\log(n-1)}\,;$$

use Taylor's formula with remainder to show that

$$\log\left(1 - \frac{1}{n}\right) = -\frac{1}{n} - \frac{A_n}{n^2},$$

where A_n is bounded as $n \to \infty$. Then show that

$$\frac{v_{n+1}}{v_n} = 1 - \frac{1}{n} - \frac{1}{n \log n} - \frac{B_n}{n^2},$$

where B_n is bounded as $n \to \infty$. From this show that

$$\left| \frac{u_{n+1}}{u_n} \right| - \frac{v_{n+1}}{v_n}$$

is positive when n is sufficiently large. Now complete the proof of Theorem XV.

19.5   The binomial series

Let us consider the Taylor's series expansion in powers of x of the function $f(x) = (1 + x)^m$, where m is any real number. We have

$$f(x) = (1 + x)^m$$
$$f'(x) = m(1 + x)^{m-1}$$
$$f''(x) = m(m - 1)(1 + x)^{m-2}$$
$$\vdots$$
$$f^{(k)}(x) = m(m - 1) \cdots (m - k + 1)(1 + x)^{m-k}$$
$$\vdots$$

Thus, $f(0) = 1$, and for $k = 1, 2, \cdots$

$$f^{(k)}(0) = m(m - 1)(m - 2) \cdots (m - k + 1).$$

Taylor's formula with remainder is

$$(1 + x)^m = 1 + mx + \frac{m(m - 1)}{2!} x^2 + \cdots$$
$$+ \frac{m(m - 1) \cdots (m - n + 1)}{n!} x^n + R_{n+1}.$$

We shall use Cauchy's formula for R_{n+1} (Theorem IV, § 4.3). From (4.3–14) with $a = 0$ and $h = x$ we have

$$R_{n+1} = \frac{f^{(n+1)}(\theta x)}{n!} x^{n+1}(1 - \theta)^n, \qquad 0 < \theta < 1.$$

If m happens to be a positive integer or zero we see that $f^{(k)}(x) \equiv 0$ when $k > m$. Hence in this case $R_{m+1} \equiv 0$ and we have

(19.5–1)

$$(1 + x)^m = 1 + mx + \frac{m(m - 1)}{2!} x^2 + \frac{m(m - 1)(m - 2)}{3!} x^3$$
$$+ \cdots + x^m.$$

This is just the ordinary binomial-expansion formula.

In the rest of this section we shall assume that m is not a positive integer or zero. The formula for R_{n+1} becomes

$$R_{n+1} = \frac{m(m-1) \cdots (m-n)(1+\theta x)^{m-n-1}}{n!} x^{n+1}(1-\theta)^n ,$$

(19.5-2)

$$R_{n+1} = \frac{m(m-1) \cdots (m-n)}{n!} \left(\frac{1-\theta}{1+\theta x}\right)^n (1+\theta x)^{m-1} x^{n+1} .$$

Now suppose that $-1 < x < 1$. Then

$$0 < \frac{1-\theta}{1+\theta x} < 1 .$$

If $m > 1$ we have $0 < (1+\theta x)^{m-1} < (1+|x|)^{m-1}$, while if $m < 1$ we have

$$(1+\theta x)^{m-1} = \frac{1}{(1+\theta x)^{1-m}} < \frac{1}{(1-|x|)^{1-m}} ,$$

so that $0 < (1+\theta x)^{m-1} < (1-|x|)^{m-1}$. Thus, if $-1 < x < 1$,

(19.5-3) $\qquad |R_{n+1}| \leqq \dfrac{m(m-1) \cdots (m-n)}{n!} (1 \pm |x|)^{m-1} |x|^{n+1} ,$

the choice of the double sign in $(1 \pm |x|)^{m-1}$ depending on the sign of $m - 1$. From (19.5-3) we shall show that $R_{n+1} \to 0$ as $n \to \infty$ if $|x| < 1$. Since $(1 \pm |x|)^{m-1}$ is independent of n, it is sufficient to show that

(19.5-4) $\qquad \lim\limits_{n \to \infty} \dfrac{m(m-1) \cdots (m-n)}{n!} x^{n+1} = 0 .$

Now let $\qquad u_n = \dfrac{m(m-1) \cdots (m-n)}{n!} x^{n+1} ,$

and consider the series $\sum u_n$. This series is convergent if $|x| < 1$, as we see by applying the ratio test of Theorem XIII, for

$$\left|\frac{u_{n+1}}{u_n}\right| = \left|\frac{m-n+1}{n+1}\right| |x| \to |x| \quad \text{as} \quad n \to \infty .$$

Since the series converges, it follows that $u_n \to 0$ as $n \to \infty$; thus (19.5-4) is established.

We have thus proved the validity of the series expansion

(19.5-5) $\quad (1+x)^m = 1 + mx + \dfrac{m(m-1)}{2!} x^2 + \cdots$

$$+ \frac{m(m-1) \cdots (m-n+1)}{n!} x^n + \cdots$$

when $|x| < 1$. This is called the *binomial series*. Except when m is a positive integer or zero it is a nonterminating series, none of the coefficients vanishing. For a discussion of the validity of (19.5–5) when $x = \pm 1$, see the Exercises. The student will note that the series (19.5–5) is the same as the series (19.4–2), which was discussed in Examples 2 and 4 of § 19.4. Observe, however, that there is a logical difference between proving merely that the series converges and proving that its sum is equal to $(1 + x)^m$.

EXERCISES

The purpose of this set of exercises is to guide the student in completing the discussion of the binomial series when $x = \pm 1$. Certain results of a general nature are needed, and these are taken up first.

1. Suppose $u_n > 0$ and

$$\frac{u_{n+1}}{u_n} = 1 - \frac{p}{n} + \frac{A_n}{n^q},$$

where $p > 0$, $q > 1$, and A_n is bounded as $n \to \infty$. Show that, if $0 < r < p$, u_n satisfies an inequality of the form $u_n \leq Cn^{-r}$, where C is a constant, for all sufficiently large values of n. As a consequence, $u_n \to 0$ as $n \to \infty$. SUGGESTION: Let $v_n = n^{-r}$ and use the formula (developed in § 19.4)

$$\frac{v_{n+1}}{v_n} = 1 - \frac{r}{n} + \frac{B_n}{n^2},$$

where B_n is bounded as $n \to \infty$, to show that

$$\frac{v_{n+1}}{v_n} - \frac{u_{n+1}}{u_n}$$

has the same sign as $p - r$ when n is sufficiently large. Then, by an argument like that in the proof of Theorem V, draw the desired conclusion.

2. Let

$$u_n = \frac{a(a + 1) \cdots (a + n - 1)b(b + 1) \cdots (b + n - 1)}{n! \, c(c + 1) \cdots (c + n - 1)},$$

where none of the numbers a, b, c is zero or a negative integer. Suppose that $1 + c - (a + b) > 0$. Show that $u_n \to 0$ as $n \to \infty$. Use the result of Exercise 1.

3. Consider the binomial series for $x = -1$. Assume throughout that m is not zero or a positive integer. (a) Explain why the terms of the series are all of the same sign for sufficiently large values of n, and show that the series converges if $m > 0$, but diverges if $m < 0$. Note Example 4, § 19.4. Actually this is a special case of Exercise 7, § 19.4. (b) Follow the suggestions given and so prove that the binomial expansion (19.5–5) is valid if $m > 0$ and $x = -1$. Start with Taylor's formula with integral remainder (Theorem II, § 4.2), taking $f(x) = (1 + x)^m$, $a = 0$, $x = -1$. In this way one finds

$$(1 - 1)^m = 1 - m + \frac{m(m - 1)}{2!} + \cdots$$

$$+ \frac{m(m - 1) \cdots (m - n + 1)}{n!} (-1)^n + R_{n+1},$$

$$R_{n+1} = (-1)^{n+1} \cdot \frac{(m - 1)(m - 2) \cdots (m - n)}{n!}.$$

Then

$$R_{n+1} = - \frac{(-m + 1)(-m + 2) \cdots (-m + n)}{n!}.$$

Now use the result of Exercise 2 to show that $\lim_{n \to \infty} R_{n+1} = 0$. This proves that if $x = -1$ the binomial series is convergent, with sum 0, when $m > 0$.

   4. Consider the binomial series for $x = 1$.
   (a) Show that, for sufficiently large values of n, the terms of the series alternate in sign. Show also that the series is divergent if $m \leq -1$ and convergent if $m > -1$. For this last part use the result of Exercise 2. The general term is

$$u_n = \frac{m(m - 1) \cdots (m - n + 1)}{n!}.$$

   (b) Show that the binomial expansion (19.5–5) is valid if $x = 1$ and $m > -1$. Start with Taylor's formula as in Exercise 3(b), this time putting $x = 1$, and getting

$$(1 + 1)^m = u_0 + u_1 + \cdots + u_n + R_{n+1},$$

$$R_{n+1} = \frac{m(m - 1) \cdots (m - n)}{n!} \int_0^1 (1 - t)^n (1 + t)^{m-n-1} \, dt.$$

If $n > m - 1$ the value of the integral is less than $1/(n + 1)$. Explain why this is so.

Thus

$$| R_{n+1} | \leq \left| \frac{m(m - 1) \cdots (m - n)}{(n + 1)!} \right| = | u_{n+1} |.$$

What is the rest of the argument?

19.6   Multiplication of series

We have already noted, at the beginning of § 19.31, that term-by-term addition (or subtraction) of convergent series is a legitimate operation. This is convenient for obtaining new series expansions from expansions already established.

Example 1.   Show that, if $| x | < 1$,

(19.6–1) $$\tfrac{1}{2} \log \frac{1 + x}{1 - x} = x + \tfrac{1}{3} x^3 + \tfrac{1}{5} x^5 + \cdots.$$

We get this series expansion by observing that

$$\log \frac{1 + x}{1 - x} = \log (1 + x) - \log (1 - x)$$

and using (19.1–1). From this latter formula we have

$$\log (1 + x) = x - \tfrac{1}{2} x^2 + \tfrac{1}{3} x^3 - \tfrac{1}{4} x^4 + \tfrac{1}{5} x^5 - \cdots$$

if $-1 < x \leq 1$. Consequently, replacing x by $-x$, we have

$$\log (1 - x) = -x - \tfrac{1}{2} x^2 - \tfrac{1}{3} x^3 - \tfrac{1}{4} x^4 - \tfrac{1}{5} x^5 - \cdots$$

if $-1 \leqq x < 1$. Combining by subtraction, which is valid if $-1 < x < 1$, since both series are then convergent, we have

$$\log (1 + x) - \log (1 - x) = 2 x + \tfrac{2}{3} x^3 + \tfrac{2}{5} x^5 + \cdots.$$

This is equivalent to (19.6–1).

It is likewise convenient to be able to find new series expansions by multiplying known series expansions. For instance, suppose we wish to expand

$$\left(\frac{1 + x}{1 + x^2} \right)^{1/2} = (1 + x)^{1/2}(1 + x^2)^{-1/2}$$

in powers of x. We might proceed as follows: The expansion of $(1 + x)^{1/2}$ is a particular case of (19.5–5); to a few terms it is

$$(19.6\text{–}2) \quad (1 + x)^{1/2} = 1 + \tfrac{1}{2} x - \tfrac{1}{8} x^2 + \tfrac{1}{16} x^3 - \tfrac{5}{128} x^4 + \cdots.$$

To get the series for $(1 + x^2)^{-1/2}$ we put x^2 in place of x in (19.5–5) and set $m = -\tfrac{1}{2}$. This is legitimate if $|x| < 1$, since then $x^2 < 1$ also. To a few terms the series is

$$(19.6\text{–}3) \qquad (1 + x^2)^{-1/2} = 1 - \tfrac{1}{2} x^2 + \tfrac{3}{8} x^4 - \tfrac{5}{16} x^6 + \cdots.$$

Thus

$$\left(\frac{1 + x}{1 + x^2} \right)^{1/2} = (1 + \tfrac{1}{2} x - \tfrac{1}{8} x^2 + \tfrac{1}{16} x^3 - \cdots)(1 - \tfrac{1}{2} x^2 + \tfrac{3}{8} x^4 - \cdots).$$

The next question is: How do we multiply the two series together to get a new series? Proceeding just as though the series were finite sums, we might write down the following scheme, which arises by multiplying the second series successively by each term of the first series:

$$
\begin{array}{lllll}
1 & -\tfrac{1}{2} x^2 & +\tfrac{3}{8} x^4 & -\tfrac{5}{16} x^6 + \cdots \\
\quad \tfrac{1}{2} x & \quad -\tfrac{1}{4} x^3 & \quad +\tfrac{3}{16} x^5 & \quad - \cdots \\
\quad\quad -\tfrac{1}{8} x^2 & \quad\quad +\tfrac{1}{16} x^4 & \quad\quad -\tfrac{3}{64} x^6 + \cdots \\
\quad\quad\quad \tfrac{1}{16} x^3 & \quad\quad\quad -\tfrac{1}{32} x^5 & \quad\quad\quad + \cdots \\
\quad\quad \cdots\cdots\cdots & \quad\quad \cdots\cdots\cdots
\end{array}
$$

There will be an infinite number of rows, each row being an infinite series. But we observe that there are only a finite number of terms of each degree, so that if we collect together terms of like degree, we obtain for the first few terms

$$(19.6\text{–}4) \qquad\qquad 1 + \tfrac{1}{2} x - \tfrac{5}{8} x^2 - \tfrac{3}{16} x^3 + \cdots.$$

It is clear that there is here a systematic process, but it remains to prove that the process gives a series which has as its sum the product of the sums of the two original series. There is a general theorem which justifies the process.

THEOREM XVII. *Suppose that each of the series $\sum u_n$, $\sum v_n$ is absolutely convergent, with sums U and V respectively:*

(19.6–5) $$U = u_0 + u_1 + u_2 + \cdots,$$

(19.6–6) $$V = v_0 + v_1 + v_2 + \cdots.$$

Let $w_0 = u_0 v_0$, $w_1 = u_0 v_1 + u_1 v_0$, and in general

(19.6–7) $$w_n = u_0 v_n + u_1 v_{n-1} + \cdots + u_n v_0.$$

Then the series $\sum w_n$ is absolutely convergent, and its sum is UV:

(19.6–8) $$UV = w_0 + w_1 + w_2 + \cdots.$$

Moreover, any infinite series which has as its terms the products $u_i v_j$ (i and $j \geq 0$) arranged in any order, each product occurring once and only once, is absolutely convergent, with sum UV.

PROOF. Let us consider the array

(I)

$$
\begin{array}{llll}
u_0 v_0 & u_0 v_1 & \cdots & u_0 v_n \quad \cdots \\
u_1 v_0 & u_1 v_1 & \cdots & u_1 v_n \quad \cdots \\
\cdots & & & \cdots \\
\cdots & & & \cdots \\
u_n v_0 & u_n v_1 & \cdots & u_n v_n \quad \cdots \\
\cdots & & & \\
\cdots & &
\end{array}
$$

and the similar array in which $u_i v_j$ is replaced by $|\,u_i\,|\,|\,v_j\,|$. Let us denote the second array by (II). Finally, let

$$A = |\,u_0\,| + |\,u_1\,| + \cdots + |\,u_n\,| + \cdots$$
$$B = |\,v_0\,| + |\,v_1\,| + \cdots + |\,v_n\,| + \cdots.$$

These last series are convergent because of the assumption that (19.6–5) and (19.6–6) are absolutely convergent. Now consider any series formed from the terms of the array (II), taken in some definite order. Such a series is convergent, since each of its partial sums is less than or equal to AB. For, any such partial sum is less than

$$(|\,u_0\,| + |\,u_1\,| + \cdots + |\,u_n\,|)(|\,v_0\,| + |\,v_1\,| + \cdots + |\,v_n\,|)$$

if n is taken sufficiently large, and this last product is certainly no larger than AB, since the first factor does not exceed A and the second factor does not exceed B. It follows that any infinite series formed by taking the terms of the array (I) in some definite order is absolutely convergent. Since, of any two such series, one is merely a rearrangement of the other, Theorem XI (§ 19.31) assures us that all such series have the same sum.

Now one possible arrangement is that in which we take $u_0 v_0$ first, then $u_0 v_1$, $u_1 v_1$, and $u_1 v_0$; then $u_0 v_2$, $u_1 v_2$, $u_2 v_2$, $u_2 v_1$, and $u_2 v_0$; and at the nth stage all terms $u_i v_j$ for which i and j do not exceed n and at least one of them is equal to n. At the nth stage our partial sum is exactly

$$(u_0 + u_1 + \cdots + u_n)(v_0 + v_1 + \cdots + v_n),$$

which is the sum of all the terms in an upper left square portion of the array (I). Hence, by (19.6–5) and (19.6–6) the limit of the partial sums is UV, and this must be the value of the series (by the fact that the limit of a product is the product of the limits).

Another arrangement is that in which we take first $u_0 v_0$, then $u_0 v_1$ and $u_1 v_0$, then $u_0 v_2$, $u_1 v_1$, and $u_2 v_0$, and at the nth stage all terms $u_i v_j$ for which $i + j = n$. This arrangement gives the same sum UV, of course. But if we group the terms according to the stage, we get

$$u_0 v_0 + (u_0 v_1 + u_1 v_0) + (u_0 v_2 + u_1 v_1 + u_2 v_0) + \cdots,$$

which is exactly the series (19.6–8). The insertion of the parentheses technically changes the series, of course, since the last series has the sum of all products in one parenthesis as a single term, whereas previously each product $u_i v_j$ was a single term. But the insertion of parentheses into a convergent series always leads to a convergent series whose sum is the same as that of the original series (see Exercise 7). Therefore (19.6–8) holds, and the proof of the theorem is complete.

As an application of the theorem let us return to (19.6–4). This series was obtained by multiplication of the series (19.6–2) and (19.6–3) by the rule of Theorem XVII. Since each of the two latter series is absolutely convergent if $|x| < 1$, the series (19.6–4) is absolutely convergent if $|x| < 1$, and its sum is the product of the sums of the other two series, namely $(1 + x)^{1/2} (1 + x^2)^{-1/2}$.

Observe that, if the series

$$a_0 + a_1 x + a_2 x^2 + \cdots + a_i x^i + \cdots,$$
$$b_0 + b_1 x + b_2 x^2 + \cdots + b_j x^j + \cdots$$

are multiplied according to Theorem XVII, the resulting series is

$$a_0 b_0 + (a_0 b_1 + a_1 b_0)x + (a_0 b_2 + a_1 b_1 + a_2 b_0)x^2 + \cdots,$$

the coefficient of x^n in the last series being

$$a_0 b_n + a_1 b_{n-1} + \cdots + a_n b_0.$$

To justify this application of Theorem XVII in any given case we need merely to check on the absolute convergence of the first two series.

Example 2.   Find the series expansion of $\dfrac{1}{1-x} \log \dfrac{1}{1-x}$ in powers of x.

We know from (19–3) that

$$(19.6\text{–}9) \qquad \frac{1}{1-x} = 1 + x + x^2 + \cdots = \sum_{n=0}^{\infty} x^n$$

if $|x| < 1$; also, from (19.1–1),

$$\log (1-x) = -x - \frac{x^2}{2} - \frac{x^3}{3} - \cdots = - \sum_{n=1}^{\infty} \frac{x^n}{n}$$

if $-1 \leq x < 1$, so that

$$(19.6\text{–}10) \qquad \log \frac{1}{1-x} = \sum_{n=1}^{\infty} \frac{x^n}{n}.$$

Each of these series is absolutely convergent if $|x| < 1$, as may be verified by Theorem XIII. The coefficients in (19.6–9) are $a_n = 1$, $n = 0, 1, 2, \cdots$; those in (19.6–10) are $b_0 = 0$, $b_n = \dfrac{1}{n}$, $n = 1, 2, \cdots$. Therefore $a_0 b_0 = 0$, and if $n \geq 1$,

$$a_0 b_n + a_1 b_{n-1} + \cdots + a_n b_0 = \frac{1}{n} + \frac{1}{n-1} + \cdots + \frac{1}{2} + 1.$$

Thus

$$\frac{1}{1-x} \log \frac{1}{1-x} = \sum_{n=1}^{\infty} \left(1 + \frac{1}{2} + \cdots + \frac{1}{n} \right) x^n$$
$$= x + (1 + \tfrac{1}{2})x^2 + (1 + \tfrac{1}{2} + \tfrac{1}{3})x^3 + \cdots.$$

This series is absolutely convergent if $|x| < 1$.

EXERCISES

1. Multiply the series for $(1-x)^{-1}$ (series (19–3), valid when $|x| < 1$) by itself and so obtain the expansion

$$(1-x)^{-2} = \sum_{n=0}^{\infty} (n+1)x^n, \quad |x| < 1.$$

2. Derive the expansion

$$(1-x)^{-3} = \sum_{n=0}^{\infty} \frac{(n+1)(n+2)}{2} x^n \qquad (|x| < 1)$$

by multiplication of series. Use the result of Exercise 1.

3. Obtain the expansion, valid when $|x| < 1$,

$$\frac{1}{1-x}\frac{\tan^{-1}\sqrt{x}}{\sqrt{x}} = 1 + (1 - \tfrac{1}{3})x + (1 - \tfrac{1}{3} + \tfrac{1}{5})x^2 + (1 - \tfrac{1}{3} + \tfrac{1}{5} - \tfrac{1}{7})x^3 + \cdots.$$

4. Prove by multiplication of series that if

$$f(x) = \sum_{n=0}^{\infty} \frac{x^n}{n!}, \quad \text{then} \quad f(x)f(y) = f(x+y)$$

for arbitrary x and y.

5. Prove that

$$2\sum_{m=0}^{\infty}(-1)^m\frac{x^{2m+1}}{(2m+1)!}\cdot\sum_{n=0}^{\infty}(-1)^n\frac{x^{2n}}{(2n)!} = \sum_{p=0}^{\infty}(-1)^p\frac{(2x)^{2p+1}}{(2p+1)!}.$$

6. Using no other properties of $\sin x$ and $\cos x$ than the fact that the series representations (19.1–8) and (19.1–9) are valid, prove that $\sin^2 x + \cos^2 x = 1$.

7. Let $\sum u_n$ be a convergent series with sum s. Let sets of parentheses be introduced into the series, e.g.:

$$(u_1 + u_2) + (u_3 + u_4 + u_5) + (u_6) + (u_7 + u_8) + \cdots.$$

Let the sum of the terms in the nth parenthesis be v_n. Prove that the series $\sum v_n$ is convergent and that its sum is s.

8. The rule for multiplying two series, as given in Theorem XVII, may give a wrong result if the series are not absolutely convergent. Verify this by taking $u_n = v_n = (-1)^n/\sqrt{n+1}$ in the theorem and showing that w_n does not approach zero as $n \rightarrow \infty$. Suggestion: Show that, for $k = 0, 1, \cdots, n$, $(k+1)(n-k+1) \leq \tfrac{1}{4}(n+2)^2$; then show that $|w_n| \geq 2(n+1)/(n+2)$.

19.7 Dirichlet's test

Thus far all our tests, with one exception, have been tests for series of positive terms, or tests for absolute convergence. The one exception is the alternating series test of § 19.32. We shall now discuss a more general test which is useful on occasion for proving that a series is convergent, though not necessarily absolutely convergent. Most tests of the character of this one depend upon an algebraic device known as *summation by parts*. The device is analogous to integration by parts.

Let $a_0, a_1, \cdots$ and $b_0, b_1, \cdots$ be two arbitrary sequences of numbers. Let

$$s_n = a_0 + a_1 + \cdots + a_n.$$

Then

(19.7–1)
$$a_0b_0 + a_1b_1 + \cdots + a_nb_n$$
$$= s_0(b_0 - b_1) + s_1(b_1 - b_2) + \cdots + s_{n-1}(b_{n-1} - b_n) + s_nb_n.$$

This is the identity of summation by parts. It is known as *Abel's summation identity*. Niels H. Abel was a Norwegian mathematician (1802–1829). The proof of the formula is very simple. We observe that $a_n = s_n - s_{n-1}$ if $n \geq 1$, and $a_0 = s_0$. Therefore

$$a_0 b_0 = s_0 b_0$$
$$a_1 b_1 = s_1 b_1 - s_0 b_1$$
$$a_2 b_2 = s_2 b_2 - s_1 b_2$$
$$\vdots \qquad \vdots \qquad \vdots$$
$$a_{n-1} b_{n-1} = s_{n-1} b_{n-1} - s_{n-2} b_{n-1}$$
$$a_n b_n = s_n b_n - s_{n-1} b_n.$$

Adding these results and grouping the terms on the right appropriately, we obtain (19.7–1).

We can now state a test known by the name of P. G. Lejeune Dirichlet, a German mathematician of the first half of the nineteenth century.

THEOREM XVIII. *(Dirichlet's test.) Consider a series of the form*

$$(19.7\text{–}2) \qquad a_0 b_0 + a_1 b_1 + a_2 b_2 + \cdots + a_n b_n + \cdots$$

which satisfies the following conditions:
(a) the terms b_n are positive, $b_{n+1} \leq b_n$, and $b_n \to 0$ as $n \to \infty$;
(b) there is some constant M independent of n such that

$$| a_0 + a_1 + \cdots + a_n | \leq M \text{ for all values of } n.$$

Then the series (19.7–2) is convergent.

PROOF.   Let $\qquad S_n = a_0 b_0 + \cdots + a_n b_n.$

We have to show that $\lim_{n \to \infty} S_n$ exists. Now by (19.7–1) we have

$$(19.7\text{–}3) \qquad S_n = T_n + s_n b_n$$

where $\qquad T_n = s_0(b_0 - b_1) + \cdots + s_{n-1}(b_{n-1} - b_n).$

Now $| s_n b_n | \leq M b_n$, by condition (b), and so $s_n b_n \to 0$, since $b_n \to 0$. Next we show that $\lim_{n \to \infty} T_n$ exists. The desired conclusion will then follow from (19.7–3). Now T_n is the partial sum of the series

$$(19.7\text{–}4) \qquad s_0(b_0 - b_1) + s_1(b_1 - b_2) + \cdots,$$

and certainly T_n will approach a limit if we show that the series (19.7–4) is absolutely convergent. Now

$$| s_0(b_0 - b_1) | + | s_1(b_1 - b_2) | + \cdots + | s_{n-1}(b_{n-1} - b_n) |$$
$$\leq M(b_0 - b_1) + M(b_1 - b_2) + \cdots + M(b_{n-1} - b_n) = M(b_0 - b_n),$$

or

$$\sum_{k=0}^{n-1} | s_k(b_k - b_{k+1}) | \leqq Mb_0.$$

Here we have used the conditions (a) and (b). Since these sums are bounded, we conclude that the series (19.7–4) is absolutely convergent. This is all that was needed to complete the proof.

Theorem XVIII is useful in connection with trigonometric series.

Example. The series

(19.7–5) $$\frac{\cos x}{1} + \frac{\cos 3 x}{3} + \frac{\cos 5 x}{5} + \cdots$$

is convergent if x is not one of the values $0, \pm \pi, \pm 2 \pi, \cdots$.

To prove this we apply Theorem XVIII, taking $a_0 = 0$, $a_1 = \cos x$, $a_2 = \cos 3 x, \cdots, a_n = \cos(2 n - 1)x$, $n = 1, 2, \cdots$, and $b_0 = 1$, $b_n = (2 n - 1)^{-1}$, $n = 1, 2, \cdots$. Condition (a) is clearly satisfied, so it remains only to show that condition (b) is satisfied. This requires a bit of ingenuity. We use the trigonometric identity

$$2 \cos A \sin B = \sin (A + B) - \sin (A - B).$$

Taking $B = x$ and A successively equal to $x, 3 x, 5 x, \cdots$, we have

$$2 \cos x \sin x = \sin 2 x - 0,$$
$$2 \cos 3 x \sin x = \sin 4 x - \sin 2 x,$$
$$2 \cos 5 x \sin x = \sin 6 x - \sin 4 x,$$
$$\vdots \qquad \qquad \vdots \qquad \qquad \vdots$$
$$2 \cos (2 n - 1)x \sin x = \sin 2 nx - \sin 2(n - 1)x.$$

Adding these results, we find

$$2 \sin x(\cos x + \cos 3 x + \cdots + \cos (2 n - 1)x) = \sin 2 nx,$$

or

(19.7–6) $$\cos x + \cos 3 x + \cdots + \cos (2 n - 1)x = \frac{\sin 2 nx}{2 \sin x}.$$

If now x is not one of the values $0, \pm \pi, \pm 2 \pi, \cdots$, then $\sin x \neq 0$, and

$$| \cos x + \cos 3 x + \cdots + \cos (2 n - 1)x | \leqq \frac{1}{2 | \sin x |}.$$

Hence condition (b) of Theorem XVIII is satisfied with $M = \dfrac{1}{2 | \sin x |}$.
Therefore the series (19.7–5) is convergent.

EXERCISES

1. Prove that the series

$$\frac{\sin x}{1} + \frac{\sin 3\,x}{\sqrt{3}} + \frac{\sin 5\,x}{\sqrt{5}} + \cdots$$

is convergent for all values of x. Use the identity

$$2 \sin A \sin B = \cos (A - B) - \cos (A + B).$$

2. Prove the identities

$$2 \sin \frac{x}{2} (\sin x + \sin 2\,x + \cdots + \sin nx) = \cos \frac{x}{2} - \cos \frac{2\,n + 1}{2}\, x,$$

$$2 \sin \frac{x}{2} (\cos x + \cos 2\,x + \cdots + \cos nx) = \sin \frac{2\,n + 1}{2}\, x - \sin \frac{x}{2}.$$

3. Suppose that $a_n > 0$, $a_{n+1} \leqq a_n$, $a_n \to 0$ as $n \to \infty$. Prove that the series $\sum_{n=1}^{\infty} a_n \sin nx$ is convergent for all values of x, and that $\sum_{n=1}^{\infty} a_n \cos nx$ is convergent with the possible exception of the cases in which $x = 2\,\pi m$, $m = 0$, ± 1, ± 2, $\cdots$. Use the identities in Exercise 2.

4. Suppose that $\sum_{n=1}^{\infty} a_n/n^p$ is convergent. Show that $\sum_{n=1}^{\infty} a_n/n^q$ is convergent if $p < q$.

5. Deduce Theorem XII as a special case of Dirichlet's test.

6. The following theorem is known as Abel's test: Suppose $\sum a_n$ is convergent, and that $b_n > 0$, $b_{n+1} \leqq b_n$. Then $\sum a_n b_n$ is convergent. Prove this.

7. Show that the series

$$\frac{1}{2 \log 2} + \frac{1}{3 \log 3} - \frac{1}{4 \log 4} - \frac{1}{5 \log 5} - \frac{1}{6 \log 6} - \frac{1}{7 \log 7} + \cdots$$

is convergent, the rule of signs being that successive terms with the same sign come in groups of 2, 4, 8, 16, $\cdots$. Begin by considering the series

$$\tfrac{1}{2} + \tfrac{1}{3} - \tfrac{1}{4} - \tfrac{1}{5} - \tfrac{1}{6} - \tfrac{1}{7} + \tfrac{1}{8} + \cdots.$$

8. Show that the series

$$1 - \tfrac{1}{2} - \tfrac{1}{3} + \tfrac{1}{4} + \tfrac{1}{5} - \tfrac{1}{6} - \tfrac{1}{7} + + - - \cdots$$

is convergent. Successive terms of like sign come in groups of two.

MISCELLANEOUS EXERCISES

1. For what values of x is $\sum n^x x^n$ convergent?

2. Find the sum of the first n terms of the series $\sum_{n=1}^{\infty} \log [1 + (1/n)]$. Is the series convergent or divergent?

3. Show that $\sum (1/n) \log [1 + (1/n)]$ is convergent.

4. Show by Theorem III, or otherwise, that $\sum_{n=2}^{\infty} 1/(\log n)^c$ is divergent for all values of c.

5. Express $(\log n)^{\log n}$ as a power of n, and use the result to show that $\sum \dfrac{1}{(\log n)^{\log n}}$ is convergent.

6. Examine each of the following series for convergence or divergence.

(a) $\sum (-1)^n \dfrac{n^n}{(n+1)^{n+1}}$,

(c) $\sum (-1)^n \left(\dfrac{n}{n+1}\right)^n$,

(b) $\sum \dfrac{(n+1)^n}{n^{n+1}}$.

(d) $\sum \dfrac{[1+(1/n)]^{n^2}}{e^n}$.

7. Show that

(a) $\dfrac{2 \cdot 4 \cdots 2n}{3 \cdot 5 \cdots (2n+1)} < \dfrac{1}{\sqrt{n+1}}$.

(b) Classify the values of x according to whether the series

$$\sum \dfrac{n!}{3 \cdot 5 \cdots (2n+1)} x^n$$

is convergent or divergent.

8. Find all values of x for which the series $\sum \dfrac{1 \cdot 3 \cdots (2n-1)}{2 \cdot 4 \cdots 2n} \dfrac{1}{n^x}$ is convergent.

9. (a) Is $\sum \sin \pi[n + (1/n)]$ absolutely convergent? Is it convergent?
(b) Show that $\sum \sin^2 \pi[n + (1/n)]$ is convergent.
(c) For what values of θ is $\sum (-1)^n(1/n) \cos(\theta/n)$ convergent?
(d) Is $\sum [1 - \cos(\pi/n)]$ convergent or divergent?

10. Discuss the convergence of each series, classifying the values of x into those for which the series converges and those for which it diverges.

(a) $\sum_{n=1}^{\infty} (-1)^n \dfrac{1 \cdot 3 \cdots (2n-1)}{2 \cdot 4 \cdots 2n} \dfrac{2}{4n+1} \left(\dfrac{x}{2}\right)^{4n+1}$.

(b) $\sum_{n=2}^{\infty} 2^{2n-2} \dfrac{((n-1)!)^2}{(2n)!} x^{2n}$.

(c) $\sum_{n=1}^{\infty} \dfrac{(n+1)^n}{n!} x^n$.

11. Show that $\sum \dfrac{\log[1+(1/n)]}{a^{\log(\log n)}}$ is convergent if $a > e$ and divergent if $0 < a \le e$.

12. (a) If $x_n = 1 - \frac{1}{2} + \frac{1}{3} - \cdots - 1/(2n)$, show that $x_n \to \log 2$ as $n \to \infty$, by using the definition of Euler's constant (see Exercise 6, §19.21). HINT: Show that $x_n = C_{2n} - C_n + \log 2$ in the notation of the exercise just mentioned.
(b) Prove that the partial sums of the series

$$1 + \tfrac{1}{2} - \tfrac{1}{3} + \tfrac{1}{4} + \tfrac{1}{5} - \tfrac{1}{6} + + - \cdots$$

approach $+\infty$ as $n \to \infty$. Make use of Euler's constant.

13. Prove that, if $u_n > 0$ and $\sum u_n$ is convergent, so is $\sum u_n^2$.

14. Show that the series $\sum \log(n \sin(1/n))$ is convergent.

15. Consider the series

$$1 - \tfrac{1}{2} - \tfrac{1}{3} + \tfrac{1}{4} + \tfrac{1}{5} + \tfrac{1}{6} + \tfrac{1}{7} - \tfrac{1}{8} - \cdots ,$$

the rule of signs being that successive terms with the same sign come in groups of 1, 2, 4, 8, 16, $\cdots$. Show that this series does not satisfy the condition of Theorem VIII, and is therefore divergent. Show also that, nevertheless, the sum s_n of the first n terms of the series satisfies the condition $0 < s_n \le 1$.

16. Consider the series

$$1 - \tfrac{1}{2} - \tfrac{1}{3} + \tfrac{1}{4} + \tfrac{1}{5} + \tfrac{1}{6} - \tfrac{1}{7} - \cdots ,$$

in which successive terms of like sign come in groups of 1, 2, 3, 4, $\cdots$. Show that this series is convergent.

17. Prove that the series $\sum_{n=2}^{\infty} (1/n) \cos (a \log n)$ is divergent by showing that it does not satisfy the condition of Theorem VIII. SUGGESTION: Suppose $0 < \delta < \pi/2$. Let n_1 denote the greatest integer $\le \exp [(2 \pi m - \delta)/a]$ and let n_2 denote the greatest integer $\le \exp [(2 \pi m + \delta)/a]$, where m is a fixed positive integer, and $\exp u = e^u$. Show that $\cos (a \log n) \ge \cos \delta > 0$ if $n_1 < n \le n_2$. Then show that

$$\frac{\cos \delta}{n_1} + \sum_{n=m_1+1}^{n_2} \frac{\cos (a \log n)}{n} > \int_{n_1}^{n_2+1} \frac{\cos \delta}{x} \, dx > \frac{2 \delta \cos \delta}{a} .$$

Now note that n_1 and $n_2 \to \infty$ if $m \to \infty$, and so finish the proof.

18. Show by an argument similar to that of Exercise 17 that the series

$$\sum_{n=4}^{\infty} \frac{\cos (\log (\log n))}{\log n} \text{ is divergent.}$$

Uniform Convergence   **20**

20. Functions defined by convergent sequences

In the more advanced parts of analysis we often deal with functions which are defined by means of infinite series. Suppose that

$$(20\text{--}1) \qquad f(x) = \sum_{n=1}^{\infty} u_n(x),$$

where it is assumed that the terms of the series are each defined on the same interval $a \leq x \leq b$, and that the series is convergent for each x of the interval. The sum of the series is then also a function of x, and we denote it by $f(x)$. The main purpose of this chapter is to deal with questions A–D below.

Assuming that we know a good deal about the functions $u_1(x)$, $u_2(x)$, $\cdots$, but nothing about the function $f(x)$ except in so far as we can draw certain conclusions from the series (20–1), we raise four questions:

A. What conditions will assure us that f is continuous?
B. What conditions will assure us that f is differentiable?
C. What conditions will justify us in writing

$$(20\text{--}2) \qquad \int_a^b f(x)dx = \int_a^b u_1(x)dx + \int_a^b u_2(x)dx + \cdots ?$$

D. What conditions will justify us in writing

$$(20\text{--}3) \qquad f'(x) = u_1'(x) + u_2'(x) + \cdots ?$$

That is, when can we integrate and differentiate an infinite series just as though it were a *finite* sum of functions?

The answers which we shall give to all these questions are dependent on the concept of *uniform convergence*. Before proceeding further, however, let us look at the problems raised by these questions in another way. Consider the partial sums

$$(20\text{--}4) \qquad s_n(x) = u_1(x) + u_2(x) + \cdots + u_n(x)$$

of the series (20–1). Then

$$(20\text{--}5) \qquad f(x) = \lim_{n \to \infty} s_n(x);$$

that is, the function $f(x)$ is the limit of a *sequence* of functions. Now we may consider the notion of the limit of a sequence of functions quite apart from the notion of a function defined by an infinite series, by presenting the sequence directly, rather than as a sum formed from the terms of a series.

644

Example 1. Consider the sequence $f_n(x) = x^n$, and define

$$(20\text{–}6) \qquad f(x) = \lim_{n \to \infty} f_n(x), \quad 0 \leq x \leq 1.$$

The sequence is convergent if $0 \leq x \leq 1$, and we see that (20–6) is equivalent to the definition

$$(20\text{–}7) \qquad \begin{cases} f(x) = 0 & \text{if } 0 \leq x < 1, \\ f(1) = 1. \end{cases}$$

The graphs of several of the functions $f_n(x)$ are shown in Fig. 179.

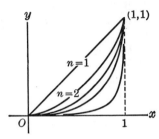

Fig. 179

We shall find that, in framing our answers to the questions A–D, the important consideration is that of functions defined as the limits of sequences. For this reason the subsequent discussion in this chapter is often phrased in terms of sequences rather than in terms of series. Let us rephrase the questions as follows:

Assuming that a function $f(x)$ is known to us entirely by

$$(20\text{–}8) \qquad f(x) = \lim_{n \to \infty} f_n(x), \quad a \leq x \leq b,$$

we ask:

A′. What conditions on the functions f_n will assure us that f is continuous?

B′. What conditions will assure us that f is differentiable?

C′. Under what conditions will it be true that

$$(20\text{–}9) \qquad \int_a^b f(x)dx = \lim_{n \to \infty} \int_a^b f_n(x)dx?$$

D′. Under what conditions will it be true that

$$(20\text{–}10) \qquad f'(x) = \lim_{n \to \infty} f_n'(x)?$$

In the particular case when $f_n(x) = s_n(x)$ and $s_n(x)$ is defined by (20–4), the questions A′–D′ are the same as the questions A–D.

We do not aim to get answers to these questions in the form of *necessary and sufficient* conditions; the conditions which we shall impose will be *sufficient* but not *necessary*.

Let us examine the situation in Example 1 with respect to question A′. Here we observe that each of the functions $f_n(x) = x^n$ is continuous on the closed interval [0, 1]; nevertheless, the limit function $f(x)$ is not continuous on the whole interval, but is discontinuous at the point $x = 1$. From this example we can conclude that in question A′ the assumption that each of the functions f_n is continuous on [a, b] is *not* sufficient to assure us that the limit function f is continuous on [a, b]. Something more is needed, not about the functions f_n themselves, but about the way in which $f_n(x)$ converges to $f(x)$.

Next we consider an example which will be instructive with respect to questions B′ and D′.

Example 2. Let $f_n(x) = x^n/n$, $n = 1, 2, \cdots$.

On the interval $0 \leq x \leq 1$, $0 \leq f_n(x) \leq 1/n$, and hence

$$\lim_{n \to \infty} f_n(x) = 0, \quad 0 \leq x \leq 1.$$

The limit function $f(x)$ has the value zero for each x in the interval [0, 1]. Thus f is differentiable. But consider the sequence of derivatives $f_n'(x) = x^{n-1}$. The successive members of this sequence are $1, x, x^2, x^3, \cdots$, and as we saw in Example 1,

$$\lim_{n \to \infty} f_n'(x) = 0 \quad \text{if} \quad 0 \leq x < 1$$

$$\lim_{n \to \infty} f_n'(1) = 1.$$

Thus $$\lim_{n \to \infty} f_n'(1) \neq f'(1).$$

This example shows us that in question D′ the relation (20–10) may fail to be true (at least for *some* values of x in the interval) even when the functions f_n and f are differentiable and the sequence of derivatives $f_n'(x)$ is convergent for *every* x in the interval.

Finally we give an example bearing on question C′.

Example 3. Let us define a sequence $f_n(x)$ so that its graph is as indicated in Fig. 180. That is, the graph of $y = f_n(x)$ for $0 \leq x \leq 1$ consists of three line segments: the line $y = 4 n^2 x$ from $x = 0$ to $x = 1/(2 n)$, the line $y = -4 n^2 x + 4 n$ from $x = 1/(2 n)$ to $x = 1/n$, and the line $y = 0$ from $x = 1/n$ to $x = 1$.

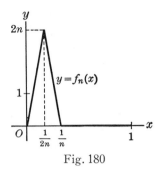

Fig. 180

The high point of the graph is at $x = 1/(2\,n)$, the height being $2\,n$. From the definition we see that, for each x in $[0, 1]$,

$$\lim_{n \to \infty} f_n(x) = 0.$$

For, $f_n(0) = 0$ for all n, and if $0 < x \le 1, f_n(x) = 0$ as soon as n is large enough to make $1/n < x$. Thus the limit function $f(x)$ has the value zero for each x, and accordingly

$$\int_0^1 f(x)dx = 0.$$

But
$$\int_0^1 f_n(x)dx = 1$$

for each value of n, because the integral is just equal to the area of a triangle of base $1/n$ and height $2\,n$. Therefore

$$\int_0^1 f(x)dx \ne \lim_{n \to \infty} \int_0^1 f_n(x)dx.$$

This shows that for the truth of (20–9) in question C′ we must have something more than the mere convergence of $f_n(x)$ to $f(x)$, even when the functions f_n and f are all continuous.

In concluding this section let us observe that any function which is defined by a convergent *sequence* of functions may also be defined by a convergent *series* of functions. For if $f(x) = \lim_{n \to \infty} f_n(x)$, let us form the series

(20–11) $f_1(x) + [f_2(x) - f_1(x)] + [f_3(x) - f_2(x)] + \cdots,$

that is, define the terms $u_n(x)$ by the formulas $u_1(x) = f_1(x)$, and $u_n(x) = [f_n(x) - f_{n-1}(x)]$ if $n > 1$. Then

$$s_n(x) = f_1(x) + [f_2(x) - f_1(x)] + \cdots + [f_n(x) - f_{n-1}(x)],$$

and when we simplify the right side of this equation we find that $s_n(x)$ = $f_n(x)$. Thus the series (20–11) has the sum $f(x)$.

For instance, the limit function $f(x)$ in Example 1 can be expressed as the infinite series

$$f(x) = x + (x^2 - x) + (x^3 - x^2) + (x^4 - x^3) + \cdots .$$

20.1   The concept of uniform convergence

Let us begin by recalling the definition of the limit of a sequence, as it applies to a sequence of functions. Suppose that

$$(20.1–1) \qquad\qquad f(x) = \lim_{n \to \infty} f_n(x)$$

for each x of the interval $a \le x \le b$. According to the definition in § 1.62 this means that if ϵ is any positive number and if x is any point of the interval, there is some integer N, the size of which will usually depend on ϵ and x, such that

$$(20.1–2) \qquad\qquad |f_n(x) - f(x)| < \epsilon$$

if $N \le n$. The inequality (20.1–2) is equivalent to the two inequalities

$$(20.1–3) \qquad\qquad f(x) - \epsilon < f_n(x) < f(x) + \epsilon .$$

Let us examine in a particular case the way in which the choice of N may depend on ϵ and x. Consider, for instance, $f_n(x) = x^n$. We saw in Example 1 of § 20 that $f_n(x) \to f(x)$ if $0 \le x \le 1$, where $f(x) = 0$ if $0 \le x < 1$ and $f(1) = 1$. Suppose that $0 < \epsilon < 1$ and $0 < x < 1$. Then (20.1–2) is equivalent to each of the following inequalities:

$$x^n < \epsilon ,$$

$$n \log x < \log \epsilon ,$$

$$\log (1/\epsilon) < n \log (1/x) ,$$

$$(20.1–4) \qquad\qquad \frac{\log (1/\epsilon)}{\log (1/x)} < n .$$

(Observe that $\log x < 0$ and $\log (1/x) > 0$ if $0 < x < 1$.) Now, to have (20.1–4) true for all $n \ge N$ it is necessary to choose the integer N large enough so that

$$(20.1–5) \qquad\qquad \frac{\log (1/\epsilon)}{\log (1/x)} < N .$$

The dependence on ϵ and x shows up clearly here. As $\epsilon \to 0$, $\log (1/\epsilon) \to + \infty$, and hence $N \to \infty$. Also, for a fixed ϵ, as $x \to 1 -$, $\log (1/x) \to 0 +$ and we see from (20.1–5) that $N \to \infty$. *There is no value of N such that (20.1–5) holds simultaneously for all values of x in the range $0 < x < 1$.*

NON-UNIFORM CONVERGENCE

The foregoing situation illustrates what we shall call *nonuniform* convergence. We shall now define what we mean by *uniform* convergence.

DEFINITION. Suppose that the sequence of functions $\{f_n(x)\}$ and the function $f(x)$ are defined on the interval $[a, b]$ and satisfy the following condition: To each $\epsilon > 0$ corresponds some integer N such that, *for every x in the interval*,

$$|f_n(x) - f(x)| < \epsilon$$

provided that $N \leq n$. Then we say that $f_n(x)$ converges uniformly to $f(x)$ on the interval $[a, b]$. The essential thing is that N is to be independent of x; it will usually depend on ϵ, however.

Example 1. If $f_n(x) = x^n/n$, then $\lim_{n \to \infty} f_n(x) = 0$, the convergence being uniform, on the interval $0 \leq x \leq 1$. For, if x is on the stated interval,

$$\left|\frac{x^n}{n}\right| \leq \frac{1}{n} < \epsilon$$

if $1/\epsilon < n$. Hence we may take N as the smallest integer greater than $1/\epsilon$. This choice is independent of x.

Example 2. The convergence in Example 3 of § 20 is *not* uniform on $0 \leq x \leq 1$. For, since $f(x) = 0$ for each value of x involved, uniform convergence would mean that $|f_n(x)| < \epsilon$ when $n \geq N$, where N depends only on ϵ, not on x. In particular, then, we should have $\left|f_n\left(\frac{1}{2n}\right)\right| < \epsilon$ if n is sufficiently large. But $\left|f_n\left(\frac{1}{2n}\right)\right| = 2n$, and certainly the inequality $2n < \epsilon$ is false if n is sufficiently large. This is a case where the choice of N depends on x, and as $x \to 0$, $N \to \infty$.

Uniform convergence may be portrayed graphically by interpreting the inequalities (20.1–3). Let the graph of $y = f(x)$ be displaced upward by the addition of ϵ. This gives us the graph of $y = f(x) + \epsilon$. Similarly, a downward displacement of amount ϵ gives us the graph of $y = f(x) - \epsilon$. Let us visualize these two displaced graphs, on the assumption that the function f is continuous. Considering the portion of the xy-plane between these two curves, for $a \leq x \leq b$, we have a ribbon-like region of width 2ϵ in the y-direction (see Fig. 181). The inequalities (20.1–3) state that the graph of $y = f_n(x)$ lies within the ribbon throughout its length. For each $\epsilon > 0$ there is

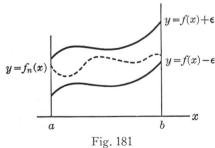

Fig. 181

a ribbon; to have uniform convergence the graph of $y = f_n(x)$ must lie entirely in the ribbon when n is large enough $(n \geq N)$, and this must be true for each ϵ.

The concept of uniform convergence relates to a whole set of values of the variable x; we have stated the definition for the case in which the set of values is a closed interval. The definition is essentially the same for any interval, whether open, closed, or neither. The crux of the matter is to have N the same for all values of x in the given interval or set.

Example 3.   Let $f_n(x) = \dfrac{nx}{1 + n^2x^2}$ · Here we have

$$\lim_{n \to \infty} f_n(x) = 0$$

for all values of x. The convergence is uniform in any closed interval which does not include $x = 0$, but it is not uniform in any interval having $x = 0$ in its interior or at one end. We shall investigate these statements about uniform convergence.

The function f_n is odd, i.e. $f_n(-x) = -f_n(x)$, so that the graph of $y = f_n(x)$ is symmetric with respect to the origin. Hence we confine our attention to values of x for which $x \geq 0$. The graph is easily constructed with the aid of information obtained from the derivative. An easy calculation shows that

$$f_n'(x) = \frac{n(1 - n^2x^2)}{(1 + n^2x^2)^2} .$$

The graph of $y = f_n(x)$, for $x > 0$, rises to a maximum at $x = 1/n$, and then diminishes toward zero as $x \to +\infty$. The maximum value is $f_n(1/n) = \frac{1}{2}$. The graphs for $n = 1, 2, 12$ are shown in Fig. 182. It is clear from

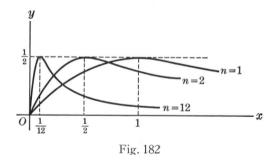

Fig. 182

this figure that the convergence is nonuniform in any interval containing the point $x = 0$, for such an interval will contain the point $x = 1/n$ if n is large enough, and so we can never have $|f_n(x)| < \epsilon$ for *all* values of x in this interval if $\epsilon < \frac{1}{2}$, no matter how large we take n. However, the convergence is uniform for all values of x such that $x \geq \delta$, where δ is any fixed positive number. For, $\epsilon > 0$ being given, if we choose N so large that

$$\frac{1}{N} < \delta \quad \text{and} \quad \frac{n\delta}{1 + n^2\delta^2} < \epsilon$$

if $N \leq n$, then it will also be true that

$$\frac{nx}{1 + n^2x^2} < \epsilon$$

if $N \leq n$ and $\delta \leq x$, so that the convergence is uniform for these latter values of x. The situation is illustrated in Fig. 183.

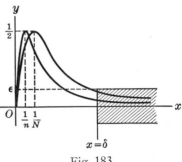

Fig. 183

When we are dealing with infinite series, we say that a series of functions, each defined on a certain fixed interval, is uniformly convergent on that interval provided that the sequence of partial sums $s_n(x)$ is uniformly convergent on the interval.

Sometimes it is convenient to express the condition for uniform convergence without reference to the limit function $f(x)$, but entirely in terms of the sequence $\{f_n(x)\}$.

THEOREM I. *Let the functions of the sequence $\{f_n(x)\}$ be defined on the interval $[a, b]$. In order that the sequence be uniformly convergent to a limit function $f(x)$ on the interval, it is necessary and sufficient that to each $\epsilon > 0$ there correspond some integer N independent of x such that*

$$(20.1\text{-}6) \qquad\qquad |f_n(x) - f_m(x)| < \epsilon$$

whenever $N \leq m$ and $N \leq n$.

The proof is a direct consequence of Cauchy's convergence criterion (Theorem VI, § 16.5). We shall prove only the sufficiency of condition (20.1-6), and leave the proof of the necessity to the student. It follows from (20.1-6) by Cauchy's criterion that the sequence $\{f_n(x)\}$ is convergent for each x of the interval, and so defines a limit function $f(x)$. We may then let $m \to \infty$ in (20.1-6) and obtain the result that $|f_n(x) - f(x)| \leq \epsilon$ if $N \leq n$. Since N is independent of x, this shows that the convergence is uniform.

Later on in this chapter we shall see that answers to questions A–D or A′–D′ may be stated in terms of the concept of uniform convergence.

EXERCISES

1. Discuss the functions $f(x) = \lim_{n \to \infty} f_n(x)$ defined by the following sequences for the values of x indicated. In each case sketch several typical graphs

$y = f_n(x)$, showing the character of the approximation of $f(x)$ by $f_n(x)$ for rather large values of n. Indicate the discontinuities, if any, of $f(x)$, and answer the particular questions in each case.

(a) $f_n(x) = (\sin x)^n$, $0 \leq x \leq \pi$. Is the convergence uniform? Explain.

(b) $f_n(x) = (\sin x)^{1/n}$, $0 \leq x \leq \pi$. For what values of a and b is the convergence uniform on $a \leq x \leq b$?

(c) $f_n(x) = (1/n)e^{-n^2x^2}$, all x. Is the convergence uniform? Prove your answer.

(d) $f_n(x) = nxe^{-nx}$, $x \geq 0$. Prove that the convergence is uniform for $x \geq \delta$, where δ is any positive number. Why is the convergence not uniform on the interval $0 \leq x \leq \delta$?

(e) $f_n(x) = \tan^{-1} nx$, all x. What conditions must a and b satisfy if the convergence is to be uniform on $a \leq x \leq b$?

(f) $f_n(x) = (x/n)e^{-x/n}$, $x \geq 0$. For given positive ϵ and A, find N such that $|f_n(x) - f(x)| < \epsilon$ if $0 \leq x \leq A$ and $N \leq n$. What is the corresponding verbal statement about uniform convergence? Is the convergence uniform for *all* x such that $x \geq 0$?

2. Follow the general directions of Exercise 1 in the case of $f_n(x) = x^n/(1+x^{2n})$, $x \geq 0$. Get what help you can from examination of $f_n'(x)$. Show that $|f_n(x)| \leq x^n$ if $x \geq 0$, and $|f_n(x)| \leq x^{-n}$ if $x > 0$. Use these inequalities to prove that the sequence converges uniformly if $0 \leq x \leq a < 1$, and also if $1 < b \leq x$, where a and b are constants. Is the convergence uniform in the interval $\frac{1}{2} \leq x \leq \frac{3}{2}$?

3. Let $f_n(x) = x^{a_n}$, $a_n = \frac{1}{2} + \frac{1}{4} + \cdots + 1/2^n$.

(a) Find $f(x) = \lim_{n \to \infty} f_n(x)$.

(b) Show that $|f_n(x) - f(x)| \leq 1 - a_n$ if $0 \leq x \leq 1$. What do you infer about uniform convergence? SUGGESTION: Consider where $f_n(x) - f(x)$ is a maximum on the interval $0 \leq x \leq 1$.

4. (a) Find a simple expression for the function $f(x) = \sum_{n=0}^{\infty} x^2/(1+x^2)^n$ when $x \neq 0$. What is $f(0)$? Does $f(x)$ have any discontinuities?

(b) If $s_n(x)$ is the sum of the first n terms of the series, show that $0 < f(x) - s_n(x) \leq 1/(1+\delta^2)^{n-1}$ if $x^2 \geq \delta^2 > 0$. What do you conclude about uniform convergence?

(c) Make a sketch showing the appearance of the curve $y = s_n(x)$ for a very large value of n. Is the series uniformly convergent when $|x| \leq \delta$?

5. Consider the function

$$f(x) = \frac{x}{x+1} + \left(\frac{2x}{2x+1} - \frac{x}{x+1}\right) + \left(\frac{3x}{3x+1} - \frac{2x}{2x+1}\right) + \cdots.$$

(a) Find the sum $s_n(x)$ of the first n terms of the series and hence find the value of $f(x)$ for each x. (b) Plot the curves $y = s_n(x)$ for several values of n, indicating how they look for large values of n. Is the convergence uniform when $0 \leq x \leq 1$? when $-1 \leq x \leq 0$? when $1 \leq x \leq 2$? (c) Show that $|s_n(x) - f(x)| \leq 1/(n\delta - 1)$ if $|x| \geq \delta > 0$ and $n\delta > 1$. What can you conclude about uniform convergence from this?

6. Give a proof of the necessity of condition (20.1-6) in Theorem I.

7. Suppose $f_0(x)$ is continuous when $0 \leq x \leq a$, and $f_n(x) = \int_0^x f_{n-1}(t)dt$. Prove that $f_n(x)$ converges uniformly to 0 when $0 \leq x \leq a$.

8. Prove that the series $\sum_{n=1}^{\infty} (-1)^{n+1}/(n + x^2)$ is uniformly convergent for all values of x. Also prove that it is not absolutely convergent for any value of x.

20.2   A comparison test for uniform convergence

We shall now give a theorem which is useful for proving that a series is uniformly convergent. While a series may be uniformly convergent without meeting the requirements of the theorem, it is nevertheless true that in a large number of important cases the theorem provides the simplest practical method for establishing the fact of uniform convergence.

THEOREM II.   *Let the terms of the series*

(20.2–1) $$u_1(x) + u_2(x) + \cdots + u_n(x) + \cdots$$

be defined on an interval. Let

(20.2–2) $$M_1 + M_2 + \cdots + M_n + \cdots$$

be a convergent series of positive constants, and let the inequality $| u_n(x) | \leq M_n$ be satisfied for all values of n and for all x on the interval. Then the series (20.2–1) is uniformly convergent on the interval. The word "interval" may be replaced by "point set" and the theorem is still true.

Example 1.   The geometric series

(20.2–3) $$1 + x + x^2 + x^3 + \cdots + x^n + \cdots$$

is uniformly convergent on the interval $-r \leq x \leq r$ if $0 < r < 1$. For, let

$$u_n(x) = x^n, \quad M_n = r^n.$$

Then, on the stated interval, $| x | \leq r$ and so $| u_n(x) | \leq M_n$. Since the series $\sum r^n$ is convergent, the uniform convergence of (20.2–3) follows by virtue of Theorem II.

We may note that although the series (20.2–3) converges at each point x of the *open* interval $-1 < x < 1$, it does not converge uniformly on that interval. For the partial sums of (20.2–3) may be written

$$1 + x + x^2 + \cdots + x^{n-1} = \frac{1 - x^n}{1 - x},$$

and the sequence $(1 - x^n)/(1 - x)$ is not uniformly convergent in the interval $0 \leq x < 1$, as may be seen by referring to the discussion in the second paragraph of § 20.1.

PROOF OF THEOREM II.   Let

$$s_n(x) = u_1(x) + \cdots + u_n(x).$$

Then, if $m < n$, $\quad s_n(x) - s_m(x) = u_{m+1}(x) + \cdots + u_n(x),$

and $\qquad\qquad |s_n(x) - s_m(x)| \leqq M_{m+1} + \cdots + M_n,$

since $|u_k(x)| \leqq M_k$. Now, if $\epsilon > 0$, there is some integer N such that

$$M_{m+1} + \cdots + M_n < \epsilon$$

if $N \leqq m < n$ (see Theorem VIII, § 19.3). Since the M's are constants, the choice of N is completely independent of x. But then we have

$$|s_n(x) - s_m(x)| < \epsilon$$

if $N \leqq m < n$; by Theorem I this establishes the uniform convergence of the series (20.2–1).

Theorem II is often called the M-test, or the Weierstrass M-test.

Example 2. The series

$$\frac{\sin x}{1^2} + \frac{\sin 2 x}{2^2} + \frac{\sin 3 x}{3^2} + \cdots + \frac{\sin nx}{n^2} + \cdots$$

is uniformly convergent on the entire x-axis. Here the comparison series may be taken to be

$$\frac{1}{1^2} + \frac{1}{2^2} + \cdots + \frac{1}{n^2} + \cdots,$$

since $|\sin nx| \leqq 1$ for all n and all x. The series of constants is known to be convergent.

EXERCISES

1. Prove what you can about the uniform convergence of each of the following series by the method of Theorem II. State specifically the range of values of x about which you make an assertion.

(a) $\displaystyle\sum_{n=1}^{\infty} \frac{n}{n+1} x^n.$

(b) $\displaystyle\frac{\cos x}{1 \cdot 2} + \frac{\cos 3 x}{3 \cdot 4} + \frac{\cos 5 x}{5 \cdot 6} + \cdots.$

(c) $\displaystyle\sum_{n=1}^{\infty} \frac{1}{n^2 + x^2}.$

(d) $\displaystyle\sum_{n=0}^{\infty} e^{-nx}.$

(e) $\displaystyle\sum_{n=1}^{\infty} \left(\frac{1}{n} - \frac{1}{x^2 + n} \right).$

2. Prove that the series $\displaystyle\sum_{n=1}^{\infty} \frac{2 x}{n^2 - x^2}$ is uniformly convergent on any finite closed interval not containing any of the points $\pm 1, \pm 2, \cdots$.

3. Show that the series $\sum_{n=1}^{\infty} 1/n^x$ converges uniformly for values of x such that $x \geq p$, if $p > 1$.

4. Suppose that $a_n \geq 0$ and that the series $\sum_{n=1}^{\infty} a_n n^{-x}$ is convergent when $x = x_0$. Show that it converges uniformly for values of x such that $x \geq x_0$.

5. Consider the series $\displaystyle\sum_{n=1}^{\infty} \frac{x^n}{n^2(1 + x^n)}$.

(a) Show that it converges uniformly when $x \geq 0$.
(b) Show that it converges uniformly if $|x| \leq a$, where $0 < a < 1$.
(c) Show that it converges uniformly if $b \leq x$, where $-1 < b$.
(d) Show that it converges uniformly if $x \leq -c$, where $c > 1$.

20.3 Continuity of the limit function

We now return to the questions A and A' of § 20.

THEOREM III. *Let the functions* $\{f_n(x)\}$ *be defined on the interval* $a \leq x \leq b$, *and let them converge uniformly on this interval to a limit function* $f(x)$. *Then, if each of the functions* f_n *is continuous at a point* x_0, *the limit function* f *is also continuous at* x_0. *In particular, if each* f_n *is continuous on the whole interval, so is* f.

PROOF. We write

$$f(x) - f(x_0) = (f(x) - f_n(x)) + (f_n(x) - f_n(x_0)) + (f_n(x_0) - f(x_0)),$$

(20.3–1) $\quad |f(x) - f(x_0)| \leq |f(x) - f_n(x)| + |f_n(x) - f_n(x_0)|$
$$+ |f_n(x_0) - f(x_0)| .$$

Now let $\epsilon > 0$ be given; x_0 is fixed, and we are to show that $|f(x) - f(x_0)| < \epsilon$ provided x is sufficiently close to x_0, for this is what is meant by saying that f is continuous at x_0. Now, by the uniform convergence, we can choose n, independent of x, so large that

$$|f_n(x) - f(x)| < \frac{\epsilon}{3}$$

for every x in the interval. If this is done, (20.3–1) shows that

(20.3–2) $\qquad |f(x) - f(x_0)| < |f_n(x) - f_n(x_0)| + \tfrac{2}{3}\,\epsilon.$

Now n has been fixed; since f_n is continuous at x_0, we shall have $|f_n(x) - f_n(x_0)| < \epsilon/3$ if x is sufficiently close to x_0. But then by (20.3–2) we shall have $|f(x) - f(x_0)| < \epsilon$. This completes the proof.

As a corollary of Theorem III we have the following assertion about series: *If the terms of an infinite series are continuous on an interval* $a \leq x \leq b$,

*and if the series is uniformly convergent on the interval, the function defined
as the sum of the series is continuous on the interval.*

EXERCISES

1. (a) Find $f(x) = \lim_{n \to \infty} x^{1/(2n-1)}$ for each value of x. From the nature of
$f(x)$ what do you conclude about uniform convergence of the sequence in intervals containing the origin?
   (b) Proceed as in (a) with $f(x) = \lim_{n \to \infty} x^{2/(2n-1)}$.

2. From an examination of the function $f(x) = \lim_{n \to \infty} 1/(1 + x^{2n})$ and use
of Theorem III, state what you conclude about lack of uniform convergence of the
sequence. Prove what you can of an affirmative nature about uniform convergence
of the sequence.

3. If $f(x) = 1 + \sum_{n=1}^{\infty} x^n/n$, find $\lim_{x \to 0} f(x)$ and justify your answer.

4. Is the function $f(x) = \sum_{n=0}^{\infty} xe^{-nx^2}$ continuous at $x = 0$? Explain your
answer.

5. Explain why the function

$$f(x) = \sum_{n=1}^{\infty} \frac{1}{2x} \left(\frac{1}{n-x} - \frac{1}{n+x} \right)$$

is continuous except at the points $x = 0, \pm 1, \pm 2, \cdots$. What is $\lim_{x \to 0} f(x)$?

20.4   Integration of sequences and series

The questions C and C' of § 20 may be answered, in one way at least,
by the following theorem.

THEOREM IV.   *Let the functions $f_n(x)$ be continuous on the closed interval
$a \le x \le b$, and let them converge uniformly on this interval to the limit
function $f(x)$. Then*

$$(20.4\text{--}1) \qquad\qquad \int_a^b f(x)dx = \lim_{n \to \infty} \int_a^b f_n(x)dx.$$

For infinite series this is equivalent to the statement that if the series

$$f(x) = \sum_{n=1}^{\infty} u_n(x)$$

*is uniformly convergent on the interval $a \le x \le b$, and if the terms $u_n(x)$
are continuous on the interval, then*

$$(20.4\text{--}2) \qquad\qquad \int_a^b f(x)dx = \sum_{n=1}^{\infty} \int_a^b u_n(x)dx.$$

PROOF.   The proof is quite easy. Consider (20.4–1).

$$\int_a^b f_n(x)dx - \int_a^b f(x)dx = \int_a^b [f_n(x) - f(x)]dx,$$

(20.4–3) $$\left| \int_a^b f_n(x)dx - \int_a^b f(x)dx \right| \leq \int_a^b \left| f_n(x) - f(x) \right| dx.$$

Now suppose $\epsilon > 0$ is given. Choose N, independent of x, so large that if $N \leq n$ we have

(20.4–4) $$|f_n(x) - f(x)| < \frac{\epsilon}{b - a}$$

if $a \leq x \leq b$. This is possible because of the assumed uniform convergence. Now from (20.4–3) and (20.4–4) we see that

(20.4–5) $$\left| \int_a^b f_n(x)dx - \int_a^b f(x)dx \right| < \epsilon$$

if $N \leq n$, for

$$\int_a^b |f_n(x) - f(x)| \, dx < \int_a^b \frac{\epsilon}{b - a} \, dx = \epsilon.$$

But (20.4–5) is just the condition which means that, by definition, (20.4–1) is true.

For the case of the infinite series, (20.4–1) becomes, with $f_n(x)$ as the partial sums,

$$\int_a^b f(x)dx = \lim_{n \to \infty} \int_a^b [u_1(x) + \cdots + u_n(x)]dx$$

$$= \lim_{n \to \infty} \left[\int_a^b u_1(x)dx + \cdots + \int_a^b u_n(x)dx \right].$$

This last relation, in turn, is just another way of expressing (20.4–2).

Example.   We saw in Example 1, § 20.2, that the geometric series

$$\frac{1}{1 - x} = 1 + x + \cdots + x^n + \cdots$$

is uniformly convergent if $-r \leq x \leq r$, where $0 < r < 1$. Let us then apply (20.4–2), integrating from $-r$ to r. The result is

(20.4–6)

$$\int_{-r}^r \frac{dx}{1 - x} = \int_{-r}^r dx + \int_{-r}^r x \, dx + \cdots + \int_{-r}^r x^n \, dx + \cdots.$$

Now
$$\int_{-r}^{r} \frac{dx}{1 - x} = - \log (1 - x) \Big|_{-r}^{r} = \log \frac{1 + r}{1 - r},$$

$$\int_{-r}^{r} x^n \, dx = \frac{r^{n+1}}{n + 1} - \frac{(- r)^{n+1}}{n + 1}.$$

The latter expression is equal to 0 if $n + 1$ is even, and equal to $2 \, r^{n+1}/(n + 1)$ if $n + 1$ is odd. Therefore (20.4–6) becomes

$$\log \frac{1 + r}{1 - r} = 2\Big(r + \frac{r^3}{3} + \frac{r^5}{5} + \cdots \Big).$$

This result was obtained by a different method in Example 1, § 19.6.

As suggested by this example, Theorem IV has important applications in deriving certain series expansions from other series expansions by integration.

The conclusions (20.4–1) and (20.4–2) of Theorem IV *may* be false if the convergence is not uniform. This is illustrated by Example 3, § 20, in which the convergence is not uniform. To have (20.4–1) and (20.4–2) it is *sufficient* to have uniform convergence; but uniform convergence is not a *necessary* condition in all cases. Suppose, for example, that we modify Example 3 of § 20 by taking the height of the triangle in the graph of $y = f_n(x)$ to be $2\sqrt{n}$ instead of $2 \, n$. The convergence is still nonuniform, and $\lim_{n \to \infty} f_n(x) = 0$ when $0 \le x \le 1$. But now

$$\int_0^1 f_n(x) dx = \frac{1}{\sqrt{n}}$$

and so (20.4–1) is true in this case.

EXERCISES

1. If $f(x) = \sum_{n=1}^{\infty} \frac{\cos nx}{n^2}$, show that $\int_0^{\pi/2} f(x) dx = \sum_{n=0}^{\infty} \frac{(-1)^n}{(2n + 1)^3}$. Justify your reasoning.

2. If $f(x) = \frac{\sin 3x}{1 \cdot 2} + \frac{\sin 5x}{3 \cdot 4} + \frac{\sin 7x}{5 \cdot 6} + \cdots$, find a series for $\int_0^{\pi/2} f(x) dx$.

3. If $f_n(x) = nxe^{-nx^2}$ and $f(x) = \lim_{n \to \infty} f_n(x)$, show that the sequence converges nonuniformly on the interval $0 \le x \le 1$, and that

$$\int_0^1 f(x) dx \ne \lim_{n \to \infty} \int_0^1 f_n(x) dx.$$

4. If $f(x) = \lim_{n \to \infty} f_n(x)$, where $f_n(x) = \frac{2 nx}{1 + n^2 x^4}$, find

$$\int_0^1 f(x)dx \quad \text{and} \quad \lim_{n \to \infty} \int_0^1 f_n(x)dx.$$

What do you conclude about uniform convergence of the sequence on the interval $0 \le x \le 1$?

5. From the series expansions of e^t, $\sin t$, and $\cos t$ in powers of t (see § 19.1), verify that $\int_a^b e^t \, dt = e^b - e^a$ and that $\int_0^x \sin t \, dt = 1 - \cos x$, $\int_0^x \cos t \, dt = \sin x$. Prove whatever you need about uniform convergence for applying Theorem IV.

20.5   Differentiation of sequences and series

We now come back to questions B, D, B', and D' of § 20. We shall answer D and D', and in so doing shall answer B and B'. That is, we shall give conditions which will not only assure us that the limit function is differentiable, but will tell us how to express the derivative itself as a limit function. Here again uniform convergence is the key to the situation. But now we must look for uniform convergence of the sequence $\{f_n'(x)\}$.

THEOREM V. *Let the functions $f_n(x)$ be defined and have continuous derivatives on the interval $a \le x \le b$. Let the sequence $\{f_n'(x)\}$ be uniformly convergent on the interval, and let the sequence $\{f_n(x)\}$ itself be convergent, with limit function $f(x)$. Then f is differentiable, and*

$$(20.5\text{--}1) \qquad\qquad f'(x) = \lim_{n \to \infty} f_n'(x).$$

For infinite series this is equivalent to the statement that if

$$f(x) = \sum_{n=1}^{\infty} u_n(x)$$

is convergent and if the series of derivatives

$$\sum_{n=1}^{\infty} u_n'(x)$$

is uniformly convergent, then

$$(20.5\text{--}2) \qquad\qquad f'(x) = \sum_{n=1}^{\infty} u_n'(x).$$

Here it is assumed that each term $u_n(x)$ has a continuous derivative.

PROOF. Let us denote the limit of the sequence $\{f_n'(x)\}$ by $g(x)$; we do not yet know that $g(x) = f'(x)$. Since $f_n'(x)$ converges uniformly to $g(x)$, Theorem IV permits us to write

$$(20.5\text{--}3) \qquad\qquad \int_a^x g(t)dt = \lim_{n \to \infty} \int_a^x f_n'(t)dt,$$

for the convergence is also uniform on any subinterval $a \leqq t \leqq x$, where x is any point of $[a, b]$. Now

$$\int_a^x f_n'(t)dt = f_n(x) - f_n(a).$$

Since $f_n(x) \to f(x)$ as $n \to \infty$, (20.5–3) becomes

$$\int_a^x g(t)dt = f(x) - f(a),$$

or

(20.5–4) $$f(x) = \int_a^x g(t)dt + f(a).$$

Now g is continuous, by Theorem III. We conclude from (20.5–4) and Theorem VII, § 1.52, that f is differentiable, with $f'(x) = g(x)$. This result is just another way of stating (20.5–1). The transition from (20.5–1) to (20.5–2) is accomplished in the same way as was done for (20.4–1) and (20.4–2) at the end of the proof of Theorem IV.

Example. Use Theorem V to aid in obtaining the series expansions, valid if $|x| < 1$,

(20.5–5)
$$\frac{1}{(1 - x)^2} = 1 + 2\,x + 3\,x^2 + 4\,x^3 + \cdots + (n + 1)x^n + \cdots,$$

(20.5–6)
$$\frac{2}{(1 - x)^3} = 2 + 3 \cdot 2\,x + 4 \cdot 3\,x^2 + \cdots + (n + 2)(n + 1)x^n + \cdots.$$

We start from the geometric series

(20.5–7) $$\frac{1}{1 - x} = 1 + x + x^2 + \cdots + x^n + \cdots,$$

which we know to be convergent if $|x| < 1$. The series (20.5–5) results from (20.5–7) according to the rule of (20.5–2), and (20.5–6) is derived from (20.5–5) in a similar manner. In order to justify this procedure by Theorem V we must prove that each of the series (20.5–5), (20.5–6) is uniformly convergent if $-r \leqq x \leqq r$ for any r such that $0 < r < 1$. If x is any fixed point such that $-1 < x < 1$, we can choose r so that $|x| < r < 1$. Then x will be in an interval of uniform convergence, and the foregoing formulas will be justified.

Suppose then that

$$0 < r < 1.$$

If $-r \leqq x \leqq r$ we have

$$|(n + 1)x^n| \leqq (n + 1)r^n.$$

Now consider the series

$$(20.5\text{-}8) \qquad 1 + 2\,r + 3\,r^2 + \cdots + (n + 1)r^n + \cdots.$$

This is a series of positive constants. It is convergent, for the ratio of successive terms is

$$\frac{(n + 2)r^{n+1}}{(n + 1)r^n} = \frac{n + 2}{n + 1}\,r,$$

and this ratio converges to r as $n \to \infty$. Since $r < 1$, the series (20.5-8) is convergent, by Theorem VII, § 19.22. The uniform convergence of the series (20.5-5) is now a consequence of the M-test (Theorem II, § 20.2); the M-series in this case is (20.5-8).

The proof that the series (20.5-6) converges uniformly if $| x | \leqq r, r < 1$, is entirely similar, and the details are left to the student.

EXERCISES

1. Which of the following functions $f(x)$ has the property that $f'(x)$ can be calculated for each x on the specified interval by differentiating the series for $f(x)$ term by term?

(a) $f(x) = \sum\limits_{n=1}^{\infty} \frac{\sin nx}{n^2}$, $0 \leqq x \leqq 2\,\pi$.

(b) $f(x) = \sum\limits_{n=1}^{\infty} (-1)^{n+1} \frac{x^n}{n}$, $0 \leqq x \leqq 1$.

(c) $f(x) = \sum\limits_{n=2}^{\infty} \frac{x^n}{n^2(\log n)^2}$, $-1 \leqq x \leqq 1$.

(d) $f(x) = \sum\limits_{n=0}^{\infty} \frac{x^{2n}}{2\,n + 1}$, $-1 < x < 1$.

(e) $f(x) = \sum\limits_{n=1}^{\infty} \left(\frac{1}{x - n\pi} + \frac{1}{n\pi} \right)$, $0 < x < \pi$.

(f) $f(x) = 1 - \frac{x^2}{2^2} + \frac{x^4}{2^2 \cdot 4^2} - \cdots + (-1)^n \frac{x^{2n}}{2^2 \cdot 4^2 \cdots (2\,n)^2} + \cdots$, all x.

2. Prove that the series (20.5-6) converges uniformly if $| x | \leqq r$, where $0 < r < 1$.

3. Let $f_n(x) = (2/\pi)x \tan^{-1} nx$, $f(x) = \lim_{n \to \infty} f_n(x)$. Show that $f(x) = | x |$, so that it is clear that f is not differentiable at $x = 0$. Show that $\lim_{n \to \infty} f_n'(x)$ exists for each x, including $x = 0$. What do you conclude about the uniformity of convergence of the sequence of derivatives?

4. Let $f(x) = \lim_{n \to \infty} f_n(x)$, $f_n(x) = (1/n)e^{-n^2x^2}$. Show that $\lim_{n \to \infty} f_n'(x) = f'(x)$ for every x, but that the convergence of the sequence of derivatives is not uniform in any interval containing the origin. The original sequence is convergent uniformly on the entire axis.

Power Series · 21

21. General remarks

The purpose of this chapter is to give a systematic exposition of some of the most important things about power series.

The representation of functions by power series is one of the most useful of mathematical techniques in a wide variety of situations. Sometimes we start from a function that is defined for us in some manner not employing series, and seek to expand the function in a power series. At other times we may form a power series, or have one presented to us, and then undertake to use this function in some way. In either of these situations we need to know something of what properties a function has if it is defined by a power series.

The general form of a power series is

$$(21\text{--}1) \quad a_0 + a_1(x - x_0) + a_2(x - x_0)^2 + \cdots + a_n(x - x_0)^n + \cdots;$$

this is called a power series in $x - x_0$. Here x_0 is fixed and x is variable. In the special case where $x_0 = 0$ the series takes the form

$$(21\text{--}2) \quad a_0 + a_1 x + a_2 x^2 + \cdots + a_n x^n + \cdots.$$

It turns out that in studying power series it is sufficient to consider (21–2), since the general case (21–1) can be reduced to (21–2) by a translation of the origin along the x-axis. For this reason all the general theory of power series will be developed for series of powers of x, of the form (21–2).

For convenience we shall often refer to the series (21–2) as the series $\sum a_n x^n$, omitting the limits $n = 0$ and ∞ for greater ease in printing.

21.1 The interval of convergence

The first important fact about a power series is expressed in the following theorem:

THEOREM I.   *Suppose a power series $\sum a_n x^n$ is convergent for $x = x_0$, where $x_0 \neq 0$. Then it is absolutely convergent when $|x| < |x_0|$. The same conclusion holds under the weaker assumption that there is some positive constant A such that $|a_n x_0^n| \leq A$ for all values of n.*

PROOF.   If $\sum a_n x_0^n$ is convergent, $a_n x_0^n \to 0$, and therefore the terms $a_n x_0^n$ are bounded. Under the hypothesis in the theorem we have

$$a_n x^n = a_n x_0^n \left(\frac{x}{x_0}\right)^n, \quad |a_n x^n| \leq A\left(\frac{|x|}{|x_0|}\right)^n.$$

662

The geometrical series $$\sum A\left(\frac{|x|}{|x_0|}\right)^n$$

is convergent if $|x| < |x_0|$; therefore the series $\sum |a_nx^n|$ is convergent, by the comparison test (Theorem II, § 19.2). This proves the theorem.

Conceivably a particular power series may be convergent for all values of x. This is the case with the series for e^x, for instance (see (19.1–6)). It is also conceivable that a series may not be convergent for any values of x except the one value $x = 0$. Such is the case with the series

$$1!x + 2!x^2 + \cdots + n!x^n + \cdots,$$

for, with $u_n = n!x^n$, if $|x| \neq 0$,

$$\left|\frac{u_{n+1}}{u_n}\right| = (n+1)|x| \to +\infty$$

as $n \to \infty$, and so the series is divergent, by Theorem XIII, § 19.4.

Let us now consider the case in which the series $\sum a_nx^n$ is convergent for at least one nonzero value of x, but is also divergent for some value of x. Let us denote by R the least upper bound of the positive numbers x for which $\sum a_nx^n$ is convergent. Denote this set of positive numbers x by S. The set S is not empty, for by Theorem I, if the series is convergent for the nonzero value x_0, it is convergent for all positive x such that $x < |x_0|$. The set S must be bounded, because of our assumption that the series is not convergent for all values of x. For if S were not bounded, then for any x we could find an x_0 in S such that $|x| < x_0$, and this would imply that $\sum a_nx^n$ is convergent. Since S is not empty and has an upper bound, it has a least upper bound (Theorem II, § 2.7). This justifies the introduction of the number R.

We now assert the following: The series $\sum a_nx^n$ is absolutely convergent if $|x| < R$ and is divergent if $|x| > R$. The proof is a simple consequence of Theorem I. Suppose $|x| < R$. Then choose x_0 so that x_0 is in the class S and $|x| < x_0 \leqq R$; this is possible, since R is the least upper bound of S. But then the series $\sum a_nx^n$ converges absolutely, by Theorem I. On the other hand, suppose that $|x| > R$. Then the series $\sum a_nx^n$ cannot converge, for if it did, Theorem I would assure us that $\sum a_ny^n$ converges if $R < y < |x|$, so that y would be in S, contrary to the fact that R is the least upper bound of S.

We sum up the foregoing conclusions in theorem form.

THEOREM II. *For a power series $\sum a_nx^n$ there are three possibilities:*

(1) *it is absolutely convergent for all values of x;*
(2) *it diverges for every $x \neq 0$;*

(3) *there is a positive number R such that the series converges absolutely if $| x | < R$ and diverges if $| x | > R$.*

In case (3) we call R the *radius of convergence* of the series. The interval $- R < x < R$ is called the *interval of convergence.* The series may or may not converge at the end points $x = \pm R$. It may converge at both, at just one, or at neither. These possibilities were illustrated in Chapter XIX.

In case (1) we say that the radius of convergence is infinite, and that the interval of convergence is the entire x-axis. It is a convenient symbolism to write $R = \infty$ in this case. In case (2) we write $R = 0$; here there is no interval of convergence.

THEOREM III. *Suppose that the series $\sum a_n x^n$ has a positive or infinite radius of convergence R. Let $0 < r < R$ (if $R = \infty$, r may be any positive number). Then the series converges uniformly on the closed interval $- r \leqq x \leqq r$.*

PROOF. The series $\sum | a_n | r^n$ is convergent, since the power series converges absolutely at $x = r$, by Theorem II. If $- r \leqq x \leqq r$ we have $| a_n x^n | \leqq | a_n | r^n$. The uniform convergence is therefore a consequence of the Weierstrass M-test (Theorem II, § 20.2), with $M_n = | a_n | r^n$.

In the cases most commonly arising in practice, the radius of convergence of a power series may be found by using d'Alembert's ratio test (Theorem XIII, § 19.4), as illustrated by Examples 1 and 2 in § 19.4. If the limit

$$(21.1–1) \qquad \lim_{n \to \infty} \left| \frac{a_{n+1}}{a_n} \right| = L$$

exists, and if we write $u_n = a_n x^n$, then

$$\lim_{n \to \infty} \left| \frac{u_{n+1}}{u_n} \right| = \lim_{n \to \infty} \left| \frac{a_{n+1}}{a_n} \right| | x | = L | x | \;;$$

the series converges if $L | x | < 1$, and diverges if $L | x | > 1$. From this we conclude that $R = 1/L$ if $L \neq 0$. Also, $R = \infty$ if $L = 0$, and $R = 0$ if $L = \infty$. As a formal statement we have

THEOREM IV. *The radius of convergence of the series $\sum a_n x^n$ is given by*

$$(21.1–2) \qquad R = \lim_{n \to \infty} \left| \frac{a_n}{a_{n+1}} \right| ,$$

provided that the limit exists or is $+ \infty$.

There are power series for which the limit (21.1–2) does not exist; there is then a means of determining R by examining the sequence $\mid a_n \mid^{1/n}$. This problem is considered in § 21.5.

Theorem III has important consequences, the chief immediate ones of which have to do with the function defined by a power series.

THEOREM V.   *Let a function f be defined by*

$$(21.1\text{–}3) \qquad\qquad f(x) = \sum_{n=0}^{\infty} a_n x^n,$$

where it is assumed that the radius of convergence of the power series is not zero. The function f so defined is continuous in the open interval of convergence of the series. Moreover, if a and b are points of this interval,

$$(21.1\text{–}4) \qquad\qquad \int_a^b f(x)dx = \sum_{n=0}^{\infty} a_n \frac{b^{n+1} - a^{n+1}}{n+1};$$

that is, the integral of the function is equal to the series obtained by integrating the original power series term by term.

PROOF.   The continuity of f is a direct consequence of Theorem III, § 20.3, for if x_0 is any point of the interval of convergence, we may choose $r > 0$ so that $\mid x_0 \mid < r < R$, and the series converges uniformly on the interval $-r \leqq x \leqq r$, which includes the point x_0. The assertion (21.1–4) is a direct application of (20.4–2) in Theorem IV, § 20.4, for the power series converges uniformly on the closed interval $[a, b]$, each point of which is interior to the interval of convergence.

Theorem V leaves something to be desired in one respect. It does not give us any information about what happens at an end of the interval of convergence. Suppose, for example, that the series (21.1–3) happens to be convergent at $x = R$, where R is the radius of convergence. Will the function f be continuous at $x = R$? That is, will

$$f(x) \rightarrow \sum_{n=0}^{\infty} a_n R^n$$

as $x \rightarrow R$ from the left? The answer to this question is in the affirmative, but the proof is not covered in Theorem V. We shall take up this matter in § 21.4. A similar question arises with regard to (21.1–4). Is it ever possible to put $b = R$ or $a = -R$? This also is considered in § 21.4.

There are many important practical applications of Theorem V, particularly of (21.1–4).

Example 1.   Derive the series expansion

$$(21.1\text{–}5)$$
$$\sin^{-1} x = x + \frac{1}{2}\frac{x^3}{3} + \frac{1 \cdot 3}{2 \cdot 4}\frac{x^5}{5} + \frac{1 \cdot 3 \cdot 5}{2 \cdot 4 \cdot 6}\frac{x^7}{7} + \cdots.$$

We start from the fact that

$$(21.1\text{-}6) \qquad\qquad \sin^{-1} x = \int_0^x \frac{dt}{\sqrt{1 - t^2}}.$$

Formula (21.1-6) is valid if $|x| \leq 1$; the integral is improper if $x = \pm 1$, since the integrand becomes infinite at $t = \pm 1$. In the binomial series (19.5-5) we replace x by $-t^2$ and set $m = -\frac{1}{2}$. The result is

$$(1 - t^2)^{-1/2} = 1 - \tfrac{1}{2}(-t^2) + \frac{(-\frac{1}{2})(-\frac{3}{2})}{2!} (-t^2)^2 +$$

$$\frac{(-\frac{1}{2})(-\frac{3}{2})(-\frac{5}{2})}{3!} (-t^2)^3 + \cdots$$

$$= 1 + \tfrac{1}{2} t^2 + \frac{1 \cdot 3}{2 \cdot 4} t^4 + \frac{1 \cdot 3 \cdot 5}{2 \cdot 4 \cdot 6} t^6 + \cdots;$$

this result is valid if $|t| < 1$, and the series has radius of convergence $R = 1$. Hence we may integrate the series from $t = 0$ to $t = x$ if $|x| < 1$. Using (21.1-6), we obtain the result (21.1-5). This argument, based on Theorem V, shows that (21.1-5) is valid if $|x| < 1$. As a matter of fact, one can easily verify that the series converges even when $x = \pm 1$, by using Raabe's test (Theorem XIV, § 19.4); it is then a consequence of the discussion in § 21.4 that (21.1-5) is valid if $-1 \leq x \leq 1$.

Example 2.   Find an expansion in powers of x of the function

$$f(x) = \int_0^1 \frac{1 - e^{-tx}}{t} \, dt,$$

and use the result to calculate $f(\frac{1}{2})$ approximately.

From the series (19.1-6) for the exponential function we have

$$e^{-tx} = 1 - tx + \frac{t^2 x^2}{2!} - \frac{t^3 x^3}{3!} + \cdots.$$

Therefore $1 - e^{-tx} = 1 - \left(1 - tx + \frac{t^2 x^2}{2!} - \frac{t^3 x^3}{3!} + \cdots\right)$

$$\frac{1 - e^{-tx}}{t} = x - \frac{tx^2}{2!} + \frac{t^2 x^3}{3!} - \cdots + (-1)^{n-1} \frac{t^{n-1} x^n}{n!} + \cdots.$$

This series representation is valid for all values of x and t; the radius of convergence of the power series in t is $R = \infty$. Integrating, we have

$$f(x) = x - \frac{x^2}{2 \cdot 2!} + \frac{x^3}{3 \cdot 3!} - \cdots + (-1)^{n-1} \frac{x^n}{n \cdot n!} + \cdots.$$

In particular,

$$f\left(\frac{1}{2}\right) = \int_0^1 \frac{1 - e^{-t/2}}{t} \, dt = \frac{1}{2} - \frac{1}{2 \cdot 2!}\left(\frac{1}{2}\right)^2 + \frac{1}{3 \cdot 3!}\left(\frac{1}{2}\right)^3 - \cdots$$

$$= 1.13 \text{ (approximately)}.$$

EXERCISES

1. Find the radius of convergence of each of the following series.

(a) $\displaystyle\sum \frac{(2\,n)!}{(n!)^2}\, x^n.$ (c) $\displaystyle\sum \frac{(3\,n)!}{(n!)^2}\, x^n.$

(b) $\displaystyle\sum \frac{(n!)^3}{(3\,n)!}\, x^n.$ (d) $\displaystyle\sum \frac{(n!)^{7/2}}{(4\,n)!}\, x^n.$

(e) $\displaystyle\sum \frac{(n+p)!}{n!(n+q)!}\, x^n,$ p and q positive integers.

(f) $\displaystyle\sum \frac{[1\cdot 3\cdots(2\,n-1)]^2}{2^{2n}(2\,n)!}\, x^n.$

2. Find the radius of convergence of the series $\displaystyle\sum \frac{(pn)!}{(n!)^q}\, x^n$, where p is a positive integer and $q > 0$. Consider all possibilities.

3. Find the radius of convergence of the following series:

$$\sum_{n=1}^{\infty} \frac{[q(q-2)\cdots(q-2\,n+2)][(q+1)(q+3)\cdots(q+2\,n-1)]}{(2\,n)!}\, x^n,$$

where q is not a positive even integer.

4. Let the radii of convergence of $\sum a_n x^n$ and $\sum b_n x^n$ be R_1 and R_2, respectively. Suppose $|\,a_n\,| \leq |\,b_n\,|$ for all sufficiently large values of n. Prove that $R_1 \geq R_2$.

5. Let $\{a_n\}$ be a bounded sequence, and let R be the radius of convergence of the series $\sum a_n x^n$. Prove that $R \geq 1$.

6. If the radius of convergence of $\sum a_n x^n$ is R, prove that the radius of convergence of $\sum a_n x^{2n}$ is $R^{1/2}$.

7. Find an expansion as a power series in x for each of the following functions. Indicate the radius of convergence in each case.

(a) $\displaystyle\int_0^x \frac{\sin t}{t}\, dt.$ (c) $\displaystyle\int_0^x e^{-t^2}\, dt.$

(b) $\displaystyle\int_0^x \cos u^2\, du.$ (d) $\displaystyle\int_0^x \frac{t^p}{\sqrt{1-t^2}}\, dt,$ p a positive integer.

8. Deduce the expansion

$$\sinh^{-1} x = x + \sum_{n=1}^{\infty}(-1)^n \frac{1\cdot 3\cdots(2\,n-1)}{2\cdot 4\cdots 2\,n}\cdot\frac{x^{2n+1}}{2\,n+1}.$$

State the justification of your procedure, and indicate for what values of x you are proving the validity of the expansion.

9. If $\displaystyle f(x) = \int_0^x \frac{\log(1+u)}{u}\, du$, find a series for $f(x)$, and calculate the approximate value of $f(\tfrac{1}{10})$.

10. Find the approximate value of $\displaystyle\int_0^1 \frac{1-\cos x}{x^2}\, dx.$

11. Find a series for the function $f(x) = \displaystyle\int_0^x \dfrac{\log(1+t)}{1+t}\, dt$, and calculate the approximate value of $f(\tfrac{1}{10})$. See Example 2, § 19.6.

12. Find the approximate value of $\displaystyle\int_0^{1/2} \dfrac{e^{-x^2}}{\sqrt{1-x^2}}\, dx$.

13. Use series (21.1–6) to write numerical series for $\pi/2$ and $\pi/6$, respectively.

21.2   Differentiation of power series

The principal fact to be established in this section is the following: If a function is defined by a power series which has a positive or infinite radius of convergence, then the function has derivatives of all orders at each point of the open interval of convergence, and these derivatives are represented by the series which are obtained by differentiation of the original series term by term. Thus, if

$$(21.2\text{–}1) \qquad f(x) = a_0 + a_1 x + a_2 x^2 + a_3 x^3 + a_4 x^4 + \cdots$$

is convergent when $|x| < R$, then

$$(21.2\text{–}2) \qquad f'(x) = a_1 + 2\,a_2 x + 3\,a_3 x^2 + 4\,a_4 x^3 + \cdots$$
$$(21.2\text{–}3) \qquad f''(x) = 2\,a_2 + 3\cdot 2\,a_3 x + 4\cdot 3\,a_4 x^2 + \cdots$$
$$(21.2\text{–}4) \qquad f'''(x) = 3!\,a_3 + 4\cdot 3\cdot 2\,a_4 x + \cdots,$$

and so on, all these series likewise being convergent if $|x| < R$. To get at these facts we begin with a consideration of the series (21.2–1) and (21.2–2).

THEOREM VI.   *In the case of any power series, the series in (21.2–1) and the series in (21.2–2) have the same radius of convergence.*

PROOF.   Let R and R' denote the radii of convergence of the series (21.2–1) and (21.2–2), respectively. Suppose $|x| < R$, and choose x_0 so that $|x| < |x_0| < R$. Then the series (21.2–1) is convergent with $x = x_0$, and consequently $a_n x_0^n \to 0$. We may therefore choose a number $A > 0$ such that $|a_n x_0^n| \le A$ for all n. Then

$$na_n x^{n-1} = \frac{n}{x_0}\,a_n x_0^n \left(\frac{x}{x_0}\right)^{n-1},$$

$$(21.2\text{–}5) \qquad |na_n x^{n-1}| \le \frac{A}{|x_0|}\,nr^{n-1},$$

where

$$r = \frac{|x|}{|x_0|} < 1.$$

The series

$$\sum \frac{A}{|x_0|}\,nr^{n-1}$$

is convergent, for the limit of the ratio of the term of index $n + 1$ to the term of index n is

$$\lim_{n \to \infty} \frac{n+1}{n} r = r < 1 .$$

Consequently, by (21.2–5), the series

$$\sum na_n x^{n-1}$$

is convergent. This is precisely the series (21.2–2), and so we have proved that this series converges if $| x | < R$. It follows that the radius of convergence of (21.2–2) is *not less* than that of (21.2–1), i.e. $R' \geq R$. If $R = \infty$ this means that $R' = \infty$ also.

To complete the proof we show that $R' > R$ is impossible, so that $R' = R$. For suppose that $R' > R$ and choose x so that $R < | x | < R'$. Then, for this x, the series (21.2–2) is absolutely convergent and the series (21.2–1) is divergent. Now

$$| a_n x^n | = | na_n x^{n-1} | \left| \frac{x}{n} \right| < | na_n x^{n-1} |$$

as soon as $n > | x |$. This comparison shows that the series (21.2–1) must be convergent, which is a contradiction. The proof of Theorem VI is now complete.

THEOREM VII. *Let $f(x)$ be defined by the power series (21.2–1), and assume that the radius of convergence is not zero. Then f is differentiable at each point inside the interval of convergence, and $f'(x)$ is represented by the series (21.2–2). Application of this conclusion to f' in place of f shows that $f''(x)$ is represented by (21.2–3), and so on.*

The proof is an immediate consequence of Theorem V, § 20.5, and Theorem III, § 21.1, in view of Theorem VI of the present section.

It is possible that the series for $f'(x)$ will diverge at an end of the interval of convergence, even though the series for $f(x)$ may be convergent at that point. An example is furnished by the series for $\log (1 + x)$ at $x = 1$ (see (19.1–1)); this is convergent, but the series for the derivative, $(1 + x)^{-1}$, is divergent at $x = 1$.

Theorem VII shows that there is a great difference between functions defined by power series and functions defined by series of less special type. Another very important class of series is the class of trigonometric series, of which

$$\frac{\sin x}{1^2} + \frac{\sin 2 x}{2^2} + \cdots + \frac{\sin nx}{n^2} + \cdots$$

is an example. We saw in Example 2, § 20.2, that this series is uniformly convergent for all values of x. If we differentiate the series term by term we get

$$\frac{\cos x}{1} + \frac{\cos 2\,x}{2} + \frac{\cos 3\,x}{3} + \cdots + \frac{\cos nx}{n} + \cdots .$$

This series is convergent for some values of x, but not for all values; for instance, it is divergent when $x = 0$. Another termwise differentiation gives the series

$$- \sin x - \sin 2\,x - \cdots - \sin nx - \cdots ,$$

which is convergent if $x = 0$, but is divergent except when x is an integral multiple of π.

Theorem VII permits us to show the relation between the general theory of power series and the expansion of a function in a power series by means of Taylor's series. In the Taylor's series expansion of a function in powers of x the coefficient of x^n is $\dfrac{f^{(n)}(0)}{n!}$ (see (19.1–4)).

THEOREM VIII.   *If a function $f(x)$ is defined by a power series (21.2–1) with positive or infinite radius of convergence, the coefficients are related to the function by the formulas*

(21.2–6) $$a_n = \frac{f^{(n)}(0)}{n!}.$$

This means that the power series is the Taylor's series of the function.

The formulas (21.2–6) are established by setting $x = 0$ in the successive series for $f(x)$, $f'(x)$, $f''(x)$, etc. (see (21.2–1)–(21.2–4)). It is clear by induction that the leading term in the series for $f^{(n)}(x)$ is $n!a_n$, and (21.2–6) is a direct consequence.

THEOREM IX.   *Suppose that two power series are convergent and have the same sum for all values of x in some interval $|x| < r$:*

$$\sum_{n=0}^{\infty} a_n x^n = \sum_{n=0}^{\infty} b_n x^n, \qquad -r < x < r.$$

Then $a_n = b_n$ for all n.

This theorem is called the *uniqueness theorem* for power series. It is a corollary of Theorem VIII. For, let $f(x)$ be the common sum of the two series. Then by (21.2–6) we see that a_n and b_n are both equal to $\dfrac{f^{(n)}(0)}{n!}$, and therefore equal to each other.

Example 1. Consider the functions $J_0(x)$, $J_1(x)$ defined as follows:

(21.2–7)

$$J_0(x) = 1 - \frac{x^2}{(1!)^2 2^2} + \frac{x^4}{(2!)^2 2^4} - \cdots + (-1)^n \frac{x^{2n}}{(n!)^2 2^{2n}} + \cdots,$$

(21.2–8)

$$J_1(x) = \frac{x}{2}\left[1 - \frac{x^2}{1!2!2^2} + \cdots + (-1)^n \frac{x^{2n}}{n!\,(n+1)!2^{-n}} + \cdots\right].$$

Show that they are defined for all values of x, and that

(21.2–9)
$$J_0'(x) = -J_1(x).$$

The function $J_0(x)$ is called the *Bessel function of order zero of first kind*; $J_1(x)$ *is called the Bessel function of order one of first kind.* These and other varieties of Bessel functions are of great importance because of the way they arise in many kinds of physical problems.

Both series are convergent for all values of x, as is readily verified by the ordinary ratio test (Theorem XIII, § 19.4). To verify (21.2–9) we write the series for $J_0(x)$ and $J_1(x)$ in the forms

$$J_0(x) = \sum_{n=0}^{\infty} (-1)^n \frac{x^{2n}}{n!n!2^{2n}}, \; J_1(x) = \sum_{n=0}^{\infty} (-1)^n \frac{x^{2n+1}}{n!(n+1)!2^{2n+1}}.$$

Then, calculating $J_0'(x)$ in the manner justified by Theorem VII, we have

$$J_0'(x) = \sum_{n=1}^{\infty} (-1)^n \frac{2nx^{2n-1}}{n!n!2^{2n}} = \sum_{n=1}^{\infty} (-1)^n \frac{x^{2n-1}}{(n-1)!n!2^{2n-1}}.$$

In the series for $J_0(x)$ the term with $n = 0$ is a constant, so that the series for $J_0'(x)$ will begin with the term for which $n = 1$. If we now write $n + 1$ instead of n in the last summation, the new index n will go from 0 to ∞, and we shall have

$$J_0'(x) = \sum_{n=0}^{\infty} (-1)^{n+1} \frac{x^{2n+1}}{n!(n+1)!2^{2n+1}}.$$

Since $(-1)^{n+1} = -(-1)^n$, we see that (21.2–9) is true, by a comparison of the summation expressions for $J_0'(x)$ and $J_1(x)$.

Example 2. Determine what can be said about solutions of the differential equation

(21.2–10)
$$(1 - x^2)\frac{d^2y}{dx^2} + 6\,y = 0$$

of the form $y = f(x) = a_0 + a_1x + a_2x^2 + \cdots + a_nx^n + \cdots,$

i.e. solutions which can be expanded in powers of x, convergent for some interval about $x = 0$.

Assuming tentatively that there is such a series solution, we differentiate it, obtaining $y = a_0 + a_1 x + a_2 x^2 + a_3 x^3 + \cdots + a_n x^n + \cdots$

$$\frac{dy}{dx} = a_1 + 2\,a_2 x + 3\,a_3 x^2 + \cdots + n a_n x^{n-1} + \cdots,$$

$$\frac{d^2 y}{dx^2} = 2\,a_2 + 3 \cdot 2\,a_3 x + \cdots + n(n-1)a_n x^{n-2} + \cdots,$$

$$x^2 \frac{d^2 y}{dx^2} = 2\,a_2 x^2 + 3 \cdot 2\,a_3 x^3 + \cdots + n(n-1)a_n x^n + \cdots.$$

Then

$$(1 - x^2)y'' + 6\,y = \sum_{n=2}^{\infty} n(n-1)a_n x^{n-2} - \sum_{n=2}^{\infty} n(n-1)a_n x^n + \sum_{n=0}^{\infty} 6\,a_n x^n$$

$$y'' - x^2 y'' + 6y \qquad = \sum_{n=0}^{\infty} [(n+2)(n+1)a_{n+2} - n(n-1)a_n + 6\,a_n]x^n \; ?$$

(observe that the term $n(n-1)a_n$ is zero for $n = 0, 1$). If now (21.2–10) is to be satisfied, we infer by Theorem IX that we must have

$$(n+2)(n+1)a_{n+2} - n(n-1)a_n + 6\,a_n = 0, \quad n = 0, 1, 2, \cdots.$$

Since $n(n-1) - 6 = (n-3)(n+2)$, this last relation is equivalent to

$$a_{n+2} = \frac{(n-3)(n+2)}{(n+2)(n+1)}\,a_n = \frac{n-3}{n+1}\,a_n.$$

From this relation we may determine $a_2, a_4, a_6, \cdots$ successively in terms of an arbitrary a_0; likewise $a_3, a_5, a_7, \cdots$ are determined in terms of an arbitrary a_1. We have

$$a_3 = \frac{1-3}{1+1}\,a_1 = -a_1, \quad a_5 = \frac{3-3}{3+1}\,a_3 = 0,$$

whence $a_7 = a_9 = a_{11} = \cdots = 0$. For the even subscripts,

$$a_2 = -\tfrac{3}{1}\,a_0$$
$$a_4 = -\tfrac{1}{3}\,a_2$$
$$a_6 = +\tfrac{1}{5}\,a_4$$
$$\vdots \qquad \vdots$$
$$a_{2n+2} = \frac{2n-3}{2n+1}\,a_{2n}.$$

Clearly none of these coefficients is zero if $a_0 \neq 0$. Now

$$a_{2n+2}a_{2n} \cdots a_4 a_2 = \frac{(2n-3)(2n-5) \cdots 1 \cdot (-1) \cdot (-3)}{(2n+1)(2n-1) \cdots 5 \cdot 3 \cdot 1}\,a_{2n} \cdots a_2 a_0 \, ;$$

when like factors are cancelled from either side of this relation we find

(21.2–11)
$$a_{2n+2} = \frac{3}{(2n+1)(2n-1)} a_0.$$

What we have done thus far shows that *if* there is a solution of (21.2–10) in the form of a series of powers of x, the solution can be written

$$y = a_1(x - x^3) + a_0 \left[1 + \frac{3}{1 \cdot (-1)} x^2 + \frac{3}{3 \cdot 1} x^4 + \cdots \right.$$
$$\left. + \frac{3}{(2n-1)(2n-3)} x^{2n} + \cdots \right].$$

Moreover, the work shows that this really is a solution within the interval of convergence of the series, provided there is an interval of convergence. Now the infinite series

(21.2–12)
$$\sum_{n=0}^{\infty} \frac{3}{(2n-1)(2n-3)} x^{2n}$$

is convergent when $|x| < 1$, as may readily be verified. Thus we have found two linearly independent solutions of the differential equation (21.2–10): the polynomial $x - x^3$ and the infinite series (21.2–12). The coefficients a_0 and a_1 are arbitrary.

EXERCISES

1. The Bessel function of order m of first kind (m a nonnegative integer) is defined as

$$J_m(x) = \sum_{n=0}^{\infty} (-1)^n \frac{1}{(n+m)!\, n!} \left(\frac{x}{2} \right)^{2n+m}$$

Show that (a) $J_0(x) = x^{-1} \dfrac{d}{dx} (x J_1(x))$, (b) $J_1(x) = x^{-2} \dfrac{d}{dx} (x^2 J_2(x))$,

(c) $J_2(x) = -x \dfrac{d}{dx} (x^{-1} J_1(x))$. (d) State and prove the rule corresponding to (a) and (b) for Bessel functions of higher orders $m - 1$ and m. (e) Of what general rule are (c) and (21.2–9) special cases? Prove the correctness of your answer.

2. Using only what you know about power series, and nothing of what you know about e^x, find the function $f(x)$, defined by a power series in x, such that $f'(x) = f(x)$ and $f(0) = 1$. What is the radius of convergence of the series?

3. Find the power series in x, denoted by $f(x)$, such that $f''(x) + f(x) = 0$ and $f(0) = 0$, $f'(0) = 1$. What is the radius of convergence of the series?

4. (a) Find $f(x)$, a power series in x, such that $(1 + x) f'(x) = mf(x)$, where m is a constant and not one of the integers $0, 1, 2, \cdots$. (b) Determine the radius of convergence of the series. (c) Let $g(x) = \dfrac{f(x)}{(1+x)^m}$, and show from the require-ment in (a) that $g'(x) = 0$, so that $g(x)$ is constant. What can you now conclude about $f(x)$ and $(1 + x)^m$ for values of x inside the interval of convergence of the series?

5. Suppose $f(x)$ is defined by a power series in x with positive radius of convergence. Let k be the smallest positive integer such that $f^{(k)}(0) \neq 0$. (We suppose $f(x)$ not a constant, so that there really is such an integer k.) **(a)** Show that f has neither a relative maximum nor a relative minimum at $x = 0$ if k is odd. **(b)** Assuming that k is even, show that f has a relative minimum at $x = 0$ if $f^{(k)}(0) > 0$ and a relative maximum at $x = 0$ if $f^{(k)}(0) < 0$.

6. (a) If the function $f(x)$ represented by a power series in x is *even*, i.e., if $f(x) = f(-x)$, show that all the coefficients of odd powers of x in the series are zero. **(b)** If the function is *odd*, i.e., if $f(-x) = -f(x)$, show that all the coefficients of even powers of x are zero.

7. Suppose $f(x)$ and $g(x)$ are power series in x, each with a positive radius of convergence. Suppose that $f(0) = f'(0) = \cdots = f^{(m-1)}(0) = 0, f^{(m)}(0) \neq 0$ and that $g(0) = g'(0) = \cdots = g^{(n-1)}(0) = 0, g^{(n)}(0) \neq 0$. **(a)** Show that neither $f(x)$ nor $g(x)$ can vanish if $0 < |x| < h$, provided h is sufficiently small. **(b)** Show that $\lim\limits_{x \to 0} \dfrac{f(x)}{g(x)} = \dfrac{f^{(n)}(0)}{g^{(n)}(0)}$ if $m = n$, and that the limit is 0 if $m > n$. Also show that $\lim\limits_{x \to 0} \left| \dfrac{f(x)}{g(x)} \right| = \infty$ if $m < n$.

8. Find a power series in x for $\dfrac{d}{dx}\left(\dfrac{e^x - 1}{x}\right)$, and deduce that

$$1 = \sum_{n=1}^{\infty} \frac{n}{(n+1)!} .$$

9. Show that $1 = \dfrac{3}{2 \cdot 1!} - \dfrac{5}{2^2 \cdot 2!} + \dfrac{7}{2^3 \cdot 3!} - \cdots$

by considering $\dfrac{d^2}{dx^2}(e^{-x^2})$.

10. Show that $4 = \sum\limits_{n=1}^{\infty} (-2)^{n+1} \dfrac{n+2}{n!}$ by considering $\dfrac{d}{dx}(x^2 e^{-x})$.

11. Find an expression for the function $f(x) = \sum\limits_{n=0}^{\infty} \dfrac{x^n}{(n+1)^2 n!}$.

SUGGESTION: Calculate successively $xf(x)$, $\dfrac{d}{dx}(xf(x))$, $x \dfrac{d}{dx}(xf(x))$ in series form, and identify the last series. Then obtain $f(x)$ by integration.

12. (a) Obtain a power-series solution $y = f(x)$ of the differential equation $xy'' + (1 - x)y' + qy = 0$, where q is an arbitrary constant. Find the radius of convergence of the series. **(b)** If q is a nonnegative integer, show that $f(x)$ is a polynomial of degree q. In this latter case, if $f(0) = q!$, $f(x)$ is denoted by $L_q(x)$ and called the Laguerre polynomial of degree q.

13. (a) Show that the differential equation $y'' - 2xy' + 2my = 0$, where

m is a constant, has two independent power-series solutions, $f_1(x) = a_0 + a_2x^2 + a_4x^4 + \cdots$, $f_2(x) = a_1x + a_3x^3 + \cdots$. Find the radii of convergence of these series. (b) If m is a nonnegative integer, show that one of these solutions is a polynomial of degree m. If the coefficient of x^m in the polynomial is 2^m, it is called the Hermite polynomial of degree m, and denoted by $H_m(x)$.

21.3   Division of power series

It is quite often advantageous to obtain the power-series expansion of a function by representing it as the quotient of two power series and then dividing one of these series by the other. We begin with an illustrative example.

Example 1. Find several terms in the expansion of tan x in powers of x. We make use of the known Taylor's series for sin x and cos x:

$$\tan x = \frac{\sin x}{\cos x} = \frac{x - \dfrac{x^3}{3!} + \dfrac{x^5}{5!} - \cdots}{1 - \dfrac{x^2}{2!} + \dfrac{x^4}{4!} - \cdots}.$$

Then we perform the long division as indicated:

$$
\begin{array}{r}
x + \tfrac{1}{3}x^3 + \tfrac{2}{15}x^5 + \cdots \\
1 - \tfrac{1}{2}x^2 + \tfrac{1}{24}x^4 - \cdots \overline{\smash{)}\; x - \tfrac{1}{6}x^3 + \tfrac{1}{120}x^5 - \cdots} \\
\underline{x - \tfrac{1}{2}x^3 + \tfrac{1}{24}x^5 - \cdots} \\
\tfrac{1}{3}x^3 - \tfrac{1}{30}x^5 + \cdots \\
\underline{\tfrac{1}{3}x^3 - \tfrac{1}{6}x^5 + \cdots} \\
\tfrac{2}{15}x^5 - \cdots \\
\underline{\tfrac{2}{15}x^5 - \cdots}
\end{array}
$$

In this way we are led to the expansion

(21.3–1) $$\tan x = x + \tfrac{1}{3}x^3 + \tfrac{2}{15}x^5 + \cdots.$$

Although we have not indicated the general term of this series, it is clear that we may compute as many terms as we please according to this systematic procedure indicated in the long-division process.

None of our previous theorems furnishes any proof of the correctness of the result (21.3–1). Note also that since we have not obtained a general formula for the coefficients in (21.3–1), we are at present unable to determine the radius of convergence of the series.

The thing of foremost practical importance is that the method of Example 1 really works. By long division we can find the quotient of two power series as another power series. *The essential condition which must be fulfilled is that the power series in the denominator must begin with a non-zero constant term.* Either of the two infinite series in the quotient may in particular cases terminate (i.e., instead of an infinite series we may have a polynomial). Experience shows that the long-division process is often the best method for obtaining a power-series representation. With a rational function, for instance, the long-division method is much more practical than the method of computing the coefficients in the Taylor's series by differentiation.

THEOREM X. *Consider a function defined as the quotient of two power series:*

$$(21.3-2) \quad f(x) = \frac{a_0 + a_1 x + a_2 x^2 + \cdots + a_n x^n + \cdots}{b_0 + b_1 x + b_2 x^2 + \cdots + b_n x^n + \cdots},$$

where $b_0 \neq 0$, and where both of the series are convergent in some interval $|x| < r$. Then for sufficiently small values of x the function f can be represented as a power series

$$(21.3-3) \quad f(x) = c_0 + c_1 x + c_2 x^2 + \cdots + c_n x^n + \cdots$$

whose coefficients may be found by the process of long division, or, what is equivalent, by solving the relations

$$
\begin{aligned}
b_0 c_0 &= a_0 \\
b_0 c_1 + b_1 c_0 &= a_1 \\
&\vdots
\end{aligned}
$$

$$(21.3-4) \qquad b_0 c_n + b_1 c_{n-1} + \cdots + b_n c_0 = a_n$$

$$\vdots$$

successively for $c_0, c_1, c_2, \cdots$.

It is impractical to give a complete proof of this theorem with the knowledge presently at our disposal, but we shall go as far as is conveniently possible. In the first place, the assumption $b_0 \neq 0$ means that the function defined by the series $\sum b_n x^n$ is not zero at $x = 0$. It is therefore different from zero throughout some interval about the origin, since the function is continuous, by Theorem V, § 21.1. If we now *assume* that a power series expansion of the form (21.3-3) is valid, we shall have the product relation

$$(21.3-5) \qquad \left(\sum_{n=0}^{\infty} b_n x^n \right) \left(\sum_{n=0}^{\infty} c_n x^n \right) = \sum_{n=0}^{\infty} a_n x^n$$

holding throughout some interval in which all three series are absolutely convergent. The rule for multiplication of absolutely convergent series (see § 19.6) gives the result

$$(21.3-6)$$

$$\left(\sum_{n=0}^{\infty} b_n x^n \right) \left(\sum_{n=0}^{\infty} c_n x^n \right) = \sum_{n=0}^{\infty} (b_0 c_n + b_1 c_{n-1} + \cdots + b_n c_0) x^n.$$

By the uniqueness theorem for power series we then conclude from (21.3–5) and (21.3–6) that the general relation (21.3–4) holds. The coefficients c_n determined successively in this way are the same as would be found by the long-division process, as the student may verify for himself. Note that to solve for c_n when $c_0, \cdots, c_{n-1}$ are known, it is essential to know that $b_0 \neq 0$.

The proof is now complete except for justifying the assumption that $f(x)$ can be expanded in a series (21.3–3). It is this part of the proof which we shall not give. It is an easy consequence of certain standard theorems in the theory of functions of a complex variable, but the development of this latter theory is beyond the scope of this text.

Example 2. The long-division method yields the series expansion

(21.3–7)
$$\frac{2 + x}{1 + x + x^2} = 2 - x - x^2 + 2\,x^3 - x^4 - x^5 + 2\,x^6 - x^7 - x^8 + \cdots.$$

Detailed verification is left to the student. It may be shown by Theorem XVI, § 19.4, that the series (21.3–7) has radius of convergence $R = 1$.

It is important to observe that the division method of obtaining the expansions (21.3–1) and (21.3–7) is much easier than using the formula (21.2–6) for computing the coefficients in the expansion.

Let us consider the situation when $b_0 = 0$ in the quotient (21.3–2). The general case may be represented by assuming that b_k is the first of the b's which is not 0, and that a_l is the first of the a's which is not 0. Then, assuming $x \neq 0$,

$$f(x) = \frac{a_l x^l + a_{l+1} x^{l+1} + \cdots}{b_k x^k + b_{k+1} x^{k+1} + \cdots},$$

(21.3–8)
$$f(x) = x^{l-k}\left(\frac{a_l + a_{l+1} x + \cdots}{b_k + b_{k+1} x + \cdots}\right).$$

This shows that, as $x \to 0$, $f(x)$ behaves like $x^{l-k}\left(\dfrac{a_l}{b_k}\right)$. In particular, $f(x)$ approaches a finite limit if $l \geq k$, but not otherwise. Let us agree to define $f(0) = \lim_{x \to 0} f(x)$ if the limit exists. The quotient in the parenthesis in (21.3–8) can be expressed as a single power series by the method of long division, since $b_k \neq 0$.

Example 3. Find a power series in x for $\dfrac{\sin x}{\sin 2\,x}$.

For $x \neq 0$ we have

$$\frac{\sin x}{\sin 2\,x} = \frac{x - \dfrac{x^3}{3!} + \dfrac{x^5}{5!} - \cdots}{2\,x - \dfrac{(2\,x)^3}{3!} + \dfrac{(2\,x)^5}{5!} - \cdots}.$$

We remove the common factor x from the two series, and deal by long division with what is left. In this way we find, for $x \neq 0$,

$$\frac{\sin x}{\sin 2 x} = \frac{1 - \dfrac{x^2}{6} + \dfrac{x^4}{120} - \cdots}{2 - \dfrac{4}{3} x^2 + \dfrac{4}{15} x^4 - \cdots} = \frac{1}{2} + \frac{1}{4} x^2 + \frac{5}{48} x^4 + \cdots .$$

EXERCISES

1. Find several terms in the power series expansions of the following quotients:

(a) $\dfrac{2 e^x}{1 + e^x}$.

(d) $\dfrac{1 - \cos x}{\sin (x^2)}$.

(b) $\dfrac{1}{\cos x}$.

(e) $\dfrac{-x}{\log (1 - x)}$.

(c) $\dfrac{1}{\cos x - \sin x}$.

(f) $\dfrac{x \cos x - \sin x}{x \sin x} \left(= \operatorname{ctn} x - \dfrac{1}{x} \right)$.

2. Show that $\displaystyle\sum_{n=0}^{\infty} \frac{a_n x^n}{1 - x} = \sum_{n=0}^{\infty} (a_0 + a_1 + \cdots + a_n) x^n$,

and use the result to find the function represented by the following series:

(a) $\displaystyle\sum_{n=0}^{\infty} \left(1 + \frac{1}{1!} + \cdots + \frac{1}{n!} \right) x^n$,

(b) $\displaystyle\sum_{n=1}^{\infty} \left(0 + 1 - \tfrac{1}{2} + \tfrac{1}{3} - \cdots + (-1)^{n+1} \frac{1}{n} \right) x^n$,

(c) $\displaystyle\sum_{n=0}^{\infty} (0 + 1 + \cdots + n) x^n$.

3. Solve the first $(n + 1)$ equations in Theorem X by determinants, and express c_n as $b_0^{-(n+1)}$ times a determinant. For the special case in which $a_0 = 1$, $a_1 = a_2 = \cdots = 0$, show that

$$c_n = \frac{(-1)^n}{b_0^{n+1}} \begin{vmatrix} b_1 & b_0 & 0 & 0 & \cdots & 0 \\ b_2 & b_1 & b_0 & 0 & \cdots & 0 \\ \multicolumn{6}{c}{\dotfill} \\ b_n & b_{n-1} & & \cdots\cdots & & b_1 \end{vmatrix} .$$

4. (a) Suppose that $\left(\sum_{n=0}^{\infty} b_n x^n \right) \left(\sum_{n=0}^{\infty} c_n x^n \right) = \sum_{n=0}^{\infty} a_n x^n$. Show that, as a consequence, $\left(\sum_{n=0}^{\infty} (-1)^n b_n x^n \right) \left(\sum_{n=0}^{\infty} (-1)^n c_n x^n \right) = \sum_{n=0}^{\infty} (-1)^n a_n x^n$. Take all questions of convergence for granted.

(b) Prove, either directly or as a consequence of (a), that if $b_0 \neq 0$ and

$$\frac{\displaystyle\sum_{n=0}^{\infty} a_{2n} x^{2n}}{\displaystyle\sum_{n=0}^{\infty} b_{2n} x^{2n}} = \sum_{n=0}^{\infty} c_{2n} x^{2n},$$

then
$$\frac{\sum_{n=0}^{\infty} (-1)^n a_{2n} x^{2n}}{\sum_{n=0}^{\infty} (-1)^n b_{2n} x^{2n}} = \sum_{n=0}^{\infty} (-1)^n c_{2n} x^{2n}.$$

(c) Use (b) to prove that if $\tanh x = \sum_{n=0}^{\infty} A_n x^{2n+1}$, then

$$\tan x = \sum_{n=0}^{\infty} (-1)^n A_n x^{2n+1}.$$

(d) Use (b) to prove that if $x \operatorname{ctnh} x = \sum_{n=0}^{\infty} A_n x^{2n}$, then

$$x \operatorname{ctn} x = \sum_{n=0}^{\infty} (-1)^n A_n x^{2n}.$$

5. Define B_n as $n!$ times the coefficient of x^n in the power-series expansion of $x/(e^x - 1)$:

$$\frac{x}{e^x - 1} = \sum_{n=0}^{\infty} \frac{B_n}{n!} x^n.$$

The numbers B_n are called Bernoulli's numbers.

(a) Show that $B_0 = 1$, $B_1 = -\frac{1}{2}$, $B_2 = \frac{1}{6}$, $B_3 = 0$.

(b) Writing
$$\frac{x}{e^x - 1} + \frac{x}{2} = 1 + \sum_{n=2}^{\infty} \frac{B_n}{n!} x^n,$$

show that the function on the left is equal to the even function $\frac{x}{2} \operatorname{ctnh} \frac{x}{2}$. Deduce from this that $B_3 = B_5 = B_7 = \cdots = 0$ (see Exercise 6, § 21.2). Then show that $x \operatorname{ctnh} x = \sum_{n=0}^{\infty} \frac{B_{2n}}{(2 n)!} (2 x)^{2n}$. The radius of convergence of this series is not readily determined by our present knowledge, but it may be shown to be π.

6. (a) Use Exercises 4(d) and 5(b) to show that

$$x \operatorname{ctn} x = \sum_{n=0}^{\infty} (-1)^n \frac{B_{2n}}{(2 n)!} (2 x)^{2n}.$$

(b) Use the identity $\tan x = \operatorname{ctn} x - 2 \operatorname{ctn} 2 x$ to obtain the expansion

$$\tan x = \sum_{n=1}^{\infty} (-1)^{n-1} \frac{B_{2n}}{(2 n)!} 2^{2n}(2^{2n} - 1)x^{2n-1}.$$

(c) Use the identity $\operatorname{ctn} x + \tan \frac{x}{2} = \frac{1}{\sin x}$ to show that

$$\frac{x}{\sin x} = \sum_{n=0}^{\infty} (-1)^{n+1} \frac{B_{2n}}{(2 n)!} (2^{2n} - 2)x^{2n}.$$

21.4   Abel's theorem

Suppose that a function is defined by a power series:

$$(21.4\text{--}1) \qquad\qquad f(x) = \sum_{n=0}^{\infty} a_n x^n,$$

and suppose that the radius of convergence R of the series is positive and finite. The series may or may not converge at $x = \pm R$. Let us suppose that the series does converge at $x = R$. In § 21.1 we raised the question as to whether, in this circumstance, $f(x) \to f(R)$ as $x \to R$ (with $x < R$). We shall now answer this question, by means of a theorem due to Abel.

THEOREM XI.   *If the series* (21.4–1) *converges at* $x = R$, *then it converges uniformly on the closed interval* $0 \leq x \leq R$. *As a consequence, the function f defined by the series is continuous on* $0 \leq x \leq R$. *A like conclusion holds for* $-R \leq x \leq 0$ *if the series converges at* $x = -R$.

The proof is made with the aid of an inequality which we state as a lemma of independent interest.

LEMMA.   *Suppose that*

$$(21.4\text{--}2) \qquad\qquad m \leq s_k \leq M, \quad k = 0, 1, \cdots, p,$$

where $\qquad\qquad s_k = u_0 + u_1 + \cdots + u_k,$

and that

$$(21.4\text{--}3) \qquad\qquad v_0 \geq v_1 \geq \cdots \geq v_p \geq 0.$$

Then

$$(21.4\text{--}4) \qquad\qquad m v_0 \leq u_0 v_0 + u_1 v_1 + \cdots + u_p v_p \leq M v_0.$$

PROOF.   According to (19.7–1) we have

$$u_0 v_0 + \cdots + u_p v_p = s_0(v_0 - v_1) + \cdots + s_{p-1}(v_{p-1} - v_p) + s_p v_p.$$

Now, because of (21.4–2) and (21.4–3) we can write

$$u_0 v_0 + \cdots + u_p v_p \leq M(v_0 - v_1) + \cdots + M(v_{p-1} - v_p) + M v_p = M v_0.$$

The other half of inequality (21.4–4) is deduced in a similar manner.

PROOF OF THEOREM XI.   Suppose ϵ is any positive number. According to Theorem I, § 20.1, our proof will be accomplished if we show that there is some integer N such that

$$(21.4\text{--}5) \qquad | a_m x^m + a_{m+1} x^{m+1} + \cdots + a_{m+p} x^{m+p} | < \epsilon$$

if $N \leq m$, $0 < p$, and $0 \leq x \leq R$. Let us set

$$u_k = a_{m+k}R^{m+k}, \quad v_k = \left(\frac{x}{R}\right)^{m+k}.$$

Then (21.4–5) is equivalent to

(21.4–6) $$-\epsilon < u_0v_0 + u_1v_1 + \cdots + u_pv_p < \epsilon.$$

The conditions (21.4–3) are fulfilled by the v's. In addition, $v_0 \leq 1$. Since the series is convergent when $x = R$, we can choose N so that

$$-\epsilon < a_mR^m + a_{m+1}R^{m+1} + \cdots + a_{m+p}R^{m+p} < \epsilon$$

if $N \leq m$ and $0 < p$ (this is just the Cauchy condition for convergence). This means, in our present notation, that (21.4–2) is satisfied with $m = -\epsilon$, $M = \epsilon$. Applying the lemma, and noting that $-\epsilon \leq -\epsilon v_0$, $\epsilon v_0 \leq \epsilon$, we see that the conclusion (21.4–4) of the lemma yields (21.4–6). This completes the proof as regards $0 \leq x \leq R$. The case of convergence at $x = -R$ is reduced to the first case by considering $g(x) = f(-x)$ at $x = R$.

Example 1. If the binomial series (19.5–5) converges at $x = 1$, its sum is 2^m. This assertion may be justified as follows: We proved the validity of the binomial series expansion for $(1 + x)^m$ when $|x| < 1$. Therefore, by Theorem XI, if the series converges at $x = 1$, its sum there is

$$\lim_{x \to 1} (1 + x)^m = 2^m.$$

This always happens if $m > 0$, by what was established in Example 4, § 19.4. It may be shown that the series converges at $x = 1$ if $-1 < m$, but diverges if $m \leq -1$ (see Exercise 4, § 19.5).

Next we show how Theorem XI permits us to extend the result of Theorem V, § 21.1, with respect to the integration of a power series. If the power series

(21.4–7) $$f(x) = \sum_{n=0}^{\infty} a_nx^n$$

is convergent when $|x| < R$, then

(21.4–8) $$\int_0^R f(x)dx = \sum_{n=0}^{\infty} \frac{a_n}{n+1} R^{n+1},$$

provided the series on the right in (21.4–8) *is convergent*, irrespective of whether or not the series in (21.4–7) is convergent at $x = R$. Of course, if (21.4–7) is not convergent at $x = R$, the integral in (21.4–8) may be improper at the upper limit.

The proof of the foregoing assertion is simple. If $0 < b < R$, we have

$$\int_0^b f(x)dx = \sum_{n=0}^{\infty} \frac{a_n}{n+1} b^{n+1}$$

by (21.1–4). Then, provided the series in (21.4–8) is convergent, we have

$$\lim_{b \to R} \int_0^b f(x)dx = \sum_{n=0}^{\infty} \frac{a_n}{n+1} R^{n+1},$$

by Theorem XI. Since it is also true that

$$\lim_{b \to R} \int_0^b f(x)dx = \int_0^R f(x)dx,$$

(21.4–8) is proved.

The remarks at the end of Example 1, § 21.1, illustrate the application of (21.4–8).

Example 2. Show that

(21.4–9)

$$\int_0^1 \frac{\log (1 - t)}{t} \, dt = - \left\{ \frac{1}{1^2} + \frac{1}{2^2} + \frac{1}{3^2} + \cdots + \frac{1}{n^2} + \cdots \right\}.$$

We start from the series (see (19.1–2))

$$\log (1 - t) = - t - \tfrac{1}{2} t^2 - \tfrac{1}{3} t^3 - \cdots - \frac{1}{n} t^n - \cdots.$$

Dividing by t, we have

$$\frac{\log (1 - t)}{t} = - 1 - \tfrac{1}{2} t - \tfrac{1}{3} t^2 - \cdots - \frac{1}{n} t^{n-1} - \cdots.$$

Observe that the series diverges at $t = 1$. But if we integrate from 0 to x we have

$$\int_0^x \frac{\log (1 - t)}{t} \, dt = - x - \frac{1}{2^2} x^2 - \frac{1}{3^2} x^3 - \cdots - \frac{1}{n^2} x^n - \cdots.$$

This series converges when $x = 1$, and so we have (21.4–9) as a special case of (21.4–8). The integral is improper at $t = 1$, but not at $t = 0$, since the integrand approaches the finite limit $- 1$ as $t \to 0$ (this may be seen by use of L'Hospital's rule).

EXERCISES

1. (a) Express $\int_0^x \tan^{-1} t \, dt$ as a power series in x, and discuss the range of validity of the series, with particular attention to the end points of the interval of convergence.

(b) Use the result in **(a)** to show that

$$\frac{\pi}{4} - \log \sqrt{2} = 1 - \tfrac{1}{2} - \tfrac{1}{3} + \tfrac{1}{4} + \tfrac{1}{5} - - + + \cdots.$$

2. Let $a_n = \left[\dfrac{1 \cdot 3 \cdots 2n - 1}{2 \cdot 4 \cdots 2n} \right]^2$. Show that $\displaystyle\sum_{n=1}^{\infty} \frac{a_n}{2n - 1} = \frac{\pi - 2}{\pi}$ by considering $\int_0^{\pi/2} (1 - x^2 \sin^2 t)^{1/2} \, dt$ as a power series in x.

3. (a) Justify the formula

$$\int_0^1 \frac{t^{p-1}}{1+t^q}\,dt = \frac{1}{p} - \frac{1}{p+q} + \frac{1}{p+2q} - \cdots,$$

where p and q are positive integers. (b) Calculate the value of the integral to two decimal places, if $p = 10$, $q = 40$.

4. Let $f(x) = \sum_{n=1}^{\infty} n\,\frac{(pn)!}{(n!)^p}\,x^{n-1}$. For what positive integral values of p is it

true that $\int_0^{p^{-p}} f(x)dx = \sum_{n=1}^{\infty} \frac{(pn)!}{(n!)^p}\,p^{-np}$?

21.5 Inferior and superior limits

The subject matter of this section is not properly a part of the theory of power series, but it is relevant to the problem of finding the radius of convergence of a power series. Inferior and superior limits have many other important uses in analysis.

Consider an arbitrary sequence $\{x_n\}$, without any assumption as to whether or not it is convergent.

DEFINITION. A sequence $\{x_n\}$ is said to *cluster*, or *accumulate*, at a point ξ if every open interval centered at ξ contains x_n for infinitely many values of n. These points x_n need not be distinct from ξ.

Example 1. Let $x_n = \frac{1}{2}$ if $n = 1, 4, 7, 10, \cdots, x_n = 1$ if $n = 2, 5, 8,$ $11, \cdots, x_n = 2$ if $n = 3, 6, 9, 12, \cdots$. This sequence clusters at the three points $\frac{1}{2}, 1, 2$.

Example 2. Let $x_n = 1 + \sin(n\pi/2)$. This sequence clusters at 0, 1, and 2.

Example 3. Let $x_n = (-1)^n (1 + (1/n))$. This sequence clusters at -1 and 1.

If we compare the definition of a cluster point of $\{x_n\}$ with the definition of the limit of a convergent sequence, *we see at once that if $\{x_n\}$ is convergent, with limit ξ, then ξ is the sole cluster point of $\{x_n\}$.*

If a sequence is bounded, it must have at least one cluster point. This is a corollary of Theorem V, § 16.31, or it may be proved by an argument much like the one used in proving the Bolzano-Weierstrass theorem (§ 16.3).

If a sequence is bounded above and has one or more cluster points, there is a cluster point farthest to the right, namely, the least upper bound of all the cluster points. This cluster point farthest to the right is called the *limit superior* of the sequence; it is denoted by

$$\overline{\lim_{n\to\infty}} x_n \quad \text{or} \quad \limsup_{n\to\infty} x_n.$$

Likewise, if the sequence is bounded below and has one or more cluster points, the cluster point farthest to the left is called the *limit inferior*, and denoted by

$$\varliminf_{n \to \infty} x_n \quad \text{or} \quad \liminf_{n \to \infty} x_n.$$

We often drop the $n \to \infty$ part of the notation, as a typographical convenience. Clearly we always have, in the case of a bounded sequence,

$$(21.5\text{–}1) \qquad\qquad \varliminf x_n \leqq \varlimsup x_n.$$

The sequence is convergent if and only if $\varliminf x_n = \varlimsup x_n$, in which case the limit of the sequence coincides with the inferior and superior limit.

If the sequence has no upper bound, we say that

$$\varlimsup_{n \to \infty} x_n = + \infty \, ;$$

if there is no lower bound, we say that

$$\varliminf_{n \to \infty} x_n = - \infty \, .$$

If $x_n \to + \infty$, we say that $\varliminf x_n = \varlimsup x_n = + \infty$, and if $x_n \to - \infty$, we say that $\varliminf x_n = \varlimsup x_n = - \infty$.

One of the main reasons for introducing the new concepts of limit superior and limit inferior at just this point in the text is that we can use them in the following theorem.

THEOREM XII. *The radius of convergence R of a power series $\sum a_n x^n$ is given by*

$$(21.5\text{–}2) \qquad\qquad \frac{1}{R} = \varlimsup_{n \to \infty} |a_n|^{1/n}.$$

The understanding is that $R = + \infty$ if the limit superior is 0, and $R = 0$ if the limit superior is $+ \infty$.

PROOF. We appeal to Cauchy's root test (Theorem XVI, § 19.4). Let $u_n = a_n x^n$. Then

$$(21.5\text{–}3)$$
$$\varlimsup |u_n|^{1/n} = \varlimsup |a_n x^n|^{1/n} = |x| \varlimsup |a_n|^{1/n} = \frac{|x|}{R} \, ,$$

where R is defined by (21.5–2). Here we have used the fact that, for any sequence $\{y_n\}$,

$$\varlimsup c \, y_n = c \varlimsup y_n \quad \text{if} \quad c > 0 \, .$$

It follows from (21.5–3) and Cauchy's root test that the series converges if $|x| < R$ and diverges if $|x| > R$. This proves Theorem XII. We have tacitly assumed R to be finite and positive. The cases $R = 0$ or $R = +\infty$ require special attention but present no difficulties. They are left to the student.

Example 4. Find the radius of convergence of the series in (21.3–7). Here $a_n = 2$ if $n = 0, 3, 6, \cdots$ and $a_n = -1$ for other values of n. Thus $|a_n|^{1/n}$ is either $2^{1/n}$ or $1^{1/n}$, depending on the value of n. We see that $|a_n|^{1/n}$ is convergent, with limit 1, so that $R = 1$, by (21.5–2). Note that R cannot be found by Theorem IV in this case, for the successive values of $\left|\dfrac{a_n}{a_{n+1}}\right|$, for $n = 0, 1, 2, \cdots$, are $2, 1, \frac{1}{2}, 2, 1, \frac{1}{2}, \cdots$, and the sequence of ratios has no limit.

It is not difficult to prove that, if $\{a_n\}$ is any sequence of positive numbers,

$$(21.5\text{–}4) \qquad \underline{\lim} \, \frac{a_{n+1}}{a_n} \leq \underline{\lim} \, a_n^{1/n}$$

and

$$(21.5\text{–}5) \qquad \overline{\lim} \, a_n^{1/n} \leq \overline{\lim} \, \frac{a_{n+1}}{a_n}.$$

It is a consequence of these inequalities that if $\lim_{n \to \infty} a_{n+1}/a_n$ exists, then $\lim_{n \to \infty} a_n^{1/n}$ also exists, and the two limits are equal. This is sometimes useful.

Example 5. Show that

$$(21.5\text{–}6) \qquad \lim_{n \to \infty} \frac{n}{(n!)^{1/n}} = e.$$

We take $a_n = \dfrac{n^n}{n!}$. Then

$$\frac{a_{n+1}}{a_n} = \frac{(n+1)^{n+1}}{(n+1)!} \frac{n!}{n^n} = \left(\frac{n+1}{n}\right)^n = \left(1 + \frac{1}{n}\right)^n \to e$$

as we know from (1.62–5). Therefore also

$$a_n^{1/n} = \frac{n}{(n!)^{1/n}} \to e.$$

This result is not so easy to show without using this method. The result (21.5–6) may also be established by using Stirling's formula (see § 22.8).

EXERCISES

1. Find the radius of convergence of $\Sigma \, a_n x^n$ in each of the following cases:
(a) $a_n = c^{\sqrt{n}}, \, c > 0$.   (b) $a_n = [1 + (1/n)]^{n^2}$.

(c) $a_n = 2^n$ if n is even.   (e) $a_n = (\sqrt{n})^{-n}$.

     $a_n = 3^n$ if n is odd.   (f) $a_n = n^{-\sqrt{n}}$.

(d) $a_n = n!/n^n$.

2. If $\sum a_n x^n$ converges for all values of x, prove that $|a_n|^{1/n} \to 0$.

3. Prove Theorem VI, § 21.2, by use of Theorem XII.

4. Find the radius of convergence of each of the following series:

(a) $\sum x^{n^2}/2^n$.   (d) $\sum r^{n^2} x^n$, $0 < r < 1$.

(b) $\sum n! \, x^{n!}$.   (e) $\sum r^n x^{n^2}$, $0 < r$.

(c) $\sum n^n x^{n^2}$.   (f) $\sum n! \, x^{n^2}$.

5. If $\sum a_n x^n$ has a finite positive radius of convergence, prove that the radius of convergence of $\sum a_n x^{n^2}$ is $R = 1$.

6. Use the method of Example 5 to find $\lim a_n^{1/n}$ in each of the following cases:

(a) $a_n = p$, a positive constant.

(b) $a_n = n$.

(c) $a_n = \dfrac{1 \cdot 3 \cdots (2n-1)}{2 \cdot 4 \cdots 2n}$.

7. If $\{x_n\}$ is bounded above, show that $\overline{\lim} \, x_n = \xi$ if and only if for each $\epsilon > 0$ it is true that $x_n < \xi + \epsilon$ for all sufficiently large values of n, and $\xi - \epsilon < x_n$ for infinitely many values of n. Devise and prove a similar statement about the limit inferior.

8. Use the formulation in Exercise 7 to prove (21.5-4) and (21.5-5). SUGGESTION: Use the relation

$$\frac{a_{N+p}}{a_N} = \frac{a_{N+1}}{a_N} \cdot \frac{a_{N+2}}{a_{N+1}} \cdots \frac{a_{N+p}}{a_{N+p-1}}.$$

Choose N suitably, depending on ϵ, and then let $p \to \infty$ after getting an appropriate inequality.

21.6   Real analytic functions

The use of power series to represent functions is so important that it is convenient to have an adjective for functions which can be so represented. This adjective is the word "analytic." There is, of course, a general meaning of this word, not solely mathematical, but we are now using the term "analytic" in a highly technical sense.

DEFINITION. Let f be a real-valued function of the real variable x, defined on the open interval $a < x < b$, and suppose that f possesses derivatives of all orders at each point of the interval. We say that f is *analytic* on the interval if, for each point x_0 in the interval, $f(x)$ is represented by the Taylor's series

$$(21.6\text{--}1) \qquad f(x) = \sum_{n=0}^{\infty} \frac{f^{(n)}(x_0)}{n!} (x - x_0)^n$$

in some subinterval $x_0 - h < x < x_0 + h$, where the size of the positive number h may vary with the point x_0.

Example 1. We see by (19.1–7) that the function e^x is analytic on the whole x-axis. Likewise one may show that $\sin x$ and $\cos x$ are analytic on the whole x-axis.

Example 2. The function $\log x$ is analytic on the whole positive x-axis. To see this we proceed as follows: Suppose $x_0 > 0$. Then

$$x = x_0 + x - x_0 = x_0 \left(1 + \frac{x - x_0}{x_0}\right),$$

$$\log x = \log x_0 + \log \left(1 + \frac{x - x_0}{x_0}\right).$$

Now use the series expansion (19.1–1), with $\dfrac{x - x_0}{x_0}$ in place of x. The result is

$$(21.6\text{--}2) \qquad \log x = \log x_0 + \sum_{n=1}^{\infty} (-1)^{n-1} \frac{1}{n} \frac{(x - x_0)^n}{x_0^n}.$$

The Taylor's series here is convergent if

$$-1 < \frac{x - x_0}{x_0} \leq 1, \quad \text{or} \quad 0 < x \leq 2 x_0.$$

Most of the basic elementary functions and the usual combinations of them are analytic on any open interval on which they are defined. We say that f is analytic at a point x_0 if it is analytic in some open interval centered at x_0. If f is not analytic at x_0, but *is* analytic at all points near x_0, we call x_0 an isolated singular point of f.

Example 3. The function $1/(1 - x^2)$ has $x = 1$ and $x = -1$ as isolated singular points. It is analytic everywhere else, as may be shown with the aid of the binomial series.

If two functions f and g are each analytic on an interval $a < x < b$, the sum $f(x) + g(x)$ and the product $f(x)g(x)$ are each analytic on that interval. The quotient $f(x)/g(x)$ is analytic at each point of the interval at which $g(x) \neq 0$. Under appropriate conditions the formation of a composite function from analytic functions yields another analytic function.

Example 4. The function $e^{\sin x}$ is analytic for all values of x, and $\log (\cos x)$ is analytic for values of x such that $\cos x > 0$.

A function defined by a power series is analytic at each point in the interior of the interval of convergence of the power series.

Some of the facts we have been stating may be proved rather easily with what we have learned in Chapter XIX and in the present chapter. The fact that a composite function of analytic functions is analytic is somewhat harder to prove. The analyticity of the quotient of two analytic functions is a special case of the result for composite functions. We omit these proofs. The theory of *real* analytic functions is much clarified and simplified by a study of the theory of analytic functions of a *complex* variable.

It is instructive to know that a function can fail to be analytic at a certain point even though it is everywhere continuous, with a continuous derivative. The function $| x |^{3/2}$ is such a function, but it is not analytic at $x = 0$. This function does not have a second derivative at $x = 0$. Much more surprising is an example of a function which has continuous derivatives of *all* orders for all values of x, yet is not analytic at $x = 0$. The function

$$f(x) = e^{-1/x^2} \quad \text{if} \quad x \neq 0, \quad f(0) = 0$$

is such a function. It is shown in Example 6, § 4.5, that f and all its derivatives have the value 0 at $x = 0$. It follows at once that $f(x)$ is not represented by its Taylor's series about the point $x = 0$, for

$$\sum_{n=0}^{\infty} \frac{f^{(n)}(0)}{n!} x^n = 0$$

for all values of x, since $f^{(n)}(0) = 0$, whereas $f(x) \neq 0$ if $x \neq 0$.

The following theorem about real analytic functions is interesting and useful:

THEOREM XIII. *Suppose $f(x)$ is defined and has derivatives of all orders when $a < x < b$, and suppose that $f^{(n)}(x) \geq 0$ for $n = 0, 1, 2, \cdots$ when $a < x < b$. Then f is analytic at each point of the interval, and the Taylor's series about the point x_0 converges to $f(x)$ for each x such that $| x - x_0 | < h$, where h is the distance from x_0 to the nearer end-point of (a, b).*

We shall not give the proof of this theorem; the proof is easily deduced from the result stated in Exercise 3.

EXERCISES

1. What does Theorem XIII imply (a) about e^x? (b) about $(1 - x)^{-m}$ if $m > 0$ and $x < 1$? (c) about $-\log (1 - x)$ if $x < 1$? (d) about $\sin x + e^{x-a}$ if $x \geq a$?

2. By induction, or otherwise, prove the formula

$$\phi(x) = \sum_{k=0}^{n} \frac{x^k \phi^{(k)}(0)}{k!} + \frac{x^{n+1}}{n!} \int_0^1 \phi^{(n+1)}(tx)(1-t)^n dt,$$

assuming that ϕ and all the derivatives occurring are continuous on the interval from 0 to x, inclusive.

3. Assume that ϕ and all its derivatives are defined and nonnegative in value when $0 \le x \le r$. Prove by the following steps that $\phi(x)$ can be expanded by Maclaurin's series if $0 \le x < r$.

(a) Use the formula in Exercise 2 to show that

$$\int_0^1 \phi^{(n+1)}(tr)(1-t)^n dt \le \frac{n!}{r^{n+1}} \phi(r).$$

(b) Use the fact that $\phi^{(n+2)}(x)$ is a nonnegative function and the result in (a) to prove that the integral remainder term in the formula of Exercise 2 is not greater than $(x/r)^{n+1}\phi(r)$ if $0 \le x < r$. Now complete the proof of the main assertion.

4. Use the result in Exercise 3 to show that $\tan x$ can be expanded by Maclaurin's series if $0 \le x < \pi/2$. What can you conclude about the expansion for $x < 0$, seeing that $\tan x$ is an odd function?

5. What can you infer about $\log \dfrac{1+x}{1-x}$ from Exercise 3?

MISCELLANEOUS EXERCISES

1. Prove the binomial coefficient relation

$$\binom{n}{0}^2 + \binom{n}{1}^2 + \cdots + \binom{n}{n}^2 = \binom{2n}{n},$$

using Theorem IX and the fact that $(1+x)^n(1+x)^n = (1+x)^{2n}$.

2. Deduce the formula

$$\frac{\sin^{-1} x}{\sqrt{1-x^2}} = x + \tfrac{2}{3} x^3 + \frac{2 \cdot 4}{3 \cdot 5} x^5 + \frac{2 \cdot 4 \cdot 6}{3 \cdot 5 \cdot 7} x^7 + \cdots,$$

and obtain a series for $\tfrac{1}{2}(\sin^{-1} x)^2$. Is the latter series valid when $x = 1$?

3. Deduce the formula

$$\frac{(\tan^{-1} x)^2}{2!} = \frac{x^2}{2} - (1 + \tfrac{1}{3})\frac{x^4}{4} + (1 + \tfrac{1}{3} + \tfrac{1}{5})\frac{x^6}{6} - \cdots.$$

4. Deduce the formulas

$$-\frac{\log(1-x)}{1-x} = x + (1 + \tfrac{1}{2})x^2 + (1 + \tfrac{1}{2} + \tfrac{1}{3})x^3 + \cdots,$$

$$\tfrac{1}{2}[\log(1-x)]^2 = \sum_{n=2}^{\infty}\left(1 + \tfrac{1}{2} + \cdots + \frac{1}{n-1}\right)\frac{x^n}{n}.$$

Prove the validity of the latter series when $x = -1$. (Use Euler's constant.)

5. Suppose that $\{x_n\}$ is a sequence with $x_n > 0$. Let

$$\overline{\lim_{n \to \infty}} \, x_n = A, \; \lim_{n \to \infty} \frac{1}{x_n} = B.$$

Show that $B = 1/A$ if A is finite and positive. Show also that $B = 0$ if $A = +\infty$, and that $B = +\infty$ if $A = 0$.

6. Apply the results of the preceding exercise to show that (21.5–2) can be replaced by the formula

$$R = \varliminf_{n \to \infty} |\, a_n \,|^{-1/n}.$$

Improper Integrals

22. Preliminary remarks

In this chapter we shall study improper integrals systematically in somewhat the same way that we studied infinite series in Chapter XIX. There are many analogies between the theory of improper integrals and the theory of infinite series. These analogies are seen most clearly in comparing

$$\int_0^\infty f(x)dx \quad \text{and} \quad \sum_{n=0}^\infty a_n.$$

In the integral we have a variable x, ranging *continuously* from 0 to ∞, while in the series we have a variable n, with the *discrete* range of values $0, 1, 2, \cdots$. The typical function value $f(x)$ is the counterpart of the typical term a_n, and the integration symbol $\int_0^\infty (\quad)dx$ is the counterpart of the summation symbol $\sum_{n=0}^\infty$. The counterpart of a partial sum $\sum_{k=0}^n a_k$ is the "partial integral"

$$\int_0^x f(t)dt.$$

DEFINITION. By an improper integral of *first kind* we mean an integral

(22–1)
$$\int_a^\infty f(x)dx$$

in which $f(x)$ is defined when $x \geq a$ and is integrable (in the sense of § 18.1) over every finite interval $[a, b]$. The integral is defined as the limit

$$\lim_{b \to \infty} \int_a^b f(x)dx$$

if the limit exists. The integral (22–1) is then said to be *convergent*. If the limit does not exist, the integral is called *divergent*. In most of the applications and illustrations $f(x)$ is continuous. The integrals

$$\int_1^\infty \frac{dx}{x^2}, \quad \int_0^\infty \sin t^2\, dt, \quad \int_0^\infty x^n e^{-x}\, dx \quad (n \geq 0)$$

are all of first kind.

DEFINITION. By an improper integral of *second kind* we mean an integral

(22–2)
$$\int_a^b f(x)dx$$

with finite limits, in which the failure to be an ordinary "proper" integral

691

arises solely from the behavior of $f(x)$ either as $x \to a$ or as $x \to b$, *but not both*. Thus, if $f(x)$ is integrable on $[a, c]$ for each c such that $a < c < b$, but is *not* integrable on $[a, b]$, we say that the integral (22–2) is improper at $x = b$. Sometimes we say that $f(x)$ has a *singularity* at $x = b$. We then define the integral (22–2) as the limit

$$\lim_{c \to b-} \int_a^c f(x)dx$$

if the limit exists. The terms *convergent* and *divergent* are applied to the integral according as the limit does or does not exist. Similar definitions are made for integrals of the second kind which are improper at the lower limit of integration.

Examples. $\displaystyle\int_0^1 \frac{dx}{\sqrt{1-x^2}}$, $\displaystyle\int_0^{\pi/3} \frac{d\theta}{\sqrt{\cos\theta - \frac{1}{2}}}$, $\displaystyle\int_{1/2}^1 \frac{\log x}{(1-x)^{3/2}}\, dx$

are improper at the upper limits, and

$$\int_0^1 \left(\log \frac{1}{x}\right)^3 dx, \quad \int_0^1 \frac{e^{-x}}{\sqrt{x}}\, dx, \quad \int_0^1 \frac{\log x}{1+x}\, dx$$

are improper at the lower limits.

As with infinite series, it is important to be able to test an improper integral for convergence or divergence. There are certain analogies between the tests for series and the tests for integrals, which we shall point out as we proceed. In practice, however, we do not need as great a variety of tests for integrals as we do for series.

Just as certain functions may be represented by infinite series whose terms depend on a variable, so certain functions may be represented by improper integrals whose integrands depend on a parameter. The principal reason for undertaking a systematic study of improper integrals at just this stage in a course in advanced calculus is to prepare the way for the study of certain special varieties of improper integrals which are important in practice. Among these we cite particularly the "gamma function" $\Gamma(x)$, defined by

$$\Gamma(x) = \int_0^\infty e^{-t} t^{x-1}\, dt, \quad 0 < x,$$

integrals of the form

$$f(s) = \int_0^\infty e^{-st} F(t)dt,$$

which are known as *Laplace transforms*, and integrals of one of the forms

$$\int_0^\infty f(t) \sin xt\, dt, \quad \int_0^\infty f(t) \cos xt\, dt,$$

which are *Fourier transforms*. Laplace and Fourier transforms are of great importance, both theoretically and practically, and they warrant extensive study. We shall be able to do no more than open the door to these subjects for the interested student.

22.1   Positive integrands.   Integrals of first kind

It is convenient to begin by studying improper integrals with positive integrands, for the same reasons that make it useful to study series of positive terms before studying more general types of series.

Let us start with an integral

$$(22.1\text{--}1) \qquad \qquad \int_a^\infty f(t)dt$$

of the first kind. This is *convergent* if the integral

$$(22.1\text{--}2) \qquad \qquad F(x) = \int_a^x f(t)dt$$

approaches a finite limit as $x \to \infty$; otherwise it is *divergent*. Now let us suppose that $f(t) \geqq 0$ when $t \geqq a$. Usually we shall have $f(t) > 0$, but zero values for $f(t)$ need not be ruled out here. Then it is clear that $F(x)$ does not decrease as x increases, for

$$F(x_2) - F(x_1) = \int_{x_1}^{x_2} f(t)dt \geqq 0$$

if $a \leqq x_1 < x_2$. There are now two possibilities: either $F(x)$ is bounded above or it is not. If it is bounded above, there is some constant M such that $F(x) \leqq M$ for all the relevant values of x. If it is not bounded above, for each M there will be some x such that $M < F(x)$, no matter how large M is, and since $F(x)$ never decreases as x increases, this means that $F(x) \to + \infty$ as $x \to \infty$. In this case the integral is divergent. If $F(x)$ is bounded, however, then $F(x)$ approaches a finite limit as $x \to \infty$, and in this case the integral is convergent. The fact that a function $F(x)$ approaches a limit as $x \to \infty$ if it is bounded and nondecreasing as x increases is analogous to a corresponding theorem about sequences (Theorem III, § 2.7), and may be proved in an analogous way (see Exercise 9). We state our conclusion about the integral formally:

THEOREM I.   *An integral* (22.1–1) *of the first kind with* $f(t) \geqq 0$ *for all t is convergent if and only if there is a constant M such that*

$$\int_a^x f(t)dt \leqq M$$

when $x > a$. *The value of the improper integral is then not greater than M.*

This theorem is the counterpart of Theorem I, § 19.2. There is also a counterpart of Theorem II, § 19.2; we call it the comparison test for integrals.

THEOREM II. *Let $\int_a^\infty f(x)dx$ and $\int_b^\infty g(x)dx$ be two integrals of first kind with nonnegative integrands, and suppose that $f(x) \leqq g(x)$ for all values of x beyond a certain point $x = c$. Then if $\int_b^\infty g(x)dx$ is convergent, so is $\int_a^\infty f(x)dx$, and if $\int_a^\infty f(x)dx$ is divergent, so is $\int_b^\infty g(x)dx$.*

For the proof we note that the convergence or divergence of the integrals is not affected if we replace both lower limits by $x = c$. We then have, if $x > c$,

$$\int_c^x f(t)dt \leqq \int_c^x g(t)dt;$$

the conclusions of the theorem now follow at once from Theorem I.

Example 1. The integral $\displaystyle\int_0^\infty \frac{dx}{\sqrt{1 + x^3}}$ is convergent, by comparison with the integral $\displaystyle\int_1^\infty \frac{dx}{x^{3/2}}$, which is convergent, as may be shown directly from the definition; for $(1 + x^3)^{-1/2} < x^{-3/2}$ when $x > 0$.

Example 2. The integral $\displaystyle\int_0^\infty \frac{dx}{(1 + x^3)^{1/3}}$ is divergent, by comparison with the integral $\displaystyle\int_0^\infty \frac{dx}{1 + x}$, which is divergent.

To see the divergence of the second integral, note that

$$\int_0^x \frac{dt}{1 + t} = \log(1 + x) \to \infty \quad \text{as} \quad x \to \infty.$$

Now
$$\frac{1}{1 + x} < \frac{1}{(1 + x^3)^{1/3}}$$

when $x > 0$, for this inequality is equivalent to $1 + x^3 < (1 + x)^3 = 1 + 3x + 3x^2 + x^3$, which is obviously correct if $x > 0$. Thus the first integral must diverge, by Theorem II.

To avoid troublesome details of working with inequalities in practice, it is often convenient to use the following theorem rather than to use the comparison test directly.

THEOREM III. *Suppose $\int_a^\infty f(x)dx$ and $\int_b^\infty g(x)dx$ are integrals of the first kind with positive integrands, and suppose that the limit*

(22.1–3) $$\lim_{x \to \infty} \frac{f(x)}{g(x)} = L$$

exists (finite) and is not zero. Then either both integrals are convergent, or both are divergent.

This is proved in exactly the same manner as we proved its counterpart for series, Theorem III, § 19.2.

Example 3. The integral $\displaystyle\int_1^{\infty} \frac{x^2 \, dx}{2 \, x^4 - x + 1}$ is convergent. To prove this, observe that for large values of x the integrand is about the size of $1/(2 \, x^2)$. More exactly, taking

$$f(x) = \frac{x^2}{2 \, x^4 - x + 1}, \qquad g(x) = \frac{1}{x^2},$$

we see that $$\lim_{x \to \infty} \frac{f(x)}{g(x)} = \tfrac{1}{2}.$$

Now the integral $\int_1^{\infty} (1/x^2)dx$ is convergent, and so the given integral is also, by Theorem III.

It is convenient to state an additional theorem about what conclusion can be drawn if $L = 0$ or $L = + \infty$ in (22.1–3).

THEOREM IV. *For the integrals described in Theorem III suppose that the limit $L = 0$. Then if $\int_b^{\infty} g(x)dx$ is convergent, so is $\int_a^{\infty} f(x)dx$. Or, alternatively, suppose that $L = + \infty$. Then if $\int_b^{\infty} g(x)dx$ is divergent, so is $\int_a^{\infty} f(x)dx$.*

The proof comes directly from Theorem II. For, in the first case we must have $f(x) < g(x)$ beyond a certain point, and in the second case it is clear that $g(x) < f(x)$ beyond a certain point. Nevertheless, it is usually easier to find L than to deal directly with the inequalities.

Example 4. The integral $\int_1^{\infty} x^{\alpha} e^{-x} \, dx$ is convergent, no matter what real number α may be. We apply the limit test, using the fact that $\int_1^{\infty} (1/x^2)dx$ is convergent.

$$\frac{x^{\alpha} e^{-x}}{x^{-2}} = x^{\alpha+2} e^{-x} = \frac{x^{\alpha+2}}{e^x},$$

and $$\lim_{x \to +\infty} \frac{x^{\alpha+2}}{e^x} = 0,$$

as we see by applying l'Hospital's rule (compare with Example 5, § 4.5). The convergence of the given integral now follows by Theorem IV.

In using Theorems III and IV the student needs to have in mind a few simple standard integrals whose convergence or divergence is known. It is easily shown that $\int_a^\infty (1/x^p)dx$ (where $a > 0$) is convergent if $p > 1$ and divergent if $p \leqq 1$. In a very large number of practical situations it will be found that the convergence or divergence of an integral $\int_a^\infty f(x)dx$ of the first kind with positive integrand can be settled by using Theorem III or Theorem IV with $g(x) = 1/x^p$, choosing an appropriate value of p as determined by trial or inspection.

Example 5. Let $f(x) = \dfrac{3\,x - 7}{(x + 1)^{5/2}}$. For large values of x this function is comparable in value to $3\,x/x^{5/2} = 3/x^{3/2}$. Thus, applying Theorem III with $g(x) = \dfrac{1}{x^{3/2}}$, we see that $\displaystyle\int_0^\infty \dfrac{3\,x - 7}{(x + 1)^{5/2}}\,dx$ is convergent.

Example 6. Consider the integral $\displaystyle\int_2^\infty \dfrac{dx}{(\log x)^p}$, where $p > 0$. In this case, we know that $\log x$ increases more slowly than any positive power of x. We apply Theorem IV with $f(x) = (\log x)^{-p}$, $g(x) = x^{-1}$. Then, using l'Hospital's rule,

$$\lim_{x \to \infty} \frac{(x)f}{g(x)} = \lim_{x \to \infty} \frac{x}{(\log x)^p} = \lim_{x \to \infty} \frac{1}{p(\log x)^{p-1} \cdot (1/x)}.$$

This is the same as $\displaystyle\lim_{x \to \infty} \frac{x}{p(\log x)^{p-1}}.$

After a certain number of applications of l'Hospital's rule we find that

$$\lim_{x \to \infty} \frac{f(x)}{g(x)} = +\infty.$$

(One must consider separately the cases in which p is or is not an integer.) Theorem IV then assures us that the given integral is divergent.

The student will note that we have not developed any analogues of the ratio tests of § 19.22. In the analogy between series and integrals there is no simple way of formulating a counterpart of a ratio test, because a typical value $f(x)$ of the integrand has no immediate successor, in the sense that a_{n+1} is the successor of a_n.

One other difference between infinite series and improper integrals of the first kind is worth noting. If a series is convergent, its typical term a_n approaches zero as $n \to \infty$. But if an integral $\int_a^\infty f(x)dx$ is convergent, it does not necessarily follow that $f(x) \to 0$ as $x \to \infty$. See Exercise 8, and Example 2, § 22.3.

EXERCISES

1. Test the following integrals for convergence or divergence, using Theorem II and the known facts about $\int_a^\infty x^{-p}\,dx$ for $a > 0$.

(a) $\displaystyle \int_1^\infty \frac{dx}{(1 + x)\sqrt{x}}.$

(b) $\displaystyle \int_0^\infty \frac{x\,dx}{(1 + x)^2(2 + \sqrt{x})}.$

(c) $\displaystyle \int_1^\infty \frac{x + 2}{x(x + 1)}\,dx.$

(d) $\displaystyle \int_2^\infty \frac{dx}{x\sqrt{1 + x^2}}.$

(e) $\displaystyle \int_4^\infty \frac{\sqrt{x + 1}}{x + \sqrt{x}}\,dx.$

(f) $\displaystyle \int_2^\infty \frac{x^2 + 4x + 4}{(\sqrt{x} - 1)^3\sqrt{x^3 - 1}}\,dx.$

2. Establish the facts about the values of the exponent p for which the integral
$$\int_a^\infty \frac{dx}{x(\log x)^p}$$
is convergent ($a > 1$). Then use either Theorem II or Theorem III to test the following integrals, using the foregoing integral as a standard, with an appropriate value of p in each case.

(a) $\displaystyle \int_1^\infty \frac{dx}{\sqrt{x^2 + 1}[\log(1 + x)]^2}.$

(b) $\displaystyle \int_6^\infty \frac{(x + 1)dx}{(x^2 - 2)\left(\log \frac{x}{2} - 1\right)}.$

3. Find whether each of the following integrals is convergent or divergent:

(a) $\displaystyle \int_0^\infty \frac{x}{(1 + x)^3}\,dx.$

(b) $\displaystyle \int_1^\infty \frac{x}{e^x - 1}\,dx.$

(c) $\displaystyle \int_0^\infty \frac{x^3}{16 + x^4}\,dx.$

(d) $\displaystyle \int_0^\infty e^{-x^2}\,dx.$

(e) $\displaystyle \int_0^\infty x^2 e^{-x^2}\,dx.$

(f) $\displaystyle \int_1^\infty \frac{\sqrt{x}}{(1 + x)^2}\,dx.$

(g) $\displaystyle \int_0^\infty \frac{\tan^{-1} x}{1 + x^2}\,dx.$

(h) $\displaystyle \int_2^\infty \frac{dx}{\sqrt{x}(\log x)^3}.$

(i) $\displaystyle \int_0^\infty e^{-x} x^2(\log x)^3\,dx.$

(j) $\displaystyle \int_1^\infty \frac{\sin \frac{1}{x}}{x}\,dx.$

(k) $\displaystyle \int_1^\infty \frac{\frac{\pi}{2} - \tan^{-1} x}{x}\,dx.$

(l) $\displaystyle \int_1^\infty \left(\frac{\pi}{2} - \tan^{-1} x\right)dx.$

(m) $\displaystyle \int_2^\infty \frac{\frac{\pi}{2} - \tan^{-1} x}{\log x}\,dx.$

(n) $\displaystyle \int_{2/\pi}^\infty \frac{\cos\left(\frac{\pi}{2} - \frac{1}{x}\right)}{1 + x}\,dx.$

(o) $\displaystyle \int_4^\infty \frac{dx}{(x - 1)\log(x - 2) \cdot \log(\log x)}.$

(p) $\displaystyle \int_3^\infty \frac{dx}{\sqrt{1 + x^2}\,\log x[\log(\log x)]^2}.$

4. For what values of α is $\displaystyle\int_1^\infty \frac{x^{\alpha-1}}{1+x}\,dx$ convergent?

5. Show that $\int_0^\infty x^n e^{-x^2}\,dx$ is convergent if $n \geq 0$. Find the value of the integral if $n = 2m + 1$, where m is a nonnegative integer.

6. Find the conditions on m and n which guarantee that $\int_1^\infty e^{-x} x^m (\log x)^n\,dx$ is a convergent integral of the first kind.

7. Let $P(x)$ and $Q(x)$ be polynomials of degrees m and n, respectively, and suppose that r is the largest real root of the equation $Q(x) = 0$. What is the necessary and sufficient condition on m and n to make the integral $\displaystyle\int_c^\infty \frac{P(x)}{Q(x)}\,dx$ converge, if $r < c$? Why is the integrand of constant sign if x is large enough?

8. Suppose $0 < a_n < \frac{1}{2}$, and $b_n > 0$, $n = 0, 1, 2, \cdots$. Define a function $f(x)$ for $x \geq 0$ as follows: $f(n) = b_n$, $n = 0, 1, 2, \cdots$, $f(x) = 0$ if $n - 1 + a_n \leq x \leq n - a_{n+1}$ and $n = 1, 2, \cdots$, $f(x)$ continuous for all $x \geq 0$, and a linear function on each of the intervals where it has not already been defined. **(a)** Sketch the graph of the function enough to show its general character. **(b)** Show that $\int_0^\infty f(x)\,dx$ is convergent if and only if the series $\sum_{n=1}^\infty a_{n+1} b_n$ is convergent. **(c)** Specialize the sequences $\{a_n\}$ and $\{b_n\}$ so as to obtain a convergent integral and yet have $b_n \to \infty$. This shows that the integral may converge even though $f(x)$ is unbounded.

9. Suppose $F(x)$ is defined when $x \geq x_0$, and that: **(a)** $F(x) < M$ for all such x, where M is a constant; **(b)** $F(x_1) \leq F(x_2)$ if $x_0 \leq x_1 < x_2$. Prove that $\lim_{x \to \infty} F(x) = A$, where A is the least upper bound of the values of $F(x)$.

10. Prove the following theorem: If $f(x) > 0$, if $f(x)$ decreases steadily as x increases, and if $\int_a^\infty f(x)\,dx$ is convergent, then $\lim_{x \to \infty} x f(x) = 0$. SUGGESTION: First prove that $\int_{x/2}^x f(t)\,dt \to 0$ as $x \to \infty$. Then use (18.1–3).

22.11 Integrals of second kind

Let us consider integrals of the second kind with positive (or non-negative) integrands. Theorems I–IV have exact analogues whose wordings differ but slightly from the statements of these theorems given in § 22.1. Suppose, for example, that we are dealing with integrals improper at the upper limit,

$$(22.11\text{–}1) \qquad\qquad \int_a^b f(t)\,dt \quad (b > a)$$

with $f(t) \geq 0$. Then, for $a \leq x < b$,

$$F(x) = \int_a^x f(t)\,dt$$

does not decrease as x increases, and the integral (22.11–1) is convergent if and only if $F(x)$ is bounded; in which case the value of the integral is $\lim_{x \to b} F(x)$. This result enables us to prove a comparison test strictly parallel to Theorem II, from which in turn we deduce limit tests in which the convergence or divergence of two integrals of the same type,

$$\int_a^b f(x)dx, \quad \int_a^b g(x)dx$$

are related by an examination of the limit

$$\lim_{x \to b-} \frac{f(x)}{g(x)}.$$

Entirely similar results obtain for integrals of the second kind improper at the lower limit of integration.

For integrals of the second kind the basic standard reference integrals are

$$(22.11–2) \qquad \int_a^b \frac{dx}{(b - x)^p}, \quad \int_a^b \frac{dx}{(x - a)^p}.$$

These integrals are convergent if $p < 1$, and divergent if $p \geqq 1$. If $p \leqq 0$ they are *proper* integrals, with no singularities of the integrands.

Example 1. The integral $\int_0^1 \frac{dx}{(1 - x^3)^{1/3}}$ is convergent. For, let

$$f(x) = \frac{1}{(1 - x^3)^{1/3}} = \frac{1}{(1 - x)^{1/3}(1 + x + x^2)^{1/3}},$$

$$g(x) = \frac{1}{(1 - x)^{1/3}}.$$

We have $\quad \lim_{x \to 1-} \frac{f(x)}{g(x)} = \lim_{x \to 1-} \frac{1}{(1 + x + x^2)^{1/3}} = (\tfrac{1}{3})^{1/3}.$

Consequently, since $\int_0^1 g(x)dx$ is of the type (22.11–2) with $p < 1$, the given integral is convergent (by the counterpart of Theorem III, § 22.1).

We often have occasion to recall the fact that $\sin x$ is approximately equal to x when x is small; more precisely,

$$\lim_{x \to 0} \frac{\sin x}{x} = 1.$$

Example 2. For what values of p is $\int_0^{\pi/2} \frac{\sin x}{x^p} dx$ convergent?

Here there is a singularity at $x = 0$ if $p > 1$; if $p = 1$ the integrand is bounded and the integral is proper. We write

$$\frac{\sin x}{x^p} = \frac{\sin x}{x} \cdot \frac{1}{x^{p-1}},$$

and take $\qquad f(x) = \dfrac{\sin x}{x^p}, \qquad g(x) = \dfrac{1}{x^{p-1}},$

so that $\qquad\qquad\qquad \lim\limits_{x \to 0} \dfrac{f(x)}{g(x)} = 1.$

The integral $\displaystyle\int_0^{\pi/2} \dfrac{dx}{x^{p-1}}$, and therefore also the integral $\displaystyle\int_0^{\pi/2} \dfrac{\sin x}{x^p}\, dx$, is convergent if $p - 1 < 1$, or $p < 2$. Both integrals are divergent if $p \geqq 2$.

It is worth observing that an integral of second kind can be transformed into an integral of first kind by a simple substitution. For an integral $\int_a^b f(x)dx$ with singularity at $x = b$ we can let

$$ y = \frac{1}{b - x}, \qquad x = b - \frac{1}{y}, \qquad dx = \frac{dy}{y^2}. $$

As $x \to b^-$, $y \to + \infty$, and we get an integral of the form $\displaystyle\int_c^\infty \phi(y)\, \dfrac{dy}{y^2}$ with $c = (b - a)^{-1}$.

Example 3. Transform $\displaystyle\int_0^1 \log\!\left(\dfrac{1}{1 - x}\right) dx.$

We set $y = (1 - x)^{-1}$ and obtain

$$ \int_0^1 \log\!\left(\frac{1}{1 - x}\right)dx = \int_1^\infty \frac{\log y}{y^2}\, dy. $$

This integral of first kind is convergent, as may be verified by the limit test of Theorem IV, with $g(y) = y^{-3/2}$. The convergence of the original integral could be established directly by a limit test with $g(x) = (1 - x)^{-1/2}$. We leave it for the student to verify by l'Hospital's rule that

$$ \lim_{x \to 1^-} \frac{\log\!\left(\dfrac{1}{1 - x}\right)}{(1 - x)^{-1/2}} = 0. $$

Other changes of variable may also be used.

Example 4. Show that

(22.11–3) $\qquad\qquad \displaystyle\int_0^1 \left(\log \frac{1}{u}\right)^3 du = \int_0^\infty t^3 e^{-t}\, dt.$

Here we set $t = \log (1/u)$, or $u = e^{-t}$, $du = - e^{-t} dt$. As $u \to 0$, $t \to + \infty$, and as $u \to 1$, $t \to 0$, whence the result follows as stated. The integral on the right in (22.11–3) has already been proved convergent (Example 4, § 22.1).

EXERCISES

1. Determine the convergence or divergence of each of the following integrals by comparison with an appropriate one of the integrals

$$\int_0^1 x^{-p}\, dx, \qquad \int_0^b (b - x)^{-q}\, dx,$$

using the known facts as to the convergence or divergence of these latter integrals.

(a) $\displaystyle\int_0^1 \frac{(1 + 2x)\sqrt{1 + x^2}}{1 - x^2}\, dx.$

(d) $\displaystyle\int_0^{\pi/2} \frac{\sqrt{x}\, dx}{(x + \sin x)(1 + x^2)}.$

(b) $\displaystyle\int_0^1 \frac{x\, dx}{\sqrt{1 - x^3}}.$

(e) $\displaystyle\int_0^1 \frac{dx}{\sqrt{x}\, (x + 2x^2)^{1/3}}.$

(c) $\displaystyle\int_1^2 \frac{4(8 - x^3)}{(2x - x^2)^2}\, dx.$

(f) $\displaystyle\int_0^1 \left(\frac{1}{x} + x\right)^{1/2} \frac{x^{1/3}}{x - \frac{1}{2}x^2}\, dx.$

2. Test the following integrals for convergence or divergence by methods analogous to those of Theorems III or IV, § 22.1.

(a) $\displaystyle\int_1^2 \frac{dx}{\sqrt{(x - 1)(3 - x)}}.$

(g) $\displaystyle\int_0^{\pi/2} \frac{x^2}{(\sin x)^3}\, dx.$

(b) $\displaystyle\int_1^{10} \frac{dx}{x\sqrt{x^2 - 1}}.$

(h) $\displaystyle\int_0^{\pi/4} \frac{x^3\, dx}{\sin (x^2)(\tan x)^3}.$

(c) $\displaystyle\int_0^2 \frac{x\, dx}{(16 - x^4)^{1/3}}.$

(i) $\displaystyle\int_0^2 \frac{\sqrt{4 - \frac{1}{4}x^2}}{\sqrt{x^3 - 2x^2 - 4x + 8}}\, dx.$

(d) $\displaystyle\int_0^1 \frac{dx}{x^{2/3}(1 + x)}.$

(j) $\displaystyle\int_0^1 \frac{-\log x}{\sqrt{x}}\, dx.$

(e) $\displaystyle\int_3^5 \frac{\sqrt{25 - x^2}}{x^2 - x - 6}\, dx.$

(k) $\displaystyle\int_1^2 \frac{\sqrt{x}}{\log x}\, dx.$

(f) $\displaystyle\int_0^1 \frac{x^3 \sin^{-1} x}{\sqrt{1 - x^2}}\, dx.$

(l) $\displaystyle\int_0^1 \left(\frac{1}{x} - x\right)^{1/2} \frac{\log (1 + x^{1/3})}{\sin x}\, dx.$

3. Give necessary and sufficient conditions on p and q for $\displaystyle\int_0^{\pi/2} \frac{x^p}{(\sin x)^q}\, dx$ to be a convergent improper integral of second kind.

4. For what values of a is the integral $\displaystyle\int_0^1 \frac{x^{a-1} + x^{-a}}{1 + x}\, dx$ a proper integral? For what values of a is the integral improper, but convergent?

5. Consider

$$\int_0^1 (1 - x^2)^{-1/2}(1 - x^3)^{-1/3} \cdots (1 - x^n)^{-1/n}\, dx, \qquad n \geqq 2.$$

Find all values of n for which the integral is convergent.

6. For what values of x is $\int_0^1 e^{-t} t^{x-1}\, dt$ (a) proper? (b) improper, but convergent?

7. Show that $\int_0^1 \dfrac{(-\log x)^p}{x^q}\, dx$ is convergent if $p \geq 0$ and $0 < q < 1$.

8. For what values of p is $\int_1^a (\log x)^{-p}\, dx$ improper, but convergent? Assume $a > 1$.

9. Prove that $\int_0^{\pi/2} \log (\sin x)\, dx$ is convergent. What can you say about $\int_0^{\pi/2} \log (\cos x)\, dx$?

10. If $f(u)$ is continuous, $0 \leq u \leq 1$, show that $\int_0^1 \dfrac{f(u)}{\sqrt{1 - u^2}}\, du$ is convergent. Show also that the substitution $u = \sin \theta$ transforms the given integral into a proper integral.

11. By a suitable change of variable transform $\int_0^1 x^{-n} e^{-1/x}\, dx$ into an improper integral of first kind, and show that it is convergent for all values of n.

12. In the integral $\int_0^1 \dfrac{x^p}{[\log (1 + x)]^q}\, dx$, use the power-series expansion of $\log (1 + x)$ to show that the integral is proper if $p \geq q$, improper but convergent if $p < q < p + 1$, and divergent if $q \geq p + 1$.

22.12   Integrals of mixed type

Many improper integrals occurring in practice are of mixed type.

Example 1.   Consider $\int_0^\infty \dfrac{dx}{(1 + x)\sqrt{x}}$ · This is of mixed type, with infinite upper limit, and a singularity of the integrand at the lower limit. There are no other singularities, so we consider the separate integrals

$$\int_0^1 \frac{dx}{(1 + x)\sqrt{x}}, \qquad \int_1^\infty \frac{dx}{(1 + x)\sqrt{x}},$$

of second and first kinds, respectively. The first integral is convergent, as may be shown by using a limit test to compare it with the convergent integral $\int_0^1 \dfrac{dx}{\sqrt{x}}$ · The second integral is also convergent, since the integrand behaves essentially like $x^{-3/2}$ as $x \to \infty$. We then write

$$\int_0^\infty \frac{dx}{(1 + x)\sqrt{x}} = \int_0^1 \frac{dx}{(1 + x)\sqrt{x}} + \int_1^\infty \frac{dx}{(1 + x)\sqrt{x}} \cdot$$

The choice of $x = 1$ as a breaking point is arbitrary. Any other positive value of x could have been used.

When an integral of mixed type is separated into its constituent "pure-type" parts, it is called divergent if any one of the constituents is divergent. If singularities occur within the interval of integration, or at both ends of a finite interval of integration, the integral must be separated into several integrals, each of which is a pure type of either first or second kind.

Integrals with $-\infty$ as a limit of integration may be treated by methods parallel to those of § 22.1, or may be reduced to integrals with $+\infty$ as a limit of integration, by the substitution $x = -u$.

Example 2.   Consider $\displaystyle\int_{-\infty}^{\infty} \frac{x\,dx}{e^x + x^4}$. We separate this into

$$(a)\int_{-\infty}^{0} \frac{x\,dx}{e^x + x^3} \quad \text{and} \quad (b)\int_{0}^{\infty} \frac{x\,dx}{e^x + x^4}\,.$$

The integral (b) is convergent, since

$$\frac{x}{e^x + x^4} < xe^{-x}$$

if $x > 0$, and $\int_0^\infty xe^{-x}\,dx$ is convergent (Example 4, § 22.1). The integral (a) is also convergent; for, as $x \to -\infty$, $e^x \to 0$, and the integrand behaves like x^{-3}.

One may set $x = -u$, and thus get

$$\int_{-\infty}^{0} \frac{x\,dx}{e^x + x^4} = \int_{\infty}^{0} \frac{u\,du}{e^{-u} + u^4} = -\int_{0}^{\infty} \frac{u\,du}{e^{-u} + u^4}\,.$$

In the transformed integral the integrand behaves like u^{-3} as $u \to +\infty$. The original integral has thus been shown to be the sum of two convergent integrals of first kind.

EXERCISES

1. Examine each of the following integrals as to convergence or divergence, giving a complete analysis of the convergence or divergence of each of the constituent pure types.

(a) $\displaystyle\int_{0}^{\infty} \frac{dx}{x\sqrt{1 + x^2}}$.

(b) $\displaystyle\int_{0}^{\infty} \frac{dx}{\sqrt{x}\sqrt{1 + x^2}}$.

(c) $\displaystyle\int_{-\infty}^{\infty} e^{-x^2}\,dx.$

(d) $\displaystyle\int_{-\infty}^{\infty} x^2 e^{-|x|}\,dx.$

(e) $\displaystyle\int_{1}^{\infty} \frac{dx}{(x-1)^{1/2}(3-x)^{2/3}}$.

(f) $\displaystyle\int_{0}^{\infty} \frac{dx}{x^{1/2}(x-1)^{4/3}}$.

(g) $\displaystyle\int_{-1}^{2} \frac{dx}{x^{1/3}}$.

(h) $\displaystyle\int_{0}^{\infty} \frac{3x-7}{(x-1)^3}\,dx.$

(i) $\displaystyle\int_{0}^{\pi} \frac{dx}{(\cos x)^{2/3}}$.

(j) $\displaystyle\int_{0}^{\pi} \frac{dx}{(\sin x)^{3/2}}$.

2. Proceed as directed in Exercise 1 with each of the following integrals:

(a) $\displaystyle\int_0^\infty \frac{(\log x)^2}{1+x^2}\,dx.$

(d) $\displaystyle\int_0^\infty \frac{(e^{-x}-1)^2}{x^3}\,dx.$

(b) $\displaystyle\int_0^\infty \frac{\log x}{(1+x^2)^2}\,dx.$

(e) $\displaystyle\int_0^\infty \frac{t^{-2/3}}{1+t}\,dt.$

(c) $\displaystyle\int_0^\infty \frac{e^{-x}-1}{x\sqrt{x}}\,dx.$

(f) $\displaystyle\int_{-\infty}^\infty \frac{\sinh\dfrac{\pi x}{2}}{\sinh \pi x}\,dx.$

3. In each of the following integrals the integrand contains a parameter. For each integral find the range of values of the parameter such that the integral is not divergent.

(a) $\displaystyle\int_0^\infty \frac{e^{-x}-1}{x^p}\,dx.$

(e) $\displaystyle\int_1^2 \left(\frac{2-x}{x-1}\right)^{a-1} \frac{dx}{x}.$

(b) $\displaystyle\int_0^\infty \frac{t^{x-1}}{1+t}\,dt.$

(f) $\displaystyle\int_0^1 \left(\log \frac{1}{u}\right)^{x-1} du.$

(c) $\displaystyle\int_0^\pi \frac{dx}{(\sin x)^p}.$

(g) $\displaystyle\int_0^{\pi/2} (\tan \theta)^{1-2x}\, d\theta.$

(d) $\displaystyle\int_0^\infty \frac{x^p}{1+x+x^2}\,dx.$

(h) $\displaystyle\int_0^\infty \left(\frac{e^{-t}}{t} - \frac{e^{-at}}{1-e^{-t}}\right) dt.$

4. Where must the point (x, y) lie in the xy-plane if the integral $\displaystyle\int_0^\infty \frac{dt}{t^x(1+t^y)}$ is to be convergent?

5. Answer the question of Exercise 4 for the integral $\displaystyle\int_0^\infty \frac{e^{-xu}-e^{-yu}}{u(1+e^{-u})}\,du.$

6. Show that the integral $\displaystyle\int_{-\infty}^\infty \frac{e^{-xu}-e^{-yu}}{1-e^{-u}}\,du$ is convergent if $0 < x < y < 1$.

22.2   The gamma function

One very interesting and important improper integral is the following, which defines what is known as the *gamma function*:

$$(22.2\text{–}1) \qquad\qquad \Gamma(x) = \int_0^\infty t^{x-1} e^{-t}\, dt.$$

If $x \geq 1$ this is an integral of the first kind, and is convergent, by Example 4, § 22.1. If $x < 1$, however, the integral is of mixed type, with a singularity of the integrand at $t = 0$, and we have to consider the integral

$$(22.2\text{-}2) \qquad\qquad \int_0^1 t^{x-1}\, e^{-t}\, dt$$

of second kind. Since $e^{-t} \to 1$ as $t \to 0$, it is clear that the integrand in (22.2-2) behaves like t^{x-1} near $t = 0$. Now

$$\int_0^1 \frac{dt}{t^{1-x}}$$

is convergent if $1 - x < 1$, i.e. if $0 < x$, and divergent if $1 - x \geqq 1$, i.e., if $x \leqq 0$. Therefore, by the analogue of Theorem III, § 22.1, for integrals of second kind, (22.2-2) is convergent if and only if $x > 0$ (it is improper only if $x < 1$).

The integral from $t = 1$ to $t = \infty$ is always convergent, as we saw in Example 4, § 22.1. The result is, then, that the integral (22.2-1) defining $\Gamma(x)$ is convergent if and only if $x > 0$. Putting $x = 1$ we have

$$\int_0^\infty e^{-t}\, dt = \lim_{T \to \infty} \int_0^T e^{-t}\, dt = \lim_{T \to \infty} [- e^{-T} + 1] = 1,$$

or

$$(22.2\text{-}3) \qquad\qquad \Gamma(1) = 1.$$

There is a very simple relation between the values of the gamma function at x and $x + 1$. This relation is found by carrying out an integration by parts. We start with

$$\Gamma(x + 1) = \int_0^\infty t^x\, e^{-t}\, dt.$$

Setting $u = t^x$, $dv = e^{-t}\, dt$, we have $du = x t^{x-1}\, dt$, $v = - e^{-t}$,

$$\int_0^T t^x e^{-t}\, dt = \left[- t^x e^{-t} \right]_0^T + \int_0^T x t^{x-1} e^{-t}\, dt;$$

letting $T \to \infty$, we see that

$$(22.2\text{-}4) \qquad \int_0^\infty t^x e^{-t}\, dt = - \lim_{T \to \infty} T^x e^{-T} + 0 + x \int_0^\infty t^{x-1} e^{-t}\, dt.$$

But

$$\lim_{T \to \infty} T^x\, e^{-T} = 0,$$

as we see by applying l'Hospital's rule n times to $\dfrac{T^x}{e^T}$, where n is the first integer greater than or equal to x. Therefore, by (22.2-4),

$$(22.2\text{-}5) \qquad\qquad \Gamma(x + 1) = x\Gamma(x).$$

From this formula and (22.2–3) we have successively,

$$\Gamma(2) = 1 \cdot \Gamma(1) = 1$$
$$\Gamma(3) = 2 \cdot \Gamma(2) = 2 \cdot 1$$
$$\Gamma(4) = 3 \cdot \Gamma(3) = 3 \cdot 2 \cdot 1$$
$$\Gamma(5) = 4 \cdot \Gamma(4) = 4 \cdot 3 \cdot 2 \cdot 1$$
$$\vdots \qquad\qquad \vdots \qquad\qquad \vdots$$

In general we can write

(22.2–6) $$\Gamma(n + 1) = n!$$

or

(22.2–7) $$\Gamma(n) = (n - 1)!.$$

In the ordinary elementary sense $n!$ is defined only if n is a positive integer. But since $\Gamma(x)$ has been defined for every positive x, we see by (22.2–7) and (22.2–3) that it is natural to make the agreement that $0! = 1$. This is customarily done.

The gamma function gives us a convenient method of interpolating between the values of the factorials $n!$, and this is one of the primary reasons for the importance of the gamma function. Just now we shall take for granted that $\Gamma(x)$ is a continuous function, though we can prove this later on (§ 22.5, following Theorem VIII). In fact, $\Gamma(x)$ has continuous derivatives of all orders, and is analytic. The derivatives are found by differentiating with respect to the parameter x *under the integral sign* in (22.2–1). Recall that

$$t^x = e^{x \log t}, \quad \frac{d}{dx}(t^x) = \log t \cdot e^{x \log t} = t^x \log t.$$

Thus

(22.2–8) $$\Gamma'(x) = \int_0^\infty t^{x-1}(\log t)e^{-t}\,dt.$$

This is the same procedure as that given in (18.5–2) for *proper* integrals dependent on a parameter. For improper integrals further justification is required; the problem is much the same as the problem of justifying the differentiation of a series term by term. We return to this problem systematically in Theorem X, § 22.5; for the present let us proceed with our study of the gamma function. We can differentiate a second time, obtaining

(22.2–9) $$\Gamma''(x) = \int_0^\infty t^{x-1}(\log t)^2 e^{-t}\,dt.$$

The integrals for $\Gamma'(x)$ and $\Gamma''(x)$ are convergent integrals of mixed type, with singularities of the integrand at $t = 0$, and the same is true for the integrals giving all the higher derivatives (see Exercise 1).

It is clear from (22.2–9) that $\Gamma'''(x) > 0$, and therefore the curve $y = \Gamma(x)$ is concave upward for all $x > 0$. We also see that $\Gamma(x) > 0$, $\Gamma(1) = \Gamma(2) = 1$. From these facts we see that $\Gamma(x)$ has just one minimum value, and that this occurs for a value of x between 1 and 2. To see how $\Gamma(x)$ behaves as $x \to 0$, we observe that if we integrate only from 0 to 1 in (22.2–1), the result is less than $\Gamma(x)$. Furthermore, e^{-t} is a decreasing function, so that $e^{-t} > e^{-1}$ if $0 \le t < 1$. Therefore

$$\Gamma(x) > \int_0^1 t^{x-1}e^{-t}\, dt > e^{-1}\int_0^1 t^{x-1}\, dt = \frac{1}{ex}.$$

It follows from this that $\Gamma(x) \to +\infty$ as $x \to 0^+$. From the information which we have now collected it is possible to show the general character of $\Gamma(x)$ on a graph. We leave it for the student to prepare such a graph for himself.

The formula (22.2–1) does not define a function if $x \le 0$. Nevertheless we can define $\Gamma(x)$ for certain negative values of x by using formula (22.2–5). If $-1 < x < 0$, then $0 < x + 1$, so that $\Gamma(x + 1)$ has a meaning already defined. We then define $\Gamma(x)$ by requiring

(22.2–10) $$\Gamma(x) = \frac{\Gamma(x + 1)}{x}.$$

Thus, for instance,

$$\Gamma(-\tfrac{1}{2}) = -2\,\Gamma(\tfrac{1}{2}).$$

Now suppose that $-2 < x < -1$; then $-1 < x + 1 < 0$, so that $\Gamma(x + 1)$ is already defined. We then define $\Gamma(x)$ by (22.2–10), e.g.,

$$\Gamma(-\tfrac{3}{2}) = -\tfrac{2}{3}\,\Gamma(-\tfrac{1}{2}).$$

This process can evidently be continued, so that we obtain a definition of $\Gamma(x)$ for all values of x except $0, -1, -2, -3, \cdots$, and the equation (22.2–10) holds for all other values of x.

It is easy to see that $\Gamma(x) < 0$ when $-1 < x < 0$, and that $\Gamma(x) \to -\infty$ as $x \to 0^-$ or $x \to -1^+$. We leave it for the student to study the situation when $-2 < x < -1$, $-3 < x < -2$, and so on. A rough graph should be constructed. It will be shown later that $\Gamma(\tfrac{1}{2}) = \sqrt{\pi}$ (see (22.41–6)); from this we may calculate $\Gamma(-\tfrac{1}{2})$, $\Gamma(-\tfrac{3}{2})$, etc.

EXERCISES

1. Show that $\int_0^\infty t^{x-1}(\log t)^n e^{-t}\, dt$ is convergent for $n = 1, 2, \cdots$ if $0 < x$.

2. Prepare a graph of $y = \Gamma(x)$, showing the general behavior of the gamma function for $x > 0$ and in the intervals $-1 < x < 0$, $-2 < x < -1$, etc.

3. Show that $\Gamma(x) = 2 \int_0^\infty u^{2x-1} e^{-u^2} \, du$ if $x > 0$.

4. Calculate the value in terms of $\sqrt{\pi}$ of

(a) $\int_0^\infty x^2 e^{-x^2} \, dx$,   (b) $\int_0^\infty x^4 e^{-x^2} \, dx$.

5. If $a > 0$, show that $\int_0^\infty x^{n-1} e^{-ax} \, dx = a^{-n} \Gamma(n)$. What is the implied restriction on n?

6. Calculate in terms of $\sqrt{\pi}$ the values of

(a) $\int_0^\infty x^{-1/2} e^{-2x} \, dx$,   (b) $\int_0^\infty x^{3/2} e^{-4x} \, dx$.

7. Show by (22.2–5) that, if $n = 1, 2, \cdots$,

$$\Gamma(n + \tfrac{1}{2}) = \frac{1 \cdot 3 \cdot 5 \cdots (2n - 1)}{2^n} \sqrt{\pi}.$$

As a consequence show that

$$\sqrt{\pi} \, \Gamma(2n + 1) = 2^{2n} \Gamma(n + \tfrac{1}{2}) \Gamma(n + 1),$$

and

$$\sqrt{\pi} \, \Gamma(2n) = 2^{2n-1} \Gamma(n) \Gamma(n + \tfrac{1}{2}).$$

These formulas suggest the conjecture that perhaps

$$\sqrt{\pi} \, \Gamma(2x) = 2^{2x-1} \Gamma(x) \Gamma(x + \tfrac{1}{2})$$

not merely for $x = n + \tfrac{1}{2}$ and $x = n$, where n is a positive integer, but for all $x > 0$. The conjecture is correct, as can be proved by later developments (see Exercise 8, § 22.7).

8. Show that $\dfrac{1 \cdot 3 \cdots (2n - 1)}{2 \cdot 4 \cdots 2n} = \dfrac{\Gamma(n + \tfrac{1}{2})}{\sqrt{\pi} \, \Gamma(n + 1)}$.

9. Derive the formula $\Gamma(x) = \displaystyle\int_0^1 \left(\log \frac{1}{u} \right)^{x-1} du$ by putting $u = e^{-t}$ in (20.2–1). Then set $u = v^a$, where $a > 0$, and so find the value of

$$\int_0^1 \left(\log \frac{1}{v} \right)^{x-1} v^{a-1} \, dv,$$

where $x > 0$.

10. Utilize the results of Exercise 9 to show that

(a) $\displaystyle\int_0^1 \left(\frac{\log (1/t)}{t} \right)^{1/2} dt = \sqrt{2\,\pi}$,

(b) $\displaystyle\int_0^1 \left(\frac{t}{\log (1/t)} \right)^{1/2} dt = \sqrt{\frac{2\,\pi}{3}}$.

22.3   Absolute convergence

An improper integral of first kind, $\int_a^\infty f(x) dx$, is called *absolutely convergent* if the integral $\int_a^\infty | f(x) | \, dx$ is convergent. Exactly the same definition is applied to integrals of second kind, and to integrals of mixed type.

The switch from $f(x)$ to $|f(x)|$ corresponds exactly to the switch from Σa_n to $\Sigma |a_n|$ in defining absolute convergence of infinite series. If an integral is convergent, but not absolutely convergent, it is called *conditionally convergent*.

The following theorem corresponds to Theorem IX, § 19.3:

THEOREM V. *If the integral $\int_a^\infty |f(x)| \, dx$ is convergent, so is $\int_a^\infty f(x)dx$.*

In other words, if an integral is absolutely convergent, it is convergent.

PROOF OF THE THEOREM. First of all we observe that

$$(22.3\text{--}1) \qquad 0 \le |f(x)| - f(x) \le 2|f(x)|.$$

Both parts of this double inequality may be checked by considering separately the cases when $f(x) \ge 0$ and $f(x) < 0$. Now let $g(x) = |f(x)| - f(x)$. Since $\int_a^\infty |f(x)| \, dx$ is assumed to be convergent, the integral with $2|f(x)|$ as integrand is also convergent. Then, by (22.3–1) and Theorem II, § 22.1, we see that $\int_a^\infty g(x)dx$ is convergent. But $f(x) = |f(x)| - g(x)$, and therefore $\int_a^\infty f(x)dx$ is convergent, for sums and differences of convergent integrals are convergent, as may be seen directly from the definition of convergence. (What theorem about limits is used at this last step in the argument?)

The theorem and its proof apply to integrals of the second kind; only the limits of integration have to be changed.

To test whether an integral is absolutely convergent, we can apply the methods of §§ 22.1, 22.11, since the integrand $|f(x)|$ is never negative. If an integral is conditionally convergent, the demonstration of its convergence is usually a more delicate matter. Many of the instances of practical importance can be handled by the following theorem, which is analogous to Dirichlet's test for series (§ 19.7).

THEOREM VI. *Consider an improper integral of first kind of the form*

$$(22.3\text{--}2) \qquad \int_a^\infty \phi(t)f(t)dt,$$

where the functions ϕ and f satisfy the conditions:

(a) $\phi'(t)$ is continuous, $\phi'(t) \le 0$, and $\lim_{t \to \infty} \phi(t) = 0$,

(b) $f(t)$ is continuous, and the integral

$$(22.3\text{--}3) \qquad F(x) = \int_a^x f(t)dt$$

is bounded for all $x \ge a$. Then the integral (22.3–2) is convergent.

PROOF. We note that $F'(x) = f(x)$. Therefore, integrating by parts and noting that $F(a) = 0$, we have

(22.3–4)

$$\int_a^x \phi(t)f(t)dt = \int_a^x \phi(t)F'(t)dt = \phi(x)F(x) - \int_a^x \phi'(t)F(t)dt.$$

Let us suppose that M is a bound for $| F(x) |$, that is, $| F(x) | \leqq M$. Then $| \phi(x)F(x) | \leqq | \phi(x) | M$, and so $\phi(x)F(x) \to 0$ as $x \to \infty$, since $\phi(x) \to 0$ by hypothesis. It then follows from (22.3–4) that (22.3–2) is convergent, provided we can show that the integral

(22.3–5)
$$\int_a^\infty \phi'(t)F(t)dt$$

is convergent. This integral is in fact absolutely convergent. For, since $\phi'(t) \leqq 0$,

$$| \phi'(t)F(t) | = - \phi'(t) | F(t) | \leqq - M\phi'(t).$$

It is then enough to show that

(22.3–6)
$$\int_a^\infty - M\phi'(t)dt$$

is convergent; then (22.3–5) will be absolutely convergent, by Theorem II, § 22.1. Now

$$\int_a^x - M\phi'(t)dt = - M\phi(x) + M\phi(a) \to M\phi(a)$$

as $x \to \infty$, by condition (a). Thus (22.3–6) is convergent, and the proof is complete.

Example 1. The integral $\int_0^\infty (1/t) \sin t\, dt$ is convergent. (There is no singularity at $t = 0$; see the remark in Example 2, § 22.11.)

Here we take $\qquad \phi(t) = \dfrac{1}{t}, \qquad f(t) = \sin t.$

Then $\qquad F(x) = \displaystyle\int_0^x \sin t\, dt = 1 - \cos x.$

The conditions of Theorem VI are fulfilled, for $\phi'(t) = - 1/t^2$ and $0 \leqq F(x) \leqq 2$. Therefore, the given integral is convergent. It is *not* absolutely convergent, however, as is not difficult to see (for a hint on this see Exercise 6).

Example 2. The integral $\int_0^\infty \sin u^2\, du$ is convergent. We make a change of variable,

$$u = \sqrt{t}, \qquad du = \frac{dt}{2\sqrt{t}}.$$

Then

(22.3–7) $$\int_0^x \sin u^2 \, du = \tfrac{1}{2} \int_0^{x^2} \frac{\sin t}{\sqrt{t}} \, dt.$$

Now the integral $\int_0^\infty \dfrac{\sin t}{\sqrt{t}} \, dt$ is convergent, by an application of Theorem VI very much the same as in Example 1. Hence, letting $x \to \infty$ in (22.3–7), we see that

$$\int_0^\infty \sin u^2 \, du = \tfrac{1}{2} \int_0^\infty \frac{\sin t}{\sqrt{t}} \, dt.$$

The integral of Example 2 illustrates the remark made at the end of § 22.1, for $\sin u^2$ does not approach zero as $u \to \infty$.

EXERCISES

1. Examine the convergence of each of the following integrals. Where possible, prove that the integral is absolutely convergent. If it is necessary to use Theorem VI, give details of the application of the theorem in the particular case. Proofs that an integral is not absolutely convergent need not be given.

(a) $\displaystyle\int_0^\infty \frac{x \cos x}{a^2 + x^2} \, dx.$

(j) $\displaystyle\int_0^\infty \frac{\sin ax}{1 + x^2} \, dx.$

(b) $\displaystyle\int_0^\infty \frac{\cos x}{(1 + x^2)(4 + x^2)} \, dx.$

(k) $\displaystyle\int_{-\infty}^\infty \frac{\cos x}{1 - x^2} \, dx.$

(c) $\displaystyle\int_0^\infty \frac{\cos x}{\sqrt{1 + x^3}} \, dx.$

(l) $\displaystyle\int_{-\infty}^\infty \frac{\cos x}{x^{2/3}} \, dx.$

(d) $\displaystyle\int_2^\infty \frac{\sin x}{\log x} \, dx.$

(m) $\displaystyle\int_0^\infty \frac{\sin x \sin mx}{x} \, dx.$

(e) $\displaystyle\int_2^\infty \frac{\log (\log x)}{\log x} \cos x \, dx.$

(n) $\displaystyle\int_0^\infty \frac{e^{-x} \sin x}{x} \, dx.$

(f) $\displaystyle\int_{-\infty}^\infty \frac{x^2 - x + 2}{x^4 + 10\,x^2 + 9} \, dx.$

(o) $\displaystyle\int_0^\infty \frac{\sin x}{e^{2x} - 1} \, dx.$

(g) $\displaystyle\int_0^\infty \frac{x(x^2 + 1) \sin x}{x^4 - x^2 + 1} \, dx.$

(p) $\displaystyle\int_0^\infty \frac{\sin (\sin x) \cos x}{x} \, dx.$

(h) $\displaystyle\int_0^\infty \frac{\sin \pi x}{x(x^2 - 1)} \, dx.$

(q) $\displaystyle\int_0^\infty \frac{x + 1}{x^{3/2}} \sin x \, dx.$

(i) $\displaystyle\int_{-\infty}^\infty \frac{x \sin x}{1 + x^2} \, dx.$

(r) $\displaystyle\int_0^\infty e^{-x} \cos 2 x \, dx.$

2. Show that $\displaystyle\int_0^\infty \frac{\sin ax}{x^p}\,dx$ is convergent if $0 < p < 2$, and absolutely convergent if $1 < p < 2$.

3. Show that $\displaystyle\int_0^\infty \frac{1 - \cos ax}{x^p}\,dx$ is convergent if $1 < p < 3$. Is it absolutely convergent for any of these values of p?

4. Show that $\displaystyle\int_0^\infty \frac{\sin x(1 - \cos x)}{x^p}\,dx$ is convergent if $0 < p < 4$. Is it absolutely convergent for any of these values of p?

5. Show that $\int_0^\infty \cos (x^2)dx$ and $\int_0^\infty x \cos (x^4)dx$ are convergent. Note that the integrand in the second integral is unbounded.

6. Show that

$$\left| \int_{n\pi}^{(n+1)\pi} \frac{\sin x}{x}\,dx \right| > \frac{2}{(n + 1)\pi}\,, \qquad n = 0, 1, \cdots.$$

Use this result to prove that $\displaystyle\int_0^\infty \frac{\sin x}{x}\,dx$ is not absolutely convergent.

22.4   Improper multiple integrals. Finite regions

Consider first the case of an integral

$$(22.4\text{--}1) \qquad\qquad \iint_R f(x, y)dA,$$

where R is a closed bounded region, f is continuous in R except at one point (x_0, y_0), and the behavior of f at that point is such that the function is not integrable over R in the sense of § 18.6. The cases of greatest practical importance are those in which $f(x, y)$ either becomes infinite or has a factor which becomes infinite as $(x, y) \to (x_0, y_0)$, e.g.

$$f(x, y) = \frac{1}{r} \quad \text{or} \quad f(x, y) = \frac{x - x_0}{r^2}\,,$$

where

$$(22.4\text{--}2) \qquad\qquad r = [(x - x_0)^2 + (y - y_0)^2]^{1/2}.$$

To define what we mean by the convergence or divergence of the integral (22.4–1) we proceed as follows: Let R' be a region derived from R by discarding a small region ΔR having the point (x_0, y_0) in its interior (R' is the shaded portion of R in Fig. 184). No restriction is placed on the shape of ΔR except that it be a Riemann region in the sense defined in § 18.6. Of course, we also assume that R is a Riemann region. Let d be the maximum diameter of ΔR, that is, the distance between two points of ΔR which are as far apart as it is possible for them to be when both are in ΔR. We then consider the integral

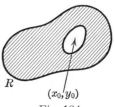

Fig. 184

$$\iint\limits_{R'} f(x, y)dA$$

over the region R'. If this integral approaches a limiting value as $d \to 0$, the value of the limit is denoted by

(22.4–3) $$\iint\limits_{R} f(x, y)dA = \lim_{d \to 0} \iint\limits_{R'} f(x, y)dA$$

and the integral (22.4–1) is said to be convergent. If the limit does not exist, the integral is called divergent.

We shall confine our attention to the case when the integral of $| f(x, y) |$ is convergent. In this case the integral of $f(x, y)$ is also convergent, as may be proved in much the same way that we proved Theorem V, § 22.3. The integral of $f(x, y)$ is then said to be *absolutely convergent*.

The comparison principle is valid in the following form for improper double integrals: *If $| f(x, y) | \leqq g(x, y)$, and if $\iint\limits_{R} g(x, y)dA$ is convergent, then $\iint\limits_{R} f(x, y)dA$ is absolutely convergent.*

If one is trying to determine whether or not the integral of a *positive* function is convergent, it is not necessary to consider regions ΔR of *arbitrary* shape in examining the limit (22.4–3). It is sufficient to confine one's attention to regions of one particular shape, say circles (or, alternatively, squares) with center at (x_0, y_0). Let δ be the radius of such a circle, and let R' be the region which is obtained by deleting from R the interior of the circle in question. Then, if $g(x, y) \geqq 0$, and if the limit of $\iint\limits_{R'} g(x, y)dA$

exists as $\delta \to 0$, the integral of g over R will be convergent and will be equal to the limit just mentioned. Proof of this fact is indicated in Exercise 8.

All the foregoing remarks apply without essential change to the case in which (x_0, y_0) is on the boundary of R instead of in the interior of R. The only difference is that we discard from R that part of R which is in a small region having (x_0, y_0) in its interior.

Example 1. Let r be defined by (22.4–2) and consider the integral

$$(22.4\text{–}4) \qquad\qquad \iint\limits_{R} \frac{1}{r^m}\, dA,$$

it being understood that (x_0, y_0) is a point of R and $m > 0$, so that the integral is improper. We shall show that it is convergent if $m < 2$. Since the only singularity is at (x_0, y_0), the typical difficulty is exposed in the case that R is a circle with center at (x_0, y_0), and of radius c. Let us delete a small concentric circle of radius δ, so that R' is the annulus between these circles. In evaluating the integral we may as well assume that (x_0, y_0) is the origin, since the value of the integral is not affected by the location of the axes. Then, with the use of polar co-ordinates,

$$\iint\limits_{R'} \frac{1}{r^m}\, dA = \int_0^{2\pi} d\theta \int_\delta^c r^{-m} r\, dr = \frac{2\,\pi}{2 - m}\, [c^{2-m} - \delta^{2-m}].$$

Since $m < 2$ we see that

$$\lim_{\delta \to 0} \iint\limits_{R'} \frac{1}{r^m}\, dA = \frac{2\,\pi}{2 - m}\, c^{2-m}.$$

Thus the integral (22.4–4) exists if R is a circle with center at (x_0, y_0) and $m < 2$. For regions of other shape, and for (x_0, y_0) located on the boundary of R, the difficulty can easily be resolved in terms of the case we have treated.

Example 2. The integral

$$(22.4\text{–}5) \qquad\qquad \iint\limits_{R} \frac{x - x_0}{r^2}\, dA$$

is absolutely convergent; for $|x - x_0| \leq r$, by (22.4–2), and so

$$\left| \frac{x - x_0}{r^2} \right| \leq \frac{1}{r},$$

whence, by the comparison-test principle and the result of Example 1, the asserted result follows.

Similar considerations apply to improper triple integrals. Improper multiple integrals in which the integrand has just one singular point in the region of integration occur typically in the theory of force fields governed by the inverse-square law, e.g., gravitational or electrostatic fields. From a purely mathematical point of view the study of such fields belongs to what is called *potential theory*. Let R be a bounded closed region in 3-space, and let it be filled with mass of density $\mu(x, y, z)$. If Q is the point (x_0, y_0, z_0), and

$$r = [(x - x_0)^2 + (y - y_0)^2 + (z - z_0)^2]^{1/2},$$

the *Newtonian potential* at Q produced by the total mass is

$$(22.4\text{-}6) \qquad \phi(x_0, y_0, z_0) = \iiint_R \frac{\mu}{r}\, dV,$$

and the *gravitational field* at Q is a vector $\mathbf{F}$ whose first component (in the x-direction) is

$$(22.4\text{-}7) \qquad F_1 = \iiint_R \mu\, \frac{x - x_0}{r^3}\, dV,$$

with similar formulas for F_2 and F_3. If Q is a point of R these are improper integrals; they are absolutely convergent if $\mu(x, y, z)$ is bounded and integrable. One can also discuss the potential and field due to distribution of mass on surfaces, and in particular on plane regions. If Q is a point in the plane region, the double integral defining the potential at Q is absolutely convergent, but the integrals defining the components of the field at Q are divergent except in very special cases.

EXERCISES

1. Let R be the circular region of radius 1 with center at the origin. Determine, for each of the following integrals, if it is convergent or divergent. Polar coordinates are denoted by r, θ.

(a) $\displaystyle\iint_R \frac{x^2}{(x^2 + y^2)^{7/4}}\, dA.$

(d) $\displaystyle\iint_R \log \frac{1}{r}\, dA.$

(b) $\displaystyle\iint_R \frac{x^2 y^2}{(x^2 + y^2)^3}\, dA.$

(e) $\displaystyle\iint_R \frac{1}{r} \log \frac{1}{r}\, dA.$

(c) $\displaystyle\iint_R \frac{\sin \frac{\theta}{2}}{r^{3/2}}\, dA.$

(f) $\displaystyle\iint_R \frac{x^2 + y^4}{(x^2 + y^2)^{3/2}}\, dA.$

2. Let $f(x, y)$ be continuous in R except at (x_0, y_0), and bounded. Let $r^2 = (x - x_0)^2 + (y - y_0)^2$. Show that $\displaystyle\iint_R \frac{f(x, y)}{r^m}\, dA$ is absolutely convergent if $m < 2$.

3. State and prove a result corresponding to that of Exercise 2, for improper triple integrals.

4. If R is a closed and bounded plane region containing the origin, for what values of p is the integral

$$\iint_R \frac{xy}{r^p}\, dA$$

certainly absolutely convergent?

5. Let R be the unit sphere $x^2 + y^2 + z^2 \leq 1$, and let $r^2 = x^2 + y^2 + z^2$. Find the values of those among the following integrals which are convergent. If a literal constant appears, indicate the restrictions you place on it to insure convergence.

(a) $\displaystyle\iiint_R \frac{1}{r}\, dV.$

(d) $\displaystyle\iiint_R \frac{z^2}{r^n}\, dV.$

(b) $\displaystyle\iiint_R r^{-n}\, dV.$

(e) $\displaystyle\iiint_R \frac{x^2 + y^4 + z^6}{r^5}\, dV.$

(c) $\displaystyle\iiint_R \frac{x^2 + y^2}{r^{9/2}}\, dV.$

(f) $\displaystyle\iiint_R \frac{x^2 y^2 z^2}{r^{17/2}}\, dV.$

6. (a) Is the integral $\displaystyle\iint_R \frac{dA}{[(x-1)^2 + y^2]^{1/2}}$ convergent or divergent, where R is the region $x^2 + y^2 \leq 1$?

(b) What if the exponent $\frac{1}{2}$ is replaced by m?

(c) For what values of p is $\displaystyle\iint_R \frac{x-1}{[(x-1)^2 + y^2]^p}\, dA$ certainly convergent?

7. Let R, f, ΔR, d have the meanings used in the discussion of (22.4–3).

(a) Let W be a subregion of R which contains (x_0, y_0) and all the points of R in some neighborhood of (x_0, y_0). Show that $\iint_R f(x, y)\, dA$ is convergent if and only if $\iint_W f(x, y)\, dA$ is convergent, and then

$$\iint_R f(x, y)\, dA = \iint_{R-W} f(x, y)\, dA + \iint_W f(x, y)\, dA,$$

where $R - W$ is the region which results by removing W from R.
Suggestion: Consider

$$\iint_{R-\Delta R} f(x, y)\, dA - \iint_{W-\Delta R} f(x, y)\, dA,$$

where ΔR is so small that it is contained in W.

(b) Deduce from the result in (a) that $\lim_{d \to 0} \iint_{\Delta R} f(x, y)\, dA = 0$ if $\iint_R f(x, y)\, dA$ is convergent.

8. Suppose $g(x, y)$ is continuous in R except at (x_0, y_0). Let $\{W_n\}$ be a sequence of subregions of the type of W in Exercise 7 **(a)**. Suppose W_1 contains W_2, W_2 contains W_3, etc., and that $d_n \to 0$ as $n \to \infty$, where d_n is the maximum diameter of W_n. Finally, if $R_n = R - W_n$, assume that

$$\lim_{n \to \infty} \iint_{R_n} g(x, y)dA \text{ exists, } = I.$$

Prove that $\iint_R g(x, y)dA$ is convergent, with value I. SUGGESTION: For a given m, choose ΔR so small that it is contained in W_m. Then choose n so that W_n is contained in ΔR. Now show that

$$\iint_{R - \Delta R} f(x, y)dA - I = \iint_{W_m - \Delta R} g(x, y)dA + \iint_{R_m} g(x, y)dA - I,$$

and that

$$\iint_{W_m - \Delta R} g(x, y)dA \leqq \iint_{R_n} g(x, y)dA - \iint_{R_m} g(x, y)dA.$$

From here it is easy to complete the proof. Write out the whole argument carefully.

9. Prove the validity of the comparison principle stated in the text.

10. Suppose $\iint_R |f(x, y)| \, dA$ is convergent. **(a)** Let $\{W_n\}$ be a sequence of regions of the type described in Exercise 8. By considering the inequality $0 \leqq |f(x, y)| - f(x, y) \leqq 2|f(x, y)|$, show that

$$\lim_{n \to \infty} \iint_{R_n} f(x, y)dA \text{ exists, } = I, \text{ say.}$$

(b) If ΔR is so small that it is contained in W_n, show that

$$\left| \iint_{R - \Delta R} f(x, y)dA - I \right| \leqq \iint_{W_n} |f(x, y)| \, dA + \left| \iint_{R_n} f(x, y)dA - I \right|,$$

and so deduce that $\iint_R f(x, y)dA = I$. Note the result of Exercise 7 **(b)**.

22.41   Improper multiple integrals.   Infinite regions

In this section we are going to consider integrals of the type

$$(22.41\text{--}1) \qquad\qquad \iint_R f(x, y)dA,$$

where R is an unbounded region, such as the first quadrant ($x \geqq 0, y \geqq 0$), an infinite strip (say between the lines $y = 0$, $y = 1$), or even the entire xy-plane. In any case we assume that the boundary of R is regular enough to cause us no trouble, and we tacitly assume the same about all other regions subsequently mentioned in this section. We also assume that f is integrable, in the ordinary proper sense, over any bounded closed subregion of R. Thus the only problem arises from the fact that R is an infinite region.

We might proceed as follows: Let us take a sequence of concentric circles with centers at O and radii becoming infinite. Let $\{R_n\}$ be the sequence of regions obtained by considering the part of R inside or on the nth circle. Then define

$$(22.41\text{–}2) \qquad \iint\limits_{R} f(x, y)dA = \lim_{n \to \infty} \iint\limits_{R_n} f(x, y)dA$$

provided this limit exists. This seems like a reasonable procedure, as indeed it is, under suitable conditions. But, one may ask, why not use squares instead of circles? Or, why not regions more general than just circles and squares? Will one get the same limit in all cases? The answer is, perhaps not, unless some further restriction is placed on the function f. Suppose, however, that we assume $f(x, y) \geqq 0$. Then it may be shown that, if the limit (22.4–2) exists when we use circles to get R_n, it also exists *and has the same value* when we use squares or any other sequence of regions subject to reasonable restrictions. To understand the essential principle of what is involved here, let $\{R_n\}$ be a sequence of regions formed by using circles as already described, and let $\{S_n\}$ be a sequence obtained in the same way, using squares (instead of circles) with center at O and length of side becoming infinite. Also, let

$$I_n = \iint\limits_{R_n} f(x, y)dA, \qquad J_n = \iint\limits_{S_n} f(x, y)dA,$$

and write

$$I = \lim_{n \to \infty} I_n.$$

Since each region R_n contains its predecessor R_{n-1}, and since $f(x, y) \geqq 0$, it is clear that

$$I_1 \leqq I_2 \leqq I_3 \leqq \cdots \leqq I.$$

Likewise

$$J_1 \leqq J_2 \leqq J_3 \leqq \cdots.$$

Now *any* one of the squares is contained in *some* one of the circles, and vice versa. Therefore, given n, there is some N such that

$$J_n \leqq I_N \leqq I,$$

and given m, there is some M such that

$$I_m \leqq J_M.$$

These inequalities show that the sequences $\{I_n\}$ and $\{J_n\}$ both have the same limit.

These arguments show that it is sufficient to define the convergence of the integral (22.41–1) in the manner already indicated if $f(x, y) \geq 0$.

We define absolute convergence of (22.41–1) in the usual way. It is then easy to see that the definition (22.41–2) is satisfactory for absolutely convergent integrals. This is all we shall deal with. In practice we shall use regions R_n obtained by using squares, circles, or other regions as convenience dictates.

Example 1. If R is the entire xy-plane, and $a > 0$,

$$(22.41–3) \qquad \iint\limits_{R} (x^2 + y^2 + a^2)^{-3/2} \, dA = \frac{2\,\pi}{a}.$$

Using polar co-ordinates, and letting R' denote an arbitrary circular region $x^2 + y^2 \leq c^2$, we have

$$\iint\limits_{R'} (x^2 + y^2 + a^2)^{-3/2} \, dA = \int_0^{2\pi} d\theta \int_0^c (r^2 + a^2)^{-3/2} \, r \, dr$$

$$= 2\,\pi \left\{ - (c^2 + a^2)^{-1/2} + \frac{1}{a} \right\}.$$

Letting $c \to \infty$, we see that there is a limit, so that the integral over all of R is convergent, and given by (22.41–3).

Example 2. If R is the entire first quadrant in the xy-plane,

$$(22.41–4) \qquad \iint\limits_{R} e^{-x^2-y^2} \, dA = \frac{\pi}{4}.$$

To obtain this result, let R' be the part of R in the circle $x^2 + y^2 \leq c^2$ (see Fig. 185).

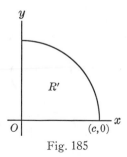

Fig. 185

Using polar co-ordinates, we have

$$\iint_{R'} e^{-x^2-y^2}\, dA = \int_0^{\pi/2} d\theta \int_0^c e^{-r^2} r\, dr$$

$$= \frac{\pi}{2}\left[-\tfrac{1}{2} e^{-r^2}\right]_0^c = \frac{\pi}{4}\left(1 - e^{-c^2}\right).$$

Hence

$$\iint_R e^{-x^2-y^2}\, dA = \lim_{c \to \infty} \iint_{R'} e^{-x^2-y^2}\, dA = \frac{\pi}{4}.$$

Let us consider what happens if we attempt to evaluate the integral in Example 2 by using squares instead of circles in the limiting process.

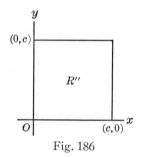

Fig. 186

Let R'' be the part of R inside the square with sides along the lines $x = \pm c$, $y = \pm c$ (see **Fig. 186**). Then

$$\iint_{R''} e^{-x^2-y^2}\, dA = \int_0^c dx \int_0^c e^{-x^2-y^2}\, dy$$

$$= \int_0^c e^{-x^2}\, dx \int_0^c e^{-y^2}\, dy = \left(\int_0^c e^{-x^2}\, dx\right)^2.$$

At the last step we have changed y to x under the integral sign, this being permitted since the value of the integral does not depend on the variable of integration.

Now we know from the earlier discussion that the value of the improper double integral is given just as well by the limiting process with squares as by that with circles. Therefore, in view of (22.41–4), we have

$$\frac{\pi}{4} = \lim_{c \to \infty}\left(\int_0^c e^{-x^2}\, dx\right)^2$$

or

(22.41–5) $$\int_0^\infty e^{-x^2}\, dx = \frac{\sqrt{\pi}}{2}.$$

This last result probably appears to the student as an unexpected by-product of our study of improper double integrals. It is a very important result, nevertheless. In fact, it was very largely to get this result that we developed this particular section of the text. Our immediate interest in the integral (22.41–5) is because of its connection with the gamma function. This connection is worked out in the next paragraph, by a change of variable. But the integral (22.41–5) is also of interest in statistics and elsewhere.

In (22.41–5) let us make the substitution

$$x = t^{1/2}, \qquad dx = \tfrac{1}{2} t^{-1/2} dt.$$

Then

$$\frac{\sqrt{\pi}}{2} = \tfrac{1}{2} \int_0^\infty t^{-1/2} e^{-1} dt = \tfrac{1}{2}\Gamma(\tfrac{1}{2}),$$

by (22.2–1). Thus we have

(22.41–6) $$\Gamma(\tfrac{1}{2}) = \sqrt{\pi}.$$

EXERCISES

1. Let the xy-plane be given a distribution of static electricity of constant density σ per unit area. Show that the resulting field at the point $(0, 0, a)$ on the positive z-axis is in the z-direction and of magnitude $2\pi\sigma$. Note that the result is independent of the distance a. The field at a point is the force which would be exerted on a unit positive charge there.

2. The electrostatic potential at $(0, 0, a)$ resulting from a charge of constant density σ on a bounded region R in the xy-plane is

$$\iint_R \frac{\sigma \, dA}{(x^2 + y^2 + a^2)^{1/2}}.$$

Would this integral be convergent if R were taken to be the whole xy-plane instead of a bounded region?

3. Suppose that, in Exercise 1, the charge is placed only on the strip of the xy-plane between the lines $x = \pm b$, where $b > 0$. Show that the electrostatic field at $(0, 0, a)$ is $4 \tan^{-1}(b/a)$. One method of solving calls for the formula

$$\int \frac{dx}{(A + Bx^2)(C + Dx^2)^{1/2}}$$
$$= \frac{1}{A}\left(\frac{A}{BC - AD}\right)^{1/2} \tan^{-1}\left[x\left(\frac{BC - AD}{A\{C + Dx^2\}}\right)^{1/2}\right] + \text{Const.},$$

which is not found in all integral tables.

4. Is the integral $\displaystyle\iint_R \frac{dA}{1 + x^2 + y^2}$ convergent, if R is the first quadrant of the xy-plane?

5. Show that, if R is the first quadrant of the xy-plane, the integral $\iint_R ye^{-y^2(1+x^2)} \, dA$ is convergent. One method is to integrate over the square $0 \leq x \leq a, 0 \leq y \leq a$, and study what happens as $a \to \infty$. It is possible to show that the integral over the square is equal to $\frac{1}{2} \tan^{-1} a - \frac{ae^{-a^2}}{2} \int_0^{a^2} \frac{e^{-t^2}dt}{a^2 + t^2}$. From this one can find the value of the double integral over R.

6. If the integrals $\int_0^\infty f(x)dx$ and $\int_0^\infty g(y)dy$ are convergent, and if $f(x)$ and $g(y)$ are never negative, show that the double integral $\iint_R f(x)g(y)dA$, over the entire first quadrant, is convergent and equal to the product

$$\left(\int_0^\infty f(x)dx \right)\left(\int_0^\infty g(y)dy \right).$$

This conclusion may *not* be correct if either or both of the functions $f(x)$, $g(y)$ can assume both positive and negative values. An illustration is afforded by $\int_0^\infty (1/x) \sin x \, dx$ and $\int_0^\infty e^{-y} \, dy$. If we attempt to compute the double integral using squares $0 \leq x \leq a, 0 \leq y \leq a$, we obtain the product of the two integrals as the limit when $a \to \infty$. But if we integrate over the region R_n defined as we shall indicate, it can be shown by estimation of the integrals over the blocks composing R_n that the double integral over R_n is greater than

$$\frac{\pi}{\log (\log n)} \log \frac{2n+1}{3} + c_n, \text{ where } c_n \to -\frac{\pi}{2} \text{ as } n \to \infty;$$ hence the integral tends to $+\infty$. The blocks composing R_n are $k\pi \leq x \leq (k+1)\pi, 0 \leq y \leq \log n$ if $k = 0, 2, \cdots, 2n$ and $k\pi \leq x \leq (k+1)\pi, 0 \leq y \leq \log [\log (\log n)]$ if $k = 1, 3, \cdots, 2n-1$.

22.5   Functions defined by improper integrals

There are many important problems in analysis in which one encounters improper integrals depending on a parameter; if such an integral is convergent, it defines a function of the parameter. A typical situation is that of the integral (22.2–1) defining the gamma function. As a general type of problem suppose we have

(22.5–1) $$F(x) = \int_c^\infty f(t, x)dt,$$

where the integral is of first kind and convergent for each value of x in some interval. In practice it is important to know whether or not F is continuous; also, under certain conditions it is possible to deal with integration and differentiation of $F(x)$ according to the formulas

$$F'(x) = \int_c^\infty \frac{\partial f}{\partial x} \, dt,$$

$$\int_a^b F(x)dx = \int_c^\infty dt \int_a^b f(t, x)dx.$$

It is important to know when these formulas are legitimate. The treatment of these matters is quite analogous to the treatment of the corresponding things in the case of functions defined by infinite series. The key concept is that of *uniform convergence*, just as it was in Chapter XX.

DEFINITION.   Suppose that the integral (22.5–1) is convergent for each x in the interval $a \leq x \leq b$. We say that it is uniformly convergent on that interval if to each $\epsilon > 0$ corresponds a number s_0, depending (possibly) on ϵ, but not on x, such that

$$\left| F(x) - \int_c^s f(t, x)dt \right| < \epsilon$$

whenever $a \leq x \leq b$ and $s_0 \leq s$.

This definition should be compared with the definition of uniform convergence of an infinite sequence of functions, in § 20.1. The $f_n(x)$ of that definition is analogous to $\int_c^s f(t, x)dt$ in our present definition, and n is analogous to s. The parallel between Chapter XX and our present work is so close that we shall in the main merely state the important theorems and omit the proofs. A student who understands Chapter XX thoroughly will have little additional difficulty in appreciating the present section. He must also be familiar with § 18.5, in which similar problems for *proper* integrals depending on a parameter are treated (see particularly Theorems XIII and XIV of § 18.5).

THEOREM VII.   *Let $f(t, x)$ be continuous in the two variables t, x when $c \leq t$ and $a \leq x \leq b$, and let the integral*

$$(22.5-2) \qquad F(x) = \int_c^\infty f(t, x)dt$$

be uniformly convergent on the interval $[a, b]$. Then F is continuous on the interval.

This corresponds to Theorem III, § 20.3.

There is a convenient practical test for uniform convergence, analogous to the M-test for series (Theorem II, § 20.2). It is the following:

THEOREM VIII.   *Suppose $g(t) \geq 0$, and that the integral of first kind*

$$\int_c^\infty g(t)dt$$

is convergent. Also suppose $|f(t, x)| \leq g(t)$ when $a \leq x \leq b$, for all values of t beyond some fixed value ($t_0 \leq t$). Then the integral (22.5–2) is uniformly convergent on the interval $[a, b]$.

There are analogous theorems for improper integrals of the second kind, with the obvious modifications in notation.

As an application, we can now prove that $\Gamma(x)$ is continuous when $x > 0$. We write

(22.5–3) $\Gamma(x) = \int_0^1 t^{x-1}e^{-1}\,dt + \int_1^\infty t^{x-1}e^{-1}\,dt.$

We prove that $\Gamma(x)$ is continuous on every interval $a \leq x \leq b$ with $0 < a < b$. Having fixed a and b, we observe that, if $a \leq x \leq b$, and $t \geq 1$, $0 < t^{x-1}e^{-t} \leq t^{b-1}e^{-t}$. Since

$$\int_1^\infty t^{b-1}e^{-t}\,dt$$

is convergent, the uniform convergence of the second integral in (22.5–3) follows by Theorem VIII. If $x > 1$, the first integral in (22.5–3) is proper, and is a continuous function of x by Theorem XIII, § 18.5. When the integral is improper, however, we use the theorem corresponding to Theorem VIII for integrals of second kind, noting that

$$0 < t^{x-1}e^{-t} \leq t^{a-1}e^{-t}$$

if $0 < t \leq 1$ and $0 < a \leq x \leq b$, and that

$$\int_0^1 t^{a-1}e^{-t}\,dt$$

is convergent. Thus $\Gamma(x)$, being the sum of two continuous functions, is continuous.

THEOREM IX.   *Under the hypothesis of Theorem VII it is true that*

(22.5–4) $\int_a^b F(x)dx = \int_c^\infty dt \int_a^b f(t,\,x)dx.$

This corresponds to Theorem IV, § 20.4.

THEOREM X.   *Let the integral*

$$F(x) = \int_c^\infty f(t,\,x)dt$$

be convergent when $a \leq x \leq b$. Let the partial derivative $\dfrac{\partial f}{\partial x}$ be continuous in the two variables t, x when $c \leq t$ and $a \leq x \leq b$, and let the integral $\displaystyle\int_c^\infty \dfrac{\partial f}{\partial x}\,dt$ converge uniformly on $[a, b]$. Then $F(x)$ has a derivative given by

(22.5–5) $$F'(x) = \int_c^\infty \frac{\partial f(t, x)}{\partial x}\, dt.$$

We shall prove this theorem, which corresponds to Theorem V, § 20.5. We set

$$G(u) = \int_c^\infty \frac{\partial f(t, u)}{\partial u}\, dt.$$

Applying Theorem IX, we have

$$\int_a^x G(u)du = \int_c^\infty dt \int_a^x \frac{\partial f(t, u)}{\partial u}\, du$$

$$= \int_c^\infty [f(t, x) - f(t, a)]dt$$

$$= F(x) - F(a).$$

Since G is a continuous function (by Theorem VII), we know from the last formula that

$$F'(x) = G(x)$$

(Theorem VII, § 1.52). This result is equivalent to (22.5–5).

With Theorem X we can justify the formula (22.2–8) for $\Gamma'(x)$, for the integrals

$$\int_0^1 t^{x-1}(\log t)e^{-t}\, dt, \quad \int_1^\infty t^{x-1}(\log t)e^{-t}\, dt$$

are uniformly convergent, as may be shown by arguments very similar to those employed in connection with (22.5–3). The procedure extends at once to the higher derivatives, so that, for $n = 1, 2, \cdots$ and $x > 0$

(22.5–6) $$\Gamma^{(n)}(x) = \int_0^\infty t^{x-1}(\log t)^n e^{-t}\, dt.$$

The theorems of this section have applications in justifying the steps in various ingenious methods by which one can sometimes find the value of an improper integral.

Example 1. Show that, if $0 < a < b$,

(22.5–7) $$\int_0^\infty \frac{e^{-at} - e^{-bt}}{t}\, dt = \log \frac{b}{a}\, .$$

The integral is of first kind, for the integrand approaches the finite limit $b - a$ as $t \to 0$. The ingenious device here consists in starting from the formula

(22.5–8) $$\frac{1}{x} = \int_0^\infty e^{-xt}\, dt,$$

which is valid if $x > 0$; this formula is obtained by direct evaluation, using the indefinite integral

$$\int e^{-xt} dt = -\frac{1}{x} e^{-xt} + C.$$

The integral (22.5-8) is uniformly convergent on the interval $a \leqq x \leqq b$, by Theorem VIII and the fact that $e^{-xt} \leqq e^{-at}$ when $t \geqq 0$. Hence, by (22.5-4),

$$\int_a^b \frac{dx}{x} = \int_0^\infty dt \int_a^b e^{-xt} dx.$$

When we calculate the integral on the left and the inner integral on the right, we obtain (22.5-7).

Example 2. Show that, if $a > 0$,

$$(22.5-9) \qquad \int_0^\infty \frac{e^{-at} \sin xt}{t} dt = \tan^{-1} \frac{x}{a}.$$

We denote the integral here by $F(x)$. We are justified in calculating $F'(x)$ by Theorem X, getting

$$(22.5-10) \qquad F'(x) = \int_0^\infty e^{-at} \cos xt \, dt,$$

for this latter integral is easily seen to converge uniformly for all values of x. Now, by an elementary integration formula

$$\int_0^s e^{-at} \cos xt \, dt = \left[\frac{e^{-at}(x \sin xt - a \cos xt)}{a^2 + x^2} \right]_0^s,$$

and hence, as $s \to \infty$ we have, after an easy calculation,

$$(22.5-11) \qquad \int_0^\infty e^{-at} \cos xt \, dt = \frac{a}{a^2 + x^2}.$$

Therefore, by integration, taking account of (22.5-10),

$$F(x) = \tan^{-1} \frac{x}{a} + C.$$

To determine the constant of integration C we observe that $F(0) = 0$ by the definition of $F(x)$. Since $\tan^{-1} 0 = 0$ also, we conclude that $C = 0$. This completes the derivation of (22.5-9).

Example 3. Show that, if $x > 0$,

$$(22.5-12) \qquad \int_0^\infty \frac{\sin xt}{t} dt = \frac{\pi}{2}.$$

This result is deduced from (22.5–9). We consider the integral in the latter formula as a function of the parameter a, writing

$$G(a) = \int_0^\infty \frac{e^{-at} \sin xt}{t} \, dt.$$

Note that $G(a) = \tan^{-1}(x/a)$ if $a > 0$, while

$$G(0) = \int_0^\infty \frac{\sin xt}{t} \, dt.$$

We regard x as fixed. This last integral is convergent, by Theorem VI, § 22.3. Now it can be shown that the integral defining $G(a)$ is uniformly convergent when $a \geq 0$; so that G is continuous for such values of a, by Theorem VII. Then $G(a) \to G(0)$ as $a \to 0^+$. Now

$$\lim_{a \to 0+} G(a) = \lim_{a \to 0+} \tan^{-1} \frac{x}{a} = \frac{\pi}{2}$$

if $x > 0$. Thus (22.5–12) is established. The assertion about uniform convergence is discussed in Exercise 20.

It is easily shown that the integral in (22.5–12) has the value $-\pi/2$ if $x < 0$, for the integral defines an odd function of x. Thus

$$(22.5\text{–}13) \qquad \int_0^\infty \frac{\sin xt}{t} \, dt = \begin{cases} \dfrac{\pi}{2} & \text{if } x > 0, \\[2mm] 0 & \text{if } x = 0, \\[2mm] -\dfrac{\pi}{2} & \text{if } x < 0. \end{cases}$$

From this it is clear that the function defined by the integral is discontinuous at $x = 0$. It must therefore fail to be uniformly convergent in any closed interval which contains $x = 0$.

EXERCISES

1. (a) Let $F(x) = \int_0^\infty e^{-t^2} \cos xt \, dt$. Assume the applicability of Theorem X, and show that $F'(x) = -\frac{1}{2} x F(x)$. Then find $F(x)$.

(b) By change of variable in the result of (a) show that

$$\int_0^\infty e^{-at^2} \cos xt \, dt = \frac{1}{2} \sqrt{\frac{\pi}{a}} \, e^{-x^2/4a}, \quad a > 0.$$

(c) By suitable use of Theorem VIII, show that the integral

$$\int_0^\infty te^{-t^2} \sin xt \, dt$$

is convergent uniformly with respect to x for all values of x, thus justifying the procedure used in (a).

2. (a) Show that, for all x,

$$\int_0^\infty e^{-t^2-(x^2/t^2)}\, dt = \frac{\sqrt{\pi}}{2}\, e^{-2|x|},$$

by denoting the integral by $F(x)$ and showing that $F'(x) = -2\,F(x)$ when $x > 0$. The substitution $u = x/t$ is useful at a certain stage in the work. Explain how you justify the answer when $x < 0$.

(b) Deduce from (a) the result

$$\int_0^\infty e^{-px^2-(q/x^2)}\, dx = \frac{1}{2}\sqrt{\frac{\pi}{p}}\, e^{-2\sqrt{pq}}, \quad p > 0,\ q \geq 0.$$

(c) Prove that the integral $F(x)$ in (a) is convergent uniformly for all values of x, and that the integral $\int_0^\infty (x/t^2)e^{-t^2-(x^2/t^2)}\, dt$ is convergent uniformly for $\delta \leq x \leq M$ if $\delta > 0$. Note that the method of calculating $F'(x)$ by differentiating under the integral sign gives a false result at $x = 0$. This is explained by the fact that the foregoing integral is not uniformly convergent in any interval including $x = 0$.

3. (a) Deduce the result

$$\int_0^\infty t^n e^{-xt}\, dt = \frac{n!}{x^{n+1}} \quad (x > 0)$$

by repeated differentiation of both sides of the formula

$$\int_0^\infty e^{-xt}\, dt = \frac{1}{x}.$$

(b) Prove that it is legitimate to differentiate under the integral sign as was done in (a).

4. (a) Deduce the result

$$\int_0^\infty \frac{dt}{(t^2 + x)^{n+1}} = \frac{1 \cdot 3 \cdots (2n - 1)}{2 \cdot 4 \cdots 2n}\, \frac{1}{x^{n+\frac{1}{2}}} \quad (x > 0)$$

by repeated differentiation of both sides of the formula

$$\int_0^\infty \frac{dt}{t^2 + x} = \frac{\pi}{2}\, x^{-1/2}.$$

(b) Prove that it is legitimate to differentiate under the integral sign as was done in (a).

5. (a) Let $F_n(x) = \int_0^\infty t^n e^{-xt^2}\, dt$, $x > 0$. Show that $F_n'(x) = -F_{n+2}(x)$. Calculate $F_0(x)$ explicitly from the result (22.41–5), and then use the foregoing relation between F_n' and F_{n+2} to prove by induction that

$$\int_0^\infty t^{2n} e^{-xt^2}\, dt = \sqrt{\pi}\, \frac{1 \cdot 3 \cdots (2n - 1)}{2^{n+1}}\, \frac{1}{x^{n+\frac{1}{2}}}.$$

(b) Prove that the integral defining $F_n(x)$ converges uniformly for $x \geq \delta$ if $\delta > 0$.

6. Show that $\int_0^\infty e^{-t^2} \sin 2\, xt\, dt = e^{-x^2} \int_0^x e^{u^2}\, du$.

7. Show that

$$\frac{d^n}{dx^n}\left(\frac{1}{1+x^2}\right) = (-1)^{n/2}\int_0^\infty t^n e^{-t}\cos xt\, dt, \quad n = 0, 2, 4, \cdots$$

$$\frac{d^n}{dx^n}\left(\frac{1}{1+x^2}\right) = (-1)^{(n+1)/2}\int_0^\infty t^n e^{-t}\sin xt\, dt, \quad n = 1, 3, 5, \cdots.$$

8. Prove that $\displaystyle\int_0^\infty \frac{\sin^2 xt}{t^2}\, dt = \frac{\pi}{2}\, x\ (x > 0)$. Assume that differentiation under the integral sign is legitimate.

9. Prove that, if $\displaystyle F(x) = \int_0^\infty \frac{1 - e^{-xt^2}}{t^2}\, dt$, where $x \geqq 0$, then $F'(x)$ can be computed by differentiation under the integral sign when $x > 0$. Deduce the value of $F(x)$.

10. (a) Start with

$$\frac{t}{1+t^2} = \int_0^\infty e^{-tx}\cos x\, dx \quad (t > 0),$$

and obtain

$$\int_0^\infty \frac{e^{-bx} - e^{-ax}}{x}\cos x\, dx = \tfrac{1}{2}\log\frac{1+a^2}{1+b^2},$$

if a and b are positive. Upon what theorem do you depend?

(b) Assume without proof that, as a function of b, the integral in **(a)** is uniformly convergent when $b \geqq 0$. Explain carefully how this implies that, if $a > 0$,

$$\int_0^\infty \frac{1 - e^{-ax}}{x}\cos x\, dx = \tfrac{1}{2}\log(1 + a^2).$$

What theorem do you use?

11. Prove that, if a and b are positive,

$$\int_a^b dx \int_0^\infty \frac{dt}{1+x^2t^2} = \int_0^\infty dt \int_a^b \frac{dx}{1+x^2t^2},$$

and in this way find the value of the integral $\displaystyle\int_0^\infty \frac{\tan^{-1}(bt) - \tan^{-1}(at)}{t}\, dt$.

12. Show that, under certain conditions

$$\int_{-\infty}^\infty \frac{f(bx) - f(ax)}{x^2}\, dx = \int_a^b dt \int_{-\infty}^\infty \frac{f'(tx)}{x}\, dx.$$

Apply this to show that

$$\int_{-\infty}^\infty \frac{\cos bx - \cos ax}{x^2}\, dx = \pi(a - b)$$

if a and b are positive. Take for granted that the "certain conditions" are satisfied.

13. Prove that $\int_0^\infty e^{-x}\dfrac{1-\cos xy}{x}\,dx = \tfrac{1}{2}\log(1+y^2)$.

Hint: $\dfrac{t}{1+t^2} = \int_0^\infty e^{-x}\sin xt\,dx$.

14. Prove that

$$\int_0^\infty e^{-x^2}\frac{\sin 2xy}{x}\,dx = \sqrt{\pi}\int_0^y e^{-t^2}\,dt.$$

15. From $\displaystyle\int_0^\infty e^{-y^2x^2}\,dx = \frac{\sqrt{\pi}}{2y}$ $(y>0)$ deduce by integration that

$$\int_0^\infty \frac{e^{-a^2x^2}-e^{-b^2x^2}}{x^2}\,dx = (b-a)\sqrt{\pi}$$

if $0 < a < b$.

16. Show that $\displaystyle\int_0^\infty \frac{e^{-t}\sin t}{t}\,dt = \frac{\pi}{4}$.

17. From (22.5–13) and trigonometric identities, deduce that

$$\int_0^\infty \frac{\sin x \cos mx}{x}\,dx = \begin{cases} \dfrac{\pi}{2} & \text{if } |m| < 1, \\[2mm] 0 & \text{if } |m| > 1, \\[2mm] \dfrac{\pi}{4} & \text{if } |m| = 1. \end{cases}$$

18. Prove that $\displaystyle\int_0^\infty \frac{\cos xy}{1+x^2}\,dx$ converges uniformly with respect to y for all y.

19. Prove that $\displaystyle\int_1^\infty \frac{dx}{x^{1+y}}$ converges uniformly with respect to y if $y \geq \delta$, where $\delta > 0$.

20. Prove that the integral $G(a) = \displaystyle\int_0^\infty e^{-at}\frac{\sin xt}{t}\,dt$, where x is fixed, is convergent uniformly with respect to a when $a \geq 0$.

Suggestion: Put $u = t^{-1}$, $dv = e^{-at}\sin xt\,dt$, and integrate by parts between $t = T$ and $t = \infty$, where $T > 0$. In this way show that

$$\left| \int_T^\infty e^{-at}\frac{\sin xt}{t}\,dt \right| \leq \frac{a+|x|}{a^2+x^2}\frac{2}{T}.$$

Explain why this inequality insures uniform convergence. (Consider the maximum of the expression on the right as a varies, assuming $x \neq 0$.)

22.51 Laplace transforms

A function $f(s)$ defined by

$$(22.51\text{–}1) \qquad\qquad f(s) = \int_0^\infty e^{-st}F(t)\,dt$$

is called the *Laplace transform* of the function $F(t)$. For instance, by (22.5–9) we see that

$$f(s) = \tan^{-1}\left(\frac{1}{s}\right) \quad \text{if} \quad F(t) = \frac{\sin t}{t},$$

and by Exercise 5, § 22.2, we see that

$$f(s) = \frac{\Gamma(n)}{s^n} \quad \text{if} \quad F(t) = t^{n-1}.$$

Laplace transforms are used a good deal in applied mathematics in solving differential equations, and they have many interesting connections with a variety of special topics in mathematics. We do not have space in this book for an extensive account of the theory and applications of Laplace transforms.

The theorems of § 22.5 are useful in the discussion of Laplace transforms. The exponential e^{-st} in (22.51–1) causes the integral to converge very nicely, provided that $F(t)$ does not grow too rapidly as $t \to \infty$. If there is some constant c such that $|F(t)| \leqq e^{ct}$ for all sufficiently large values of t, the integral defining $f(s)$ converges when $s > c$, and f has derivatives of all orders, which can be calculated by differentiating under the integral sign.

One of the most important properties of the Laplace transform is revealed when we integrate by parts in (22.51–1), taking $u = F(t)$, $dv = e^{-st}\,dt$. We first obtain

$$\int_0^b e^{-st}F(t)dt = -\frac{1}{s}\,e^{-st}F(t)\,\Big|_0^b + \frac{1}{s}\int_0^b e^{-st}F'(t)dt.$$

Then, letting $b \to \infty$, we get

$$(22.51\text{–}2) \qquad \int_0^\infty e^{-st}F(t)dt = \frac{1}{s}\,F(0) + \frac{1}{s}\int_0^\infty e^{-st}F'(t)dt,$$

provided certain conditions are satisfied. These conditions must be such as to guarantee the convergence of the integrals, and also such as to guarantee that

$$e^{-st}F(t) \to 0 \quad \text{when} \quad t \to \infty.$$

It is convenient to denote the Laplace transform of a function by using a symbol L in this way: $f(s) = L\{F(t)\}$. Formula (22.51–2) can now be written

$$(22.51\text{–}3) \qquad L\{F'(t)\} = sL\{F(t)\} - F(0).$$

If the procedure can be repeated with F' in place of F, we obtain

$$L\{F''(t)\} = sL\{F'(t)\} - F'(0).$$

If in this result we substitute for $L\{F'(t)\}$ from (22.51–3), we obtain

(22.51–4) $L\{F''(t)\} = s^2 L\{F(t)\} - sF(0) - F'(0).$

Formulas for the Laplace transform of higher-order derivatives can be found by carrying on the process.

The foregoing formulas provide the basis for using Laplace transforms to solve differential equations. As a preliminary to further discussion of this question we observe the following formulas:

(22.51–5) $L\{\sin at\} = \int_0^\infty e^{-st} \sin at \, dt = \dfrac{a}{s^2 + a^2} \, ,$

(22.51–6) $L\{\cos at\} = \int_0^\infty e^{-st} \cos at \, dt = \dfrac{s}{s^2 + a^2} \cdot$

These may be proved by direct use of elementary integration formulas (see the derivation of (22.5–11)).

Now we consider a differential equation problem.

Example. Find the function $y = F(t)$ such that

(22.51–7) $\dfrac{d^2 y}{dt^2} + 4 y = 3 \sin t$

and $F(0) = 1, \quad F'(0) = 0.$

We assume that the problem has a solution, and we let $f(s) = L\{F(t)\}$. From (22.51–7) we see that

$$L\{F''(t)\} + 4 L\{F(t)\} = 3 L\{\sin t\}.$$

We use (22.51–4) and (22.51–5) to obtain

$$s^2 f(s) - s \cdot 1 - 0 + 4 f(s) = \frac{3}{s^2 + 1} \cdot$$

We solve this equation to find $f(s)$:

$$f(s) = \frac{3}{(s^2 + 4)(s^2 + 1)} + \frac{s}{s^2 + 4} \cdot$$

The first fraction on the right can be expressed in terms of simpler fractions, and we find

$$f(s) = \frac{1}{s^2 + 1} - \frac{1}{s^2 + 4} + \frac{s}{s^2 + 4} \cdot$$

We now observe by (22.51–5) and (22.51–6) that

$$f(s) = L\{\sin t\} - \tfrac{1}{2} L\{\sin 2 t\} + L\{\cos 2 t\}$$
$$= L\{\sin t - \tfrac{1}{2} \sin 2 t + \cos 2 t\}.$$

Since $f(s) = L\{F(t)\}$, this suggests that

$$F(t) = \sin t - \tfrac{1}{2} \sin 2t + \cos 2t.$$

A check shows that this function satisfies the differential equation and the initial conditions.

This example illustrates a procedure which can be developed into a systematic technique for solving differential equations of certain types. Of course, the foregoing problem could also have been solved by a variety of standard elementary methods.

EXERCISES

1. Verify that $L\{1\} = s^{-1}$ and $L\{e^{ct}\} = (s - c)^{-1}$.

2. Derive the formulas

$$L\{\sinh at\} = \frac{a}{s^2 - a^2}, \quad L\{\cosh at\} = \frac{s}{s^2 - a^2}.$$

3. Find a function $y = F(t)$ such that $y'' + 2y' = e^t$ and $F(0) = 1$, $F'(0) = 0$. Use the results in Exercise 1.

4. Find a function $y = F(t)$ such that (a) $y'' + 9y = \cos 2t$, $F(0) = -1$, $F'(0) = 1$; (b) $y'' - 4y = -3 \sinh t$, $F(0) = 0$, $F'(0) = 5$.

5. If $L\{F(t)\} = f(s)$, show that $L\{e^{ct} F(t)\} = f(s - c)$. Use this to find Laplace transforms of (a) $t^{n-1}e^{-t}$; (b) $e^{-t} \sin t$; (c) $e^{2t} \cos 3t$.

22.6   Repeated improper integrals

We sometimes encounter improper integrals whose integrands are improper integrals, e.g.,

$$\int_0^\infty dy \int_0^\infty e^{-xy} \sin x \, dx, \quad \int_0^1 dy \int_0^\infty y^{-1/4} x^{-1/2} e^{-y(1+x)} \, dx.$$

It is often highly useful, as a technique in solving problems, to be able to invert the order of integration in such integrals. The legitimacy of such an inversion of order is not guaranteed by Theorem IX of § 22.5. We do not intend to take up theorems which deal with this problem. Our main intent is to point out that the inversion of the order of integration requires proof. Such proofs are sometimes rather delicate, and a systematic treatment of them would be too long and difficult for this book. We shall nevertheless give some examples of the use of inversions of the order of integration.

Example 1. Assuming the correctness of

$$(22.6\text{--}1) \qquad \int_0^\infty dy \int_0^\infty e^{-xy} \sin x \, dx = \int_0^\infty dx \int_0^\infty e^{-xy} \sin x \, dy,$$

we can easily show that

$$(22.6\text{-}2) \qquad \int_0^\infty \frac{\sin x}{x}\, dx = \frac{\pi}{2} \; ;$$

for the right side of (22.6–1) is $\int_0^\infty \frac{\sin x}{x}\, dx$, by (22.5–8). The inner integral on the left is handled in a manner similar to that of deriving (22.5–11). The result is

$$(22.6\text{-}3) \qquad \int_0^\infty e^{-xy} \sin x \, dx = \frac{1}{1 + y^2}, \quad y > 0.$$

Thus (22.6–1) becomes

$$\int_0^\infty \frac{dy}{1 + y^2} = \int_0^\infty \frac{\sin x}{x}\, dx.$$

The left side here is $\lim_{y \to \infty} \tan^{-1} y = \pi/2$. Thus (22.6–2) is established. We thus have an alternative to the method of Example 3, § 22.5.

Example 2. Show that

$$(22.6\text{-}4) \qquad \int_0^\infty \frac{\sin t}{\sqrt{t}}\, dt = \frac{2}{\sqrt{\pi}} \int_0^\infty \frac{ds}{1 + s^4} \; .$$

To get started, substitute $x = s\sqrt{t}$ in (22.41–5), taking s as the new variable of integration. The result is

$$\frac{2}{\sqrt{\pi}} \int_0^\infty e^{-s^2 t}\, ds = \frac{1}{\sqrt{t}} \; .$$

Thus $\qquad \displaystyle\int_0^\infty \frac{\sin t}{\sqrt{t}}\, dt = \frac{2}{\sqrt{\pi}} \int_0^\infty dt \int_0^\infty e^{-s^2 t} \sin t \, ds.$

If we now assume that it is legitimate to invert the order of integration on the right, we obtain the repeated integral

$$\frac{2}{\sqrt{\pi}} \int_0^\infty ds \int_0^\infty e^{-s^2 t} \sin t \, dt.$$

The inner integral here is evaluated by (22.6–3), and we have

$$(22.6\text{-}5) \qquad \frac{2}{\sqrt{\pi}} \int_0^\infty \frac{ds}{1 + s^4} \; ;$$

thus (22.6–4) has been derived. The integral (22.6–5) can be calculated by elementary methods, and is found to have the value $\dfrac{\pi}{2\sqrt{2}}$ (see Exercise 9). Thus we arrive at the formula

$$(22.6\text{--}6) \qquad \int_0^\infty \frac{\sin t}{\sqrt{t}}\, dt = \left(\frac{\pi}{2}\right)^{1/2}.$$

EXERCISES

In the following exercises it is to be assumed by the student that reversal of the order of integration in the repeated integrals is legitimate.

1. Let $I = \int_0^\infty e^{-x^2}\, dx$. If $x > 0$, show that $\int_0^\infty xe^{-x^2y^2}\, dy = I$. Thus

$$I^2 = \int_0^\infty e^{-x^2}\, dx \int_0^\infty xe^{-x^2y^2}\, dy.$$

Now reverse the order of integration and deduce the value of I. This is an alternative derivation of (22.41–5).

2. (a) Verify that $2\int_0^\infty ye^{-(1+x^2)y^2}\, dy = 1/(1 + x^2)$. From this deduce that

$$\int_0^\infty \frac{\cos ax}{1 + x^2}\, dx = \frac{\pi}{2}\, e^{-|a|}.$$

Use the result of Exercise 1 **(b)**, § 22.5.

(b) Use the method of **(a)** to show that $\displaystyle\int_0^\infty \frac{x \sin ax}{1 + x^2}\, dx = \frac{\pi}{2}\, e^{-a}$, $a > 0$. In the process you will have to deal with $\int_0^\infty xe^{-x^2y^2} \sin ax\, dx$ by integration by parts, and with $\int_0^\infty y^{-2}e^{-y^2-(a^2/4y^2)}\, dy$ by an appropriate substitution of a new variable of integration.

3. By reversing the order of integration in the repeated integral

$$\int_0^\infty dy \int_0^\infty y^{-1/4}x^{-1/2}e^{-y(1+x)}\, dx,$$

show that

$$\int_0^\infty \frac{dx}{x^{1/2}(1 + x)^{3/4}} = \frac{\Gamma(\tfrac14)\Gamma(\tfrac12)}{\Gamma(\tfrac34)}.$$

4. Use the formula $1/(1 + x^2) = \int_0^\infty e^{-xt} \sin t\, dt$ to show that

$$\int_0^\infty \frac{x^{a-1}}{1 + x^2}\, dx = \Gamma(a)\int_0^\infty \frac{\sin t}{t^a}\, dt$$

if $0 < a < 2$.

5. Show that, if $a \geq 0$, $\displaystyle\int_0^\infty \frac{e^{-ax}}{1 + x^2}\, dx = \int_0^\infty \frac{\sin t}{a + t}\, dt.$

6. Show that $\displaystyle\int_0^\infty \frac{\cos ax}{1 + x^2}\, dx = \int_0^\infty \frac{t \sin t}{a^2 + t^2}\, dt$, and that, if $a > 0$, these integrals are equal to $\displaystyle\int_0^\infty \frac{x \sin ax}{1 + x^2}\, dx$. From this result deduce that the value of the first integral is $\dfrac{\pi}{2}\, e^{-|a|}$. Assume the legitimacy of all procedures of reversing order of integration and differentiation under the integral sign.

7. Show that $\displaystyle\int_0^\infty t^{a-1} e^{-x\sqrt{t}}\, dt = 2\,\frac{\Gamma(2\,a)}{x^{2a}}$ if $a > 0$. Use this and the result

$\displaystyle\int_0^\infty e^{-x\sqrt{t}} \sin x\, dx = \frac{1}{1+t}$ to show that $\displaystyle\int_0^\infty \frac{t^{a-1}}{1+t}\, dt = 2\,\Gamma(2\,a)\int_0^\infty \frac{\sin x}{x^{2a}}\, dx$

if $0 < a < 1$.

8. Use the formula $\displaystyle\int_0^\infty t^{a-1} e^{-xt}\, dt = \frac{\Gamma(a)}{x^a}$ $(x > 0,\ a > 0)$ to show that

$\displaystyle\int_0^\infty \frac{\cos x}{x^a}\, dx = \frac{1}{\Gamma(a)}\int_0^\infty \frac{t^a}{1+t^2}\, dt$ if $0 < a < 1$.

9. Prove that $\displaystyle\int_0^\infty \frac{dx}{1+x^4} = \frac{\pi\sqrt{2}}{4}$ by expressing the integrand in terms of

partial fractions with quadratic denominators and deriving the integration formula

$$\int \frac{dx}{1+x^4} =$$

$$\frac{\sqrt{2}}{8} \log \frac{x^2 + x\sqrt{2} + 1}{x^2 - x\sqrt{2} + 1} + \frac{\sqrt{2}}{4}\{\tan^{-1}(x\sqrt{2} - 1) + \tan^{-1}(x\sqrt{2} + 1)\}.$$

10. (a) Prove that $\displaystyle\int_0^\infty \frac{x^2\, dx}{1+x^4} = \int_0^\infty \frac{dx}{1+x^4}$ by substituting $y = \frac{1}{x}$ in the

integral.

(b) Prove that $\displaystyle\int_0^\infty \frac{\cos t}{\sqrt{t}}\, dt = \left(\frac{\pi}{2}\right)^{1/2}$ by the method of illustrative Example 2.

22.7   The beta function

The beta function is defined by

$$(22.7\text{–}1) \qquad\qquad B(x, y) = \int_0^1 t^{x-1}(1 - t)^{y-1}\, dt.$$

If $x \geqq 1$ and $y \geqq 1$, the integral is proper. If $x > 0$ and $y > 0$, and either $x < 1$ or $y < 1$ (or both), the integral is improper, but convergent.

We shall show that there is an important connection between the beta function and the gamma function. In fact, if x and y are positive

$$(22.7\text{–}2) \qquad\qquad B(x, y) = \frac{\Gamma(x)\Gamma(y)}{\Gamma(x + y)}\,.$$

We shall verify this relation without trying to explain how it might be discovered.

To begin with, let us substitute

$$t = \frac{u}{1 + u}, \quad dt = \frac{du}{(1 + u)^2}$$

in (22.7–1), thus getting

$$(22.7\text{–}3) \qquad B(x, y) = \int_0^\infty \left(\frac{u}{1+u}\right)^{x-1}\left(\frac{1}{1+u}\right)^{y-1}\frac{du}{(1+u)^2}$$

$$= \int_0^\infty \frac{u^{x-1}}{(1+u)^{x+y}}\,du.$$

Next, in the formula (22.2–1) let us substitute $t = vu$, $dt = v\,du$, $(v > 0)$. Then

$$\Gamma(x) = \int_0^\infty (vu)^{x-1}e^{-vu}v\,du = \int_0^\infty v^x u^{x-1}e^{-vu}\,du.$$

Multiply this by $v^{y-1}e^{-v}$ and integrate with respect to v:

$$\Gamma(x)\int_0^\infty v^{y-1}e^{-v}\,dv = \int_0^\infty dv \int_0^\infty v^{x+y-1}u^{x-1}e^{-v(1+u)}\,du.$$

The left side here is $\Gamma(x)\Gamma(y)$. We assume that we can reverse the order of integration on the right. Then we have

$$\Gamma(x)\Gamma(y) = \int_0^\infty u^{x-1}\,du \int_0^\infty v^{x+y-1}e^{-v(1+u)}\,dv.$$

In the inner integral set

$$v = \frac{w}{1+u}, \qquad dv = \frac{dw}{1+u}.$$

Then

$$\int_0^\infty v^{x+y-1}e^{-v(1+u)}\,dv = \frac{1}{(1+u)^{x+y}}\int_0^\infty w^{x+y-1}e^{-w}\,dw$$

$$= \frac{\Gamma(x+y)}{(1+u)^{x+y}},$$

and so

$$\Gamma(x)\Gamma(y) = \Gamma(x+y)\int_0^\infty \frac{u^{x-1}}{(1+u)^{x+y}}\,du = \Gamma(x+y)\,B(x, y),$$

by (22.7–3). Thus (22.7–2) is established. The justification of the reversal of order of integration is a problem of the type referred to in § 22.6. There are various methods for proving (22.7–2). Some of the methods use the theory of improper double integrals, as sketched in § 22.41.

A useful alternative form for the beta function comes from putting

$$t = \sin^2 \theta, \quad dt = 2 \sin \theta \cos \theta \, d\theta$$

in (22.7–1). The new formula is

$$(22.7\text{–}4) \qquad B(x, y) = 2\int_0^{\pi/2} \sin^{2x-1} \theta \cos^{2y-1} \theta \, d\theta.$$

In particular, with $x = n$, $y = \frac{1}{2}$, we have

(22.7-5) $\qquad \displaystyle\int_0^{\pi/2} \sin^{2n-1} \theta\, d\theta = \tfrac{1}{2} B(n, \tfrac{1}{2}) = \frac{1}{2} \cdot \frac{\Gamma(n)\Gamma(\frac{1}{2})}{\Gamma(n + \frac{1}{2})}.$

By repeated use of (22.2-5) we can write

$$\Gamma(n + \tfrac{1}{2}) = (n - \tfrac{1}{2})\Gamma(n - \tfrac{1}{2}) = (n - \tfrac{1}{2})(n - \tfrac{3}{2})\Gamma(n - \tfrac{3}{2})$$

$$\cdot \quad \cdot \quad \cdot \quad \cdot \quad \cdot \quad \cdot$$

$$= (n - \tfrac{1}{2})(n - \tfrac{3}{2}) \cdots \tfrac{1}{2}\, \Gamma(\tfrac{1}{2}),$$

or

(22.7-6) $\qquad \displaystyle\Gamma(n + \tfrac{1}{2}) = \frac{1 \cdot 3 \cdot 5 \cdots (2n - 1)}{2^n}\, \Gamma(\tfrac{1}{2}).$

When this is combined with (22.7-5) we have

(22.7-7)
$$\int_0^{\pi/2} \sin^{2n-1} \theta\, d\theta = \frac{2^{n-1}\Gamma(n)}{1 \cdot 3 \cdot 5 \cdots (2n - 1)} = \frac{2 \cdot 4 \cdots (2n - 2)}{1 \cdot 3 \cdots (2n - 1)}.$$

In a similar manner it may be shown that

(22.7-8) $\qquad \displaystyle\int_0^{\pi/2} \sin^{2n} \theta\, d\theta = \frac{1 \cdot 3 \cdots (2n - 1)}{2 \cdot 4 \cdots (2n)} \cdot \frac{\pi}{2}.$

These two formulas are valid for $n = 1, 2, \cdots$. They may be derived in a completely elementary fashion by using the reduction formula

$$\int_0^{\pi/2} \sin^m \theta\, d\theta = \frac{m - 1}{m} \int_0^{\pi/2} \sin^{m-2} \theta\, d\theta.$$

Our purpose in deriving (22.7-7) and (22.7-8) here is so that we may prove the formula

(22.7-9) $\qquad \displaystyle\frac{\pi}{2} = \lim_{n \to \infty} \left[\frac{2 \cdot 4 \cdots (2n)}{1 \cdot 3 \cdots (2n - 1)} \right]^2 \frac{1}{2n + 1}.$

This formula is called Wallis's formula, after a 17th century English mathematician. To prove (22.7-9) we observe that, if $0 < \theta < \dfrac{\pi}{2}$,

$$\sin^{2n+1} \theta < \sin^{2n} \theta < \sin^{2n-1} \theta,$$

and so

$$\int_0^{\pi/2} \sin^{2n+1} \theta\, d\theta < \int_0^{\pi/2} \sin^{2n} \theta\, d\theta < \int_0^{\pi/2} \sin^{2n-1} \theta\, d\theta.$$

Therefore

$$\frac{2 \cdot 4 \cdots (2n)}{1 \cdot 3 \cdots (2n + 1)} < \frac{1 \cdot 3 \cdots (2n - 1)}{2 \cdot 4 \cdots (2n)} \cdot \frac{\pi}{2} < \frac{2 \cdot 4 \cdots (2n - 2)}{1 \cdot 3 \cdots (2n - 1)}.$$

These inequalities can be rewritten in the form

$$\frac{2\,n}{2\,n+1}\cdot\frac{\pi}{2} < \left[\frac{2\cdot 4\cdots(2\,n)}{1\cdot 3\cdots(2\,n-1)}\right]^2 \frac{1}{2\,n+1} < \frac{\pi}{2},$$

and (22.7–9) follows at once.

EXERCISES

1. Show directly from (22.7–1) that $B(y, x) = B(x, y)$ (make a simple change of the variable of integration).

2. Show that $B(x, 1) = x^{-1}$.

3. Using integration by parts, show directly from (22.7–1) that $yB(x + 1, y) = xB(x, y + 1)$. Hence, if $m + 1$ and $n + 1$ are positive integers, show that $B(m + 1, n + 1) = \dfrac{m!\,n!}{(m + n + 1)!}$ without recourse to (22.7–2).

4. Show directly from (22.7–1) that $B(x + 1, y) + B(x, y + 1) = B(x, y)$. Combine this with the result in Exercise 3 to show that

$$B(x + 1, y) = \frac{x}{x + y}\,B(x, y).$$

Hence, if n is a positive integer, show that

$$B(n, y) = \frac{(n - 1)!}{y(y + 1)\cdots(y + n - 1)}$$

without recourse to (22.7–2).

5. Show that

$$\int_0^1 \frac{x^{2a}}{\sqrt{1 - x^2}}\,dx = \frac{\sqrt{\pi}\,\Gamma(a + \tfrac{1}{2})}{2\,\Gamma(a + 1)}\,.$$

What is the restriction on a?

6. Show that $$\int_0^1 \frac{dx}{\sqrt{1 - x^n}} = \frac{\sqrt{\pi}\,\Gamma\!\left(\dfrac{1}{n}\right)}{n\Gamma\!\left(\dfrac{n + 2}{2\,n}\right)}\,.$$

7. Express $\int_0^1 x^m(1 - x^p)^n\,dx$ in terms of the beta function, and hence in terms of values of the gamma function. As particular cases, calculate

(a) $\int_0^1 (1 - \sqrt{x})^2\,dx$,   (b) $\int_0^1 x^{3/2}(1 - \sqrt{x})^{1/2}\,dx$,

(c) the total area enclosed by the curve $x^{2/3} + y^{2/3} = 1$, (d) $\displaystyle\int_0^1 \frac{x^4}{\sqrt{1 - x^4}}\,dx$.

8. Show that $B(x, x) = 2\int_0^{1/2} (t - t^2)^{x-1}\,dt$. Make the change of variable $u = 4(t - t^2)$, and deduce the relation $B(x, x) = 2^{1-2x}\,B(x, \tfrac{1}{2})$. From this deduce that $\sqrt{\pi}\,\Gamma(2\,x) = 2^{2x-1}\,\Gamma(x)\Gamma(x + \tfrac{1}{2})$. Compare with Exercise 7, § 22.2.

9. The gamma function can be expressed in the form

$$\Gamma(x) = 2\int_0^\infty s^{2x-1}e^{-s^2}\, ds,$$

as we see by putting $t = s^2$ in (22.2–1). Show that, as a consequence,

$$\tfrac{1}{4}\,\Gamma(x)\Gamma(y) = \iint_R e^{-s^2-t^2}s^{2x-1}t^{2y-1}\, dA,$$

where R is the entire first quadrant of the st-plane. Using a method similar to that of Example 2, § 22.41, evaluate this improper double integral with the aid of polar co-ordinates, and show that it has the value $\tfrac{1}{4}\,B(x, y)\Gamma(x + y)$. This gives us a new derivation of the formula (22.7–2).

10. Show that $\displaystyle\int_1^\infty \frac{y^{a-1}}{1+y}\, dy = \int_0^1 \frac{x^{-a}}{1+x}\, dx$ if $0 < a < 1$, and hence (see

(22.7–3)) that $B(a, 1 - a) = \displaystyle\int_0^1 \frac{x^{a-1} + x^{-a}}{1+x}\, dx.$

11. Use the geometric series for $(1 + x)^{-1}$ in the last integral in Exercise 10, and integrate the resulting series term by term. Show that the result can be

put in the form $\Gamma(a)\Gamma(1 - a) = \dfrac{1}{a} + 2a \displaystyle\sum_{n=1}^\infty (-1)^{n+1}\,\dfrac{1}{n^2 - a^2}\cdot$ It can be

shown that the series on the right has the value $\pi/(\sin a\pi)$. Thus

$$B(a, 1 - a) = \int_0^\infty \frac{t^{a-1}}{1+t}\, dt = \frac{\pi}{\sin a\pi} \text{ if } 0 < a < 1.$$

12. If $0 < a < 1$, show, using (22.7–3), that $B(a, 1 - a) = \dfrac{1}{a}\displaystyle\int_0^\infty \dfrac{dx}{1 + x^{1/a}}\cdot$

Hence show that, if $n > 1$, $\displaystyle\int_0^\infty \frac{dx}{1 + x^n} = \frac{\pi}{n}\csc\frac{\pi}{n}\cdot$ Use the result stated in Exercise 11.

22.8   Stirling's formula

In this section we shall develop an approximate formula for $n!$ when n is large. We start by considering

$$\log(n!) = \log 1 + \log 2 + \cdots + \log n.$$

It is natural to try to estimate this sum by considering the integral of $\log t$, the idea being much the same as in the proof of the integral test (Theorem IV, § 19.21). Since $\log t$ increases as t increases, it is clear that when $k \geq 2$,

$$\int_{k-1}^k \log t\, dt < \log k < \int_k^{k+1} \log t\, dt.$$

Adding these inequalities for $k = 2, 3, \cdots, n$, and noting that $\log 1 = 0$, we have

$$\int_1^n \log t \, dt < \log n! < \int_2^{n+1} \log t \, dt.$$

The right member of this inequality is increased if the lower limit is changed from 2 to 1. Evaluation of the integrals then leads to the inequality

$$(22.8\text{-}1) \quad n \log n - n + 1 < \log n! < (n + 1) \log (n + 1) - n.$$

As our first estimate of $\log n!$ we now take the arithmetic mean of the extremes of the inequality (22.8-1). This arithmetic mean is

$$M_n = \frac{n \log n - n + 1 + (n + 1) \log (n + 1) - n}{2.}$$

By writing $\log (n + 1) = \log n + \log\left(1 + \dfrac{1}{n}\right)$, it is easy to see that M_n can be expressed in the form

$$M_n = n \log n - n + \tfrac{1}{2} \log n + a_n,$$

where $\quad a_n = \tfrac{1}{2} \log\left(1 + \dfrac{1}{n}\right)^n + \tfrac{1}{2} \log\left(1 + \dfrac{1}{n}\right) + \tfrac{1}{2}.$

Observe that, as $n \to \infty$, $a_n \to 1$, because $\left(1 + \dfrac{1}{n}\right)^n \to e$.

The foregoing considerations suggest that, for large values of n, the order of magnitude of $\log n!$ may be about the same as that of $n \log n - n + \tfrac{1}{2} \log n$. To test the validity of this plausible assumption, we set

$$(22.8\text{-}2) \quad S_n = \log n! - (n \log n - n + \tfrac{1}{2} \log n).$$

We shall prove that S_n approaches a limit, and that this limit is $\log \sqrt{2\,\pi}$. Moreover, we shall obtain an estimate of the amount by which S_n differs from its limit. When we have done this, formula (22.8-2) will give us a controlled estimate of $\log n!$ and hence of $n!$ itself.

Before beginning this work, we recall the expansion

$$\log(1 - x) = -(x + \tfrac{1}{2} x^2 + \tfrac{1}{3} x^3 + \cdots),$$

valid if $| x | < 1$.

An easy calculation shows that

$$S_n - S_{n-1} = 1 + (n - \tfrac{1}{2}) \log \left(1 - \dfrac{1}{n}\right).$$

We leave it to the student to verify this calculation. If $n \geq 2$,

$$\log\left(1 - \dfrac{1}{n}\right) = -\left(\dfrac{1}{n} + \dfrac{1}{2\,n^2} + \dfrac{1}{3\,n^3} + \cdots\right),$$

and so $S_n - S_{n-1} = 1 - \left(1 + \dfrac{1}{2\,n} + \dfrac{1}{3\,n^2} + \cdots\right)$

$$+ \left(\dfrac{1}{2\,n} + \dfrac{1}{4\,n^2} + \dfrac{1}{6\,n^3} + \cdots\right),$$

$$S_n - S_{n-1} = -(\tfrac{1}{3} - \tfrac{1}{4})\dfrac{1}{n^2} - (\tfrac{1}{4} - \tfrac{1}{6})\dfrac{1}{n^3} - (\tfrac{1}{5} - \tfrac{1}{8})\dfrac{1}{n^4} - \cdots,$$

(22.8–3) $\qquad S_n - S_{n-1} = -\dfrac{1}{3 \cdot 4}\dfrac{1}{n^2} - \dfrac{2}{4 \cdot 6}\dfrac{1}{n^3} - \cdots$

$$- \dfrac{k}{(k+2)(2\,k+2)}\dfrac{1}{n^{k+1}} - \cdots.$$

From this we see that

$$n^2(S_n - S_{n-1}) \to -\tfrac{1}{12}$$

as $n \to \infty$, and hence that the series

$$S_1 + (S_2 - S_1) + (S_3 - S_2) + \cdots + (S_n - S_{n-1}) + \cdots$$

is convergent. Since the sum of the first n terms of this series is S_n:

$$S_n = S_1 + (S_2 - S_1) + \cdots + (S_n - S_{n-1}),$$

we have proved that the sequence $\{S_n\}$ is convergent. Let

$$S = \lim_{n \to \infty} S_n, \quad S = S_n + R_n,$$

so that

(22.8–4) $\qquad R_n = (S_{n+1} - S_n) + (S_{n+2} - S_{n+1}) + \cdots.$

We shall estimate R_n presently. First we must find the value of S. For this purpose we write

$$S_n = \log T_n, \quad T_n = e^{S_n}, \quad T_n \to e^S.$$

By (22.8–2) $\qquad \log T_n = \log n! - \log[(n/e)^n \sqrt{n}\,],$

so $\qquad T_n = \dfrac{n!\,e^n}{n^n\sqrt{n}}.$

To find the limit of T_n we make use of the Wallis limit formula (22.7–9). The method is an ingenious one:

$$\dfrac{T_n^2}{T_{2n}} = \dfrac{(n!)^2 e^{2n}}{n^{2n} \cdot n} \cdot \dfrac{(2\,n)^{2n}\sqrt{2\,n}}{(2\,n)!\,e^{2n}} = \dfrac{(n!)^2 2^{2n}\sqrt{2\,n}}{(2\,n)!\,n}.$$

$$= \dfrac{2 \cdot 4 \cdots (2\,n)}{1 \cdot 3 \cdots (2\,n-1)} \dfrac{\sqrt{2\,n}}{n}.$$

To get this expression in a form for using Wallis's formula we write the expression as

$$\frac{2 \cdot 4 \cdots (2\,n)}{1 \cdot 3 \cdots (2\,n - 1)} \frac{1}{\sqrt{2\,n + 1}} \cdot \left[\frac{2\,n(2\,n + 1)}{n^2}\right]^{1/2};$$

the limit is then seen to be

$$\lim_{n \to \infty} \frac{T_n^2}{T_{2n}} = \left(\frac{\pi}{2}\right)^{1/2} \cdot 2 = \sqrt{2\,\pi}.$$

But, since T_n and T_{2n} have the same limit, it is clear that

$$\lim_{n \to \infty} \frac{T_n^2}{T_{2n}} = \frac{e^{2S}}{e^S} = e^S.$$

Therefore $e^S = \sqrt{2\,\pi}$,   or $S = \log \sqrt{2\,\pi}$.

If we now write $S_n = S - R_n$, and use the value just found for S, (22.8-2) may be written in the form

$$(22.8\text{-}5) \qquad \log n! = \log\left[\left(\frac{n}{e}\right)^n \sqrt{2\,\pi n}\right] - R_n.$$

We know that R_n is given by (22.8-4), and that $R_n \to 0$ as $n \to \infty$ (because R_n is the remainder after n terms in a convergent series). The mere fact that $R_n \to 0$ enables us to conclude from (22.8-5) that

$$(22.8\text{-}6) \qquad \lim_{n \to \infty} \frac{n!}{\left(\dfrac{n}{e}\right)^n \sqrt{2\,\pi n}} = 1.$$

This means that, for large values of n, $n!$ is approximately equal to $(n/e)^n \sqrt{2\,\pi n}$. The percentage error in the approximation approaches zero as $n \to \infty$. This approximate formula for $n!$ is called Stirling's formula. It dates from the 18th century.

Now we turn to an estimation of the magnitude of R_n. From (22.8-3) and (22.8-4) we see that $R_n < 0$. It is more convenient to write $r_n = - R_n$, and deal with r_n. We can write (for $n \geqq 1$)

$$r_n = - \sum_{j=n+1}^{\infty} (S_j - S_{j-1}),$$

$$S_j - S_{j-1} = - \sum_{k=1}^{\infty} \frac{k}{2(k+1)(k+2)} \frac{1}{j^{k+1}},$$

and so $r_n = \sum_{j=n+1}^{\infty}\left(\sum_{k=1}^{\infty} \frac{k}{2(k+1)(k+2)} \frac{1}{j^{k+1}}\right).$

We have here a repeated summation. Each of the infinite series involved is convergent, and all the terms are positive. Under these conditions it is easy to prove that the order of writing the summation signs may be reversed without affecting the value of the whole expression, so that

$$(22.8\text{–}7) \qquad r_n = \sum_{k=1}^{\infty}\left(\sum_{j=n+1}^{\infty} \frac{k}{2(k+1)(k+2)} \frac{1}{j^{k+1}}\right).$$

In this new form it is easier for us to estimate the size of r_n.

If $k \geq 1$, it is readily seen that

$$\int_{j}^{j+1} \frac{dx}{x^{k+1}} < \frac{1}{j^{k+1}} < \int_{j-1}^{j} \frac{dx}{x^{k+1}}, \quad j \geq 2,$$

and so

$$\int_{n+1}^{\infty} \frac{dx}{x^{k+1}} < \sum_{j=n+1}^{\infty} \frac{1}{j^{k+1}} < \int_{n}^{\infty} \frac{dx}{x^{k+1}}.$$

Evaluating the integrals, we obtain

$$\frac{1}{k(n+1)^k} < \sum_{j=n+1}^{\infty} \frac{1}{j^{k+1}} < \frac{1}{kn^k}.$$

These inequalities may now be used in (22.8–7) to obtain the inequalities

$$(22.8\text{–}8)$$

$$\sum_{k=1}^{\infty} \frac{1}{2(k+1)(k+2)(n+1)^k} < r_n < \sum_{n=1}^{\infty} \frac{1}{2(k+1)(k+2)n^k}.$$

Written out to a few terms, these inequalities read

$$\frac{1}{12(n+1)} + \frac{1}{24(n+1)^2} + \cdots < r_n < \frac{1}{12\,n} + \frac{1}{24\,n^2} + \cdots.$$

The series in the left member of the inequality exceeds $\dfrac{1}{12(n+1)}$ · That on the right is smaller than the geometric series

$$\frac{1}{12\,n} + \frac{1}{12\,n^2} + \frac{1}{12\,n^3} + \cdots,$$

whose value is $\dfrac{1}{12(n-1)}$. Thus a simpler inequality which is true is

$$(22.8\text{–}9) \qquad \frac{1}{12(n+1)} < r_n < \frac{1}{12(n-1)}.$$

Still more exact estimates of r_n can be made, but we shall stop with this last one.

Stirling's formula may now be written in the more precise form

$$(22.8\text{–}10) \qquad n! = \left(\frac{n}{e}\right)^n \sqrt{2\,\pi n}\ e^{r_n},$$

where r_n satisfies (22.8–9). This formula is an immediate consequence of (22.8–5).

Example 1. Find $\lim\limits_{n \to \infty} \dfrac{(n!)^{1/n}}{n}$.

By (22.8–10),

$$(n!)^{1/n} = \frac{n}{e}(2\,\pi)^{1/2n}n^{1/2n}e^{r_n/n}.$$

Now, as $n \to \infty$,

$$(2\,\pi)^{1/n} \to 1, \quad n^{1/2n} \to 1, \quad \frac{r_n}{n} \to 0.$$

Hence $$\lim_{n \to \infty} \frac{(n!)^{1/n}}{n} = \frac{1}{e}.$$

Example 2. If $(n/e)^n\sqrt{2\,\pi n}$ is used as an approximation for $n!$, show that the relative error is less than r_n.

By definition, the relative error is

$$\frac{n! - \left(\dfrac{n}{e}\right)^n\sqrt{2\,\pi n}}{n!}.$$

By (22.8–10), this expression is equal to

$$1 - e^{-r_n} = r_n - \frac{(r_n)^2}{2!} + \frac{(r_n)^3}{3!} - \cdots.$$

Since $r_n > 0$, this quantity is positive, and less than r_n.

By (22.8–9) the relative error is then certainly less than $\dfrac{1}{12(n-1)}$. For example, if $n = 10$, the error is less than 1%, and if $n = 52$ the error is less than 0.17%.

MISCELLANEOUS EXERCISES

1. Prove the following:

(a) $\displaystyle\int_0^\infty \frac{x}{(1+x)^3}\,dx = \frac{1}{2}\int_0^\infty \frac{dx}{(1+x)^2} = \frac{1}{2}.$

(b) $\displaystyle\int_1^\infty \frac{\sqrt{x}}{(1+x)^2}\,dx = \frac{1}{2} + \frac{\pi}{4}.$

(c) $\displaystyle\int_a^\infty \frac{dx}{x^4\sqrt{a^2+x^2}} = \frac{2-\sqrt{2}}{3\,a^4}\ (a > 0).$

(d) $\displaystyle\int_0^\infty \frac{dx}{(a^2+x^2)(b^2+x^2)} = \frac{\pi}{2\,ab(a+b)}.$

(e) $\displaystyle\int_0^\infty \frac{x^2\,dx}{(a^2 + x^2)(b^2 + x^2)} = \frac{\pi}{2(a + b)}$.

(f) $\displaystyle\int_0^\infty \frac{x^2\,dx}{(x^2 - a^2)^2 + b^2x^2} = \frac{\pi}{2b}\ (b > 0).$

(g) $\displaystyle\int_0^\infty \frac{x^4\,dx}{[(x^2 - a^2)^2 + b^2x^2]^2} = \frac{\pi}{4b^3}\ (b > 0).$

2. Examine the following integrals as to convergence or divergence:

(a) $\displaystyle\int_1^\infty \frac{\log [1 + (1/x)]}{x}\,dx.$
(b) $\displaystyle\int_1^\infty \log (1 + e^{-x})dx.$

3. Suppose $f(x) \geqq 0$, $a_n = \int_{x_n}^{x_{n+1}} f(x)dx$, where $c = x_0 < x_1 < x_2 < \cdots$, and $x_n \to \infty$. Prove that $\int_c^\infty f(x)dx$ is convergent if and only if $\sum a_n$ is convergent.

4. Let $F(x) = \int_0^{\pi/2} \log (1 - x^2 \cos^2 \theta)d\theta$, $0 \leqq x^2 \leqq 1$.

(a) Find $F'(x)$ when $0 \leqq x < 1$, by differentiation under the integral sign (this is justified by Theorem XIV, § 18.5); evaluate the resulting integral by use of standard tables. Then integrate and find $F(x) = \pi \log \dfrac{1 + \sqrt{1 - x^2}}{2}$, at least if $0 \leqq x^2 < 1$.

(b) Prove that F is continuous at $x = 1$, by use of Theorem VII, and hence deduce from (a) that $\int_0^{\pi/2} \log \sin \theta\, d\theta = (\pi/2) \log \tfrac{1}{2}$.

5. Let $I = \int_0^\pi \log \sin \theta\, d\theta$. Show that

$$I = 2\int_0^{\pi/2} \log \sin \theta\, d\theta = 2\int_0^{\pi/2} \log \cos \theta\, d\theta.$$

Then use the formula $\sin \theta = 2 \sin (\theta/2) \cos (\theta/2)$ to deduce that $I = -\pi \log 2$.

6. Show that $\displaystyle\int_0^\infty xe^{-ax} \cos bx\, dx = \frac{a^2 - b^2}{(a^2 + b^2)^2}\ (a > 0).$

7. In the integral $F(x) = \int_0^\infty e^{-t^2 - (x^2/t^2)}\, dt$, assume that $x > 0$, and make the change of variable $u = t - (x/t)$. The resulting integral will be of the form $\int_{-\infty}^\infty \phi(x, u)du$. Express $\int_{-\infty}^0 \phi(x, u)du$ in the form $\int_0^\infty \phi(x, -u)du$ and in this way deduce the value of $F(x)$.

8. Suppose $\int_a^\infty f(x)dx$ is convergent. Suppose also that $\phi(x) > 0$, that $\phi'(x)$ is continuous, and $\phi'(x) \leqq 0$. Show that $\int_a^\infty \phi(x) f(x)dx$ is convergent.

9. Use the result of Exercise 8 to show that $\displaystyle\int_1^\infty \frac{\cos x^2}{\tan^{-1} x}\, dx$ is convergent.

10. Let $F(x) = \dfrac{2}{\pi}\displaystyle\int_0^\infty \frac{x\,dt}{x^2 + t^2}$ · Find the value of $F(x)$ for each x. Is F continuous at $x = 0$?

11. Let $F(y) = \int_0^\infty y^3 e^{-xy^2}\, dx$. Show that $F(y) = y$ for all values of y. Verify that

$$F'(y) = \int_0^\infty \left[\frac{\partial}{\partial y} (y^3 e^{-xy^2}) \right] dx$$ if $y \neq 0$, but that this is false if $y = 0$. What do you conclude from this about the uniform convergence of the integral defining $F(y)$?

12. Is the equation

$$\int_0^a dy \int_0^\infty (2y - 2xy^3) e^{-xy^2} dx = \int_0^\infty dx \int_0^a (2y - 2xy^3) e^{-xy^2} dy$$

true or false when $a \neq 0$?

13. Let $f(x, y) = \sin(x^2 + y^2)$. Let R be the square region $0 \leq x \leq a$, $0 \leq y \leq a$, and let T be the portion of the circular region $x^2 + y^2 \leq a^2$ which lies in the first quadrant. Show that, if $I = \iint_R f(x, y) dA$ and $J = \iint_T f(x, y) dA$, then $I \to \pi/4$ as $a \to \infty$, but J does not approach any limit.

14. (a) Let $F(x) = \int_0^\infty \frac{\sin xt}{t(1 + t^2)} dt$. Show that $F''(x) - F(x) = -\pi/2$ if $x > 0$. Take for granted that $F''(x)$ can be calculated by differentiating under the integral. Solve the differential equation and evaluate $F(0)$, $F'(0)$, and thus show that, if $x \geq 0$,

$$\int_0^\infty \frac{\sin xt}{t(1 + t^2)} dt = \frac{\pi}{2} (1 - e^{-x}),$$

$$\int_0^\infty \frac{\cos xt}{1 + t^2} dt = \frac{\pi}{2} e^{-x}.$$

(b) Show how to deduce the value of $\int_0^\infty \frac{t \sin xt}{1 + t^2} dt$ from the results in (a). State what must be proved to be legitimate for the success of your method, but do not prove it.

15. Suppose $f(x)$ is continuous when $x \geq 0$, that $\lim\limits_{x \to 0+} \dfrac{f(x) - f(0)}{x}$ exists, and that $\int_1^\infty \dfrac{f(u)}{u} du$ is convergent. From these assumptions it may be proved that

$$\int_0^\infty \frac{f(bx) - f(ax)}{x} dx = f(0) \log \frac{a}{b}$$

if a and b are positive. Write out a proof with the aid of the following suggestions:
(a) Show that, if $y > 0$,

$$\int_0^y \frac{f(ax) - f(0)}{x} dx = \int_0^{ay} \frac{f(u) - f(0)}{u} du,$$

explaining why the integrals are not improper at the lower limit.
(b) Explain why

$$\lim_{y \to \infty} \int_{ay}^{by} \frac{f(u)}{u} du = 0.$$

(c) Complete the proof.

16. Show that the functions $\cos x$ and $(\pi/2) - \tan^{-1} x$ satisfy the conditions on $f(x)$ in Exercise 15, and so find the values of the integrals

$$\int_0^\infty \frac{\cos bx - \cos ax}{x} \, dx, \quad \int_0^\infty \frac{\tan^{-1} bx - \tan^{-1} ax}{x} \, dx.$$

17. Suppose that $f(x)$ is continuous when $x \geq 0$, and that $f'(x)$ is continuous when $x > 0$. Suppose also that $\lim_{x \to \infty} f(x)$ exists, and denote the limit by $f(\infty)$. From these assumptions it may be proved that, if a and b are positive,

$$\int_0^\infty \frac{f(bx) - f(ax)}{x} \, dx = (f(\infty) - f(0)) \log (b/a).$$

Write out a proof with the aid of the following suggestions:
(a) Apply Theorem IX to $\int_a^b dt \int_0^\infty f'(tx)dx$.
(b) Show that $\int_1^\infty f'(tx)dx$ is convergent uniformly for $t \geq a$, and the same for $\int_0^1 f'(tx)dx$ in case this integral is improper at $x = 0$.

18. Apply the result of Exercise 17 to each of the functions e^{-x}, $\tan^{-1} x$.

19. Show that, if $0 < a < 1$, $\displaystyle\int_{-\infty}^\infty \frac{e^{at}}{1 + e^t} \, dt = \Gamma(a)\Gamma(1 - a)$.

20. (a) Let $F(\theta) = \int_0^\infty e^{-t\cos\theta} t^{a-1} \cos (t \sin \theta) \, dt$,

$\qquad G(\theta) = \int_0^\infty e^{-t\cos\theta} t^{a-1} \sin (t \sin \theta) \, dt$,

where $-\pi/2 < \theta < \pi/2$ and $a > 0$. Show that $F'(\theta) = -aG(\theta)$ and $G'(\theta) = aF(\theta)$, whence $F''(\theta) + a^2 F(\theta) = 0$. From this deduce that $F(\theta) = \Gamma(a) \cos a\theta$ and $G(\theta) = \Gamma(a) \sin a\theta$.
(b) Show how to obtain the formulas

$$\int_0^\infty \frac{\cos t}{t^{1-a}} \, dt = \Gamma(a) \cos \frac{\pi a}{2}, \quad \int_0^\infty \frac{\sin t}{t^{1-a}} \, dt = \Gamma(a) \sin \frac{\pi a}{2}$$

from the results in (a). State without proof the facts about uniform convergence which would be sufficient to justify the procedure.
(c) The first formula in (b) is valid if $0 < a < 1$. The second is actually valid if $-1 < a < 1$, $a \neq 0$, $\Gamma(a)$ for $a < 0$ being defined as in § 22.2 (see (22.2–10)). Deduce that $\displaystyle\int_0^\infty \frac{\sin t}{t^a} \, dt = \Gamma(1 - a) \cos \frac{\pi a}{2}$ if $0 < a < 2$, $a \neq 1$.

21. Some improper integrals can be evaluated by expanding the integrand in a series and integrating term by term. For improper integrals the justification of this procedure is not covered by Theorem IV, § 20.4. In the following, assume the legitimacy of the procedure, and deduce the results as listed.

(a) $\displaystyle\int_0^1 \frac{\log x}{1 + x} \, dx = -\left(\frac{1}{1^2} - \frac{1}{2^2} + \frac{1}{3^2} - \frac{1}{4^2} + \cdots\right).$

(b) $\displaystyle\int_0^1 \frac{1}{x} \log \frac{1 + x}{1 - x} \, dx = 2\left(\frac{1}{1^2} + \frac{1}{3^2} + \frac{1}{5^2} + \frac{1}{7^2} + \cdots\right).$

(c) $\displaystyle\int_0^1 \frac{\log x}{1 - x^2}\, dx = -\left(\frac{1}{1^2} + \frac{1}{3^2} + \frac{1}{5^2} + \frac{1}{7^2} + \cdots\right).$

(d) $\displaystyle\int_0^\infty \log \frac{e^x + 1}{e^x - 1}\, dx = 2\left(\frac{1}{1^2} + \frac{1}{3^2} + \frac{1}{5^2} + \frac{1}{7^2} + \cdots\right).$

(e) $\displaystyle\int_0^\infty \frac{x^2}{e^x - 1}\, dx = 2\left(\frac{1}{1^3} + \frac{1}{2^3} + \frac{1}{3^3} + \frac{1}{4^3} + \cdots\right).$

22. Deduce the formula $\int_0^\infty e^{-ax^2} \cos 2bx\, dx = \frac{1}{2}\sqrt{\pi/a}\; e^{-b^2/a}$ (assuming $a > 0$) by expressing the cosine as a power series and integrating term by term.

Answers to Selected Exercises

CHAPTER I

§ 1.1 Pages 9–11

1. (a) 8; (c) n; (e) $-\frac{1}{4}$; (h) 2. 2. (a) 0; (b) 1; (d) $\frac{\pi}{2}$.

3. (a) Yes; (c) $\lim\limits_{x \to 0} f(x) = 0$; (d) yes; (e) no.

4. (a) 2; (b) 1; (c) 12. 9. (a) 0; (b) 1; (c) -1.

10. (a) Approaches 0; (b) approaches ∞; (c) does not exist.

13. (a) and (c). 14. (c) 0.

15. (b) $f(x) = -1 - x$ if $0 < x < 1$; $f(x) = 1 + x$ if $1 < x$.

16. (a) $f(-1) = 2$; (c) the limit does not exist.

17. (a) $f(\frac{3}{2}) = 1$; (b) yes; (d) no.

18. (a) $f(\pi/4) = 0$; (b) no; (c) 0; (d) -1.

19. Take δ to be the smaller of the numbers 1, $\epsilon/5$.

21. Take δ to be the smaller of the numbers 1, $\epsilon/7$.

§ 1.11 Pages 18–20

9. $200\,\pi$. 10. $x = ct^3$. 11. 0. 14. (c) No. 15. (c) 0.

18. (a) $n \geq 2$; (b) $n \geq 4$; (c) $n \geq 5$.

§ 1.12 Pages 25–27

1. (b) Max. 0 at $x = -5$, min. -226 at $x = -6$.

2. Max. 7000, min. $200(35 - 27\sqrt{5})$. 3. 10. 5. $\frac{32}{27}$.

6. Abs. min. $48\sqrt[3]{4}$; no abs. max. 7. Abs. min. 125.

8. (a) 2, -2; (b) $\frac{3}{2}$, -3. 9. $\dfrac{4\,c}{3}$. 11. (c) $2\,r < h$.

12. (b) Max. 4, min. π. HINT: Use symmetry about line $x = \frac{1}{2}$, and study the graph of $F(x)$, where $f'(x) = \dfrac{F(x)}{x^2(1 - x)^2}$.

13. (d) 25. 14. (c) $c_1 > c_2$ and $c_2 a < b\sqrt{c_1{}^2 - c_2{}^2}$. 15. Max. $\frac{7}{2}$.

17. $\dfrac{a^2 - b^2}{2\,ab}$.

§ 1.2 Pages 31–33

3. $2 - \frac{1}{3}\sqrt{21}$. 4. $c \geq 1$.

§ 1.3 Pages 36–37

1. (a) $dy = -2$; (b) $dx = 2$. 5. $d\psi = \dfrac{r'^2 - rr''}{r^2 + r'^2}\, d\theta$.

§ 1.4 Page 40

3. If $F(x) = 0$ when $0 \leq x \leq 1$, $F(x) = x - 1$ when $1 \leq x \leq 2$, and

751

$F(x) = 2x - 3$ when $2 \leq x \leq 3$, then $F'(x) = [x]$ except at $x = 1, 2, 3$, but $F'(x)$ is not defined at these points.

§ 1.5 Pages 45–47

1. (a) 10, −2;   (b) 7, 1;   (c) 4.
2. (a) 1.09458;   (b) $s = 0.99564, S = 1.21786$.
6. (b) $A_3(n) = \frac{1}{4}n^2(n + 1)^2$.   9. Low 1.3339, high 1.4052.

§ 1.51 Page 48

2. $\frac{1}{4}\pi a$.   3. $\frac{1}{3}\pi R^2$.

§ 1.52 Page 51

2. (a) $1/\sqrt{3}$.   3. (b) $2/\pi$;   (c) 0.   4. (a) 1;   (b) 0.
5. (a) $x = 0, 1$;   (b) 0.   6. 0.

§ 1.53 Pages 54–56

3. (a) $\log 4$;   (b) $\log 2$;   (c) $\log \frac{3}{4}$;   (d) $\log \frac{4}{3}$;   (e) $\frac{1}{6}\log 10$.
5. (a) $\pi/3$;   (b) $\frac{7}{12}\pi$;   (c) $5\pi/12\sqrt{3}$;   (d) $\pi/2$.   8. (d) $(\pi/2) - \frac{1}{2}\tan^{-1} 2$.
15. $\frac{1}{8}\pi(b - a)^2$.   18. $\pi^2/4$.

§ 1.61 Pages 61–62

6. Take M the larger of the numbers $\frac{1}{3}$, $(7 + \epsilon)/(3\epsilon)$.
11. (a) $f'(0) = +\infty$;   (b) $f'_+(0) = +\infty, f'_-(0) = -\infty$.

§ 1.62 Pages 69–71

1. (b), (e), and (f) convergent.   2. $N > 1/(3\epsilon)$ will do.   3. $A = 6$.
4. Take N greater than 10 and $10^{11}/(10!\,\epsilon)$.   7. (b) 2^r.   8. $\frac{4}{3}$.
9. (a) 0;   (b) $\frac{1}{2}$.   11. $f(x) = 0$ if $x \neq 1, f(1) = \frac{1}{2}$.   14. (a) $\frac{1}{2}$.
20. (b) e^2.

Miscellaneous Exercises Pages 74–76

3. $\frac{1}{2}(a + b)$.   5. 2.   7. (b) $(1 - \cos \pi\alpha)/(\pi\alpha)$.   9. $a + b$.
10. Continuous, but not differentiable.
12. $F'(0) = 2$.   13. Max. 8, min. −4.

CHAPTER II

§ 2.7 Page 88

5. The limit is 2.   6. The limit is $\frac{1}{2}(1 + \sqrt{1 + 4c})$.

Miscellaneous Exercises Pages 89–90

4. $N = 3$.   6. The l.u.b. is 2.
7. The l.u.b. is $\frac{3}{2}$; the g.l.b. is −1.   8. (b) 1.

CHAPTER III

§ 3 Page 92

1. No.
3. The answer is "yes" in both cases. In the second case, consider $f(x) = -x$ if $x < 0$, $f(x) = 1$ if $x \geq 0$, and $g(x) = 1$ if $x < 0$, $g(x) = x$ if $x \geq 0$.
4. The answers are "no" and "yes," respectively.
5. No.

§ 3.1 Page 94

1. Yes. 2. Yes. 3. (b), (c), (d). 4. (a) 14; (b) 6; (c) 13.

§ 3.2 Pages 96–97

2. $M = 1$, $m = 0$; M not attained.
3. $M = \pi/2$, $m = -\pi/2$; neither attained.
4. $M = 1$, $m = 0$; m not attained.

§ 3.3 Page 98

5. Consider the function $F(x)$ defined by $F(x)(\int_a^b f(t)dt) = \int_a^x f(t)dt$.

Miscellaneous Exercises Pages 98–101

1. (ii). (a) g undefined at 1, -1; (b) continuous elsewhere; (c) may be defined at 1 by $g(1) = \frac{1}{2}$, and is then continuous there.
(iv). (a) F undefined at 4, 9, and for $x < 0$; (b) continuous elsewhere; (c) continuous at 4 if one defines $F(4) = -32$.
(vi). (a) H undefined at 2, and for $x < \frac{5}{3}$; (b) continuous elsewhere; (c) no.
7. There must be at least one positive root and at least one negative root.
16. (a) Only at $x = \frac{1}{2}$. 17. (c) Yes.

CHAPTER IV

§ 4.3 Pages 111–113

1. $x^4 = 81 + 108(x - 3) + 54(x - 3)^2 + 12(x - 3)^3 + (x - 3)^4$.
3. (a) $\sin^2 x = x^2 - \frac{1}{3}x^4 \cos 2X$.

(e) $\dfrac{\log(1 - x)}{1 - x} = -x + x^2 \left[\dfrac{2\log(1 - X) - 3}{2(1 - X)^3} \right].$

4. $\dfrac{\cos X}{5!} \left(x - \dfrac{\pi}{2} \right)^5.$ 9. $x = \dfrac{\pi}{2(1 + \lambda)}.$

§ 4.5 Pages 121–123

1. (a) $+\infty$; (b) $\frac{1}{2}$; (g) 1; (h) $\log 2$; (i) $\pi/2$.
2. (b) 0; (e) 2; (f) 0; (g) $+\infty$, (h) 1.

3. (b) 1; (d) e^{-2}; (f) 1; (h) e^{-1}; (j) $e^{3/2}$.
4. (a) 0; (c) $+\infty$; (e) $\frac{2}{3}$; (h) $\frac{1}{2}$; (j) $-\frac{1}{2}$.
7. (b) 0; (d) 0; (e) $\dfrac{e}{2}$; (g) 0.

Miscellaneous Exercises   Pages 123–125

6. (a) $-\pi/4$; (c) 0; (e) e^{-1}; (f) -2.
7. The limit is a_n in the general case.   **8.** 1.   **11.** $\theta \to 1/(n+2)$.

CHAPTER V

§ 5.1   Pages 131–132

1. S is open. $B(S)$ is the set described in Exercise 2.
2. S is closed. It has no interior points.
3. S is closed. $B(S)$ is composed of the line segment $y = 1$, $-1 \le x \le 1$, and the parabolic arc $y = x^2$, $-1 \le x \le 1$.
4. S is neither open nor closed. $B(S)$ is composed of the part of the curve $xy = 1$ in the first quadrant, and the nonnegative portion of each co-ordinate axis.
5. S has no interior points. It is not closed. $B(S)$ consists of S and the line segment $x = 0$, $-1 \le y \le 1$.
6. S is not open. $B(S)$ consists of the semicircular arc $y = \sqrt{4 - x^2}$, the segment $y = 0$, $-2 \le x \le 2$, the segment $x = 1/n$, $0 < y \le 1$, and the segment $x = 0$, $0 < y \le 1$. The points of this last segment are in S. S is a region, but $C(S)$ is not.
7. This set is open, and therefore a region. Its boundary consists of the y-axis and the curve $y = \sin(1/x)$.
8. This set is open. Its boundary consists of the half-lines $x = n\pi$, $y \ge 0$, and the segments $y = 0$, $2n\pi \le x \le (2n+1)\pi$, $n = 0$, ± 1, ± 2, $\cdots$.

§ 5.2   Page 134

2. No.   **8.** $\delta = 2\sqrt{\epsilon}$.   **10.** $\delta = \sqrt{\epsilon/2}$ will do.   **12.** Yes.

§ 5.3   Page 137

2. No.   **4.** (a) Yes;   (b) no.   **5.** No discontinuities.
7. Define $f(x, x) = 0$.   **8.** (a) No;   (b) continuous elsewhere.
9. No.   **10.** Yes. Define $f(0, 0, 0) = 0$.

CHAPTER VI

§ 6.1   Pages 144–145

2. $\dfrac{\partial u}{\partial x} = \dfrac{yu + 4xv}{2(u^2 + v^2)}$, $\dfrac{\partial v}{\partial x} = \dfrac{4xu - yv}{2(u^2 + v^2)}$.

3. $\dfrac{\partial z}{\partial x} = 2$, $\dfrac{\partial z}{\partial y} = -1$.   **4.** $\dfrac{\partial z}{\partial x} = \dfrac{yz - x^2}{z^2 - xy}$.

§ 6.2   Page 148

1. (a) $x + z = 1$;   (c) $y = z$.   2. $5x + 7y - 21z + 9 = 0$.
4. The angle is $\cos^{-1}(1/\sqrt{3})$.

§ 6.3   Pages 154–155

1. $(3, 4, 5)$.   3. $\frac{1}{3}$, at $(\frac{1}{2}, \frac{1}{3})$.
4. Min. at $(6, 6)$; no max.   5. $4A^2(a^2 + b^2 + c^2)^{-1}$.
6. Max. $\frac{5}{4}$; min. -1.   7. Max. $\frac{1}{2}$; min. -1.   10. $\frac{3}{8}\sqrt{3}$.
11. Max. $2e^{-1}$ at $(0, \pm 1)$.
12. Max. 16 at $(0, 2)$ and $(2, 0)$; min. 0 at $(0, 0)$.
13. Max. 45 at $(3, 3)$; min. -30 at $(2, 1)$.
14. 18.   15. $\frac{1}{864}$.
18. (a) Along x-axis and at $(0, 1)$;   (b) $x = 0, 0 < y < 1$;
(c) min. 1 at $x = 0, 0 \leqq y \leqq 1$.
19. 1 at $(0, 1)$.   20. 1 at $(0, 0, 0)$.

§ 6.4   Page 166

4. $dz = \pi^2$.   5. 0.02.

§ 6.5   Pages 172–175

2. $\dfrac{\partial G}{\partial y} = -2y\,\dfrac{\partial F}{\partial u} + 2x\,\dfrac{\partial F}{\partial v}$ (partial answer).

4. $\dfrac{\partial G}{\partial u} = \dfrac{\partial F}{\partial x} + \dfrac{\partial F}{\partial y} + \dfrac{\partial F}{\partial z}$, $\dfrac{\partial G}{\partial w} = \dfrac{\partial F}{\partial z}$ (partial answer).

11. $\left(\dfrac{\partial w}{\partial x}\right)^2 - \left(\dfrac{\partial w}{\partial y}\right)^2$.   22. (a) 3.

§ 6.52   Pages 180–182

5. $-4c^2\,\dfrac{\partial^2 u}{\partial \xi\,\partial \eta}$.   9. (a) $F(r) = Ar^{-1} + B$, A and B constant.

§ 6.8   Pages 200–204

1. $abc/27$.   4. $A^2/(3abc)$.   5. $(\sqrt{a} + \sqrt{b} + \sqrt{c})^2$.   7. a^2b^2.
8. $(a^{3/2} + b^{3/2} + c^{3/2})^{-2}$.   10. (c) Max. $\frac{8}{3}$, min. 1;   (d) 9.
11. $8A^3/(27abc)$.   13. $bc/(1 + b^2)^{1/2}$.   14. $|a - b|$.
19. Max. 54; min. $153 - 45\sqrt{5}$.   20. $\left(\sum a_i^2\right)^{-1}$.   21. $\sum a_i^2$.
22. $\frac{4}{5}$.   23. $(13 - 4\sqrt{3})^{1/2} = 2\sqrt{3} - 1$.

§ 6.9   Pages 209–210

1. Semi-axes 1, $\sqrt{2}/2$.   2. Max. at $(2/\sqrt{5}, -1/\sqrt{5})$.
3. Max. 10.   4. Max. 2; min. -1.
5. (a) $\lambda_1 = \sqrt{2} + 1$, $\lambda_2 = 0$, $\lambda_3 = 1 - \sqrt{2}$;   (b) $\lambda_1 = 18$, $\lambda_2 = \lambda_3 = 9$.

6. (a) $\lambda_1 = 1$, $\lambda_2 = \lambda_3 = -\frac{1}{2}$; **(b)** 1; **(c)** $x = y = z = \pm \dfrac{\sqrt{3}}{3}$.

7. (a) and **(b)** hyperboloids of one sheet; **(c)** ellipsoid.

8. Max. 16; min. 4. **9.** $\frac{27}{10}$.

Miscellaneous Exercises Pages 210–211

1. $a = \frac{4}{15}$, $b = \frac{4}{5}$.

3. (b) $(0, a)$, (a, a), $(a, 0)$, $(\frac{2}{3}\,a, \frac{2}{3}a)$; **(c)** rel. max. at $(\frac{2}{3}\,a, \frac{2}{3}\,a)$, others neither max. nor min.; **(d)** no.

4. $24\sqrt{3}$. **5.** $(8, 4, 12)$. **6.** Max. 25; min $\frac{675}{43}$.

7. $\dfrac{y}{x} = 1 + 1/\sqrt{5}$, $\dfrac{x}{z} = \sqrt{5}/2$. **8.** Max. 4; min. 3.

9. $(a^q + b^q + c^q)^{1/q}$, where $q = 1/(1 - p)$.

11. Max. at $(0, 1)$; min. at $(\frac{1}{3}, 0)$.

12. (a) $5\sqrt{5}$; **(b)** $\sqrt{5}$; **(c)** no in **(a)**, yes in **(b)**.

CHAPTER VII

§ 7.1 Page 214

1. Yes.

§ 7.4 Pages 221–222

1. (a) $1/\sqrt{3}$; **(b)** $1/\sqrt{3}$; **(c)** $\frac{5}{9}$ (approx.). **7. (a)** $1/\sqrt{3}$; **(b)** $\frac{2}{3}$; **(c)** $\frac{1}{2}$.

§ 7.5 Pages 224–225

1. $\sin h \sin k = hk - \frac{1}{6}h(h^2 + 3\,k^2)\cos\theta h \sin\theta k - \frac{1}{6}k(3\,h^2 + k^2)\sin\theta h\cos\theta k$.

3. $F(3 + h, 3 + k) = \dfrac{1}{2!}(18\,h^2 - 18\,hk + 18\,k^2) + \dfrac{1}{3!}(6\,h^3 + 6\,k^3)$.

4. $\log 2 + \frac{1}{2}[(x - 1) + y] - \frac{1}{8}[(x - 1)^2 + 2(x - 1)y - y^2] + \cdots$.

5. $2(x - 1) + y + \frac{1}{2}[2(x - 1)^2 + y^2] + \cdots$.

7. $1 + (2\,xy - y^2) + (2\,x^2y^2 - 2\,xy^3 + \frac{1}{2}\,y^4) + \cdots$.

10. (a) $x - y$; **(b)** $\frac{1}{4}\pi + \frac{1}{10}(x - 3) - \frac{4}{5}(y - \frac{1}{2})$.

11. (a) $h^4 - h^2k^2 + k^4$; **(b)** $(\pi^2/96)(h + k)^4 - \frac{1}{2}h^2k^2$.

§ 7.6 Pages 230–232

1. (b) Saddle at $(0, 0)$; min. at $(0, 1)$ and $(0, -2)$. **(d)** Max. at $(0, 0)$; four saddle points. **(f)** Saddle at $(0, 0)$, $(0, 3)$, and $(4, 0)$; max. at $(\frac{4}{3}, 1)$. **(g)** Max. at $(a/2, a/3)$; degenerate if $x = 0$ or $y = 0$. **(h)** Saddle at $(1, 1)$, $(0, 1)$, $(1, 0)$; max. at $(\frac{2}{3}, \frac{2}{3})$. **(k)** Min. at $(6, 8)$. **(l)** Min. at $(6, 8)$; saddle at $(-\frac{6}{5}, -\frac{8}{5})$.

3. 5 critical points.

4. Shortest distance $= \sqrt{2}$. Saddle point at $(x, -x)$, where $x^3 + 2\,x - 1 = 0$.

5. Min. $= a$ if $0 < a \le 1$; min. $= \sqrt{2\,a - 1}$ if $a > 1$.

6. Min. at $(-2, 0, 1)$; max. at $(\frac{16}{7}, 0, -\frac{8}{7})$.

7. Max. at $(1, 1, 1)$; min. at $(-1, -1, -1)$. **9.** Min. at $x = y = z = \frac{1}{4}$.

CHAPTER VIII

§ 8.2 Pages 240–242

1. Yes. 2. $f(1 + h, k) \sim -2h - k$. 3. Yes. No. 4. Yes.
7. (b) No. (c) A spherical surface and its center. 11. No.
12. An infinite number of isolated points, among them $(0, \pi/2)$, $(\pi, 5\pi/2)$, $(-\pi, \pi/2)$.
14. (a) $f(x) = x\sqrt{x - 1}$. (c) $(0, 0)$ and $(1, 0)$.
15. (b) $f(x) = x^2 + x^{5/2}$. 16. $a^2 \neq 3b^2$. 17. Yes. 20. No.

§ 8.3 Pages 246–248

1. Yes. 2. (a) $u_0 v_0 w_0 \neq 0$. 3. $u_0 v_0 \neq 0$, $x_0^2 \neq y_0^2$, $x_0 y_0 \neq 0$.
4. (b) $x_0 \neq y_0$, $y_0 \neq z_0$, $z_0 \neq x_0$. 5. $y_0 \neq z_0$.
6. (a) $x_0^2 + y_0^2 \neq 0$. (b) $z_0 \neq 0$, $y_0 \neq. 2 x_0$.

CHAPTER IX

§ 9.3 Pages 260–261

1. (b) Yes; all points where $u = 0$ are singular. (c) The regions where $y^2 > x^2$.
4. $u = [\tfrac{1}{2}(\sqrt{x^2 + 4y^2} + x)]^{1/2}$, $v = [\tfrac{1}{2}(\sqrt{x^2 + 4y^2} - x)]^{1/2}$.
6. (a) The u-curves are hyperbolas. The v-curves are ellipses. (b) Singular points correspond to the foci $x = \pm 1$, $y = 0$.
8. (a) $r^2 \sin \phi$. (b) A sphere. A nappe of a cone. A half plane.
9. The singular points correspond to the circle $x^2 + y^2 = 1$, $z = 0$. The u-surfaces are tori. The v-surfaces are spheres.

§ 9.4 Pages 266–267

1. The circle $u^2 + v^2 \leq 1$.
3. The triangle bounded by $u = v$, $u = -v$, $u = 1$.
5. Regions in the first and third quadrants, bounded by $y = x + 3$, $y = x - 3$, $xy = 4$, $xy = 16$.
6. $u = x + y$, $v = y/(x + y)$. The u-curves are parallel lines of slope -1. The v-curves are straight lines through the origin. The x-curves are hyperbolas with asymptotes $u = 0$, $v = 1$. The y-curves are hyperbolas with asymptotes $u = 0$, $v = 0$.
8. (a) $(1 - x - y)^{-3}$. (b) $x = u(1 + u + v)^{-1}$, $y = v(1 + u + v)^{-1}$. (c) The u-curves are straight lines through $(0, 1)$. The v-curves are straight lines through $(1, 0)$. (e) A quadrangle with vertices at $(\tfrac{2}{3}, 1)$, $(1, 1)$, $(1, 2)$, $(\tfrac{1}{2}, \tfrac{3}{2})$.
9. (a) Ellipses if $u > 0$ or $u < -1$; hyperbolas if $-1 < u < 0$. (c) Along the lines $x = 0$ and $y = 0$.

§ 9.6 Page 272

1. No. 2. $ad - bc = 0$.

4. (a) $w = u^2 - 2v$;   **(b)** $w = -(1 + uv)/(u + v)$.

Miscellaneous Exercises   Pages 272–273

   1. (c) $-r^{-6}$.   **2.** $r^3 \sin^2 \phi \sin \psi$.

CHAPTER X

§ 10.21   Pages 279–281

   1. (a) Neither;   **(b)** perpendicular;   **(c)** neither;   **(d)** collinear.
   2. (a) 7;   **(b)** 6;   **(c)** 3;   **(d)** 5.   **3.** $\frac{4}{9}$.   **7.** $2 \sin (\theta/2)$.

§ 10.22   Pages 283–284

   1. (a) $i - 2j + k$;   **(c)** $-11i - 5j - k$.   **2.** $\sqrt{110}$.   **3. (b)** $\frac{1}{2}\sqrt{69}$.
   4. (a) $-79, 79$.   **5. (e)** 2;   **(f)** both zero.   **6.** $(A \times B) \cdot (C \times D) = 0$.

§ 10.3   Pages 286–287

   5. (a) $(1, 26, 3)$; $(-16, -31, 29)$.
   6. (a) $-i' + j' - \sqrt{3}\,k'$;   **(b)** $-2i - \sqrt{2}(j - k)$.

§ 10.51   Pages 294–295

   1. (a) $\frac{1}{2}(x'^2 - 6x'y' + y'^2 - 2z'^2)$;   **(b)** $F(x, y, z) = 2y$;   **(c)** $-x' + \sqrt{2}z', y'$
$\dfrac{1}{\sqrt{2}}(x' - y') + z'$;   **(d)** $-\left(\dfrac{x}{\sqrt{2}} + z\right)i + \dfrac{3y - z}{2\sqrt{2}}j + \left(x - \dfrac{y + z}{2\sqrt{2}}\right)k$.
   2. (b) $\frac{1}{49}(-23x^2 + 41y^2 + 31z^2 - 48xy + 72xz + 24yz)$;
(c) $\frac{1}{7}(3y' + 2z')i' - \frac{1}{7}(3x' + 6z')j' - \frac{1}{7}(2x' - 6y')k'$.

§ 10.6   Pages 298–299

   1. (a) $(4x - 3y - 4z)i + (2y - 3x)j + (12z - 4x)k$;
(d) $r^{-5}[-3xzi - 3yzj + (r^2 - 3z^2)k]$, where $r^2 = x^2 + y^2 + z^2$.
   2. $-t : 11t : -t$, where $t = (123)^{-1/2}$.   **3.** $-i - 2j - 2k$; rate = 3.
   4. (a) 0;   **(b)** $14\sqrt{2}$.   **7. (e)** $nr^{n-2}R$;   **(f)** $-2e^{-r^2}R$.   **8. (c)** $r^{-1} + C$.

§ 10.7   Pages 304–305

   4. (b) $Ar^{-1} + B$.   **5. (b)** Cr^{-3}.   **6. (b)** $\nabla \cdot V = 0$.

§ 10.8   Page 308

   3. (a) 0;   **(c)** $-i - j - k$;   **(e)** 0.

Miscellaneous Exercises   Page 308

   2. (a) $nr^{n-2}R \cdot A$ and $nr^{n-2}R \times A$;   **(b)** 0.

CHAPTER XII

§ 12.1–§ 12.8   Pages 373–375

2. If T_1 and T_2 are both differentials of $\mathbf{f}$ at $\mathbf{a}$, then their difference is a linear transformation L with the property that

$$\lim_{\mathbf{h} \to 0} \frac{||\, L(\mathbf{h}) \,||}{||\, \mathbf{h} \,||} = 0.$$

Show that the only transformation in $\operatorname{Hom}(R^n, R^m)$ which has this property is the zero element.

17. Let a continuously differentiable function $\mathbf{F}$ from R^3 to R^2 be defined by $u = f(x, y, z)$, $v = g(x, y, z)$, and let

$$\Delta_1 = \begin{vmatrix} \dfrac{\partial f}{\partial x} & \dfrac{\partial f}{\partial y} \\[2mm] \dfrac{\partial g}{\partial x} & \dfrac{\partial g}{\partial y} \end{vmatrix}, \quad \Delta_2 = \begin{vmatrix} \dfrac{\partial f}{\partial x} & \dfrac{\partial f}{\partial z} \\[2mm] \dfrac{\partial g}{\partial x} & \dfrac{\partial g}{\partial z} \end{vmatrix}, \quad \Delta_3 = \begin{vmatrix} \dfrac{\partial f}{\partial y} & \dfrac{\partial f}{\partial z} \\[2mm] \dfrac{\partial g}{\partial y} & \dfrac{\partial g}{\partial z} \end{vmatrix}.$$

Suppose that there exists a point (x_0, y_0, z_0) where at least one of these determinants is different from 0, and let (u_0, v_0) denote the image of this point in R^2. Then the equations $f(x, y, z) = u_0$, $g(x, y, z) = v_0$ determine two of the three variables as a function of the third near (x_0, y_0, z_0) and hence the mapping cannot be one-to-one.

From now on, assume that $\Delta_1{}^2 + \Delta_2{}^2 + \Delta_3{}^2 = 0$ everywhere.

If there is an open ball in R^3 throughout which all six of the first partial derivatives of f and g vanish identically, then the mapping is constant in B and hence not one-to-one.

Suppose that there is some open ball B centered at (x_0, y_0, z_0) such that one, but not both, of the two gradients $\mathbf{grad}\, f$ and $\mathbf{grad}\, g$ vanishes identically. Assume that $\mathbf{grad}\, g \equiv 0$ in B and put $u_0 = f(x_0, y_0, z_0)$, $v_0 = g(x_0, y_0, z_0)$. Then $g(x, y, z) \equiv v_0$ in B and the equation $f(x, y, z) = u_0$ determines one of the three variables as a function of the other two and the mapping $\mathbf{F}$ cannot be one-to-one.

If there is no open set throughout which either $\mathbf{grad}\, f$ or $\mathbf{grad}\, g$ is always zero, there must be an open ball at each point of which $\mathbf{grad}\, f \neq 0$ and $\mathbf{grad}\, g \neq 0$. Then, from the fact that $\Delta_1 = \Delta_2 = \Delta_3 = 0$, it follows that $\mathbf{grad}\, g$ is a multiple of $\mathbf{grad}\, f$ at each point of the ball. If (x_0, y_0, z_0) in the ball has the image (u_0, v_0), the equation $f(x, y, z) = u_0$ can be solved for one of the variables (say, z) in terms of the other two, and this solution $z = \phi(x, y)$ will also be a solution of $g(x, y, z) = v_0$. Then the mapping is not one-to-one, for many points $(x, y, \phi(x, y))$ map into the point (u_0, v_0).

CHAPTER XIII

§ 13.3   Pages 389–390

1. (b) $\frac{1}{3} a^3$;   (d) $\frac{1}{2} \pi H a^2$;   (e) $\frac{4}{3} \pi a b c$;   (f) $\frac{1}{12} \pi a^2 h$;   (g) $\frac{8}{15} a b c$.

3. $\frac{2}{3} a^3$.

4. (a) $(\frac{2}{3} a, \frac{1}{3} b)$;   (b) $(\frac{1}{3}(a + b), \frac{1}{3} c)$;   (c) $((4 a/3 \pi), 0)$;   (e) $(\frac{3}{5} B, \frac{3}{8} H)$;   (g) $(a/2, \frac{2}{5} b)$.

7. πa^3.

8. (a) $\left(\dfrac{187 + 4\sqrt{2}}{310}, \dfrac{29 + 116\sqrt{2}}{310} \right)$;   (c) $(\frac{5}{14}, \frac{32}{35})$.

§ 13.4   Pages 395–396

1. (a) $\frac{2}{3} a^3$;   (c) $\frac{1}{2} \pi a^2 h$;   (e) $\frac{16}{9} a^3(3\pi - 4)$;   (f) $\frac{32}{9} a^3$;

(g) $\dfrac{a^2 h}{12} \left[1 + \dfrac{3\sqrt{2}}{2} \log(1 + \sqrt{2}) \right]$.

2. (c) $\bar{y} = \dfrac{5a}{6}$;   (e) $\bar{x} = \dfrac{a}{2} \dfrac{3\sqrt{3} + 8\pi}{3\sqrt{3} + 2\pi}$, $\bar{y} = \dfrac{11a}{2(3\sqrt{3} + 2\pi)}$;

(f) $\bar{x} = \dfrac{a\pi\sqrt{2}}{8}$.

3. One part is $\frac{1}{9} a^3(12\sqrt{3}\,\pi - 20)$.   5. $\pi/32$.

§ 13.5   Pages 400–402

1. (a) $M = \frac{1}{3} ka^2 b$, $\bar{x} = \frac{3}{4} a$, $\bar{y} = \frac{3}{8} b$;   (c) $M = \frac{1}{3} ka^3$, $\bar{x} = \frac{3}{16} \pi a$, $\bar{y} = \frac{3}{8} a$;
(e) $M = \frac{1}{6} k\pi a^3$, $\bar{x} = \bar{y} = \frac{3}{2} a/\pi$;   (g) $M = \frac{4}{15} kab^2$, $\bar{x} = \frac{5}{16} a$, $\bar{y} = \frac{3}{7} b$;
(i) $M = \frac{16}{9} ka^3$, $\bar{x} = \frac{6}{5} a$, $\bar{y} = \frac{9}{20} a$.

3. (b) $I_x = \frac{1}{4} Ma^2$;   (d) $I_x = \frac{1}{6} Mb^2$, $I_y = \frac{1}{2} Ma^2$;
(f) $I_x = \frac{4}{5} Ma^2$, $I_y = \frac{12}{7} Ma^2$;

(h) $I_x = \dfrac{Mb^2}{4} \left(1 - \dfrac{2\sin^3 \beta \cos \beta}{3\beta - 3\sin \beta \cos \beta} \right)$,

$I_y = \dfrac{Mb^2}{4} \left(1 + \dfrac{2\sin^3 \beta \cos \beta}{\beta - \sin \beta \cos \beta} \right)$.

4. (b) $\dfrac{\sigma a^4}{8} = \dfrac{Ma^2}{2\pi}$;   (d) $-\dfrac{Ma^2}{4}$;   (f) 0.

7. (a) $\dfrac{\sigma a^3 b^3}{6(a^2 + b^2)}$;   (c) $\dfrac{656 \sigma a^4}{35}$.

8. (a) $(1 + \sqrt{2})y = x, y = -(1 + \sqrt{2})x$;   (c) $y = \pm x$;   (e) $y = 0, x = 0$.

9. $\frac{1}{4} \pi a^6$ about $y = -x$, $\frac{1}{12} \pi a^6$ about $y = x$.

§ 13.51   Pages 405–406

4. $2 \sigma b \log(1 + \sqrt{2})$.   5. $4 \sigma b$.

6. $u = 2\sigma \left(b \tan^{-1} \dfrac{b}{\sqrt{b^2 + 2a^2}} + \dfrac{a}{2} \log \dfrac{\sqrt{b^2 + 2a^2} + a}{\sqrt{b^2 + 2a^2} - a} \right) - \dfrac{\pi \sigma b}{2}$.

§ 13.7   Pages 413–414

2. $(\frac{1}{4}, \frac{1}{4}, \frac{3}{4})$.   3. (a) $(\frac{1}{4} a, \frac{1}{4} b, \frac{1}{4} c)$;   (b) $\frac{1}{10} M(a^2 + c^2)$.
4. $\bar{x} = \frac{3}{16} \pi a$, $\bar{y} = \frac{3}{8} a$.   5. (b) $2\sqrt{2} - \sqrt{3} - 1$.
7. (a) $(\frac{3}{8} a, \frac{3}{8} b, \frac{3}{8} c)$;   (b) $\frac{1}{5} M(b^2 + c^2)$.
8. (a) $M = 8 a^3$, avg. density $= 1$;   (b) $I_z = \frac{38}{45} Ma^2$;   (c) $\frac{19}{15}$.

§ 13.8   Pages 415–416

2. $\frac{2}{5} Ma^2$.   3. (a) $\frac{1}{2} Ma^2$; $\frac{1}{12} M(3a^2 + h^2)$.

§ 13.9   Pages 418–419

4. (b) $1 + 2\sqrt{2}$;   (c) $\frac{4}{3} \pi \mu a$.

CHAPTER XIV

§ 14.2   Pages 425–426

2. Avg. speed $= \frac{1}{2}\sqrt{2 + \omega^2} + \frac{1}{\omega} \log \frac{\omega + \sqrt{2 + \omega^2}}{\sqrt{2}}$.

3. $4\pi a$.   9. $a\sqrt{2} \sinh t_1$.

§ 14.32   Pages 430–432

5. (a) $T = i$, $N = j$, $B = k$, $\kappa = 2$, $\tau = 3$;   (b) $\frac{2}{27}\sqrt{5}$;

(c) $\kappa = \dfrac{216\sqrt{334}}{(1621)^{3/2} a}$, $\tau = -\dfrac{9}{167 a}$;

(d) $T = i$, $N = -\frac{1}{2}\sqrt{2}(j + k)$, $B = \frac{1}{2}\sqrt{2}(j - k)$, $\kappa = \frac{1}{4}\sqrt{2}$, $\tau = 0$;

(e) $\kappa = (1 + 2 e^{2t})^{1/2}(1 + e^{2t})^{-3/2}$, $\tau = 2 e^t(1 + 2 e^{2t})^{-1}$;

(g) $T = \frac{3}{5} i + \frac{4}{5} k$, $B = -\frac{4}{5} i + \frac{3}{5} k$, $N = j$, $\kappa = \tau = \frac{6}{25}$.

8. (a) $T = -i$, $N = \dfrac{1}{\sqrt{5}} (2 j - k)$, $B = -\dfrac{1}{\sqrt{5}} (j + 2 k)$, $\kappa = \dfrac{\sqrt{5}}{2 a}$,

$\tau = 0$;   (b) $T = -\dfrac{\sqrt{3}}{3} (\sqrt{2} j - k)$, $N = -\dfrac{1}{\sqrt{39}} (6 i + j + \sqrt{2} k)$,

$B = \dfrac{1}{\sqrt{13}} (i - 2 j - 2\sqrt{2} k)$, $\kappa = \dfrac{\sqrt{39}}{9 a}$, $\tau = -\dfrac{3\sqrt{2}}{13 a}$;

(c) $T = \dfrac{1}{\sqrt{2}} (i + k)$, $N = -j$, $B = \dfrac{1}{\sqrt{2}} (i - k)$, $\kappa = \dfrac{1}{2 a}$, $\tau = -\dfrac{3}{8 a}$.

§ 14.4   Pages 436–437

3. (a) $(x^2/a^2) + (y^2/b^2) = z^2$;   (c) $(x^2/a^2) + (y^2/b^2) - (z^2/c^2) = 1$;

(e) $(x^2/a^2) - (y^2/b^2) = z$;

(f) The part of the cone $(x^2/a^2) = (y^2/b^2) + (z^2/c^2)$ on which $x \geqq a$.

6. Cylinder, $x^2 + y^2 = a^2$.   7. Plane, $x + y = z$.

§ 14.5   Pages 440–441

1. Circles in which planes through z-axis intersect the torus; $ds = b \, d\phi$.

3. The u-curves are circles $y^2 + z^2 = a^2$ in planes perpendicular to the x-axis; the v-curves are straight lines parallel to the x-axis. $ds^2 = du^2 + a^2 \, dv^2$.

5. (a) $ds^2 = (1 + 4 k^2 u^2) du^2 + u^2 \, dv^2$;

(c) $ds^2 = (1 + r^2 \sin^2 2 \, \theta) dr^2 + r^3 \sin 4 \, \theta \, dr \, d\theta + r^2(1 + r^2 \cos^2 2 \, \theta) d\theta^2$;

(e) $ds^2 = (1 + 4 u^2 + 9 u^4) du^2 + 2(1 + 4 uv + 9 u^2 v^2) du \, dv + (1 + 4 v^2 + 9 v^4) dv^2$.

6. (b) $\theta \tan \omega = \log \tan (\frac{1}{2} \lambda + \frac{1}{4} \pi)$.

§ 14.6   Pages 447–449

2. $8\sqrt{2} \, a^2$.   3. $2 \pi a^2 \sqrt{2}$.   4. $\frac{2}{3} \pi[(1 + a^2)^{3/2} - 1]$.

5. $\frac{2}{3} \pi ab(2\sqrt{2} - 1)$.   6. (a) $\frac{2}{3}\sqrt{2}(a + b)\sqrt{ab}$;   (b) $\frac{1}{3}\sqrt{2}$.

7. $\pi[\sqrt{2} + \log (1 + \sqrt{2})]$.   9. $8 \, a^2(\pi - 2)$.   11. $\frac{14}{9} \pi a^2$.

15. $2 \, a^2$.

§ 14.8   Pages 460–462

1. (a) $y^2 = 4x$;   (b) $y = \pm x$;   (c) $y^3 = 27x$;   (d) $27y^2 = 4x^3$;
(e) $x^2 - y^2 = a^2$;   (f) $x^4 + y^4 = a^4$;   (g) $27x^4 + 16y^3 = 0$;
(h) $x^2 = 16 - 8y$.

2. $x = 0$ and $3x - 4y = 0$.

3. (a) $4xy = c^2$;   (b) $x^{1/2} + y^{1/2} = c^{1/2}$;   (c) $x^{2/3} + y^{2/3} = c^{2/3}$.

4. (a) $2\pi xy = \pm A$;   (b) $x^{2/3} + y^{2/3} = c^{2/3}$.   5. $\dfrac{x^2}{a^2 + b^2} + \dfrac{y^2}{b^2} = 1$.

6. A cycloid with arches half as high as the original cycloid.

8. $(x^2 + y^2)^2 = 16xy$.   9. $y^2 = 4ax$.   10. $\dfrac{x^2}{a^2} - \dfrac{y^2}{b^2 - a^2} = 1$.

15. $4(x - 2a)^3 = 27ay^2$.

§ 14.9   Pages 466–467

1. (a) $x^2 + y^2 = 1$;   (b) $x^2 + y^2 = p^2$;   (c) a torus;   (d) a right circular cylinder of radius 4 with axis along the line $x = \alpha$, $y = 2\alpha$, $z = 3\alpha$.

2. (a) $(z - 1)^2 = 4xy$;   (b) $x^2 + y^2 = z^2$;   (c) $x^2 + 4xy + 4y^2 + 8z = 0$;
(d) $4z = x^2 + y^2 - 2xy + 2x + 2y + 1$.

3. The cylinder $(x + y + z)^2 + 4(ax + by + cz) = 0$. No envelope if $a = b = c$.

4. (a) $27xyz = p^3$;   (b) $x^{2/3} + y^{2/3} + z^{2/3} = p^{2/3}$.   5. $z = -xy$.

7. $(c^2 - kz)^2 = c^2(c^2 - k^2)\left(\dfrac{x^2}{a^2} + \dfrac{y^2}{b^2}\right)$.   9. $z(x^2 + y^2 + z^2) = 2ax^2$.

10. $cz(x^2 + y^2 + z^2) = a^2x^2 + b^2y^2$.

11. $(x^2 + y^2 + z^2)^2 = 4(a^2x^2 + b^2y^2 + c^2z^2)$.

CHAPTER XV

§ 15.12   Pages 473–475

1. (a) $\frac{4}{3}$;   (b) $\frac{16}{3}$;   (c) $-\frac{32}{3}$;   (d) $\frac{256}{3}$;   (e) $\frac{8}{15}$;   (f) -2;   (g) $\frac{4}{5}$;
(h) $-\frac{8}{3} - \pi$.

2. (a) $\frac{4}{15}$;   (b) 16;   (c) $\frac{1}{4}\pi a^2$;   (d) $-\frac{1}{4}\pi a^4$;   (e) 5π;   (f) 4π;   (g) 2π;
(h) $\frac{4}{3}ab^2$.

3. (a) $\frac{3}{4}\pi$;   (b) $\frac{1}{3}\pi$;   (c) $-\frac{1}{2}\pi$.   4. $\frac{1}{2}\pi(a^2 + 1)$.   6. $\frac{1}{4}$.   7. $\frac{184}{3} + 2\pi$.

8. $\dfrac{2}{3} - \dfrac{\pi}{2}$.   9. (a) $-\frac{16}{9}$;   (b) $\dfrac{32}{3\sqrt{3}}$;   (c) $\dfrac{32}{9\sqrt{3}}$.

14. $2\pi\displaystyle\int_C xy\,dy$.

§ 15.13   Pages 479–480

1. $2a^2$.   2. $2\sqrt{2}\,a$.   3. $2w + \frac{8}{3}$.   4. (a) $2a$;   (b) none.

5. 50 ft. lb.

§ 15.3   Pages 487–489

1. (a) 56;   (b) 0;   (c) 4π;   (d) 0;   (e) $-\frac{1}{4}\pi a^4$;   (f) 0.

§ 15.32 Pages 493–495

1. $53\frac{1}{3}$. **2.** $(a^2 + b^2)^{-1}$. **3.** $\frac{1}{30}$. **4.** $\bar{x} = 5\frac{1}{2}$. **5.** $\frac{1}{2}\pi ab$.
6. 3. **7.** $\frac{1}{4}(e - e^{-1})$. **8.** $\frac{1}{2}(1 - \log 2)$. **9.** $\frac{21}{8}$. **11.** $\frac{1}{32}$.
12. 1. **14.** $2 \log 2 - \frac{1}{2}$.

§ 15.41 Pages 504–505

1. (b) and (e) not exact. **2.** (b) $-\frac{8}{5}$; $-\frac{148}{9}$. **3.** (b) $2 \log \frac{5}{3}$.
4. $w = e^x$; $e^x(x \sin y - \sin y + y \cos y)$.
5. (b) Along $y = 0$, $x > -1$; (c) along $y = 0$, $x^2 < 1$.

§ 15.51 Pages 510–511

1. $\bar{z} = \dfrac{a}{2}$. **2.** (a) 9π; (b) 8π; (c) $\frac{20}{3}\pi$.

3. $\sqrt{2}$. **4.** 0. **5.** $\frac{1}{2}\pi^2 a$. **6.** 2.

7. $A = 2 a^2(\pi - 2)$, $A\bar{x} = \frac{8}{3} a^3$, $A\bar{y} = \frac{2}{3} a^3(3\pi - 8)$, $A\bar{z} = \pi a^3$.

8. $16 a^4 \sigma \dfrac{3\pi - 7}{9}$. **10.** $\dfrac{e}{2}\left(1 - \dfrac{1}{\sqrt{5}}\right)$, where e is the total charge.

15. $-\dfrac{4\pi}{3}, \dfrac{8\pi a^3}{3 c^3}, \dfrac{2\pi}{3}$ in the corresponding cases.

16. $\dfrac{4\pi\sigma a^2}{\sqrt{a^2 + c^2} + |a - c|}$.

§ 15.6 Pages 518–520

4. $3 a^4$. **6.** $4\pi b^5$. **8.** $8\pi b^2$.

§ 15.62 Pages 525–527

2. (a) 900; (b) $(5, 5, -2)$. **3.** $\frac{4}{15}\pi(a^2 + b^2)abc$.
4. $\frac{315}{32}\pi$. **5.** $\frac{1}{4}$. **6.** $\frac{1}{24}$ and $\frac{1}{2}$. **7.** $\frac{21}{8}\pi$.

§ 15.7 Pages 532–533

1. $-2\sqrt{2}\pi b^2$. **2.** $-\pi$. **3.** $-\pi a^2$. **4.** (d) $-4\pi a^2$. **5.** 0.
6. (a) $\frac{2}{3}(3\pi - 8)a^2$; (b) πa^2; (c) $\frac{8}{3} a^2$.

§ 15.8 Pages 537–538

1. (b) $a^2 b^2 c$; (d) 1. **2.** $\dfrac{a}{a^2 + b^2 + c^2} - 1$.

3. $x^2 y + y \log z$. **4.** $\dfrac{x^2 + y^2 + z^2}{z} - 3$. **5.** $(xy - z)e^{-xy}$.
6. (b) $x^2 + y^2 < a^2$, $z = 0$; (c) $x^2 + y^2 = 0$, $z \neq 0$.

Miscellaneous Exercises Pages 538–539

2. $e^x \cos y - 3 \sin^{-1} x + 7 \tan y$. **4.** Discontinuous along $y = 0$, $x^2 > 1$.

5. $\sqrt{3}$.   **7.** $8/(3\,\pi)$.

9. $4\,\pi\sigma(b/a)\{(a+b)E(k) + (a-b)K(k)\}$.   **10.** $\pi^2/16$.

CHAPTER XVI

§ 16.2   Pages 544–545

1. (a) Open; (b) closed; (c) neither; (d) neither.   **2.** 0 and 1.

3. (a) No; (b) yes, 0 and 1.

6. (a) 1. Not closed.   (b) 0. Closed.   (c) 1 and -1. Not closed.

§ 16.41   Page 550

1. Yes.   **2.** (c) $x = 0$, $-1 \leqq y \leqq 1$.

3. Neither. All points of the square $0 \leqq x \leqq 1$, $0 \leqq y \leqq 1$.

CHAPTER XVIII

§ 18.1   Pages 568–569

2. (b) 1.   **3.** $I = 2$, $J = 4$.

§ 18.21   Page 578

1. $\int_a^b F(x)\, f'(x)dx$.

§ 18.4   Page 580

4. $\frac{1}{3}$.

7. (b) $F(x) = -\frac{1}{2}x^2 + \frac{1}{2}$ if $x < 0$, $F(x) = \frac{1}{2} + \frac{1}{2}x^2$ if $x > 0$.

8. (a) $F(x) = -x - 1$ if $x < 0$, $F(x) = -1 + x$ if $x > 0$.

§ 18.5   Page 585

1. $F'(u) = \displaystyle\int_0^4 \frac{-2\,ux^2}{1 - u^2x^2}\, dx.$ **2.** $(1 - a^2)^{-1/2}$.

3. $e^{-x^4} - 2x\displaystyle\int_0^x t^2 e^{-x^2t^2}\, dt.$

4. $(x^2 - \sin^2 x)^n \cos x - (x^2 - x^4)^n\, 2x + \displaystyle\int_{x^2}^{\sin x} 2\,xn(x^2 - t^2)^{n-1}\, dt.$

7. $\dfrac{\partial^2 u}{\partial x^2} = \dfrac{1}{x^2} f\!\left(x, \dfrac{1}{x}\right) + \displaystyle\int_{1/x}^{y} f_1(x,\, t)dt.$

$\dfrac{\partial^2 u}{\partial y^2} = \dfrac{1}{y^2} f\!\left(\dfrac{1}{y},\, y\right) + \displaystyle\int_{1/y}^{x} f_2(s,\, y)ds.$

§ 18.6   Page 588

1. Yes.   **2.** Yes.   **3.** No.

4. No. Use the fact that f is uniformly continuous.

§ 18.9   Page 597

1. 3 in each case.   2. $-\frac{1}{2}$, $\frac{7}{2}$, 3 respectively.
3. 5, -15, 35 respectively.   4. (a) $2e^{-1} - 1$;   (b) $\frac{1}{2}(e^{-2} - 1)$.
5. (a) $-\frac{1}{2}\log 2$;   (b) $\frac{7}{288}\pi^2$.
6. (a) 0, 1, 1, 3 respectively;   (b) $4\sqrt{7} - \sqrt{2} - 3$.
7. (a) $w(0) = 0$, $w(x) = 1 + \frac{1}{2}x$ if $0 < x < 5$, $w(5) = \frac{9}{2}$,
$w(x) = 2 + \frac{1}{2}x$ if $5 < x < 10$, $w(10) = 8$;   (b) 40 and $\frac{275}{3}$ respectively.
8. (a) Second moment about y-axis is
$\int_a^b x^2\, dA(x)$;   (b) $\bar{y}\int_a^b dA(x) = \frac{1}{2}\int_a^b \phi(x)dA(x)$.
9. $\bar{x}\int_a^b dV(x) = \int_a^b x\, dV(x)$.

CHAPTER XIX

§ 19   Pages 600–601

1. (b) $|x| < 1$, $\dfrac{8x}{1 - x^2}$;   (d) $|x| > 1$, $\dfrac{10}{x - 1}$;   (f) $x > 0$, $\dfrac{1 + x}{x}$;
(h) $\dfrac{1}{e} < x < e$, $\dfrac{\log x}{1 - \log x}$.
4. (a), (b), (d) are divergent; (c) is convergent.   5. Divergent.

§ 19.2   Pages 608–609

1. (a) and (b) divergent; (c) and (d) convergent.
5. (b), (c), and (f) divergent; (a), (d), (e), and (g) convergent.

§ 19.21   Pages 611–612

1. (a), (d), and (g) divergent; (b), (c), (e), (f), and (h) convergent.
3. $p > 1$.   5. $p > 1$ and any q. Also $p = 1$ and $q < -1$.

§ 19.22   Pages 613–614

1. (c) and (e) divergent; (a), (b), (d), (f), and (g) convergent.
3. (a) and (d).

§ 19.32   Pages 622–623

2. (b) and (d) divergent; (a) and (c) convergent.

§ 19.4   Pages 628–630

1. (b) Convergent if $|x| \leq 1$, divergent if $|x| > 1$;   (d) convergent if
$|x| < \frac{5}{2}$, divergent if $|x| \geq \frac{5}{2}$;   (f) convergent if $|x| < 2$, divergent if $|x| > 2$,
indecisive if $|x| = 2$;   (h) convergent if $|x| < \frac{1}{2}$, divergent if $|x| > \frac{1}{2}$, inde-
cisive if $|x| = \frac{1}{2}$.
2. (b) Convergent if $|x| < 1$, divergent if $|x| \geq 1$;   (d) indecisive;
(f) convergent if $|x| < \sqrt{\frac{3}{2}}$, divergent if $|x| \geq \sqrt{\frac{3}{2}}$;   (h) convergent if
$|x| < \frac{2}{5}$, divergent if $|x| \geq \frac{2}{5}$.

3. (b) Convergent;   **(d)** divergent;   **(f)** convergent if $p > \frac{2}{3}$, divergent if $p < \frac{2}{3}$, indecisive if $p = \frac{2}{3}$.

4. (a) Divergent;   **(c)** convergent;   **(e)** convergent if $p > 2$, divergent if $p \leq 2$;   **(f)** divergent if $p = \frac{2}{3}$ (see answer to 3(f));   **(g)** divergent.

8. (b) The first two series are convergent; the third is divergent.

Miscellaneous Exercises   Pages 641–643

1. Convergent if and only if $-1 \leq x < 1$.

6. (a) Convergent, others divergent.

7. (b) Convergent if and only if $-2 \leq x < 2$.   **8.** $x > \frac{1}{2}$.

9. (a) No. Yes.   **(c)** All values.   **(d)** Convergent.

10. (a) Convergent if and only if $|x| \leq 2$.   **(b)** Convergent if and only if $|x| \leq 1$.   **(c)** Convergent if and only if $-e^{-1} \leq x < e^{-1}$.

CHAPTER XX

§ 20.1   Pages 651–653

1. (a) No.   **(b)** $0 < a < b < \pi$.   **(e)** $ab > 0$.

4. (a) $1 + x^2$. 0.   Yes.   **(c)** No.   **5. (b)** No. No. Yes.

§ 20.2   Pages 654–655

1. Uniform convergence for the ranges of x as indicated: **(a)** $|x| \leq c$ if $0 < c < 1$.   **(b)** All x.   **(c)** All x.   **(d)** $x \geq c$ if $c > 0$.   **(e)** All x on any finite interval.

§ 20.3   Page 656

1. (a) Not uniformly convergent on any interval having 0 inside or at one end.

3. The function is continuous at 0.

§ 20.4   Pages 658–659

4. $\int_0^1 f(x)dx = 0$; $\int_0^1 f_n(x)dx \to \pi/2$.

§ 20.5   Page 661

1. All but **(a)** and **(b)**.

CHAPTER XXI

§ 21.1   Pages 667–668

1. (a) $\frac{1}{4}$.   **(b)** 27.   **(c)** 0.   **(d)** ∞.   **(e)** ∞.   **(f)** 4.

2. ∞ if $p < q$; 0 if $p > q$; p^{-p} if $p = q$.

3. 1.   **9.** 0.0976.   **10.** 0.4864.   **11.** 0.0045.   **12.** 0.482.

§ 21.2   Pages 673–675

11. $\dfrac{1}{x} \displaystyle\int_0^x \dfrac{e^t - 1}{t}\, dt.$

12. $L_q(x) = q! \left[1 + \sum_{n=1}^{q} (-1)^n \dfrac{q(q-1) \cdots (q-n+1)}{(n!)^2} x^n \right].$

13. $H_m(x) = (2x)^m - \dfrac{m(m-1)}{1!} (2x)^{m-2}$

$\qquad + \dfrac{m(m-1)(m-2)(m-3)}{2!} (2x)^{m-4} - \cdots$

(breaking off with the term of exponent 0 or 1 according as m is even or odd).

§ 21.3 Pages 678–679

1. (a) $1 + \frac{1}{2} x - \frac{1}{24} x^3 + \frac{1}{240} x^5 - \cdots.$
(c) $1 + x + \frac{3}{2} x^2 + \frac{11}{6} x^3 + \cdots.$
(e) $1 - \frac{1}{2} x - \frac{1}{12} x^2 - \frac{1}{24} x^3 - \cdots.$
(f) $-\frac{1}{3} x - \frac{1}{45} x^3 - \frac{2}{945} x^5 - \cdots.$

2. (a) $\dfrac{e^x}{1-x}.$ **(b)** $\dfrac{\log(1+x)}{1-x}.$ **(c)** $\dfrac{x}{(1-x)^3}.$

§ 21.4 Pages 682–683

3. (b) 0.09. **4.** $p > 3.$

§ 21.5 Pages 685–686

1. (a) 1. **(b)** $e^{-1}.$ **(c)** $\frac{1}{3}.$ **(d)** $e.$ **(e)** $\infty.$ **(f)** 1.
4. (a) 1. **(b)** 1. **(c)** 1. **(d)** $\infty.$ **(e)** 1. **(f)** 1.

CHAPTER XXII

§ 22.1 Pages 697–698

1. (a), (b), (d) convergent; **(c), (e), (f)** divergent.
2. (a) Convergent. **(b)** Divergent.
3. (a), (b), (d), (e), (f), (g), (i), (j), (k), (n), (p) convergent; others divergent.
5. $\frac{1}{2} m!.$ **7.** $m + 1 < n.$

§ 22.11 Pages 701–702

1. (a) Divergent. **(c)** Divergent. **(e)** Convergent.
2. (a), (b), (c), (d), (f), (j) convergent; others divergent.
4. Improper but convergent if $0 < a < 1.$
5. $n = 2, 3.$ **6. (b)** $0 < x < 1.$ **7.** $0 < p < 1.$

§ 22.12 Pages 703–704

1. (b), (c), (d), (e), (g), (i) convergent; others divergent.
2. (d) Divergent; others convergent.
3. (a) $1 < p < 2.$ **(b)** $0 < x < 1.$ **(c)** $p < 1.$ **(d)** $-1 < p < 1.$
(e) $0 < a < 2.$ **(f)** $0 < x.$ **(g)** $0 < x < 1.$ **(h)** $0 < a.$
4. To the left of $x = 1$ and above $x + y = 1.$

§ 22.2   Pages 707–708

4. (a) $\frac{1}{4}\sqrt{\pi}$.   (b) $\frac{3}{8}\sqrt{\pi}$.   6. (a) $\sqrt{\pi}/2$.   (b) $\frac{3}{128}\sqrt{\pi}$.

§ 22.3   Pages 711–712

1. (b) Abs. convergent.   (d) Convergent.   (f) Abs. convergent.
(h) Abs. convergent.   (j) Abs. convergent.   (l) Convergent.
(n) Abs. convergent.   (p) Convergent.   (r) Abs. convergent.
   3. Yes, whenever $1 < p < 3$.   4. Yes, whenever $1 < p < 4$.

§ 22.4   Pages 715–717

1. (b) Divergent; others convergent.   4. $p < 4$.

5. (a) 2π.   (b) $\dfrac{4\pi}{3-n}$, $(n < 3)$.   (c) Convergent.

(d) $\dfrac{4\pi}{3(5-n)}$, $(n < 5)$.   (e) Divergent.   (f) $\dfrac{8\pi}{105}$.

6. (a) Convergent.   (c) $p < \frac{3}{2}$.

§ 22.41   Pages 721–722

2. No.   5. The value is $\pi/4$.

§ 22.5   Pages 727–730

9. $\sqrt{\pi x}$.   11. $\frac{1}{2}\pi \log (b/a)$.

§ 22.51   Page 733

3. $\frac{1}{2} + \frac{1}{6} e^{-2t} + \frac{1}{3} e^{t}$.
4. (a) $-\frac{6}{5}\cos 3t + \frac{1}{3}\sin 3t + \frac{1}{5}\cos 2t$.   (b) $2\sinh 2t + \sinh t$.
5. (a) $\Gamma(n)(s+1)^{-n}$.   (b) $(s^2 + 2s + 2)^{-1}$.   (c) $3(s^2 - 4s + 13)^{-1}$.

§ 22.7   Pages 739–740

5. $-\frac{1}{2} < a$.

7. (a) $\frac{1}{6}$.   (b) $\dfrac{2^{11}\, 4!\, 5!}{11!}$.   (c) $\dfrac{3\pi}{8}$.   (d) $\dfrac{\sqrt{\pi}\,\Gamma(\frac{1}{4})}{12\,\Gamma(\frac{3}{4})}$.

Miscellaneous Exercises   Pages 745–749

2. Both convergent.

Index